A BRIEF GUID THE MOST FF K

1 Read the Book

FEATURE	DESCRIPTION	BENEFIT
Applications Using Real-World Data	From the chapter and section openers through the examples and exercises, interesting applications from nearly every discipline, supported by up-to-date real-world data, are included in every section.	Ever wondered how you'll use algebra? This feature will show you how algebra can solve real problems.
Detailed Worked-Out Examples	Examples are clearly written and provide step-by-step solutions. No steps are omitted, and key steps are thoroughly explained to the right of the mathematics.	The blue annotations will help you to understand the solutions by providing the reason why the algebraic steps are true.
Explanatory Voice Balloons	Voice balloons help to demystify algebra. They translate algebraic language into plain English, clarify problem-solving procedures, and present alternative ways of understanding.	Does math ever look foreign to you? This feature translates math into everyday English.
Study Tips	Study Tip boxes offer suggestions for problem solving, point out common errors to avoid, and provide informal hints and suggestions.	By seeing common mistakes, you'll be able to avoid them.

2 Work the Problems

FEATURE	DESCRIPTION	BENEFIT
Check Point Examples	Each example is followed by a similar problem, called a Check Point, that offers you the opportunity to work a similar exercise. Answers to all Check Points are provided in the answer section.	You learn best by doing. You'll solidify your understanding of worked examples if you try a similar problem right away to be sure you understand what you've just read.
Extensive and Varied Exercise Sets	An abundant collection of exercises is included in an exercise set at the end of each section. Exercises are organized within several categories. Practice Exercises follow the same order as the section's worked examples. Practice PLUS exercises contain more challenging problems that often require you to combine several skills or concepts.	The parallel order of the Practice Exercises lets you refer to the worked examples and use them as models for solving these problems. Practice PLUS provides you with ample opportunity to dig in and develop your problem-solving skills.

3 Review for Quizzes and Tests

FEATURE	DESCRIPTION	BENEFIT
Mid-Chapter Check Points	Near the midway point in the chapter, an integrated set of review exercises allows you to review the skills and concepts you learned separately over several sections.	Combining exercises from the first half of the chapter gives you a comprehensive review before you continue on.
Chapter Review Chart	Each chapter contains a review chart that summarizes the definitions and concepts in every section of the chapter, complete with examples.	Review this chart and you'll know the most important material in the chapter!
Chapter Tests	Each chapter contains a practice test with problems that cover the important concepts in the chapter. Take the test, check your answers, and then watch the Chapter Test Prep Video CD to see worked-out solutions for any exercises you miss.	You can use the chapter test to determine whether you have mastered the material covered in the chapter.
Chapter Test Prep Video CD	This video CD found at the front of your text contains worked-out solutions to every exercise in each chapter test.	The video lets you review any exercises you miss on the chapter test.

Licensing Agreement

Robert Blitzer

Chapter Test Prep Video CD to accompany Introductory Algebra for College Students, 5e

ISBN: 0-13-602926-4/978-0-13-602926-4

Pearson Prentice Hall
Pearson Education, Inc.
Upper Saddle River, NJ 07458

YOU SHOULD CAREFULLY READ THE TERMS AND CONDITIONS BEFORE USING THE CD-ROM PACKAGE. USING THIS CD-ROM PACKAGE INDICATES YOUR ACCEPTANCE OF THESE TERMS AND CONDITIONS.
Pearson Education, Inc. provides this program and licenses its use. You assume responsibility for the selection of the program to achieve your intended results, and for the installation, use, and results obtained from the program. This license extends only to use of the program in the United States or countries in which the program is marketed by authorized distributors.

LICENSE GRANT
You hereby accept a nonexclusive, nontransferable, permanent license to install and use the program ON A SINGLE COMPUTER at any given time. You may copy the program solely for backup or archival purposes in support of your use of the program on the single computer. You may not modify, translate, disassemble, decompile, or reverse engineer the program, in whole or in part.

TERM
The License is effective until terminated. Pearson Education, Inc. reserves the right to terminate this License automatically if any provision of the License is violated. You may terminate the License at any time. To terminate this License, you must return the program, including documentation, along with a written warranty stating that all copies in your possession have been returned or destroyed.

LIMITED WARRANTY
THE PROGRAM IS PROVIDED "AS IS" WITHOUT WARRANTY OF ANY KIND, EITHER EXPRESSED OR IMPLIED, INCLUDING, BUT NOT LIMITED TO, THE IMPLIED WARRANTIES OF MERCHANTABILITY AND FITNESS FOR A PARTICULAR PURPOSE. THE ENTIRE RISK AS TO THE QUALITY AND PERFORMANCE OF THE PROGRAM IS WITH YOU. SHOULD THE PROGRAM PROVE DEFECTIVE, YOU (AND NOT PEARSON EDUCATION, INC. OR ANY AUTHORIZED DEALER) ASSUME THE ENTIRE COST OF ALL NECESSARY SERVICING, REPAIR, OR CORRECTION. NO ORAL OR WRITTEN INFORMATION OR ADVICE GIVEN BY PEARSON EDUCATION, INC., ITS DEALERS, DISTRIBUTORS, OR AGENTS SHALL CREATE A WARRANTY OR INCREASE THE SCOPE OF THIS WARRANTY. SOME STATES DO NOT ALLOW THE EXCLUSION OF IMPLIED WARRANTIES, SO THE ABOVE EXCLUSION MAY NOT APPLY TO YOU. THIS WARRANTY GIVES YOU SPECIFIC LEGAL RIGHTS AND YOU MAY ALSO HAVE OTHER LEGAL RIGHTS THAT VARY FROM STATE TO STATE.
Pearson Education, Inc. does not warrant that the functions contained in the program will meet your requirements or that the operation of the program will be uninterrupted or error-free. However, Pearson Education, Inc. warrants the CD-ROM(s) on which the program is furnished to be free from defects in material and workmanship under normal use for a period of ninety (90) days from the date of delivery to you as evidenced by a copy of your receipt. The program should not be relied on as the sole basis to solve a problem whose incorrect solution could result in injury to person or property. If the program is employed in such a manner, it is at the user's own risk and Pearson Education, Inc. explicitly disclaims all liability for such misuse.

LIMITATION OF REMEDIES
Pearson Education, Inc.'s entire liability and your exclusive remedy shall be:
1. the replacement of any CD-ROM not meeting Pearson Education, Inc.'s "LIMITED WARRANTY" and that is returned to Pearson Education, or
2. if Pearson Education is unable to deliver a replacement CD-ROM that is free of defects in materials or workmanship, you may terminate this agreement by returning the program.

IN NO EVENT WILL PEARSON EDUCATION, INC. BE LIABLE TO YOU FOR ANY DAMAGES, INCLUDING ANY LOST PROFITS, LOST SAVINGS, OR OTHER INCIDENTAL OR CONSEQUENTIAL DAMAGES ARISING OUT OF THE USE OR INABILITY TO USE SUCH PROGRAM EVEN IF PEARSON EDUCATION, INC. OR AN AUTHORIZED DISTRIBUTOR HAS BEEN ADVISED OF THE POSSIBILITY OF SUCH DAMAGES, OR FOR ANY CLAIM BY ANY OTHER PARTY.
SOME STATES DO NOT ALLOW FOR THE LIMITATION OR EXCLUSION OF LIABILITY FOR INCIDENTAL OR CONSEQUENTIAL DAMAGES, SO THE ABOVE LIMITATION OR EXCLUSION MAY NOT APPLY TO YOU.

GENERAL
You may not sublicense, assign, or transfer the license of the program. Any attempt to sublicense, assign or transfer any of the rights, duties, or obligations hereunder is void.

This Agreement will be governed by the laws of the State of New York.
Should you have any questions concerning this Agreement, you may contact Pearson Education, Inc. by writing to:
ESM Media Development
Higher Education Division
Pearson Education, Inc.
1 Lake Street
Upper Saddle River, NJ 07458
Should you have any questions concerning technical support, you may write to:
New Media Production
Higher Education Division
Pearson Education, Inc.
1 Lake Street, Upper Saddle River, NJ 07458

YOU ACKNOWLEDGE THAT YOU HAVE READ THIS AGREEMENT, UNDERSTAND IT, AND AGREE TO BE BOUND BY ITS TERMS AND CONDITIONS. YOU FURTHER AGREE THAT IT IS THE COMPLETE AND EXCLUSIVE STATEMENT OF THE AGREEMENT BETWEEN US THAT SUPERSEDES ANY PROPOSAL OR PRIOR AGREEMENT, ORAL OR WRITTEN, AND ANY OTHER COMMUNICATIONS BETWEEN US RELATING TO THE SUBJECT MATTER OF THIS AGREEMENT.

System Requirements
Macintosh:
Power PC G3 233 MHz or better
Mac OS 10.x
64 MB RAM
19 MB of additional hard drive space (if QuickTime installation is needed)
800 x 600 resolution
8x CD-ROM drive
QuickTime 7.x
Internet browser

Windows:
Pentium II 300 MHz processor-based computer
Windows 2000, XP, Vista Ultimate
64 MB RAM (128 MB RAM required for Windows XP)
7.2 MB additional hard drive space (if QuickTime installation is needed)
800 x 600 resolution
8x CD-ROM drive
QuickTime 7.x
Sound Card
Internet browser

Support Information
If you are having problems with this software, you can get support by filling out the web form located at: http://247pearsoned.custhelp.com/
Our technical staff will need to know certain things about your system in order to help us solve your problems more quickly and efficiently. If possible, please be at your computer when you call for support. You should have the following information ready:

- ISBN
- CD-ROM ISBN
- corresponding product and title
- computer make and model
- Operating System (Windows or Macintosh) and Version
- RAM available
- hard disk space available
- Sound card? Yes or No
- printer make and model
- network connection
- detailed description of the problem, including the exact wording of any error messages.

NOTE: Pearson does not support and/or assist with the following:

- third-party software (i.e. Microsoft including Microsoft Office suite, Apple, Borland, etc.)
- homework assistance
- Textbooks and CD-ROMs purchased used are not supported and are non-replaceable. To purchase a new CD-ROM, contact Pearson Individual Order Copies at 1-800-282-0693.

PEARSON ALWAYS LEARNING

Review of Introductory Algebra

A Custom Edition for Portland Community College

Taken from:
Intermediate Algebra for College Students, Fifth Edition
by Robert Blitzer

Introductory Algebra for College Students, Fifth Edition
by Robert Blitzer

Cover photograph courtesy of Will Freeman.

Taken from:

Intermediate Algebra for College Students, Fifth Edition
by Robert Blitzer

Published by Prentice Hall
Upper Saddle River, New Jersey 07458

Introductory Algebra for College Students, Fifth Edition
by Robert Blitzer

Published by Prentice Hall

Pearson Learning Solutions, 501 Boylston Street, Suite 900, Boston, MA 02116
A Pearson Education Company
www.pearsoned.com

Printed in the United States of America

2 3 4 5 6 7 8 9 10 V0ZN 17 16 15 14 13

000200010271680358

CB

PEARSON

ISBN 10: 1-256-84045-9
ISBN 13: 978-1-256-84045-9

NOTE TO STUDENTS: The material in this book has been specifically selected by your instructors. They have chosen to eliminate some material which is not critical to your course, and in doing so the price of this book has been adjusted to reflect the content removed. Please note that because content has been removed from your text, you will see some breaks in the page numbering.

CONTENTS

Taken from: *Intermediate Algebra for College Students*, Fifth Edition
by Robert Blitzer

CHAPTER 3

Systems of Linear Equations 165

CHAPTER 4

Inequalities and Problem Solving 239

CHAPTER 5

Polynomials, Polynomial Functions, and Factoring 303

PREFACE

Intermediate Algebra for College Students, Fifth Edition, provides comprehensive, in-depth coverage of the topics required in a one-term course in intermediate algebra. The book is written for college students who have had a course in introductory algebra. I wrote the book to help diverse students, with different backgrounds and career plans, to succeed in intermediate algebra. *Intermediate Algebra for College Students,* Fifth Edition, has two primary goals:

1. To help students acquire a solid foundation in the skills and concepts of intermediate algebra.
2. To show students how algebra can model and solve authentic real-world problems.

One major obstacle in the way of achieving these goals is the fact that very few students actually read their textbook. This has been a regular source of frustration for me and for my colleagues in the classroom. Anecdotal evidence gathered over years highlights two basic reasons why students do not take advantage of their textbook:

- "I'll never use this information."
- "I can't follow the explanations."

I've written every page of the Fifth Edition with the intent of eliminating these two objections. The ideas and tools I've used to do so are described in the features that follow. These features and their benefits are highlighted for the student in "A Brief Guide to Getting the Most from This Book," which appears inside the front cover.

What's New in the Fifth Edition?

- **New Applications and Real World Data.** I'm on a constant search for real-world data that can be used to illustrate unique algebraic applications. I researched hundreds of books, magazines, newspapers, almanacs, and online sites to prepare the Fifth Edition. With 223 worked-out examples and application exercises based on new data sets, the Fifth Edition contains a greater array of applications than any previous revision of this book.
- **"Make Sense?" Classroom Discussion Exercises.** Each exercise set contains four Critical Thinking exercises intended for classroom discussion in order to engage participation in the learning process. These items test conceptual understanding by asking students to determine whether statements are sensible, and to explain why or why not. Although sample answers are provided, students have skills and perspectives that frequently differ from those of math teachers, so answers and explanations may vary. The important part of this new feature is to let you ask students what they think about selected statements, determine whether they understand the concepts, and give them feedback to clarify any misunderstandings.
- **New Directions for the True/False Critical Thinking Exercises.** The Fifth Edition asks students to determine whether each statement in an itemized list is true or false. If the statement is false, students are then asked to make the necessary change or changes to produce a true statement.
- **Preview Exercises.** Each exercise set concludes with three problems to help students prepare for the next section. Some of these problems review previously covered material that students will need to be successful in the forthcoming section. Other problems are designed to get students thinking about concepts they will soon encounter.
- **More Detailed Directions When Comparing Mathematical Models with Actual Data.** The Fifth Edition asks students if values obtained from mathematical models underestimate or overestimate data displayed by graphs, and, if so, by how much.
- **Increased Study Tip Boxes.** The book's Study Tip boxes offer suggestions for problem solving, point out common errors to avoid, and provide informal hints and suggestions. These invaluable hints appear in greater abundance in the Fifth Edition.

- **New Chapter-Opening and Section-Opening Scenarios.** Every chapter and every section open with a scenario based on an application, the majority of which are unique to the Fifth Edition. These scenarios are revisited in the course of the chapter or section in one of the book's new examples, exercises, or discussions. The often-humorous tone of these openers is intended to help fearful and reluctant students overcome their negative perceptions about math.
- **745 New Examples and Exercises.** The Fifth Edition contains 27 detailed worked-out examples involving new data, 196 new application exercises, 256 "make sense" discussion exercises, 189 preview exercises, and 77 new exercises that appear in the various other categories of the exercise sets.

What Content and Organizational Changes Have Been Made to the Fifth Edition?

- **Section 2.1** (Introduction to Functions) and **Section 2.2** (Graphs of Functions) introduce functions over two sections, rather than only one section, as in the previous edition. This gradual approach allows for a discussion of functions represented by tables in Section 2.1.
- **Section 2.5** (The Point-Slope Form of the Equation of a Line) contains a more thoroughly developed example on writing equations of lines perpendicular to a given line.
- **Section 4.3** (Equations and Inequalities Involving Absolute Value) uses boundary points to show students how solving inequalities involving absolute value is connected to the graph of $f(x) = |x|$. Instructors are given the option of solving absolute value inequalities using boundary points or by the more traditional method of rewriting as equivalent compound inequalities.
- **Section 9.5** (Exponential and Logarithmic Equations) has been reorganized into four categories:

  Solving exponential equations using like bases
  Solving exponential equations using logarithms and logarithmic properties
  Solving logarithmic equations using the definition of a logarithm
  Solving logarithmic equations using the one-to-one property of logarithms.

  New examples appear throughout the section to ensure adequate coverage of each category.

What Familiar Features Have Been Retained in the Fifth Edition?

- **Detailed Worked-Out Examples.** Each worked example is titled, making clear the purpose of the example. Examples are clearly written and provide students with detailed step-by-step solutions. No steps are omitted and key steps are thoroughly explained to the right of the mathematics.
- **Explanatory Voice Balloons.** Voice balloons are used in a variety of ways to demystify mathematics. They translate algebraic ideas into everyday English, help clarify problem-solving procedures, present alternative ways of understanding concepts, and connect problem solving to concepts students have already learned.
- **Check Point Examples.** Each example is followed by a similar matched problem, called a Check Point, offering students the opportunity to test their understanding of the example by working a similar exercise. The answers to the Check Points are provided in the answer section.
- **Extensive and Varied Exercise Sets.** An abundant collection of exercises is included in an exercise set at the end of each section. Exercises are organized within eight category types: Practice Exercises, Practice Plus Exercises, Application Exercises, Writing in Mathematics, Technology Exercises, Critical Thinking Exercises,

Review Exercises, and Preview Exercises. This format makes it easy to create well-rounded homework assignments. The order of the practice exercises is exactly the same as the order of the section's worked examples. This parallel order enables students to refer to the titled examples and their detailed explanations to achieve success working the practice exercises.

- **Practice Plus Problems.** This category of exercises contains more challenging practice problems that often require students to combine several skills or concepts. With an average of ten practice plus problems per exercise set, instructors are provided with the option of creating assignments that take practice exercises to a more challenging level.
- **Mid-Chapter Check Points.** At approximately the midway point in each chapter, an integrated set of review exercises allows students to review and assimilate the skills and concepts they learned separately over several sections.
- **Graphing and Functions.** Graphing is introduced in Chapter 1 and functions are introduced in Chapter 2, with an integrated graphing functional approach emphasized throughout the book. Graphs and functions that model data appear in nearly every section and exercise set. Examples and exercises use graphs of functions to explore relationships between data and to provide ways of visualizing a problem's solution. Because functions are the core of this course, students are repeatedly shown how functions relate to equations and graphs.
- **Section Objectives.** Learning objectives are clearly stated at the beginning of each section. These objectives help students recognize and focus on the section's most important ideas. The objectives are restated in the margin at their point of use.
- **Integration of Technology Using Graphical and Numerical Approaches to Problems.** Side-by-side features in the technology boxes connect algebraic solutions to graphical and numerical approaches to problems. Although the use of graphing utilities is optional, students can use the explanatory voice balloons to understand different approaches to problems even if they are not using a graphing utility in the course.
- **Chapter Review Grids.** Each chapter contains a review chart that summarizes the definitions and concepts in every section of the chapter. Examples that illustrate these key concepts are also included in the chart.
- **End-of-Chapter Materials.** A comprehensive collection of review exercises for each of the chapter's sections follows the review grid. This is followed by a chapter test that enables students to test their understanding of the material covered in the chapter. Beginning with Chapter 2, each chapter concludes with a comprehensive collection of mixed cumulative review exercises.
- **Chapter Test Prep Video CD.** Packaged at the front of the text, this video CD provides students with step-by-step solutions for each of the exercises in the book's chapter tests.
- **Blitzer Bonuses.** These enrichment essays provide historical, interdisciplinary, and otherwise interesting connections to the algebra under study, showing students that math is an interesting and dynamic discipline.
- **Discovery.** Discover for Yourself boxes, found throughout the text, encourage students to further explore algebraic concepts. These explorations are optional and their omission does not interfere with the continuity of the topic under consideration.
- **Chapter Projects.** At the end of each chapter is a collaborative activity that gives students the opportunity to work cooperatively as they think and talk about mathematics. Additional group projects can be found in the *Instructor's Resource Manual*. Many of these exercises should result in interesting group discussions.

I hope that my passion for teaching, as well as my respect for the diversity of students I have taught and learned from over the years, is apparent throughout this new edition. By connecting algebra to the whole spectrum of learning, it is my intent to show students that their world is profoundly mathematical, and indeed, π is in the sky.

Robert Blitzer

Resources for the Fifth Edition

FOR STUDENTS

Student Solutions Manual Fully worked solutions to the odd-numbered section exercises plus all Check Points, Review/Preview Exercises, Mid-Chapter Check Points, Chapter Reviews, Chapter Tests, and Cumulative Reviews.

Worksheets Provide a ready-to-use lesson and exercise set for every section of the text with ample student work space.

CD Lecture Series A comprehensive set of CD-ROMS, tied to the textbook, containing short video clips of an instructor working key text examples/exercises. (Also available separately on DVD.)

MathXL® Tutorials on CD This interactive tutorial CD-ROM provides algorithmically generated practice exercises that are correlated at the objective level to the exercises in the textbook. Every practice exercise is accompanied by an example and a guided solution designed to involve students in the solution process.

Chapter Test Prep Video CD Provides step-by-step video solutions to each problem in each Chapter Test in the textbook. Packaged with a new text, inside the front cover.

Pearson Tutor Services Tutors provide one-on-one tutoring for any problem with an answer in the back of the text. Access via toll-free phone, fax, or email.

FOR INSTRUCTORS

Instructor Resource Distribution Most instructor resources can be downloaded from the Web site, *www.prenhall.com*. Select "Browse our catalog," then click on "mathematics," select your course and choose your text. Under "resources," on the left side, select "instructor" and choose the supplement you need to download. You will be required to run through a one time registration before you can complete this process.

Instructor's Solutions Manual Fully worked solutions to every exercise in the text.

Instructor's Resource Manual with Tests Includes a Mini-Lecture, Skill Builder, and Additional Exercises for ever section of the text; two short group Activities per chapter, several chapter test forms, both free-response and multiple-choice, as well as cumulative tests and final exams. Answers to all items also included.

TestGen® Easily create tests from section objectives. Questions are algorithmically generated allowing for unlimited versions. Edit problems or create your own. There's a chapter test file for each Chapter Test in the text.

Annotated Instructor's Edition Answers to exercises are printed on the same text page with graphing answers in a special Graphing Answer Section in the back of the text.

FOR BOTH

MyMathLab — MathXL PRACTICE · WATCH · DOWNLOAD · READ · REVIEW

MathXL®

MathXL® is a powerful online homework, tutorial, and assessment system that accompanies this textbook. Instructors can create, edit, and assign online homework and tests using algorithmically generated exercises correlated at the objective level to the textbook. Student work is tracked in an online gradebook. Students can take chapter tests and receive personalized study plans based on their results. The study plan diagnoses weaknesses and links students directly to tutorial exercises for objectives they need to study. Students can also access video clips directly from selected exercises. MathXL is available to qualified adopters. For more information, visit our website at *www.mathxl.com* or contact your Pearson sales representative.

MyMathLab®

MyMathLab® is a series of text-specific, customizable online courses for your textbooks. Powered by CourseCompass™ (Pearson Education's online teaching and learning environment) and MathXL® (our online homework, tutorial, and assessment system), MyMathLab gives you the tools you need to deliver all or a portion of your course online, whether students are in a lab setting or working from home. MyMathLab provides a rich and flexible set of course materials, featuring free-response exercises that are algorithmically generated for unlimited practice. Students can also use online tools, such as video lectures and a multimedia textbook, to improve their performance. Instructors can use MyMathLab's homework and test managers to select and assign online exercises correlated to the textbook, and can import TestGen tests for added flexibility. The online gradebook—designed specifically for mathematics—automatically tracks students' homework and test results and gives the instructor control over how to calculate final grades. MyMathLab is available to qualified adopters. For more information, visit our website at *www.mymathlab.com* or contact your Pearson sales representative.

Acknowledgments

An enormous benefit of authoring a successful series is the broad-based feedback I receive from the students, dedicated users, and reviewers. Every change to this edition is the result of their thoughtful comments and suggestions. I would like to express my appreciation to all the reviewers, whose collective insights form the backbone of this revision. In particular, I would like to thank the following people for reviewing *Intermediate Algebra for College Students*.

Gwen P. Aldridge	*Northwest Mississippi Community College*
Howard Anderson	*Skagit Valley College*
John Anderson	*Illinois Valley Community College*
Michael H. Andreoli	*Miami Dade College–North Campus*
Jan Archibald	*Ventura College*
Donna Beatty	*Ventura College*
Michael S. Bowen	*Ventura College*
Gale Brewer	*Amarillo College*
Hien Bui	*Hillsborough Community College*
Warren J. Burch	*Brevard Community College*
Alice Burstein	*Middlesex Community College*
Edie Carter	*Amarillo College*
Thomas B. Clark	*Trident Technical College*
Sandra Pryor Clarkson	*Hunter College*
Bettyann Daley	*University of Delaware*
Robert A. Davies	*Cuyahoga Community College*
Paige Davis	*Lurleen B. Wallace Community College*
Ben Divers, Jr.	*Ferrum College*
Irene Doo	*Austin Community College*
Charles C. Edgar	*Onondaga Community College*
Rhoderick Fleming	*Wake Technical Community College*
Susan Forman	*Bronx Community College*
Donna Gerken	*Miami-Dade College*
Marion K. Glasby	*Anne Arundel Community College*
Sue Glascoe	*Mesa Community College*
Jay Graening	*University of Arkansas*
Robert B. Hafer	*Brevard Community College*
Mary Lou Hammond	*Spokane Community College*
Donald Herrick	*Northern Illinois University*
Beth Hooper	*Golden West College*
Tracy Hoy	*College of Lake County*
Judy Kasabian	*Lansing Community College*
Gary Kersting	*North Central Michigan College*
Gary Knippenberg	*Lansing Community College*
Mary Kochler	*Cuyahoga Community College*
Mary A. Koehler	*Cuyahoga Community College*
Kristi Laird	*Jackson Community College*
Jennifer Lempke	*North Central Michigan College*
Sandy Lofstock	*St. Petersburg College*
Hank Martel	*Broward Community College*
Diana Martelly	*Miami-Dade College*
John Robert Martin	*Tarrant County Junior College*

Mikal McDowell	*Cedar Valley College*
Irwin Metviner	*State University of New York at Old Westbury*
Terri Moser	*Austin Community College*
Robert Musselman	*California State University, Fresno*
Kamilia Nemri	*Spokane Community College*
Allen R. Newhart	*Parkersburg Community College*
Steve O'Donnell	*Rogue Community College*
Jeff Parent	*Oakland Community College*
Tian Ren	*Queensborough Community College*
Kate Rozsa	*Mesa Community College*
Scott W. Satake	*Eastern Washington University*
Mike Schramm	*Indian River Community College*
Terri Seiver	*San Jacinto College*
Kathy Shepard	*Monroe County Community College*
Gayle Smith	*Lane Community College*
Linda Smoke	*Central Michigan University*
Dick Spangler	*Tacoma Community College*
Janette Summers	*University of Arkansas*
Kory Swart	*Kirkwood Community College*
Robert Thornton	*Loyola University*
Lucy C. Thrower	*Francis Marion College*
Andrew Walker	*North Seattle Community College*
Kathryn Wetzel	*Amarillo College*
Margaret Williamson	*Milwaukee Area Technical College*
Roberta Yellott	*McNeese State University*
Marilyn Zopp	*McHenry County College*

Additional acknowledgments are extended to Dan Miller and Kelly Barber, for preparing the solutions manuals, Brad Davis, for preparing the answer section and serving as accuracy checker, the Preparè, Inc. formatting team, for the book's brilliant paging, Aaron Darnall at Scientific Illustrators, for superbly illustrating the book, Diane Austin, photo researcher, for obtaining the book's new photographs, and Barbara Mack, whose talents as production editor kept every aspect of this complex project moving through its many stages.

I would like to thank my editors at Prentice Hall, Paul Murphy and Chris Hoag, and Project Manager, Dawn Nuttall, who guided and coordinated the book from manuscript through production. Thanks to Maureen Eide and John Christiana for the beautiful cover, Juan López for the wonderful interior design, Kate Valentine and Patrice Jones for their innovative marketing efforts, and to the entire Pearson Education sales force for their confidence and enthusiasm about the book.

Robert Blitzer

TO THE STUDENT

I have written this book so that you can learn about the power of algebra and how it relates directly to your life outside the classroom. All concepts are carefully explained, important definitions and procedures are set off in boxes, and worked-out examples that present solutions in a step-by-step manner appear in every section. Each example is followed by a similar matched problem, called a Check Point, for you to try so that you can actively participate in the learning process as you read the book. (Answers to all Check Points appear in the back of the book.) Study Tips offer hints and suggestions and often point out common errors to avoid. A great deal of attention has been given to applying algebra to your life to make your learning experience both interesting and relevant.

As you begin your studies, I would like to offer some specific suggestions for using this book and for being successful in this course:

- **Attend all lectures.** No book is intended to be a substitute for valuable insights and interactions that occur in the classroom. In addition to arriving for lecture on time and being prepared, you will find it useful to read the section before it is covered in lecture. This will give you a clear idea of the new material that will be discussed.
- **Read the book.** Read each section with pen (or pencil) in hand. Move through the worked-out examples with great care. These examples provide a model for doing exercises in the exercise sets. As you proceed through the reading, do not give up if you do not understand every single word. Things will become clearer as you read on and see how various procedures are applied to specific worked-out examples.
- **Work problems every day and check your answers.** The way to learn mathematics is by doing mathematics, which means working the Check Points and assigned exercises in the exercise sets. The more exercises you work, the better you will understand the material.
- **Review for quizzes and tests.** After completing a chapter, study the chapter review chart, work the exercises in the Chapter Review, and work the exercises in the Chapter Test. Answers to all these exercises are given in the back of the book.

  The methods that I've used to help you read the book, work the problems, and review for tests are described in "A Brief Guide to Getting the Most from This Book" that appears inside the front cover. Spend a few minutes reviewing the guide to familiarize yourself with the book's features and their benefits.

- **Use the resources available with this book.** Additional resources to aid your study are described on page xii. These resources include a Solutions Manual; a Chapter Test Prep Video CD; MyMathLab®, an online version of the book with links to multimedia resources; MathXL®, an online homework, tutorial, and assessment system of the text; and tutorial support at the Pearson Tutor Services.

I wrote this book in Point Reyes National Seashore, 40 miles north of San Francisco. The park consists of 75,000 acres with miles of pristine surf-washed beaches, forested ridges, and bays bordered by white cliffs. It was my hope to convey the beauty and excitement of mathematics using nature's unspoiled beauty as a source of inspiration and creativity. Enjoy the pages that follow as you empower yourself with the algebra needed to succeed in college, your career, and in your life.

Regards,
Bob
Robert Blitzer

Bob and his horse Jerid

ABOUT THE AUTHOR

Bob Blitzer is a native of Manhattan and received a Bachelor of Arts degree with dual majors in mathematics and psychology (minor: English literature) from the City College of New York. His unusual combination of academic interests led him toward a Master of Arts in mathematics from the University of Miami and a doctorate in behavioral sciences from Nova University. Bob's love for teaching mathematics was nourished for nearly 30 years at Miami Dade College, where he received numerous teaching awards, including Innovator of the Year from the League for Innovations in the Community College and an endowed chair based on excellence in the classroom. In addition to *Intermediate Algebra for College Students,* Bob has written textbooks covering introductory algebra, college algebra, algebra and trigonometry, precalculus, and liberal arts mathematics, all published by Prentice Hall. When not secluded in his Northern California writer's cabin, Bob can be found hiking the beaches and trails of Point Reyes National Seashore, and tending to the chores required by his beloved entourage of horses, chickens, and irritable roosters.

CHAPTER 1

Algebra, Mathematical Models, and Problem Solving

Tuition and Fees at Four-Year Colleges

	School Year Ending 2000	School Year Ending 2007
Public	\$3362	\$5836
Private	\$15,518	\$22,218

Source: The College Board

The cost of a college education is skyrocketing. If these trends continue, what can we expect by the 2010s and beyond? We can answer this question by representing the data mathematically. With such representations, called *mathematical models*, we can gain insights and predict what might occur in the future on a variety of issues, ranging from college costs to a possible Social Security doomsday, and even the changes that occur as we age.

Mathematical models involving college costs appear as Exercises 67–68 in Exercise Set 1.4. The insecurities of Social Security are explored in Exercise 75 in the Review Exercises. Some surprising changes that occur with aging appear as Example 2 in Section 1.1, Exercises 75–78 in Exercise Set 1.1, and Exercises 53–56 in Exercise Set 1.3.

SECTION 1.1

Algebraic Expressions and Real Numbers

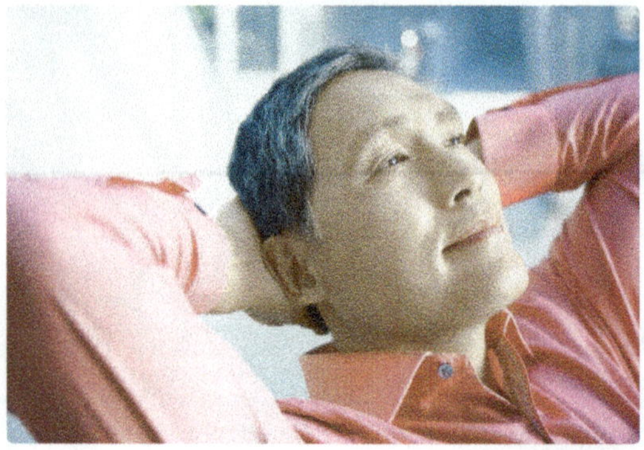

Objectives

1. Translate English phrases into algebraic expressions.
2. Evaluate algebraic expressions.
3. Use mathematical models.
4. Recognize the sets that make up the real numbers.
5. Use set-builder notation.
6. Use the symbols $\in$ and $\notin$.
7. Use inequality symbols.

As we get older, do we mellow out or become more neurotic? In this section, you will learn how the special language of algebra describes your world, including our improving emotional health with age.

Algebraic Expressions

Algebra uses letters, such as x and y, to represent numbers. If a letter is used to represent various numbers, it is called a **variable**. For example, imagine that you are basking in the sun on the beach. We can let x represent the number of minutes that you can stay in the sun without burning with no sunscreen. With a number 6 sunscreen, exposure time without burning is six times as long, or 6 times x. This can be written $6 \cdot x$, but it is usually expressed as $6x$. Placing a number and a letter next to one another indicates multiplication.

Notice that $6x$ combines the number 6 and the variable x using the operation of multiplication. A combination of variables and numbers using the operations of addition, subtraction, multiplication, or division, as well as powers or roots, is called an **algebraic expression**. Here are some examples of algebraic expressions:

$$x + 6, \quad x - 6, \quad 6x, \quad \frac{x}{6}, \quad 3x + 5, \quad x^2 - 3, \quad \sqrt{x} + 7.$$

Is every letter in algebra a variable? No. Some letters stand for a particular number. Such a letter is called a **constant**. For example, let $d =$ the number of days in a week. The letter d represents just one number, namely 7, and is a constant.

1. Translate English phrases into algebraic expressions.

Translating English Phrases into Algebraic Expressions

Problem solving in algebra involves translating English phrases into algebraic expressions. Here is a list of words and phrases for the four basic operations:

Addition	Subtraction	Multiplication	Division
sum	difference	product	quotient
plus	minus	times	divide
increased by	decreased by	of (used with fractions)	per
more than	less than	twice	ratio

EXAMPLE 1 Translating English Phrases into Algebraic Expressions

Write each English phrase as an algebraic expression. Let x represent the number.

a. Nine less than six times a number

b. The quotient of five and a number, increased by twice the number

Solution

a. Nine less than | six times a number

$$6x - 9$$

b. The quotient of five and a number, | increased by | twice the number

$$\frac{5}{x} + 2x$$

Study Tip

After working each Check Point, check your answer in the answer section before continuing your reading.

✓ **CHECK POINT 1** Write each English phrase as an algebraic expression. Let x represent the number.

a. Five more than 8 times a number

b. The quotient of a number and 7, decreased by twice the number

2 Evaluate algebraic expressions.

Evaluating Algebraic Expressions

Evaluating an algebraic expression means to find the value of the expression for a given value of the variable.

EXAMPLE 2 Evaluating an Algebraic Expression

A test measuring neurotic traits, such as anxiety and hostility, indicates that people may become less neurotic as they get older. **Figure 1.1** shows the average level of neuroticism, on a scale of 0 to 50, for persons at various ages.

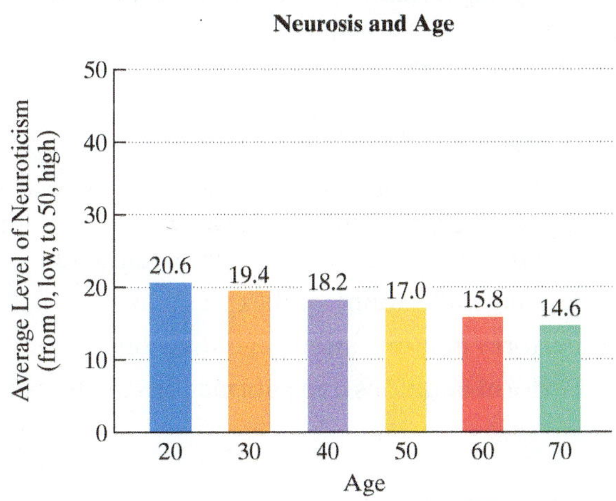

FIGURE 1.1

Source: L. M. Williams, "The Mellow Years? Neural Basis of Improving Emotional Stability over Age," *The Journal of Neuroscience*, June 14, 2006.

The algebraic expression $23 - 0.12x$ describes the average neurotic level for people who are x years old. Evaluate the expression for $x = 80$. Describe what the answer means in practical terms.

Solution We begin by substituting 80 for x. Because $x = 80$, we will be finding the average neurotic level at age 80.

$$23 - 0.12x$$

Replace x with 80.

$$= 23 - 0.12(80) = 23 - 9.6 = 13.4$$

Thus, at age 80, the average level of neuroticism on a scale of 0 to 50 is 13.4.

✓ **CHECK POINT 2** Evaluate the expression from Example 2, $23 - 0.12x$, for $x = 10$. Describe what the answer means in practical terms.

Many algebraic expressions involve *exponents*. For example, the algebraic expression

$$233x^2 + 2296x + 16{,}197$$

approximates the number of registered lobbyists in Washington x years after 2000. The expression x^2 means $x \cdot x$, and is read "x to the second power" or "x squared." The exponent, 2, indicates that the base, x, appears as a factor two times.

Exponential Notation

If n is a counting number (1, 2, 3, and so on),

Exponent or Power

$$b^n = \underbrace{b \cdot b \cdot b \cdot \cdots \cdot b}_{b \text{ appears as a factor } n \text{ times.}}.$$

Base

b^n is read "the nth power of b" or "b to the nth power." Thus, the nth power of b is defined as the product of n factors of b. The expression b^n is called an **exponential expression**. Furthermore, $b^1 = b$.

Using Technology

You can use a calculator to evaluate exponential expressions. For example, to evaluate 2^4, press the following keys:

Many Scientific Calculators

2 $\boxed{y^x}$ 4 $\boxed{=}$

Many Graphing Calculators

2 $\boxed{\wedge}$ 4 $\boxed{\text{ENTER}}$

Although calculators have special keys to evaluate powers of ten and to square bases, you can always use one of the sequences shown here.

For example,

$$8^2 = 8 \cdot 8 = 64, \quad 5^3 = 5 \cdot 5 \cdot 5 = 125, \quad \text{and} \quad 2^4 = 2 \cdot 2 \cdot 2 \cdot 2 = 16.$$

Many algebraic expressions involve more than one operation. Evaluating an algebraic expression without a calculator involves carefully applying the following order of operations agreement:

The Order of Operations Agreement

1. Perform operations within the innermost parentheses and work outward. If the algebraic expression involves a fraction, treat the numerator and the denominator as if they were each enclosed in parentheses.
2. Evaluate all exponential expressions.
3. Perform multiplications and divisions as they occur, working from left to right.
4. Perform additions and subtractions as they occur, working from left to right.

EXAMPLE 3 Evaluating an Algebraic Expression

Evaluate $7 + 5(x - 4)^3$ for $x = 6$.

Solution

$7 + 5(x - 4)^3 = 7 + 5(6 - 4)^3$ Replace x with 6.

$= 7 + 5(2)^3$ First work inside parentheses: $6 - 4 = 2$.

$= 7 + 5(8)$ Evaluate the exponential expression: $2^3 = 2 \cdot 2 \cdot 2 = 8$.

$= 7 + 40$ Multiply: $5(8) = 40$.

$= 47$ Add.

✓ **CHECK POINT 3** Evaluate $8 + 6(x - 3)^2$ for $x = 13$.

3 Use mathematical models.

Formulas and Mathematical Models

An **equation** is formed when an equal sign is placed between two algebraic expressions. One aim of algebra is to provide a compact, symbolic description of the world. These descriptions involve the use of *formulas*. A **formula** is an equation that uses variables to express a relationship between two or more quantities. Here is an example of a formula:

$$C = \frac{5}{9}(F - 32).$$

Celsius temperature | is | $\frac{5}{9}$ of | the difference between Fahrenheit temperature and 32°.

The process of finding formulas to describe real-world phenomena is called **mathematical modeling**. Such formulas, together with the meaning assigned to the variables, are called **mathematical models**. We often say that these formulas model, or describe, the relationships among the variables.

EXAMPLE 4 Modeling the Number of Lobbyists in Washington

In 2006, the ease with which conniving lobbyists bought members of Congress put pressure on lawmakers to mend a broken system. The formula

$$L = 233x^2 + 2296x + 16{,}197$$

models the number of registered lobbyists, L, in Washington, x years after 2000.

a. Use the formula to find the number of lobbyists in Washington in 2004.

b. By how much is the model value for 2004 greater than or less than the actual data value shown in **Figure 1.2**?

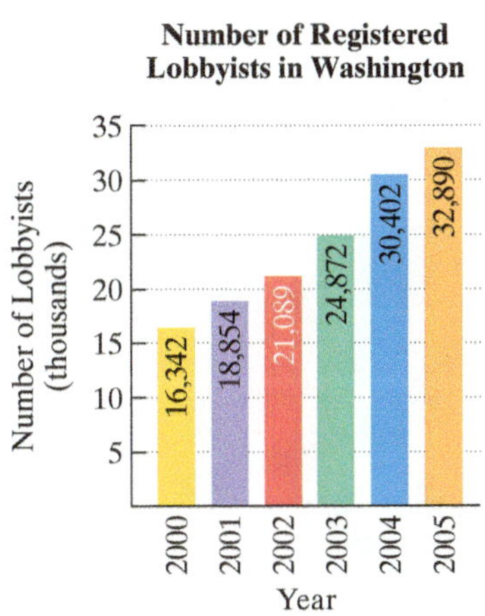

FIGURE 1.2

Source: Senate Office of Public Records

Solution

a. Because 2004 is 4 years after 2000, we substitute 4 for x in the given formula. Then we use the order of operations to find L, the number of lobbyists in 2004.

$$L = 233x^2 + 2296x + 16{,}197 \quad \text{This is the given mathematical model.}$$

$$L = 233(4)^2 + 2296(4) + 16{,}197 \quad \text{Replace each occurrence of } x \text{ with 4.}$$

$$L = 233(16) + 2296(4) + 16{,}197 \quad \text{Evaluate the exponential expression: } 4^2 = 4 \cdot 4 = 16.$$

$$L = 3728 + 9184 + 16{,}197 \quad \text{Multiply from left to right: } 233(16) = 3728 \text{ and } 2296(4) = 9184.$$

$$L = 29{,}109 \quad \text{Add.}$$

The formula indicates that in 2004, there were 29,109 lobbyists in Washington.

b. The number of lobbyists for 2004 given in **Figure 1.2** is 30,402. The model value, 29,109, is less than the actual data value by 30,402 − 29,109, or by 1293 lobbyists.

✓ CHECK POINT 4

a. Use the formula in Example 4 to find the number of lobbyists in Washington in 2005.

b. By how much is the model value for 2005 greater than or less than the actual data value shown in **Figure 1.2**?

Sometimes a mathematical model gives an estimate that is not a good approximation or is extended to include values of the variable that do not make sense. In these cases, we say that **model breakdown** has occurred. Models that accurately describe data for the past ten years might not serve as reliable predictions for what can reasonably be expected to occur in the future. Model breakdown can occur when formulas are extended too far into the future.

4 Recognize the sets that make up the real numbers.

The Set of Real Numbers

Before we describe the set of real numbers, let's be sure you are familiar with some basic ideas about sets. A **set** is a collection of objects whose contents can be clearly determined. The objects in a set are called the **elements** of the set. For example, the set of numbers used for counting can be represented by

$$\{1, 2, 3, 4, 5, \ldots\}.$$

The braces, { }, indicate that we are representing a set. This form of representation, called the **roster method**, uses commas to separate the elements of the set. The three dots after the 5, called an *ellipsis*, indicate that there is no final element and that the listing goes on forever.

Three common sets of numbers are the *natural numbers*, the *whole numbers*, and the *integers*.

Study Tip

Grouping symbols such as parentheses, ( ), and square brackets, [ ], are not used to represent sets. Only commas are used to separate the elements of a set. Separators such as colons or semicolons are not used.

Natural Numbers, Whole Numbers, and Integers

The Set of Natural Numbers

$$\{1, 2, 3, 4, 5, \ldots\}$$

These are the numbers that we use for counting.

The Set of Whole Numbers

$$\{0, 1, 2, 3, 4, 5, \ldots\}$$

The set of whole numbers includes 0 and the natural numbers.

The Set of Integers

$$\{\ldots, -5, -4, -3, -2, -1, 0, 1, 2, 3, 4, 5, \ldots\}$$

The set of integers includes the negatives of the natural numbers and the whole numbers.

5 Use set-builder notation.

A set can also be written in **set-builder notation**. In this notation, the elements of the set are described, but not listed. Here is an example:

$$\{x \mid x \text{ is a natural number less than } 6\}.$$

The set of all x | such that | x is a natural number less than 6.

The same set is written using the roster method as

$$\{1, 2, 3, 4, 5\}.$$

6 Use the symbols $\in$ and $\notin$.

The symbol $\in$ is used to indicate that a number or object is in a particular set. The symbol $\in$ is read "is an element of." Here is an example:

$$7 \in \{1, 2, 3, 4, 5, \ldots\}.$$

7 | is an element of | the set of natural numbers.

The symbol $\notin$ is used to indicate that a number or object is not in a particular set. The symbol $\notin$ is read "is not an element of." Here is an example:

$$\frac{1}{2} \notin \{1, 2, 3, 4, 5, \ldots\}.$$

$\frac{1}{2}$ is not an element of the set of natural numbers.

EXAMPLE 5 Using the Symbols $\in$ and $\notin$

Determine whether each statement is true or false:

a. $100 \in \{x \mid x \text{ is an integer}\}$ **b.** $20 \notin \{5, 10, 15\}$.

Solution

a. Because 100 is an integer, the statement

$$100 \in \{x \mid x \text{ is an integer}\}$$

is true. The number 100 is an element of the set of integers.

b. Because 20 is not an element of $\{5, 10, 15\}$, the statement $20 \notin \{5, 10, 15\}$ is true.

✓ CHECK POINT 5 Determine whether each statement is true or false:

a. $13 \in \{x \mid x \text{ is an integer}\}$ **b.** $6 \notin \{7, 8, 9, 10\}$.

Another common set is the set of *rational numbers*. Each of these numbers can be expressed as an integer divided by a nonzero integer.

Rational Numbers

The set of **rational numbers** is the set of all numbers that can be expressed as a quotient of two integers, with the denominator not 0.

$$\left\{\frac{a}{b} \,\middle|\, a \text{ and } b \text{ are integers and } b \neq 0\right\}$$

This means that b is not equal to zero.

Three examples of rational numbers are

$$\frac{1}{4}, \quad \frac{-2}{3}, \quad \text{and } 5 = \frac{5}{1}.$$

($a = 1$, $b = 4$; $a = -2$, $b = 3$; $a = 5$, $b = 1$)

Can you see that integers are also rational numbers because they can be written in terms of division by 1?

Rational numbers can be expressed in fraction or decimal notation. To express the fraction $\frac{a}{b}$ as a decimal, divide the denominator, b, into the numerator, a. In decimal notation, rational numbers either terminate (stop) or have a digit, or block of digits, that repeats. For example,

$$\frac{3}{8} = 3 \div 8 = 0.375 \quad \text{and} \quad \frac{7}{11} = 7 \div 11 = 0.6363\ldots = 0.\overline{63}.$$

The decimal stops: it is a terminating decimal.

This is a repeating decimal. The bar is written over the repeating part.

Some numbers cannot be expressed as terminating or repeating decimals. An example of such a number is $\sqrt{2}$, the square root of 2. The number $\sqrt{2}$ is a number that can be squared to give 2. No terminating or repeating decimal can be squared to get 2. However, some approximations have squares that come close to 2. We use the symbol $\approx$, which means "is approximately equal to."

- $\sqrt{2} \approx 1.4$ because $(1.4)^2 = (1.4)(1.4) = 1.96$.
- $\sqrt{2} \approx 1.41$ because $(1.41)^2 = (1.41)(1.41) = 1.9881$.
- $\sqrt{2} \approx 1.4142$ because $(1.4142)^2 = (1.4142)(1.4142) = 1.99996164$.

$\sqrt{2}$ is an example of an *irrational number.*

Irrational Numbers

The set of **irrational numbers** is the set of numbers whose decimal representations neither terminate nor repeat. Irrational numbers cannot be expressed as quotients of integers.

Examples of irrational numbers include

$$\sqrt{3} \approx 1.73205 \qquad \text{and} \qquad \pi(\text{pi}) \approx 3.141593.$$

Not all square roots are irrational. For example, $\sqrt{25} = 5$ because $5^2 = 5 \cdot 5 = 25$. Thus, $\sqrt{25}$ is a natural number, a whole number, an integer, and a rational number $\left(\sqrt{25} = \frac{5}{1}\right)$.

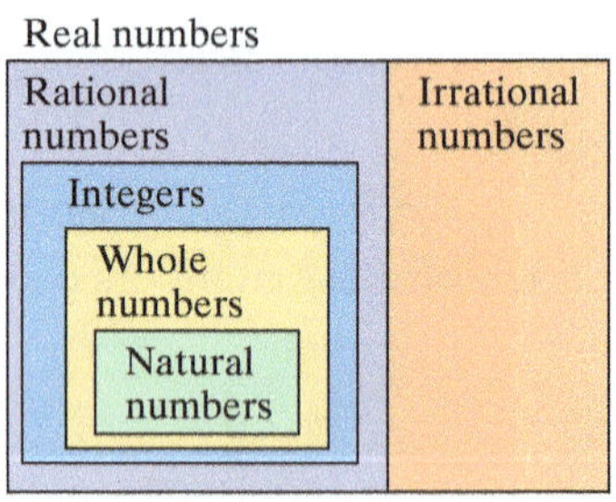

FIGURE 1.3 Every real number is either rational or irrational.

The set of *real numbers* is formed by combining the sets of rational numbers and irrational numbers. Thus, every real number is either rational or irrational, as shown in **Figure 1.3**.

Real Numbers

The set of **real numbers** is the set of numbers that are either rational or irrational:

$$\{x \mid x \text{ is rational or } x \text{ is irrational}\}.$$

The Real Number Line

The **real number line** is a graph used to represent the set of real numbers. An arbitrary point, called the **origin**, is labeled 0. Select a point to the right of 0 and label it 1. The distance from 0 to 1 is called the **unit distance**. Numbers to the right of the origin are **positive** and numbers to the left of the origin are **negative**. The real number line is shown in **Figure 1.4**.

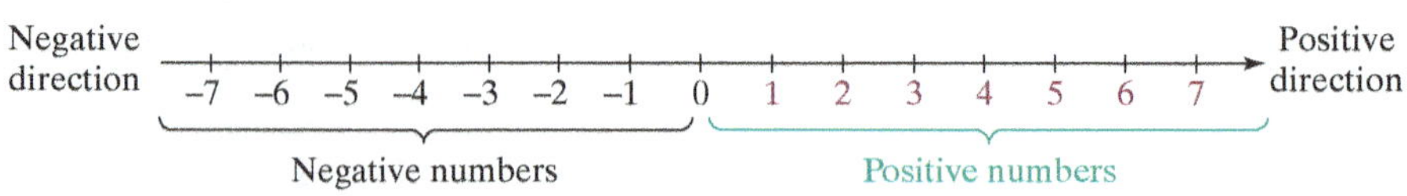

FIGURE 1.4 The real number line

Real numbers are **graphed** on a number line by placing a dot at the correct location for each number. The integers are easiest to locate. In **Figure 1.5**, we've graphed six rational numbers and three irrational numbers on a real number line.

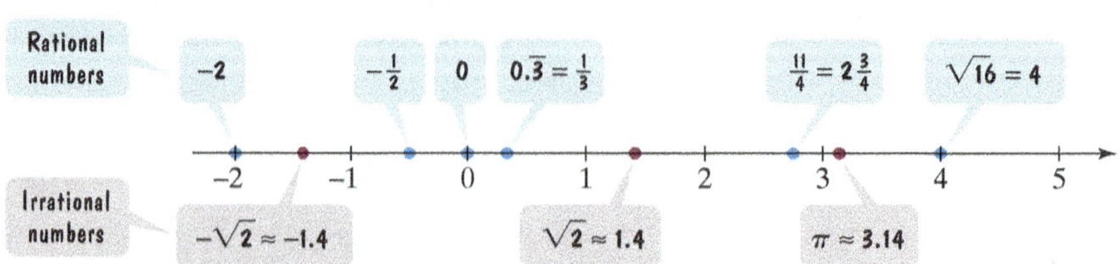

FIGURE 1.5 Graphing numbers on a real number line

Every real number corresponds to a point on the number line and every point on the number line corresponds to a real number. We say that there is a **one-to-one correspondence** between all the real numbers and all points on a real number line.

Study Tip

Wondering how we located $\sqrt{2}$ as a precise point on the number line in **Figure 1.5**? We used a right triangle with two legs of length 1. The remaining side has a length measuring $\sqrt{2}$.

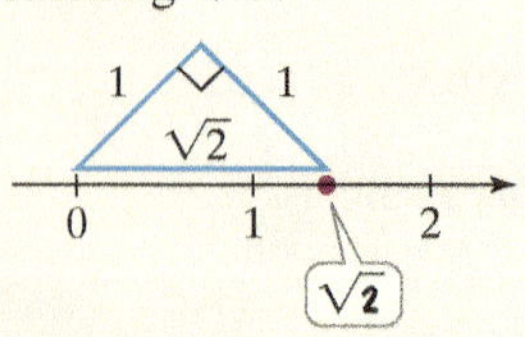

We'll have lots more to say about right triangles later in the book.

7 Use inequality symbols.

Ordering the Real Numbers

On the real number line, the real numbers increase from left to right. The lesser of two real numbers is the one farther to the left on a number line. The greater of two real numbers is the one farther to the right on a number line.

Look at the number line in **Figure 1.6**. The integers -4 and -1 are graphed.

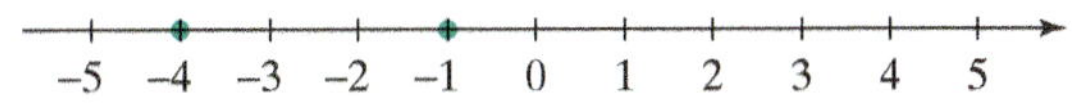

FIGURE 1.6

Observe that -4 is to the left of -1 on the number line. This means that -4 is less than -1.

$$-4 < -1$$

-4 is less than -1 because -4 is to the **left** of -1 on the number line.

In **Figure 1.6**, we can also observe that -1 is to the right of -4 on the number line. This means that -1 is greater than -4.

$$-1 > -4$$

-1 is greater than -4 because -1 is to the **right** of -4 on the number line.

The symbols $<$ and $>$ are called **inequality symbols**. These symbols always point to the lesser of the two real numbers when the inequality statement is true.

-4 is less than -1. $\quad -4 < -1 \quad$ The symbol points to -4, the lesser number.

-1 is greater than -4. $\quad -1 > -4 \quad$ The symbol still points to -4, the lesser number.

The symbols $<$ and $>$ may be combined with an equal sign, as shown in the following table:

Symbol	Meaning	Examples	Explanation
$a \le b$	a is less than or equal to b.	$2 \le 9$ $9 \le 9$	Because $2 < 9$ Because $9 = 9$
$b \ge a$	b is greater than or equal to a.	$9 \ge 2$ $2 \ge 2$	Because $9 > 2$ Because $2 = 2$

This inequality is true if either the $<$ part or the $=$ part is true.

This inequality is true if either the $>$ part or the $=$ part is true.

EXAMPLE 6 Using Inequality Symbols

Write out the meaning of each inequality. Then determine whether the inequality is true or false.

a. $-5 < -1$ **b.** $6 > -2$ **c.** $-6 \le 3$ **d.** $10 \ge 10$ **e.** $-9 \ge 6$

Solution The solution is illustrated by the number line in **Figure 1.7**.

FIGURE 1.7

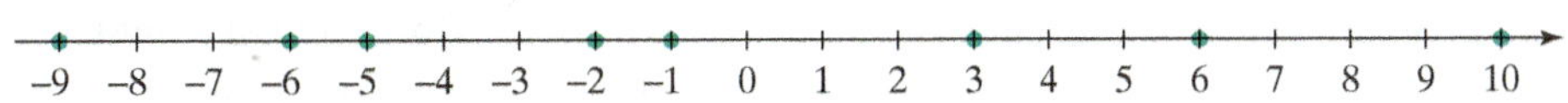

Inequality	Meaning
a. $-5 < -1$	"-5 is less than -1." Because -5 is to the left of -1 on the number line, the inequality is true.
b. $6 > -2$	"6 is greater than -2." Because 6 is to the right of -2 on the number line, the inequality is true.
c. $-6 \le 3$	"-6 is less than or equal to 3." Because $-6 < 3$ is true (-6 is to the left of 3 on the number line), the inequality is true.
d. $10 \ge 10$	"10 is greater than or equal to 10." Because $10 = 10$ is true, the inequality is true.
e. $-9 \ge 6$	"-9 is greater than or equal to 6." Because neither $-9 > 6$ nor $-9 = 6$ is true, the inequality is false.

Study Tip

Here are three similar English phrases that have very different translations:

- 7 minus a number: $7 - x$
- 7 less than a number: $x - 7$
- 7 is less than a number: $7 < x$

Think carefully about what is expressed in English before you translate into the language of algebra.

CHECK POINT 6 Write out the meaning of each inequality. Then determine whether the inequality is true or false.

a. $-8 < -2$

b. $7 > -3$

c. $-1 \leq -4$

d. $5 \geq 5$

e. $2 \geq -14$

1.1 EXERCISE SET

MyMathLab MathXL PRACTICE WATCH DOWNLOAD READ REVIEW

Study Tip

Researchers say the mind can be strengthened, just like your muscles, with regular training and rigorous practice. Think of the book's exercise sets as brain calisthenics. If you're feeling a bit sluggish before any of your mental workouts, try this warmup:

In the list below, say the color the word is printed in, not the word itself. Once you can do this in 15 seconds without an error, the warmup is over and it's time to move on to the assigned exercises.

Blue Yellow Red Green Yellow Green Blue Red Yellow Red

Practice Exercises

In Exercises 1–14, write each English phrase as an algebraic expression. Let x represent the number.

1. Five more than a number
2. A number increased by six
3. Four less than a number
4. Nine less than a number
5. Four times a number
6. Twice a number
7. Ten more than twice a number
8. Four more than five times a number
9. The difference of six and half of a number
10. The difference of three and half of a number
11. Two less than the quotient of four and a number
12. Three less than the quotient of five and a number
13. The quotient of three and the difference of five and a number
14. The quotient of six and the difference of ten and a number

In Exercises 15–26, evaluate each algebraic expression for the given value or values of the variable(s).

15. $7 + 5x$, for $x = 10$
16. $8 + 6x$, for $x = 5$
17. $6x - y$, for $x = 3$ and $y = 8$
18. $8x - y$, for $x = 3$ and $y = 4$
19. $x^2 + 3x$, for $x = \frac{1}{3}$
20. $x^2 + 2x$ for $x = \frac{1}{2}$
21. $x^2 - 6x + 3$, for $x = 7$
22. $x^2 - 7x + 4$, for $x = 8$
23. $4 + 5(x - 7)^3$, for $x = 9$
24. $6 + 5(x - 6)^3$, for $x = 8$
25. $x^2 - 3(x - y)$, for $x = 8$ and $y = 2$
26. $x^2 - 4(x - y)$, for $x = 8$ and $y = 3$

In Exercises 27–34, use the roster method to list the elements in each set.

27. $\{x|x \text{ is a natural number less than } 5\}$
28. $\{x|x \text{ is a natural number less than } 4\}$
29. $\{x|x \text{ is an integer between } -8 \text{ and } -3\}$
30. $\{x|x \text{ is an integer between } -7 \text{ and } -2\}$
31. $\{x|x \text{ is a natural number greater than } 7\}$
32. $\{x|x \text{ is a natural number greater than } 9\}$
33. $\{x|x \text{ is an odd whole number less than } 11\}$
34. $\{x|x \text{ is an odd whole number less than } 9\}$

In Exercises 35–48, use the meaning of the symbols $\in$ and $\notin$ to determine whether each statement is true or false.

35. $7 \in \{x|x \text{ is an integer}\}$
36. $9 \in \{x|x \text{ is an integer}\}$
37. $7 \in \{x|x \text{ is a rational number}\}$
38. $9 \in \{x|x \text{ is a rational number}\}$

39. $7 \in \{x|x \text{ is an irrational number}\}$

40. $9 \in \{x|x \text{ is an irrational number}\}$

41. $3 \notin \{x|x \text{ is an irrational number}\}$

42. $5 \notin \{x|x \text{ is an irrational number}\}$

43. $\frac{1}{2} \notin \{x|x \text{ is a rational number}\}$

44. $\frac{1}{4} \notin \{x|x \text{ is a rational number}\}$

45. $\sqrt{2} \notin \{x|x \text{ is a rational number}\}$

46. $\pi \notin \{x|x \text{ is a rational number}\}$

47. $\sqrt{2} \notin \{x|x \text{ is a real number}\}$

48. $\pi \notin \{x|x \text{ is a real number}\}$

In Exercises 49–64, write out the meaning of each inequality. Then determine whether the inequality is true or false.

49. $-6 < -2$

50. $-7 < -3$

51. $5 > -7$

52. $3 > -8$

53. $0 < -4$

54. $0 < -5$

55. $-4 \le 1$

56. $-5 \le 1$

57. $-2 \le -6$

58. $-3 \le -7$

59. $-2 \le -2$

60. $-3 \le -3$

61. $-2 \ge -2$

62. $-3 \ge -3$

63. $2 \le -\frac{1}{2}$

64. $4 \le -\frac{1}{2}$

Practice PLUS

By definition, an "and" statement is true only when the statements before and after the "and" connective are both true. Use this definition to determine whether each statement in Exercises 65–74 is true or false.

65. $0.\overline{3} > 0.3$ and $-10 < 4 + 6$.

66. $0.6 < 0.\overline{6}$ and $2 \cdot 5 \le 4 + 6$.

67. $12 \in \{1, 2, 3, \dots\}$ and $\{3\} \in \{1, 2, 3, 4\}$.

68. $17 \in \{1, 2, 3, \dots\}$ and $\{4\} \in \{1, 2, 3, 4, 5\}$.

69. $\left(\frac{2}{5} + \frac{3}{5}\right) \in \{x|x \text{ is a natural number}\}$ and the value of $9x^2(x + 11) - 9(x + 11)x^2$, for $x = 100$, is 0.

70. $\left(\frac{14}{19} + \frac{5}{19}\right) \in \{x|x \text{ is a natural number}\}$ and the value of $12x^2(x + 10) - 12(x + 10)x^2$, for $x = 50$, is 0.

71. $\{x|x \text{ is an integer between } -3 \text{ and } 0\} = \{-3, -2, -1, 0\}$ and $-\pi > -3.5$.

72. $\{x|x \text{ is an integer between } -4 \text{ and } 0\} = \{-4, -3, -2, -1, 0\}$ and $-\frac{\pi}{2} > -2.3$.

73. Twice the sum of a number and three is represented by $2x + 3$ and $-1{,}100{,}000 \in \{x|x \text{ is an integer}\}$.

74. Three times the sum of a number and five is represented by $3x + 5$ and $-4{,}500{,}000 \in \{x|x \text{ is an integer}\}$.

Application Exercises

Why do people become less neurotic as they get older? One theory is that key centers of the brain tend to create less resistance to feelings of happiness as we age. The graph shows the average resistance to happiness, on a scale of 0 (no resistance) to 8 (completely resistant), for persons at various ages.

Resistance to Happiness and Age

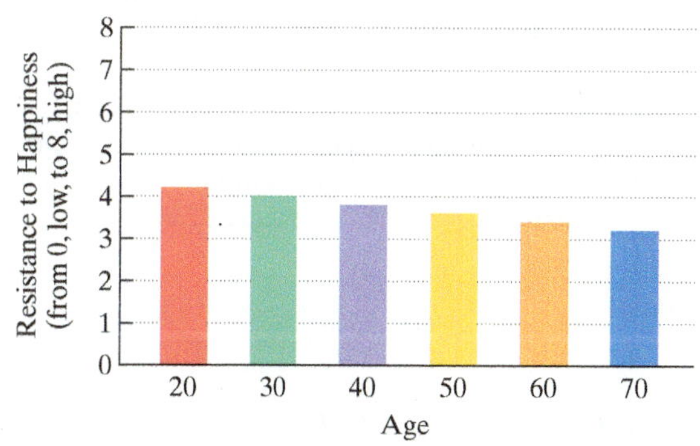

Source: L. M. Williams, *Ibid.*

The data in the graph can be modeled by the formula

$$R = 4.6 - 0.02x,$$

where R represents the average resistance to happiness, on a scale of 0 to 8, for a person who is x years old. Use this formula to solve Exercises 75–78.

75. According to the formula, what is the average resistance to happiness at age 20?

76. According to the formula, what is the average resistance to happiness at age 30?

77. What is the difference between the average resistance to happiness at age 30 and at age 50?

78. What is the difference between the average resistance to happiness at age 20 and at age 70?

In 2005, the average CEO was paid 821 times as much per hour as a full-time minimum-wage earner, who earned \$5.15 per hour. The graph shows the ratio of CEO compensation to minimum-wage salary from 2001 through 2005.

Ratio of CEO Hourly Earnings to Minimum Hourly Wage

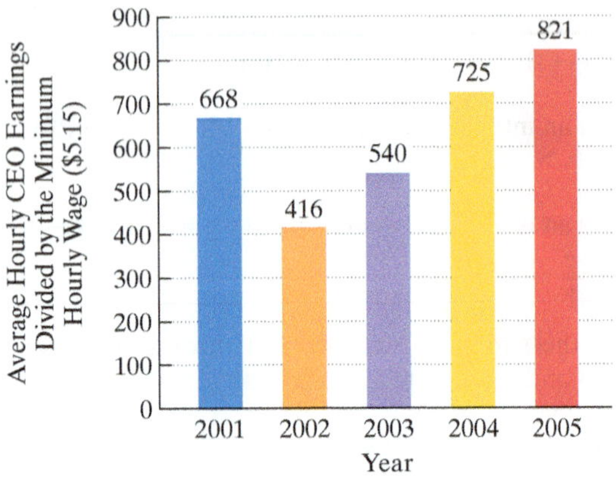

Source: Economic Policy Institute

The data in the graph can be modeled by the formula

$$R = 54x^2 - 263x + 828,$$

where R represents the ratio of CEO compensation to minimum-wage salary (\$5.15 per hour) x years after 2000. Use the formula to solve Exercises 79–80.

79. According to the formula, what was the ratio of CEO compensation to minimum-wage salary in 2005? Does the formula underestimate or overestimate the actual ratio shown by the bar graph? By how much?

80. According to the formula, what was the ratio of CEO compensation to minimum-wage salary in 2003? Does the formula underestimate or overestimate the actual ratio shown by the bar graph? By how much?

The formula

$$C = \frac{5}{9}(F - 32)$$

expresses the relationship between Fahrenheit temperature, F, and Celsius temperature, C. In Exercises 81–82, use the formula to convert the given Fahrenheit temperature to its equivalent temperature on the Celsius scale.

81. 50°F **82.** 86°F

A football is kicked vertically upward from a height of 4 feet with an initial speed of 60 feet per second. The formula

$$h = 4 + 60t - 16t^2$$

describes the ball's height above the ground, h, in feet, t seconds after it was kicked. Use this formula to solve Exercises 83–84.

83. What is the ball's height 2 seconds after it was kicked?

84. What is the ball's height 3 seconds after it was kicked?

Writing in Mathematics

Writing about mathematics will help you to learn mathematics. For all writing exercises in this book, use complete sentences to respond to the question. Some writing exercises can be answered in a sentence. Others require a paragraph or two. You can decide how much you need to write as long as your writing clearly and directly answers the question in the exercise. Standard references such as a dictionary and a thesaurus should be helpful.

85. What is a variable?

86. What is an algebraic expression? Give an example with your explanation.

87. If n is a natural number, what does b^n mean? Give an example with your explanation.

88. What does it mean when we say that a formula models real-world phenomena?

89. What is model breakdown?

90. What is a set?

91. Describe the roster method for representing a set.

92. What are the natural numbers?

93. What are the whole numbers?

94. What are the integers?

95. Describe the rational numbers.

96. Describe the difference between a rational number and an irrational number.

97. What are the real numbers?

98. What is set-builder notation?

99. Describe the meanings of the symbols $\in$ and $\notin$. Provide an example showing the correct use of each symbol.

100. What is the real number line?

101. If you are given two real numbers, explain how to determine which one is the lesser.

Critical Thinking Exercises

Make Sense? *In Exercises 102–105, determine whether each statement "makes sense" or "does not make sense" and explain your reasoning.*

102. My mathematical model describes the data for the past ten years extremely well, so it will serve as an accurate prediction for what will occur in 2050.

103. My calculator will not display the value of 13^{1500}, so the algebraic expression $4x^{1500} - 3x + 7$ cannot be evaluated for $x = 13$ even without a calculator.

104. Regardless of what real numbers I substitute for x and y, I will always obtain zero when evaluating $2x^2y - 2yx^2$.

105. A model that describes the number of lobbyists x years after 2000 cannot be used to estimate the number in 2000.

In Exercises 106–109, determine whether each statement is true or false. If the statement is false, make the necessary change(s) to produce a true statement.

106. Every rational number is an integer.

107. Some whole numbers are not integers.

108. Some rational numbers are not positive.

109. Some irrational numbers are negative.

110. A bird lover visited a pet shop where there were twice 4 and 20 parrots. The bird lover purchased $\frac{1}{7}$ of the birds. English, of course, can be ambiguous and "twice 4 and 20" can mean $2(4 + 20)$ or $2 \cdot 4 + 20$. Explain how the conditions of the situation determine if "twice 4 and 20" means $2(4 + 20)$ or $2 \cdot 4 + 20$.

In Exercises 111–112, insert parentheses to make each statement true.

111. $2 \cdot 3 + 3 \cdot 5 = 45$

112. $8 + 2 \cdot 4 - 3 = 10$

113. Between which two consecutive integers is $-\sqrt{26}$? Do not use a calculator.

Preview Exercises

Exercises 114–116 will help you prepare for the material covered in the next section.

114. There are two real numbers whose distance is five units from zero on a real number line. What are these numbers?

115. Simplify: $\frac{16 + 3(2)^4}{12 - (10 - 6)}$.

116. Evaluate $2(3x + 5)$ and $6x + 10$ for $x = 4$.

SECTION 1.2

Operations with Real Numbers and Simplifying Algebraic Expressions

Objectives

1. Find a number's absolute value.
2. Add real numbers.
3. Find opposites.
4. Subtract real numbers.
5. Multiply real numbers.
6. Evaluate exponential expressions.
7. Divide real numbers.
8. Use the order of operations.
9. Use commutative, associative, and distributive properties.
10. Simplify algebraic expressions.

From both the political left and the right, watchdogs of the federal budget are warning of fiscal trouble. David Walker, Comptroller General of the United States and the nation's top auditor, admits to being "terrified" about the budget deficit in coming decades. To hear him tell it, the United States can be likened to Rome before the fall of the empire. America's financial condition is "worse than advertised," he says. It has a "broken business model. It faces deficits in its budget, its balance of payments, and its savings." In this section, we review operations with real numbers so that you can make numerical sense of the nation's finances.

Absolute Value

Absolute value is used to describe how to operate with positive and negative numbers.

Geometric Meaning of Absolute Value

The **absolute value** of a real number a, denoted by $|a|$, is the distance from 0 to a on the number line. This distance is always taken to be nonnegative.

1 Find a number's absolute value.

EXAMPLE 1 Finding Absolute Value

Find the absolute value:

a. $|-4|$ b. $|3.5|$ c. $|0|$.

Solution The solution is illustrated in **Figure 1.8**.

a. $|-4| = 4$ The absolute value of -4 is 4 because -4 is 4 units from 0.

b. $|3.5| = 3.5$ The absolute value of 3.5 is 3.5 because 3.5 is 3.5 units from 0.

c. $|0| = 0$ The absolute value of 0 is 0 because 0 is 0 units from itself.

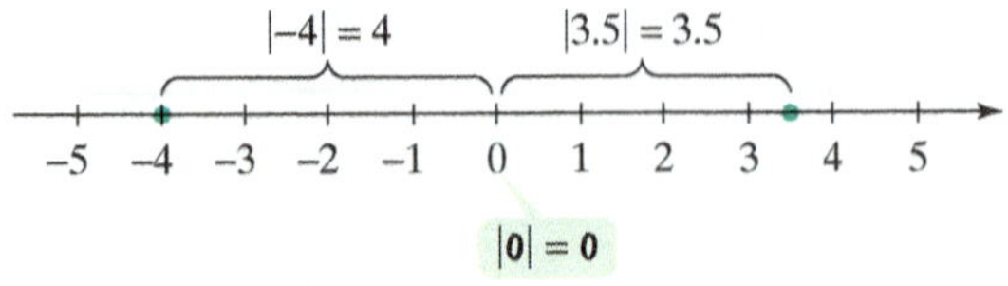

FIGURE 1.8

Can you see that the absolute value of a real number is either positive or zero? Zero is the only real number whose absolute value is 0:

$$|0| = 0.$$

The absolute value of a real number is never negative.

CHECK POINT 1 Find the absolute value:

a. $|-6|$ b. $|4.5|$ c. $|0|$.

2 Add real numbers.

Adding Real Numbers

Table 1.1 reviews how to add real numbers.

Table 1.1 Adding Real Numbers

Rule	Example
To add two real numbers with the same sign, add their absolute values. Use the common sign as the sign of the sum.	$(-7) + (-4) = -(\lvert-7\rvert + \lvert-4\rvert)$ $= -(7 + 4)$ $= -11$
To add two real numbers with different signs, subtract the smaller absolute value from the greater absolute value. Use the sign of the number with the greater absolute value as the sign of the sum.	$7 + (-15) = -(\lvert-15\rvert - \lvert7\rvert)$ $= -(15 - 7)$ $= -8$

EXAMPLE 2 Adding Real Numbers

Add: a. $-12 + (-14)$ b. $-0.3 + 0.7$ c. $-\frac{3}{4} + \frac{1}{2}$.

Solution

a. $-12 + (-14) = -26$

Add absolute values: $12 + 14 = 26$.

Use the common sign.

b. $-0.3 + 0.7 = 0.4$

Subtract absolute values: $0.7 - 0.3 = 0.4$.

Use the sign of the number with the greater absolute value. The sign of the sum is assumed to be positive.

c. $-\frac{3}{4} + \frac{1}{2} = -\frac{1}{4}$

Subtract absolute values: $\frac{3}{4} - \frac{1}{2} = \frac{3}{4} - \frac{2}{4} = \frac{1}{4}$.

Use the sign of the number with the greater absolute value.

CHECK POINT 2 Add:

a. $-10 + (-18)$ b. $-0.2 + 0.9$ c. $-\frac{3}{5} + \frac{1}{2}$.

If one of two numbers being added is zero, the sum is the other number. For example,

$$-3 + 0 = -3 \qquad \text{and} \qquad 0 + 2 = 2.$$

In general,

$$a + 0 = a \qquad \text{and} \qquad 0 + a = a.$$

We call 0 the **identity element of addition** or the **additive identity**. Thus, the additive identity can be deleted from a sum.

3 Find opposites.

Numbers with different signs but the same absolute value are called **opposites** or **additive inverses**. For example, 3 and −3 are additive inverses. When additive inverses are added, their sum is 0. For example,

$$3 + (-3) = 0 \qquad \text{and} \qquad -3 + 3 = 0.$$

Inverse Property of Addition

The sum of a real number and its additive inverse is 0, the additive identity.

$$a + (-a) = 0 \qquad \text{and} \qquad (-a) + a = 0$$

The symbol "−" is used to name the opposite, or additive inverse, of a. When a is a negative number, $-a$, its opposite, is positive. For example, if a is −4, its opposite is 4. Thus,

$$-(-4) = 4.$$

The opposite of −4 is 4.

In general, if a is any real number,

$$-(-a) = a.$$

EXAMPLE 3 Finding Opposites

Find $-x$ if **a.** $x = -6$ **b.** $x = \frac{1}{2}$.

Solution

a. If $x = -6$, then $-x = -(-6) = 6$. The opposite of −6 is 6.

b. If $x = \frac{1}{2}$, then $-x = -\frac{1}{2}$. The opposite of $\frac{1}{2}$ is $-\frac{1}{2}$.

✓ CHECK POINT 3 Find $-x$ if

a. $x = -8$ **b.** $x = \frac{1}{3}$.

We can define the absolute value of the real number a using opposites, without referring to a number line. The algebraic definition of the absolute value of a is given as follows:

Definition of Absolute Value

$$|a| = \begin{cases} a & \text{if } a \geq 0 \\ -a & \text{if } a < 0 \end{cases}$$

If a is nonnegative (that is, $a \geq 0$), the absolute value of a is the number itself: $|a| = a$. For example,

$$|5| = 5 \qquad |\pi| = \pi \qquad \left|\frac{1}{3}\right| = \frac{1}{3} \qquad |0| = 0$$

Zero is the only number whose absolute value is 0.

If a is a negative number (that is, $a < 0$), the absolute value of a is the opposite of a: $|a| = -a$. This makes the absolute value positive. For example,

$$|-3| = -(-3) = 3 \qquad |-\pi| = -(-\pi) = \pi \qquad \left|-\frac{1}{3}\right| = -\left(-\frac{1}{3}\right) = \frac{1}{3}.$$

This middle step is usually omitted.

4 Subtract real numbers.

Subtracting Real Numbers

Subtraction of real numbers is defined in terms of addition.

Definition of Subtraction

If a and b are real numbers,

$$a - b = a + (-b).$$

To subtract a real number, add its opposite or additive inverse.

Thus, to subtract real numbers,

1. Change the subtraction to addition.
2. Change the sign of the number being subtracted.
3. Add, using one of the rules for adding numbers with the same sign or different signs.

EXAMPLE 4 Subtracting Real Numbers

Subtract: **a.** $6 - 13$ **b.** $5.1 - (-4.2)$ **c.** $-\frac{11}{3} - \left(-\frac{4}{3}\right)$.

Solution

a. $6 - 13 = 6 + (-13) = -7$

Change the subtraction to addition. Replace 13 with its opposite.

b. $5.1 - (-4.2) = 5.1 + 4.2 = 9.3$

Change the subtraction to addition. Replace −4.2 with its opposite.

c. $-\frac{11}{3} - \left(-\frac{4}{3}\right) = -\frac{11}{3} + \frac{4}{3} = -\frac{7}{3}$

Change the subtraction to addition. Replace $-\frac{4}{3}$ with its opposite.

CHECK POINT 4 Subtract:

a. $7 - 10$ **b.** $4.3 - (-6.2)$ **c.** $-\frac{4}{5} - \left(-\frac{1}{5}\right)$.

5 Multiply real numbers.

Multiplying Real Numbers

You can think of multiplication as repeated addition or subtraction that starts at 0. For example,

$$3(-4) = 0 + (-4) + (-4) + (-4) = -12$$

The numbers have different signs and the product is negative.

and

$$(-3)(-4) = 0 - (-4) - (-4) - (-4) = 0 + 4 + 4 + 4 = 12$$

The numbers have the same sign and the product is positive.

Table 1.2 reviews how to multiply real numbers.

Table 1.2 Multiplying Real Numbers

Rule	Example
The product of two real numbers with different signs is found by multiplying their absolute values. The product is negative.	$7(-5) = -35$
The product of two real numbers with the same sign is found by multiplying their absolute values. The product is positive.	$(-6)(-11) = 66$
The product of 0 and any real number is 0: $a \cdot 0 = 0$ and $0 \cdot a = 0$.	$-17(0) = 0$
If no number is 0, a product with an odd number of negative factors is found by multiplying absolute values. The product is negative.	$-2(-3)(-5) = -30$ Three (odd) negative factors
If no number is 0, a product with an even number of negative factors is found by multiplying absolute values. The product is positive.	$-2(3)(-5) = 30$ Two (even) negative factors

6 Evaluate exponential expressions.

Because exponents indicate repeated multiplication, rules for multiplying real numbers can be used to evaluate exponential expressions.

EXAMPLE 5 Evaluating Exponential Expressions

Evaluate: **a.** $(-6)^2$ **b.** -6^2 **c.** $(-5)^3$ **d.** $\left(-\frac{2}{3}\right)^4$.

Solution

a. $(-6)^2 = (-6)(-6) = 36$

Base is −6. Same signs give positive product.

b. $-6^2 = -(6 \cdot 6) = -36$

Base is 6. The negative is not inside parentheses and is not taken to the second power.

c. $(-5)^3 = (-5)(-5)(-5) = -125$

An odd number of negative factors gives a negative product.

d. $\left(-\frac{2}{3}\right)^4 = \left(-\frac{2}{3}\right)\left(-\frac{2}{3}\right)\left(-\frac{2}{3}\right)\left(-\frac{2}{3}\right) = \frac{16}{81}$

Base is $-\frac{2}{3}$.

An even number of negative factors gives a positive product.

CHECK POINT 5 Evaluate:

a. $(-5)^2$ **b.** -5^2 **c.** $(-4)^3$ **d.** $\left(-\frac{3}{5}\right)^4$.

7 Divide real numbers.

Dividing Real Numbers

If a and b are real numbers and b is not 0, then the quotient of a and b is defined as follows:

$$a \div b = a \cdot \frac{1}{b} \quad \text{or} \quad \frac{a}{b} = a \cdot \frac{1}{b}.$$

Thus, to find the quotient of a and b, we can divide by b or multiply by $\frac{1}{b}$. The nonzero real numbers b and $\frac{1}{b}$ are called **reciprocals**, or **multiplicative inverses**, of one another. When reciprocals are multiplied, their product is 1:

$$b \cdot \frac{1}{b} = 1.$$

Because division is defined in terms of multiplication, the sign rules for dividing numbers are the same as the sign rules for multiplying them.

Dividing Real Numbers

The quotient of two numbers with different signs is negative. The quotient of two numbers with the same sign is positive. The quotient is found by dividing absolute values.

EXAMPLE 6 Dividing Real Numbers

Divide: **a.** $\frac{20}{-5}$ **b.** $-\frac{3}{4} \div \left(-\frac{5}{9}\right)$.

Solution

a. $\frac{20}{-5} = -4$

Divide absolute values: $\frac{20}{5} = 4$.

Different signs: negative quotient

b. $-\frac{3}{4} \div \left(-\frac{5}{9}\right) = \frac{27}{20}$

Divide absolute values: $\frac{3}{4} \div \frac{5}{9} = \frac{3}{4} \cdot \frac{9}{5} = \frac{27}{20}$.

Same sign: positive quotient

CHECK POINT 6 Divide:

a. $\frac{32}{-4}$ **b.** $-\frac{2}{3} \div \left(-\frac{5}{4}\right)$.

We must be careful with division when 0 is involved. Zero divided by any nonzero real number is 0. For example,

$$\frac{0}{-5} = 0.$$

Can you see why $\frac{0}{-5}$ must be 0? The definition of division tells us that

$$\frac{0}{-5} = 0 \cdot \left(-\frac{1}{5}\right)$$

and the product of 0 and any real number is 0. By contrast, what happens if we divide -5 by 0. The answer must be a number that, when multiplied by 0, gives -5. However, any number multiplied by 0 is 0. Thus, we cannot divide -5, or any other real number, by 0.

Division by Zero

Division by zero is not allowed; it is undefined. A real number can never have a denominator of 0.

8 Use the order of operations.

Order of Operations

The rules for order of operations can be applied to positive and negative real numbers. Recall that if no grouping symbols are present, we

- Evaluate exponential expressions.
- Multiply and divide, from left to right.
- Add and subtract, from left to right.

EXAMPLE 7 Using the Order of Operations

Simplify: $4 - 7^2 + 8 \div 2(-3)^2$.

Solution

$4 - 7^2 + 8 \div 2(-3)^2$

$= 4 - 49 + 8 \div 2(9)$ Evaluate exponential expressions: $7^2 = 7 \cdot 7 = 49$ and $(-3)^2 = (-3)(-3) = 9$.

$= 4 - 49 + 4(9)$ Divide: $8 \div 2 = 4$.

$= 4 - 49 + 36$ Multiply: $4(9) = 36$.

$= -45 + 36$ Subtract: $4 - 49 = 4 + (-49) = -45$.

$= -9$ Add.

✓ CHECK POINT 7 Simplify: $3 - 5^2 + 12 \div 2(-4)^2$.

If an expression contains grouping symbols, we perform operations within these symbols first. Common grouping symbols are parentheses, brackets, and braces. Other grouping symbols include fraction bars, absolute value symbols, and radical symbols, such as square root signs ($\sqrt{}$).

EXAMPLE 8 Using the Order of Operations

Simplify: $\frac{13 - 3(-2)^4}{3 - (6 - 10)}$.

Solution Simplify the numerator and the denominator separately. Then divide.

$$\frac{13 - 3(-2)^4}{3 - (6 - 10)}$$

$$= \frac{13 - 3(16)}{3 - (-4)}$$ Evaluate the exponential expression in the numerator: $(-2)^4 = (-2)(-2)(-2)(-2) = 16$. Subtract inside parentheses in the denominator: $6 - 10 = 6 + (-10) = -4$.

$$= \frac{13 - 48}{7}$$ Multiply in the numerator: $3(16) = 48$. Subtract in the denominator: $3 - (-4) = 3 + 4 = 7$.

$$= \frac{-35}{7}$$ Subtract in the numerator: $13 - 48 = 13 + (-48) = -35$.

$$= -5$$ Divide.

✓ **CHECK POINT 8** Simplify: $\frac{4 + 3(-2)^3}{2 - (6 - 9)}$.

9 Use commutative, associative, and distributive properties.

The Commutative, Associative, and Distributive Properties

Basic algebraic properties enable us to write *equivalent algebraic expressions.* Two algebraic expressions that have the same value for all replacements are called **equivalent algebraic expressions**. In Section 1.4, you will use such expressions to solve equations.

In arithmetic, when two numbers are added or multiplied, the order in which the numbers are written does not affect the answer. These facts are called **commutative properties**.

The Commutative Properties

Let a and b represent real numbers, variables, or algebraic expressions.

Addition: $a + b = b + a$

Multiplication: $ab = ba$

Changing order when adding or multiplying does not affect a sum or product.

EXAMPLE 9 Using the Commutative Properties

Write an algebraic expression equivalent to $3x + 7$ using each of the commutative properties.

Solution

Commutative of Addition

$3x + 7 = 7 + 3x$

Change the order of the addition.

Commutative of Multiplication

$3x + 7 = x \cdot 3 + 7$

Change the order of the multiplication.

✓ **CHECK POINT 9** Write an algebraic expression equivalent to $4x + 9$ using each of the commutative properties.

The **associative properties** enable us to form equivalent expressions by regrouping.

The Associative Properties

Let a, b, and c represent real numbers, variables, or algebraic expressions.

Addition: $(a + b) + c = a + (b + c)$

Multiplication: $(ab)c = a(bc)$

Changing grouping when adding or multiplying does not affect a sum or product.

Blitzer Bonus

The Associative Property and the English Language

In the English language, phrases can take on different meanings depending on the way the words are associated with commas. Here are three examples.

- Woman, without her man, is nothing.
  Woman, without her, man is nothing.
- What's the latest dope?
  What's the latest, dope?
- Population of Amsterdam broken down by age and sex
  Population of Amsterdam, broken down by age and sex

EXAMPLE 10 Using the Associative Properties

Use an associative property to write an equivalent expression and simplify:

a. $7 + (3 + x)$ **b.** $-6(5x)$.

Solution

a. $7 + (3 + x) = (7 + 3) + x = 10 + x$

b. $-6(5x) = (-6 \cdot 5)x = -30x$

CHECK POINT 10 Use an associative property to write an equivalent expression and simplify:

a. $6 + (12 + x)$ **b.** $-7(4x)$.

The **distributive property** allows us to rewrite the product of a number and a sum as the sum of two products.

The Distributive Property

Let a, b, and c represent real numbers, variables, or algebraic expressions.

$$a(b + c) = ab + ac$$

Multiplication distributes over addition.

EXAMPLE 11 Using the Distributive Property

Use the distributive property to write an equivalent expression:

$$-2(3x + 5).$$

Solution

$$-2(3x + 5) = -2 \cdot 3x + (-2) \cdot 5 = -6x + (-10) = -6x - 10$$

CHECK POINT 11 Use the distributive property to write an equivalent expression: $-4(7x + 2)$.

Study Tip

When using a distributive property to remove parentheses, be sure to multiply *each term* inside the parentheses by the factor outside.

Incorrect

$5(3y + 7) = 5 \cdot 3y + 7$
$= 15y + 7$

Table 1.3 shows a number of other forms of the distributive property.

Table 1.3 Other Forms of the Distributive Property

Property	Meaning	Example
$a(b - c) = ab - ac$	Multiplication distributes over subtraction.	$6(4x - 5) = 6 \cdot 4x - 6 \cdot 5$ $= 24x - 30$
$a(b + c + d) = ab + ac + ad$	Multiplication distributes over three or more terms in parentheses.	$5(x + 4 + 7y)$ $= 5x + 5 \cdot 4 + 5 \cdot 7y$ $= 5x + 20 + 35y$
$(b + c)a = ba + ca$	Multiplication on the right distributes over addition (or subtraction).	$(x + 10)8 = x \cdot 8 + 10 \cdot 8$ $= 8x + 80$

10 Simplify algebraic expressions.

Combining Like Terms and Simplifying Algebraic Expressions

The **terms** of an algebraic expression are those parts that are separated by addition. For example, consider the algebraic expression

$$7x - 9y + z - 3,$$

which can be expressed as

$$7x + (-9y) + z + (-3).$$

This expression contains four terms, namely $7x$, $-9y$, z, and -3.

The numerical part of a term is called its **coefficient**. In the term $7x$, the 7 is the coefficient. If a term containing one or more variables is written without a coefficient, the coefficient is understood to be 1. Thus, z means $1z$. If a term is a constant, its coefficient is that constant. Thus, the coefficient of the constant term -3 is -3.

$$7x + (-9y) + z + (-3)$$

Coefficient is 7. | Coefficient is −9. | Coefficient is 1; z means $1z$. | Coefficient is −3.

The parts of each term that are multiplied are called the **factors** of the term. The factors of the term $7x$ are 7 and x.

Like terms are terms that have exactly the same variable factors. For example, $3x$ and $7x$ are like terms. The distributive property in the form

$$ba + ca = (b + c)a$$

enables us to add or subtract like terms. For example,

$$3x + 7x = (3 + 7)x = 10x$$
$$7y^2 - y^2 = 7y^2 - 1y^2 = (7 - 1)y^2 = 6y^2.$$

This process is called **combining like terms**.

An algebraic expression is **simplified** when grouping symbols have been removed and like terms have been combined.

Study Tip

To combine like terms mentally, add or subtract the coefficients of the terms. Use this result as the coefficient of the terms' variable factor(s).

EXAMPLE 12 Simplifying an Algebraic Expression

Simplify: $7x + 12x^2 + 3x + x^2$.

Solution

$$7x + 12x^2 + 3x + x^2$$

$= (7x + 3x) + (12x^2 + x^2)$ Rearrange terms and group like terms using commutative and associative properties. This step is often done mentally.

$x^2 = 1x^2$

$= (7 + 3)x + (12 + 1)x^2$ Apply the distributive property.

$= 10x + 13x^2$ Simplify. Because $10x$ and $13x^2$ are not like terms, this is the final answer.

Using the commutative property of addition, we can write this simplified expression as $13x^2 + 10x$.

✓ **CHECK POINT 12** Simplify: $3x + 14x^2 + 11x + x^2$.

EXAMPLE 13 Simplifying an Algebraic Expression

Simplify: $4(7x - 3) - 10x$.

Solution

$$4(7x - 3) - 10x$$

$= 4 \cdot 7x - 4 \cdot 3 - 10x$ Use the distributive property to remove the parentheses.

$= 28x - 12 - 10x$ Multiply.

$= (28x - 10x) - 12$ Group like terms.

$= (28 - 10)x - 12$ Apply the distributive property.

$= 18x - 12$ Simplify.

CHECK POINT 13 Simplify: $8(2x - 5) - 4x$.

It is not uncommon to see algebraic expressions with parentheses preceded by a negative sign or subtraction. An expression of the form $-(b + c)$ can be simplified as follows:

$$-(b + c) = -1(b + c) = (-1)b + (-1)c = -b + (-c) = -b - c.$$

Do you see a fast way to obtain the simplified expression on the right? **If a negative sign or a subtraction symbol appears outside parentheses, drop the parentheses and change the sign of every term within the parentheses.** For example,

$$-(3x^2 - 7x - 4) = -3x^2 + 7x + 4.$$

EXAMPLE 14 Simplifying an Algebraic Expression

Simplify: $8x + 2[5 - (x - 3)]$.

Solution

$$8x + 2[5 - (x - 3)]$$

$= 8x + 2[5 - x + 3]$ Drop parentheses and change the sign of each term in parentheses: $-(x - 3) = -x + 3$.

$= 8x + 2[8 - x]$ Simplify inside brackets: $5 + 3 = 8$.

$= 8x + 16 - 2x$ Apply the distributive property: $2[8 - x] = 2 \cdot 8 - 2x = 16 - 2x$.

$= (8x - 2x) + 16$ Group like terms.

$= (8 - 2)x + 16$ Apply the distributive property.

$= 6x + 16$ Simplify.

CHECK POINT 14 Simplify: $6 + 4[7 - (x - 2)]$.

1.2 EXERCISE SET

MyMathLab MathXL PRACTICE WATCH DOWNLOAD READ REVIEW

Practice Exercises

In Exercises 1–12, find each absolute value.

1. $|-7|$
2. $|-10|$
3. $|4|$
4. $|13|$
5. $|-7.6|$
6. $|-8.3|$
7. $\left|\frac{\pi}{2}\right|$
8. $\left|\frac{\pi}{3}\right|$
9. $|-\sqrt{2}|$
10. $|-\sqrt{3}|$
11. $-\left|-\frac{2}{5}\right|$
12. $-\left|-\frac{7}{10}\right|$

In Exercises 13–28, add as indicated.

13. $-3 + (-8)$
14. $-5 + (-10)$
15. $-14 + 10$
16. $-15 + 6$
17. $-6.8 + 2.3$
18. $-7.9 + 2.4$
19. $\frac{11}{15} + \left(-\frac{3}{5}\right)$
20. $\frac{7}{10} + \left(-\frac{4}{5}\right)$
21. $-\frac{2}{9} - \frac{3}{4}$
22. $-\frac{3}{5} - \frac{4}{7}$
23. $-3.7 + (-4.5)$
24. $-6.2 + (-5.9)$
25. $0 + (-12.4)$
26. $0 + (-15.3)$
27. $12.4 + (-12.4)$
28. $15.3 + (-15.3)$

In Exercises 29–34, find $-x$ for the given value of x.

29. $x = 11$
30. $x = 13$
31. $x = -5$
32. $x = -9$
33. $x = 0$
34. $x = -\sqrt{2}$

In Exercises 35–46, subtract as indicated.

35. $3 - 15$
36. $4 - 20$
37. $8 - (-10)$
38. $7 - (-13)$
39. $-20 - (-5)$
40. $-30 - (-10)$
41. $\frac{1}{4} - \frac{1}{2}$
42. $\frac{1}{10} - \frac{2}{5}$
43. $-2.3 - (-7.8)$
44. $-4.3 - (-8.7)$
45. $0 - (-\sqrt{2})$
46. $0 - (-\sqrt{3})$

In Exercises 47–58, multiply as indicated.

47. $9(-10)$
48. $8(-10)$
49. $(-3)(-11)$
50. $(-7)(-11)$
51. $\frac{15}{13}(-1)$
52. $\frac{11}{13}(-1)$
53. $-\sqrt{2}\cdot 0$
54. $-\sqrt{3}\cdot 0$
55. $(-4)(-2)(-1)$
56. $(-5)(-3)(-2)$
57. $2(-3)(-1)(-2)(-4)$
58. $3(-2)(-1)(-5)(-3)$

In Exercises 59–70, evaluate each exponential expression.

59. $(-10)^2$
60. $(-8)^2$
61. -10^2
62. -8^2
63. $(-2)^3$
64. $(-3)^3$
65. $(-1)^4$
66. $(-4)^4$
67. $(-1)^{33}$
68. $(-1)^{35}$
69. $-\left(-\frac{1}{2}\right)^3$
70. $-\left(-\frac{1}{4}\right)^3$

In Exercises 71–82, divide as indicated or state that the division is undefined.

71. $\frac{12}{-4}$
72. $\frac{30}{-5}$
73. $\frac{-90}{-2}$
74. $\frac{-55}{-5}$
75. $\frac{0}{-4.6}$
76. $\frac{0}{-5.3}$
77. $-\frac{4.6}{0}$
78. $-\frac{5.3}{0}$
79. $-\frac{1}{2} \div \left(-\frac{7}{9}\right)$
80. $-\frac{1}{2} \div \left(-\frac{3}{5}\right)$
81. $6 \div \left(-\frac{2}{5}\right)$
82. $8 \div \left(-\frac{2}{9}\right)$

In Exercises 83–100, use the order of operations to simplify each expression.

83. $4(-5) - 6(-3)$
84. $8(-3) - 5(-6)$
85. $3(-2)^2 - 4(-3)^2$
86. $5(-3)^2 - 2(-2)^2$
87. $8^2 - 16 \div 2^2 \cdot 4 - 3$
88. $10^2 - 100 \div 5^2 \cdot 2 - 3$
89. $\frac{5 \cdot 2 - 3^2}{[3^2 - (-2)]^2}$
90. $\frac{10 \div 2 + 3 \cdot 4}{(12 - 3 \cdot 2)^2}$
91. $8 - 3[-2(2 - 5) - 4(8 - 6)]$
92. $8 - 3[-2(5 - 7) - 5(4 - 2)]$
93. $\frac{2(-2) - 4(-3)}{5 - 8}$
94. $\frac{6(-4) - 5(-3)}{9 - 10}$
95. $\frac{(5 - 6)^2 - 2|3 - 7|}{89 - 3 \cdot 5^2}$
96. $\frac{12 \div 3 \cdot 5|2^2 + 3^2|}{7 + 3 - 6^2}$
97. $15 - \sqrt{3 - (-1)} + 12 \div 2 \cdot 3$
98. $17 - |5 - (-2)| + 12 \div 2 \cdot 3$
99. $20 + 1 - \sqrt{10^2 - (5 + 1)^2}\,(-2)$
100. $24 \div \sqrt{3 \cdot (5 - 2)} \div [-1 - (-3)]^2$

In Exercises 101–104, write an algebraic expression equivalent to the given expression using each of the commutative properties.

101. $4x + 10$
102. $5x + 30$
103. $7x - 5$
104. $3x - 7$

In Exercises 105–110, use an associative property to write an algebraic expression equivalent to each expression and simplify.

105. $4 + (6 + x)$
106. $12 + (3 + x)$
107. $-7(3x)$
108. $-10(5x)$
109. $-\frac{1}{3}(-3y)$
110. $-\frac{1}{4}(-4y)$

In Exercises 111–116, use the distributive property to write an equivalent expression.

111. $3(2x + 5)$
112. $5(4x + 7)$
113. $-7(2x + 3)$
114. $-9(3x + 2)$
115. $-(3x - 6)$
116. $-(6x - 3)$

In Exercises 117–130, simplify each algebraic expression.

117. $7x + 5x$
118. $8x + 10x$
119. $6x^2 - x^2$
120. $9x^2 - x^2$
121. $6x + 10x^2 + 4x + 2x^2$
122. $9x + 5x^2 + 3x + 4x^2$
123. $8(3x - 5) - 6x$
124. $7(4x - 5) - 8x$
125. $5(3y - 2) - (7y + 2)$
126. $4(5y - 3) - (6y + 3)$
127. $7 - 4[3 - (4y - 5)]$
128. $6 - 5[8 - (2y - 4)]$
129. $18x^2 + 4 - [6(x^2 - 2) + 5]$
130. $14x^2 + 5 - [7(x^2 - 2) + 4]$

Practice PLUS

In Exercises 131–138, write each English phrase as an algebraic expression. Then simplify the expression. Let x represent the number.

131. A number decreased by the sum of the number and four
132. A number decreased by the difference between eight and the number
133. Six times the product of negative five and a number
134. Ten times the product of negative four and a number
135. The difference between the product of five times a number and twice the number
136. The difference between the product of six and a number and negative two times the number
137. The difference between eight times a number and six more than three times the number
138. Eight decreased by three times the sum of a number and six

Application Exercises

The bar graph shows the U.S. trade balance in goods and services, in billions of dollars, from 2000 through 2005. The most complete scorecard of the U.S. international trade performance deteriorated to a record $805 billion deficit in 2005. Use the information shown by the graph to solve Exercises 139–142. Express answers in billions of dollars.

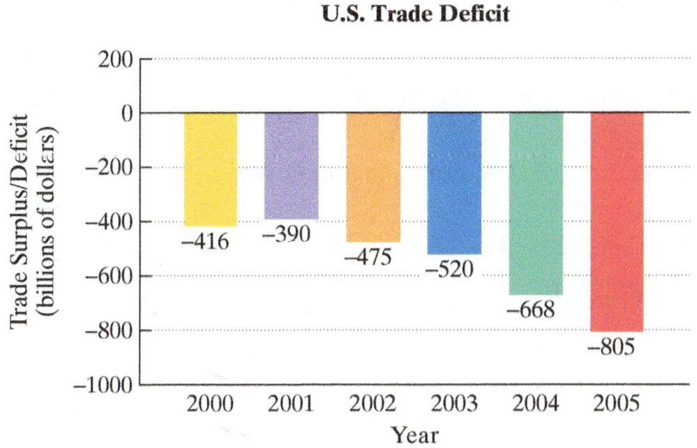

Source: Bureau of Economic Analysis

139. Find the difference between the 2000 trade deficit and the 2005 trade deficit.
140. Find the difference between the 2001 trade deficit and the 2005 trade deficit.
141. By how much did the 2005 deficit exceed twice the 2001 deficit?
142. Find the average trade deficit for 2003 and 2004 combined. By how much did the 2005 deficit exceed this average?

The bar graph shows the amount of money, in billions of dollars, collected and spent by the U.S. government from 2001 through 2006. Use the information from the graph to solve Exercises 143–146. Express answers in billions of dollars.

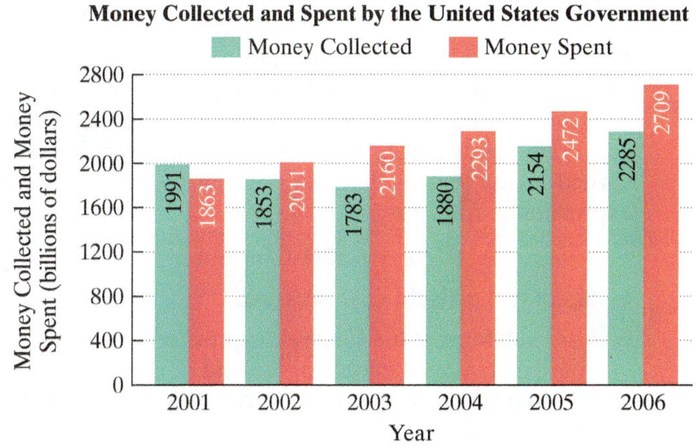

Source: Budget of the U.S. Government

143. In 2003, what was the difference between the amount of money collected and the amount spent? Was there a budget surplus or deficit in 2003?
144. In 2004, what was the difference between the amount of money collected and the amount spent? Was there a budget surplus or deficit in 2004?
145. What is the difference between the 2001 surplus and the 2006 deficit?
146. What is the difference between the 2001 surplus and the 2005 deficit?

The bar graph shows the percentage of U.S. adults who have been tested for the HIV virus, by age.

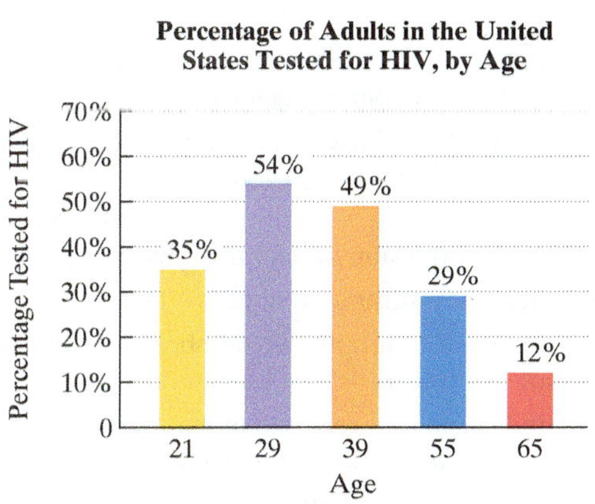

Source: National Center for Health Statistics

The data in the graph at the bottom of the previous page can be modeled by the formula

$$P = -0.05x^2 + 3.6x - 15,$$

where P represents the percentage of U.S. adults tested for HIV at age x. Use this formula to solve Exercises 147–148.

147. According to the formula, what percentage of U.S. adults who are 21 years old have been tested for HIV? Does the model underestimate or overestimate the percent displayed by the bar graph? By how much?

148. According to the formula, what percentage of U.S. adults who are 39 years old have been tested for HIV? Does the model underestimate or overestimate the percent displayed by the bar graph? By how much?

149. You had \$10,000 to invest. You put x dollars in a safe, government-insured certificate of deposit paying 5% per year. You invested the remainder of the money in noninsured corporate bonds paying 12% per year. Your total interest earned at the end of the year is given by the algebraic expression

$$0.05x + 0.12(10{,}000 - x).$$

a. Simplify the algebraic expression.

b. Use each form of the algebraic expression to determine your total interest earned at the end of the year if you invested \$6000 in the safe, government-insured certificate of deposit.

150. It takes you 50 minutes to get to campus. You spend t minutes walking to the bus stop and the rest of the time riding the bus. Your walking rate is 0.06 mile per minute and the bus travels at a rate of 0.5 mile per minute. The total distance walking and traveling by bus is given by the algebraic expression

$$0.06t + 0.5(50 - t).$$

a. Simplify the algebraic expression.

b. Use each form of the algebraic expression to determine the total distance that you travel if you spend 20 minutes walking to the bus stop.

Writing in Mathematics

151. What is the meaning of $|a|$ in terms of a number line?

152. Explain how to add two numbers with the same sign. Give an example with your explanation.

153. Explain how to add two numbers with different signs. Give an example with your explanation.

154. What are opposites, or additive inverses? What happens when finding the sum of a number and its opposite?

155. Explain how to subtract real numbers.

156. Explain how to multiply two numbers with different signs. Give an example with your explanation.

157. Explain how to multiply two numbers with the same sign. Give an example with your explanation.

158. Explain how to determine the sign of a product that involves more than two numbers.

159. Explain how to divide real numbers.

160. Why is $\frac{0}{4} = 0$, although $\frac{4}{0}$ is undefined?

161. What are equivalent algebraic expressions?

162. State a commutative property and give an example of how it is used to write equivalent algebraic expressions.

163. State an associative property and give an example of how it is used to write equivalent algebraic expressions.

164. State a distributive property and give an example of how it is used to write equivalent algebraic expressions.

165. What are the terms of an algebraic expression? How can you tell if terms are like terms?

166. What does it mean to simplify an algebraic expression?

167. If a negative sign appears outside parentheses, explain how to simplify the expression. Give an example.

168. What explanations can you offer for the trend in the percentage of U.S. adults tested for HIV, by age, shown in the bar graph in Exercises 147–148?

Critical Thinking Exercises

Make Sense? *In Exercises 169–172, determine whether each statement "makes sense" or "does not make sense" and explain your reasoning.*

169. My mathematical model, although it contains an algebraic expression that is not simplified, describes the data perfectly well, so it will describe the data equally well when simplified.

170. Subtraction actually means the addition of an additive inverse.

171. The terms $13x^2$ and $10x$ both contain the variable x, so I can combine them to obtain $23x^3$.

172. There is no number in front of the term x, so this means that the term has no coefficient.

In Exercises 173–177, determine whether each statement is true or false. If the statement is false, make the necessary change(s) to produce a true statement.

173. $16 \div 4 \cdot 2 = 16 \div 8 = 2$

174. $6 - 2(4 + 3) = 4(4 + 3) = 4(7) = 28$

175. $5 + 3(x - 4) = 8(x - 4) = 8x - 32$

176. $-x - x = -x + (-x) = 0$

177. $x - 0.02(x + 200) = 0.98x - 4$

In Exercises 178–179, insert parentheses to make each statement true.

178. $8 - 2 \cdot 3 - 4 = 14$

179. $2 \cdot 5 - \frac{1}{2} \cdot 10 \cdot 9 = 45$

180. Simplify: $\dfrac{9[4 - (1 + 6)] - (3 - 9)^2}{5 + \dfrac{12}{5 - \dfrac{6}{2 + 1}}}$.

Review Exercises

From here on, each exercise set will contain three review exercises. It is important to review previously covered topics to improve your understanding of the topics and to help maintain your mastery of the material. If you are not certain how to solve a review exercise, turn to the section and the worked-out example given in parentheses at the end of each exercise.

181. Write the following English phrase as an algebraic expression: "The quotient of ten and a number, decreased by four times the number." Let x represent the number. (Section 1.1, Example 1)

182. Evaluate $10 + 2(x - 5)^4$ for $x = 7$. (Section 1.1, Example 3)

183. Determine whether the following statement is true or false: $\frac{1}{2} \notin \{x \mid x \text{ is an irrational number}\}$. (Section 1.1, Example 5)

Preview Exercises

Exercises 184–186 will help you prepare for the material covered in the next section.

184. If $y = 4 - x^2$, find the value of y that corresponds to values of x for each integer starting with -3 and ending with 3.

185. If $y = 1 - x^2$, find the value of y that corresponds to values of x for each integer starting with -3 and ending with 3.

186. If $y = |x + 1|$, find the value of y that corresponds to values of x for each integer starting with -4 and ending with 2.

SECTION 1.3

Graphing Equations

Objectives

1. Plot points in the rectangular coordinate system.
2. Graph equations in the rectangular coordinate system.
3. Use the rectangular coordinate system to visualize relationships between variables.
4. Interpret information about a graphing utility's viewing rectangle or table.

The beginning of the seventeenth century was a time of innovative ideas and enormous intellectual progress in Europe. English theatergoers enjoyed a succession of exciting new plays by Shakespeare. William Harvey proposed the radical notion that the heart was a pump for blood rather than the center of emotion. Galileo, with his new-fangled invention called the telescope, supported the theory of Polish astronomer Copernicus that the sun, not the Earth, was the center of the solar system. Monteverdi was writing the world's first grand operas. French mathematicians Pascal and Fermat invented a new field of mathematics called probability theory.

Into this arena of intellectual electricity stepped French aristocrat René Descartes (1596–1650). Descartes (pronounced "day cart"), propelled by the creativity surrounding him, developed a new branch of mathematics that brought together algebra and geometry in a unified way—a way that visualized numbers as points on a graph, equations as geometric figures, and geometric figures as equations. This new branch of mathematics, called *analytic geometry*, established Descartes as one of the founders of modern thought and among the most original mathematicians and philosophers of any age. We begin this section by looking at Descartes's deceptively simple idea, called the **rectangular coordinate system** or (in his honor) the **Cartesian coordinate system**.

1 Plot points in the rectangular coordinate system.

Points and Ordered Pairs

Descartes used two number lines that intersect at right angles at their zero points, as shown in **Figure 1.9**. The horizontal number line is the ***x*-axis**. The vertical number line is the ***y*-axis**. The point of intersection of these axes is their zero points, called the **origin**. Positive numbers are shown to the right and above the origin. Negative numbers are shown to the left and below the origin. The axes divide the plane into four quarters, called **quadrants**. The points located on the axes are not in any quadrant.

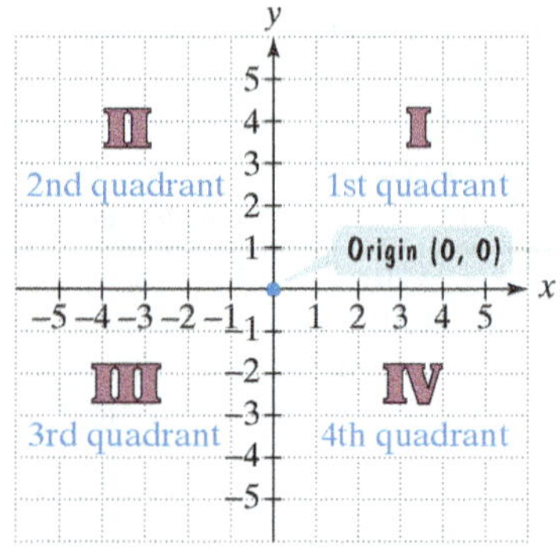

FIGURE 1.9 The rectangular coordinate system

Each point in the rectangular coordinate system corresponds to an **ordered pair** of real numbers, (x, y). Examples of such pairs are $(-5, 3)$ and $(3, -5)$. The first number in each pair, called the ***x*-coordinate**, denotes the distance and direction from the origin along the x-axis. The second number, called the ***y*-coordinate**, denotes vertical distance and direction along a line parallel to the y-axis or along the y-axis itself.

Figure 1.10 shows how we **plot**, or locate, the points corresponding to the ordered pairs $(-5, 3)$ and $(3, -5)$. We plot $(-5, 3)$ by going 5 units from 0 to the left along the x-axis. Then we go 3 units up parallel to the y-axis. We plot $(3, -5)$ by going 3 units from 0 to the right along the x-axis and 5 units down parallel to the y-axis. The phrase "the points corresponding to the ordered pairs $(-5, 3)$ and $(3, -5)$" are often abbreviated as "the points $(-5, 3)$ and $(3, -5)$."

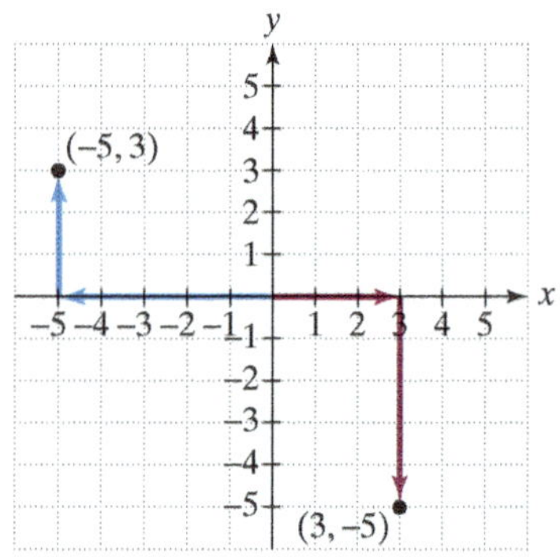

FIGURE 1.10 Plotting $(-5, 3)$ and $(3, -5)$

Study Tip

The phrase *ordered pair* is used because order is important. The order in which coordinates appear makes a difference in a point's location. This is illustrated in **Figure 1.10**.

EXAMPLE 1 Plotting Points in the Rectangular Coordinate System

Plot the points: $A(-4, 5)$, $B(3, -4)$, $C(-5, 0)$, $D(-4, -2)$, $E(0, 3.5)$, and $F(0, 0)$.

Solution See **Figure 1.11**. We move from the origin and plot the points in the following way:

$A(-4, 5)$: 4 units left, 5 units up

$B(3, -4)$: 3 units right, 4 units down

$C(-5, 0)$: 5 units left, 0 units up or down

$D(-4, -2)$: 4 units left, 2 units down

$E(0, 3.5)$: 0 units right or left, 3.5 units up

$F(0, 0)$: 0 units right or left, 0 units up or down

Notice that the origin is represented by (0, 0).

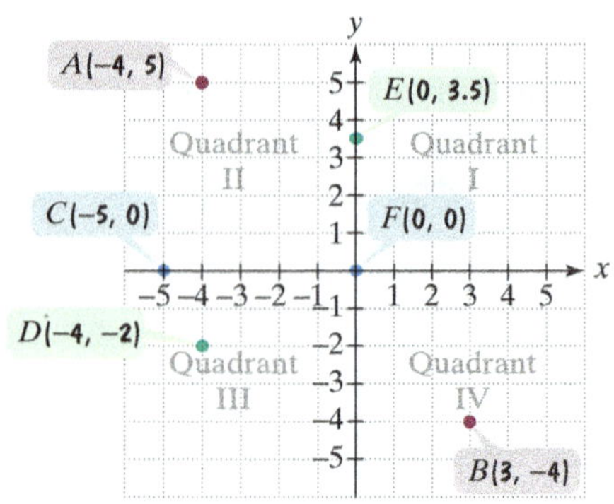

FIGURE 1.11 Plotting points

CHECK POINT 1 Plot the points:

$A(2, 5)$, $B(-1, 3)$, $C(-1.5, -4.5)$, and $D(0, -2)$.

2 Graph equations in the rectangular coordinate system.

Graphs of Equations

A relationship between two quantities can be expressed as an **equation in two variables**, such as

$$y = 4 - x^2.$$

A **solution of an equation in two variables**, x and y, is an ordered pair of real numbers with the following property: When the x-coordinate is substituted for x and the y-coordinate is substituted for y in the equation, we obtain a true statement. For example, consider the equation $y = 4 - x^2$ and the ordered pair $(3, -5)$. When 3 is substituted for x and -5 is substituted for y, we obtain the statement $-5 = 4 - 3^2$, or $-5 = 4 - 9$, or $-5 = -5$. Because this statement is true, the ordered pair $(3, -5)$ is a solution of the equation $y = 4 - x^2$. We also say that $(3, -5)$ **satisfies** the equation.

We can generate as many ordered-pair solutions as desired to $y = 4 - x^2$ by substituting numbers for x and then finding the corresponding values for y. For example, suppose we let $x = 3$:

Start with x. Compute y. Form the ordered pair (x, y).

x	$y = 4 - x^2$	Ordered Pair (x, y)
3	$y = 4 - 3^2 = 4 - 9 = -5$	$(3, -5)$

Let $x = 3$. $(3, -5)$ is a solution of $y = 4 - x^2$.

The **graph of an equation in two variables** is the set of all points whose coordinates satisfy the equation. One method for graphing such equations is the **point-plotting method.** First, we find several ordered pairs that are solutions of the equation. Next, we plot these ordered pairs as points in the rectangular coordinate system. Finally, we connect the points with a smooth curve or line. This often gives us a picture of all ordered pairs that satisfy the equation.

EXAMPLE 2 Graphing an Equation Using the Point-Plotting Method

Graph $y = 4 - x^2$. Select integers for x, starting with -3 and ending with 3.

Solution For each value of x, we find the corresponding value for y.

Start with x. Compute y. Form the ordered pair (x, y).

We selected integers from -3 to 3, inclusive, to include three negative numbers, 0, and three positive numbers. We also wanted to keep the resulting computations for y relatively simple.

x	$y = 4 - x^2$	Ordered Pair (x, y)
-3	$y = 4 - (-3)^2 = 4 - 9 = -5$	$(-3, -5)$
-2	$y = 4 - (-2)^2 = 4 - 4 = 0$	$(-2, 0)$
-1	$y = 4 - (-1)^2 = 4 - 1 = 3$	$(-1, 3)$
0	$y = 4 - 0^2 = 4 - 0 = 4$	$(0, 4)$
1	$y = 4 - 1^2 = 4 - 1 = 3$	$(1, 3)$
2	$y = 4 - 2^2 = 4 - 4 = 0$	$(2, 0)$
3	$y = 4 - 3^2 = 4 - 9 = -5$	$(3, -5)$

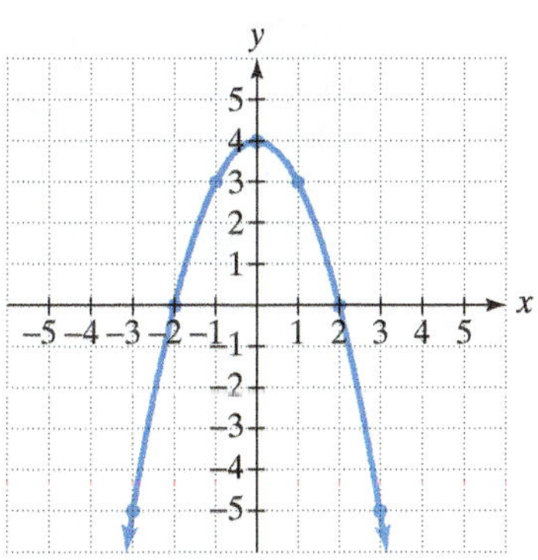

FIGURE 1.12 The graph of $y = 4 - x^2$

Now we plot the seven points and join them with a smooth curve, as shown in **Figure 1.12**. The graph of $y = 4 - x^2$ is a curve where the part of the graph to the right of the y-axis is a reflection of the part to the left of it and vice versa. The arrows on the left and the right of the curve indicate that it extends indefinitely in both directions.

CHECK POINT 2 Graph $y = 1 - x^2$. Select integers for x, starting with -3 and ending with 3.

EXAMPLE 3 Graphing an Equation Using the Point-Plotting Method

Graph $y = |x|$. Select integers for x, starting with -3 and ending with 3.

Solution For each value of x, we find the corresponding value for y.

x	$y = \|x\|$	Ordered Pair (x, y)
-3	$y = \|-3\| = 3$	$(-3, 3)$
-2	$y = \|-2\| = 3$	$(-2, 2)$
-1	$y = \|-1\| = 3$	$(-1, 1)$
0	$y = \|0\| = 0$	$(0, 0)$
1	$y = \|1\| = 1$	$(1, 1)$
3	$y = \|2\| = 2$	$(2, 2)$
3	$y = \|3\| = 3$	$(3, 3)$

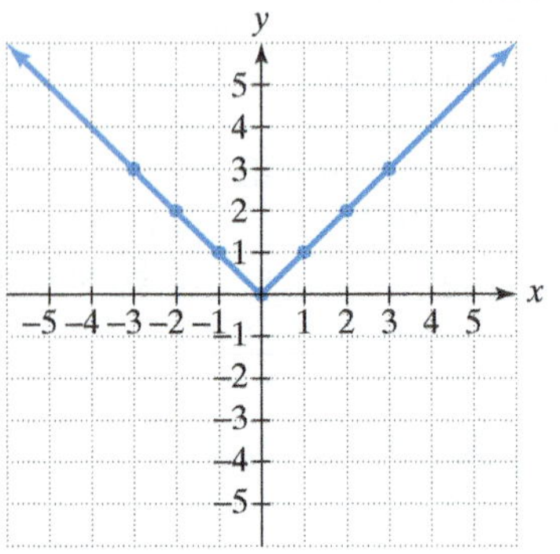

FIGURE 1.13 The graph of $y = |x|$

We plot the points and connect them, resulting in the graph shown in **Figure 1.13**. The graph is V-shaped and centered at the origin. For every point (x, y) on the graph, the point $(-x, y)$ is also on the graph. This shows that the absolute value of a positive number is the same as the absolute value of its opposite.

CHECK POINT 3 Graph $y = |x + 1|$. Select integers for x, starting with -4 and ending with 2.

3 Use the rectangular coordinate system to visualize relationships between variables.

Applications

The rectangular coordinate system allows us to visualize relationships between two variables by associating any equation in two variables with a graph. Graphs in the rectangular coordinate system can also be used to tell a story.

EXAMPLE 4 Telling a Story with a Graph

Too late for that flu shot now! It's only 8 A.M. and you're feeling lousy. Fascinated by the way that algebra models the world (your author is projecting a bit here), you construct a graph showing your body temperature from 8 A.M. through 3 P.M. You decide to let x represent the number of hours after 8 A.M. and y your body temperature at time x. The graph is shown in **Figure 1.14**. The symbol ⌇ on the y-axis shows that there is a break in values between 0 and 98. Thus, the first tick mark on the y-axis represents a temperature of 98°F.

a. What is your temperature at 8 A.M.?

b. During which period of time is your temperature decreasing?

c. Estimate your minimum temperature during the time period shown. How many hours after 8 A.M. does this occur? At what time does this occur?

d. During which period of time is your temperature increasing?

e. Part of the graph is shown as a horizontal line segment. What does this mean about your temperature and when does this occur?

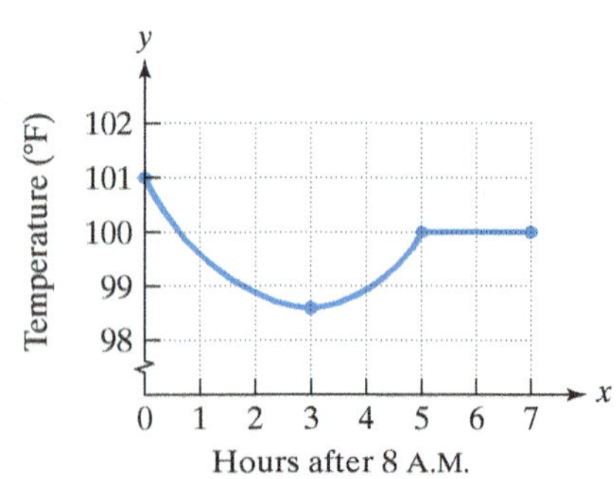

FIGURE 1.14 Body temperature from 8 A.M. through 3 P.M.

Solution

a. Because x is the number of hours after 8 A.M., your temperature at 8 A.M. corresponds to $x = 0$. Locate 0 on the horizontal axis and look at the point on the graph above 0. The first figure on the right shows that your temperature at 8 A.M. is 101°F.

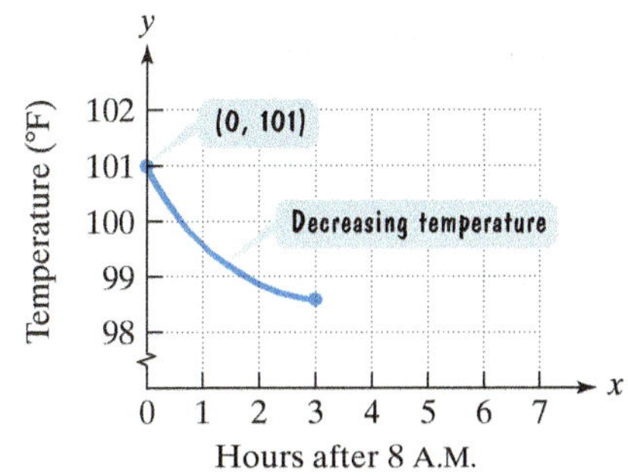

b. Your temperature is decreasing when the graph falls from left to right. This occurs between $x = 0$ and $x = 3$, also shown in the first figure on the right. Because x represents the number of hours after 8 A.M., your temperature is decreasing between 8 A.M. and 11 A.M.

c. Your minimum temperature can be found by locating the lowest point on the graph. This point lies above 3 on the horizontal axis, shown in the second figure on the right. The y-coordinate of this point falls more than midway between 98 and 99, at approximately 98.6. The lowest point on the graph, (3, 98.6), shows that your minimum temperature, 98.6°F, occurs 3 hours after 8 A.M., at 11 A.M.

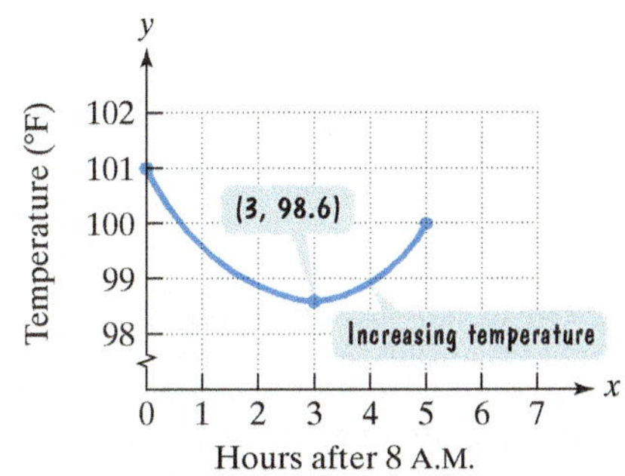

d. Your temperature is increasing when the graph rises from left to right. This occurs between $x = 3$ and $x = 5$, shown in the second figure. Because x represents the number of hours after 8 A.M., your temperature is increasing between 11 A.M. and 1 P.M.

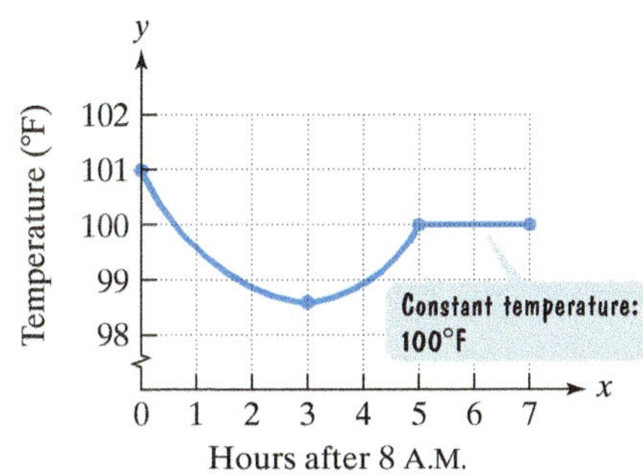

e. The horizontal line segment shown in the figure on the right indicates that your temperature is neither increasing nor decreasing. Your temperature remains the same, 100°F, between $x = 5$ and $x = 7$. Thus, your temperature is at a constant 100°F between 1 P.M. and 3 P.M.

✓ CHECK POINT 4 When a person receives a drug injected into a muscle, the concentration of the drug in the body, measured in milligrams per 100 milliliters, depends on the time elapsed after the injection, measured in hours. The figure shows the graph of drug concentration over time, where x represents hours after the injection and y represents the drug concentration at time x.

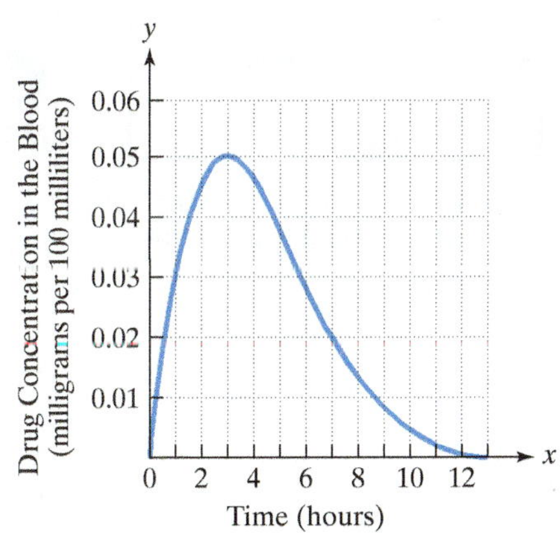

a. During which period of time is the drug concentration increasing?

b. During which period of time is the drug concentration decreasing?

c. What is the drug's maximum concentration and when does this occur?

d. What happens by the end of 13 hours?

4 Interpret information about a graphing utility's viewing rectangle or table.

Graphing Equations and Creating Tables Using a Graphing Utility

Graphing calculators and graphing software packages for computers are referred to as **graphing utilities** or graphers. A graphing utility is a powerful tool that quickly generates the graph of an equation in two variables. **Figures 1.15(a)** and **1.15(b)** show two such graphs for the equations in Examples 2 and 3.

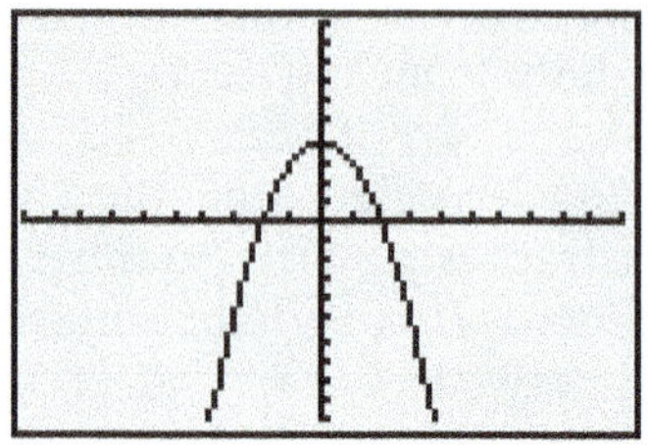

FIGURE 1.15(a) The graph of $y = 4 - x^2$

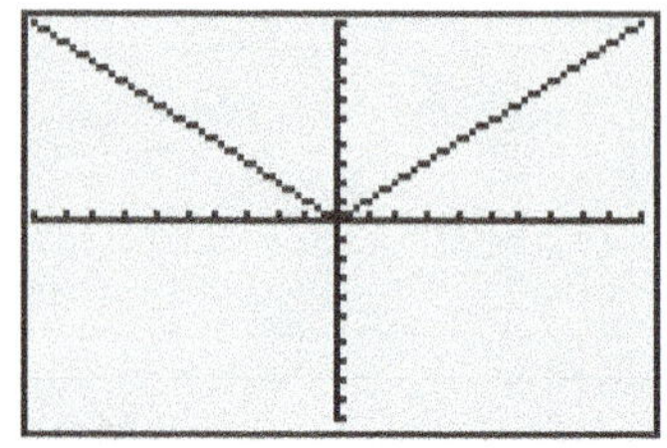

FIGURE 1.15(b) The graph of $y = |x|$

Study Tip

Even if you are not using a graphing utility in the course, read this part of the section. Knowing about viewing rectangles will enable you to understand the graphs that we display in the technology boxes throughout the book.

What differences do you notice between these graphs and the graphs that we drew by hand? They do seem a bit "jittery." Arrows do not appear on the left and right ends of the graphs. Furthermore, numbers are not given along the axes. For both graphs in **Figure 1.15**, the x-axis extends from -10 to 10 and the y-axis also extends from -10 to 10. The distance represented by each consecutive tick mark is one unit. We say that the **viewing rectangle**, or the **viewing window**, is $[-10, 10, 1]$ by $[-10, 10, 1]$.

[−10,	10,	1]	by	[−10,	10,	1]
The minimum x-value along the x-axis is −10.	The maximum x-value along the x-axis is 10.	Distance between consecutive tick marks on the x-axis is one unit.		The minimum y-value along the y-axis is −10.	The maximum y-value along the y-axis is 10.	Distance between consecutive tick marks on the y-axis is one unit.

To graph an equation in x and y using a graphing utility, enter the equation and specify the size of the viewing rectangle. The size of the viewing rectangle sets minimum and maximum values for both the x- and y-axes. Enter these values, as well as the values between consecutive tick marks, on the respective axes. The $[-10, 10, 1]$ by $[-10, 10, 1]$ viewing rectangle used in **Figure 1.15** is called the **standard viewing rectangle**.

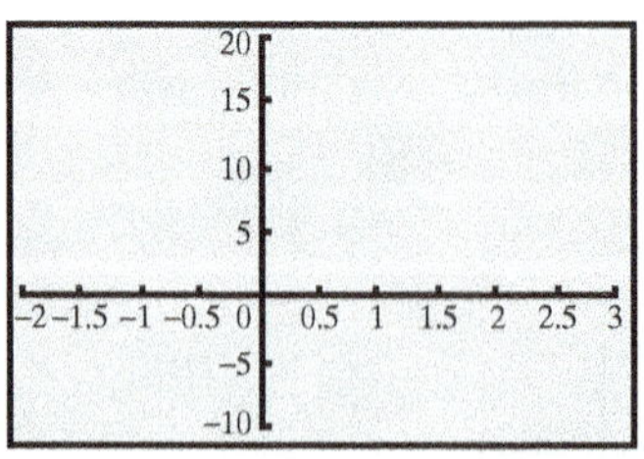

FIGURE 1.16 A $[-2, 3, 0.5]$ by $[-10, 20, 5]$ viewing rectangle

EXAMPLE 5 Understanding the Viewing Rectangle

What is the meaning of a $[-2, 3, 0.5]$ by $[-10, 20, 5]$ viewing rectangle?

Solution We begin with $[-2, 3, 0.5]$, which describes the x-axis. The minimum x-value is -2 and the maximum x-value is 3. The distance between consecutive tick marks is 0.5.

Next, consider $[-10, 20, 5]$, which describes the y-axis. The minimum y-value is -10 and the maximum y-value is 20. The distance between consecutive tick marks is 5.

Figure 1.16 illustrates a $[-2, 3, 0.5]$ by $[-10, 20, 5]$ viewing rectangle. To make things clearer, we've placed numbers by each tick mark. These numbers do not appear on the axes when you use a graphing utility to graph an equation.

✓ **CHECK POINT 5** What is the meaning of a $[-100, 100, 50]$ by $[-100, 100, 10]$ viewing rectangle? Create a figure like the one in **Figure 1.16** that illustrates this viewing rectangle.

On most graphing utilities, the display screen is two-thirds as high as it is wide. By using a square setting, you can equally space the x and y tick marks. (This does not occur in the standard viewing rectangle.) Graphing utilities can also *zoom in* and *zoom out*. When you zoom in, you see a smaller portion of the graph, but you do so in greater detail. When you zoom out, you see a larger portion of the graph. Thus, zooming out may help you to develop a better understanding of the overall character of the graph. With practice, you will become more comfortable with graphing equations in two variables using your graphing utility. You will also develop a better sense of the size of the viewing rectangle that will reveal needed information about a particular graph.

Graphing utilities can also be used to create tables showing solutions of equations in two variables. Use the Table Setup function to choose the starting value of x and to input the increment, or change, between the consecutive x-values. The corresponding y-values are calculated based on the equation(s) in two variables in the [Y=] screen. In **Figure 1.17**, we used a TI-84 Plus to create a table for $y = 4 - x^2$ and $y = |x|$, the equations in Examples 2 and 3.

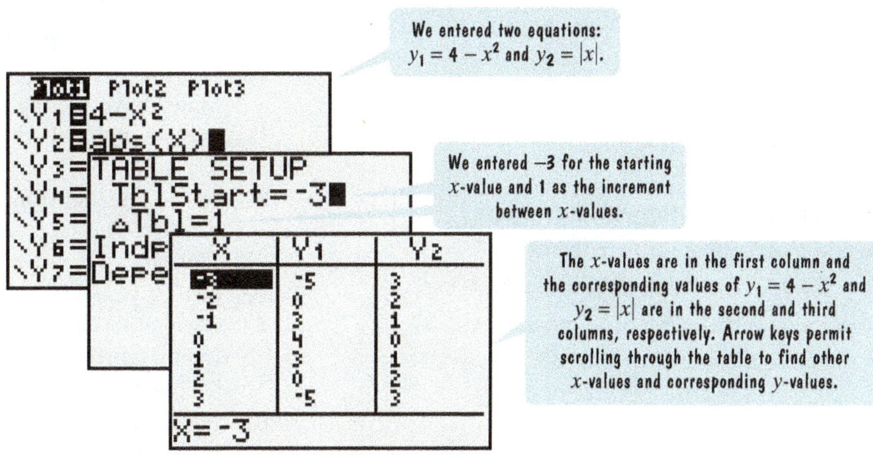

FIGURE 1.17 Creating a table for $y_1 = 4 - x^2$ and $y_2 = |x|$

1.3 EXERCISE SET

MyMathLab PRACTICE WATCH DOWNLOAD READ REVIEW

Practice Exercises

In Exercises 1–10, plot the given point in a rectangular coordinate system.

1. (1, 4)
2. (2, 5)
3. (−2, 3)
4. (−1, 4)
5. (−3, −5)
6. (−4, −2)
7. (4, −1)
8. (3, −2)
9. (−4, 0)
10. (0, −3)

Graph each equation in Exercises 11–26. Let $x = -3, -2, -1, 0,$ 1, 2, and 3.

11. $y = x^2 - 4$
12. $y = x^2 - 9$
13. $y = x - 2$
14. $y = x + 2$
15. $y = 2x + 1$
16. $y = 2x - 4$
17. $y = -\frac{1}{2}x$
18. $y = -\frac{1}{2}x + 2$
19. $y = |x| + 1$
20. $y = |x| - 1$
21. $y = 2|x|$
22. $y = -2|x|$
23. $y = -x^2$
24. $y = -\frac{1}{2}x^2$
25. $y = x^3$
26. $y = x^3 - 1$

In Exercises 27–30, match the viewing rectangle with the correct figure. Then label the tick marks in the figure to illustrate this viewing rectangle.

27. [−5, 5, 1] by [−5, 5, 1]
28. [−10, 10, 2] by [−4, 4, 2]
29. [−20, 80, 10] by [−30, 70, 10]
30. [−40, 40, 20] by [−1000, 1000, 100]

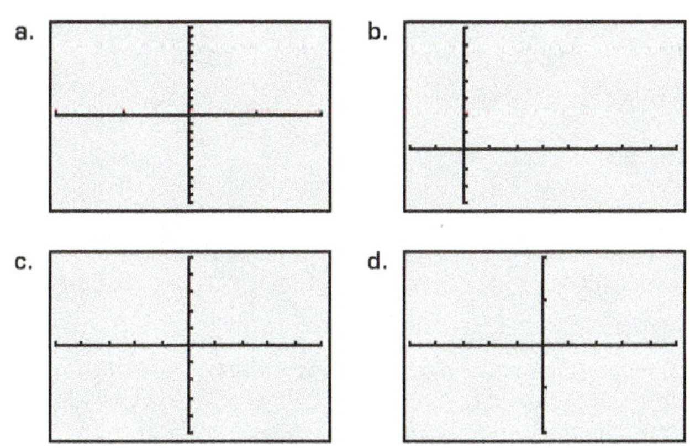

The table of values was generated by a graphing utility with a TABLE feature. Use the table to solve Exercises 31–38.

X	Y1	Y2
-3	9	5
-2	4	4
-1	1	3
0	0	2
1	1	1
2	4	0
3	9	-1
X=-3		

31. Which equation corresponds to Y_2 in the table?

a. $y_2 = x + 8$

b. $y_2 = x - 2$

c. $y_2 = 2 - x$

d. $y_2 = 1 - 2x$

32. Which equation corresponds to Y_1 in the table?

a. $y_1 = -3x$

b. $y_1 = x^2$

c. $y_1 = -x^2$

d. $y_1 = 2 - x$

33. Does the graph of Y_2 pass through the origin?

34. Does the graph of Y_1 pass through the origin?

35. At which point does the graph of Y_2 cross the x-axis?

36. At which point does the graph of Y_2 cross the y-axis?

37. At which points do the graphs of Y_1 and Y_2 intersect?

38. For which values of x is $Y_1 = Y_2$?

Practice PLUS

In Exercises 39–42, write each English sentence as an equation in two variables. Then graph the equation.

39. The y-value is four more than twice the x-value.

40. The y-value is the difference between four and twice the x-value.

41. The y-value is three decreased by the square of the x-value.

42. The y-value is two more than the square of the x-value.

In Exercises 43–46, graph each equation.

43. $y = 5$ (Let $x = -3, -2, -1, 0, 1, 2,$ and 3.)

44. $y = -1$ (Let $x = -3, -2, -1, 0, 1, 2,$ and 3.)

45. $y = \frac{1}{x}$ (Let $x = -2, -1, -\frac{1}{2}, -\frac{1}{3}, \frac{1}{3}, \frac{1}{2}, 1,$ and 2.)

46. $y = -\frac{1}{x}$ (Let $x = -2, -1, -\frac{1}{2}, -\frac{1}{3}, \frac{1}{3}, \frac{1}{2}, 1,$ and 2.)

Application Exercises

The line graph at the top of the next column shows the top marginal income tax rates in the United States from 1925 through 2005. Use the graph to solve Exercises 47–52.

Top United States Marginal Tax Rates, 1925–2005

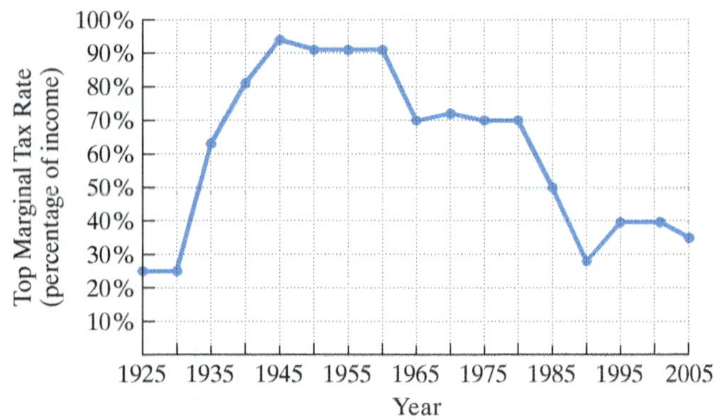

Source: National Taxpayers Union

47. Estimate the top marginal tax rate in 2005.

48. Estimate the top marginal tax rate in 1925.

49. For the period shown, during which year did the United States have the highest marginal tax rate? Estimate, to the nearest percent, the tax rate for that year.

50. For the period from 1950 through 2005, during which year did the United States have the lowest marginal tax rate? Estimate, to the nearest percent, the tax rate for that year.

51. For the period shown, during which ten-year period did the top marginal tax rate remain constant? Estimate, to the nearest percent, the tax rate for that period.

52. For the period shown, during which five-year period did the top marginal tax rate increase most rapidly? Estimate, to the nearest percent, the increase in the top tax rate for that period.

Contrary to popular belief, older people do not need less sleep than younger adults. However, the line graphs show that they awaken more often during the night. The numerous awakenings are one reason why some elderly individuals report that sleep is less restful than it had been in the past. Use the line graphs to solve Exercises 53–56.

Average Number of Awakenings During the Night, by Age and Gender

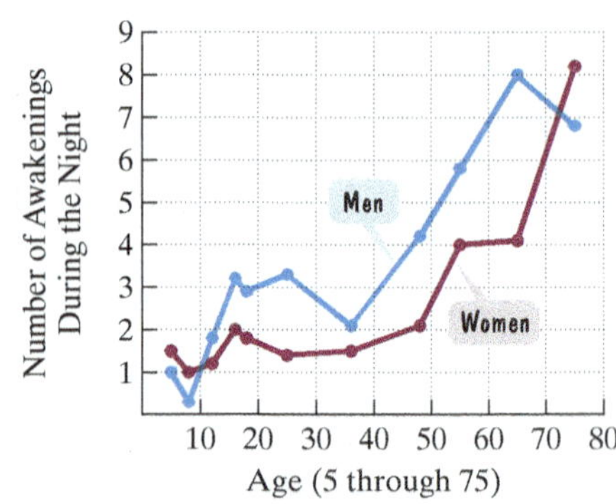

Source: Stephen Davis and Joseph Palladino, *Psychology*, 5th Edition, Prentice Hall, 2007

53. At which age, estimated to the nearest year, do women have the least number of awakenings during the night? What is the average number of awakenings at that age?

54. At which age do men have the greatest number of awakenings during the night? What is the average number of awakenings at that age?

55. Estimate, to the nearest tenth, the difference between the average number of awakenings during the night between 25-year-old men and 25-year-old women.

56. Estimate, to the nearest tenth, the difference between the average number of awakenings during the night between 18-year-old men and 18-year-old women.

In Exercises 57–60, match the story with the correct figure. The figures are labeled (a), (b), (c), and (d).

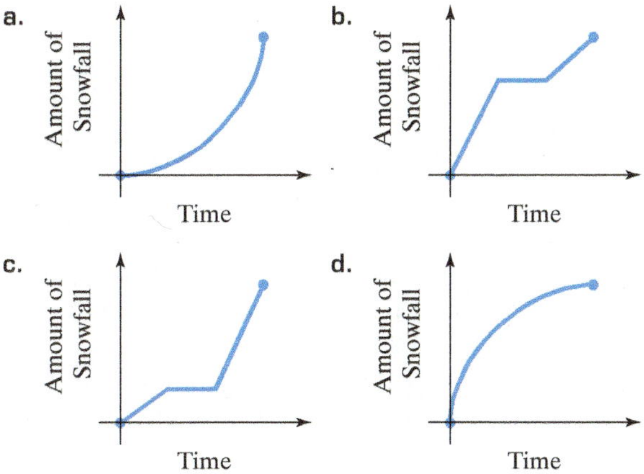

57. As the blizzard got worse, the snow fell harder and harder.

58. The snow fell more and more softly.

59. It snowed hard, but then it stopped. After a short time, the snow started falling softly.

60. It snowed softly, and then it stopped. After a short time, the snow started falling hard.

In Exercises 61–64, select the graph that best illustrates each story.

61. An airplane flew from Miami to San Francisco.

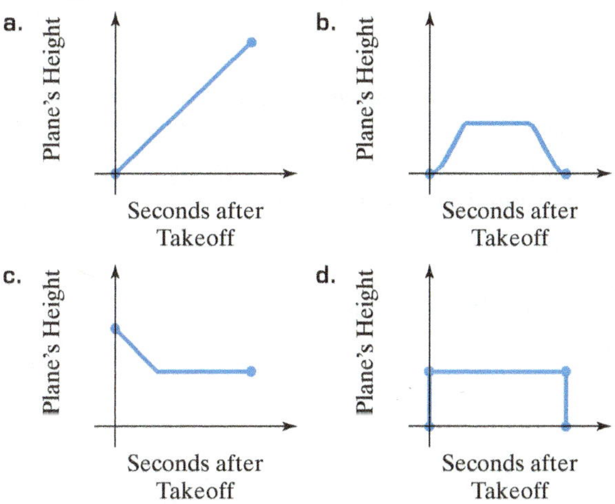

62. At noon, you begin to breathe in.

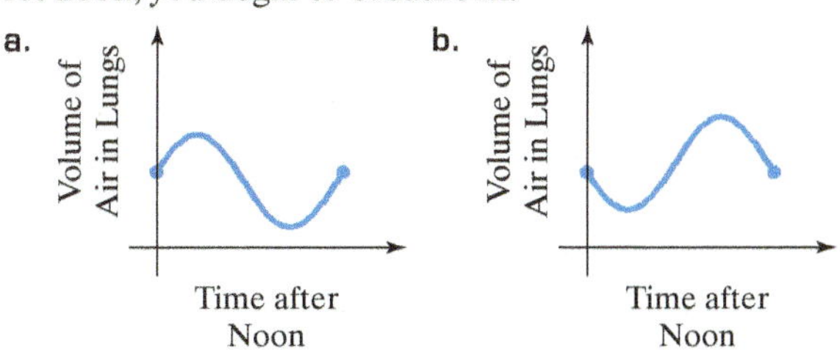

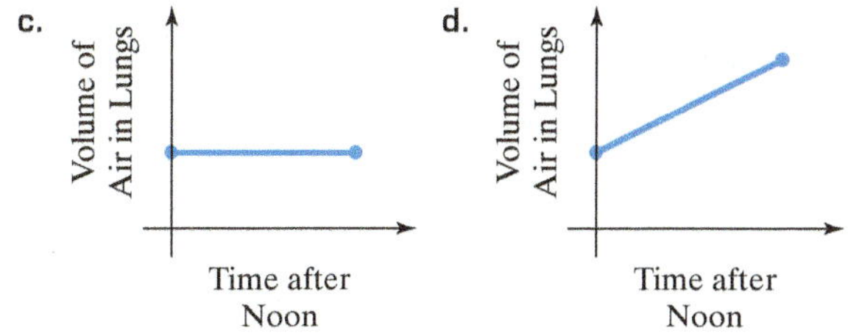

63. Measurements are taken of a person's height from birth to age 100.

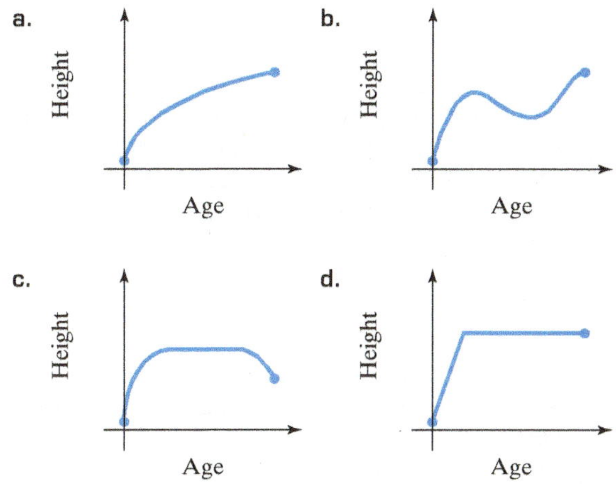

64. You begin your bike ride by riding down a hill. Then you ride up another hill. Finally, you ride along a level surface before coming to a stop.

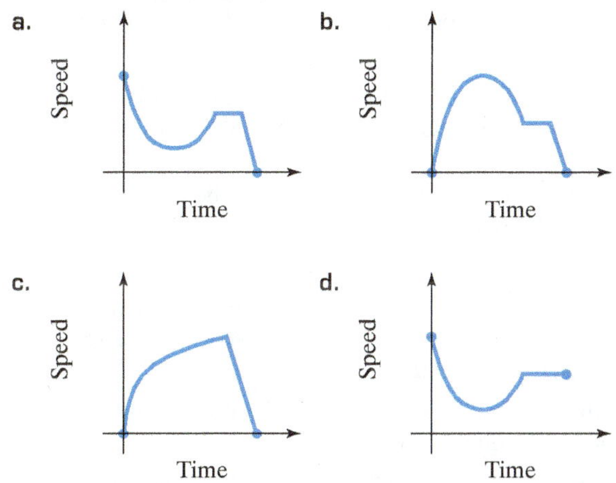

Writing in Mathematics

65. What is the rectangular coordinate system?

66. Explain how to plot a point in the rectangular coordinate system. Give an example with your explanation.

67. Explain why $(5, -2)$ and $(-2, 5)$ do not represent the same point.

68. Explain how to graph an equation in the rectangular coordinate system.

69. What does a $[-20, 2, 1]$ by $[-4, 5, 0.5]$ viewing rectangle mean?

In Exercises 70–71, write a story, or description, to match each title and graph.

70. **Checking Account Balance**

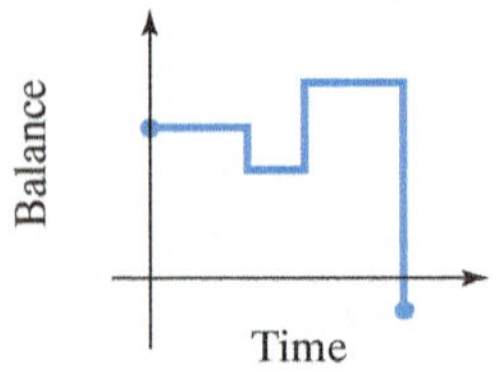

71. **Hair Length**

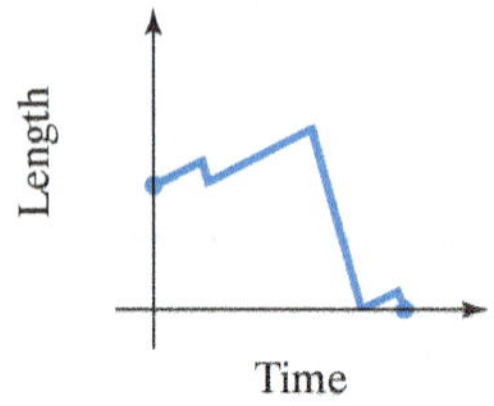

Technology Exercise

72. Use a graphing utility to verify each of your hand-drawn graphs in Exercises 11–26. Experiment with the viewing rectangle to make the graph displayed by the graphing utility resemble your hand-drawn graph as much as possible.

Critical Thinking Exercises

Make Sense? *In Exercises 73–76, determine whether each statement "makes sense" or "does not make sense" and explain your reasoning.*

73. The rectangular coordinate system provides a geometric picture of what an equation in two variables looks like.

74. There is something wrong with my graphing utility because it is not displaying numbers along the x- and y-axes.

75. A horizontal line is not a graph that tells the story of the number of calories that I burn throughout the day.

76. I told my story with a graph, so I can be confident that there is a mathematical model that perfectly describes the graph's data.

In Exercises 77–80, determine whether each statement is true or false. If the statement is false, make the necessary change(s) to produce a true statement.

77. If the product of a point's coordinates is positive, the point must be in quadrant I.

78. If a point is on the x-axis, it is neither up nor down, so $x = 0$.

79. If a point is on the y-axis, its x-coordinate must be 0.

80. The ordered pair (2, 5) satisfies $3y - 2x = -4$.

The graph shows the costs at a parking garage that allows cars to be parked for up to ten hours per day. Closed dots indicate that points belong to the graph and open dots indicate that points are not part of the graph. Use the graph to solve Exercises 81–82.

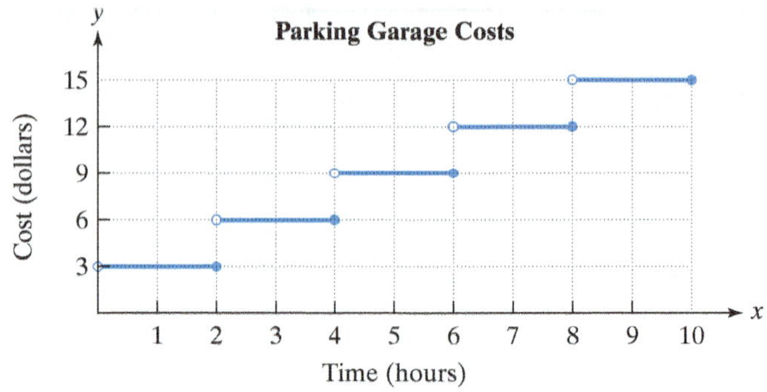

81. You park your car at the garage for four hours on Tuesday and five hours on Wednesday. What are the total parking garage costs for the two days?

82. On Thursday, you paid \$12 for parking at the garage. Describe how long your car was parked.

Review Exercises

83. Find the absolute value: $|-14.3|$. (Section 1.2, Example 1)

84. Simplify: $[12 - (13 - 17)] - [9 - (6 - 10)]$. (Section 1.2, Examples 7 and 8)

85. Simplify: $6x - 5(4x + 3) - 10$. (Section 1.2, Example 13)

Preview Exercises

Exercises 86–88 will help you prepare for the material covered in the next section.

86. If -9 is substituted for x in the equation $4x - 3 = 5x + 6$, is the resulting statement true or false?

87. Simplify: $13 - 3(x + 2)$.

88. Simplify: $10\left(\frac{3x + 1}{2}\right)$.

SECTION 1.4

Objectives

1. Solve linear equations.
2. Recognize identities, conditional equations, and inconsistent equations.
3. Solve applied problems using mathematical models.

Solving Linear Equations

Teutul Time: Jr. and Sr. in a rare peaceful moment

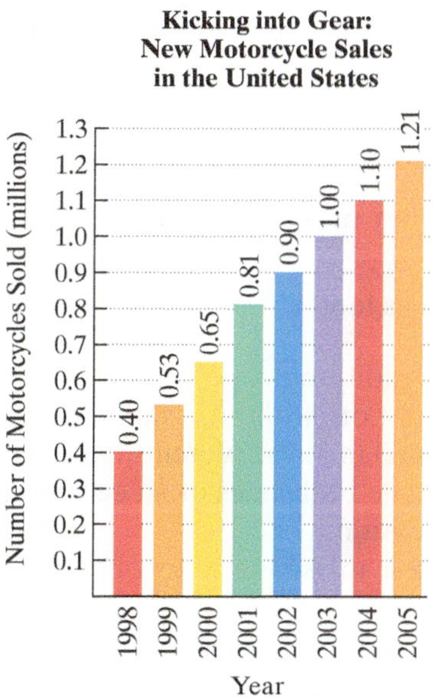

FIGURE 1.18

Source: Motorcycle Industry Council

Talk about a booming business cycle: TV's "American Chopper" is the hottest thing since *Easy Rider* and motorcycles are selling faster than they have in a generation. The 10.5 million weekly viewers of "American Chopper" are drawn as much by the titanic tussels between superstar bike builders Paul Senior and Paul Junior Teutul as by the craftsmanship in their elaborate theme bikes.

Although "American Chopper" is bringing more people into showrooms, the bar graph in **Figure 1.18** indicates that motorcycle sales were rising long before the Teutuls showed up. The rising sales from 1998 through 2005 can be modeled by the formula

$$N = 0.12x + 0.4,$$

where N represents the number of new motorcycles sold in the United States, in millions, x years after 1998. In 2004, cycle sales reached 1.1 million, topping 1 million for the first time since the post–*Easy Rider* days of the early 1970s. So, when will motorcycle sales double the 2004 level and reach 2.2 million? Substitute 2.2 for N in the formula $N = 0.12x + 0.4$:

$$2.2 = 0.12x + 0.4.$$

Our goal is to determine the value of x, the number of years after 1998, when sales will reach 2.2 million. Notice that the exponent on the variable in this equation is 1. In this section, we will study how to determine the value of x in such equations. With this skill, you will be able to use certain mathematical models, such as the model for motorcycle sales, to project what might occur in the future.

1 Solve linear equations.

Solving Linear Equations in One Variable

We begin with a general definition of a linear equation in one variable.

Definition of a Linear Equation

A **linear equation in one variable x** is an equation that can be written in the form

$$ax + b = 0,$$

where a and b are real numbers, and $a \neq 0$ (a is not equal to 0).

An example of a linear equation in one variable is

$$4x + 12 = 0.$$

Solving an equation in x involves determining all values of x that result in a true statement when substituted into the equation. Such values are **solutions**, or **roots**, of the equation. For example, substitute -3 for x in $4x + 12 = 0$. We obtain

$$4(-3) + 12 = 0, \quad \text{or} \quad -12 + 12 = 0.$$

This simplifies to the true statement $0 = 0$. Thus, -3 is a solution of the equation $4x + 12 = 0$. We also say that -3 **satisfies** the equation $4x + 12 = 0$, because when we substitute -3 for x, a true statement results. The set of all such solutions is called the equation's **solution set**. For example, the solution set of the equation $4x + 12 = 0$ is $\{-3\}$.

Two or more equations that have the same solution set are called **equivalent equations**. For example, the equations

$$4x + 12 = 0 \quad \text{and} \quad 4x = -12 \quad \text{and} \quad x = -3$$

are equivalent equations because the solution set for each is $\{-3\}$. To solve a linear equation in x, we transform the equation into an equivalent equation one or more times. Our final equivalent equation should be of the form

$$x = \text{a number}.$$

The solution set of this equation is the set consisting of the number.

To generate equivalent equations, we will use the following properties:

The Addition and Multiplication Properties of Equality

The Addition Property of Equality

The same real number or algebraic expression may be added to both sides of an equation without changing the equation's solution set.

$$a = b \text{ and } a + c = b + c \text{ are equivalent equations.}$$

The Multiplication Property of Equality

The same nonzero real number may multiply both sides of an equation without changing the equation's solution set.

$$a = b \text{ and } ac = bc \text{ are equivalent equations as long as } c \neq 0.$$

Because subtraction is defined in terms of addition, the addition property also lets us subtract the same number from both sides of an equation without changing the equation's solution set. Similarly, because division is defined in terms of multiplication, the multiplication property of equality can be used to divide both sides of an equation by the same nonzero number to obtain an equivalent equation.

Table 1.4 illustrates how these properties are used to isolate x to obtain an equation of the form x = a number.

Study Tip

Your final equivalent equation should not be of the form

$$-x = \text{a number}.$$

We're not finished. A negative sign should not precede the variable.

Isolate x by multiplying or dividing both sides of this equation by -1.

Table 1.4 Using Properties of Equality to Solve Linear Equations

	Equation	How to Isolate x	Solving the Equation	The Equation's Solution Set
These equations are solved using the Addition Property of Equality.	$x - 3 = 8$	Add 3 to both sides.	$x - 3 + 3 = 8 + 3$ $x = 11$	$\{11\}$
	$x + 7 = -15$	Subtract 7 from both sides.	$x + 7 - 7 = -15 - 7$ $x = -22$	$\{-22\}$
These equations are solved using the Multiplication Property of Equality.	$6x = 30$	Divide both sides by 6 (or multiply both sides by $\frac{1}{6}$).	$\frac{6x}{6} = \frac{30}{6}$ $x = 5$	$\{5\}$
	$\frac{x}{5} = 9$	Multiply both sides by 5.	$5 \cdot \frac{x}{5} = 5 \cdot 9$ $x = 45$	$\{45\}$

EXAMPLE 1 Solving a Linear Equation

Solve and check: $2x + 3 = 17$.

Solution Our goal is to obtain an equivalent equation with x isolated on one side and a number on the other side.

$2x + 3 = 17$	This is the given equation.
$2x + 3 - 3 = 17 - 3$	Subtract 3 from both sides.
$2x = 14$	Simplify.
$\frac{2x}{2} = \frac{14}{2}$	Divide both sides by 2.
$x = 7$	Simplify.

Now we check the proposed solution, 7, by replacing x with 7 in the original equation.

$2x + 3 = 17$	This is the original equation.
$2 \cdot 7 + 3 \stackrel{?}{=} 17$	Substitute 7 for x. The question mark indicates that we do not yet know if the two sides are equal.
$14 + 3 \stackrel{?}{=} 17$	Multiply: $2 \cdot 7 = 14$.
This statement is true. $17 = 17$	Add: $14 + 3 = 17$.

Because the check results in a true statement, we conclude that the solution of the given equation is 7, or the solution set is $\{7\}$.

CHECK POINT 1 Solve and check: $4x + 5 = 29$.

Study Tip

We simplify algebraic expressions. We solve algebraic equations. Notice the differences between the procedures:

Simplifying an Algebraic Expression

Simplify: $3(x - 7) - (5x - 11)$.

This is not an equation. There is no equal sign.

Solution
$$\begin{aligned} &3(x - 7) - (5x - 11)\\ &= 3x - 21 - 5x + 11\\ &= (3x - 5x) + (-21 + 11)\\ &= -2x + (-10)\\ &= -2x - 10 \end{aligned}$$

Stop! Further simplification is not possible. Avoid the common error of setting $-2x - 10$ equal to 0.

Solving an Algebraic Equation

Solve: $3(x - 7) - (5x - 11) = 14$.

This is an equation. There is an equal sign.

Solution
$$\begin{aligned} 3(x - 7) - (5x - 11) &= 14\\ 3x - 21 - 5x + 11 &= 14\\ -2x - 10 &= 14\\ -2x - 10 + 10 &= 14 + 10\\ -2x &= 24\\ \frac{-2x}{-2} &= \frac{24}{-2}\\ x &= -12 \end{aligned}$$

Add 10 to both sides.

Divide both sides by -2.

The solution set is $\{-12\}$.

Here is a step-by-step procedure for solving a linear equation in one variable. Not all of these steps are necessary to solve every equation.

Solving a Linear Equation

1. Simplify the algebraic expression on each side by removing grouping symbols and combining like terms.
2. Collect all the variable terms on one side and all the numbers, or constant terms, on the other side.
3. Isolate the variable and solve.
4. Check the proposed solution in the original equation.

EXAMPLE 2 Solving a Linear Equation

Solve and check: $2x - 7 + x = 3x + 1 + 2x$.

Solution

Step 1. Simplify the algebraic expression on each side.

$2x - 7 + x = 3x + 1 + 2x$	This is the given equation.
$3x - 7 = 5x + 1$	Combine like terms: $2x + x = 3x$ and $3x + 2x = 5x$.

Step 2. Collect variable terms on one side and constant terms on the other side. We will collect variable terms on the left by subtracting $5x$ from both sides. We will collect the numbers on the right by adding 7 to both sides.

$3x - 5x - 7 = 5x - 5x + 1$	Subtract $5x$ from both sides.
$-2x - 7 = 1$	Simplify.
$-2x - 7 + 7 = 1 + 7$	Add 7 to both sides.
$-2x = 8$	Simplify.

Step 3. Isolate the variable and solve. We isolate x by dividing both sides by -2.

$\dfrac{-2x}{-2} = \dfrac{8}{-2}$	Divide both sides by -2.
$x = -4$	Simplify.

Step 4. Check the proposed solution in the original equation. Substitute -4 for x in the original equation.

$2x - 7 + x = 3x + 1 + 2x$	This is the original equation.
$2(-4) - 7 + (-4) \stackrel{?}{=} 3(-4) + 1 + 2(-4)$	Substitute -4 for x.
$-8 - 7 + (-4) \stackrel{?}{=} -12 + 1 + (-8)$	Multiply: $2(-4) = -8$, $3(-4) = -12$, and $2(-4) = -8$.
$-15 + (-4) \stackrel{?}{=} -11 + (-8)$	Add or subtract from left to right: $-8 - 7 = -15$ and $-12 + 1 = -11$.
$-19 = -19$	Add.

The true statement $-19 = -19$ verifies that -4 is the solution, or the solution set is $\{-4\}$.

Discover for Yourself

Solve the equation in Example 2 by collecting terms with the variable on the right and constant terms on the left. What do you observe?

✓ **CHECK POINT 2** Solve and check: $2x - 12 + x = 6x - 4 + 5x$.

Solving a Linear Equation

Solve and check: $4(2x + 1) - 29 = 3(2x - 5)$.

Solution

Step 1. Simplify the algebraic expression on each side.

$4(2x + 1) - 29 = 3(2x - 5)$	This is the given equation.
$8x + 4 - 29 = 6x - 15$	Use the distributive property.
$8x - 25 = 6x - 15$	Simplify.

Step 2. Collect variable terms on one side and constant terms on the other side. We will collect the variable terms on the left by subtracting $6x$ from both sides. We will collect the numbers on the right by adding 25 to both sides.

$8x - 6x - 25 = 6x - 6x - 15$	Subtract $6x$ from both sides.
$2x - 25 = -15$	Simplify.
$2x - 25 + 25 = -15 + 25$	Add 25 to both sides.
$2x = 10$	Simplify.

Step 3. Isolate the variable and solve. We isolate x by dividing both sides by 2.

$\frac{2x}{2} = \frac{10}{2}$	Divide both sides by 2.
$x = 5$	Simplify.

(The example continues on page 42.)

Using Technology

Numeric and Graphic Connections

In many algebraic situations, technology provides numeric and visual insights into problem solving. For example, you can use a graphing utility to check the solution of a linear equation, giving numeric and geometric meaning to the solution. Enter each side of the equation separately under y_1 and y_2. Then use the table or the graphs to locate the x-value for which the y-values are the same. This x-value is the solution.

Let's verify our work in Example 3 and show that 5 is the solution of

$$4(2x + 1) - 29 = 3(2x - 5).$$

Enter $y_1 = 4(2x + 1) - 29$ in the [y=] screen.

Enter $y_2 = 3(2x - 5)$ in the [y=] screen.

Numeric Check

Display a table for y_1 and y_2.

$y_1 = 4(2x + 1) - 29$ | $y_2 = 3(2x - 5)$

X	Y1	Y2
1	-17	-9
2	-9	-3
3	-1	3
4	7	9
5	15	15
6	23	21
7	31	27

X=5

When $x = 5$, y_1 and y_2 have the same value, namely 15. This verifies 5 is the solution of $4(2x + 1) - 29 = 3(2x - 5)$.

Graphic Check

Display graphs for y_1 and y_2 and use the intersection feature. The solution is the x-coordinate of the intersection point.

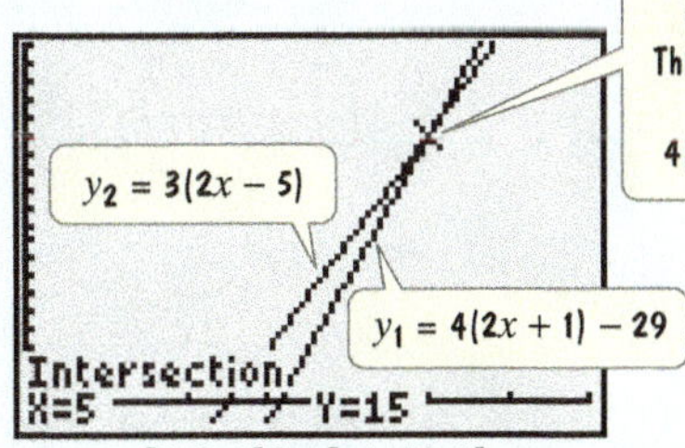

Graphs intersect at (5, 15). When $x = 5$, y_1 and y_2 have the same value, namely, 15. This graphically verifies 5 is the solution of $4(2x + 1) - 29 = 3(2x - 5)$.

$[0, 7, 1]$ by $[-1, 20, 1]$

Choose a large enough viewing rectangle so that you can see the intersection point.

Step 4. Check the proposed solution in the original equation. Substitute 5 for x in the original equation.

$4(2x + 1) - 29 = 3(2x - 5)$	This is the original equation.
$4(2 \cdot 5 + 1) - 29 \stackrel{?}{=} 3(2 \cdot 5 - 5)$	Substitute 5 for x.
$4(11) - 29 \stackrel{?}{=} 3(5)$	Simplify inside parentheses: $2 \cdot 5 + 1 = 10 + 1 = 11$ and $2 \cdot 5 - 5 = 10 - 5 = 5$.
$44 - 29 \stackrel{?}{=} 15$	Multiply: $4(11) = 44$ and $3(5) = 15$.
$15 = 15$	Subtract.

The true statement $15 = 15$ verifies that 5 is the solution, or the solution set is $\{5\}$.

☑ **CHECK POINT 3** Solve and check: $2(x - 3) - 17 = 13 - 3(x + 2)$.

Linear Equations with Fractions

Equations are easier to solve when they do not contain fractions. How do we remove fractions from an equation? We begin by multiplying both sides of the equation by the least common denominator (LCD) of any fractions in the equation. The least common denominator is the smallest number that all denominators will divide into. Multiplying every term on both sides of the equation by the least common denominator will eliminate the fractions in the equation. Example 4 shows how we "clear an equation of fractions."

EXAMPLE 4 Solving a Linear Equation Involving Fractions

Solve: $\frac{2x + 5}{5} + \frac{x - 7}{2} = \frac{3x + 1}{2}$.

Solution The denominators are 5, 2, and 2. The smallest number that is divisible by 5, 2, and 2 is 10. We begin by multiplying both sides of the equation by 10, the least common denominator.

$\frac{2x + 5}{5} + \frac{x - 7}{2} = \frac{3x + 1}{2}$	This is the given equation.
$10\left(\frac{2x + 5}{5} + \frac{x - 7}{2}\right) = 10\left(\frac{3x + 1}{2}\right)$	Multiply both sides by 10.
$\frac{10}{1} \cdot \left(\frac{2x + 5}{5}\right) + \frac{10}{1} \cdot \left(\frac{x - 7}{2}\right) = \frac{10}{1} \cdot \left(\frac{3x + 1}{2}\right)$	Use the distributive property and multiply each term by 10.
$\frac{\overset{2}{\cancel{10}}}{1} \cdot \left(\frac{2x + 5}{\underset{1}{\cancel{5}}}\right) + \frac{\overset{5}{\cancel{10}}}{1} \cdot \left(\frac{x - 7}{\underset{1}{\cancel{2}}}\right) = \frac{\overset{5}{\cancel{10}}}{1} \cdot \left(\frac{3x + 1}{\underset{1}{\cancel{2}}}\right)$	Divide out common factors in each multiplication.
$2(2x + 5) + 5(x - 7) = 5(3x + 1)$	The fractions are now cleared.

At this point, we have an equation similar to those we have previously solved. Use the distributive property to begin simplifying each side.

$4x + 10 + 5x - 35 = 15x + 5$	Use the distributive property.
$9x - 25 = 15x + 5$	Combine like terms on the left side: $4x + 5x = 9x$ and $10 - 35 = -25$.

For variety, let's collect variable terms on the right and constant terms on the left.

$$9x - 9x - 25 = 15x - 9x + 5 \quad \text{Subtract } 9x \text{ from both sides.}$$
$$-25 = 6x + 5 \quad \text{Simplify.}$$
$$-25 - 5 = 6x + 5 - 5 \quad \text{Subtract 5 from both sides.}$$
$$-30 = 6x \quad \text{Simplify.}$$

Isolate x on the right side by dividing both sides by 6.

$$\frac{-30}{6} = \frac{6x}{6} \quad \text{Divide both sides by 6.}$$
$$-5 = x \quad \text{Simplify.}$$

Check the proposed solution in the original equation. Substitute -5 for x in the original equation. You should obtain $-7 = -7$. This true statement verifies that -5 is the solution, or the solution set is $\{-5\}$.

CHECK POINT 4 Solve: $\frac{x+5}{7} + \frac{x-3}{4} = \frac{5}{14}$.

2 Recognize identities, conditional equations, and inconsistent equations.

Types of Equations

Equations can be placed into categories that depend on their solution sets.

An equation that is true for all real numbers for which both sides are defined is called an **identity**. An example of an identity is

$$x + 3 = x + 2 + 1.$$

Every number plus 3 is equal to that number plus 2 plus 1. Therefore, the solution set to this equation is the set of all real numbers. This set is written either as

$$\{x | x \text{ is a real number}\} \quad \text{or} \quad \mathbb{R}.$$

An equation that is not an identity, but that is true for at least one real number, is called a **conditional equation**. The equation $2x + 3 = 17$ is an example of a conditional equation. The equation is not an identity and is true only if x is 7.

An **inconsistent equation** is an equation that is not true for even one real number. An example of an inconsistent equation is

$$x = x + 7.$$

There is no number that is equal to itself plus 7. The equation $x = x + 7$ has no solution. Its solution set is written either as

$$\{\ \} \quad \text{or} \quad \varnothing.$$

These symbols stand for the empty set, a set with no elements.

Study Tip

If you are concerned by the vocabulary of equation types, keep in mind that there are three possible situations. We can state these situations informally as follows:

1. x = a real number
   Conditional equation
2. x = all real numbers
   Identity
3. x = no real numbers.
   Inconsistent equation

If you attempt to solve an identity or an inconsistent equation, you will eliminate the variable. A true statement such as $6 = 6$ or a false statement such as $2 = 3$ will be the result. **If a true statement results, the equation is an identity that is true for all real numbers. If a false statement results, the equation is an inconsistent equation with no solution.**

EXAMPLE 5 Categorizing an Equation

Solve and determine whether the equation

$$2(x + 1) = 2x + 3$$

is an identity, a conditional equation, or an inconsistent equation.

Solution Begin by applying the distributive property on the left side. We obtain

$$2x + 2 = 2x + 3.$$

Does something look strange about $2x + 2 = 2x + 3$? Can doubling a number and increasing the product by 2 give the same result as doubling the same number and increasing the product by 3? No. Let's continue solving the equation by subtracting $2x$ from both sides of $2x + 2 = 2x + 3$.

$$2x - 2x + 2 = 2x - 2x + 3$$

Keep reading. 2 = 3 is not the solution.

$$2 = 3$$

The original equation is equivalent to the statement $2 = 3$, which is false for every value of x. The equation is inconsistent and has no solution. You can express this by writing "no solution" or using one of the symbols for the empty set, $\{\ \}$ or $\varnothing$.

Using Technology

Graphic Connections

How can technology visually reinforce the fact that the equation

$$2(x + 1) = 2x + 3$$

has no solution? Enter $y_1 = 2(x + 1)$ and $y_2 = 2x + 3$. The graphs of y_1 and y_2 appear to be parallel lines with no intersection point. This supports our conclusion that $2(x + 1) = 2x + 3$ is an inconsistent equation with no solution.

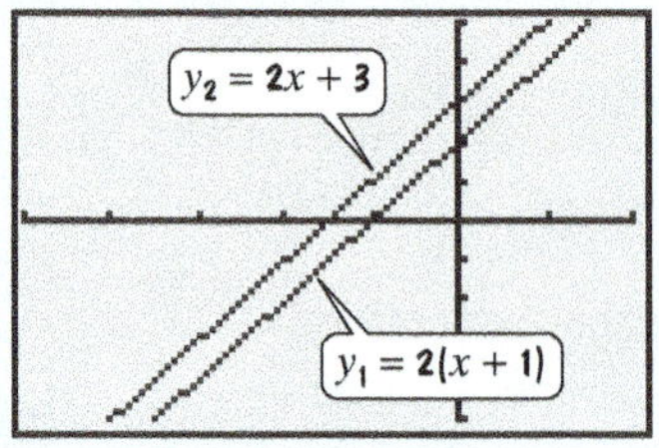

[−5, 2, 1] by [−5, 5, 1]

CHECK POINT 5 Solve and determine whether the equation

$$4x - 7 = 4(x - 1) + 3$$

is an identity, a conditional equation, or an inconsistent equation.

EXAMPLE 6 Categorizing an Equation

Solve and determine whether the equation

$$4x + 6 = 6(x + 1) - 2x$$

is an identity, a conditional equation, or an inconsistent equation.

Solution

$$4x + 6 = 6(x + 1) - 2x$$ This is the given equation.

$$4x + 6 = 6x + 6 - 2x$$ Apply the distributive property on the right side.

$$4x + 6 = 4x + 6$$ Combine like terms on the right side: $6x - 2x = 4x$.

Can you see that the equation $4x + 6 = 4x + 6$ is true for every value of x? Let's continue solving the equation by subtracting $4x$ from both sides.

$$4x - 4x + 6 = 4x - 4x + 6$$

Keep reading. 6 = 6 is not the solution.

$$6 = 6$$

The original equation is equivalent to the statement $6 = 6$, which is true for every value of x. The equation is an identity, and all real numbers are solutions. You can express this by writing "all real numbers" or using one of the following notations:

$$\{x \mid x \text{ is a real number}\} \quad \text{or} \quad \mathbb{R}.$$

Using Technology

Numeric Connections

A graphing utility's TABLE feature can be used to numerically verify that the solution set of

$$4x + 6 = 6(x + 1) - 2x$$

is the set of all real numbers.

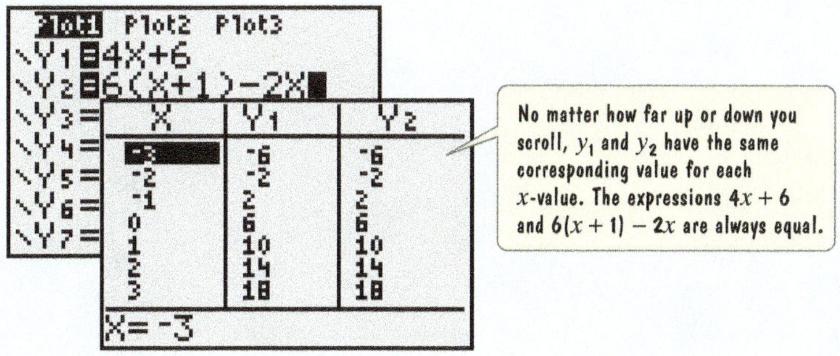

CHECK POINT 6 Solve and determine whether the equation

$$7x + 9 = 9(x + 1) - 2x$$

is an identity, a conditional equation, or an inconsistent equation.

---

3 Solve applied problems using mathematical models.

Applications

Our next example shows how the procedure for solving linear equations can be used to find the value of a variable in a mathematical model.

EXAMPLE 7 Motorcycle Sales in the United States

The formula

$$N = 0.12x + 0.4$$

models the number of new motorcycles sold in the United States, N, in millions, x years after 1998. When will new motorcycle sales reach 2.2 million?

Solution We are interested in when sales will reach 2.2 million, so substitute 2.2 for N in the formula and solve for x, the number of years after 1998.

$$N = 0.12x + 0.4 \quad \text{This is the given formula.}$$

$$2.2 = 0.12x + 0.4 \quad \text{Replace } N \text{ with 2.2.}$$

$$2.2 - 0.4 = 0.12x + 0.4 - 0.4 \quad \text{Subtract 0.4 from both sides.}$$

$$1.8 = 0.12x \quad \text{Simplify.}$$

$$\frac{1.8}{0.12} = \frac{0.12x}{0.12} \quad \text{Divide both sides by 0.12.}$$

$$15 = x \quad \text{Simplify.}$$

The model indicates that 15 years after 1998, or in 2013, new motorcycle sales will reach 2.2 million.

CHECK POINT 7 Use the formula in Example 7 to find when new motorcycle sales reached 1.6 million.

1.4 EXERCISE SET

MyMathLab MathXL PRACTICE WATCH DOWNLOAD READ REVIEW

Practice Exercises

In Exercises 1–24, solve and check each linear equation.

1. $5x + 3 = 18$
2. $3x + 8 = 50$
3. $6x - 3 = 63$
4. $5x - 8 = 72$
5. $14 - 5x = -41$
6. $25 - 6x = -83$
7. $11x - (6x - 5) = 40$
8. $5x - (2x - 8) = 35$
9. $2x - 7 = 6 + x$
10. $3x + 5 = 2x + 13$
11. $7x + 4 = x + 16$
12. $8x + 1 = x + 43$
13. $8y - 3 = 11y + 9$
14. $5y - 2 = 9y + 2$
15. $3(x - 2) + 7 = 2(x + 5)$
16. $2(x - 1) + 3 = x - 3(x + 1)$
17. $3(x - 4) - 4(x - 3) = x + 3 - (x - 2)$
18. $2 - (7x + 5) = 13 - 3x$
19. $16 = 3(x - 1) - (x - 7)$
20. $5x - (2x + 2) = x + (3x - 5)$
21. $7(x + 1) = 4[x - (3 - x)]$
22. $2[3x - (4x - 6)] = 5(x - 6)$
23. $\frac{1}{2}(4z + 8) - 16 = -\frac{2}{3}(9z - 12)$
24. $\frac{3}{4}(24 - 8z) - 16 = -\frac{2}{3}(6z - 9)$

In Exercises 25–38, solve each equation.

25. $\frac{x}{3} = \frac{x}{2} - 2$
26. $\frac{x}{5} = \frac{x}{6} + 1$
27. $20 - \frac{x}{3} = \frac{x}{2}$
28. $\frac{x}{5} - \frac{1}{2} = \frac{x}{6}$
29. $\frac{3x}{5} = \frac{2x}{3} + 1$
30. $\frac{x}{2} = \frac{3x}{4} + 5$
31. $\frac{3x}{5} - x = \frac{x}{10} - \frac{5}{2}$
32. $2x - \frac{2x}{7} = \frac{x}{2} + \frac{17}{2}$
33. $\frac{x + 3}{6} = \frac{2}{3} + \frac{x - 5}{4}$
34. $\frac{x + 1}{4} = \frac{1}{6} + \frac{2 - x}{3}$
35. $\frac{x}{4} = 2 + \frac{x - 3}{3}$
36. $5 + \frac{x - 2}{3} = \frac{x + 3}{8}$
37. $\frac{x + 1}{3} = 5 - \frac{x + 2}{7}$
38. $\frac{3x}{5} - \frac{x - 3}{2} = \frac{x + 2}{3}$

In Exercises 39–50, solve each equation. Then state whether the equation is an identity, a conditional equation, or an inconsistent equation.

39. $5x + 9 = 9(x + 1) - 4x$
40. $4x + 7 = 7(x + 1) - 3x$
41. $3(y + 2) = 7 + 3y$
42. $4(y + 5) = 21 + 4y$
43. $10x + 3 = 8x + 3$
44. $5x + 7 = 2x + 7$
45. $\frac{1}{2}(6z + 20) - 8 = 2(z - 4)$
46. $\frac{1}{3}(6z + 12) = \frac{1}{5}(20z + 30) - 8$
47. $-4x - 3(2 - 2x) = 7 + 2x$
48. $3x - 3(2 - x) = 6(x - 1)$
49. $y + 3(4y + 2) = 6(y + 1) + 5y$
50. $9y - 3(6 - 5y) = y - 2(3y + 9)$

In Exercises 51–54, use the Y= *screen to write the equation being solved. Then use the table to solve the equation.*

51.

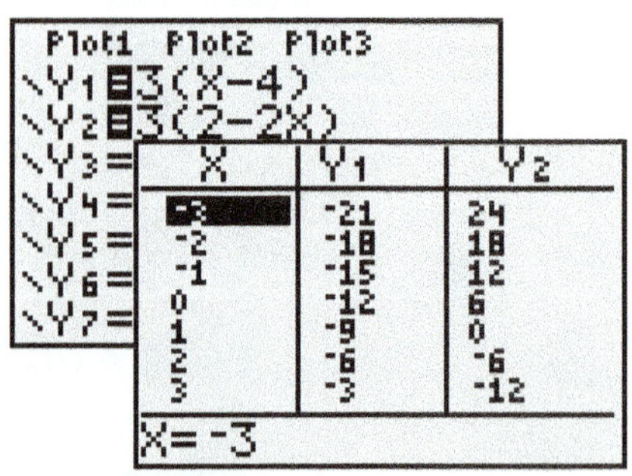

52.

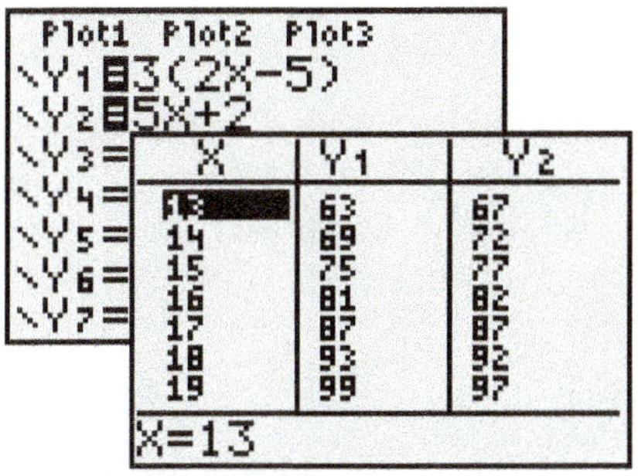

53.

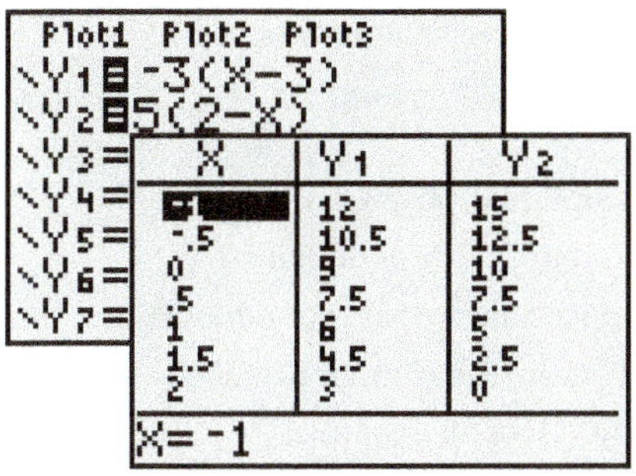

54.

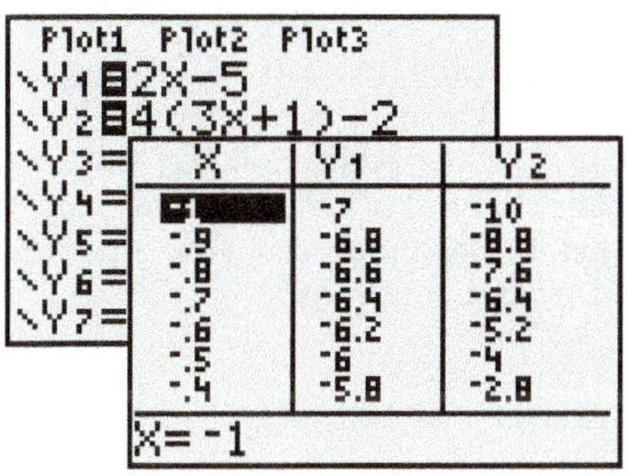

Practice PLUS

55. Evaluate $x^2 - x$ for the value of x satisfying $4(x - 2) + 2 = 4x - 2(2 - x)$.

56. Evaluate $x^2 - x$ for the value of x satisfying $2(x - 6) = 3x + 2(2x - 1)$.

57. Evaluate $x^2 - (xy - y)$ for x satisfying $\frac{3(x + 3)}{5} = 2x + 6$ and y satisfying $-2y - 10 = 5y + 18$.

58. Evaluate $x^2 - (xy - y)$ for x satisfying $\frac{13x - 6}{4} = 5x + 2$ and y satisfying $5 - y = 7(y + 4) + 1$.

In Exercises 59–66, solve each equation.

59. $[(3 + 6)^2 \div 3] \cdot 4 = -54x$

60. $2^3 - [4(5 - 3)^3] = -8x$

61. $5 - 12x = 8 - 7x - [6 \div 3(2 + 5^3) + 5x]$

62. $2(5x + 58) = 10x + 4(21 \div 3.5 - 11)$

63. $0.7x + 0.4(20) = 0.5(x + 20)$

64. $0.5(x + 2) = 0.1 + 3(0.1x + 0.3)$

65. $4x + 13 - \{2x - [4(x - 3) - 5]\} = 2(x - 6)$

66. $-2\{7 - [4 - 2(1 - x) + 3]\} = 10 - [4x - 2(x - 3)]$

Application Exercises

67. The bar graph shows the average cost of tuition and fees at private four-year colleges in the United States.

Average Cost of Tuition and Fees at Private Four-Year United States Colleges

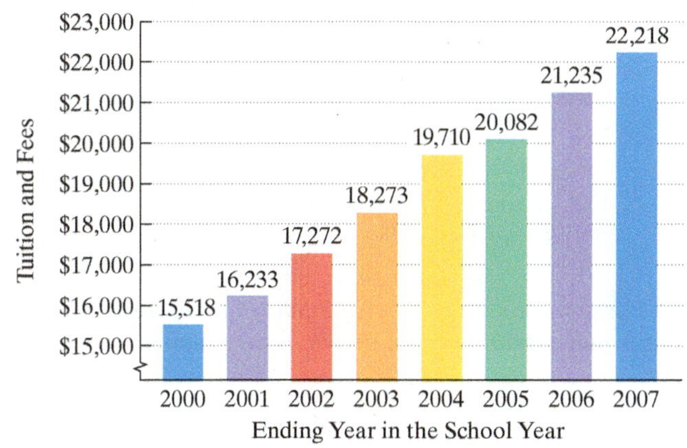

Source: The College Board

Here are two mathematical models for the data shown in the graph. In each formula, T represents the average cost of tuition and fees at private U.S. colleges for the school year ending x years after 2000.

Model 1 $T = 974x + 15{,}410$

Model 2 $T = -2.1x^2 + 988x + 15{,}395$

a. Use each model to find the average cost, to the nearest dollar, of tuition and fees at private U.S. colleges for the school year ending in 2007. By how much does each model underestimate or overestimate the actual cost shown for the school year ending in 2007?

b. Use model 1 to determine when tuition and fees at private four-year colleges will average $27,098.

68. The bar graph shows the average cost of tuition and fees at public four-year colleges in the United States.

Average Cost of Tuition and Fees at Public Four-Year United States Colleges

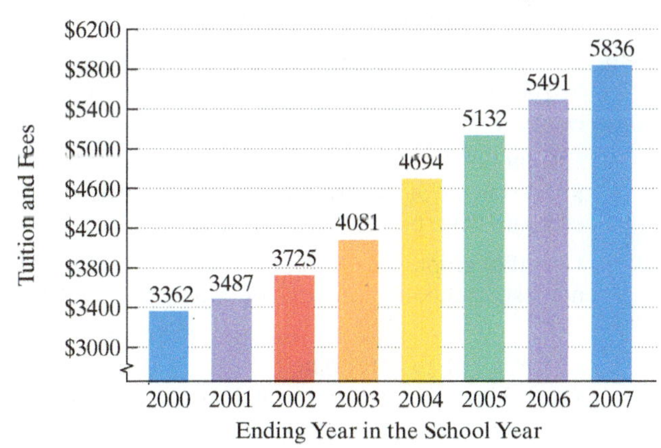

Source: The College Board

At the top of the next page are two mathematical models for the data shown in the graph. In each formula, T represents the average cost of tuition and fees at public U.S. colleges for the school year ending x years after 2000.

Model 1 $T = 383x + 3136$

Model 2 $T = 17x^2 + 261x + 3257$

a. Use each model to find the average cost of tuition and fees at public U.S. colleges for the school year ending in 2007. Which model provides the better description for the actual cost shown by the bar graph?

b. Use model 1 to determine when tuition and fees at public four-year colleges will average \$8498.

The line graph shows the cost of inflation. What cost \$10,000 in 1975 would cost the amount shown by the graph in subsequent years.

The Cost of Inflation

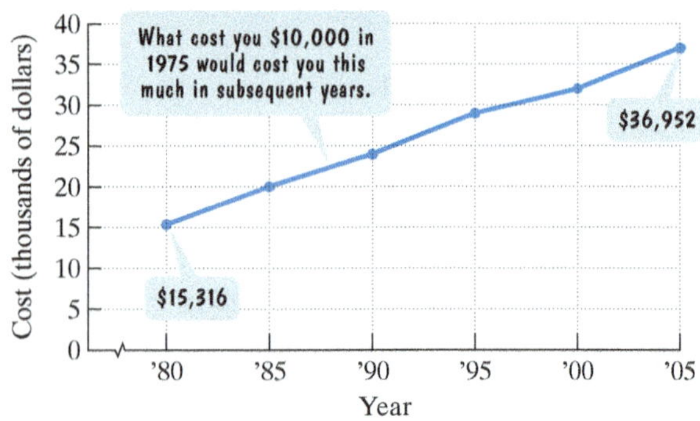

Source: Bureau of Labor Statistics

Here are two mathematical models for the data shown in the graph. In each formula, C represents the cost x years after 1980 of what cost \$10,000 in 1975.

Model 1 $C = 865x + 15{,}316$

Model 2 $C = -2x^2 + 900x + 15{,}397$

Use these models to solve Exercises 69–73.

69. a. Use the graph to estimate the cost in 2000, to the nearest thousand dollars, of what cost \$10,000 in 1975.

b. Use model 1 to determine the cost in 2000 of what cost \$10,000 in 1975. By how much does this differ from your estimate from part (a)?

c. Use model 2 to determine the cost in 2000 of what cost \$10,000 in 1975. By how much does this differ from your estimate from part (a)?

70. a. Use the graph to estimate the cost in 1990, to the nearest thousand dollars, of what cost \$10,000 in 1975.

b. Use model 1 to determine the cost in 1990 of what cost \$10,000 in 1975. By how much does this differ from your estimate from part (a)?

c. Use model 2 to determine the cost in 1990 of what cost \$10,000 in 1975. By how much does this differ from your estimate from part (a)?

71. Which model is a better description for the cost in 2005 of what cost \$10,000 in 1975?

72. Use model 1 to determine in which year the cost will be \$43,861 for what cost \$10,000 in 1975.

73. Use model 1 to determine in which year the cost will be \$54,241 for what cost \$10,000 in 1975.

Writing in Mathematics

74. What is a linear equation in one variable? Give an example of this type of equation.

75. What does it mean to solve an equation?

76. How do you determine if a number is a solution of an equation?

77. What are equivalent equations? Give an example.

78. What is the addition property of equality?

79. What is the multiplication property of equality?

80. Explain how to clear an equation of fractions.

81. What is an identity? Give an example.

82. What is a conditional equation? Give an example.

83. What is an inconsistent equation? Give an example.

84. Despite low rates of inflation, the cost of a college education continues to skyrocket. This is a departure from the trend during the 1970s: In constant dollars (which negate the effect of inflation), the cost of college actually decreased several times. What explanations can you offer for the increasing cost of a college education?

Technology Exercises

In Exercises 85–88, use your graphing utility to enter each side of the equation separately under y_1 and y_2. Then use the utility's TABLE *or* GRAPH *feature to solve the equation.*

85. $5x + 2(x - 1) = 3x + 10$

86. $2x + 3(x - 4) = 4x - 7$

87. $3(2x + 11) = 3(5 + x)$

88. $\frac{2x - 1}{3} - \frac{x - 5}{6} = \frac{x - 3}{4}$

Critical Thinking Exercises

Make Sense? *In Exercises 89–92, determine whether each statement "makes sense" or "does not make sense" and explain your reasoning*

89. Because $x = x + 5$ is an inconsistent equation, the graphs of $y = x$ and $y = x + 5$ should not intersect.

90. Because subtraction is defined in terms of addition, it's not necessary to state a separate subtraction property of equality to generate equivalent equations.

91. The number 3 satisfies the equation $7x + 9 = 9(x + 1) - 2x$, so $\{3\}$ is the equation's solution set.

92. I can solve $-2x = 10$ using the addition property of equality.

In Exercises 93–96, determine whether each statement is true or false. If the statement is false, make the necessary change(s) to produce a true statement.

93. The equation $-7x = x$ has no solution.

94. The equations $\frac{x}{x-4} = \frac{4}{x-4}$ and $x = 4$ are equivalent.

95. The equations $3y - 1 = 11$ and $3y - 7 = 5$ are equivalent.

96. If a and b are any real numbers, then $ax + b = 0$ always has only one number in its solution set.

97. Solve for x: $ax + b = c$.

98. Write three equations that are equivalent to $x = 5$.

99. If x represents a number, write an English sentence about the number that results in an inconsistent equation.

100. Find b such that $\frac{7x+4}{b} + 13 = x$ will have a solution set given by $\{-6\}$.

Review Exercises

In Exercises 101–102, perform the indicated operations.

101. $-\frac{1}{5} - \left(-\frac{1}{2}\right)$ (Section 1.2, Example 4)

102. $4(-3)(-1)(-5)$ (Section 1.2, Examples in **Table 1.2**)

103. Graph $y = x^2 - 4$. Let $x = -3, -2, -1, 0, 1, 2,$ and 3. (Section 1.3, Example 2)

Preview Exercises

Exercises 104–106 will help you prepare for the material covered in the next section.

104. Let x represent a number.

a. Write an equation in x that describes the following conditions:

Four less than three times the number is 32.

b. Solve the equation and determine the number.

105. Let x represent the number of countries in the world that are not free. The number of free countries exceeds the number of not-free countries by 44. Write an algebraic expression that represents the number of free countries.

106. You purchase a new car for $20,000. Each year the value of the car decreases by $2500. Write an algebraic expression that represents the car's value, in dollars, after x years.

MID-CHAPTER CHECK POINT Section 1.1–Section 1.4

✓ What You Know: We reviewed a number of topics from introductory algebra, including the real numbers and their representations on number lines. We performed operations with real numbers and applied the order-of-operations agreement to expressions containing more than one operation. We used commutative, associative, and distributive properties to simplify algebraic expressions. We used the rectangular coordinate system to represent ordered pairs of real numbers and graph equations in two variables. Finally, we solved linear equations, including equations with fractions. We saw that some equations have no solution, whereas others have all real numbers as solutions.

In Exercises 1–14, simplify the expression or solve the equation, whichever is appropriate.

1. $-5 + 3(x + 5)$
2. $-5 + 3(x + 5) = 2(3x - 4)$
3. $3[7 - 4(5 - 2)]$
4. $\frac{x-3}{5} - 1 = \frac{x-5}{4}$
5. $\frac{-2^4 + (-2)^2}{-4 - (2 - 2)}$
6. $7x - [8 - 3(2x - 5)]$
7. $3(2x - 5) - 2(4x + 1) = -5(x + 3) - 2$
8. $3(2x - 5) - 2(4x + 1) - 5(x + 3) - 2$
9. $-4^2 \div 2 + (-3)(-5)$
10. $3x + 1 - (x - 5) = 2x - 4$
11. $\frac{3x}{4} - \frac{x}{3} + 1 = \frac{4x}{5} - \frac{3}{20}$
12. $(6 - 9)(8 - 12) \div \frac{5^2 + 4 \div 2}{8^2 - 9^2 + 8}$
13. $4x - 2(1 - x) = 3(2x + 1) - 5$
14. $\frac{3[4 - 3(-2)^2]}{2^2 - 2^4}$

In Exercises 15–17, graph each equation in a rectangular coordinate system.

15. $y = 2x - 1$ 16. $y = 1 - |x|$ 17. $y = x^2 + 2$

In Exercises 18–21, determine whether each statement is true or false.

18. $-\left|-\frac{\sqrt{3}}{5}\right| = -\frac{\sqrt{3}}{5}$
19. $\{x | x \text{ is a negative integer greater than } -4\} = \{-4, -3, -2, -1\}$
20. $-17 \notin \{x | x \text{ is a rational number}\}$
21. $-128 \div (2 \cdot 4) > (-128 \div 2) \cdot 4$

SECTION 1.5

Problem Solving and Using Formulas

Objectives

1. Solve algebraic word problems using linear equations.
2. Solve a formula for a variable.

The human race is undeniably becoming a faster race. Since the beginning of the past century, track-and-field records have fallen in everything from sprints to miles to marathons. The performance arc is clearly rising, but no one knows how much higher it can climb. At some point, even the best-trained body simply has to up and quit. The question is, just where is that point, and is it possible for athletes, trainers, and genetic engineers to push it higher? In this section, you will learn a problem-solving strategy that uses linear equations to determine if anyone will ever run a 3-minute mile.

1 Solve algebraic word problems using linear equations.

Problem Solving with Linear Equations

We have seen that a model is a mathematical representation of a real-world situation. In this section, we will be solving problems that are presented in English. This means that we must obtain models by translating from the ordinary language of English into the language of algebraic equations. To translate, however, we must understand the English prose and be familiar with the forms of algebraic language. Here are some general steps we will follow in solving word problems:

Study Tip

When solving word problems, particularly problems involving geometric figures, drawing a picture of the situation is often helpful. Label x on your drawing and, where appropriate, label other parts of the drawing in terms of x.

Strategy for Solving Word Problems

Step 1. Read the problem carefully. Attempt to state the problem in your own words and state what the problem is looking for. Let x (or any variable) represent one of the unknown quantities in the problem.

Step 2. If necessary, write expressions for any other unknown quantities in the problem in terms of x.

Step 3. Write an equation in x that models the verbal conditions of the problem.

Step 4. Solve the equation and answer the problem's question.

Step 5. Check the solution *in the original wording* of the problem, not in the equation obtained from the words.

World's Countries by Status of Freedom

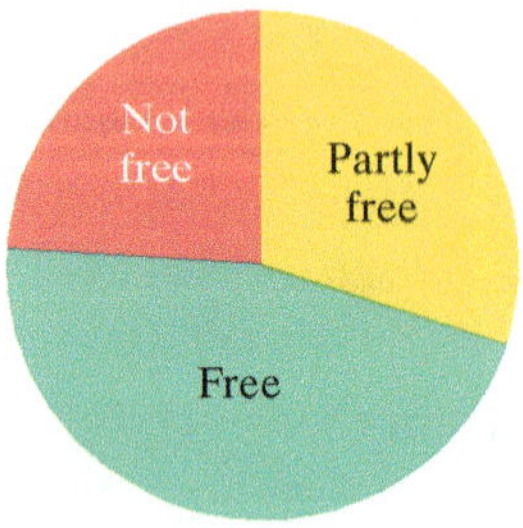

FIGURE 1.19

Source: Larry Berman and Bruce Murphy, *Approaching Democracy*, 5th Edition, Prentice Hall, 2007

EXAMPLE 1 Status of Freedom in the World

The circle graph in **Figure 1.19** represents the 2006 breakdown of the world's 192 countries by free, partly free, or not free. The number of free countries exceeds the number of not-free countries by 44. The number of partly free countries exceeds the number of not-free countries by 13. Determine the number of countries in the world that fall into each of the categories in the circle graph.

Solution We must determine the numbers of countries that are free, partly free, and not free.

Step 1. Let x represent one of the quantities. We know something about free countries and partly free countries: The numbers exceed the number of not-free countries by 44 and 13, respectively. We will let

$$x = \text{the number of not-free countries.}$$

Step 2. Represent other unknown quantities in terms of x. Because the number of free countries exceeds the number of not-free countries by 44, let

$$x + 44 = \text{the number of free countries.}$$

Because the number of partly free countries exceeds the number of not-free countries by 13, let

$$x + 13 = \text{the number of partly free countries.}$$

Step 3. Write an equation in x that models the conditions. **Figure 1.19** categorizes the world's 192 countries by free, partly free, or not free.

The number of not-free countries	plus	the number of free countries	plus	the number of partly free countries	equals	192 countries.
x	$+$	$(x + 44)$	$+$	$(x + 13)$	$=$	192

Step 4. Solve the equation and answer the question.

$x + (x + 44) + (x + 13) = 192$ This is the equation that models the problem's conditions.

$3x + 57 = 192$ Remove parentheses, regroup, and combine like terms.

$3x = 135$ Subtract 57 from both sides.

$x = 45$ Divide both sides by 3.

Thus,

the number of not-free countries $= x = 45$,

the number of free countries $= x + 44 = 45 + 44 = 89$,

and the number of partly free countries $= x + 13 = 45 + 13 = 58$.

There are 45 countries that are not free, 89 countries that are free, and 58 countries that are partly free.

Step 5. Check the proposed solution in the original wording of the problem. The problem states that we are categorizing the world's 192 countries by status of freedom. By adding 45, 89, and 58, the numbers that we found in each category, we obtain

$$45 + 89 + 58 = 192,$$

as specified by the problem's conditions.

Study Tip

Modeling with the word *exceeds* can be a bit tricky. It's helpful to identify the smaller quantity. Then add to this quantity to represent the larger quantity. For example, suppose that Tim's height exceeds Tom's height by a inches. Tom is the shorter person. If Tom's height is represented by x, then Tim's height is represented by $x + a$.

✓ CHECK POINT 1 There are 46 countries in the world with Muslim majorities. Of these 46 countries, the number that are not free exceeds the number that are free by 20. The number that are partly free exceeds the number that are free by 17. Determine the numbers of countries with Muslim majorities that are free, not free, and partly free. (*Source*: 2006 data from Larry Berman and Bruce Murphy, *Ibid.*)

Mile Records

1886	4:12.3	1958	3:54.5
1923	4:10.4	1966	3:51.3
1933	4:07.6	1979	3:48.9
1945	4:01.3	1985	3:46.3
1954	3:59.4	1999	3:43.1

Source: U.S.A. Track and Field

EXAMPLE 2 Will Anyone Ever Run a Three-Minute Mile?

One yardstick for measuring how steadily—if slowly—athletic performance has improved is the mile run. In 1923, the record for the mile was a comparatively sleepy 4 minutes, 10.4 seconds. In 1954, Roger Bannister of Britain cracked the 4-minute mark, coming in at 3 minutes, 59.4 seconds. In the half-century since, about 0.3 second per year has been shaved off Bannister's record. If this trend continues, by which year will someone run a 3-minute mile?

Solution In solving this problem, we will express time for the mile run in seconds. Our interest is in a time of 3 minutes, or 180 seconds.

Step 1. Let x represent one of the quantities. Here is the critical information in the problem:

- In 1954, the record was 3 minutes, 59.4 seconds, or 239.4 seconds.
- The record has decreased by 0.3 second per year since then.

We are interested in when the record will be 180 seconds. Let

x = the number of years after 1954 when someone will run a 3-minute mile.

Step 2. Represent other unknown quantities in terms of x. There are no other unknown quantities to find, so we can skip this step.

Step 3. Write an equation in x that models the conditions.

The 1954 record time	decreased by	0.3 second per year for x years	equals	the 3-minute, or 180-second, mile.
239.4	$-$	$0.3x$	$=$	180

Step 4. Solve the equation and answer the question.

$$239.4 - 0.3x = 180 \quad \text{This is the equation that models the problem's conditions.}$$

$$239.4 - 239.4 - 0.3x = 180 - 239.4 \quad \text{Subtract 239.4 from both sides.}$$

$$-0.3x = -59.4 \quad \text{Simplify.}$$

$$\frac{-0.3x}{-0.3} = \frac{-59.4}{-0.3} \quad \text{Divide both sides by } -0.3.$$

$$x = 198 \quad \text{Simplify.}$$

Using current trends, by 198 years (gasp!) after 1954, or in 2152, someone will run a 3-minute mile.

Step 5. Check the proposed solution in the original wording of the problem. The problem states that the record time should be 180 seconds. Do we obtain 180 seconds if we decrease the 1954 record time, 239.4 seconds, by 0.3 second per year for 198 years, our proposed solution?

$$239.4 - 0.3(198) = 239.4 - 59.4 = 180$$

This verifies that, using current trends, the 3-minute mile will be run 198 years after 1954.

✓ CHECK POINT 2 Cars in the United States are being driven longer. In 2005, the average age of a U.S. automobile was 9 years. Between 1990 and 2005, this average age in operation had increased by 0.17 year per year. If this trend continues, by which year will automobiles in the United States have an average age of 10.7 years? (*Source*: Bureau of Transportation Statistics)

EXAMPLE 3 Selecting a Long-Distance Carrier

You are choosing between two long-distance telephone plans. Plan A has a monthly fee of \$20 with a charge of \$0.05 per minute for all long-distance calls. Plan B has a monthly fee of \$5 with a charge of \$0.10 per minute for all long-distance calls. For how many minutes of long-distance calls will the costs for the two plans be the same?

Solution

Step 1. Let x represent one of the quantities. Let

x = the number of minutes of long-distance calls for which the two plans cost the same.

Step 2. Represent other unknown quantities in terms of x. There are no other unknown quantities, so we can skip this step.

Step 3. Write an equation in x that models the conditions. The monthly cost for plan A is the monthly fee, \$20, plus the per minute charge, \$0.05, times the number of minutes of long-distance calls, x. The monthly cost for plan B is the monthly fee, \$5, plus the per-minute charge, \$0.10, times the number of minutes of long-distance calls, x.

The monthly cost for plan A | must equal | the monthly cost for plan B.

$$20 + 0.05x \quad = \quad 5 + 0.10x$$

Step 4. Solve the equation and answer the question.

$$20 + 0.05x = 5 + 0.10x$$ This is the equation that models the problem's conditions.

$$20 = 5 + 0.05x$$ Subtract 0.05x from both sides.

$$15 = 0.05x$$ Subtract 5 from both sides.

$$\frac{15}{0.05} = \frac{0.05x}{0.05}$$ Divide both sides by 0.05.

$$300 = x$$ Simplify.

Because x represents the number of minutes of long-distance calls for which the two plans cost the same, the costs will be the same for 300 minutes of long-distance calls.

Step 5. Check the proposed solution in the original wording of the problem. The problem states that the costs for the two plans should be the same. Let's see if they are with 300 minutes of long-distance calls:

$$\text{Cost for plan A} = \$20 + \$0.05(300) = \$20 + \$15 = \$35$$

Monthly fee | Per-minute charge

$$\text{Cost for plan B} = \$5 + \$0.10(300) = \$5 + \$30 = \$35.$$

With 300 minutes, or 5 hours, of long-distance chatting, both plans cost \$35 for the month. Thus, the proposed solution, 300 minutes, satisfies the problem's conditions.

Using Technology

Numeric and Graphic Connections

We can use a graphing utility to numerically or graphically verify our work in Example 3.

The monthly cost for plan A | must equal | the monthly cost for plan B.

$$20 + 0.05x \quad = \quad 5 + 0.10x$$

Enter $y_1 = 20 + .05x$. | Enter $y_2 = 5 + .10x$.

Numeric Check

Display a table for y_1 and y_2.

X	Y1	Y2
100	25	15
150	27.5	20
200	30	25
250	32.5	30
300	35	35
350	37.5	40
400	40	45

X=300

When $x = 300$, y_1 and y_2 have the same value, 35. With 300 minutes of calls, costs are the same, \$35, for both plans.

Graphic Check

Display graphs for y_1 and y_2. Use the intersection feature.

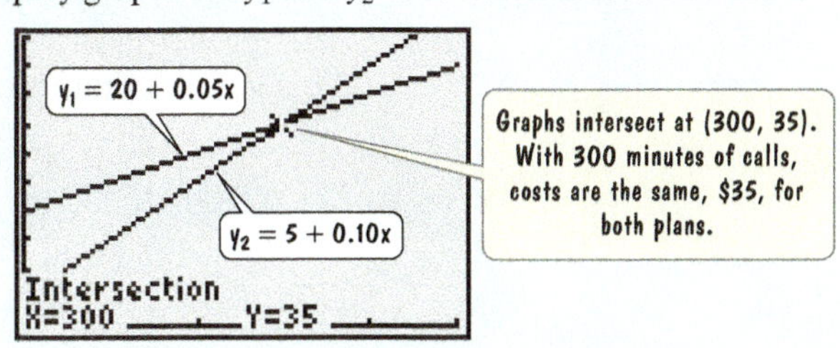

[0, 500, 100] by [0, 50, 5]

☑ **CHECK POINT 3** You are choosing between two long-distance telephone plans. Plan A has a monthly fee of \$15 with a charge of \$0.08 per minute for all long-distance calls. Plan B has a monthly fee of \$3 with a charge of \$0.12 per minute for all long-distance calls. For how many minutes of long-distance calls will the costs for the two plans be the same?

EXAMPLE 4 A Price Reduction on a Digital Camera

Your local computer store is having a terrific sale on digital cameras. After a 40% price reduction, you purchase a digital camera for \$276. What was the camera's price before the reduction?

Solution

Step 1. Let x represent one of the quantities. We will let

x = the original price of the digital camera prior to the reduction.

Step 2. Represent other unknown quantities in terms of x. There are no other unknown quantities to find, so we can skip this step.

Step 3. Write an equation in x that models the conditions. The camera's original price minus the 40% reduction is the reduced price, \$276.

Study Tip

Observe that the original price, x, reduced by 40% is $x - 0.4x$ and *not* $x - 0.4$.

Original price	minus	the reduction (40% of the original price)	is	the reduced price, \$276.
x	$-$	$0.4x$	$=$	276

Step 4. Solve the equation and answer the question.

$$x - 0.4x = 276$$ This is the equation that models the problem's conditions.

$$0.6x = 276$$ Combine like terms: $x - 0.4x = 1x - 0.4x = 0.6x$.

$$\frac{0.6x}{0.6} = \frac{276}{0.6}$$ Divide both sides by 0.6.

$$x = 460$$ Simplify: $0.6\overline{)276.0}$ = 460.

The digital camera's price before the reduction was \$460.

Step 5. Check the proposed solution in the original wording of the problem. The price before the reduction, \$460, minus the 40% reduction should equal the reduced price given in the original wording, \$276:

$$460 - 40\% \text{ of } 460 = 460 - 0.4(460) = 460 - 184 = 276.$$

This verifies that the digital camera's price before the reduction was \$460.

☑ **CHECK POINT 4** After a 30% price reduction, you purchase a new computer for \$840. What was the computer's price before the reduction?

Solving geometry problems usually requires a knowledge of basic geometric ideas and formulas. Formulas for area, perimeter, and volume are given in **Table 1.5**.

Table 1.5 Common Formulas for Area, Perimeter, and Volume

Square	Rectangle	Circle	Triangle	Trapezoid
$A = s^2$	$A = lw$	$A = \pi r^2$	$A = \frac{1}{2}bh$	$A = \frac{1}{2}h(a+b)$
$P = 4s$	$P = 2l + 2w$	$C = 2\pi r$		

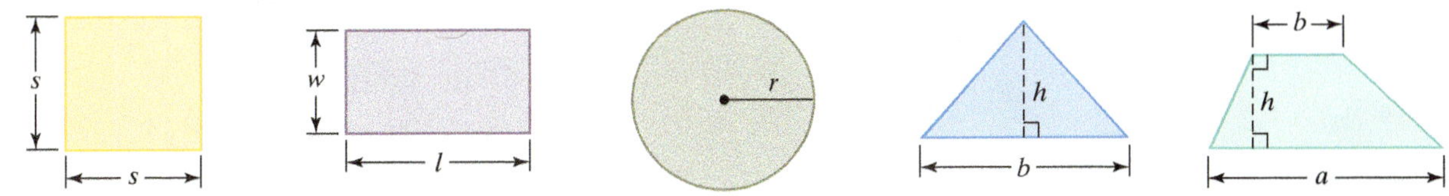

Cube	Rectangular Solid	Circular Cylinder	Sphere	Cone
$V = s^3$	$V = lwh$	$V = \pi r^2 h$	$V = \frac{4}{3}\pi r^3$	$V = \frac{1}{3}\pi r^2 h$

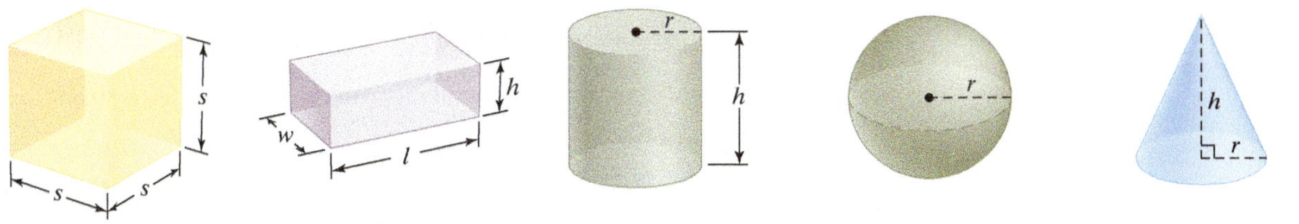

We will be using the formula for the perimeter of a rectangle, $P = 2l + 2w$, in our next example. The formula states that a rectangle's perimeter is the sum of twice its length and twice its width.

EXAMPLE 5 Finding the Dimensions of an American Football Field

The length of an American football field is 200 feet more than the width. If the perimeter of the field is 1040 feet, what are its dimensions?

Solution

Step 1. Let x represent one of the quantities. We know something about the length; the length is 200 feet more than the width. We will let

$$x = \text{the width.}$$

Step 2. Represent other unknown quantities in terms of x. Because the length is 200 feet more than the width, let

$$x + 200 = \text{the length.}$$

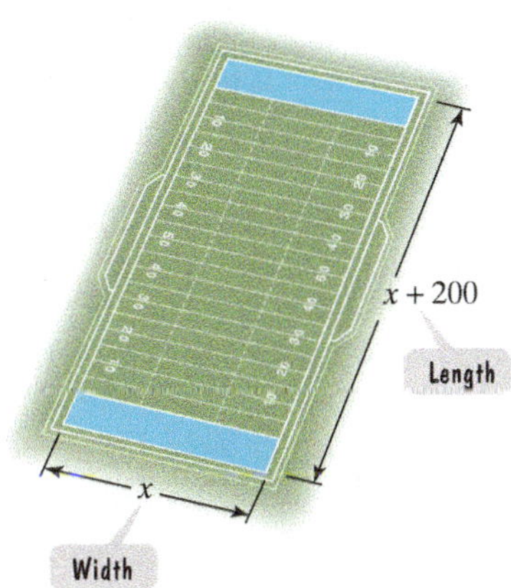

FIGURE 1.20
An American football field

Figure 1.20 illustrates an American football field and its dimensions.

Step 3. Write an equation in x that models the conditions. Because the perimeter of the field is 1040 feet,

Twice the length	plus	twice the width	is	the perimeter.
$2(x + 200)$	$+$	$2x$	$=$	$1040.$

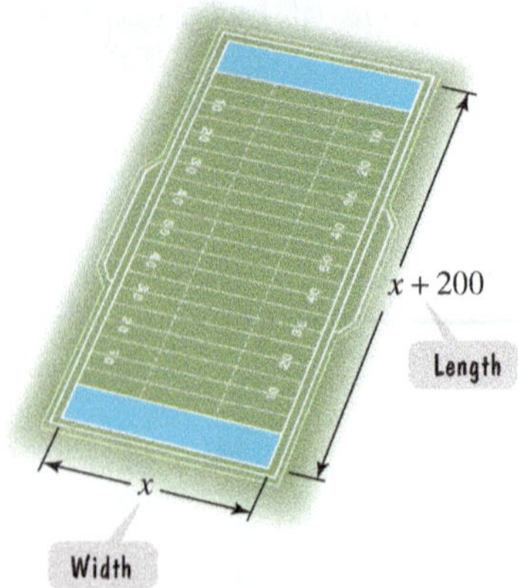

FIGURE 1.20 (repeated)

Step 4. Solve the equation and answer the question.

$$2(x + 200) + 2x = 1040 \quad \text{This is the equation that models the problem's conditions.}$$
$$2x + 400 + 2x = 1040 \quad \text{Apply the distributive property.}$$
$$4x + 400 = 1040 \quad \text{Combine like terms: } 2x + 2x = 4x.$$
$$4x = 640 \quad \text{Subtract 400 from both sides.}$$
$$x = 160 \quad \text{Divide both sides by 4.}$$

Thus,

$$\text{width} = x = 160.$$
$$\text{length} = x + 200 = 160 + 200 = 360.$$

The dimensions of an American football field are 160 feet by 360 feet. (The 360-foot length is usually described as 120 yards.)

Step 5. Check the proposed solution in the original wording of the problem. The perimeter of the football field using the dimensions that we found is

$$2(360 \text{ feet}) + 2(160 \text{ feet}) = 720 \text{ feet} + 320 \text{ feet} = 1040 \text{ feet}.$$

Because the problem's wording tells us that the perimeter is 1040 feet, our dimensions are correct.

✓ **CHECK POINT 5** The length of a rectangular basketball court is 44 feet more than the width. If the perimeter of the basketball court is 288 feet, what are its dimensions?

---

2 Solve a formula for a variable.

Solving a Formula for One of its Variables

We know that solving an equation is the process of finding the number (or numbers) that make the equation a true statement. All of the equations we have solved contained only one letter, x.

By contrast, formulas contain two or more letters, representing two or more variables. An example is the formula for the perimeter of a rectangle:

$$2l + 2w = P.$$

We say that this formula is solved for the variable P because P is alone on one side of the equation and the other side does not contain a P.

Solving a formula for a variable means using the addition and multiplication properties of equality to rewrite the formula so that the variable is isolated on one side of the equation. It does not mean obtaining a numerical value for that variable.

To solve a formula for one of its variables, treat that variable as if it were the only variable in the equation. Think of the other variables as if they were numbers. Use the addition property of equality to isolate all terms with the specified variable on one side of the equation and all terms without the specified variable on the other side. Then use the multiplication property of equality to get the specified variable alone. The next example shows how to do this.

EXAMPLE 6 Solving a Formula for a Variable

Solve the formula $2l + 2w = P$ for l.

Solution First, isolate $2l$ on the left by subtracting $2w$ from both sides. Then solve for l by dividing both sides by 2.

We need to isolate l.

$$2l + 2w = P \quad \text{This is the given formula.}$$

$$2l + 2w - 2w = P - 2w \quad \text{Isolate } 2l \text{ by subtracting } 2w \text{ from both sides.}$$

$$2l = P - 2w \quad \text{Simplify.}$$

$$\frac{2l}{2} = \frac{P - 2w}{2} \quad \text{Solve for } l \text{ by dividing both sides by 2.}$$

$$l = \frac{P - 2w}{2} \quad \text{Simplify.}$$

CHECK POINT 6 Solve the formula $2l + 2w = P$ for w.

EXAMPLE 7 Solving a Formula for a Variable

Circular Cylinder

$V = \pi r^2 h$

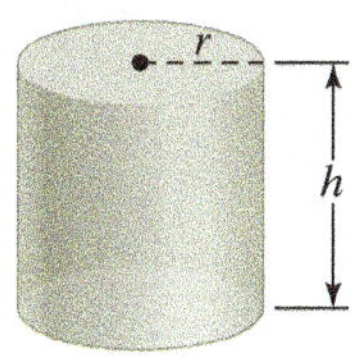

Table 1.5 on page 55 shows that the volume of a circular cylinder is given by the formula

$$V = \pi r^2 h,$$

where r is the radius of the circle at either end and h is the height. Solve this formula for h.

Solution Our goal is to get h by itself on one side of the formula. There is only one term with h, $\pi r^2 h$, and it is already isolated on the right side. We isolate h on the right by dividing both sides by πr^2.

We need to isolate h.

$$V = \pi r^2 h \quad \text{This is the given formula.}$$

$$\frac{V}{\pi r^2} = \frac{\pi r^2 h}{\pi r^2} \quad \text{Isolate } h \text{ by dividing both sides by } \pi r^2.$$

$$\frac{V}{\pi r^2} = h \quad \text{Simplify: } \frac{\pi r^2 h}{\pi r^2} = \frac{\pi r^2}{\pi r^2} \cdot h = 1h = h.$$

Equivalently,

$$h = \frac{V}{\pi r^2}.$$

CHECK POINT 7 The volume of a rectangular solid is the product of its length, width, and height:

$$V = lwh.$$

Solve this formula for h.

The Celsius scale is on the left and the Fahrenheit scale is on the right.

You'll be leaving the cold of winter for a vacation to Hawaii. CNN International reports a temperature in Hawaii of 30°C. Should you pack a winter coat? You can convert from Celsius temperature, C, to Fahrenheit temperature, F, using the formula

$$F = \frac{9}{5}C + 32.$$

A temperature of 30°C corresponds to a Fahrenheit temperature of

$$F = \frac{9}{5} \cdot 30 + 32 = \frac{9}{\cancel{5}_1} \cdot \frac{\cancel{30}^6}{1} + 32 = 54 + 32 = 86,$$

or a balmy 86°F. (Don't pack the coat.)

Visitors to the United States are more likely to be familiar with the Celsius temperature scale. For them, a useful formula is one that can be used to convert from Fahrenheit to Celsius. In Example 8, you will see how to obtain such a formula.

EXAMPLE 8 Solving a Formula for a Variable

Solve the formula

$$F = \frac{9}{5}C + 32$$

for C.

Solution We begin by multiplying both sides of the formula by 5 to clear the fraction. Then we isolate the variable C.

$$F = \frac{9}{5}C + 32 \quad \text{This is the given formula.}$$

$$5F = 5\left(\frac{9}{5}C + 32\right) \quad \text{Multiply both sides by 5.}$$

$$5F = 5 \cdot \frac{9}{5}C + 5 \cdot 32 \quad \text{Apply the distributive property.}$$

We need to isolate C.

$$5F = 9C + 160 \quad \text{Simplify.}$$

$$5F - 160 = 9C + 160 - 160 \quad \text{Subtract 160 from both sides.}$$

$$5F - 160 = 9C \quad \text{Simplify.}$$

$$\frac{5F - 160}{9} = \frac{9C}{9} \quad \text{Divide both sides by 9.}$$

$$\frac{5F - 160}{9} = C \quad \text{Simplify.}$$

Using the distributive property, we can express $5F - 160$ as $5(F - 32)$. Thus,

$$C = \frac{5F - 160}{9} = \frac{5(F - 32)}{9}.$$

This formula, used to convert from Fahrenheit to Celsius, is usually given as

$$C = \frac{5}{9}(F - 32).$$

CHECK POINT 8 The formula

$$\frac{W}{2} - 3H = 53$$

models the recommended weight, W, in pounds, for a male, where H represents his height, in inches, over 5 feet. Solve this formula for W.

EXAMPLE 9 Solving a Formula for a Variable That Occurs Twice

The formula

$$A = P + Prt$$

describes the amount, A, that a principal of P dollars is worth after t years when invested at a simple annual interest rate, r. Solve this formula for P.

Solution Notice that all the terms with P already occur on the right side of the formula.

We need to isolate P.

$$A = P + Prt$$

Study Tip

You cannot solve $A = P + Prt$ for P by subtracting Prt from both sides and writing

$$A - Prt = P.$$

When a formula is solved for a specified variable, that variable must be isolated on one side. The variable P occurs on both sides of

$$A - Prt = P.$$

We can use the distributive property in the form $ab + ac = a(b + c)$ to convert the two occurrences of P into one.

$$A = P + Prt \quad \text{This is the given formula.}$$

$$A = P(1 + rt) \quad \text{Use the distributive property to obtain a single occurrence of } P.$$

$$\frac{A}{1 + rt} = \frac{P(1 + rt)}{1 + rt} \quad \text{Divide both sides by } 1 + rt.$$

$$\frac{A}{1 + rt} = P \quad \text{Simplify: } \frac{P(1 + rt)}{1(1 + rt)} = \frac{P}{1} = P.$$

Equivalently,

$$P = \frac{A}{1 + rt}.$$

CHECK POINT 9 Solve the formula $P = C + MC$ for C.

Blitzer Bonus

Einstein's Famous Formula: $E = mc^2$

One of the most famous formulas in the world is $E = mc^2$, formulated by Albert Einstein. Einstein showed that any form of energy has mass and that mass itself is a form of energy. In this formula, E represents energy, in ergs, m represents mass, in grams, and c represents the speed of light. Because light travels at 30 billion centimeters per second, the formula indicates that 1 gram of mass will produce 900 billion billion ergs of energy.

Einstein's formula implies that the mass of a golf ball could provide the daily energy needs of the metropolitan Boston area. Mass and energy are equivalent, and the transformation of even a tiny amount of mass releases an enormous amount of energy. If this energy is released suddenly, a destructive force is unleashed, as in an atom bomb. When the release is gradual and controlled, the energy can be used to generate power.

The theoretical results implied by Einstein's formula $E = mc^2$ have not been realized because scientists have not yet developed a way of converting a mass completely to energy.

1.5 EXERCISE SET

MyMathLab — MathXL PRACTICE — WATCH — DOWNLOAD — READ — REVIEW

Practice Exercises

Use the five-step strategy for solving word problems to find the number or numbers described in Exercises 1–10.

1. When five times a number is decreased by 4, the result is 26. What is the number?
2. When two times a number is decreased by 3, the result is 11. What is the number?
3. When a number is decreased by 20% of itself, the result is 20. What is the number?
4. When a number is decreased by 30% of itself, the result is 28. What is the number?
5. When 60% of a number is added to the number, the result is 192. What is the number?
6. When 80% of a number is added to the number, the result is 252. What is the number?
7. 70% of what number is 224?
8. 70% of what number is 252?
9. One number exceeds another by 26. The sum of the numbers is 64. What are the numbers?
10. One number exceeds another by 24. The sum of the numbers is 58. What are the numbers?

Practice PLUS

In Exercises 11–16, find all values of x satisfying the given conditions.

11. $y_1 = 13x - 4$, $y_2 = 5x + 10$, and y_1 exceeds y_2 by 2.

12. $y_1 = 10x + 6$, $y_2 = 12x - 7$, and y_1 exceeds y_2 by 3.

13. $y_1 = 10(2x - 1)$, $y_2 = 2x + 1$, and y_1 is 14 more than 8 times y_2.

14. $y_1 = 9(3x - 5)$, $y_2 = 3x - 1$, and y_1 is 51 less than 12 times y_2.

15. $y_1 = 2x + 6$, $y_2 = x + 8$, $y_3 = x$, and the difference between 3 times y_1 and 5 times y_2 is 22 less than y_3.

16. $y_1 = 2.5$, $y_2 = 2x + 1$, $y_3 = x$, and the difference between 2 times y_1 and 3 times y_2 is 8 less than 4 times y_3.

Application Exercises

In Exercises 17–48, use the five-step strategy for solving word problems.

17. The bar graph represents the average credit card debt for U.S. college students. The average credit card debt for juniors exceeds the debt for sophomores by $421, and the average credit card debt for seniors exceeds the debt for sophomores by $1265. The combined credit card debt for a sophomore, a junior, and a senior is $6429. Determine the average credit card debt for a sophomore, a junior, and a senior.

Average Credit Card Debt for United States College Students

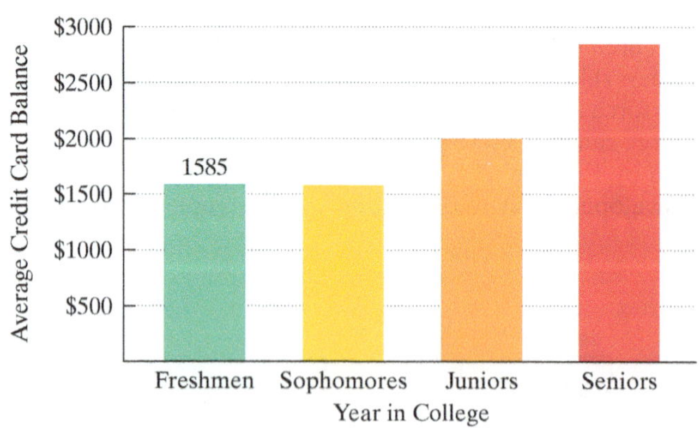

Source: Nellie Mae

18. The bar graph at the top of the next column represents the millions of barrels of oil consumed each day by the countries with the greatest oil consumption. Oil consumption in China exceeds Japan's by 0.8 million barrels per day, and oil consumption in the United States exceeds Japan's by 15 million barrels per day. Of the 82 million barrels of oil used by the world every day, the combined consumption for the United States, China, and Japan is 32.3 million barrels. Determine the daily oil consumption, in millions of barrels, for the United States, China, and Japan.

Countries with the Greatest Oil Consumption

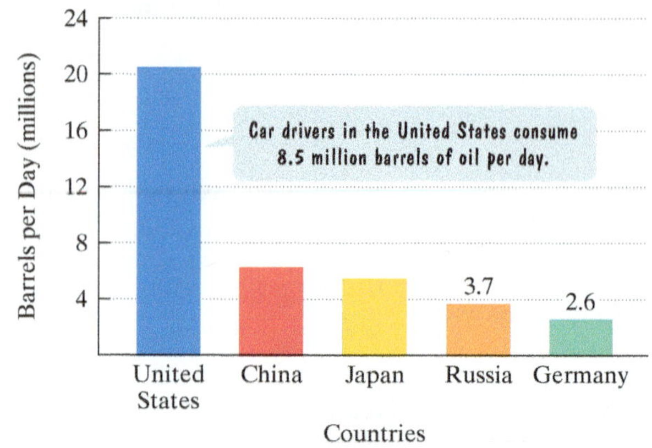

Source: U.S. Energy Information Administration

Solve Exercises 19–22 using the fact that the sum of the measures of the three angles of a triangle is 180°.

19. In a triangle, the measure of the first angle is twice the measure of the second angle. The measure of the third angle is 8° less than the measure of the second angle. What is the measure of each angle?

20. In a triangle, the measure of the first angle is three times the measure of the second angle. The measure of the third angle is 35° less than the measure of the second angle. What is the measure of each angle?

21. In a triangle, the measures of the three angles are consecutive integers. What is the measure of each angle?

22. In a triangle, the measures of the three angles are consecutive even integers. What is the measure of each angle?

According to one mathematical model, the average life expectancy for American men born in 1900 was 55 years. Life expectancy has increased by about 0.2 year for each birth year after 1900. Use this information to solve Exercises 23–24.

23. If this trend continues, for which birth year will the average life expectancy be 85 years?

24. If this trend continues, for which birth year will the average life expectancy be 91 years?

The line graph indicates that in 2005, 19.4% of people in the United States spoke a language other than English at home. For the period from 2000 through 2005, this had been increasing by approximately 0.4% per year. Use this information to solve Exercises 25–26.

Percentage of People in the United States Speaking a Language Other Than English at Home

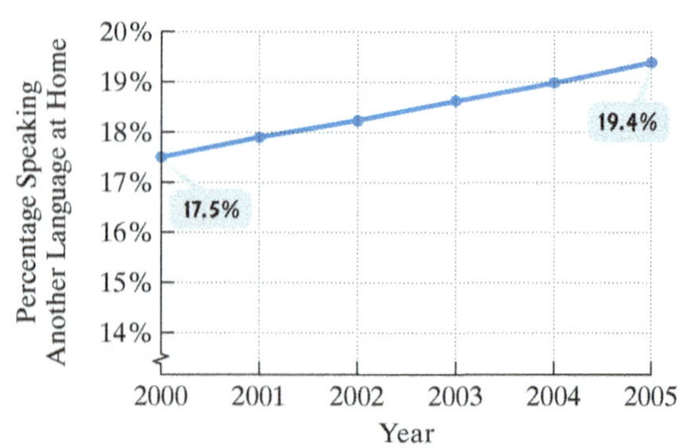

Source: U.S. Census Bureau

25. If this trend continues, by which year will 23% of people in the United States speak a language other than English at home?

26. If this trend continues, by which year will 25% of people in the United States speak a language other than English at home?

27. You are choosing between two health clubs. Club A offers membership for a fee of \$40 plus a monthly fee of \$25. Club B offers membership for a fee of \$15 plus a monthly fee of \$30. After how many months will the total cost at each health club be the same? What will be the total cost for each club?

28. Video Store A charges \$9 to rent a video game for one week. Although only members can rent from the store, membership is free. Video Store B charges only \$4 to rent a video game for one week. Only members can rent from the store and membership is \$50 per year. After how many video-game rentals will the total amount spent at each store be the same? What will be the total amount spent at each store?

29. The bus fare in a city is \$1.25. People who use the bus have the option of purchasing a monthly coupon book for \$15.00. With the coupon book, the fare is reduced to \$0.75. Determine the number of times in a month the bus must be used so that the total monthly cost without the coupon book is the same as the total monthly cost with the coupon book.

30. A coupon book for a bridge costs \$30 per month. The toll for the bridge is normally \$5.00, but it is reduced to \$3.50 for people who have purchased the coupon book. Determine the number of times in a month the bridge must be crossed so that the total monthly cost without the coupon book is the same as the total monthly cost with the coupon book.

31. In 2005, there were 13,300 students at college A, with a projected enrollment increase of 1000 students per year. In the same year, there were 26,800 students at college B, with a projected enrollment decline of 500 students per year.

a. According to these projections, when will the colleges have the same enrollment? What will be the enrollment in each college at that time?

b. Use the following table to numerically check your work in part (a). What equations were entered for y_1 and y_2 to obtain this table?

X	Y1	Y2
7	20300	23300
8	21300	22800
9	22300	22300
10	23300	21800
11	24300	21300
12	25300	20800
13	26300	20300

X=7

32. In 2000, the population of Greece was 10,600,000, with projections of a population decrease of 28,000 people per year. In the same year, the population of Belgium was 10,200,000, with projections of a population decrease of 12,000 people per year. (*Source:* United Nations) According to these projections, when will the two countries have the same population? What will be the population at that time?

33. After a 20% reduction, you purchase a television for \$336. What was the television's price before the reduction?

34. After a 30% reduction, you purchase a dictionary for \$30.80. What was the dictionary's price before the reduction?

35. Including 8% sales tax, an inn charges \$162 per night. Find the inn's nightly cost before the tax is added.

36. Including 5% sales tax, an inn charges \$252 per night. Find the inn's nightly cost before the tax is added.

The graph shows average yearly earnings in the United States by highest educational attainment. Use the relevant information shown in the graph to solve Exercises 37–38.

Average Earnings of Full-Time Workers in the United States, by Highest Educational Attainment

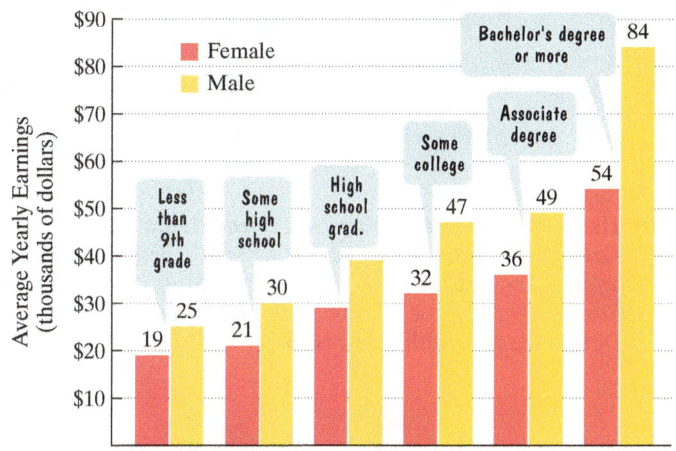

Source: U.S. Census Bureau

37. The annual salary for men with some college is an increase of 20.5% over the annual salary for men whose highest educational attainment is a high school degree. What is the annual salary, to the nearest thousand dollars, for men whose highest educational attainment is a high school degree?

38. The annual salary for women with an associate degree is an increase of 24.1% over the annual salary for women whose highest educational attainment is a high school degree. What is the annual salary, to the nearest thousand dollars, for women whose highest educational attainment is a high school degree?

Exercises 39–40 involve markup, the amount added to the dealer's cost of an item to arrive at the selling price of that item.

39. The selling price of a refrigerator is \$584. If the markup is 25% of the dealer's cost, what is the dealer's cost of the refrigerator?

40. The selling price of a scientific calculator is \$15. If the markup is 25% of the dealer's cost, what is the dealer's cost of the calculator?

41. A rectangular soccer field is twice as long as it is wide. If the perimeter of the soccer field is 300 yards, what are its dimensions?

42. A rectangular swimming pool is three times as long as it is wide. If the perimeter of the pool is 320 feet, what are its dimensions?

43. The length of the rectangular tennis court at Wimbledon is 6 feet longer than twice the width. If the court's perimeter is 228 feet, what are the court's dimensions?

44. The length of a rectangular pool is 6 meters less than twice the width. If the pool's perimeter is 126 meters, what are its dimensions?

45. The rectangular painting in the figure shown measures 12 inches by 16 inches and contains a frame of uniform width around the four edges. The perimeter of the rectangle formed by the painting and its frame is 72 inches. Determine the width of the frame.

46. The rectangular swimming pool in the figure shown measures 40 feet by 60 feet and contains a path of uniform width around the four edges. The perimeter of the rectangle formed by the pool and the surrounding path is 248 feet. Determine the width of the path.

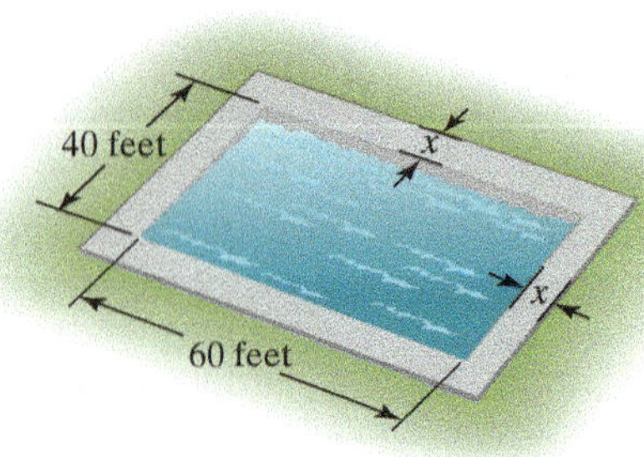

47. For a long-distance person-to-person telephone call, a telephone company charges \$0.43 for the first minute, \$0.32 for each additional minute, and a \$2.10 service charge. If the cost of a call is \$5.73, how long did the person talk?

48. A job pays an annual salary of \$33,150, which includes a holiday bonus of \$750. If paychecks are issued twice a month, what is the gross amount for each paycheck?

In Exercises 49–74, solve each formula for the specified variable.

49. $A = lw$ for l

50. $A = lw$ for w

51. $A = \frac{1}{2}bh$ for b

52. $A = \frac{1}{2}bh$ for h

53. $I = Prt$ for P

54. $I = Prt$ for t

55. $T = D + pm$ for p

56. $P = C + MC$ for M

57. $A = \frac{1}{2}h(a + b)$ for a

58. $A = \frac{1}{2}h(a + b)$ for b

59. $V = \frac{1}{3}\pi r^2 h$ for h

60. $V = \frac{1}{3}\pi r^2 h$ for r^2

61. $y - y_1 = m(x - x_1)$ for m

62. $y_2 - y_1 = m(x_2 - x_1)$ for m

63. $V = \frac{d_1 - d_2}{t}$ for d_1

64. $z = \frac{x - u}{s}$ for x

65. $Ax + By = C$ for x

66. $Ax + By = C$ for y

67. $s = \frac{1}{2}at^2 + vt$ for v

68. $s = \frac{1}{2}at^2 + vt$ for a

69. $L = a + (n - 1)d$ for n

70. $L = a + (n - 1)d$ for d

71. $A = 2lw + 2lh + 2wh$ for l

72. $A = 2lw + 2lh + 2wh$ for h

73. $IR + Ir = E$ for I

74. $A = \frac{x_1 + x_2 + x_3}{n}$ for n

Writing in Mathematics

75. In your own words, describe a step-by-step approach for solving algebraic word problems.

76. Write an original word problem that can be solved using a linear equation. Then solve the problem.

77. Explain what it means to solve a formula for a variable.

78. Did you have difficulties solving some of the problems that were assigned in this exercise set? Discuss what you did if this happened to you. Did your course of action enhance your ability to solve algebraic word problems?

79. The mile records in Example 2 on page 52 are a yardstick for measuring how athletes are getting better and better. Do you think that there is a limit to human performance? Explain your answer. If so, when might we reach it?

80. The bar graph in Exercises 37–38 shows average earnings of U.S. men and women, by highest educational attainment. Describe the trend shown by the graph. Discuss any aspects of the data that surprise you.

Technology Exercises

81. Use a graphing utility to numerically or graphically verify your work in any one exercise from Exercises 27–30. For assistance on how to do this, refer to the Using Technology box on page 53.

82. The formula $y = 55 + 0.2x$ models the average life expectancy, y, of American men born x years after 1900. Graph the formula in a [0, 200, 20] by [0, 100, 10] viewing rectangle. Then use the TRACE or ZOOM feature to verify your answer in Exercise 23 or 24.

83. In Exercises 25–26, we saw that in 2005, 19.4% of people in the United States spoke a language other than English at home, increasing by approximately 0.4% per year.

a. Write a formula that models the percentage of people in the United States speaking a language other than English at home, y, x years after 2005.

b. Enter the formula from part (a) as y_1 in your graphing utility. Then use either a table for y_1 or a graph of y_1 to numerically or graphically verify your answer to Exercise 25 or 26.

Critical Thinking Exercises

Make Sense? *In Exercises 84–87, determine whether each statement "makes sense" or "does not make sense" and explain your reasoning.*

84. I solved the formula for one of its variables, so now I have a numerical value for that variable.

85. Reasoning through a word problem often increases problem-solving skills in general.

86. The hardest part in solving a word problem is writing the equation that models the verbal conditions.

87. When traveling in Europe, the most useful form of the two Celsius-Fahrenheit conversion formulas is the formula used to convert from Fahrenheit to Celsius.

In Exercises 88–91, determine whether each statement is true or false. If the statement is false, make the necessary change(s) to produce a true statement.

88. If $I = prt$, then $t = I - pr$.

89. If $y = \frac{kx}{z}$, then $z = \frac{kx}{y}$.

90. It is not necessary to use the distributive property to solve $P = C + MC$ for C.

91. An item's price, x, reduced by $\frac{1}{3}$ is modeled by $x - \frac{1}{3}$.

92. The price of a dress is reduced by 40%. When the dress still does not sell, it is reduced by 40% of the reduced price. If the price of the dress after both reductions is \$72, what was the original price?

93. Suppose that we agree to pay you 8¢ for every problem in this chapter that you solve correctly and fine you 5¢ for every problem done incorrectly. If at the end of 26 problems we do not owe each other any money, how many problems did you solve correctly?

94. It was wartime when the Ricardos found out Mrs. Ricardo was pregnant. Ricky Ricardo was drafted and made out a will, deciding that \$14,000 in a savings account was to be divided between his wife and his child-to-be. Rather strangely, and certainly with gender bias, Ricky stipulated that if the child were a boy, he would get twice the amount of the mother's portion. If it were a girl, the mother would get twice the amount the girl was to receive. We'll never know what Ricky was thinking, for (as fate would have it) he did not return from war. Mrs. Ricardo gave birth to twins—a boy and a girl. How was the money divided?

95. A thief steals a number of rare plants from a nursery. On the way out, the thief meets three security guards, one after another. To each security guard, the thief is forced to give one-half the plants that he still has, plus 2 more. Finally, the thief leaves the nursery with 1 lone palm. How many plants were originally stolen?

96. Solve for C: $V = C - \frac{C - S}{L}N$.

Review Exercises

97. What does $-6 \le -6$ mean? Is the statement true or false? (Section 1.1, Example 6)

98. Simplify: $\frac{(2 + 4)^2 + (-1)^5}{12 \div 2 \cdot 3 - 3}$. (Section 1.2, Example 8)

99. Solve: $\frac{2x}{3} - \frac{8}{3} = x$. (Section 1.4, Example 4)

Preview Exercises

Exercises 100–102 will help you prepare for the material covered in the next section.

100. In parts (a) and (b), complete each statement.

a. $b^4 \cdot b^3 = (b \cdot b \cdot b \cdot b)(b \cdot b \cdot b) = b^?$

b. $b^5 \cdot b^5 = (b \cdot b \cdot b \cdot b \cdot b)(b \cdot b \cdot b \cdot b \cdot b) = b^?$

c. Generalizing from parts (a) and (b), what should be done with the exponents when multiplying exponential expressions with the same base?

101. In parts (a) and (b), complete each statement.

a. $\frac{b^7}{b^3} = \frac{\cancel{b} \cdot \cancel{b} \cdot \cancel{b} \cdot b \cdot b \cdot b \cdot b}{\cancel{b} \cdot \cancel{b} \cdot \cancel{b}} = b^?$

b. $\frac{b^8}{b^2} = \frac{\cancel{b} \cdot \cancel{b} \cdot b \cdot b \cdot b \cdot b \cdot b \cdot b}{\cancel{b} \cdot \cancel{b}} = b^?$

c. Generalizing from parts (a) and (b), what should be done with the exponents when dividing exponential expressions with the same base?

102. Simplify: $\frac{1}{\left(-\frac{1}{2}\right)^3}$.

SECTION 1.6

Properties of Integral Exponents

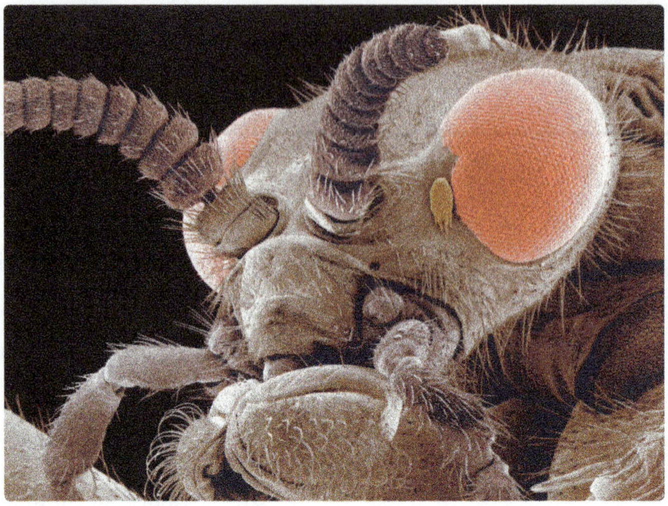

Objectives

1 Use the product rule.
2 Use the quotient rule.
3 Use the zero-exponent rule.
4 Use the negative-exponent rule.
5 Use the power rule.
6 Find the power of a product.
7 Find the power of a quotient.
8 Simplify exponential expressions.

Our opening photo shows the head of a fly as seen under an electronic microscope. Some electronic microscopes can view objects that are less than 10^{-4} meter, or 0.0001 meter, in size. In this section, we'll make sense of the negative exponent in 10^{-4}, as we turn to integral exponents and their properties.

The Product and Quotient Rules

1 Use the product rule.

We have seen that exponents are used to indicate repeated multiplication. Now consider the multiplication of two exponential expressions, such as $b^4 \cdot b^3$. We are multiplying 4 factors of b and 3 factors of b. We have a total of 7 factors of b:

4 factors of b | 3 factors of b

$$b^4 \cdot b^3 = (b \cdot b \cdot b \cdot b)(b \cdot b \cdot b) = b^7.$$

Total: 7 factors of b

The product is exactly the same if we add the exponents:

$$b^4 \cdot b^3 = b^{4+3} = b^7.$$

This suggests the following rule:

The Product Rule

$$b^m \cdot b^n = b^{m+n}$$

When multiplying exponential expressions with the same base, add the exponents. Use this sum as the exponent of the common base.

EXAMPLE 1 Using the Product Rule

Multiply each expression using the product rule:

a. $b^8 \cdot b^{10}$ b. $(6x^4y^3)(5x^2y^7)$.

Solution

a. $b^8 \cdot b^{10} = b^{8+10} = b^{18}$

b. $(6x^4y^3)(5x^2y^7)$

$= 6 \cdot 5 \cdot x^4 \cdot x^2 \cdot y^3 \cdot y^7$ Use the associative and commutative properties. This step can be done mentally.

$= 30x^{4+2}y^{3+7}$

$= 30x^6y^{10}$

CHECK POINT 1 Multiply each expression using the product rule:

a. $b^6 \cdot b^5$ b. $(4x^3y^4)(10x^2y^6)$.

2 Use the quotient rule.

Now, consider the division of two exponential expressions, such as the quotient of b^7 and b^3. We are dividing 7 factors of b by 3 factors of b.

$$\frac{b^7}{b^3} = \frac{b \cdot b \cdot b \cdot b \cdot b \cdot b \cdot b}{b \cdot b \cdot b} = \boxed{\frac{b \cdot b \cdot b}{b \cdot b \cdot b}} \cdot b \cdot b \cdot b \cdot b = 1 \cdot b \cdot b \cdot b \cdot b = b^4$$

This factor is equal to 1.

The quotient is exactly the same if we subtract the exponents:

$$\frac{b^7}{b^3} = b^{7-3} = b^4.$$

This suggests the following rule:

The Quotient Rule

$$\frac{b^m}{b^n} = b^{m-n}, \quad b \neq 0$$

When dividing exponential expressions with the same nonzero base, subtract the exponent in the denominator from the exponent in the numerator. Use this difference as the exponent of the common base.

EXAMPLE 2 Using the Quotient Rule

Divide each expression using the quotient rule:

a. $\frac{(-2)^7}{(-2)^4}$ b. $\frac{30x^{12}y^9}{5x^3y^7}$.

Solution

a. $\frac{(-2)^7}{(-2)^4} = (-2)^{7-4} = (-2)^3$ or -8 $(-2)^3 = (-2)(-2)(-2) = -8$

b. $\frac{30x^{12}y^9}{5x^3y^7} = \frac{30}{5} \cdot \frac{x^{12}}{x^3} \cdot \frac{y^9}{y^7} = 6x^{12-3}y^{9-7} = 6x^9y^2$

CHECK POINT 2 Divide each expression using the quotient rule:

a. $\frac{(-3)^6}{(-3)^3}$ b. $\frac{27x^{14}y^8}{3x^3y^5}$.

3 Use the zero-exponent rule.

Zero as an Exponent

A nonzero base can be raised to the 0 power. The quotient rule can be used to help determine what zero as an exponent should mean. Consider the quotient of b^4 and b^4, where b is not zero. We can determine this quotient in two ways.

$$\frac{b^4}{b^4} = 1 \qquad \frac{b^4}{b^4} = b^{4-4} = b^0$$

Any nonzero expression divided by itself is 1.

Use the quotient rule and subtract exponents.

This means that b^0 must equal 1.

The Zero-Exponent Rule

If b is any real number other than 0,

$$b^0 = 1.$$

EXAMPLE 3 Using the Zero-Exponent Rule

Use the zero-exponent rule to simplify each expression:

a. 8^0 **b.** $(-6)^0$ **c.** -6^0 **d.** $5x^0$ **e.** $(5x)^0$.

Solution

a. $8^0 = 1$ Any nonzero number raised to the 0 power is 1.

b. $(-6)^0 = 1$

c. $-6^0 = -(6^0) = -1$

Only 6 is raised to the 0 power.

d. $5x^0 = 5 \cdot 1 = 5$ Only x is raised to the 0 power.

e. $(5x)^0 = 1$ The entire expression, 5x, is raised to the 0 power.

CHECK POINT 3 Use the zero-exponent rule to simplify each expression:

a. 7^0 **b.** $(-5)^0$ **c.** -5^0 **d.** $10x^0$ **e.** $(10x)^0$.

4 Use the negative-exponent rule.

Negative Integers as Exponents

A nonzero base can be raised to a negative power. The quotient rule can be used to help determine what a negative integer as an exponent should mean. Consider the quotient of b^3 and b^5, where b is not zero. We can determine this quotient in two ways.

$$\frac{b^3}{b^5} = \frac{\cancel{b} \cdot \cancel{b} \cdot \cancel{b}}{\cancel{b} \cdot \cancel{b} \cdot \cancel{b} \cdot b \cdot b} = \frac{1}{b^2} \qquad \frac{b^3}{b^5} = b^{3-5} = b^{-2}$$

After dividing common factors, we have two factors of b in the denominator.

Use the quotient rule and subtract exponents.

Notice that $\frac{b^3}{b^5}$ equals both b^{-2} and $\frac{1}{b^2}$. This means that b^{-2} must equal $\frac{1}{b^2}$. This example is a special case of the **negative-exponent rule.**

The Negative-Exponent Rule

If b is any real number other than 0 and n is a natural number, then

$$b^{-n} = \frac{1}{b^n}.$$

EXAMPLE 4 Using the Negative-Exponent Rule

Use the negative-exponent rule to write each expression with a positive exponent. Simplify, if possible:

a. 9^{-2} b. $(-2)^{-5}$ c. $\frac{1}{6^{-2}}$ d. $7x^{-5}y^2$.

Solution

a. $9^{-2} = \frac{1}{9^2} = \frac{1}{81}$

b. $(-2)^{-5} = \frac{1}{(-2)^5} = \frac{1}{(-2)(-2)(-2)(-2)(-2)} = \frac{1}{-32} = -\frac{1}{32}$

Only the sign of the exponent, −5, changes. The base, −2, does not change sign.

c. $\frac{1}{6^{-2}} = \frac{1}{\frac{1}{6^2}} = 1 \cdot \frac{6^2}{1} = 6^2 = 36$

d. $7x^{-5}y^2 = 7 \cdot \frac{1}{x^5} \cdot y^2 = \frac{7y^2}{x^5}$

CHECK POINT 4 Use the negative-exponent rule to write each expression with a positive exponent. Simplify, if possible:

a. 5^{-2} b. $(-3)^{-3}$ c. $\frac{1}{4^{-2}}$ d. $3x^{-6}y^4$.

In Example 4 and Check Point 4, did you notice that

$$\frac{1}{6^{-2}} = 6^2 \quad \text{and} \quad \frac{1}{4^{-2}} = 4^2?$$

In general, if a negative exponent appears in a denominator, an expression can be written with a positive exponent using

$$\frac{1}{b^{-n}} = b^n.$$

Negative Exponents in Numerators and Denominators

If b is any real number other than 0 and n is a natural number, then

$$b^{-n} = \frac{1}{b^n} \quad \text{and} \quad \frac{1}{b^{-n}} = b^n.$$

When a negative number appears as an exponent, switch the position of the base (from numerator to denominator or from denominator to numerator) and make the exponent positive. The sign of the base does not change.

EXAMPLE 5 Using Negative Exponents

Write each expression with positive exponents only. Then simplify, if possible:

a. $\frac{5^{-3}}{4^{-2}}$ b. $\frac{1}{6x^{-4}}$.

Solution

a. $\frac{5^{-3}}{4^{-2}} = \frac{4^2}{5^3} = \frac{4 \cdot 4}{5 \cdot 5 \cdot 5} = \frac{16}{125}$ Switch the position of each base to the other side of the fraction bar and change the sign of the exponent.

b. $\frac{1}{6x^{-4}} = \frac{x^4}{6}$ Switch the position of x to the other side of the fraction bar and change −4 to 4.

Don't switch the position of 6. It is not affected by a negative exponent.

CHECK POINT 5 Write each expression with positive exponents only. Then simplify, if possible:

a. $\frac{7^{-2}}{4^{-3}}$ b. $\frac{1}{5x^{-2}}$.

5 Use the power rule.

The Power Rule for Exponents (Powers to Powers)

The next property of exponents applies when an exponential expression is raised to a power. Here is an example:

$$(b^2)^4.$$

The exponential expression b^2 is raised to the fourth power.

There are 4 factors of b^2. Thus,

$$(b^2)^4 = b^2 \cdot b^2 \cdot b^2 \cdot b^2 = b^{2+2+2+2} = b^8.$$

Add exponents when multiplying with the same base.

We can obtain the answer, b^8, by multiplying the exponents:

$$(b^2)^4 = b^{2 \cdot 4} = b^8.$$

This suggests the following rule:

The Power Rule (Powers to Powers)

$$(b^m)^n = b^{mn}$$

When an exponential expression is raised to a power, multiply the exponents. Place the product of the exponents on the base and remove the parentheses.

EXAMPLE 6 Using the Power Rule (Powers to Powers)

Simplify each expression using the power rule:

a. $(x^6)^4$ b. $(y^5)^{-3}$ c. $(b^{-4})^{-2}$.

Solution

a. $(x^6)^4 = x^{6 \cdot 4} = x^{24}$

b. $(y^5)^{-3} = y^{5(-3)} = y^{-15} = \frac{1}{y^{15}}$

c. $(b^{-4})^{-2} = b^{(-4)(-2)} = b^8$

CHECK POINT 6 Simplify each expression using the power rule:

a. $(x^5)^3$ b. $(y^7)^{-2}$ c. $(b^{-3})^{-4}$.

6 Find the power of a product.

The Products-to-Powers Rule for Exponents

The next property of exponents applies when we are raising a product to a power. Here is an example:

$$(2x)^4.$$

The product $2x$ is raised to the fourth power.

There are four factors of $2x$. Thus,

$$(2x)^4 = 2x \cdot 2x \cdot 2x \cdot 2x = 2 \cdot 2 \cdot 2 \cdot 2 \cdot x \cdot x \cdot x \cdot x = 2^4x^4.$$

We can obtain the answer, 2^4x^4, by raising each factor within the parentheses to the fourth power:

$$(2x)^4 = 2^4x^4.$$

This suggests the following rule:

Products to Powers

$$(ab)^n = a^nb^n$$

When a product is raised to a power, raise each factor to that power.

EXAMPLE 7 Using the Products-to-Powers Rule

Simplify each expression using the products-to-powers rule:

a. $(6x)^3$ b. $(-2y^2)^4$ c. $(-3x^{-1}y^3)^{-2}$.

Solution

a. $(6x)^3 = 6^3x^3$ Raise each factor to the third power.

$= 216x^3$ Simplify: $6^3 = 6 \cdot 6 \cdot 6 = 216$.

b. $(-2y^2)^4 = (-2)^4(y^2)^4$ Raise each factor to the fourth power.

$= (-2)^4y^{2 \cdot 4}$ To raise an exponential expression to a power, multiply exponents: $(b^m)^n = b^{mn}$.

$= 16y^8$ Simplify: $(-2)^4 = (-2)(-2)(-2)(-2) = 16$.

c. $(-3x^{-1}y^3)^{-2} = (-3)^{-2}(x^{-1})^{-2}(y^3)^{-2}$ Raise each factor to the -2 power.

$= (-3)^{-2}x^{(-1)(-2)}y^{3(-2)}$ Use $(b^m)^n = b^{mn}$ on the second and third factors.

$= (-3)^{-2}x^2y^{-6}$ Simplify.

$= \frac{1}{(-3)^2} \cdot x^2 \cdot \frac{1}{y^6}$ Apply $b^{-n} = \frac{1}{b^n}$ to the first and last factors.

$= \frac{x^2}{9y^6}$ Simplify: $(-3)^2 = (-3)(-3) = 9$.

✓ CHECK POINT 7 Simplify each expression using the products-to-powers rule:

a. $(2x)^4$ b. $(-3y^2)^3$ c. $(-4x^5y^{-1})^{-2}$.

7 Find the power of a quotient.

The Quotients-to-Powers Rule for Exponents

The following rule is used to raise a quotient to a power:

Quotients to Powers

If b is a nonzero real number, then

$$\left(\frac{a}{b}\right)^n = \frac{a^n}{b^n}.$$

When a quotient is raised to a power, raise the numerator to that power and divide by the denominator to that power.

EXAMPLE 8 Using the Quotients-to-Powers Rule

Simplify each expression using the quotients-to-powers rule:

a. $\left(\frac{x^2}{4}\right)^3$ **b.** $\left(\frac{2x^3}{y^{-4}}\right)^5$ **c.** $\left(\frac{x^3}{y^2}\right)^{-4}$.

Study Tip

When simplifying exponential expressions, the first step should be to simplify inside parentheses. In Example 8, the expressions inside parentheses have different bases and cannot be simplified. This is why we begin with the quotients-to-powers rule.

Solution

a. $\left(\frac{x^2}{4}\right)^3 = \frac{(x^2)^3}{4^3} = \frac{x^{2\cdot 3}}{4\cdot 4\cdot 4} = \frac{x^6}{64}$ Cube the numerator and the denominator.

b. $\left(\frac{2x^3}{y^{-4}}\right)^5 = \frac{(2x^3)^5}{(y^{-4})^5}$ Raise the numerator and the denominator to the fifth power.

$= \frac{2^5(x^3)^5}{(y^{-4})^5}$ Raise each factor in the numerator to the fifth power.

$= \frac{2^5\cdot x^{3\cdot 5}}{y^{(-4)(5)}}$ Multiply exponents in both powers-to-powers expressions: $(b^m)^n = b^{mn}$.

$= \frac{32x^{15}}{y^{-20}}$ Simplify.

$= 32x^{15}y^{20}$ Move y to the other side of the fraction bar and change -20 to 20: $\frac{1}{b^{-n}} = b^n$.

c. $\left(\frac{x^3}{y^2}\right)^{-4} = \frac{(x^3)^{-4}}{(y^2)^{-4}}$ Raise the numerator and the denominator to the -4 power.

$= \frac{x^{3(-4)}}{y^{2(-4)}}$ Multiply exponents in both powers-to-powers expressions: $(b^m)^n = b^{mn}$.

$= \frac{x^{-12}}{y^{-8}}$ Simplify.

$= \frac{y^8}{x^{12}}$ Move each base to the other side of the fraction bar and make each exponent positive.

CHECK POINT 8 Simplify each expression using the quotients-to-powers rule:

a. $\left(\frac{x^5}{4}\right)^3$ **b.** $\left(\frac{2x^{-3}}{y^2}\right)^4$ **c.** $\left(\frac{x^{-3}}{y^4}\right)^{-5}$.

8 Simplify exponential expressions.

Simplifying Exponential Expressions

Properties of exponents are used to simplify exponential expressions. An exponential expression is **simplified** when

- No parentheses appear.
- No powers are raised to powers.
- Each base occurs only once.
- No negative or zero exponents appear.

Simplifying Exponential Expressions

	Example
1. If necessary, remove parentheses by using $(ab)^n = a^n b^n$ or $\left(\frac{a}{b}\right)^n = \frac{a^n}{b^n}$.	$(xy)^3 = x^3y^3$
2. If necessary, simplify powers to powers by using $(b^m)^n = b^{mn}$.	$(x^4)^3 = x^{4\cdot 3} = x^{12}$
3. If necessary, be sure that each base appears only once by using $b^m \cdot b^n = b^{m+n}$ or $\frac{b^m}{b^n} = b^{m-n}$.	$x^4 \cdot x^3 = x^{4+3} = x^7$
4. If necessary, rewrite exponential expressions with zero powers as 1 $(b^0 = 1)$. Furthermore, write the answer with positive exponents by using $b^{-n} = \frac{1}{b^n}$ or $\frac{1}{b^{-n}} = b^n$.	$\frac{x^5}{x^8} = x^{5-8} = x^{-3} = \frac{1}{x^3}$

The following example shows how to simplify exponential expressions. Throughout the example, assume that no variable in a denominator is equal to zero.

EXAMPLE 9 Simplifying Exponential Expressions

Simplify:

a. $(-2xy^{-14})(-3x^4y^5)^3$ **b.** $\left(\frac{25x^2y^4}{-5x^6y^{-8}}\right)^2$ **c.** $\left(\frac{x^{-4}y^7}{2}\right)^{-5}$.

Solution

a. $(-2xy^{-14})(-3x^4y^5)^3$

$= (-2xy^{-14})(-3)^3(x^4)^3(y^5)^3$ Cube each factor in the second parentheses.

$= (-2xy^{-14})(-27)x^{12}y^{15}$ Multiply the exponents when raising a power to a power: $(x^4)^3 = x^{4\cdot 3} = x^{12}$ and $(y^5)^3 = y^{5\cdot 3} = y^{15}$.

$= (-2)(-27)x^{1+12}y^{-14+15}$ Mentally rearrange factors and multiply like bases by adding the exponents.

$= 54x^{13}y$ Simplify.

b. $\left(\dfrac{25x^2y^4}{-5x^6y^{-8}}\right)^2$ The expression inside parentheses contains the same bases and can be simplified. Begin with this simplification.

$= (-5x^{2-6}y^{4-(-8)})^2$ Simplify inside the parentheses. Subtract the exponents when dividing.

$= (-5x^{-4}y^{12})^2$ Simplify.

$= (-5)^2(x^{-4})^2(y^{12})^2$ Square each factor in parentheses.

$= 25x^{-8}y^{24}$ Multiply the exponents when raising a power to a power: $(x^{-4})^2 = x^{-4(2)} = x^{-8}$ and $(y^{12})^2 = y^{12\cdot 2} = y^{24}$.

$= \dfrac{25y^{24}}{x^8}$ Simplify x^{-8} using $b^{-n} = \dfrac{1}{b^n}$.

c. $\left(\dfrac{x^{-4}y^7}{2}\right)^{-5}$

$= \dfrac{(x^{-4}y^7)^{-5}}{2^{-5}}$ Raise the numerator and the denominator to the -5 power.

$= \dfrac{(x^{-4})^{-5}(y^7)^{-5}}{2^{-5}}$ Raise each factor in the numerator to the -5 power.

$= \dfrac{x^{20}y^{-35}}{2^{-5}}$ Multiply the exponents when raising a power to a power: $(x^{-4})^{-5} = x^{-4(-5)} = x^{20}$ and $(y^7)^{-5} = y^{7(-5)} = y^{-35}$.

$= \dfrac{2^5x^{20}}{y^{35}}$ Move each base with a negative exponent to the other side of the fraction bar and make each negative exponent positive.

$= \dfrac{32x^{20}}{y^{35}}$ Simplify: $2^5 = 2\cdot2\cdot2\cdot2\cdot2 = 32$.

CHECK POINT 9 Simplify:

a. $(-3x^{-6}y)(-2x^3y^4)^2$ b. $\left(\dfrac{10x^3y^5}{5x^6y^{-2}}\right)^2$ c. $\left(\dfrac{x^3y^5}{4}\right)^{-3}$.

Study Tip

Try to avoid the following common errors that can occur when simplifying exponential expressions.

Correct	Incorrect	Description of Error
$b^3\cdot b^4 = b^7$	$b^3\cdot b^4 = b^{12}$	The exponents should be added, not multiplied.
$3^2\cdot 3^4 = 3^6$	$3^2\cdot 3^4 = 9^6$	The common base should be retained, not multiplied.
$\dfrac{5^{16}}{5^4} = 5^{12}$	$\dfrac{5^{16}}{5^4} = 5^4$	The exponents should be subtracted, not divided.
$(4a)^3 = 64a^3$	$(4a)^3 = 4a^3$	Both factors should be cubed.
$b^{-n} = \dfrac{1}{b^n}$	$b^{-n} = -\dfrac{1}{b^n}$	Only the exponent should change sign.
$(a+b)^{-1} = \dfrac{1}{a+b}$	$(a+b)^{-1} = \dfrac{1}{a} + \dfrac{1}{b}$	The exponent applies to the entire expression $a + b$.

1.6 EXERCISE SET

MyMathLab | MathXL PRACTICE | WATCH | DOWNLOAD | READ | REVIEW

Practice Exercises

In Exercises 1–14, multiply using the product rule.

1. $b^4 \cdot b^7$
2. $b^5 \cdot b^9$
3. $x \cdot x^3$
4. $x \cdot x^4$
5. $2^3 \cdot 2^2$
6. $2^4 \cdot 2^2$
7. $3x^4 \cdot 2x^2$
8. $5x^3 \cdot 3x^2$
9. $(-2y^{10})(-10y^2)$
10. $(-4y^8)(-8y^4)$
11. $(5x^3y^4)(20x^7y^8)$
12. $(4x^5y^6)(20x^7y^4)$
13. $(-3x^4y^0z)(-7xyz^3)$
14. $(-9x^3yz^4)(-5xy^0z^2)$

In Exercises 15–24, divide using the quotient rule.

15. $\frac{b^{12}}{b^3}$
16. $\frac{b^{25}}{b^5}$
17. $\frac{15x^9}{3x^4}$
18. $\frac{18x^{11}}{3x^4}$
19. $\frac{x^9y^7}{x^4y^2}$
20. $\frac{x^9y^{12}}{x^2y^6}$
21. $\frac{50x^2y^7}{5xy^4}$
22. $\frac{36x^{12}y^4}{4xy^2}$
23. $\frac{-56a^{12}b^{10}c^8}{7ab^2c^4}$
24. $\frac{-66a^9b^7c^6}{6a^3bc^2}$

In Exercises 25–34, use the zero-exponent rule to simplify each expression.

25. 6^0
26. 9^0
27. $(-4)^0$
28. $(-2)^0$
29. -4^0
30. -2^0
31. $13y^0$
32. $17y^0$
33. $(13y)^0$
34. $(17y)^0$

In Exercises 35–52, write each expression with positive exponents only. Then simplify, if possible.

35. 3^{-2}
36. 4^{-2}
37. $(-5)^{-2}$
38. $(-7)^{-2}$
39. -5^{-2}
40. -7^{-2}
41. x^2y^{-3}
42. x^3y^{-4}
43. $8x^{-7}y^3$
44. $9x^{-8}y^4$
45. $\frac{1}{5^{-3}}$
46. $\frac{1}{2^{-5}}$
47. $\frac{1}{(-3)^{-4}}$
48. $\frac{1}{(-2)^{-4}}$
49. $\frac{x^{-2}}{y^{-5}}$
50. $\frac{x^{-3}}{y^{-7}}$
51. $\frac{a^{-4}b^7}{c^{-3}}$
52. $\frac{a^{-3}b^8}{c^{-2}}$

In Exercises 53–58, simplify each expression using the power rule.

53. $(x^6)^{10}$
54. $(x^3)^2$
55. $(b^4)^{-3}$
56. $(b^8)^{-3}$
57. $(7^{-4})^{-5}$
58. $(9^{-4})^{-5}$

In Exercises 59–72, simplify each expression using the products-to-powers rule.

59. $(4x)^3$
60. $(2x)^5$
61. $(-3x^7)^2$
62. $(-4x^9)^2$
63. $(2xy^2)^3$
64. $(3x^2y)^4$
65. $(-3x^2y^5)^2$
66. $(-3x^4y^6)^2$
67. $(-3x^{-2})^{-3}$
68. $(-2x^{-4})^{-3}$
69. $(5x^3y^{-4})^{-2}$
70. $(7x^2y^{-5})^{-2}$
71. $(-2x^{-5}y^4z^2)^{-4}$
72. $(-2x^{-4}y^5z^3)^{-4}$

In Exercises 73–84, simplify each expression using the quotients-to-powers rule.

73. $\left(\frac{2}{x}\right)^4$
74. $\left(\frac{y}{2}\right)^5$
75. $\left(\frac{x^3}{5}\right)^2$
76. $\left(\frac{x^4}{6}\right)^2$
77. $\left(-\frac{3x}{y}\right)^4$
78. $\left(-\frac{2x}{y}\right)^5$
79. $\left(\frac{x^4}{y^2}\right)^6$
80. $\left(\frac{x^5}{y^3}\right)^6$
81. $\left(\frac{x^3}{y^{-4}}\right)^3$
82. $\left(\frac{x^4}{y^{-2}}\right)^3$
83. $\left(\frac{a^{-2}}{b^3}\right)^{-4}$
84. $\left(\frac{a^{-3}}{b^5}\right)^{-4}$

In Exercises 85–116, simplify each exponential expression.

85. $\frac{x^3}{x^9}$
86. $\frac{x^6}{x^{10}}$
87. $\frac{20x^3}{-5x^4}$
88. $\frac{10x^5}{-2x^6}$
89. $\frac{16x^3}{8x^{10}}$
90. $\frac{15x^2}{3x^{11}}$

91. $\dfrac{20a^3b^8}{2ab^{13}}$

92. $\dfrac{72a^5b^{11}}{9ab^{17}}$

93. $x^3 \cdot x^{-12}$

94. $x^4 \cdot x^{-12}$

95. $(2a^5)(-3a^{-7})$

96. $(4a^2)(-2a^{-5})$

97. $\left(-\frac{1}{4}x^{-4}y^5z^{-1}\right)(-12x^{-3}y^{-1}z^4)$

98. $\left(-\frac{1}{3}x^{-5}y^4z^6\right)(-18x^{-2}y^{-1}z^{-7})$

99. $\dfrac{6x^2}{2x^{-8}}$

100. $\dfrac{12x^5}{3x^{-10}}$

101. $\dfrac{x^{-7}}{x^3}$

102. $\dfrac{x^{-10}}{x^4}$

103. $\dfrac{30x^2y^5}{-6x^8y^{-3}}$

104. $\dfrac{24x^2y^{13}}{-2x^5y^{-2}}$

105. $\dfrac{-24a^3b^{-5}c^5}{-3a^{-6}b^{-4}c^{-7}}$

106. $\dfrac{-24a^2b^{-2}c^8}{-8a^{-5}b^{-1}c^{-3}}$

107. $\left(\dfrac{x^3}{x^{-5}}\right)^2$

108. $\left(\dfrac{x^4}{x^{-11}}\right)^3$

109. $\left(\dfrac{-15a^4b^2}{5a^{10}b^{-3}}\right)^3$

110. $\left(\dfrac{-30a^{14}b^8}{10a^{17}b^{-2}}\right)^3$

111. $\left(\dfrac{3a^{-5}b^2}{12a^3b^{-4}}\right)^0$

112. $\left(\dfrac{4a^{-5}b^3}{12a^3b^{-5}}\right)^0$

113. $\left(\dfrac{x^{-5}y^8}{3}\right)^{-4}$

114. $\left(\dfrac{x^6y^{-7}}{2}\right)^{-3}$

115. $\left(\dfrac{20a^{-3}b^4c^5}{-2a^{-5}b^{-2}c}\right)^{-2}$

116. $\left(\dfrac{-2a^{-4}b^3c^{-1}}{3a^{-2}b^{-5}c^{-2}}\right)^{-4}$

Practice PLUS

In Exercises 117–124, simplify each exponential expression.

117. $\dfrac{9y^4}{x^{-2}} + \left(\dfrac{x^{-1}}{y^2}\right)^{-2}$

118. $\dfrac{7x^3}{y^{-9}} + \left(\dfrac{x^{-1}}{y^3}\right)^{-3}$

119. $\left(\dfrac{3x^4}{y^{-4}}\right)^{-1}\left(\dfrac{2x}{y^2}\right)^3$

120. $\left(\dfrac{2^{-1}x^{-2}y}{x^4y^{-1}}\right)^{-2}\left(\dfrac{xy^{-3}}{x^{-3}y}\right)^3$

121. $(-4x^3y^{-5})^{-2}(2x^{-8}y^{-5})$

122. $(-4x^{-4}y^5)^{-2}(-2x^5y^{-6})$

123. $\dfrac{(2x^2y^4)^{-1}(4xy^3)^{-3}}{(x^2y)^{-5}(x^3y^2)^4}$

124. $\dfrac{(3x^3y^2)^{-1}(2x^2y)^{-2}}{(xy^2)^{-5}(x^2y^3)^3}$

Application Exercises

The formula

$$A = 1000 \cdot 2^t$$

models the population, A, of aphids in a field of potato plants after t weeks. Use this formula to solve Exercises 125–126.

125. a. What is the present aphid population?
   b. What will the aphid population be in 4 weeks?
   c. What was the aphid population 3 weeks ago?

126. a. What is the present aphid population?
   b. What will the aphid population be in 3 weeks?
   c. What was the aphid population 2 weeks ago?

A rumor about algebra CDs that you can listen to as you sleep, allowing you to awaken refreshed and algebraically empowered, is spreading among the students in your math class. The formula

$$N = \frac{25}{1 + 24 \cdot 2^{-t}}$$

models the number of people in the class, N, who have heard the rumor after t minutes. Use this formula to solve Exercises 127–128.

127. a. How many people in the class started the rumor?
   b. How many people in the class have heard the rumor after 4 minutes?

128. a. How many people in the class started the rumor?
   b. How many people in the class, rounded to the nearest whole number, have heard the rumor after 6 minutes?

Use the graph of the rumor model to solve Exercises 129–132.

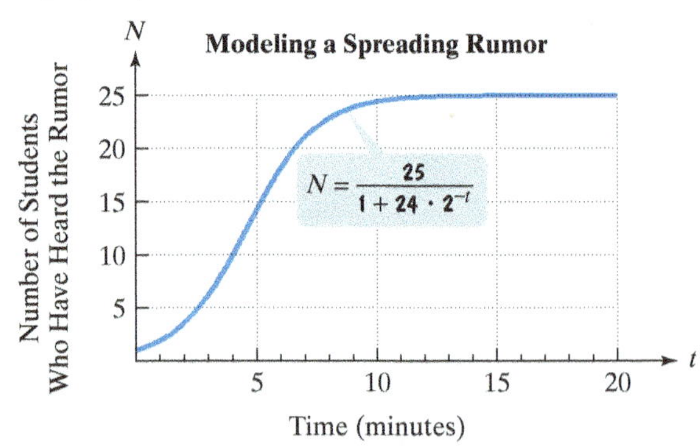

129. Identify your answers to Exercise 127, parts (a) and (b), as points on the graph.

130. Identify your answers to Exercise 128, parts (a) and (b), as points on the graph.

131. Which one of the following best describes the rate of growth of the rumor as shown by the graph?
   a. The number of people in the class who heard the rumor grew steadily over time.
   b. The number of people in the class who heard the rumor remained constant over time.
   c. The number of people in the class who heard the rumor increased slowly at the beginning, but this rate of increase continued to escalate over time.
   d. The number of people in the class who heard the rumor increased quite rapidly at the beginning, but this rate of increase eventually slowed down, ultimately limited by the number of students in the class.

132. Use the graph to determine how many people in the class eventually heard the rumor.

The astronomical unit (AU) is often used to measure distances within the solar system. One AU is equal to the average distance between Earth and the sun, or 92,955,630 miles. The distance, d, of the nth planet from the sun is modeled by the formula

$$d = \frac{3(2^{n-2}) + 4}{10},$$

where d is measured in astronomical units. Use this formula to solve Exercises 133–136.

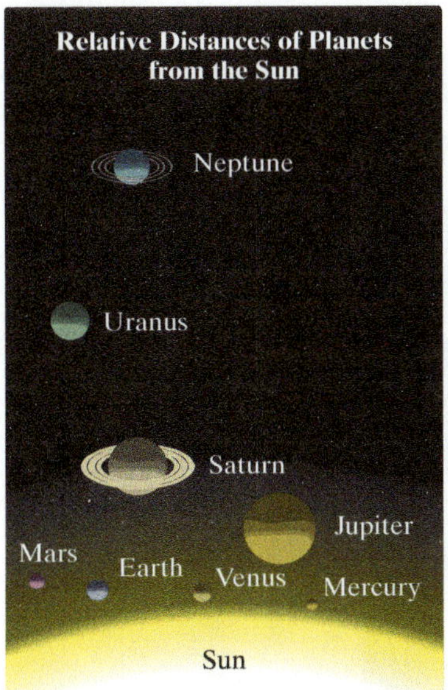

133. Substitute 1 for n and find the distance between Mercury and the sun.

134. Substitute 2 for n and find the distance between Venus and the sun.

135. How much farther from the sun is Jupiter than Earth?

136. How much farther from the sun is Uranus than Earth?

Writing in Mathematics

137. Explain the product rule for exponents. Use $b^2 \cdot b^3$ in your explanation.

138. Explain the quotient rule for exponents. Use $\frac{b^8}{b^2}$ in your explanation.

139. Explain how to find any nonzero number to the 0 power.

140. Explain the negative-exponent rule and give an example.

141. Explain the power rule for exponents. Use $(b^2)^3$ in your explanation.

142. Explain how to simplify an expression that involves a product raised to a power. Give an example.

143. Explain how to simplify an expression that involves a quotient raised to a power. Give an example.

144. How do you know if an exponential expression is simplified?

Technology Exercise

145. Enter the rumor formula

$$N = \frac{25}{1 + 24 \cdot 2^{-t}}$$

in your graphing utility as

$y_1 = 25$ [÷] [(] 1 [+] 24 [×] 2 [∧] [(−)] [x] [)].

Then use a table for y_1 to numerically verify your answers to Exercise 127 or 128.

Critical Thinking Exercises

Make Sense? *In Exercises 146–149, determine whether each statement "makes sense" or "does not make sense" and explain your reasoning.*

146. The properties $(ab)^n = a^n b^n$ and $\left(\frac{a}{b}\right)^n = \frac{a^n}{b^n}$ are like distributive properties of powers over multiplication and division.

147. If 7^{-2} is raised to the third power, the result is a number between 0 and 1.

148. There are many exponential expressions that are equal to $25x^{12}$, such as $(5x^6)^2$, $(5x^3)(5x^9)$, $25(x^3)^9$, and $5^2(x^2)^6$.

149. The expression $\frac{a^n}{b^0}$ is undefined because division by 0 is undefined.

In Exercises 150–157, determine whether each statement is true or false. If the statement is false, make the necessary change(s) to produce a true statement.

150. $2^2 \cdot 2^4 = 2^8$

151. $5^6 \cdot 5^2 = 25^8$

152. $2^3 \cdot 3^2 = 6^5$

153. $\frac{1}{(-2)^3} = 2^{-3}$

154. $\frac{2^8}{2^{-3}} = 2^5$

155. $2^4 + 2^5 = 2^9$

156. $2000.002 = (2 \times 10^3) + (2 \times 10^{-3})$

157. $40{,}000.04 = (4 \times 10^4) + (4 \times 10^{-2})$

In Exercises 158–161, simplify the expression. Assume that all variables used as exponents represent integers and that all other variables represent nonzero real numbers.

158. $x^{n-1} \cdot x^{3n+4}$

159. $(x^{-4n} \cdot x^n)^{-3}$

160. $\left(\frac{x^{3-n}}{x^{6-n}}\right)^{-2}$

161. $\left(\frac{x^n y^{3n+1}}{y^n}\right)^3$

Review Exercises

162. Graph $y = 2x - 1$ in a rectangular coordinate system. Let $x = -3, -2, -1, 0, 1, 2,$ and 3. (Section 1.3, Example 2)

163. Solve $Ax + By = C$ for y. (Section 1.5, Example 6)

164. The length of a rectangular playing field is 5 meters less than twice its width. If 230 meters of fencing enclose the field, what are its dimensions? (Section 1.5, Example 5)

Preview Exercises

Exercises 165–167 will help you prepare for the material covered in the next section.

165. If 6.2 is multiplied by 10^3, what does this multiplication do to the decimal point in 6.2?

166. If 8.5 is multiplied by 10^{-2}, what does this multiplication do to the decimal point in 8.5?

167. Write each computation as a single power of 10. Then evaluate this exponential expression.

a. $10^9 \times 10^{-4}$ **b.** $\dfrac{10^4}{10^{-2}}$

SECTION 1.7 Scientific Notation

Objectives

1. Convert from scientific to decimal notation.
2. Convert from decimal to scientific notation.
3. Perform computations with scientific notation.
4. Use scientific notation to solve problems.

People who complain about paying their income tax can be divided into two types: men and women. Perhaps we can quantify the complaining by examining the data in **Figure 1.21**. The bar graphs show the U.S. population, in millions, and the total amount we paid in federal taxes, in trillions of dollars, for nine selected years.

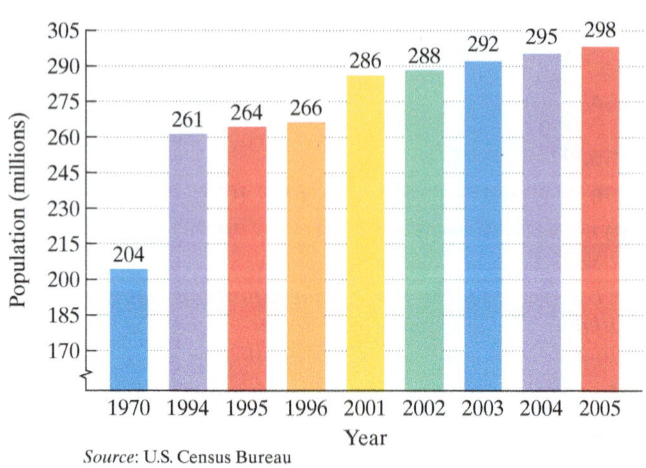

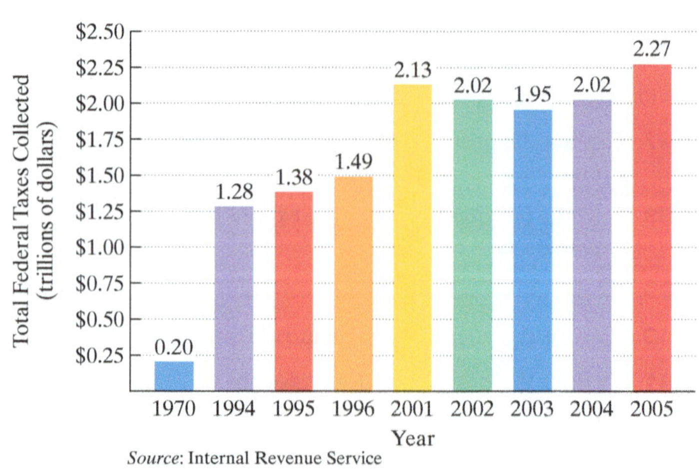

FIGURE 1.21 Population and total tax collections in the United States

The bar graph on the right shows that in 2005 total tax collections were $2.27 trillion. How can we place this amount in the proper perspective? If the total tax collections were evenly divided among all Americans, how much would each citizen pay in taxes?

In this section, you will learn to use exponents to provide a way of putting large and small numbers in perspective. Using this skill, we will explore the per capita tax for some of the years shown in **Figure 1.21**.

1 Convert from scientific to decimal notation.

Scientific Notation

We have seen that in 2005 total tax collections were \$2.27 trillion. Because a trillion is 10^{12} (see **Table 1.6**), this amount can be expressed as

$$2.27 \times 10^{12}.$$

The number 2.27×10^{12} is written in a form called *scientific notation.*

Table 1.6 Names of Large Numbers

10^2	hundred
10^3	thousand
10^6	million
10^9	billion
10^{12}	trillion
10^{15}	quadrillion
10^{18}	quintillion
10^{21}	sextillion
10^{24}	septillion
10^{27}	octillion
10^{30}	nonillion
10^{100}	googol

Scientific Notation

A number is written in **scientific notation** when it is expressed in the form

$$a \times 10^n,$$

where the absolute value of a is greater than or equal to 1 and less than 10 $(1 \leq |a| < 10)$, and n is an integer.

It is customary to use the multiplication symbol, $\times$, rather than a dot, when writing a number in scientific notation.

Converting from Scientific to Decimal Notation

Here are two examples of numbers in scientific notation:

$$6.4 \times 10^5 \quad \text{means} \quad 640{,}000.$$
$$2.17 \times 10^{-3} \quad \text{means} \quad 0.00217.$$

Do you see that the number with the positive exponent is relatively large and the number with the negative exponent is relatively small?

We can use n, the exponent on the 10 in $a \times 10^n$, to change a number in scientific notation to decimal notation. If n is **positive**, move the decimal point in a to the **right** n places. If n is **negative**, move the decimal point in a to the **left** $|n|$ places.

EXAMPLE 1 Converting from Scientific to Decimal Notation

Write each number in decimal notation:

a. 6.2×10^7 **b.** -6.2×10^7 **c.** 2.019×10^{-3} **d.** -2.019×10^{-3}.

Solution In each case, we use the exponent on the 10 to move the decimal point. In parts (a) and (b), the exponent is positive, so we move the decimal point to the right. In parts (c) and (d), the exponent is negative, so we move the decimal point to the left.

a. $6.2 \times 10^7 = 62{,}000{,}000$

$n = 7$ — Move the decimal point 7 places to the right.

b. $-6.2 \times 10^7 = -62{,}000{,}000$

$n = 7$ — Move the decimal point 7 places to the right.

c. $2.019 \times 10^{-3} = 0.002019$

$n = -3$ — Move the decimal point $|-3|$ places, or 3 places, to the left.

d. $-2.019 \times 10^{-3} = -0.002019$

$n = -3$ — Move the decimal point $|-3|$ places, or 3 places, to the left.

CHECK POINT 1 Write each number in decimal notation:

a. -2.6×10^9 **b.** 3.017×10^{-6}.

2 Convert from decimal to scientific notation.

Converting from Decimal to Scientific Notation

To convert from decimal notation to scientific notation, we reverse the procedure of Example 1.

Converting from Decimal to Scientific Notation

Write the number in the form $a \times 10^n$.

- Determine a, the numerical factor. Move the decimal point in the given number to obtain a number whose absolute value is between 1 and 10, including 1.
- Determine n, the exponent on 10^n. The absolute value of n is the number of places the decimal point was moved. The exponent n is positive if the decimal point was moved to the left, negative if the decimal point was moved to the right, and 0 if the decimal point was not moved.

Using Technology

You can use your calculator's [EE] (enter exponent) or [EXP] key to convert from decimal to scientific notation. Here is how it's done for 0.0000000000802.

Many Scientific Calculators

Keystrokes

.0000000000802 [EE] [=]

Display

8.02 − 11

Many Graphing Calculators

Use the mode setting for scientific notation.

Keystrokes

.0000000000802 [ENTER]

Display

8.02E − 11

EXAMPLE 2 Converting from Decimal Notation to Scientific Notation

Write each number in scientific notation:

a. 34,970,000,000,000 b. −34,970,000,000,000
c. 0.0000000000802 d. −0.0000000000802.

Solution

a. $34{,}970{,}000{,}000{,}000 = 3.497 \times 10^{13}$

Move the decimal point to get a number whose absolute value is between 1 and 10.

The decimal point was moved 13 places to the left, so $n = 13$.

b. $-34{,}970{,}000{,}000{,}000 = -3.497 \times 10^{13}$

c. $0.0000000000802 = 8.02 \times 10^{-11}$

Move the decimal point to get a number whose absolute value is between 1 and 10.

The decimal point was moved 11 places to the right, so $n = -11$.

d. $-0.0000000000802 = -8.02 \times 10^{-11}$

Study Tip

If the absolute value of a number is greater than 10, it will have a positive exponent in scientific notation. If the absolute value of a number is less than 1, it will have a negative exponent in scientific notation.

✓ CHECK POINT 2 Write each number in scientific notation:

a. 5,210,000,000 b. −0.00000006893.

EXAMPLE 3 Expressing the Number of Cellphone Spam Messages in Scientific Notation

As feature-rich cellphones function more like PCs, digital intruders are targeting them with viruses, spam, and phishing scams. The bar graph in **Figure 1.22** on the next page shows the number of cellphone spam messages, in millions, from 2002 through 2005. Express the number of spam messages in 2005 in scientific notation.

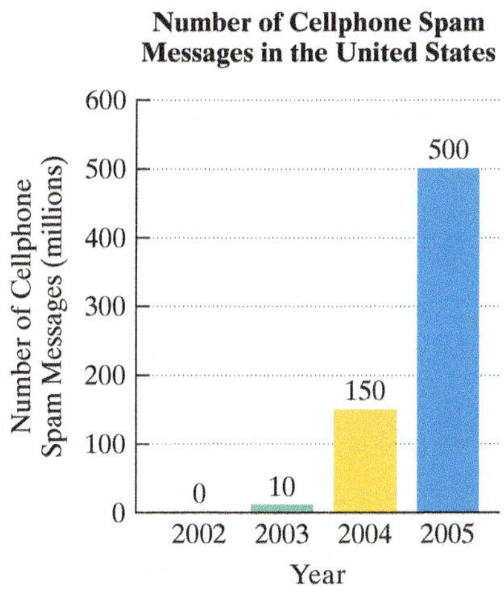

FIGURE 1.22

Source: Ferris Research

Solution Because a million is 10^6, the number of cellphone spam messages in 2005 can be expressed as

$$500 \times 10^6.$$

This factor is not between 1 and 10, so the number is not in scientific notation.

The voice balloon indicates that we need to convert 500 to scientific notation.

$$500 \times 10^6 = (5 \times 10^2) \times 10^6 = 5 \times (10^2 \times 10^6) = 5 \times 10^{2+6} = 5 \times 10^8$$

There were 5×10^8 cellphone spam messages in 2005.

Study Tip

Many of the large numbers we encounter in newspapers, magazines, and almanacs are expressed in millions (10^6), billions (10^9), and trillions (10^{12}). We can use exponential properties to describe the number of cellphone spam messages in 2005 using millions or billions.

$$500 \times 10^6 = 5 \times 10^8 = 0.5 \times 10^9$$

- 500×10^6: There were 500 million messages.
- 5×10^8: This expresses the number of messages in scientific notation.
- 0.5×10^9: There were half a billion messages.

CHECK POINT 3 In 2005, the federal cost of social security was \$519 billion. Express this amount in scientific notation.

3 Perform computations with scientific notation.

Computations with Scientific Notation

Properties of exponents are used to perform computations with numbers that are expressed in scientific notation.

Computations with Numbers in Scientific Notation

Multiplication

$$(a \times 10^n)(b \times 10^m) = (a \times b) \times 10^{n+m}$$

Add the exponents on 10 and multiply the other parts of the numbers separately.

Division

$$\frac{a \times 10^n}{b \times 10^m} = \left(\frac{a}{b}\right) \times 10^{n-m}$$

Subtract the exponents on 10 and divide the other parts of the numbers separately.

After the computation is completed, the answer may require an adjustment before it is back in scientific notation.

EXAMPLE 4 Computations with Scientific Notation

Perform the indicated computations, writing the answers in scientific notation:

a. $(6.1 \times 10^5)(4 \times 10^{-9})$ b. $\dfrac{1.8 \times 10^4}{3 \times 10^{-2}}$.

Solution

a. $(6.1 \times 10^5)(4 \times 10^{-9})$

$= (6.1 \times 4) \times (10^5 \times 10^{-9})$ Regroup factors.

$= 24.4 \times 10^{5+(-9)}$ Add the exponents on 10 and multiply the other parts.

$= 24.4 \times 10^{-4}$ Simplify.

$= (2.44 \times 10^1) \times 10^{-4}$ Convert 24.4 to scientific notation: $24.4 = 2.44 \times 10^1$.

$= 2.44 \times 10^{-3}$ $10^1 \times 10^{-4} = 10^{1+(-4)} = 10^{-3}$

b. $\dfrac{1.8 \times 10^4}{3 \times 10^{-2}} = \left(\dfrac{1.8}{3}\right) \times \left(\dfrac{10^4}{10^{-2}}\right)$ Regroup factors.

$= 0.6 \times 10^{4-(-2)}$ Subtract the exponents on 10 and divide the other parts.

$= 0.6 \times 10^6$ Simplify: $4 - (-2) = 4 + 2 = 6$.

$= (6 \times 10^{-1}) \times 10^6$ Convert 0.6 to scientific notation: $0.6 = 6 \times 10^{-1}$.

$= 6 \times 10^5$ $10^{-1} \times 10^6 = 10^{-1+6} = 10^5$

Using Technology

$(6.1 \times 10^5)(4 \times 10^{-9})$
on a Calculator:

Many Scientific Calculators

6.1 [EE] 5 [×] 4 [EE] 9 [+/−] [=]

Display

2.44 − 03

Many Graphing Calculators

6.1 [EE] 5 [×] 4 [EE] [(−)] 9 [ENTER]

Display (in scientific notation mode)

2.44 E − 3

✓ **CHECK POINT 4** Perform the indicated computations, writing the answers in scientific notation:

a. $(7.1 \times 10^5)(5 \times 10^{-7})$ b. $\dfrac{1.2 \times 10^6}{3 \times 10^{-3}}$.

4 Use scientific notation to solve problems.

Applications: Putting Numbers in Perspective

We have seen that in 2005 the U.S. government collected $2.27 trillion in taxes. Example 5 shows how we can use scientific notation to comprehend the meaning of a number such as 2.27 trillion.

EXAMPLE 5 Tax per Capita

In 2005, the U.S. government collected 2.27×10^{12} dollars in taxes. At that time, the U.S. population was approximately 298 million, or 2.98×10^8. If the total tax collections were evenly divided among all Americans, how much would each citizen pay? Express the answer in decimal notation, rounded to the nearest dollar.

Solution The amount that we would each pay, or the tax per capita, is the total amount collected, 2.27×10^{12}, divided by the number of Americans, 2.98×10^8.

$$\frac{2.27 \times 10^{12}}{2.98 \times 10^8} = \left(\frac{2.27}{2.98}\right) \times \left(\frac{10^{12}}{10^8}\right) \approx 0.7617 \times 10^{12-8} = 0.7617 \times 10^4 = 7617$$

To obtain an answer in decimal notation, it is not necessary to express this number in scientific notation.

Move the decimal point 4 places to the right.

If total tax collections were evenly divided, we would each pay approximately \$7617 in taxes.

CHECK POINT 5 In 2004, the U. S. government collected 2.02×10^{12} dollars in taxes. At that time, the U.S. population was approximately 295 million or 2.95×10^8. Find the per capita tax, rounded to the nearest dollar, in 2004.

Many problems in algebra involve motion. Suppose that you ride your bike at an average speed of 12 miles per hour. What distance do you cover in 2 hours? Your distance is the product of your speed and the time that you travel:

$$\frac{12 \text{ miles}}{\text{hour}} \times 2 \text{ hours} = 24 \text{ miles}.$$

Your distance is 24 miles. Notice how the hour units cancel. The distance is expressed in miles.

In general, the distance covered by any moving body is the product of its average speed, or rate, and its time in motion.

A Formula for Motion

$$d = rt$$

Distance equals rate times time.

EXAMPLE 6 Using the Motion Formula

Light travels at a rate of approximately 1.86×10^5 miles per second. It takes light 5×10^2 seconds to travel from the sun to Earth. What is the distance between Earth and the sun?

Solution

$d = rt$ — Use the motion formula.

$d = (1.86 \times 10^5) \times (5 \times 10^2)$ — Substitute the given values.

$d = (1.86 \times 5) \times (10^5 \times 10^2)$ — Rearrange factors.

$d = 9.3 \times 10^7$ — Add the exponents on 10 and multiply the other parts.

The distance between Earth and the sun is approximately 9.3×10^7 miles, or 93 million miles.

CHECK POINT 6 A futuristic spacecraft traveling at 1.55×10^3 miles per hour takes 20,000 hours (about 833 days) to travel from Venus to Mercury. What is the distance from Venus to Mercury?

1.7 EXERCISE SET

MyMathLab MathXL PRACTICE WATCH DOWNLOAD READ REVIEW

Practice Exercises

In Exercises 1–14, write each number in decimal notation without the use of exponents.

1. 3.8×10^2
2. 9.2×10^2
3. 6×10^{-4}
4. 7×10^{-5}
5. -7.16×10^6
6. -8.17×10^6
7. 1.4×10^0
8. 2.4×10^0
9. 7.9×10^{-1}
10. 6.8×10^{-1}
11. -4.15×10^{-3}
12. -3.14×10^{-3}
13. -6.00001×10^{10}
14. -7.00001×10^{10}

In Exercises 15–30, write each number in scientific notation.

15. 32,000
16. 64,000
17. 638,000,000,000,000,000
18. 579,000,000,000,000,000
19. −317
20. −326
21. −5716
22. −3829
23. 0.0027
24. 0.0083
25. −0.00000000504
26. −0.00000000405
27. 0.007
28. 0.005
29. 3.14159
30. 2.71828

In Exercises 31–50, perform the indicated computations. Write the answers in scientific notation. If necessary, round the decimal factor in your scientific notation answer to two decimal places.

31. $(3 \times 10^4)(2.1 \times 10^3)$
32. $(2 \times 10^4)(4.1 \times 10^3)$
33. $(1.6 \times 10^{15})(4 \times 10^{-11})$
34. $(1.4 \times 10^{15})(3 \times 10^{-11})$
35. $(6.1 \times 10^{-8})(2 \times 10^{-4})$
36. $(5.1 \times 10^{-8})(3 \times 10^{-4})$
37. $(4.3 \times 10^8)(6.2 \times 10^4)$
38. $(8.2 \times 10^8)(4.6 \times 10^4)$
39. $\dfrac{8.4 \times 10^8}{4 \times 10^5}$
40. $\dfrac{6.9 \times 10^8}{3 \times 10^5}$
41. $\dfrac{3.6 \times 10^4}{9 \times 10^{-2}}$
42. $\dfrac{1.2 \times 10^4}{2 \times 10^{-2}}$
43. $\dfrac{4.8 \times 10^{-2}}{2.4 \times 10^6}$
44. $\dfrac{7.5 \times 10^{-2}}{2.5 \times 10^6}$
45. $\dfrac{2.4 \times 10^{-2}}{4.8 \times 10^{-6}}$
46. $\dfrac{1.5 \times 10^{-2}}{3 \times 10^{-6}}$
47. $\dfrac{480{,}000{,}000{,}000}{0.00012}$
48. $\dfrac{282{,}000{,}000{,}000}{0.00141}$
49. $\dfrac{0.00072 \times 0.003}{0.00024}$
50. $\dfrac{66{,}000 \times 0.001}{0.003 \times 0.002}$

Practice PLUS

In Exercises 51–58, solve each equation. Express the solution in scientific notation.

51. $(2 \times 10^{-5})x = 1.2 \times 10^9$
52. $(3 \times 10^{-2})x = 1.2 \times 10^4$
53. $\dfrac{x}{2 \times 10^8} = -3.1 \times 10^{-5}$
54. $\dfrac{x}{5 \times 10^{11}} = -2.9 \times 10^{-3}$
55. $x - (7.2 \times 10^{18}) = 9.1 \times 10^{18}$
56. $x - (5.3 \times 10^{-16}) = 8.4 \times 10^{-16}$
57. $(-1.2 \times 10^{-3})x = (1.8 \times 10^{-4})(2.4 \times 10^6)$
58. $(-7.8 \times 10^{-4})x = (3.9 \times 10^{-7})(6.8 \times 10^5)$

Application Exercises

The graph shows the net worth, in billions of dollars, of the five richest Americans. Use 10^9 for one billion and the figures shown to solve Exercises 59–62. Express all answers in scientific notation.

Net Worth of the Five Richest Americans

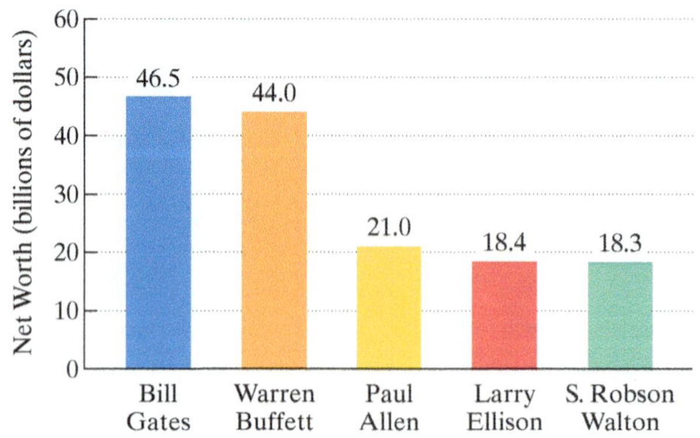

Source: *Forbes* Billionaires List, 2005

59. How much is Bill Gates worth?
60. How much is Warren Buffett worth?
61. By how much does Larry Ellison's worth exceed that of S. Robson Walton?
62. If each person doubled his net worth, by how much would Larry Ellison's worth exceed that of S. Robson Walton?

Our ancient ancestors hunted for their meat and expended a great deal of energy chasing it down. Today, our animal protein is raised in cages and on feedlots, delivered in great abundance nearly to our door. Use the numbers shown below to solve Exercises 63–66. Use 10^6 for one million and 10^9 for one billion.

Source: Time, October 20, 2003

In Exercises 63–64, use 300 million, or 3×10^8, for the U.S. population. Express answers in decimal notation, rounded, if necessary, to the nearest whole number.

63. Find the number of hot dogs consumed by each American in a year.

64. If the consumption of Big Macs was divided evenly among all Americans, how many Big Macs would we each consume in a year?

In Exercises 65–66, use the fact that there are approximately 3.2×10^7 seconds in a year.

65. How many chickens are raised for food each second in the United States? Express the answer in scientific and decimal notations.

66. How many chickens are eaten per year in the United States? Express the answer in scientific notation.

The graph shows the cost, in billions of dollars, and the enrollment, in millions of people, for various federal social programs in 2005. Use the numbers shown to solve Exercises 67–69.

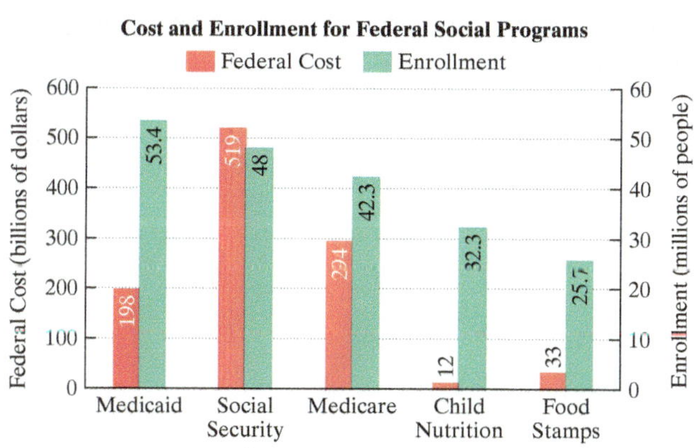

Source: Office of Management and Budget

67. a. What was the average per person benefit for Social Security? Express the answer in scientific notation and in decimal notation, rounded to the nearest dollar.

b. What was the average monthly per person benefit, rounded to the nearest dollar, for Social Security?

68. a. What was the average per person benefit for the food stamps program? Express the answer in scientific notation and in decimal notation, rounded to the nearest dollar.

b. What was the average monthly per person benefit, rounded to the nearest dollar, for the food stamps program?

69. Medicaid provides health insurance for the poor. Medicare provides health insurance for people 65 and older, as well as younger people who are disabled. Which program provides the greater per person benefit? By how much, rounded to the nearest dollar?

70. The area of Alaska is approximately 3.66×10^8 acres. The state was purchased in 1867 from Russia for \$7.2 million. What price per acre, to the nearest cent, did the United States pay Russia?

71. The mass of one oxygen molecule is 5.3×10^{-23} gram. Find the mass of 20,000 molecules of oxygen. Express the answer in scientific notation.

72. The mass of one hydrogen atom is 1.67×10^{-24} gram. Find the mass of 80,000 hydrogen atoms. Express the answer in scientific notation.

73. In Exercises 65–66, we used 3.2×10^7 as an approximation for the number of seconds in a year. Convert 365 days (one year) to hours, to minutes, and, finally, to seconds, to determine precisely how many seconds there are in a year. Express the answer in scientific notation.

Writing in Mathematics

74. How do you know if a number is written in scientific notation?

75. Explain how to convert from scientific to decimal notation and give an example.

76. Explain how to convert from decimal to scientific notation and give an example.

77. Describe one advantage of expressing a number in scientific notation over decimal notation.

Technology Exercises

78. Use a calculator to check any three of your answers in Exercises 1–14.

79. Use a calculator to check any three of your answers in Exercises 15–30.

80. Use a calculator with an EE or EXP key to check any four of your computations in Exercises 31–50. Display the result of the computation in scientific notation.

Critical Thinking Exercises

Make Sense? *In Exercises 81–84, determine whether each statement "makes sense" or "does not make sense" and explain your reasoning.*

81. For a recent year, total tax collections in the United States were $\$2.02 \times 10^7$.

82. I just finished reading a book that contained approximately 1.04×10^5 words.

83. If numbers in the form $a \times 10^n$ are listed from least to greatest, values of a need not appear from least to greatest.

84. When expressed in scientific notation, 58 million and 58 millionths have exponents on 10 with the same absolute value.

In Exercises 85–89, determine whether each statement is true or false. If the statement is false, make the necessary change(s) to produce a true statement.

85. $534.7 = 5.347 \times 10^3$

86. $\frac{8 \times 10^{30}}{4 \times 10^{-5}} = 2 \times 10^{25}$

87. $(7 \times 10^5) + (2 \times 10^{-3}) = 9 \times 10^2$

88. $(4 \times 10^3) + (3 \times 10^2) = 43 \times 10^2$

89. The numbers 8.7×10^{25}, 1.0×10^{26}, 5.7×10^{26}, and 3.7×10^{27} are listed from least to greatest.

In Exercises 90–91, perform the indicated additions. Write the answers in scientific notation.

90. $5.6 \times 10^{13} + 3.1 \times 10^{13}$

91. $8.2 \times 10^{-16} + 4.3 \times 10^{-16}$

92. Our hearts beat approximately 70 times per minute. Express in scientific notation how many times the heart beats over a lifetime of 80 years. Round the decimal factor in your scientific notation answer to two decimal places.

93. Give an example of a number where there is no advantage in using scientific notation over decimal notation.

Review Exercises

94. Simplify: $9(10x - 4) - (5x - 10)$. (Section 1.2, Example 14)

95. Solve: $\frac{4x - 1}{10} = \frac{5x + 2}{4} - 4$. (Section 1.4, Example 4)

96. Simplify: $(8x^4y^{-3})^{-2}$. (Section 1.6, Example 7)

Preview Exercises

Exercises 97–99 will help you prepare for the material covered in the first section of the next chapter.

97. Here are two sets of ordered pairs:

set 1: $\{(1, 5), (2, 5)\}$

set 2: $\{(5, 1), (5, 2)\}$

In which set is each x-coordinate paired with only one y-coordinate?

98. Evaluate $r^3 - 2r^2 + 5$ for $r = -5$.

99. Evaluate $5x + 7$ for $x = a + h$.

GROUP PROJECT

CHAPTER 1

One of the best ways to learn how to *solve* a word problem in algebra is to *design* word problems of your own. Creating a word problem makes you very aware of precisely how much information is needed to solve the problem. You must also focus on the best way to present information to a reader and on how much information to give. As you write your problem, you gain skills that will help you solve problems created by others.

The group should design five different word problems that can be solved using an algebraic equation. All of the problems should be on different topics. For example, the group should not have more than one problem on a price reduction. The group should turn in both the problems and their algebraic solutions.

Chapter 1 Summary

Definitions and Concepts / Examples

Section 1.1 Algebraic Expressions and Real Numbers

Letters that represent numbers are called variables.
An algebraic expression is a combination of variables, numbers, and operation symbols. English phrases can be translated into algebraic expressions:

- Addition: sum, plus, increased by, more than
- Subtraction: difference, minus, decreased by, less than
- Multiplication: product, times, of, twice
- Division: quotient, divide, per, ratio

Translate: Six less than the product of a number and five.

6 less than — the product of a number and 5

$$5x - 6$$

or $x \cdot 5 - 6$

Many algebraic expressions contain exponents. If b is a natural number, b^n, the nth power of b, is the product of n factors of b. Furthermore, $b^1 = b$.
Evaluating an algebraic expression means to find the value of the expression for a given value of the variable.

Evaluate $6 + 5(x - 10)^3$ for $x = 12$.

$$6 + 5(12 - 10)^3$$
$$= 6 + 5 \cdot 2^3$$
$$= 6 + 5 \cdot 8$$
$$= 6 + 40 = 46$$

An equation is a statement that two expressions are equal. Formulas are equations that express relationships among two or more variables. Mathematical modeling is the process of finding formulas to describe real-world phenomena. Such formulas, together with the meaning assigned to the variables, are called mathematical models. The formulas are said to model, or describe, the relationships among the variables.

The formula

$$h = -16t^2 + 200t + 4$$

models the height, h, in feet, of fireworks t seconds after launch. What is the height after 2 seconds?

$$h = -16(2)^2 + 200(2) + 4$$
$$= -16(4) + 200(2) + 4$$
$$= -64 + 400 + 4 = 340$$

The height is 340 feet.

A set is a collection of objects, called elements, enclosed in braces. The roster method uses commas to separate the elements of the set. Set-builder notation describes the elements of a set, but does not list them. The symbol $\in$ means that a number or object is in a set; $\notin$ means that a number or object is not in a set. The set of real numbers is the set of all numbers that can be represented by points on the number line. Sets that make up the real numbers include

Natural numbers: $\{1, 2, 3, 4, \ldots\}$

Whole numbers: $\{0, 1, 2, 3, 4, \ldots\}$

Integers: $\{\ldots, -4, -3, -2, -1, 0, 1, 2, 3, 4, \ldots\}$

Rational numbers: $\left\{\frac{a}{b} \middle| a \text{ and } b \text{ are integers and } b \neq 0\right\}$

Irrational numbers:
$\{x | x \text{ is a real number and } x \text{ is not a rational number}\}$.

In decimal form, rational numbers terminate or repeat.
In decimal form, irrational numbers do neither.

- Use the roster method to list the elements of

$$\{x | x \text{ is a natural number less than } 6\}.$$

Solution

$$\{1, 2, 3, 4, 5\}$$

- True or false:

$$\sqrt{2} \notin \{x | x \text{ is a rational number}\}.$$

Solution

The statement is true:

$$\sqrt{2} \text{ is not a rational number.}$$

The decimal form of $\sqrt{2}$ neither terminates nor repeats. Thus, $\sqrt{2}$ is an irrational number.

For any two real numbers, a and b, a is less than b if a is to the left of b on the number line.

Inequality Symbols

$<$: is less than
$>$: is greater than
$\leq$: is less than or equal to
$\geq$: is greater than or equal to

- $-1 < 5$, or -1 is less than 5, is true because -1 is to the left of 5 on a number line.
- $-3 \geq 7$, -3 is greater than or equal to 7, is false. Neither $-3 > 7$ nor $-3 = 7$ is true.

Definitions and Concepts	Examples

Section 1.2 Operations with Real Numbers and Simplifying Algebraic Expressions

Absolute Value

$$|a| = \begin{cases} a & \text{if } a \geq 0 \\ -a & \text{if } a < 0 \end{cases}$$

The opposite, or additive inverse, of a is $-a$. When a is a negative number, $-a$ is positive.

- $|6.03| = 6.03$
- $|0| = 0$
- $|-4.9| = -(-4.9) = 4.9$

Adding Real Numbers

To add two numbers with the same sign, add their absolute values and use their common sign. To add two numbers with different signs, subtract the smaller absolute value from the greater absolute value and use the sign of the number with the greater absolute value.

- $-4.1 + (-6.2) = -10.3$
- $-30 + 25 = -5$
- $12 + (-8) = 4$

Subtracting Real Numbers

$$a - b = a + (-b)$$

$$-\frac{3}{4} - \left(-\frac{1}{2}\right) = -\frac{3}{4} + \frac{1}{2} = -\frac{3}{4} + \frac{2}{4} = -\frac{1}{4}$$

Multiplying and Dividing Real Numbers

The product or quotient of two numbers with the same sign is positive and with different signs is negative. If no number is 0, a product with an even number of negative factors is positive and a product with an odd number of negative factors is negative. Division by 0 is undefined.

- $2(-6)(-1)(-5) = -60$

  Three (odd) negative factors give a negative product.
- $(-2)^3 = (-2)(-2)(-2) = -8$
- $-\frac{1}{3}\left(-\frac{2}{5}\right) = \frac{2}{15}$
- $\frac{-14}{2} = -7$

Order of Operations

1. Perform operations within grouping symbols, starting with the innermost grouping symbols. Grouping symbols include parentheses, brackets, fraction bars, absolute value symbols, and square root signs.
2. Evaluate exponential expressions.
3. Multiply and divide from left to right.
4. Add and subtract from left to right.

Simplify: $\frac{6(8-10)^3 + (-2)}{(-5)^2(-2)}$.

$$= \frac{6(-2)^3 + (-2)}{(-5)^2(-2)} = \frac{6(-8) + (-2)}{25(-2)}$$

$$= \frac{-48 + (-2)}{-50} = \frac{-50}{-50} = 1$$

Basic Algebraic Properties

Commutative: $a + b = b + a$
$ab = ba$

Associative: $(a + b) + c = a + (b + c)$
$(ab)c = a(bc)$

Distributive: $a(b + c) = ab + ac$
$a(b - c) = ab - ac$
$(b + c)a = ba + ca$

- Commutative

  $3x + 5 = 5 + 3x = 5 + x \cdot 3$
- Associative

  $-4(6x) = (-4 \cdot 6)x = -24x$
- Distributive

  $-4(9x + 3) = -4(9x) + (-4) \cdot 3$
  $= -36x + (-12)$
  $= -36x - 12$

Definitions and Concepts	Examples

Section 1.2 Operations with Real Numbers and Simplifying Algebraic Expressions (continued)

Simplifying Algebraic Expressions

Terms are separated by addition. Like terms have the same variable factors and are combined using the distributive property. An algebraic expression is simplified when grouping symbols have been removed and like terms have been combined.

Simplify: $7(3x - 4) - (10x - 5)$.

$$= 21x - 28 - 10x + 5$$
$$= 21x - 10x - 28 + 5$$
$$= 11x - 23$$

Section 1.3 Graphing Equations

The rectangular coordinate system consists of a horizontal number line, the x-axis, and a vertical number line, the y-axis, intersecting at their zero points, the origin. Each point in the system corresponds to an ordered pair of real numbers (x, y). The first number in the pair is the x-coordinate; the second number is the y-coordinate.

Plot: $(4, 2)$, $(-3, 4)$, $(-5, -4)$, and $(4, -3)$.

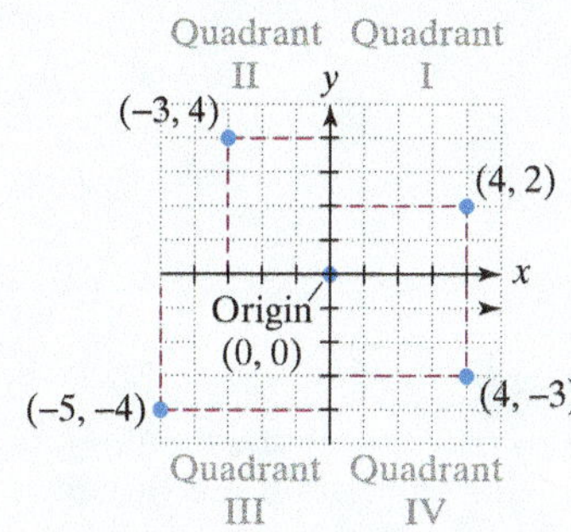

An ordered pair is a solution of an equation in two variables if replacing the variables by the corresponding coordinates results in a true statement. The ordered pair is said to satisfy the equation. The graph of the equation is the set of all points whose coordinates satisfy the equation. One method for graphing an equation is to plot ordered-pair solutions and connect them with a smooth curve or line.

Graph: $y = x^2 - 1$.

x	$y = x^2 - 1$
-2	$(-2)^2 - 1 = 3$
-1	$(-1)^2 - 1 = 0$
0	$0^2 - 1 = -1$
1	$1^2 - 1 = 0$
2	$2^2 - 1 = 3$

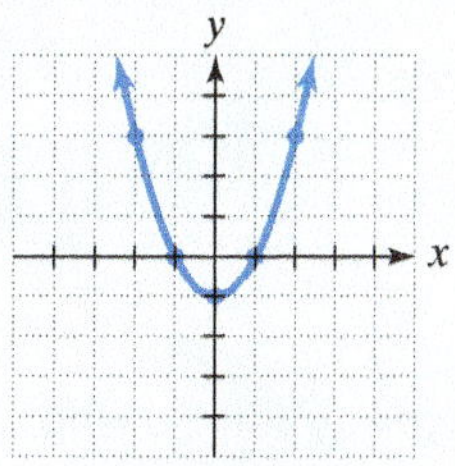

Section 1.4 Solving Linear Equations

A linear equation in one variable can be written in the form $ax + b = 0, a \neq 0$. A solution is a value of the variable that makes the equation a true statement. The set of all such solutions is the equation's solution set. Equivalent equations have the same solution set. To solve a linear equation,

1. Simplify each side.
2. Collect variable terms on one side and constant terms on the other side.
3. Isolate the variable and solve.
4. Check the proposed solution in the original equation.

Solve: $4(x - 5) = 2x - 14$.

$$4x - 20 = 2x - 14$$
$$4x - 2x - 20 = 2x - 2x - 14$$
$$2x - 20 = -14$$
$$2x - 20 + 20 = -14 + 20$$
$$2x = 6$$
$$\frac{2x}{2} = \frac{6}{2}$$
$$x = 3$$

Checking gives $-8 = -8$, so 3 is the solution, or $\{3\}$ is the solution set.

Definitions and Concepts	Examples

Section 1.4 Solving Linear Equations (continued)

Equations Containing Fractions

Multiply both sides (all terms) by the least common denominator. This clears the equation of fractions.

Solve: $\frac{x-2}{5} + \frac{x+2}{2} = \frac{x+4}{3}$.

$$30\left(\frac{x-2}{5} + \frac{x+2}{2}\right) = 30\left(\frac{x+4}{3}\right)$$

$$6(x-2) + 15(x+2) = 10(x+4)$$

$$6x - 12 + 15x + 30 = 10x + 40$$

$$21x + 18 = 10x + 40$$

$$11x = 22$$

$$x = 2$$

Checking gives $2 = 2$, so 2 is the solution, or $\{2\}$ is the solution set.

Types of Equations

An equation that is true for all real numbers, $\mathbb{R}$, is called an identity. When solving an identity, the variable is eliminated and a true statement, such as $3 = 3$, results. An equation that is not true for even one real number is called an inconsistent equation. A false statement, such as $3 = 7$, results when solving such an equation, whose solution set is $\varnothing$, the empty set. A conditional equation is not an identity, but is true for at least one real number.

Solve: $4x + 5 = 4(x + 2)$.

$$4x + 5 = 4x + 8$$

$$5 = 8, \quad \text{false}$$

The inconsistent equation has no solution: $\varnothing$.

Solve: $5x - 4 = 5(x + 1) - 9$.

$$5x - 4 = 5x + 5 - 9$$

$$5x - 4 = 5x - 4$$

$$-4 = -4, \quad \text{true}$$

All real numbers satisfy the identity: $\mathbb{R}$.

Section 1.5 Problem Solving and Using Formulas

Strategy for Solving Algebraic Word Problems

1. Let x represent one of the quantities.
2. Represent other unknown quantities in terms of x.
3. Write an equation that models the conditions.
4. Solve the equation and answer the question.
5. Check the proposed solution in the original wording of the problem.

After a 60% reduction, a suit sold for \$32. What was the original price?

Let x = the original price.

Original price	minus	60% reduction	=	reduced price
x	$-$	$0.6x$	$=$	32

$$0.4x = 32$$

$$\frac{0.4x}{0.4} = \frac{32}{0.4}$$

$$x = 80$$

The original price was \$80. Check this amount using the first sentence in the problem's conditions.

To solve a formula for a variable, use the steps for solving a linear equation and isolate that variable on one side of the equation.

Solve for r: $E = I(R + r)$.

$$E = IR + Ir$$ We need to isolate r.

$$E - IR = Ir$$

$$\frac{E - IR}{I} = r$$

Definitions and Concepts	Examples

Section 1.6 Properties of Integral Exponents

The Product Rule

$$b^m \cdot b^n = b^{m+n}$$

$$(-3x^{10})(5x^{20}) = -3 \cdot 5x^{10+20} = -15x^{30}$$

The Quotient Rule

$$\frac{b^m}{b^n} = b^{m-n}, b \neq 0$$

$$\frac{5x^{20}}{10x^{10}} = \frac{5}{10} \cdot x^{20-10} = \frac{x^{10}}{2}$$

Zero and Negative Exponents

$$b^0 = 1, b \neq 0$$

$$b^{-n} = \frac{1}{b^n} \text{ and } \frac{1}{b^{-n}} = b^n$$

- $(3x)^0 = 1$
- $3x^0 = 3 \cdot 1 = 3$
- $\frac{2^{-3}}{4^{-2}} = \frac{4^2}{2^3} = \frac{16}{8} = 2$

Power Rule

$$(b^m)^n = b^{mn}$$

$$(x^5)^{-4} = x^{5(-4)} = x^{-20} = \frac{1}{x^{20}}$$

Products to Powers

$$(ab)^n = a^n b^n$$

$$(5x^3y^{-4})^{-2} = 5^{-2} \cdot (x^3)^{-2} \cdot (y^{-4})^{-2} = 5^{-2}x^{-6}y^8 = \frac{y^8}{5^2x^6} = \frac{y^8}{25x^6}$$

Quotients to Powers

$$\left(\frac{a}{b}\right)^n = \frac{a^n}{b^n}$$

$$\left(\frac{2}{x^3}\right)^{-4} = \frac{2^{-4}}{(x^3)^{-4}} = \frac{2^{-4}}{x^{-12}} = \frac{x^{12}}{2^4} = \frac{x^{12}}{16}$$

An exponential expression is simplified when

- No parentheses appear.
- No powers are raised to powers.
- Each base occurs only once.
- No negative or zero exponents appear.

Simplify: $\frac{-5x^{-3}y^2}{-20x^2y^{-6}}$.

$$= \frac{-5}{-20} \cdot x^{-3-2} \cdot y^{2-(-6)} = \frac{1}{4}x^{-5}y^8 = \frac{y^8}{4x^5}$$

Section 1.7 Scientific Notation

A number in scientific notation is expressed in the form

$$a \times 10^n,$$

where $|a|$ is greater than or equal to 1 and less than 10, and n is an integer.

Write in decimal notation: 3.8×10^{-3}.

$$3.8 \times 10^{-3} = .0038 = 0.0038$$

Write in scientific notation: 26,000.

$$26{,}000 = 2.6 \times 10^4$$

Computations with Numbers in Scientific Notation

$$(a \times 10^n)(b \times 10^m) = (a \times b) \times 10^{n+m}$$

$$\frac{a \times 10^n}{b \times 10^m} = \left(\frac{a}{b}\right) \times 10^{n-m}$$

$$(8 \times 10^3)(5 \times 10^{-8}) = 8 \cdot 5 \times 10^{3+(-8)} = 40 \times 10^{-5} = (4 \times 10^1) \times 10^{-5} = 4 \times 10^{-4}$$

CHAPTER 1 REVIEW EXERCISES

1.1

In Exercises 1–3, write each English phrase as an algebraic expression. Let x represent the number.

1. Ten less than twice a number
2. Four more than the product of six and a number
3. The quotient of nine and a number, increased by half of the number

In Exercises 4–6, evaluate each algebraic expression for the given value or values of the variable.

4. $x^2 - 7x + 4$, for $x = 10$
5. $6 + 2(x - 8)^3$, for $x = 11$
6. $x^4 - (x - y)$, for $x = 2$ and $y = 1$

In Exercises 7–8, use the roster method to list the elements in each set.

7. $\{x|x$ is a natural number less than $3\}$
8. $\{x|x$ is an integer greater than -4 and less than $2\}$

In Exercises 9–11, determine whether each statement is true or false.

9. $0 \in \{x|x$ is a natural number$\}$
10. $-2 \in \{x|x$ is a rational number$\}$
11. $\frac{1}{3} \notin \{x|x$ is an irrational number$\}$

In Exercises 12–14, write out the meaning of each inequality. Then determine whether the inequality is true or false.

12. $-5 < 2$
13. $-7 \geq -3$
14. $-7 \leq -7$
15. You are riding along an expressway traveling x miles per hour. The formula

$$S = 0.015x^2 + x + 10$$

models the recommended safe distance, S, in feet, between your car and other cars on the expressway. What is the recommended safe distance when your speed is 60 miles per hour?

1.2

In Exercises 16–18, find each absolute value.

16. $|-9.7|$
17. $|5.003|$
18. $|0|$

In Exercises 19–30, perform the indicated operation.

19. $-2.4 + (-5.2)$
20. $-6.8 + 2.4$
21. $-7 - (-20)$
22. $(-3)(-20)$
23. $-\frac{3}{5} - \left(-\frac{1}{2}\right)$
24. $\left(\frac{2}{7}\right)\left(-\frac{3}{10}\right)$
25. $4(-3)(-2)(-10)$
26. $(-2)^4$
27. -2^5
28. $-\frac{2}{3} \div \frac{8}{5}$
29. $\frac{-35}{-5}$
30. $\frac{54.6}{-6}$
31. Find $-x$ if $x = -7$.

In Exercises 32–38, simplify each expression.

32. $-11 - [-17 + (-3)]$
33. $\left(-\frac{1}{2}\right)^3 \cdot 2^4$
34. $-3[4 - (6 - 8)]$
35. $8^2 - 36 \div 3^2 \cdot 4 - (-7)$
36. $\frac{(-2)^4 + (-3)^2}{2^2 - (-21)}$
37. $\frac{(7 - 9)^3 - (-4)^2}{2 + 2(8) \div 4}$
38. $4 - (3 - 8)^2 + 3 \div 6 \cdot 4^2$

In Exercises 39–43, simplify each algebraic expression.

39. $5(2x - 3) + 7x$
40. $5x + 7x^2 - 4x + 2x^2$
41. $3(4y - 5) - (7y + 2)$
42. $8 - 2[3 - (5x - 1)]$
43. $6(2x - 3) - 5(3x - 2)$

1.3

In Exercises 44–46, plot the given point in a rectangular coordinate system.

44. $(-1, 3)$
45. $(2, -5)$
46. $(0, -6)$

In Exercises 47–50, graph each equation. Let $x = -3, -2, -1, 0, 1, 2,$ and 3.

47. $y = 2x - 2$
48. $y = x^2 - 3$
49. $y = x$
50. $y = |x| - 2$
51. What does a $[-20, 40, 10]$ by $[-5, 5, 1]$ viewing rectangle mean? Draw axes with tick marks and label the tick marks to illustrate this viewing rectangle.

The caseload of Alzheimer's disease in the United States is expected to explode as baby boomers head into their later years. The graph shows the percentage of Americans with the disease, by age. Use the graph to solve Exercises 52–54.

Alzheimer's Prevalence in the United States, by Age

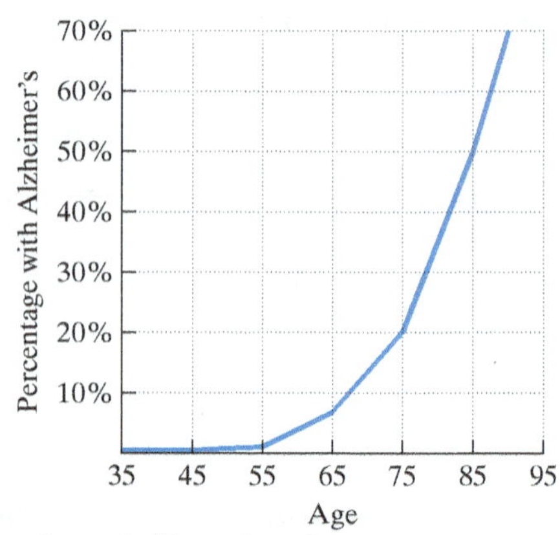

Source: Centers for Disease Control

52. What percentage of Americans who are 75 have Alzheimer's disease?
53. What age represents 50% prevalence of Alzheimer's disease?

54. Describe the trend shown by the graph.

55. Select the graph that best illustrates the following description: A train pulls into a station and lets off its passengers.

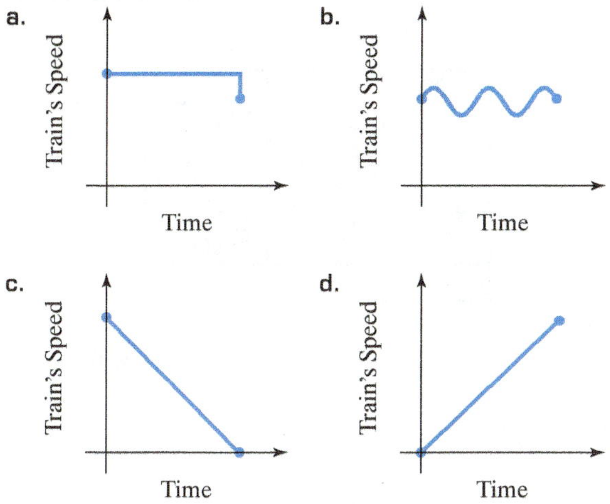

1.4 *In Exercises 56–61, solve and check each linear equation.*

56. $2x - 5 = 7$

57. $5x + 20 = 3x$

58. $7(x - 4) = x + 2$

59. $1 - 2(6 - x) = 3x + 2$

60. $2(x - 4) + 3(x + 5) = 2x - 2$

61. $2x - 4(5x + 1) = 3x + 17$

In Exercises 62–66, solve each equation.

62. $\frac{2x}{3} = \frac{x}{6} + 1$

63. $\frac{x}{2} - \frac{1}{10} = \frac{x}{5} + \frac{1}{2}$

64. $\frac{2x}{3} = 6 - \frac{x}{4}$

65. $\frac{x}{4} = 2 + \frac{x - 3}{3}$

66. $\frac{3x + 1}{3} - \frac{13}{2} = \frac{1 - x}{4}$

In Exercises 67–71, solve each equation. Then state whether the equation is an identity, a conditional equation, or an inconsistent equation.

67. $7x + 5 = 5(x + 3) + 2x$

68. $7x + 13 = 4x - 10 + 3x + 23$

69. $7x + 13 = 3x - 10 + 2x + 23$

70. $4(x - 3) + 5 = x + 5(x - 2)$

71. $(2x - 3)2 - 3(x + 1) = (x - 2)4 - 3(x + 5)$

72. The bar graph shows the number of corporations that owned the majority of the media industry in the United States for selected years. Through mergers and buyouts, by 2004, a majority of American newspapers, magazines, TV and radio stations, book publishers, and movie studios were owned by just five corporations: Time Warner, Disney, News Corp., Bertelsmann, and Viacom.

Number of Corporations Owning the Majority of United States Media

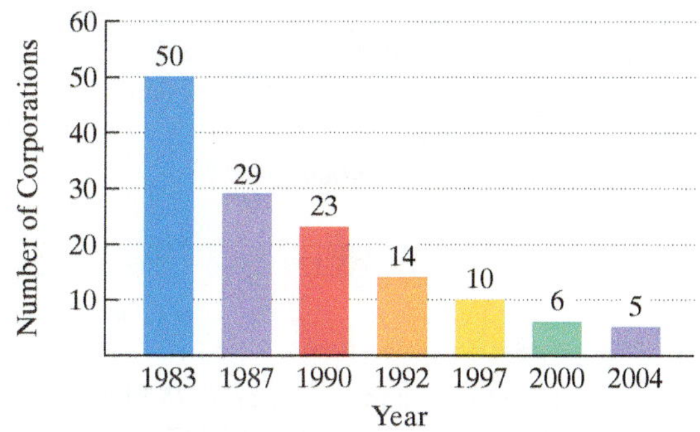

Source: Jonathan Teller-Elsberg et al., *Field Guide to the U.S. Economy*, The New Press, 2006

The data can be modeled by the formula $N = -2x + 40$, in which N represents the number of corporations owning the majority of U.S. media x years after 1983.

a. According to the model, in which year were there 12 corporations that owned the majority of media available to Americans?

b. Does the information provided by the model in part (a) overestimate or underestimate the number of corporations indicated by the graph? By how much?

1.5 *In Exercises 73–79, use the five-step strategy for solving word problems.*

73. The bar graph represents money earned, in millions of dollars, by each of the top five concert tours in 2005. Gross earnings by U2 exceeded earnings by The Eagles by \$143 million. Gross earnings by The Rolling Stones exceeded The Eagles' earnings by \$24 million. Combined, these three groups earned \$518 million on their concert tours. Determine the gross earnings on concert tours, in millions of dollars, for each of the three groups.

Top United States Concert Tours, 2005

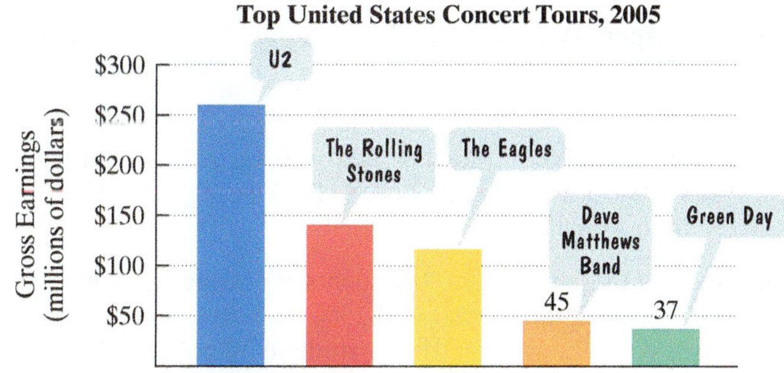

Source: Billboard magazine

74. One angle of a triangle measures 10° more than the second angle. The measure of the third angle is twice the sum of the measures of the first two angles. Determine the measure of each angle.

75. Without changes, the graphs show projections for the amount being paid in Social Security benefits and the amount going into the system. All data are expressed in billions of dollars.

Social Insecurity: Projected Income and Outflow of the Social Security System

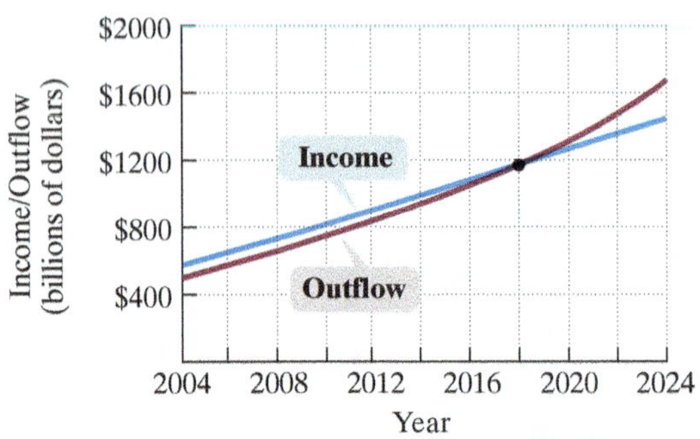

Source: 2004 Social Security Trustees Report

a. In 2004, the system's income was \$575 billion, projected to increase at an average rate of \$43 billion per year. In which year will the system's income be \$1177 billion?

b. The data for the system's outflow can be modeled by the formula

$$B = 0.07x^2 + 47.4x + 500,$$

where B represents the amount paid in benefits, in billions of dollars, x years after 2004. According to this model, what will be the amount paid in benefits for the year you determined in part (a)? Round to the nearest billion dollar.

c. How are your answers to parts (a) and (b) shown by the graphs?

76. You are choosing between two long-distance telephone plans. One plan has a monthly fee of \$15 with a charge of \$0.05 per minute. The other plan has a monthly fee of \$5 with a charge of \$0.07 per minute. For how many minutes of long-distance calls will the costs for the two plans be the same?

77. After a 20% price reduction, a cordless phone sold for \$48. What was the phone's price before the reduction?

78. A salesperson earns \$300 per week plus 5% commission of sales. How much must be sold to earn \$800 in a week?

79. The length of a rectangular field is 6 yards less than triple the width. If the perimeter of the field is 340 yards, what are its dimensions?

80. In 2005, there were 14,100 students at college A, with a projected enrollment increase of 1500 students per year. In the same year, there were 41,700 students at college B, with a projected enrollment decline of 800 students per year.

a. Let x represent the number of years after 2005. Write, but do not solve, an equation that can be used to find how many years after 2005 the colleges will have the same enrollment.

b. The following table is based on your equation in part (a). Y_1 represents one side of the equation and Y_2 represents the other side of the equation. Use the table to answer these questions: In which year will the colleges have the same enrollment? What will be the enrollment in each college at that time?

X	Y1	Y2
7	24600	36100
8	26100	35300
9	27600	34500
10	29100	33700
11	30600	32900
12	32100	32100
13	33600	31300

X=7

In Exercises 81–86, solve each formula for the specified variable.

81. $V = \frac{1}{3}Bh$ for h

82. $y - y_1 = m(x - x_1)$ for x

83. $E = I(R + r)$ for R

84. $C = \frac{5F - 160}{9}$ for F

85. $s = vt + gt^2$ for g

86. $T = gr + gvt$ for g

1.6 *In Exercises 87–101, simplify each exponential expression. Assume that no denominators are 0.*

87. $(-3x^7)(-5x^6)$

88. x^2y^{-5}

89. $\frac{3^{-2}x^4}{y^{-7}}$

90. $(x^3)^{-6}$

91. $(7x^3y)^2$

92. $\frac{16y^3}{-2y^{10}}$

93. $(-3x^4)(4x^{-11})$

94. $\frac{12x^7}{4x^{-3}}$

95. $\frac{-10a^5b^6}{20a^{-3}b^{11}}$

96. $(-3xy^4)(2x^2)^3$

97. $2^{-2} + \frac{1}{2}x^0$

98. $(5x^2y^{-4})^{-3}$

99. $(3x^4y^{-2})(-2x^5y^{-3})$

100. $\left(\frac{3xy^3}{5x^{-3}y^{-4}}\right)^2$

101. $\left(\frac{-20x^{-2}y^3}{10x^5y^{-6}}\right)^{-3}$

1.7 *In Exercises 102–103, write each number in decimal notation.*

102. 7.16×10^6

103. 1.07×10^{-4}

In Exercises 104–105, write each number in scientific notation.

104. −41,000,000,000,000

105. 0.00809

In Exercises 106–107, perform the indicated computations. Write the answers in scientific notation.

106. $(4.2 \times 10^{13})(3 \times 10^{-6})$

107. $\dfrac{5 \times 10^{-6}}{20 \times 10^{-8}}$

108. The human body contains approximately 3.2×10^4 microliters of blood for every pound of body weight. Each microliter of blood contains approximately 5×10^6 red blood cells. Express in scientific notation the approximate number of red blood cells in the body of a 180-pound person.

CHAPTER 1 TEST

Remember to use your Chapter Test Prep Video CD to see the worked-out solutions to the test questions you want to review.

1. Write the following English phrase as an algebraic expression:

Five less than the product of a number and four.

Let x represent the number.

2. Evaluate $8 + 2(x - 7)^4$, for $x = 10$.

3. Use the roster method to list the elements in the set:

$$\{x \mid x \text{ is a negative integer greater than } -5\}.$$

4. Determine whether the following statement is true or false:

$$\frac{1}{4} \notin \{x \mid x \text{ is a natural number}\}.$$

5. Write out the meaning of the inequality $-3 > -1$. Then determine whether the inequality is true or false.

6. The bar graph shows the number of billionaires in the United States from 2000 through 2004.

A Growing Club: U.S. Billionaires

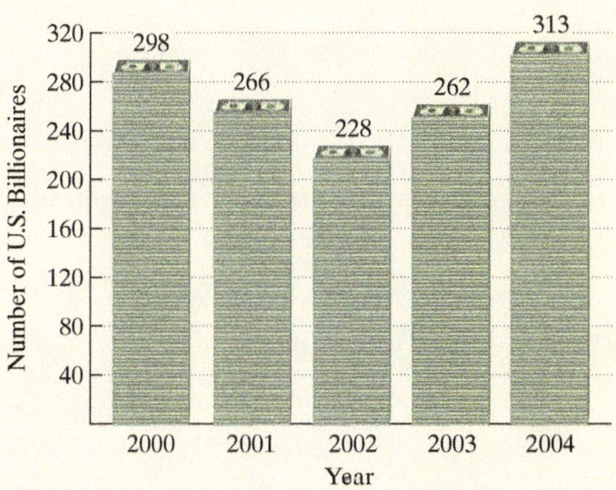

Source: Forbes magazine

The formula

$$N = 17x^2 - 65.4x + 302.2$$

models the number of billionaires, N, in the United States, x years after 2000. According to the formula, how many U.S. billionaires, to the nearest whole number, were there in 2003? Does the formula overestimate or underestimate the actual number shown by the bar graph? By how much?

7. Find the absolute value: $|-17.9|$.

In Exercises 8–12, perform the indicated operation.

8. $-10.8 + 3.2$

9. $-\frac{1}{4} - \left(-\frac{1}{2}\right)$

10. $2(-3)(-1)(-10)$

11. $-\frac{1}{4}\left(-\frac{1}{2}\right)$

12. $\dfrac{-27.9}{-9}$

In Exercises 13–18, simplify each expression.

13. $24 - 36 \div 4 \cdot 3$

14. $(5^2 - 2^4) + [9 \div (-3)]$

15. $\dfrac{(8 - 10)^3 - (-4)^2}{2 + 8(2) \div 4}$

16. $7x - 4(3x + 2) - 10$

17. $5(2y - 6) - (4y - 3)$

18. $9x - [10 - 4(2x - 3)]$

19. Plot $(-2, -4)$ in a rectangular coordinate system.

20. Graph $y = x^2 - 4$ in a rectangular coordinate system.

In Exercises 21–23, solve each equation. If the solution set is $\varnothing$ or $\mathbb{R}$, classify the equation as an inconsistent equation or an identity.

21. $3(2x - 4) = 9 - 3(x + 1)$

22. $\dfrac{2x - 3}{4} = \dfrac{x - 4}{2} - \dfrac{x + 1}{4}$

23. $3(x - 4) + x = 2(6 + 2x)$

In Exercises 24–28, use the five-step strategy for solving word problems.

24. Find two numbers such that the second number is 3 more than twice the first number and the sum of the two numbers is 72.

25. You bought a new car for \$13,805. Its value is decreasing by \$1820 per year. After how many years will its value be \$4705?

26. Photo Shop A charges \$1.60 to develop a roll of film plus \$0.11 for each print. Photo Shop B charges \$1.20 to develop a roll of film plus \$0.13 per print. For how many prints will the amount spent at each photo shop be the same? What will be that amount?

27. After a 60% reduction, a jacket sold for \$20. What was the jacket's price before the reduction?

28. The length of a rectangular field exceeds the width by 260 yards. If the perimeter of the field is 1000 yards, what are its dimensions?

In Exercises 29–30, solve each formula for the specified variable.

29. $V = \frac{1}{3}lwh$ for h

30. $Ax + By = C$ for y

In Exercises 31–35, simplify each exponential expression.

31. $(-2x^5)(7x^{-10})$

32. $(-8x^{-5}y^{-3})(-5x^2y^{-5})$

33. $\frac{-10x^4y^3}{-40x^{-2}y^6}$

34. $(4x^{-5}y^2)^{-3}$

35. $\left(\frac{-6x^{-5}y}{2x^3y^{-4}}\right)^{-2}$

36. Write in decimal notation: 3.8×10^{-6}.

37. Write in scientific notation: 407,000,000,000.

38. Divide and write the answer in scientific notation:

$$\frac{4 \times 10^{-3}}{8 \times 10^{-7}}.$$

39. In 2006, world population was approximately 6.5×10^9. By some projections, world population will double by 2080. Express the population at that time in scientific notation.

CHAPTER 2

Functions and Linear Functions

A vast expanse of open water at the top of our world was once covered with ice. The melting of the Arctic ice caps has forced polar bears to swim as far as 40 miles, causing them to drown in significant numbers. Such deaths were rare in the past.

There is strong scientific consensus that human activities are changing the Earth's climate. Scientists now believe that there is a striking correlation between atmospheric carbon dioxide concentration and global temperature. As both of these variables increase at significant rates, there are warnings of a planetary emergency that threatens to condemn coming generations to a catastrophically diminished future.*

In this chapter, you'll learn to approach our climate crisis mathematically by creating formulas, called functions, that model data for average global temperature and carbon dioxide concentration over time. Understanding the concept of a function will give you a new perspective on many situations, ranging from global warming to using mathematics in a way that is similar to making a movie.

*Sources: Al Gore, *An Inconvenient Truth*, Rodale, 2006; *Time*, April 3, 2006

Mathematical models involving global warming are developed in Exercises 69–70 in Exercise Set 2.5. Using mathematics in a way that is similar to making a movie is discussed in the Blitzer Bonus on page 138.

SECTION 2.1

Introduction to Functions

Objectives

1. Find the domain and range of a relation.
2. Determine whether a relation is a function.
3. Evaluate a function.

Actors Tommy Lee Jones and Will Smith

Top U.S. Last Names

Name	% of All Names
Smith	1.006%
Johnson	0.810%
Williams	0.699%
Brown	0.621%
Jones	0.621%

Source: Russell Ash, *The Top 10 of Everything 2006*

The top five U.S. last names shown above account for nearly 4% of the entire population. The table indicates a correspondence between a last name and the percentage of Americans who share that name. We can write this correspondence using a set of ordered pairs:

{(Smith, 1.006%), (Johnson, 0.810%), (Williams, 0.699%), (Brown, 0.621%), (Jones, 0.621%)}.

These braces indicate we are representing a set.

The mathematical term for a set of ordered pairs is a *relation*.

Definition of a Relation

A **relation** is any set of ordered pairs. The set of all first components of the ordered pairs is called the **domain** of the relation and the set of all second components is called the **range** of the relation.

1 Find the domain and range of a relation.

EXAMPLE 1 Finding the Domain and Range of a Relation

Find the domain and range of the relation:

{(Smith, 1.006%), (Johnson, 0.810%), (Williams, 0.699%), (Brown, 0.621%), (Jones, 0.621%)}.

Solution The domain is the set of all first components. Thus, the domain is

{Smith, Johnson, Williams, Brown, Jones}.

The range is the set of all second components. Thus, the range is

{1.006%, 0.810%, 0.699%, 0.621%}.

Although Brown and Jones are both shared by 0.621% of the U.S. population, it is not necessary to list 0.621% twice.

✓ CHECK POINT 1 Find the domain and the range of the relation:

$$\{(0, 9.1), (10, 6.7), (20, 10.7), (30, 13.2), (34, 15.5)\}.$$

As you worked Check Point 1, did you wonder if there was a rule that assigned the "inputs" in the domain to the "outputs" in the range? For example, for the ordered pair (30, 13.2), how does the output 13.2 depend on the input 30? The ordered pair is based on the data in **Figure 2.1(a)**, which shows the percentage of first-year U.S. college students claiming no religious affiliation.

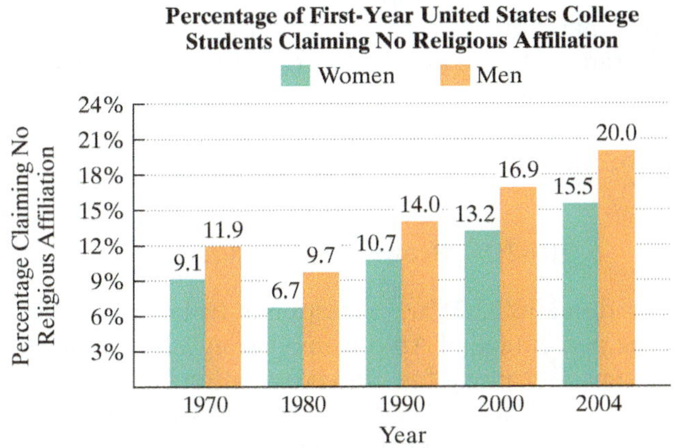

FIGURE 2.1(a) Data for women and men

Source: John Macionis, *Sociology, 11th Edition*, Prentice Hall, 2007

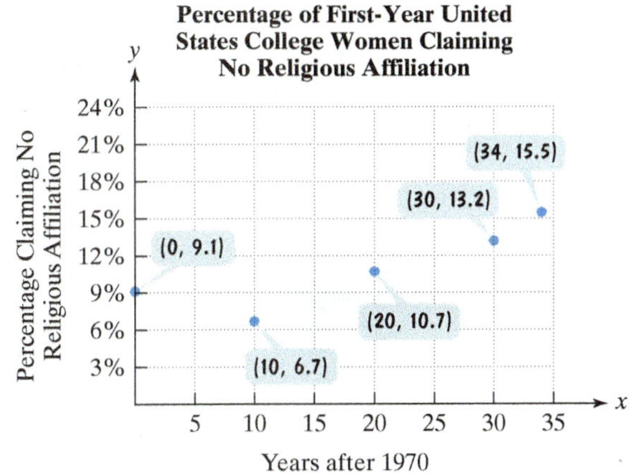

FIGURE 2.1(b) Visually representing the relation for women's data

In **Figure 2.1(b)**, we used the data for college women to create the following ordered pairs:

$$\left(\text{years after 1970},\ \begin{array}{l}\text{percentage of first-year college}\\ \text{women claiming no religious}\\ \text{affiliation}\end{array}\right).$$

Consider, for example, the ordered pair (30, 13.2).

(30, 13.2)

30 years after 1970, or in 2000,

13.2% of first-year college women claimed no religious affiliation.

The five points in **Figure 2.1(b)** visually represent the relation formed from the women's data. Another way to visually represent the relation is as follows:

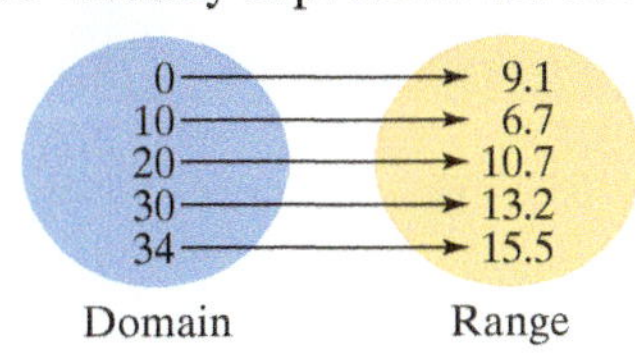

2 Determine whether a relation is a function.

Functions

Shown, again, in the margin are the top five U.S. last names and the percentage of Americans who share those names. We've used this information to define two relations. **Figure 2.2(a)** shows a correspondence between last names and percents sharing those names. **Figure 2.2(b)** shows a correspondence between percents sharing last names and those last names.

Top U.S. Last Names

Name	% of All Names
Smith	1.006%
Johnson	0.810%
Williams	0.699%
Brown	0.621%
Jones	0.621%

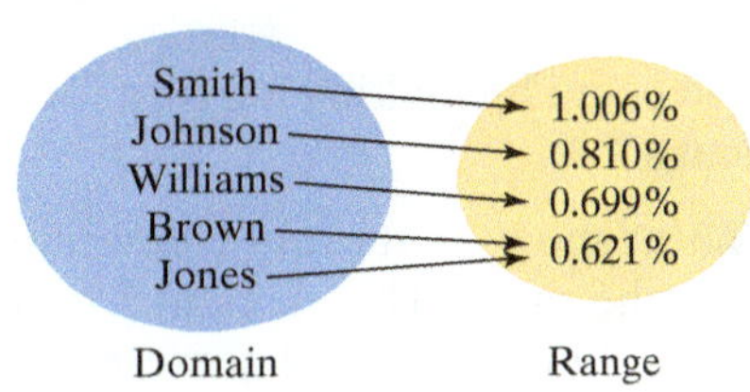

FIGURE 2.2(a) Names correspond to percents.

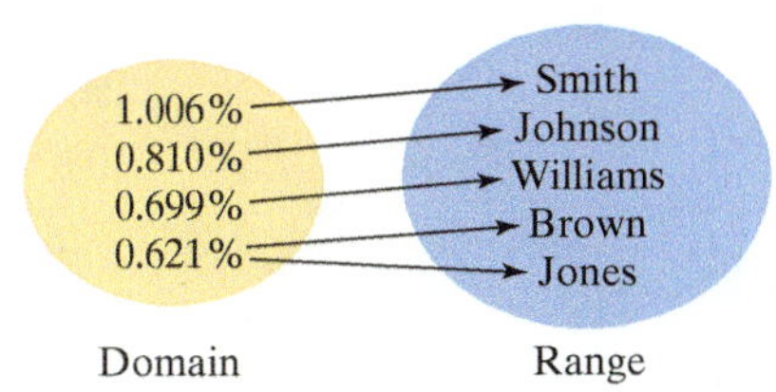

FIGURE 2.2(b) Percents correspond to names.

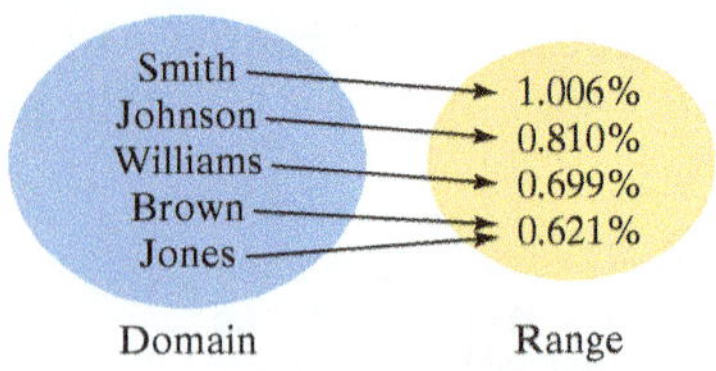

FIGURE 2.2(a) (repeated)

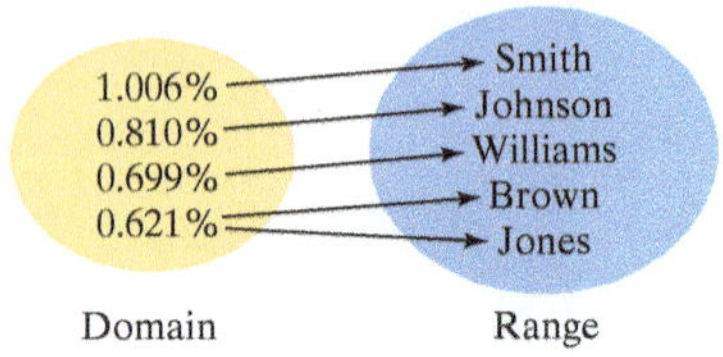

FIGURE 2.2(b) (repeated)

A relation in which each member of the domain corresponds to exactly one member of the range is a **function**. Can you see that the relation in **Figure 2.2(a)** is a function? Each last name in the domain corresponds to exactly one peercent in the range. If we know the last name, we can be sure of the percentage of Americans sharing that name. Notice that more than one element in the domain can correspond to the same element in the range: Brown and Jones are both shared by 0.621% of Americans.

Is the relation in **Figure 2.2(b)** a function? Does each member of the domain correspond to precisely one member of the range? This relation is not a function because there is a member of the domain that corresponds to two different members of the range:

$$(0.621\%, \text{Brown}) \quad (0.621\%, \text{Jones}).$$

The member of the domain, 0.621%, corresponds to both Brown and Jones in the range. If we know the percentage of Americans sharing a last name, 0.621%, we cannot be sure of that last name. Because **a function is a relation in which no two ordered pairs have the same first component and different second components**, the ordered pairs (0.621%, Brown) and (0.621%, Jones) are not ordered pairs of a function.

Same first component

$$(0.621\%, \text{Brown}) \quad (0.621\%, \text{Jones})$$

Different second components

Definition of a Function

A **function** is a correspondence from a first set, called the **domain**, to a second set, called the **range**, such that each element in the domain corresponds to *exactly one* element in the range.

In Check Point 1, we considered a relation that gave a correspondence between years after 1970 and the percentage of first-year college women claiming no religious affiliation. Can you see that this relation is a function?

Each element in the domain

$$\{(0, 9.1), (10, 6.7), (20, 10.7), (30, 13.2), (34, 15.5)\}$$

corresponds to exactly one element in the range.

However, Example 2 illustrates that not every correspondence between sets is a function.

EXAMPLE 2 Determining Whether a Relation Is a Function

Determine whether each relation is a function:

a. $\{(1, 5), (2, 5), (3, 7), (4, 8)\}$ **b.** $\{(5, 1), (5, 2), (7, 3), (8, 4)\}$.

Solution We begin by making a figure for each relation that shows the domain and the range (**Figure 2.3**).

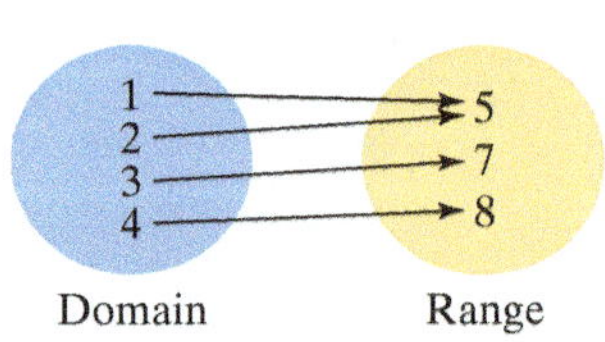

FIGURE 2.3(a)

a. **Figure 2.3(a)** shows that every element in the domain corresponds to exactly one element in the range. The element 1 in the domain corresponds to the element 5 in the range. Furthermore, 2 corresponds to 5, 3 corresponds to 7, and 4 corresponds to 8. No two ordered pairs in the given relation have the same first component and different second components. Thus, the relation is a function.

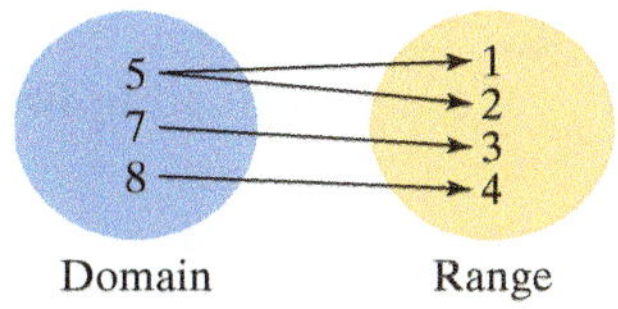

FIGURE 2.3(b)

b. **Figure 2.3(b)** shows that 5 corresponds to both 1 and 2. If any element in the domain corresponds to more than one element in the range, the relation is not a function. This relation is not a function because two ordered pairs have the same first component and different second components.

Same first component

(5, 1) (5, 2)

Different second components

Study Tip

If a relation is a function, reversing the components in each of its ordered pairs may result in a relation that is not a function.

Look at **Figure 2.3(a)** again. The fact that 1 and 2 in the domain correspond to the same number, 5, in the range does not violate the definition of a function. **A function can have two different first components with the same second component.** By contrast, a relation is not a function when two different ordered pairs have the same first component and different second components. Thus, the relation in Example 2(b) is not a function.

✓ **CHECK POINT 2** Determine whether each relation is a function:

a. $\{(1, 2), (3, 4), (5, 6), (5, 7)\}$

b. $\{(1, 2), (3, 4), (6, 5), (7, 5)\}$.

3 Evaluate a function.

Functions as Equations and Function Notation

Functions are usually given in terms of equations rather than as sets of ordered pairs. For example, here is an equation that models the percentage of first-year college women claiming no religious affiliation as a function of time:

$$y = 0.012x^2 - 0.19x + 8.7.$$

The variable x represents the number of years after 1970. The variable y represents the percentage of first-year college women claiming no religious affiliation. The variable y is a function of the variable x. For each value of x, there is one and only one value of y. The variable x is called the **independent variable** because it can be assigned any value from the domain. Thus, x can be assigned any nonnegative integer representing the number of years after 1970. The variable y is called the **dependent variable** because its value depends on x. The percentage claiming no religious affiliation depends on the number of years after 1970. The value of the dependent variable, y, is calculated after selecting a value for the independent variable, x.

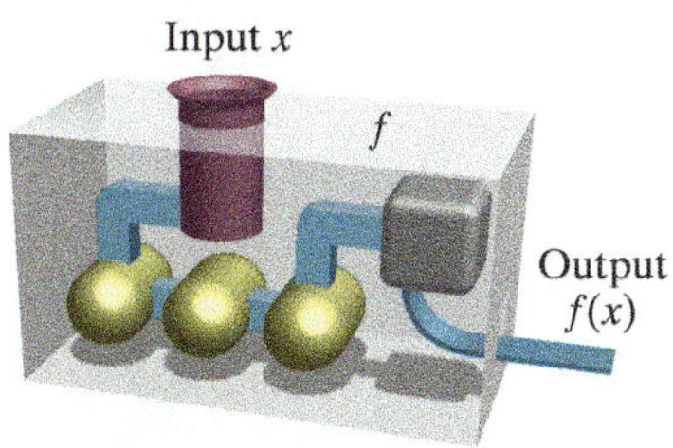

FIGURE 2.4 A "function machine" with inputs and outputs

If an equation in x and y gives one and only one value of y for each value of x, then the variable y is a function of the variable x. When an equation represents a function, the function is often named by a letter such as f, g, h, F, G, or H. Any letter can be used to name a function. Suppose that f names a function. Think of the domain as the set of the function's inputs and the range as the set of the function's outputs. As shown in **Figure 2.4**, the input is represented by x and the output by $f(x)$. The special notation $f(x)$, read "f of x" or "f at x," represents the **value of the function at the number x.**

Let's make this clearer by considering a specific example. We know that the equation

$$y = 0.012x^2 - 0.19x + 8.7$$

defines y as a function of x. We'll name the function f. Now, we can apply our new function notation.

Input	Output	Equation
x	$f(x)$	$f(x) = 0.012x^2 - 0.19x + 8.7$

We read this equation as "f of x equals $0.012x^2 - 0.19x + 8.7$."

Study Tip

The notation $f(x)$ does *not* mean "f times x." The notation describes the value of the function at x.

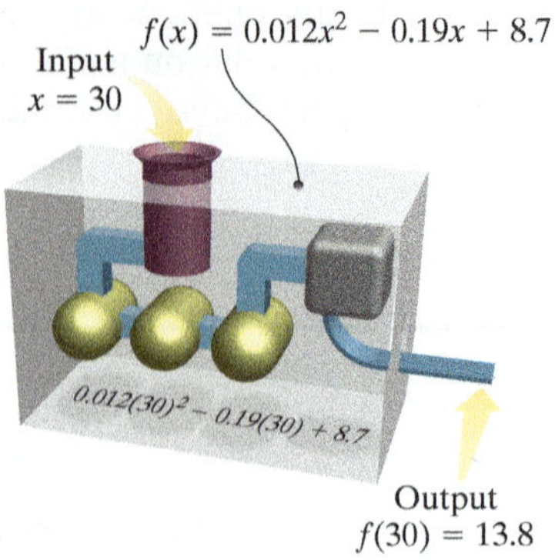

FIGURE 2.5 A function machine at work

Suppose we are interested in finding $f(30)$, the function's output when the input is 30. To find the value of the function at 30, we substitute 30 for x. We are **evaluating the function** at 30.

$$f(x) = 0.012x^2 - 0.19x + 8.7 \quad \text{This is the given function.}$$

$$f(30) = 0.012(30)^2 - 0.19(30) + 8.7 \quad \text{Replace each occurrence of } x \text{ with 30.}$$

$$= 0.012(900) - 0.19(30) + 8.7 \quad \text{Evaluate the exponential expression: } 30^2 = 30 \cdot 30 = 900.$$

$$= 10.8 - 5.7 + 8.7 \quad \text{Perform the multiplications.}$$

$$f(30) = 13.8 \quad \text{Subtract and add from left to right.}$$

The statement $f(30) = 13.8$, read "f of 30 equals 13.8," tells us that the value of the function at 30 is 13.8. When the function's input is 30, its output is 13.8. **Figure 2.5** illustrates the input and output in terms of a function machine.

$$f(30) = 13.8$$

30 years after 1970, or in 2000, | 13.8% of first-year college women claimed no religious affiliation.

We have seen that in 2000, 13.2% actually claimed nonaffiliation, so our function that models the data slightly overestimates the percent for 2000.

Using Technology

Graphing utilities can be used to evaluate functions. The screens on the right show the evaluation of

$$f(x) = 0.012x^2 - 0.19x + 8.7$$

at 30 on a TI-84 Plus graphing calculator. The function f is named Y_1.

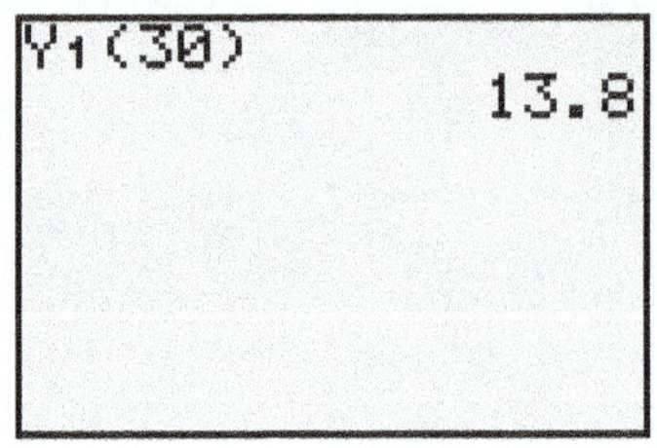

We used $f(x) = 0.012x^2 - 0.19x + 8.7$ to find $f(30)$. To find other function values, such as $f(40)$ or $f(55)$, substitute the specified input value, 40 or 55, for x in the function's equation.

If a function is named f and x represents the independent variable, the notation $f(x)$ corresponds to the y-value for a given x. Thus,

$$f(x) = 0.012x^2 - 0.19x + 8.7 \quad \text{and} \quad y = 0.012x^2 - 0.19x + 8.7$$

define the same function. This function may be written as

$$y = f(x) = 0.012x^2 - 0.19x + 8.7.$$

EXAMPLE 3 Using Function Notation

Find the indicated function value:

a. $f(4)$ for $f(x) = 2x + 3$

b. $g(-2)$ for $g(x) = 2x^2 - 1$

c. $h(-5)$ for $h(r) = r^3 - 2r^2 + 5$

d. $F(a + h)$ for $F(x) = 5x + 7$.

Solution

a. $f(x) = 2x + 3$ This is the given function.

$f(4) = 2 \cdot 4 + 3$ To find f of 4, replace x with 4.

$= 8 + 3$ Multiply: $2 \cdot 4 = 8$.

$f(4) = 11$ (f of 4 is 11.) Add.

b.

$g(x) = 2x^2 - 1$ — This is the given function.

$g(-2) = 2(-2)^2 - 1$ — To find g of -2, replace x with -2.

$= 2(4) - 1$ — Evaluate the exponential expression: $(-2)^2 = 4$.

$= 8 - 1$ — Multiply: $2(4) = 8$.

$g(-2) = 7$ — Subtract.

g of −2 is 7.

c.

$h(r) = r^3 - 2r^2 + 5$ — The function's name is h and r represents the independent variable.

$h(-5) = (-5)^3 - 2(-5)^2 + 5$ — To find h of -5, replace each occurrence of r with -5.

$= -125 - 2(25) + 5$ — Evaluate exponential expressions.

$= -125 - 50 + 5$ — Multiply.

$h(-5) = -170$ — $-125 - 50 = -175$ and $-175 + 5 = -170$.

h of −5 is −170.

d.

$F(x) = 5x + 7$ — This is the given function.

$F(a + h) = 5(a + h) + 7$ — Replace x with $a + h$.

$F(a + h) = 5a + 5h + 7$ — Apply the distributive property.

F of *a* + *h* is 5*a* + 5*h* + 7.

✓ CHECK POINT 3 Find the indicated function value:

a. $f(6)$ for $f(x) = 4x + 5$

b. $g(-5)$ for $g(x) = 3x^2 - 10$

c. $h(-4)$ for $h(r) = r^2 - 7r + 2$

d. $F(a + h)$ for $F(x) = 6x + 9$.

Functions Represented by Tables and Function Notation

Function notation can be applied to functions that are represented by tables.

EXAMPLE 4 Using Function Notation

Function f is defined by the following table:

x	$f(x)$
−2	5
−1	0
0	3
1	1
2	4

a. Explain why the table defines a function.

b. Find the domain and the range of the function.

Find the indicated function value:

c. $f(-1)$

d. $f(0)$.

e. Find x such that $f(x) = 4$.

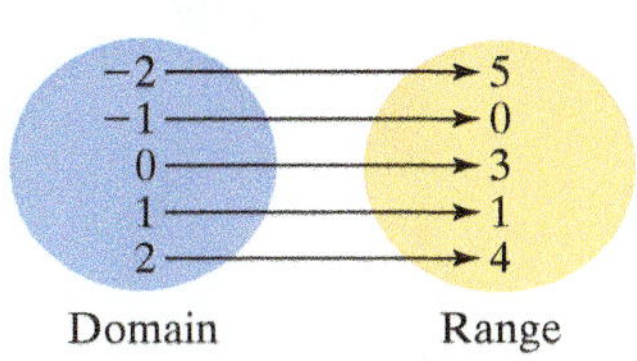

FIGURE 2.6

Solution

a. Values in the first column of the table make up the domain, or input values. Values in the second column of the table make up the range, or output values. We see that every element in the domain corresponds to exactly one element in the range, shown in **Figure 2.6**. Therefore, the relation given by the table is a function.

The voice balloons pointing to appropriate parts of the table illustrate the solution to parts (b)-(e).

x	$f(x)$
-2	5
-1	0
0	3
1	1
2	4

c. $f(-1) = 0$: When the input is -1, the output is 0.

d. $f(0) = 3$: When the input is 0, the output is 3.

e. $f(x) = 4$ when $x = 2$: The output, $f(x)$, is 4 when the input, x, is 2.

b. The domain is the set of inputs: $\{-2, -1, 0, 1, 2\}$.

b. The range is the set of outputs: $\{5, 0, 3, 1, 4\}$.

CHECK POINT 4 Function g is defined by the following table:

x	$g(x)$
0	3
1	0
2	1
3	2
4	3

a. Explain why the table defines a function.

b. Find the domain and the range of the function.

Find the indicated function value:

c. $g(1)$

d. $g(3)$

e. Find x such that $g(x) = 3$.

2.1 EXERCISE SET

MyMathLab — MathXL PRACTICE · WATCH · DOWNLOAD · READ · REVIEW

Practice Exercises

In Exercises 1–8, determine whether each relation is a function. Give the domain and range for each relation.

1. $\{(1, 2), (3, 4), (5, 5)\}$
2. $\{(4, 5), (6, 7), (8, 8)\}$
3. $\{(3, 4), (3, 5), (4, 4), (4, 5)\}$
4. $\{(5, 6), (5, 7), (6, 6), (6, 7)\}$
5. $\{(-3, -3), (-2, -2), (-1, -1), (0, 0)\}$
6. $\{(-7, -7), (-5, -5), (-3, -3), (0, 0)\}$
7. $\{(1, 4), (1, 5), (1, 6)\}$
8. $\{(4, 1), (5, 1), (6, 1)\}$

In Exercises 9–22, find the indicated function values.

9. $f(x) = x + 1$
   a. $f(0)$ b. $f(5)$ c. $f(-8)$
   d. $f(2a)$ e. $f(a + 2)$

10. $f(x) = x + 3$
    a. $f(0)$ b. $f(5)$ c. $f(-8)$
    d. $f(2a)$ e. $f(a + 2)$

11. $g(x) = 3x - 2$
    a. $g(0)$ b. $g(-5)$ c. $g\left(\frac{2}{3}\right)$
    d. $g(4b)$ e. $g(b + 4)$

12. $g(x) = 4x - 3$
    a. $g(0)$ b. $g(-5)$ c. $g\left(\frac{3}{4}\right)$
    d. $g(5b)$ e. $g(b + 5)$

13. $h(x) = 3x^2 + 5$
    a. $h(0)$ b. $h(-1)$ c. $h(4)$
    d. $h(-3)$ e. $h(4b)$

14. $h(x) = 2x^2 - 4$
    a. $h(0)$ b. $h(-1)$ c. $h(5)$
    d. $h(-3)$ e. $h(5b)$

15. $f(x) = 2x^2 + 3x - 1$
    a. $f(0)$ b. $f(3)$ c. $f(-4)$
    d. $f(b)$
    e. $f(5a)$

16. $f(x) = 3x^2 + 4x - 2$

a. $f(0)$ b. $f(3)$ c. $f(-5)$
d. $f(b)$
e. $f(5a)$

17. $f(x) = \dfrac{2x - 3}{x - 4}$

a. $f(0)$ b. $f(3)$ c. $f(-4)$
d. $f(-5)$ e. $f(a + h)$
f. Why must 4 be excluded from the domain of f?

18. $f(x) = \dfrac{3x - 1}{x - 5}$

a. $f(0)$ b. $f(3)$ c. $f(-3)$
d. $f(10)$ e. $f(a + h)$
f. Why must 5 be excluded from the domain of f?

19.

x	f(x)
−4	3
−2	6
0	9
2	12
4	15

a. $f(-2)$
b. $f(2)$
c. For what value of x is $f(x) = 9$?

20.

x	f(x)
−5	4
−3	8
0	12
3	16
5	20

a. $f(-3)$
b. $f(3)$
c. For what value of x is $f(x) = 12$?

21.

x	h(x)
−2	2
−1	1
0	0
1	1
2	2

a. $h(-2)$
b. $h(1)$
c. For what values of x is $h(x) = 1$?

22.

x	h(x)
−2	−2
−1	−1
0	0
1	−1
2	−2

a. $h(-2)$
b. $h(1)$
c. For what values of x is $h(x) = -1$?

Practice PLUS

In Exercises 23–24, let $f(x) = x^2 - x + 4$ and $g(x) = 3x - 5$.

23. Find $g(1)$ and $f(g(1))$.

24. Find $g(-1)$ and $f(g(-1))$.

In Exercises 25–26, let f and g be defined by the following table:

x	f(x)	g(x)
−2	6	0
−1	3	4
0	−1	1
1	−4	−3
2	0	−6

25. Find $\sqrt{f(-1) - f(0)} - [g(2)]^2 + f(-2) \div g(2) \cdot g(-1)$.

26. Find $|f(1) - f(0)| - [g(1)]^2 + g(1) \div f(-1) \cdot g(2)$.

In Exercises 27–28, find $f(-x) - f(x)$ for the given function f. Then simplify the expression.

27. $f(x) = x^3 + x - 5$

28. $f(x) = x^2 - 3x + 7$

In Exercises 29–30, each function is defined by two equations. The equation in the first row gives the output for negative numbers in the domain. The equation in the second row gives the output for non-negative numbers in the domain. Find the indicated function values.

29. $f(x) = \begin{cases} 3x + 5 & \text{if } x < 0 \\ 4x + 7 & \text{if } x \ge 0 \end{cases}$

a. $f(-2)$ b. $f(0)$
c. $f(3)$ d. $f(-100) + f(100)$

30. $f(x) = \begin{cases} 6x - 1 & \text{if } x < 0 \\ 7x + 3 & \text{if } x \ge 0 \end{cases}$

a. $f(-3)$ b. $f(0)$
c. $f(4)$ d. $f(-100) + f(100)$

Application Exercises

31. The bar graph shows the breakdown of political ideologies in the United States.

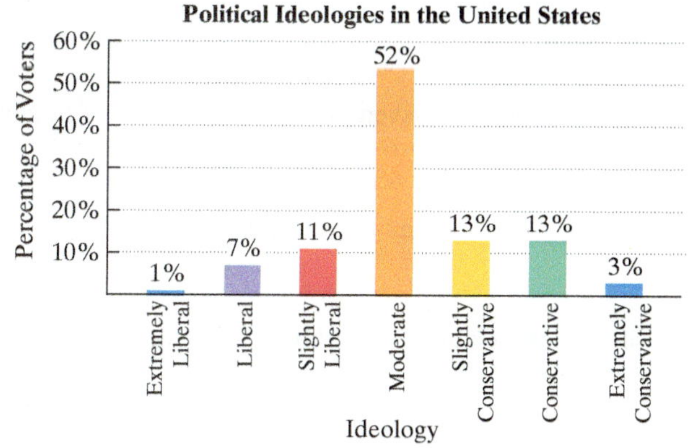

Source: Center for Political Studies, University of Michigan

(Refer to the graph at the bottom of page 103 as you work this exercise.)

a. Write a set of seven ordered pairs in which political ideologies correspond to percentages. Each ordered pair should be in the form

(ideology, percent).

Use EL, L, SL, M, SC, C, and EC to represent the respective ideologies from left to right.

b. Is the relation in part (a) a function? Explain your answer.

c. Write a set of seven ordered pairs in which percentages correspond to political ideologies. Each ordered pair should be in the form

(percent, ideology).

d. Is the relation in part (c) a function? Explain your answer.

32. Actors with the most Oscar nominations include Jack Nicholson, Laurence Olivier, Paul Newman, Spencer Tracy, Marlon Brando, Jack Lemmon, and Al Pacino. The bar graph shows the number of nominations for each of these men.

Actors with the Most Oscar Nominations

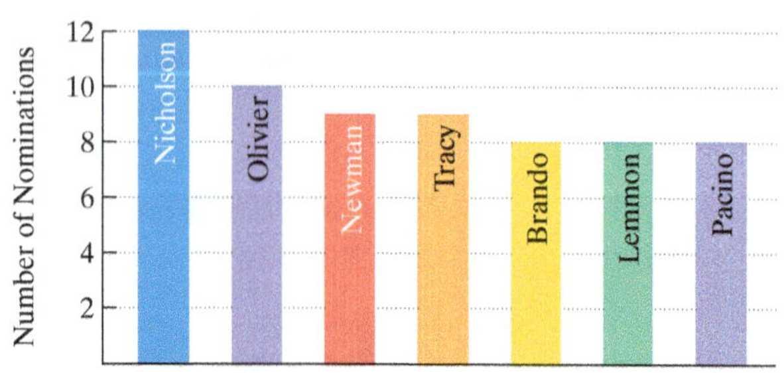

Source: Russell Ash, *The Top 10 of Everything 2007*

a. Write a set of seven ordered pairs in which actors correspond to numbers of Oscar nominations. Each ordered pair should be in the form

(actor, number of nominations).

b. Is the relation in part (a) a function? Explain your answer.

c. Write a set of seven ordered pairs in which numbers of nominations correspond to actors. Each ordered pair should be in the form

(number of nominations, actor).

d. Is the relation in part (c) a function? Explain your answer.

Writing in Mathematics

33. What is a relationed? Describe what is meant by its domain and its range.

34. Explain how to determine whether a relation is a function. What is a function?

35. Does $f(x)$ mean f times x when referring to function f? If not, what does $f(x)$ mean? Provide an example with your explanation.

36. For people filing a single return, federal income tax is a function of adjusted gross income because for each value of adjusted gross income there is a specific tax to be paid. By contrast, the price of a house is not a function of the lot size on which the house sits because houses on same-sized lots can sell for many different prices.

a. Describe an everyday situation between variables that is a function.

b. Describe an everyday situation between variables that is not a function.

Critical Thinking Exercises

Make Sense? *In Exercises 37–40, determine whether each statement "makes sense" or "does not make sense" and explain your reasoning.*

37. Today's temperature is a function of the time of day.

38. My height is a function of my age.

39. Although I presented my function as a set of ordered pairs, I could have shown the correspondences using a table or using points plotted in a rectangular coordinate system.

40. My function models how the chance of divorce depends on the number of years of marriage, so the range is $\{x | x \text{ is the number of years of marriage}\}$.

In Exercises 41–46, determine whether each statement is true or false. If the statement is false, make the necessary change(s) to produce a true statement.

41. All relations are functions.

42. No two ordered pairs of a function can have the same second components and different first components.

Using the tables that define f and g, determine whether each statement in Exercises 43–46 is true or false.

x	$f(x)$
−4	−1
−3	−2
−2	−3
−1	−4

x	$g(x)$
−1	−4
−2	−3
−3	−2
−4	−1

43. The domain of f = the range of f

44. The range of f = the domain of g

45. $f(-4) - f(-2) = 2$

46. $g(-4) + f(-4) = 0$

47. If $f(x) = 3x + 7$, find $\dfrac{f(a + h) - f(a)}{h}$.

48. Give an example of a relation with the following characteristics: The relation is a function containing two ordered pairs. Reversing the components in each ordered pair results in a relation that is not a function.

49. If $f(x + y) = f(x) + f(y)$ and $f(1) = 3$, find $f(2)$, $f(3)$, and $f(4)$. Is $f(x + y) = f(x) + f(y)$ for all functions?

Review Exercises

50. Simplify: $24 \div 4[2 - (5 - 2)]^2 - 6$. (Section 1.2, Example 7)

51. Simplify: $\left(\dfrac{3x^2y^{-2}}{y^3}\right)^{-2}$. (Section 1.6, Example 9)

52. Solve: $\dfrac{x}{3} = \dfrac{3x}{5} + 4$. (Section 1.4, Example 4)

Preview Exercises

Exercises 53–55 will help you prepare for the material covered in the next section.

53. Graph $y = 2x$. Select integers for x, starting with -2 and ending with 2.

54. Graph $y = 2x + 4$. Select integers for x, starting with -2 and ending with 2.

55. Use the following graph to solve this exercise.

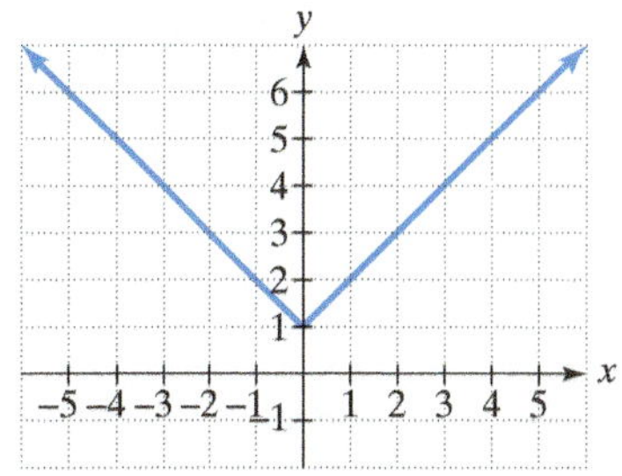

a. What is the y-coordinate when the x-coordinate is 2?

b. What are the x-coordinates when the y-coordinate is 4?

c. Describe the x-coordinates of all points on the graph.

d. Describe the y-coordinates of all points on the graph.

SECTION 2.2

Graphs of Functions

Objectives

1. Graph functions by plotting points.
2. Use the vertical line test to identify functions.
3. Obtain information about a function from its graph.
4. Identify the domain and range of a function from its graph.

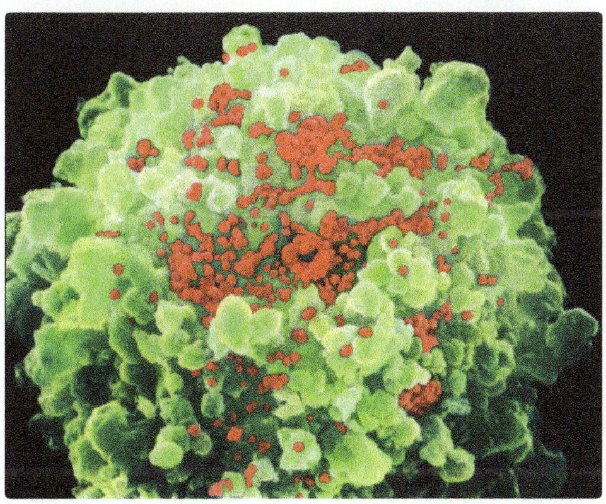

Magnified 6000 times, this color-scanned image shows a T-lymphocyte blood cell (green) infected with the HIV virus (red). Depletion of the number of T cells causes destruction of the immune system.

The number of T cells in a person with HIV is a function of time after infection. In this section, we'll analyze the graph of this function, using the rectangular coordinate system to visualize what functions look like.

1 Graph functions by plotting points.

Graphs of Functions

The **graph of a function** is the graph of its ordered pairs. For example, the graph of $f(x) = 2x$ is the set of points (x, y) in the rectangular coordinate system satisfying $y = 2x$. Similarly, the graph of $g(x) = 2x + 4$ is the set of points (x, y) in the rectangular coordinate system satisfying the equation $y = 2x + 4$. In the next example, we graph both of these functions in the same rectangular coordinate system.

EXAMPLE 1 Graphing Functions

Graph the functions $f(x) = 2x$ and $g(x) = 2x + 4$ in the same rectangular coordinate system. Select integers for x, starting with -2 and ending with 2.

Solution We begin by setting up a partial table of coordinates for each function. Then, we plot the five points in each table and connect them, as shown in **Figure 2.7**. The graph of each function is a straight line. Do you see a relationship between the two graphs? The graph of g is the graph of f shifted vertically up by 4 units.

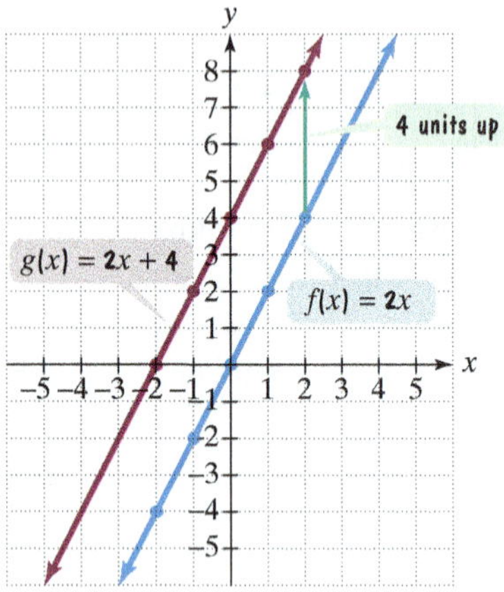

FIGURE 2.7

x	$f(x) = 2x$	(x, y) or $(x, f(x))$
-2	$f(-2) = 2(-2) = -4$	$(-2, -4)$
-1	$f(-1) = 2(-1) = -2$	$(-1, -2)$
0	$f(0) = 2 \cdot 0 = 0$	$(0, 0)$
1	$f(1) = 2 \cdot 1 = 2$	$(1, 2)$
2	$f(2) = 2 \cdot 2 = 4$	$(2, 4)$
Choose x.	Compute $f(x)$ by evaluating f at x.	Form the ordered pair.

x	$g(x) = 2x + 4$	(x, y) or $(x, g(x))$
-2	$g(-2) = 2(-2) + 4 = 0$	$(-2, 0)$
-1	$g(-1) = 2(-1) + 4 = 2$	$(-1, 2)$
0	$g(0) = 2 \cdot 0 + 4 = 4$	$(0, 4)$
1	$g(1) = 2 \cdot 1 + 4 = 6$	$(1, 6)$
2	$g(2) = 2 \cdot 2 + 4 = 8$	$(2, 8)$
Choose x.	Compute $g(x)$ by evaluating g at x.	Form the ordered pair.

The graphs in Example 1 are straight lines. All functions with equations of the form $f(x) = mx + b$ graph as straight lines. Such functions, called **linear functions**, will be discussed in detail in Sections 2.4–2.5.

Using Technology

We can use a graphing utility to check the tables and the graphs in Example 1 for the functions

$$f(x) = 2x \quad \text{and} \quad g(x) = 2x + 4.$$

Enter $y_1 = 2x$ in the $\boxed{y=}$ screen.

Enter $y_2 = 2x + 4$ in the $\boxed{y=}$ screen.

Checking Tables

We entered -2 for the starting x-value and 1 as an increment between x-values to check our tables in Example 1.

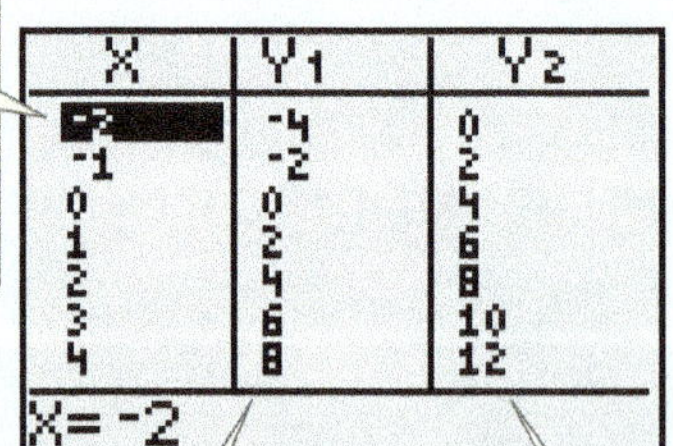

Use the first five ordered pairs (x, y_1) to check the first table.

Use the first five ordered pairs (x, y_2) to check the second table.

Checking Graphs

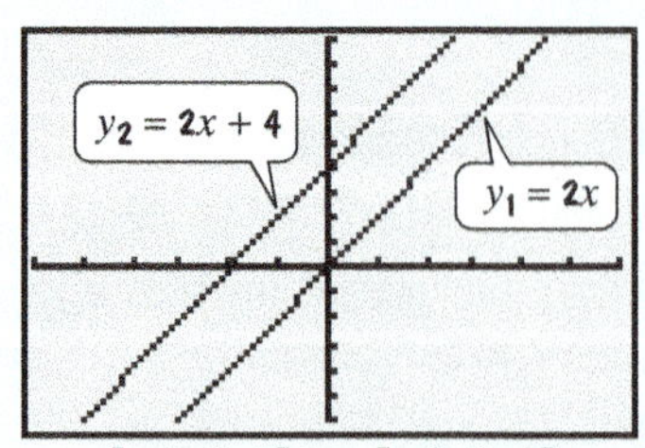

[−6, 6, 1] by [−6, 9, 1]

We selected this viewing rectangle, or window, to match **Figure 2.7.**

CHECK POINT 1 Graph the functions $f(x) = 2x$ and $g(x) = 2x - 3$ in the same rectangular coordinate system. Select integers for x, starting with -2 and ending with 2. How is the graph of g related to the graph of f?

2 Use the vertical line test to identify functions.

The Vertical Line Test

Not every graph in the rectangular coordinate system is the graph of a function. The definition of a function specifies that no value of x can be paired with two or more different values of y. Consequently, if a graph contains two or more different points

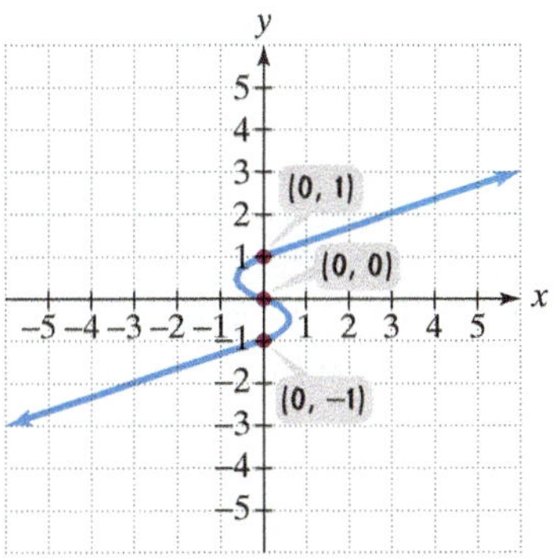

FIGURE 2.8 y is not a function of x because 0 is paired with three values of y, namely, 1, 0, and −1.

with the same first coordinate, the graph cannot represent a function. This is illustrated in **Figure 2.8**. Observe that points sharing a common first coordinate are vertically above or below each other.

This observation is the basis of a useful test for determining whether a graph defines y as a function of x. The test is called the **vertical line test**.

The Vertical Line Test for Functions

If any vertical line intersects a graph in more than one point, the graph does not define y as a function of x.

EXAMPLE 2 Using the Vertical Line Test

Use the vertical line test to identify graphs in which y is a function of x.

a. b. c. d.

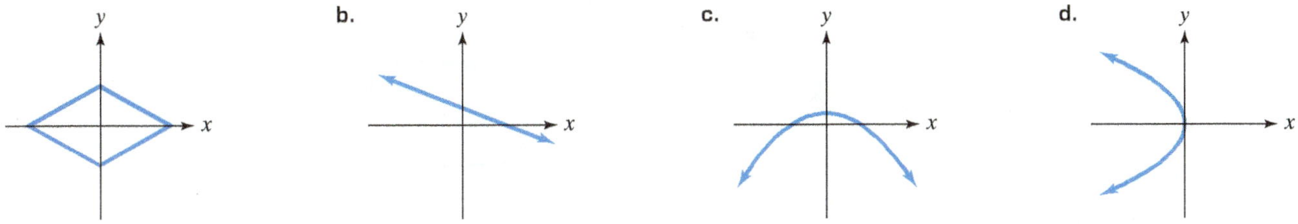

Solution y is a function of x for the graphs in (b) and (c).

a. b. c. d.

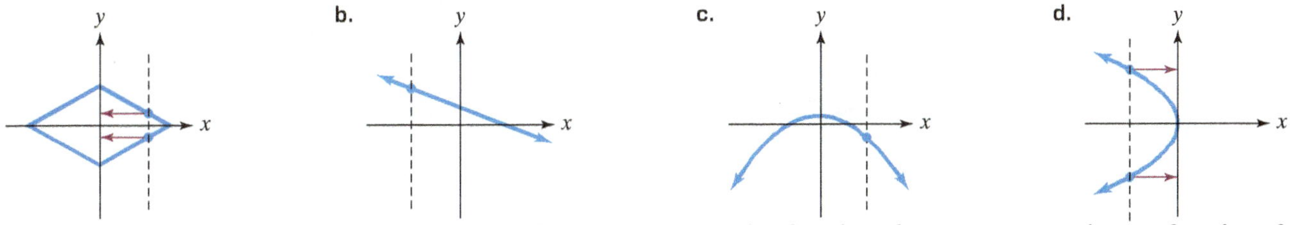

a. y **is not a function** of x. Two values of y correspond to one x-value.

b. y **is a function** of x.

c. y **is a function** of x.

d. y **is not a function** of x. Two values of y correspond to one x-value.

CHECK POINT 2 Use the vertical line test to identify graphs in which y is a function of x.

a. b. c.

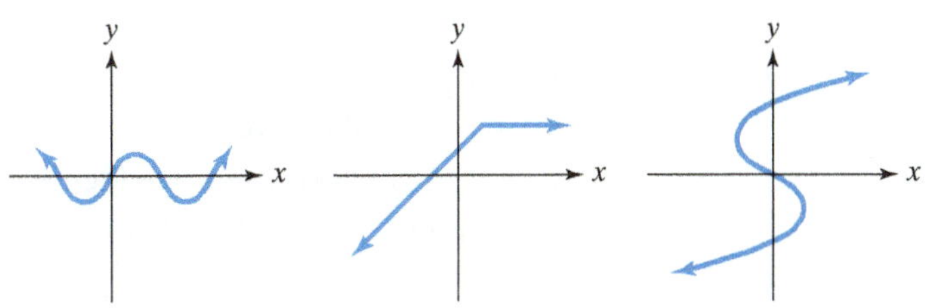

3 Obtain information about a function from its graph.

Obtaining Information from Graphs

You can obtain information about a function from its graph. At the right or left of a graph, you will often find closed dots, open dots, or arrows.

- A closed dot indicates that the graph does not extend beyond this point and the point belongs to the graph.
- An open dot indicates that the graph does not extend beyond this point and the point does not belong to the graph.
- An arrow indicates that the graph extends indefinitely in the direction in which the arrow points.

EXAMPLE 3 Analyzing the Graph of a Function

The human immunodeficiency virus, or HIV, infects and kills helper T cells. Because T cells stimulate the immune system to produce antibodies, their destruction disables the body's defenses against other pathogens. By counting the number of T cells that remain active in the body, the progression of HIV can be monitored. The fewer helper T cells, the more advanced the disease. **Figure 2.9** shows a graph that is used to monitor the average progression of the disease. The number of T cells, $f(x)$, is a function of time after infection, x.

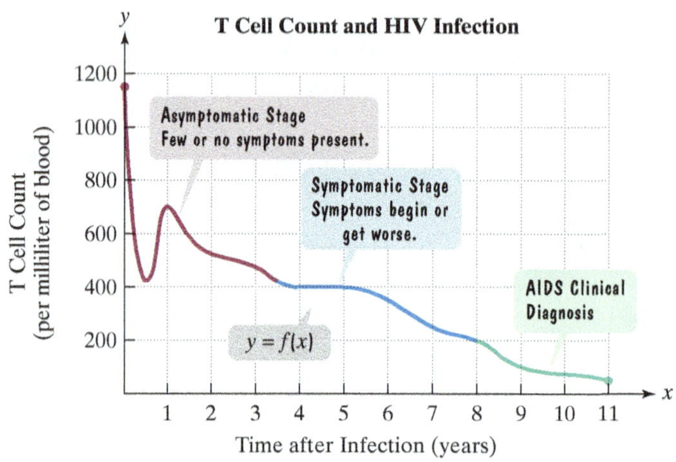

FIGURE 2.9

Source: B.E. Pruitt et al., *Human Sexuality*, Prentice Hall, 2007

a. Explain why f represents the graph of a function.
b. Use the graph to find $f(8)$.
c. For what value of x is $f(x) = 350$?
d. Describe the general trend shown by the graph.

Solution

a. No vertical line can be drawn that intersects the graph of f more than once. By the vertical line test, f represents the graph of a function.

b. To find $f(8)$, or f of 8, we locate 8 on the x-axis. **Figure 2.10** shows the point on the graph of f for which 8 is the first coordinate. From this point, we look to the y-axis to find the corresponding y-coordinate. We see that the y-coordinate is 200. Thus,

$$f(8) = 200.$$

When the time after infection is 8 years, the T cell count is 200 cells per milliliter of blood. (AIDS clinical diagnosis is given at a T cell count of 200 or below.)

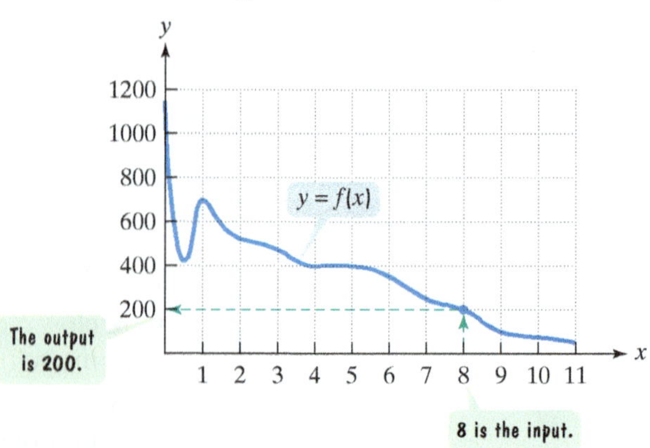

FIGURE 2.10 Finding $f(8)$

c. To find the value of x for which $f(x) = 350$, we approximately locate 350 on the y-axis. **Figure 2.11** shows that there is one point on the graph of f for which 350 is the second coordinate. From this point, we look to the x-axis to find the corresponding x-coordinate. We see that the x-coordinate is 6. Thus,

$$f(x) = 350 \text{ for } x = 6.$$

A T cell count of 350 occurs 6 years after infection.

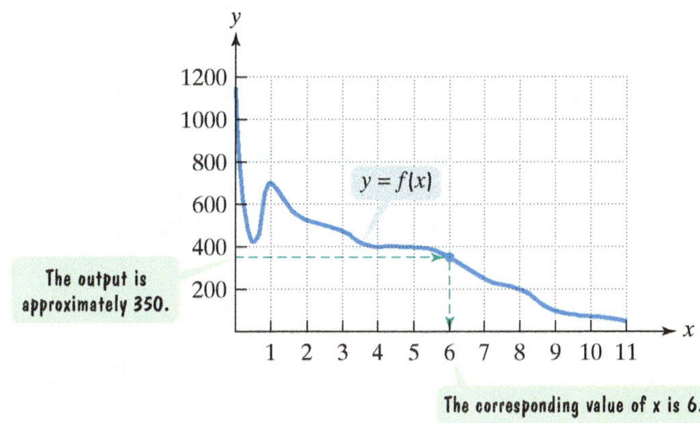

FIGURE 2.11 Finding x for which $f(x) = 350$

d. **Figure 2.12** uses voice balloons to describe the general trend shown by the graph.

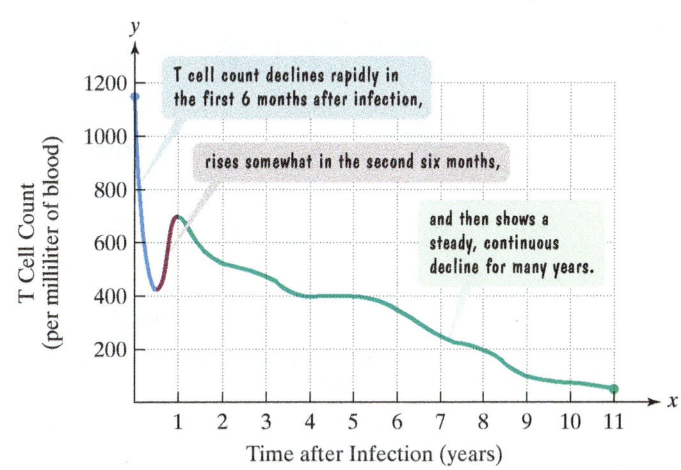

FIGURE 2.12 Describing changing T cell count over time in a person infected with HIV

CHECK POINT 3

a. Use the graph of f in **Figure 2.9** on page 108 to find $f(5)$.

b. For what value of x is $f(x) = 100$?

c. Estimate the minimum T cell count during the asymptomatic stage.

4 Identify the domain and range of a function from its graph.

Identifying Domain and Range from a Function's Graph

Figure 2.13 illustrates how the graph of a function is used to determine the function's domain and its range.

Domain: set of inputs

Found on the x-axis

Range: set of outputs

Found on the y-axis

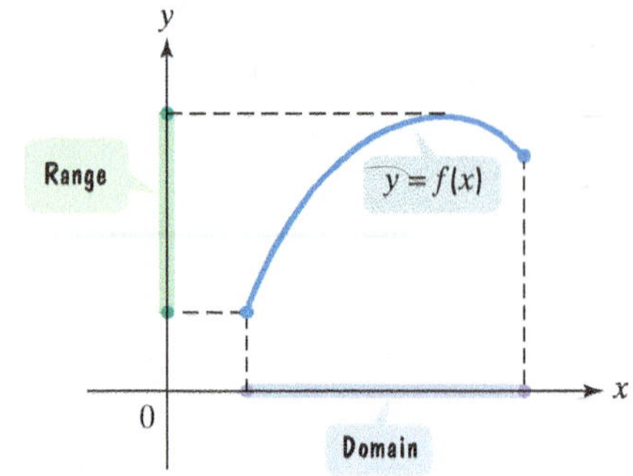

FIGURE 2.13 Domain and range of f

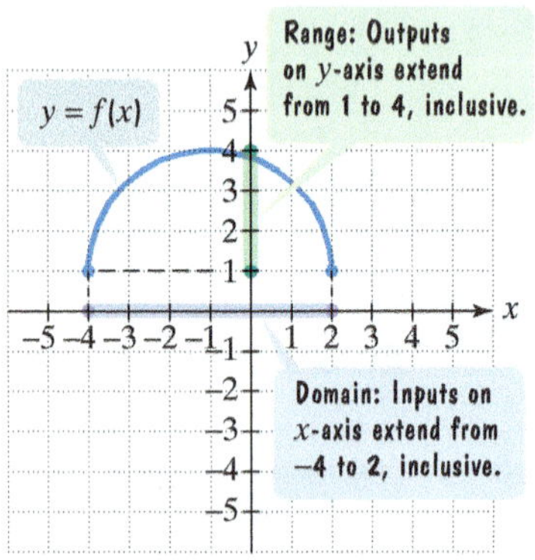

FIGURE 2.14 Domain and range of f

Let's apply these ideas to the graph of the function shown in **Figure 2.14.** To find the domain, look for all the inputs on the x-axis that correspond to points on the graph. Can you see that they extend from -4 to 2, inclusive? Using set-builder notation, the function's domain can be represented as follows:

$$\{x \mid -4 \leq x \leq 2\}.$$

The set of all x | such that | x is greater than or equal to -4 and less than or equal to 2.

To find the range, look for all the outputs on the y-axis that correspond to points on the graph. They extend from 1 to 4, inclusive. Using set-builder notation, the function's range can be represented as follows:

$$\{y \mid 1 \leq y \leq 4\}$$

The set of all y | such that | y is greater than or equal to 1 and less than or equal to 4.

EXAMPLE 4 Identifying the Domain and Range of a Function from Its Graph

Use the graph of each function to identify its domain and its range.

a.

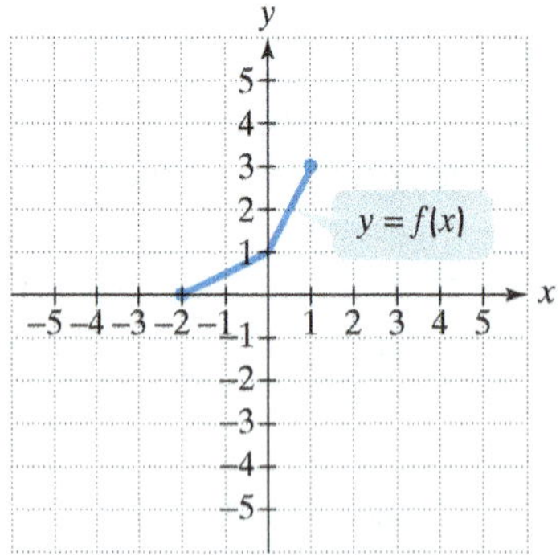

b.

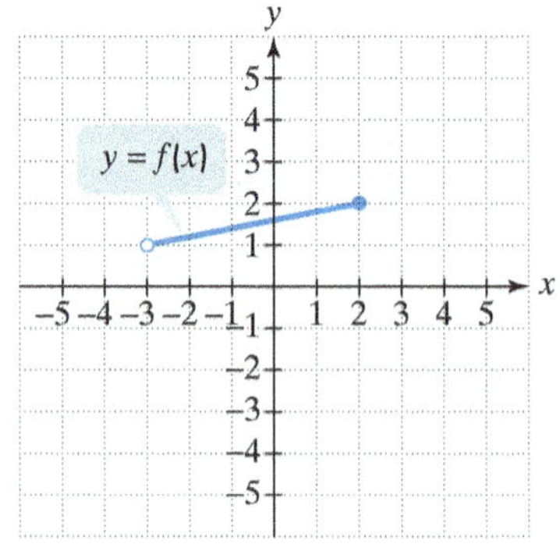

c.

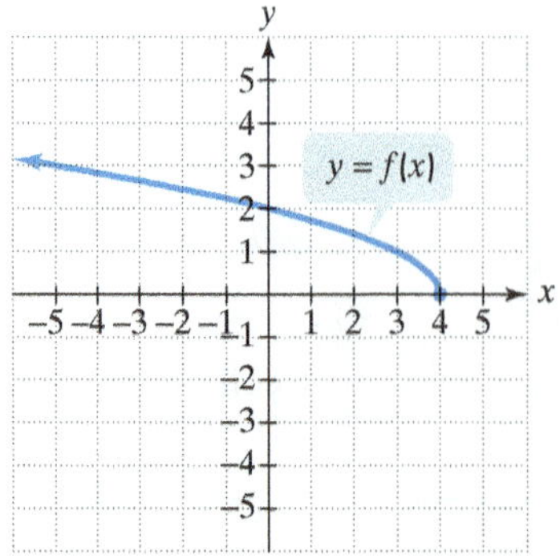

d.

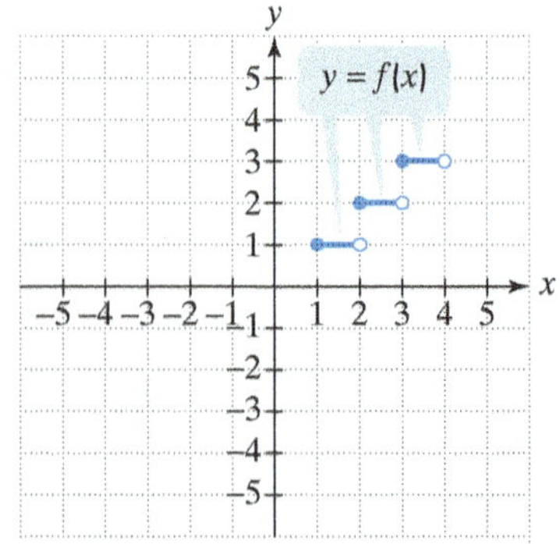

Solution For the graph of each function, the domain is highlighted in purple on the x-axis and the range is highlighted in green on the y-axis.

a.

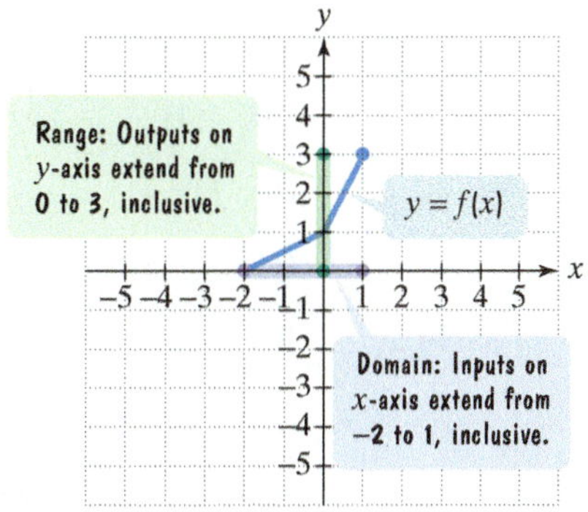

$$\text{Domain} = \{x \mid -2 \le x \le 1\}$$
$$\text{Range} = \{y \mid 0 \le y \le 3\}$$

b.

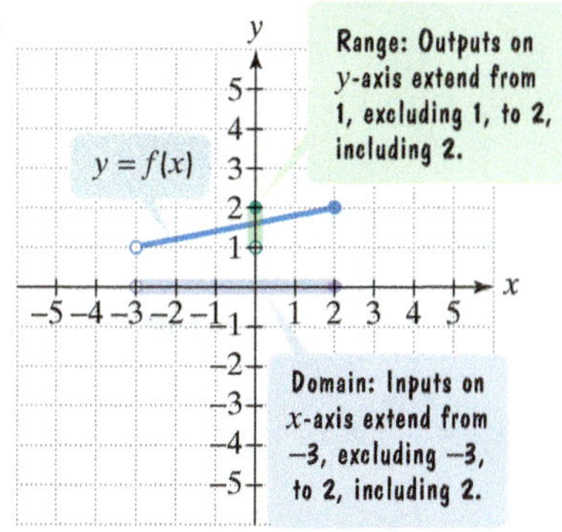

$$\text{Domain} = \{x \mid -3 < x \le 2\}$$
$$\text{Range} = \{y \mid 1 < y \le 2\}$$

c.

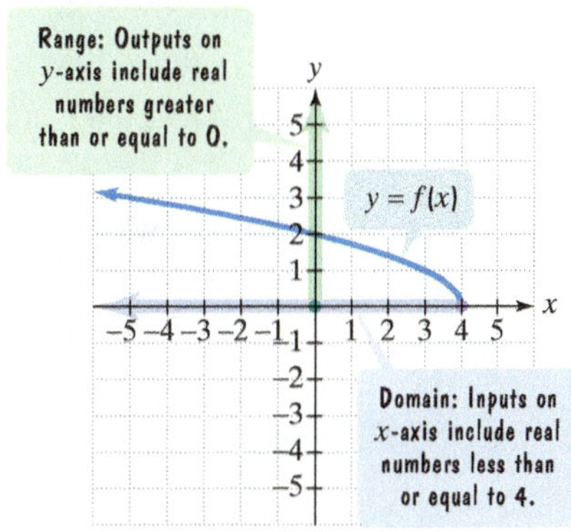

$$\text{Domain} = \{x \mid x \le 4\}$$
$$\text{Range} = \{y \mid y \ge 0\}$$

d.

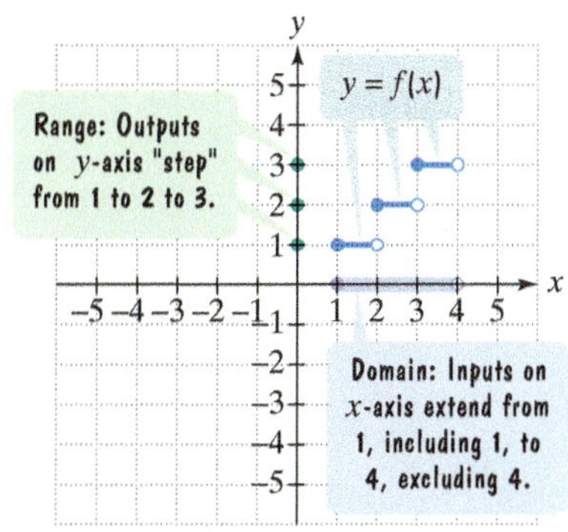

$$\text{Domain} = \{x \mid 1 \le x < 4\}$$
$$\text{Range} = \{y \mid y = 1, 2, 3\}$$

☑ **CHECK POINT 4** Use the graph of each function to identify its domain and its range.

a.

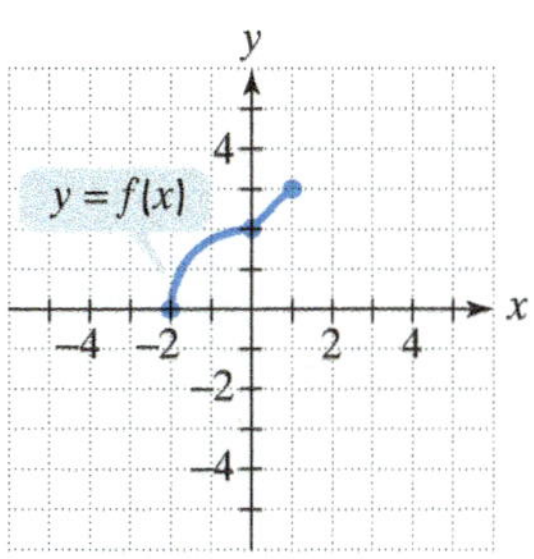

b.

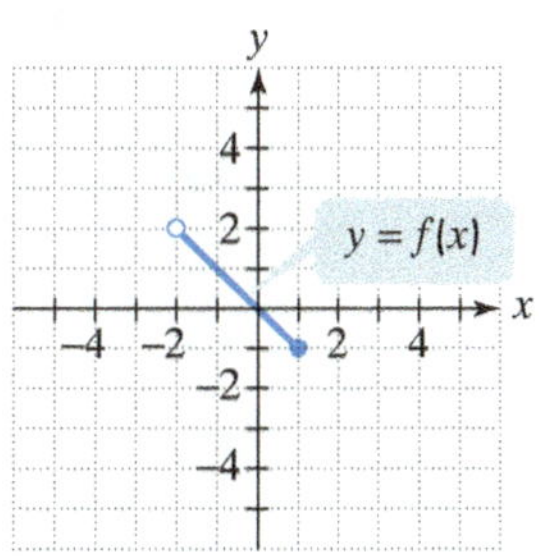

c.

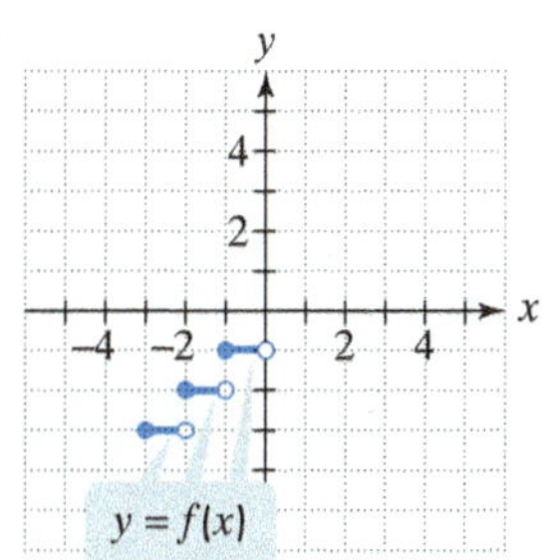

2.2 EXERCISE SET

MyMathLab — MathXL PRACTICE | WATCH | DOWNLOAD | READ | REVIEW

Practice Exercises

In Exercises 1–10, graph the given functions, f and g, in the same rectangular coordinate system. Select integers for x, starting with −2 and ending with 2. Once you have obtained your graphs, describe how the graph of g is related to the graph of f.

1. $f(x) = x, g(x) = x + 3$

2. $f(x) = x, g(x) = x - 4$

3. $f(x) = -2x, g(x) = -2x - 1$

4. $f(x) = -2x,\ g(x) = -2x + 3$

5. $f(x) = x^2,\ g(x) = x^2 + 1$

6. $f(x) = x^2,\ g(x) = x^2 - 2$

7. $f(x) = |x|,\ g(x) = |x| - 2$

8. $f(x) = |x|,\ g(x) = |x| + 1$

9. $f(x) = x^3,\ g(x) = x^3 + 2$

10. $f(x) = x^3,\ g(x) = x^3 - 1$

In Exercises 11–18, use the vertical line test to identify graphs in which y is a function of x.

11.
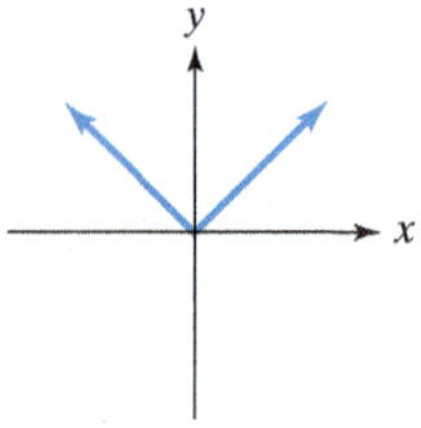

12.
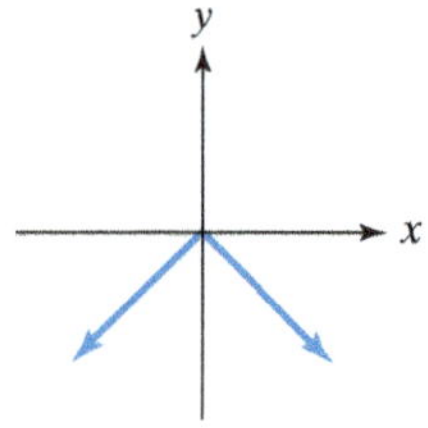

13.
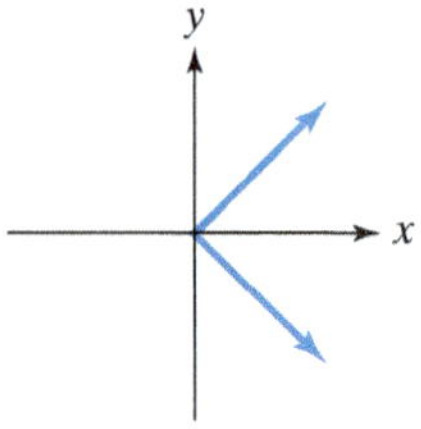

14.
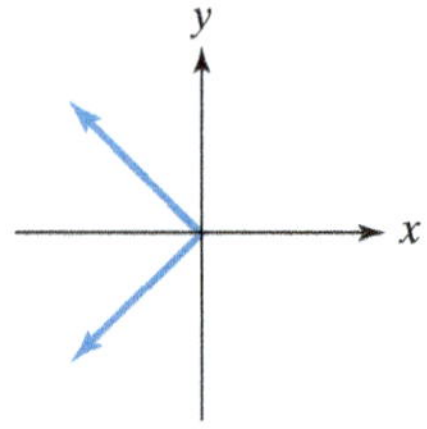

15.
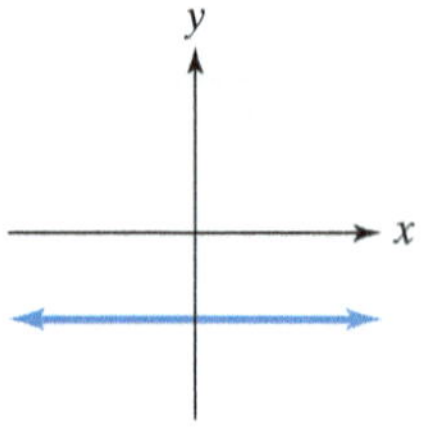

16.
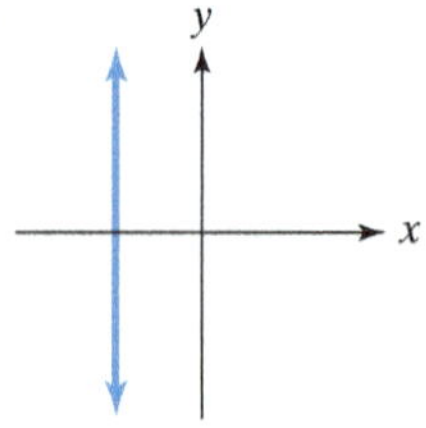

17.
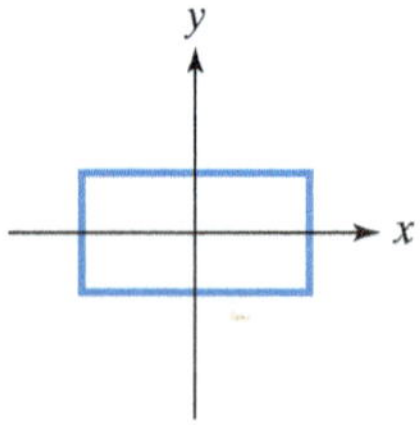

18.
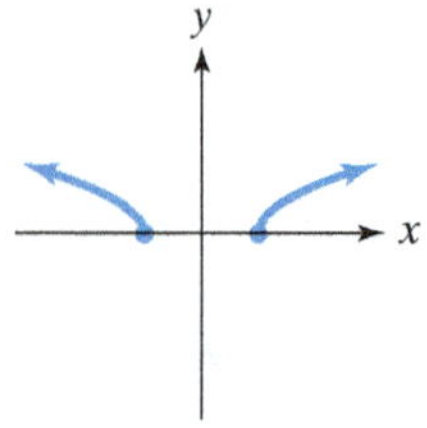

In Exercises 19–24, use the graph of f to find each indicated function value.

19. $f(-2)$

20. $f(2)$

21. $f(4)$

22. $f(-4)$

23. $f(-3)$

24. $f(-1)$

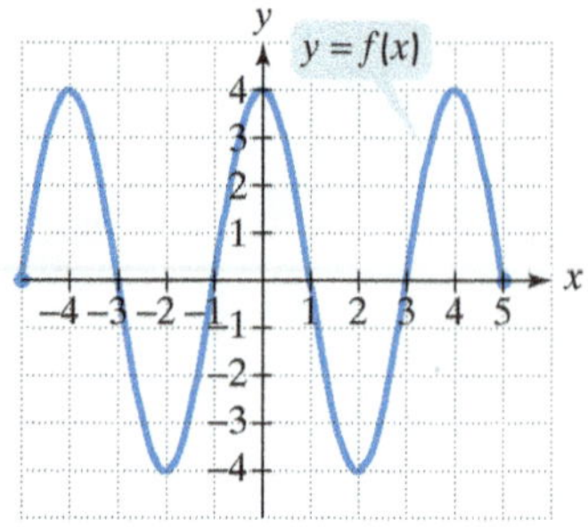

Use the graph of g to solve Exercises 25–30.

25. Find $g(-4)$.

26. Find $g(2)$.

27. Find $g(-10)$.

28. Find $g(10)$.

29. For what value of x is $g(x) = 1$?

30. For what value of x is $g(x) = -1$?

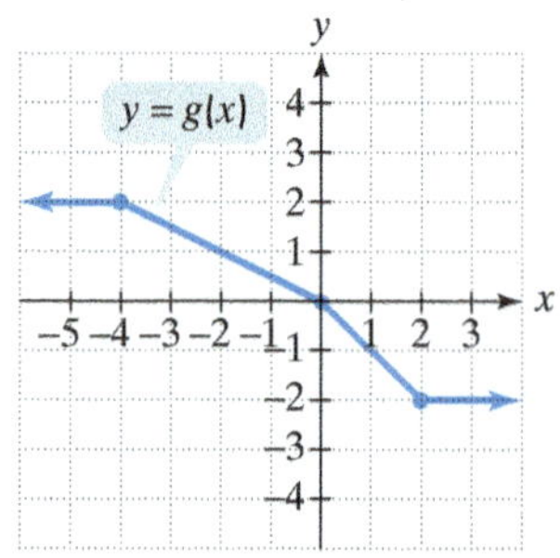

In Exercises 31–40, use the graph of each function to identify its domain and its range.

31.
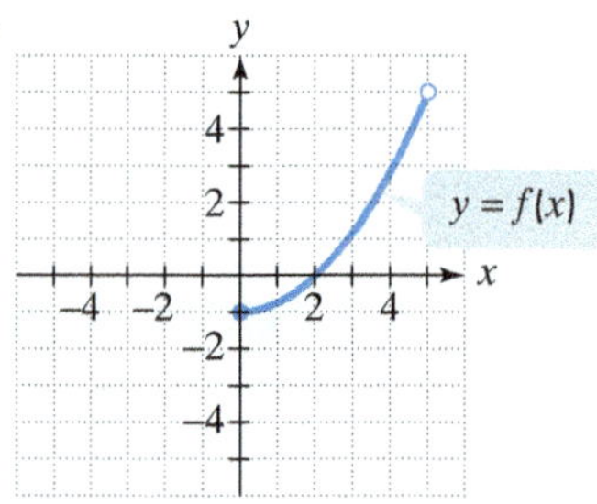

32.
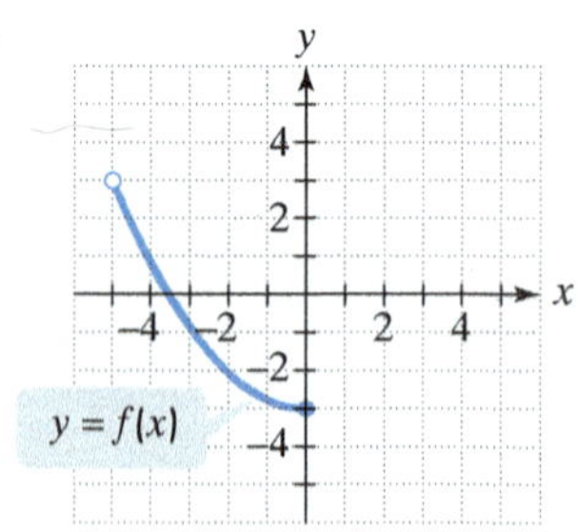

33.
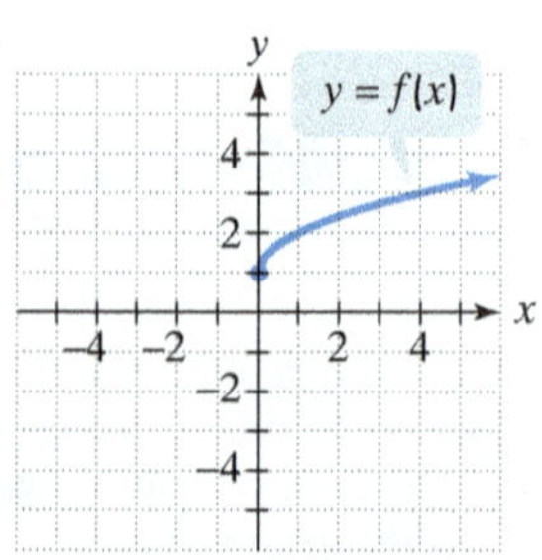

34.

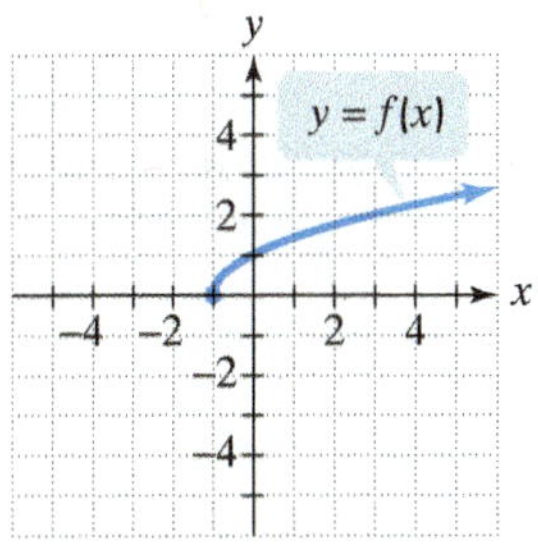

35.

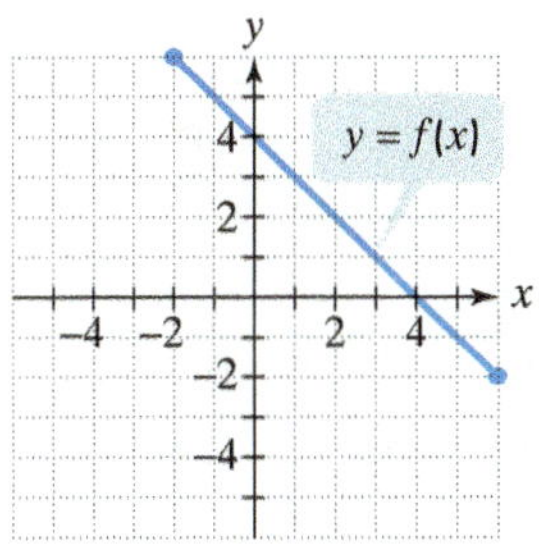

36.

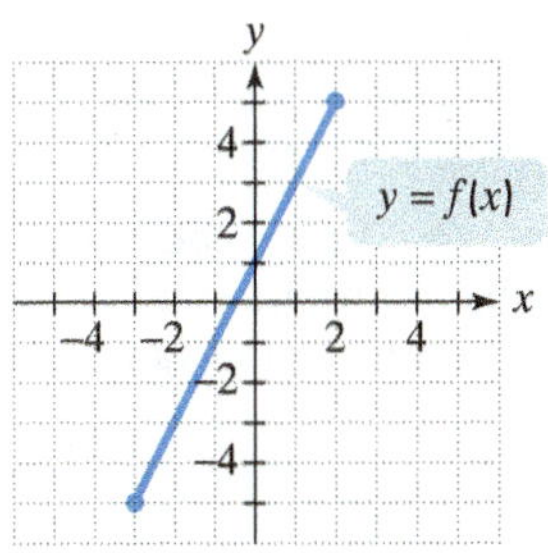

37.

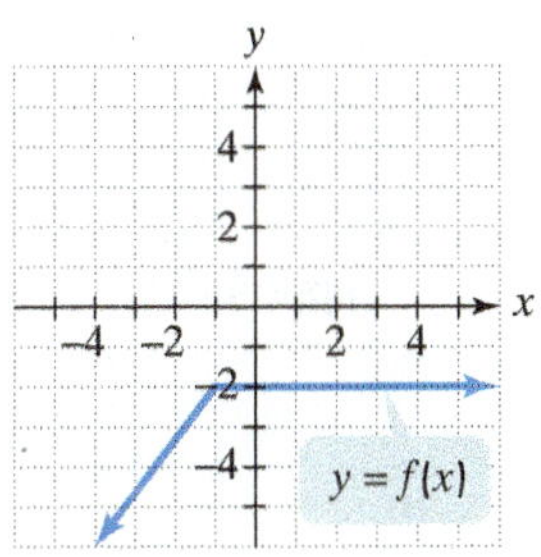

38.

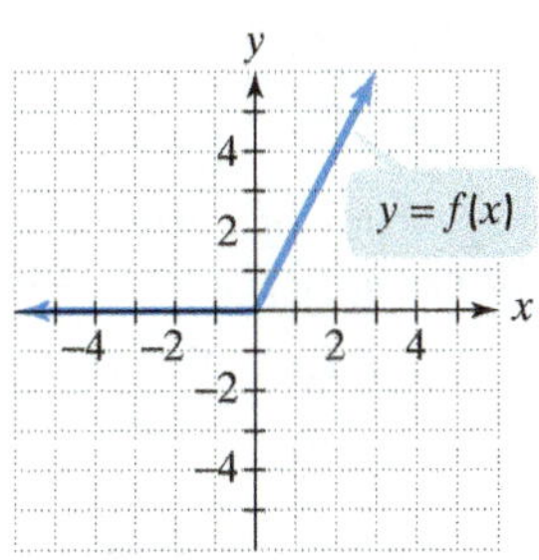

39.

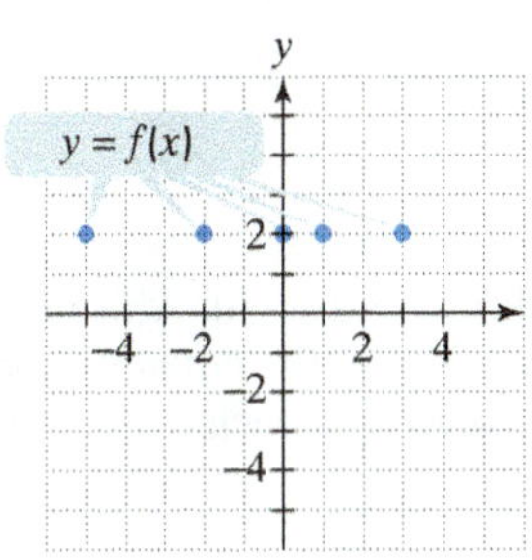

40.

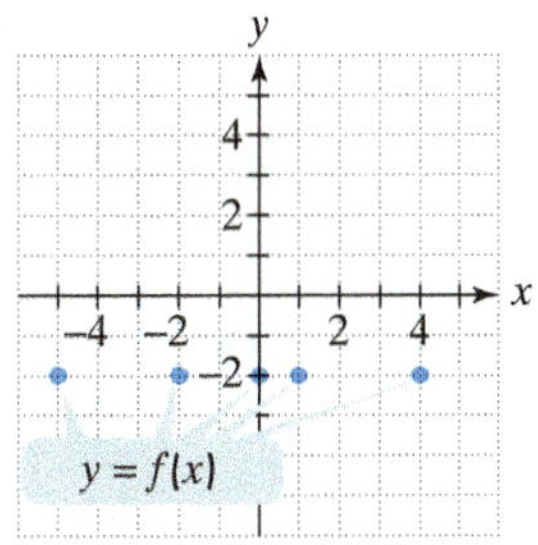

Practice PLUS

41. Use the graph of f to determine each of the following. Where applicable, use set-builder notation.

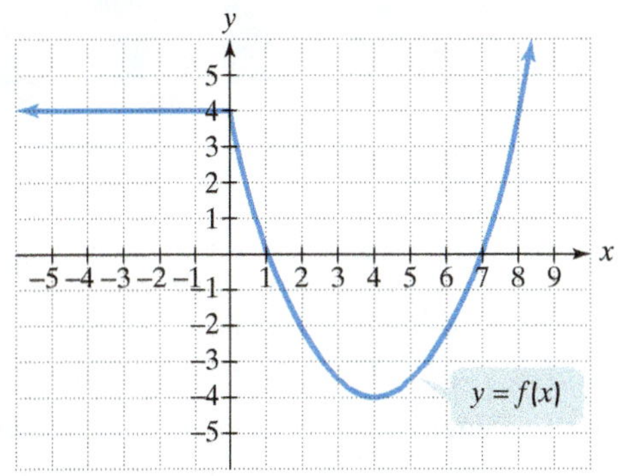

a. the domain of f

b. the range of f

c. $f(-3)$

d. the values of x for which $f(x) = -2$

e. the points where the graph of f crosses the x-axis

f. the point where the graph of f crosses the y-axis

g. values of x for which $f(x) < 0$

h. Is $f(-8)$ positive or negative?

42. Use the graph of f to determine each of the following. Where applicable, use set-builder notation.

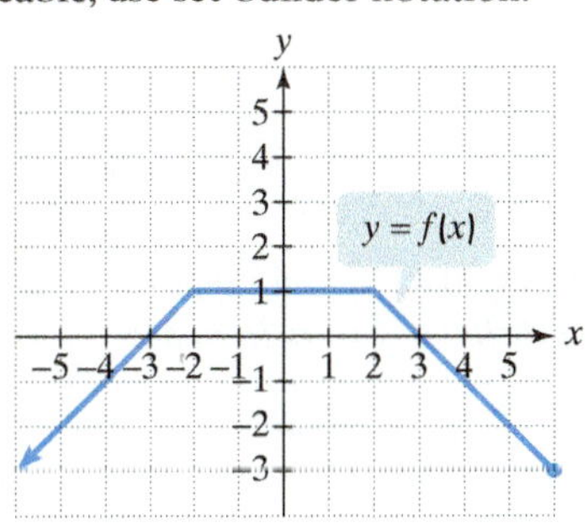

a. the domain of f

b. the range of f

c. $f(-4)$

d. the values of x for which $f(x) = -3$

e. the points where the graph of f crosses the x-axis

f. the point where the graph of f crosses the y-axis

g. values of x for which $f(x) > 0$

h. Is $f(-2)$ positive or negative?

Application Exercises

The male minority? The bar graph shows the number of bachelor's degrees, in thousands, awarded to men and women in the United States for selected years, with projections for 2010. The trend indicated by the graphs is among the hottest topics of debate among college-admissions officers. Some private liberal arts colleges have quietly begun special efforts to recruit men–including admissions preferences for them.

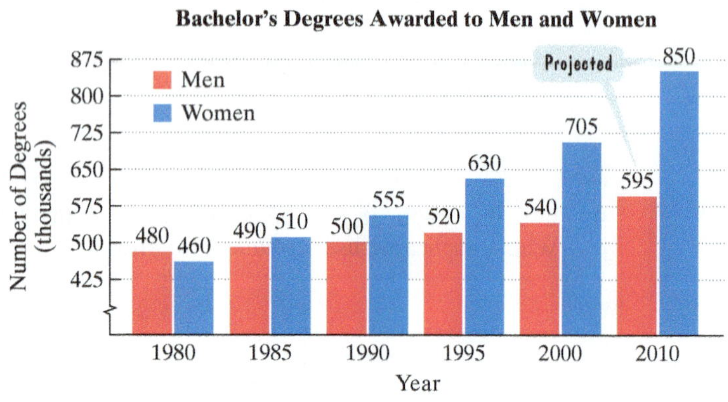

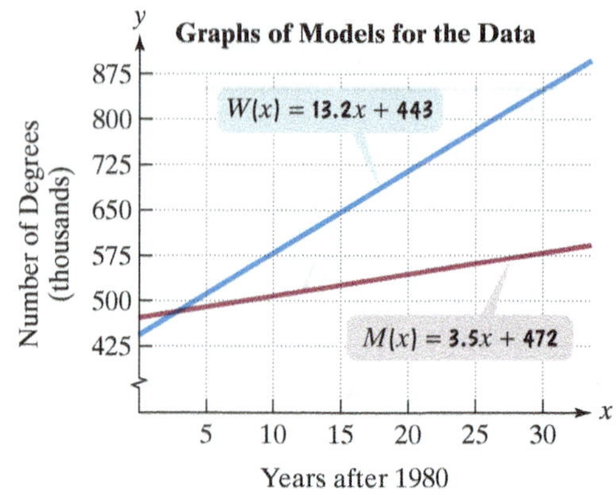

Source: Department of Education

The function $M(x) = 3.5x + 472$ *models the number of bachelor's degrees, in thousands, awarded to men x years after 1980. The function* $W(x) = 13.2x + 443$ *models the number of bachelor's degrees, in thousands, awarded to women x years after 1980. The graphs of functions M and W are shown to the right of the actual data. Use this information to solve Exercises 43–46.*

43. a. Find and interpret $W(20)$. Identify this information as a point on the graph of the function for women.

b. Does $W(20)$ overestimate or underestimate the actual data shown by the bar graph? By how much?

44. a. Find and interpret $M(20)$. Identify this information as a point on the graph of the function for men.

b. Does $M(20)$ overestimate or underestimate the actual data shown by the bar graph? By how much?

45. a. Use the two functions to find and interpret $W(10) - M(10)$. Identify this information as an appropriate distance between the graphs of the functions for women and men.

b. Does $W(10) - M(10)$ overestimate or underestimate the actual data shown by the bar graph? By how much?

46. a. Use the two functions to find and interpret $W(5) - M(5)$. Identify this information as an appropriate distance between the graphs of the functions for women and men.

b. Does $W(5) - M(5)$ overestimate or underestimate the actual data shown by the bar graph? By how much?

The function $f(x) = 0.4x^2 - 36x + 1000$ *models the number of accidents, f(x), per 50 million miles driven as a function of a driver's age, x, in years, where x includes drivers from ages 16 through 74, inclusive. The graph of f is shown. Use the equation for f to solve Exercises 47–50.*

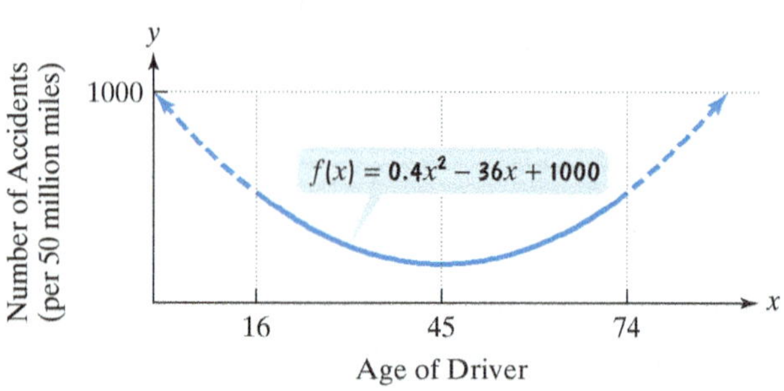

47. Find and interpret $f(20)$. Identify this information as a point on the graph of f.

48. Find and interpret $f(50)$. Identify this information as a point on the graph of f.

49. For what value of x does the graph reach its lowest point? Use the equation for f to find the minimum value of y. Describe the practical significance of this minimum value.

50. Use the graph to identify two different ages for which drivers have the same number of accidents. Use the equation for f to find the number of accidents for drivers at each of these ages.

The figure shows the percentage of Jewish Americans in the U.S. population, f(x), x years after 1900. Use the graph to solve Exercises 51–58.

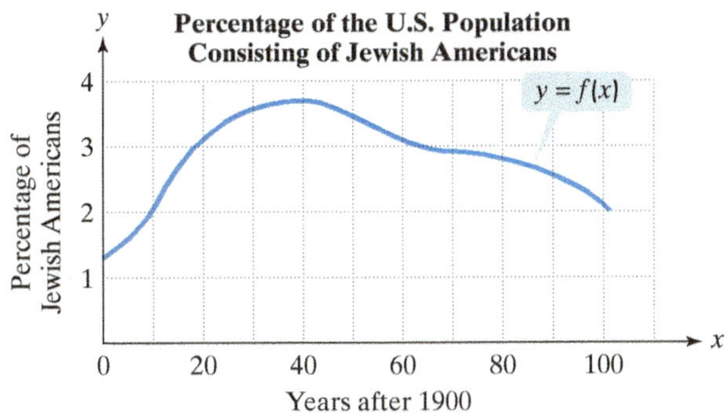

Source: American Jewish Yearbook

51. Use the graph to find a reasonable estimate of $f(60)$. What does this mean in terms of the variables in this situation?

52. Use the graph to find a reasonable estimate of $f(100)$. What does this mean in terms of the variables in this situation?

53. For what value or values of x is $f(x) = 3$? Round to the nearest year. What does this mean in terms of the variables in this situation?

54. For what value or values of x is $f(x) = 2.5$? Round to the nearest year. What does this mean in terms of the variables in this situation?

55. In which year did the percentage of Jewish Americans in the U.S. population reach a maximum? What is a reasonable estimate of the percentage for that year?

56. In which year was the percentage of Jewish Americans in the U.S. population at a minimum? What is a reasonable estimate of the percentage for that year?

57. Explain why f represents the graph of a function.

58. Describe the general trend shown by the graph.

The figure shows the cost of mailing a first-class letter, f(x), as a function of its weight, x, in ounces, for weights not exceeding 3.5 ounces. Use the graph to solve Exercises 59–62.

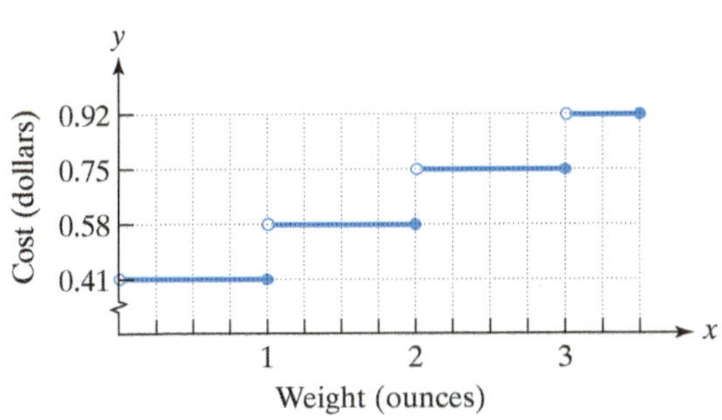

59. Find $f(3)$. What does this mean in terms of the variables in this situation?

60. Find $f(3.5)$. What does this mean in terms of the variables in this situation?

61. What is the cost of mailing a letter that weighs 1.5 ounces?

62. What is the cost of mailing a letter that weighs 1.8 ounces?

Writing in Mathematics

63. What is the graph of a function?

64. Explain how the vertical line test is used to determine whether a graph represents a function.

65. Explain how to identify the domain and range of a function from its graph.

66. Do you believe that the trend shown by the graphs for Exercises 43–46 should be reversed by providing admissions preferences for men? Explain your position on this issue.

Technology Exercises

67. Use a graphing utility to verify the pairs of graphs that you drew by hand in Exercises 1–10.

68. The function

$$f(x) = -0.00002x^3 + 0.008x^2 - 0.3x + 6.95$$

models the number of annual physician visits, $f(x)$, by a person of age x. Graph the function in a [0, 100, 5] by [0, 40, 2] viewing rectangle. What does the shape of the graph indicate about the relationship between one's age and the number of annual physician visits? Use the TRACE or minimum function capability to find the coordinates of the minimum point on the graph of the function. What does this mean?

Critical Thinking Exercises

Make Sense? *In Exercises 69–72, determine whether each statement "makes sense" or "does not make sense" and explain your reasoning.*

69. I knew how to use point plotting to graph the equation $y = x^2 - 1$, so there was really nothing new to learn when I used the same technique to graph the function $f(x) = x^2 - 1$.

70. The graph of my function revealed aspects of its behavior that were not obvious by just looking at its equation.

71. I graphed a function showing how paid vacation days depend on the number of years a person works for a company. The domain was the number of paid vacation days.

72. I graphed a function showing how the number of annual physician visits depends on a person's age. The domain was the number of annual physician visits.

In Exercises 73–78, determine whether each statement is true or false. If the statement is false, make the necessary change(s) to produce a true statement.

73. The graph of every line is a function.

74. A horizontal line can intersect the graph of a function in more than one point.

Use the graph of f to determine whether each statement in Exercises 75–78 is true or false.

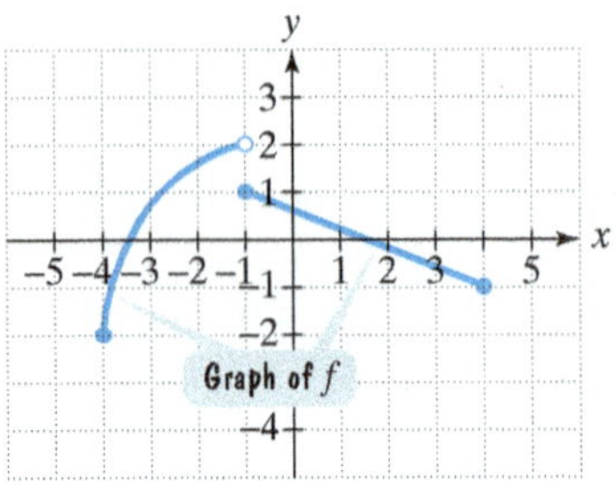

75. The domain of f is $\{x|-4 \leq x \leq 4\}$.

76. The range of f is $\{y|-2 \leq y \leq 2\}$.

77. $f(-1) - f(4) = 2$

78. $f(0) = 2.1$

In Exercises 79–80, let f be defined by the following graph:

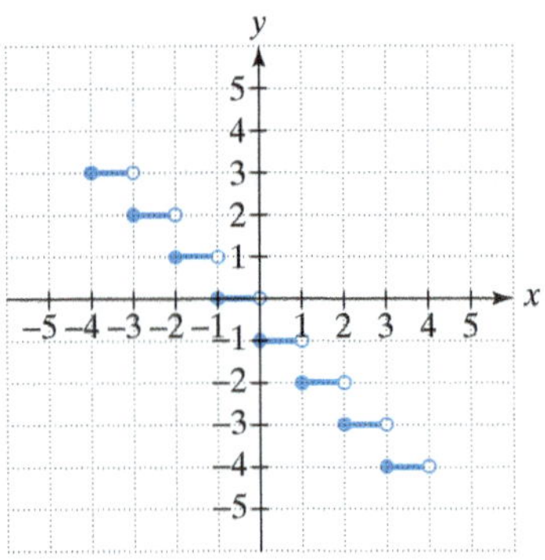

79. Find
$$\sqrt{f(-1.5) + f(-0.9)} - [f(\pi)]^2 + f(-3) \div f(1) \cdot f(-\pi).$$

80. Find
$$\sqrt{f(-2.5) - f(1.9)} - [f(-\pi)]^2 + f(-3) \div f(1) \cdot f(\pi).$$

Review Exercises

81. Is $\{(1, 1), (2, 2), (3, 3), (4, 4)\}$ a function? (Section 2.1, Example 2)

82. Solve: $12 - 2(3x + 1) = 4x - 5$. (Section 1.4, Example 3)

83. The length of a rectangle exceeds 3 times the width by 8 yards. If the perimeter of the rectangle is 624 yards, what are its dimensions? (Section 1.5, Example 5)

Preview Exercises

Exercises 84–86 will help you prepare for the material covered in the next section.

84. If $f(x) = \dfrac{4}{x - 3}$, why must 3 be excluded from the domain of f?

85. If $f(x) = x^2 + x$ and $g(x) = x - 5$, find $f(4) + g(4)$.

86. Simplify: $7.4x^2 - 15x + 4046 - (-3.5x^2 + 20x + 2405)$.

SECTION 2.4

Linear Functions and Slope

Objectives

1. Use intercepts to graph a linear function in standard form.
2. Compute a line's slope.
3. Find a line's slope and y-intercept from its equation.
4. Graph linear functions in slope-intercept form.
5. Graph horizontal or vertical lines.
6. Interpret slope as rate of change.
7. Find a function's average rate of change.
8. Use slope and y-intercept to model data.

Is there a relationship between literacy and child mortality? As the percentage of adult females who are literate increases, does the mortality of children under age five decrease? **Figure 2.16**, based on data from the United Nations, indicates that this is, indeed, the case. Each point in the figure represents one country.

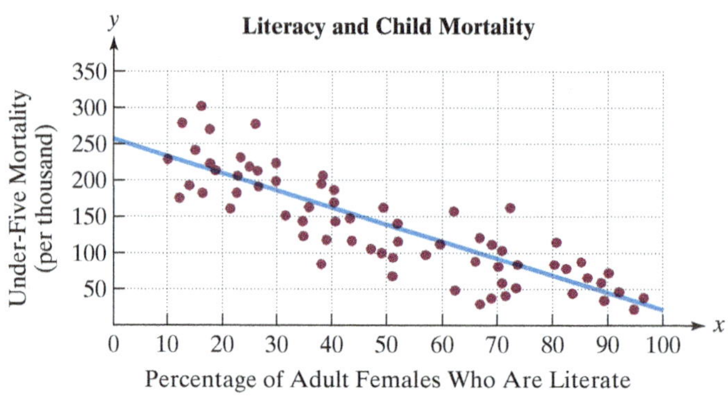

FIGURE 2.16
Source: United Nations

Data presented in a visual form as a set of points is called a **scatter plot**. Also shown in **Figure 2.16** is a line that passes through or near the points. A line that best fits the data points in a scatter plot is called a **regression line**. By writing the equation of this line, we can obtain a model of the data and make predictions about child mortality based on the percentage of literate adult females in a country.

Data often fall on or near a line. In the remainder of this chapter, we will use equations to model such data and make predictions. We begin with a discussion of graphing linear functions using intercepts.

1 Use intercepts to graph a linear function in standard form.

Graphing Using Intercepts

The equation of the regression line in **Figure 2.16** is

$$y = -2.39x + 254.47.$$

The variable x represents the percentage of adult females in a country who are literate. The variable y represents child mortality, per thousand, for children under five in that country. Using function notation, we can rewrite the equation as

$$f(x) = -2.39x + 254.47.$$

A function such as this, whose graph is a straight line, is called a **linear function**. There is another way that we can write the function's equation

$$y = -2.39x + 254.47.$$

We will collect the x- and y-terms on the left side. This is done by adding $2.39x$ to both sides:

$$2.39x + y = 254.47.$$

The form of this equation is $Ax + By = C$.

$$2.39x \quad + \quad y = 254.47$$

A, the coefficient of x, is 2.39.

B, the coefficient of y, is 1.

C, the constant on the right, is 254.47.

All equations of the form $Ax + By = C$ are straight lines when graphed, as long as A and B are not both zero. Such an equation is called the **standard form of the equation of a line**. To graph equations of this form, we will use two important points: the **intercepts**.

An ***x*-intercept** of a graph is a point where the graph intersects the x-axis. For example, look at the graph of $2x - 4y = 8$ in **Figure 2.17**. The graph crosses the x-axis at $(4, 0)$. **The *y*-coordinate corresponding to an *x*-intercept is always zero.**

A ***y*-intercept** of a graph is a point where the graph intersects the y-axis. The graph of $2x - 4y = 8$ in **Figure 2.17** shows that the graph crosses the y-axis at $(0, -2)$. **The *x*-coordinate corresponding to a *y*-intercept is always zero.**

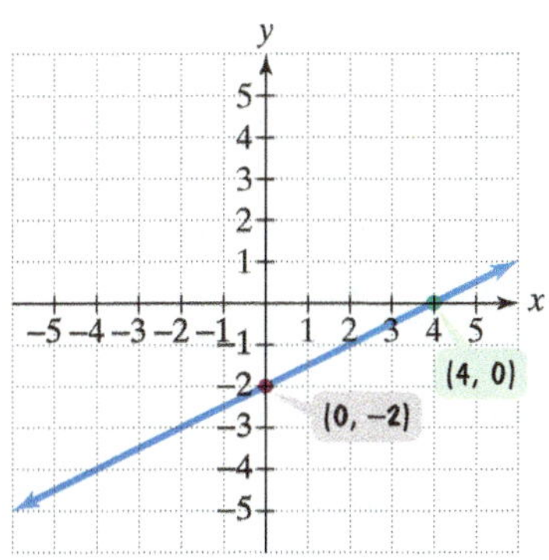

FIGURE 2.17 The graph of $2x - 4y = 8$

Study Tip

Here's the important thing to keep in mind:

x-intercept: The corresponding y-coordinate is 0.
y-intercept: The corresponding x-coordinate is 0.

When graphing using intercepts, it is a good idea to use a third point, a check-point, before drawing the line. A checkpoint can be obtained by selecting a value for either variable, other than 0, and finding the corresponding value for the other variable. The checkpoint should lie on the same line as the x- and y-intercepts. If it does not, recheck your work and find the error.

Using Intercepts to Graph $Ax + By = C$

1. Find the x-intercept. Let $y = 0$ and solve for x.
2. Find the y-intercept. Let $x = 0$ and solve for y.
3. Find a checkpoint, a third ordered-pair solution.
4. Graph the equation by drawing a line through the three points.

EXAMPLE 1 Using Intercepts to Graph a Linear Equation

Graph: $4x - 3y = 6$.

Solution

Step 1. Find the x-intercept. Let $y = 0$ and solve for x.

$$4x - 3 \cdot 0 = 6 \quad \text{Replace } y \text{ with 0 in } 4x - 3y = 6.$$

$$4x = 6 \quad \text{Simplify.}$$

$$x = \frac{6}{4} = \frac{3}{2} \quad \text{Divide both sides by 4.}$$

The x-intercept is $\left(\frac{3}{2}, 0\right)$ or $(1.5, 0)$.

Step 2. Find the y-intercept. Let $x = 0$ and solve for y.

$$4 \cdot 0 - 3y = 6 \quad \text{Replace } x \text{ with 0 in } 4x - 3y = 6.$$

$$-3y = 6 \quad \text{Simplify.}$$

$$y = -2 \quad \text{Divide both sides by } -3.$$

The y-intercept is $(0, -2)$.

Step 3. Find a checkpoint, a third ordered-pair solution. For our checkpoint, we will let $x = 1$ and find the corresponding value for y.

$$4x - 3y = 6 \quad \text{This is the given equation.}$$

$$4 \cdot 1 - 3y = 6 \quad \text{Substitute 1 for } x.$$

$$4 - 3y = 6 \quad \text{Simplify.}$$

$$-3y = 2 \quad \text{Subtract 4 from both sides.}$$

$$y = -\frac{2}{3} \quad \text{Divide both sides by } -3.$$

The checkpoint is the ordered pair $\left(1, -\frac{2}{3}\right)$.

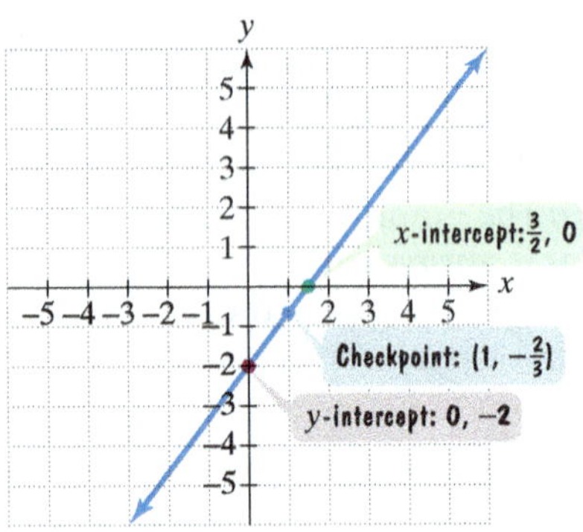

FIGURE 2.18 The graph of $4x - 3y = 6$

Step 4. Graph the equation by drawing a line through the three points. The three points in **Figure 2.18** lie along the same line. Drawing a line through the three points results in the graph of $4x - 3y = 6$.

Using Technology

You can use a graphing utility to graph equations of the form $Ax + By = C$. Begin by solving the equation for y. For example, to graph $4x - 3y = 6$, solve the equation for y.

$$4x - 3y = 6 \quad \text{This is the equation to be graphed.}$$

$$4x - 4x - 3y = -4x + 6 \quad \text{Add } -4x \text{ to both sides.}$$

$$-3y = -4x + 6 \quad \text{Simplify.}$$

$$\frac{-3y}{-3} = \frac{-4x + 6}{-3} \quad \text{Divide both sides by } -3.$$

$$y = \frac{4}{3}x - 2 \quad \text{Simplify.}$$

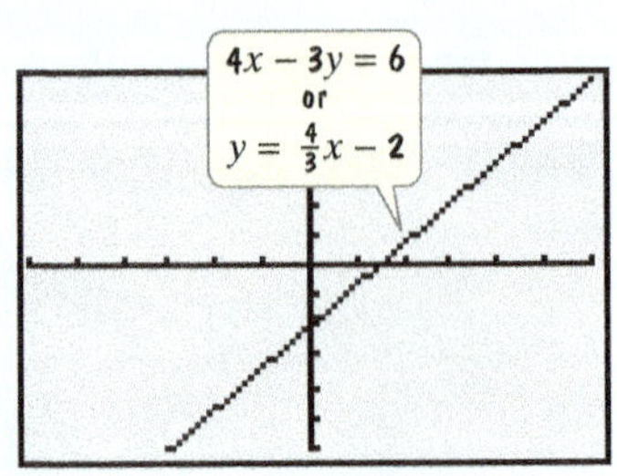

This is the equation to enter in your graphing utility. The graph of $y = \frac{4}{3}x - 2$, or, equivalently, $4x - 3y = 6$ is shown above in a $[-6, 6, 1]$ by $[-6, 6, 1]$ viewing rectangle.

✓ CHECK POINT 1 Graph: $3x - 2y = 6$.

2 Compute a line's slope.

The Slope of a Line

Mathematicians have developed a useful measure of the steepness of a line, called the **slope** of the line. Slope compares the vertical change (the **rise**) to the horizontal change (the **run**) when moving from one fixed point to another along the line. To calculate the slope of a line, we use a ratio that compares the change in y (the rise) to the change in x (the run).

Definition of Slope

The **slope** of the line through the distinct points (x_1, y_1) and (x_2, y_2) is

$$\frac{\text{Change in } y}{\text{Change in } x} = \frac{\text{Rise}}{\text{Run}} = \frac{y_2 - y_1}{x_2 - x_1},$$

Vertical change (Rise); Horizontal change (Run)

where $x_2 - x_1 \neq 0$.

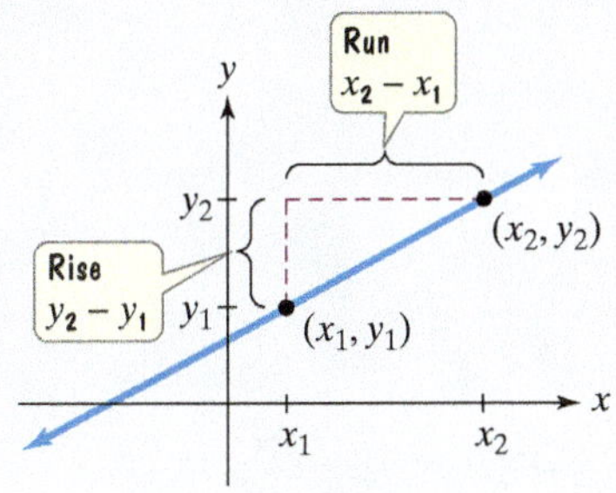

It is common notation to let the letter m represent the slope of a line. The letter m is used because it is the first letter of the French verb *monter*, meaning "to rise," or "to ascend."

EXAMPLE 2 Using the Definition of Slope

Find the slope of the line passing through each pair of points:

a. $(-3, -4)$ and $(-1, 6)$ **b.** $(-1, 3)$ and $(-4, 5)$.

Solution

a. Let $(x_1, y_1) = (-3, -4)$ and $(x_2, y_2) = (-1, 6)$. The slope is obtained as follows:

$$m = \frac{\text{Change in } y}{\text{Change in } x} = \frac{y_2 - y_1}{x_2 - x_1} = \frac{6 - (-4)}{-1 - (-3)} = \frac{6 + 4}{-1 + 3} = \frac{10}{2} = 5.$$

The situation is illustrated in **Figure 2.19**. The slope of the line is 5, or $\frac{10}{2}$. For every vertical change, or rise, of 10 units, there is a corresponding horizontal change, or run, of 2 units. The slope is positive and the line rises from left to right.

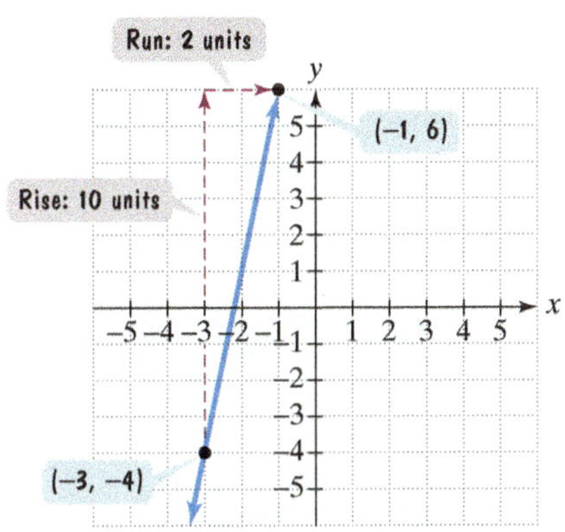

FIGURE 2.19 Visualizing a slope of 5

Study Tip

When computing slope, it makes no difference which point you call (x_1, y_1) and which point you call (x_2, y_2). If we let $(x_1, y_1) = (-1, 6)$ and $(x_2, y_2) = (-3, -4)$, the slope is still 5:

$$m = \frac{y_2 - y_1}{x_2 - x_1} = \frac{-4 - 6}{-3 - (-1)} = \frac{-10}{-2} = 5.$$

However, you should not subtract in one order in the numerator $(y_2 - y_1)$ and then in a different order in the denominator $(x_1 - x_2)$.

$$\frac{-4 - 6}{-1 - (-3)} = \frac{-10}{2} = -5.$$ (crossed out)

Incorrect! The slope is not −5.

b. To find the slope of the line passing through $(-1, 3)$ and $(-4, 5)$, we can let $(x_1, y_1) = (-1, 3)$ and $(x_2, y_2) = (-4, 5)$. The slope is computed as follows:

$$m = \frac{\text{Change in } y}{\text{Change in } x} = \frac{y_2 - y_1}{x_2 - x_1} = \frac{5 - 3}{-4 - (-1)} = \frac{2}{-3} = -\frac{2}{3}.$$

The situation is illustrated in **Figure 2.20**. The slope of the line is $-\frac{2}{3}$. For every vertical change of -2 units (2 units down), there is a corresponding horizontal change of 3 units. The slope is negative and the line falls from left to right.

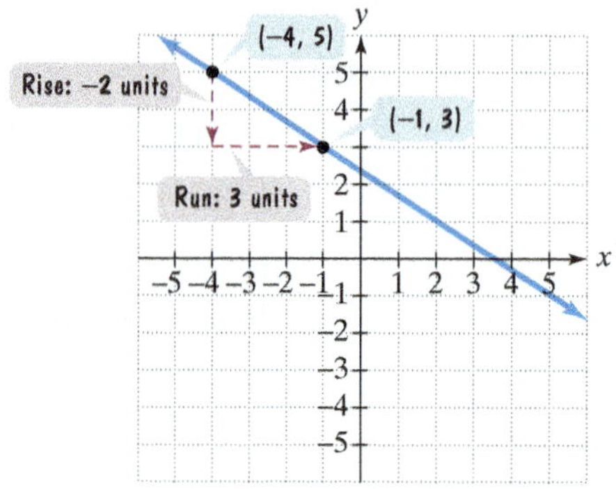

FIGURE 2.20 Visualizing a slope of $-\frac{2}{3}$

✓ CHECK POINT 2 Find the slope of the line passing through each pair of points:

a. $(-3, 4)$ and $(-4, -2)$ **b.** $(4, -2)$ and $(-1, 5)$.

Example 2 illustrates that a line with a positive slope is rising from left to right and a line with a negative slope is falling from left to right. By contrast, a horizontal line neither rises nor falls and has a slope of zero. A vertical line has no horizontal change, so $x_2 - x_1 = 0$ in the formula for slope. Because we cannot divide by zero, the slope of a vertical line is undefined. This discussion is summarized in **Table 2.1.**

Table 2.1 Possibilities for a Line's Slope

Positive Slope	Negative Slope	Zero Slope	Undefined Slope
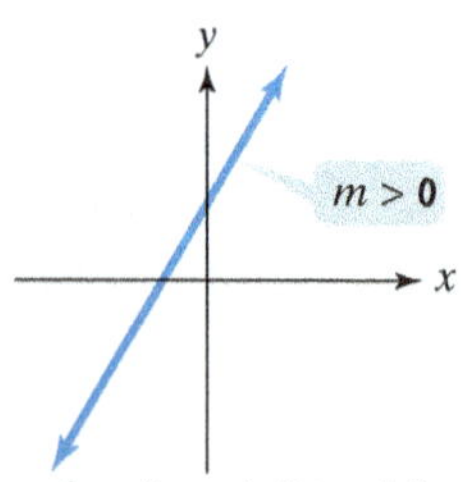	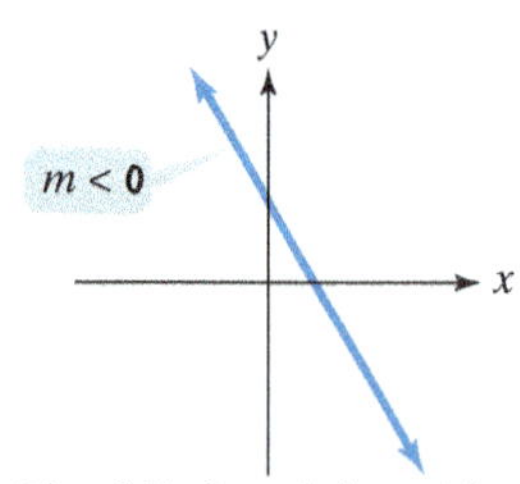	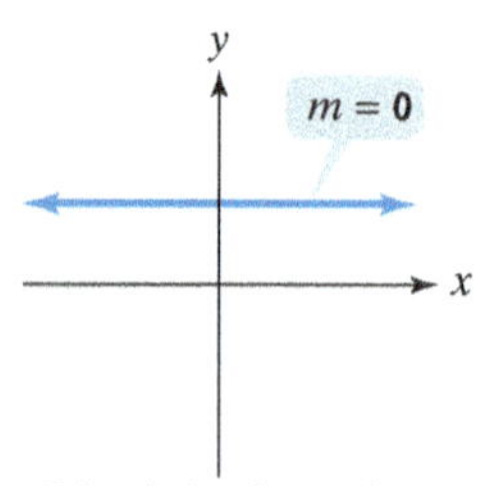	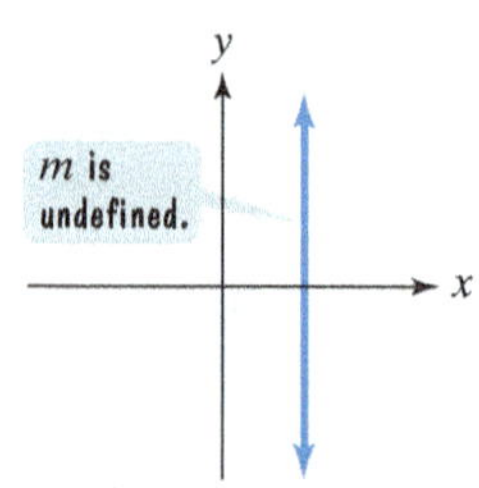
Line rises from left to right.	Line falls from left to right.	Line is horizontal.	Line is vertical.

Study Tip

Always be clear in the way you use language, especially in mathematics. For example, it's not a good idea to say that a line has "no slope." This could mean that the slope is zero or that the slope is undefined.

The Slope-Intercept Form of the Equation of a Line

We opened this section with a linear function that modeled child mortality as a function of literacy. The function's equation can be expressed as

$$y = -2.39x + 254.47 \quad \text{or} \quad f(x) = -2.39x + 254.47.$$

What is the significance of −2.39, the x-coefficient, or of 254.47, the constant term? To answer this question, let's look at an equation in the same form with simpler numbers. In particular, consider the equation $y = 2x + 4$.

Figure 2.21 shows the graph of $y = 2x + 4$. Verify that the x-intercept is $(-2, 0)$ by setting y equal to 0 and solving for x. Similarly, verify that the y-intercept is $(0, 4)$ by setting x equal to 0 and solving for y.

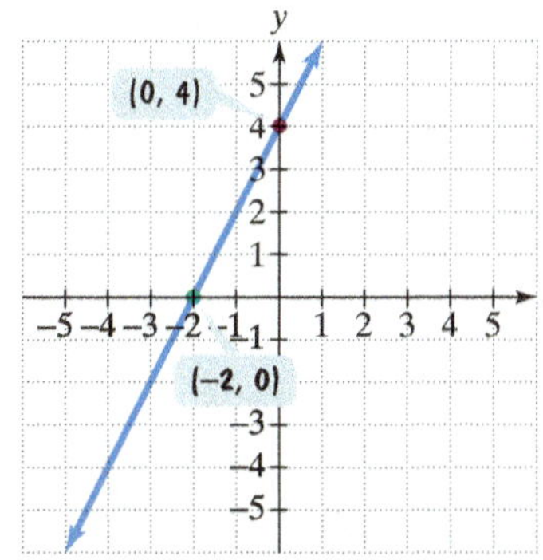

FIGURE 2.21 The graph of $y = 2x + 4$

Now that we have two points on the line, we can calculate the slope of the graph of $y = 2x + 4$.

$$\text{Slope} = \frac{\text{Change in } y}{\text{Change in } x} = \frac{4 - 0}{0 - (-2)} = \frac{4}{2} = 2$$

We see that the slope of the line is 2, the same as the coefficient of x in the equation $y = 2x + 4$. The y-intercept is (0, 4), the same as the constant in the equation $y = 2x + 4$.

$$y = 2x + 4$$

The slope is 2. The y-intercept is (0, 4).

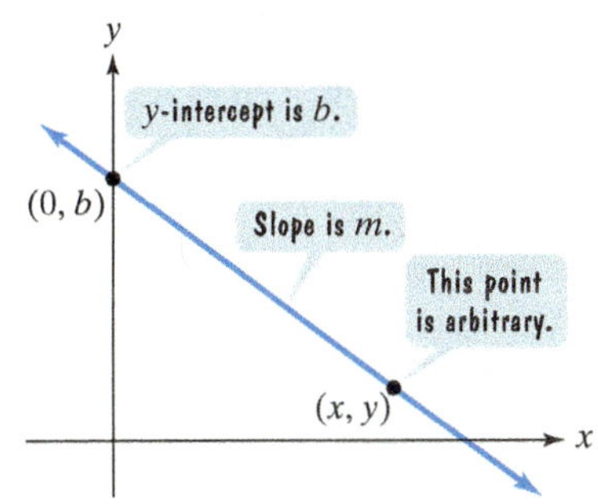

FIGURE 2.22 A line with slope m and y-intercept b

It is not merely a coincidence that the x-coefficient is the line's slope and the constant term is the y-cordinate of the y-intercept. Let's find the equation of any nonvertical line with slope m and y-intercept $(0, b)$. Now, let (x, y) represent any other point on the line, shown in **Figure 2.22**. Keep in mind that the point (x, y) is arbitrary and is not in one fixed position. By contrast, the point $(0, b)$ is fixed.

Regardless of where the point (x, y) is located, the steepness of the line in **Figure 2.22** remains the same. Thus, the ratio for slope stays a constant m. This means that for all points along the line,

$$m = \frac{\text{Change in } y}{\text{Change in } x} = \frac{y - b}{x - 0} = \frac{y - b}{x}.$$

We can clear the fraction by multiplying both sides by x, the least common denominator.

$$m = \frac{y - b}{x}$$ This is the slope of the line in Figure 2.22.

$$mx = \frac{y - b}{x} \cdot x$$ Multiply both sides by x.

$$mx = y - b$$ Simplify: $\frac{y - b}{\cancel{x}} \cdot \cancel{x} = y - b$.

$$mx + b = y - b + b$$ Add b to both sides and solve for y.

$$mx + b = y$$ Simplify.

Now, if we reverse the two sides, we obtain the slope-intercept form of the equation of a line.

Slope-Intercept Form of the Equation of a Line

The **slope-intercept form of the equation** of a nonvertical line with slope m and y-intercept $(0, b)$ is

$$y = mx + b.$$

3 Find a line's slope and y-intercept from its equation.

The slope-intercept form of a line's equation, $y = mx + b$, can be expressed in function notation by replacing y with $f(x)$:

$$f(x) = mx + b.$$

We have seen that functions in this form are called **linear functions**. Thus, in the equation of a linear function, the x-coefficient is the line's slope and the constant term is the y-intercept. Here are two examples:

$$y = 2x - 4 \qquad\qquad f(x) = \frac{1}{2}x + 2.$$

The slope is 2. The y-intercept is $(0, -4)$. The slope is $\frac{1}{2}$. The y-intercept is $(0, 2)$.

Study Tip

The variables in $y = mx + b$ vary in different ways. The values for slope, m, and y-intercept, b, vary from one line's equation to another. However, they remain constant in the equation of a single line. By contrast, the variables x and y represent the infinitely many points, (x, y), on a single line. Thus, these variables vary in both the equation of a single line, as well as from one equation to another.

If a linear function's equation is in standard form, $Ax + By = C$, do you see how we can identify the line's slope and y-intercept? Solve the equation for y and convert to slope-intercept form.

EXAMPLE 3 Converting from Standard Form to Slope-Intercept Form

Give the slope and the y-intercept for the line whose equation is

$$5x + 3y = -12.$$

Solution We convert $5x + 3y = -12$ to slope-intercept form by solving the equation for y. In this form, the coefficient of x is the line's slope and the constant term is the y-intercept.

$$5x + 3y = -12$$ This is the given equation in standard form, $Ax + By = C$.

Our goal is to isolate y.

$$5x - 5x + 3y = -5x - 12$$ Add $-5x$ to both sides.

$$3y = -5x - 12$$ Simplify.

$$\frac{3y}{3} = \frac{-5x - 12}{3}$$ Divide both sides by 3.

$$y = -\frac{5}{3}x - 4$$ Divide each term in the numerator by 3.

The slope is $-\frac{5}{3}$. The y-intercept is $(0, -4)$.

CHECK POINT 3 Give the slope and the y-intercept for the line whose equation is $8x - 4y = 20$.

4 Graph linear functions in slope-intercept form.

If a linear function's equation is in slope-intercept form, we can use the y-intercept and the slope to obtain its graph.

Graphing $y = mx + b$ Using the Slope and y-Intercept

1. Plot the y-intercept on the y-axis. This is the point $(0, b)$.
2. Obtain a second point using the slope, m. Write m as a fraction, and use rise over run, starting at the point containing the y-intercept, to plot this point.
3. Use a straightedge to draw a line through the two points. Draw arrowheads at the ends of the line to show that the line continues indefinitely in both directions.

Study Tip

Writing the slope, m, as a fraction allows you to identify the rise (the fraction's numerator) and the run (the fraction's denominator).

EXAMPLE 4 Graphing by Using the Slope and *y*-Intercept

Graph the line whose equation is $y = 3x - 4$.

Solution The equation $y = 3x - 4$ is in the form $y = mx + b$. The slope, m, is the coefficient of x. The y-coordinate of the y-intercept, b, is the constant term.

$$y = 3x + (-4)$$

The slope is 3. The y-intercept is $(0, -4)$.

Now that we have identified the slope and the y-intercept, we use the three-step procedure to graph the equation.

Step 1. Plot the *y*-intercept on the *y*-axis. The y-intercept is the point $(0, -4)$, shown in **Figure 2.23(a)**.

Step 2. Obtain a second point using the slope, *m*. Write *m* as a fraction, and use rise over run, starting at the point containing the *y*-intercept, to plot this point. We express the slope, 3, as a fraction.

$$m = \frac{3}{1} = \frac{\text{Rise}}{\text{Run}}$$

We plot the second point on the line by starting at $(0, -4)$, the first point. Based on the slope, we move 3 units *up* (the rise) and 1 unit to the *right* (the run). This puts us at a second point on the line, $(1, -1)$, shown in **Figure 2.23(b)**.

Step 3. Use a straightedge to draw a line through the two points. The graph of $y = 3x - 4$ is shown in **Figure 2.23(c)**.

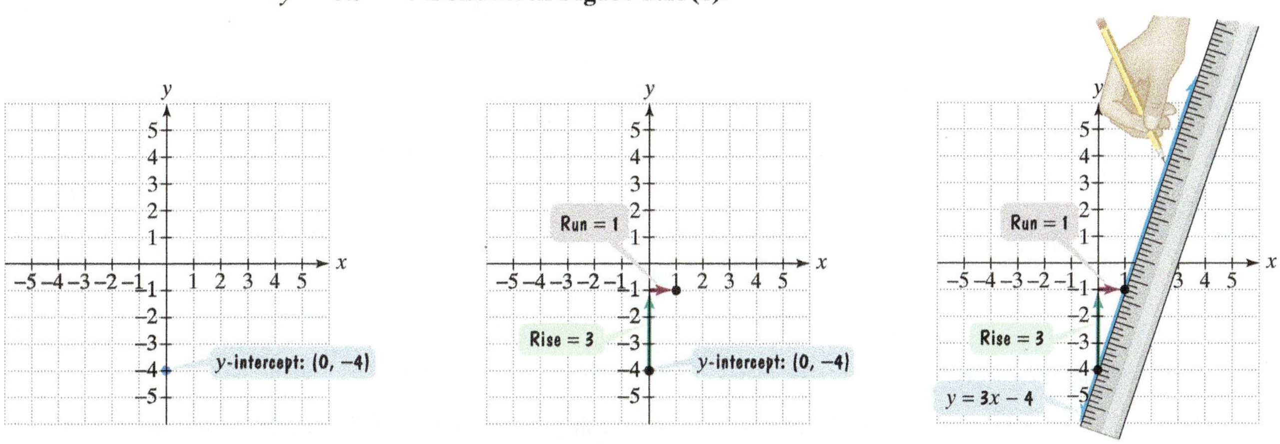

(a) The y-intercept is -4, so $(0, -4)$ is a point on the line.

(b) The slope is 3.

(c) The graph of $y = 3x - 4$.

FIGURE 2.23 Graphing $y = 3x - 4$ using the y-intercept and slope

✓ **CHECK POINT 4** Graph the line whose equation is $y = 4x - 3$.

EXAMPLE 5 Graphing by Using the Slope and *y*-Intercept

Graph the linear function: $f(x) = -\frac{3}{2}x + 2$.

Solution The equation of the line is in the form $f(x) = mx + b$. We can find the slope, m, by identifying the coefficient of x. We can find the y-intercept, b, by identifying the constant term.

$$f(x) = -\frac{3}{2}x + 2$$

The slope is $-\frac{3}{2}$. The y-intercept is $(0, 2)$.

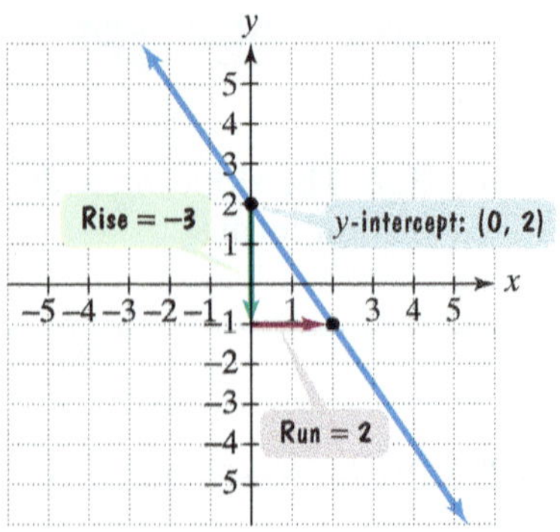

FIGURE 2.24 The graph of $f(x) = -\frac{3}{2}x + 2$

Now that we have identified the slope $-\frac{3}{2}$, and the y-intercept, $(0, 2)$, we use the three-step procedure to graph the equation.

Step 1. Plot the y-intercept on the y-axis. The y-intercept is $(0, 2)$. We plot $(0, 2)$, shown in **Figure 2.24**.

Step 2. Obtain a second point using the slope, m. Write m as a fraction, and use rise over run, starting at the point containing the y-intercept, to plot this point. The slope, $-\frac{3}{2}$, is already written as a fraction.

$$m = -\frac{3}{2} = \frac{-3}{2} = \frac{\text{Rise}}{\text{Run}}$$

We plot the second point on the line by starting at $(0, 2)$, the first point. Based on the slope, we move 3 units *down* (the rise) and 2 units to the *right* (the run). This puts us at a second point on the line, $(2, -1)$, shown in **Figure 2.24**.

Step 3. Use a straightedge to draw a line through the two points. The graph of the linear function $f(x) = -\frac{3}{2}x + 2$ is shown as a blue line in **Figure 2.24**.

Discover for Yourself

Obtain a second point in Example 5 by writing the slope as follows:

$$m = \frac{3}{-2} = \frac{\text{Rise}}{\text{Run}}.$$

$-\frac{3}{2}$ can be expressed as $\frac{-3}{2}$ or $\frac{3}{-2}$.

Obtain a second point in **Figure 2.24** by moving *up* 3 units and to the *left* 2 units, starting at $(0, 2)$. What do you observe once you draw the line?

✓ CHECK POINT 5 Graph the linear function: $f(x) = -\frac{2}{3}x$.

5 Graph horizontal or vertical lines.

Equations of Horizontal and Vertical Lines

If a line is horizontal, its slope is zero: $m = 0$. Thus, the equation $y = mx + b$ becomes $y = b$, where b is the y-intercept. All horizontal lines have equations of the form $y = b$.

EXAMPLE 6 Graphing a Horizontal Line

Graph $y = -4$ in the rectangular coordinate system.

Solution All ordered pairs that are solutions of $y = -4$ have a value of y that is always -4. Any value can be used for x. In the table on the right, we have selected three of the possible values for x -2, 0, and 3. The table shows that three ordered pairs that are solutions of $y = -4$ are $(-2, -4)$, $(0, -4)$, and $(3, -4)$. Drawing a line that passes through the three points gives the horizontal line shown in **Figure 2.25**.

x	$y = -4$	(x, y)
−2	−4	(−2, −4)
0	−4	(0, −4)
3	−4	(3, −4)

For all choices of x,

y is a constant −4.

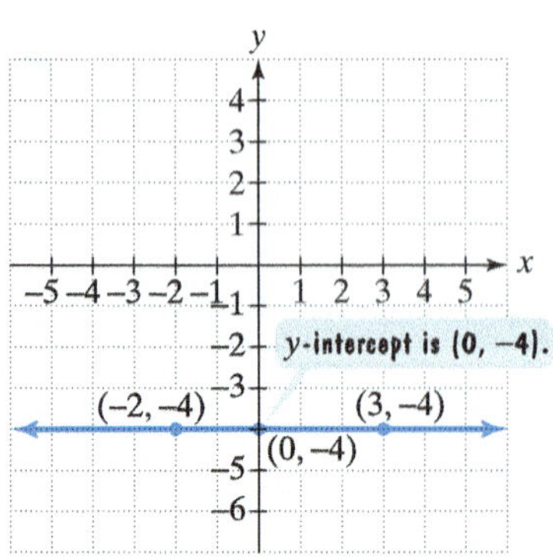

FIGURE 2.25 The graph of $y = -4$ or $f(x) = -4$

✓ CHECK POINT 6 Graph $y = 3$ in the rectangular coordinate system.

Equation of a Horizontal Line

A horizontal line is given by an equation of the form

$$y = b,$$

where $(0, b)$ is the y-intercept.

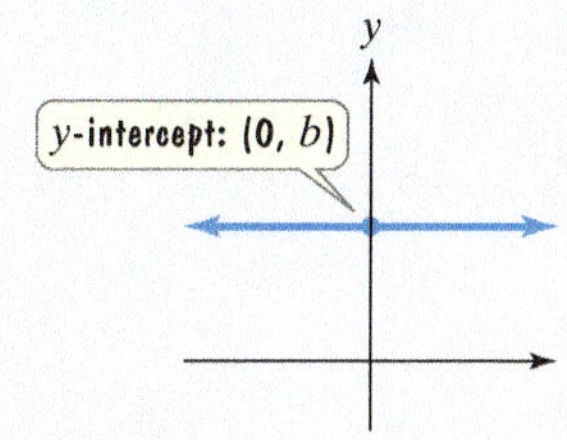

Because any vertical line can intersect the graph of a horizontal line $y = b$ only once, a horizontal line is the graph of a function. Thus, we can express the equation $y = b$ as $f(x) = b$. This linear function is often called a **constant function.**

Next, let's see what we can discover about the graph of an equation of the form $x = a$ by looking at an example.

EXAMPLE 7 Graphing a Vertical Line

Graph the linear equation: $x = 2$.

Solution All ordered pairs that are solutions of $x = 2$ have a value of x that is always 2. Any value can be used for y. In the table on the right, we have selected three of the possible values for y: -2, 0, and 3. The table shows that three ordered pairs that are solutions of $x = 2$ are $(2, -2)$, $(2, 0)$, and $(2, 3)$. Drawing a line that passes through the three points gives the vertical line shown in **Figure 2.26**.

For all choices of y,

$x = 2$	y	(x, y)
2	-2	$(2, -2)$
2	0	$(2, 0)$
2	3	$(2, 3)$

x is always 2.

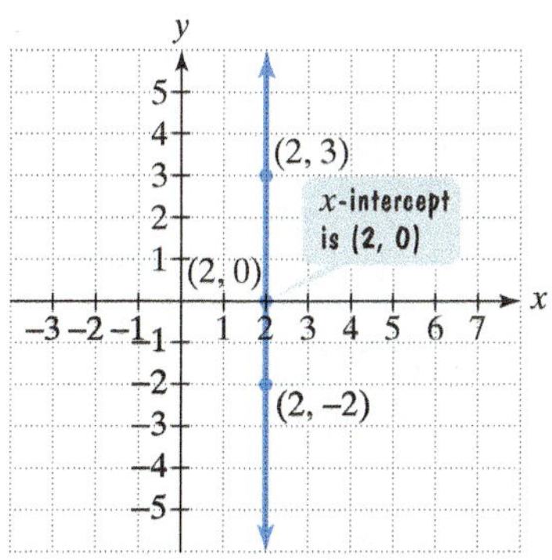

FIGURE 2.26 The graph of $x = 2$

Equation of a Vertical Line

A vertical line is given by an equation of the form

$$x = a,$$

where $(a, 0)$ is the x-intercept.

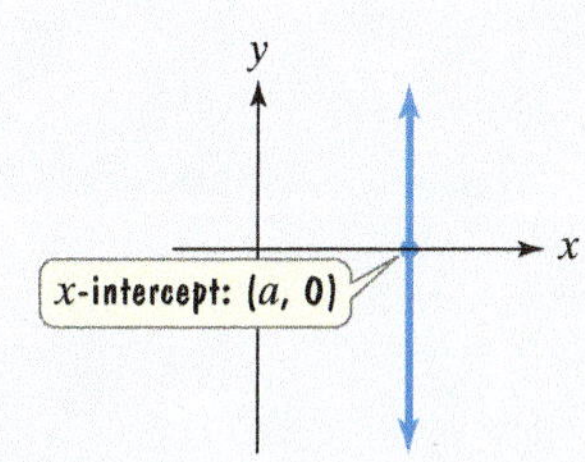

Does a vertical line represent the graph of a linear function? No. Look at the graph of $x = 2$ in **Figure 2.26**. A vertical line drawn through $(2, 0)$ intersects the graph infinitely many times. This shows that infinitely many outputs are associated with the input 2. **No vertical line is a linear function.**

CHECK POINT 7 Graph the linear equation: $x = -3$.

See graphing answer section.

Study Tip

The linear equations in Examples 6 and 7, $y = -4$ and $x = 2$, each show only one variable. However, these are equations in two variables in the standard form $Ax + By = C$:

- $y = -4$ means $0x + 1y = -4$.
- $x = 2$ means $1x + 0y = 2$.

6 Interpret slope as rate of change.

Slope as Rate of Change

Slope is defined as the ratio of a change in y to a corresponding change in x. It describes how fast y is changing with respect to x. For a linear function, slope may be interpreted as the rate of change of the dependent variable per unit change in the independent variable.

Our next example shows how slope can be interpreted as a rate of change in an applied situation. When calculating slope in applied problems, keep track of the units in the numerator and the denominator.

EXAMPLE 8 Slope as a Rate of Change

The line graphs in **Figure 2.27** show the percentage of Americans in two age groups who reported using illegal drugs in the previous month. Find the slope of the line segment for the 12–17 age group. Describe what this slope represents.

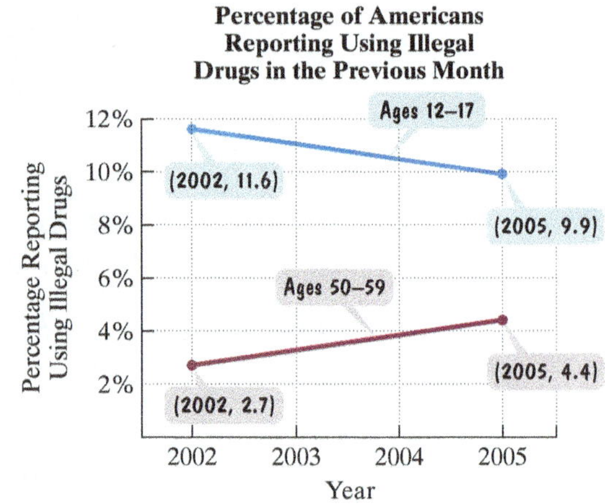

FIGURE 2.27

Source: Substance Abuse and Mental Health Services Administration

Solution We will let x represent a year and y the percentage reporting illegal drug use in the previous month for that year. The two points shown on the line segment for the 12–17 age group have the following coordinates:

$$(2002, 11.6) \quad \text{and} \quad (2005, 9.9).$$

In 2002, 11.6% reported using illegal drugs in the previous month.

In 2005, 9.9% reported using illegal drugs in the previous month.

Now we compute the slope:

$$m = \frac{\text{Change in } y}{\text{Change in } x} = \frac{9.9 - 11.6}{2005 - 2002} = \frac{-1.7}{3} \approx -0.57$$

The unit in the numerator is *percent*.

The unit in the denominator is *year*.

The slope indicates that for the 12–17 age group, the percentage reporting using illegal drugs in the previous month decreased by approximately 0.57% each year. The rate of change is a decrease of approximately 0.57% per year.

CHECK POINT 8 Use the graph in **Figure 2.27** to find the slope of the line segment for the 50–59 age group. Express the slope correct to two decimal places and describe what it represents.

---

7 Find a function's average rate of change.

The Average Rate of Change of a Function

If the graph of a function is not a straight line, the **average rate of change** between any two points is the slope of the line containing the two points. For example, **Figure 2.28** shows the graph of a particular man's height, in inches, as a function of his age, in years. Two points on the graph are labeled (13, 57) and (18, 76). At age 13, this man was 57 inches tall, and at age 18, he was 76 inches tall.

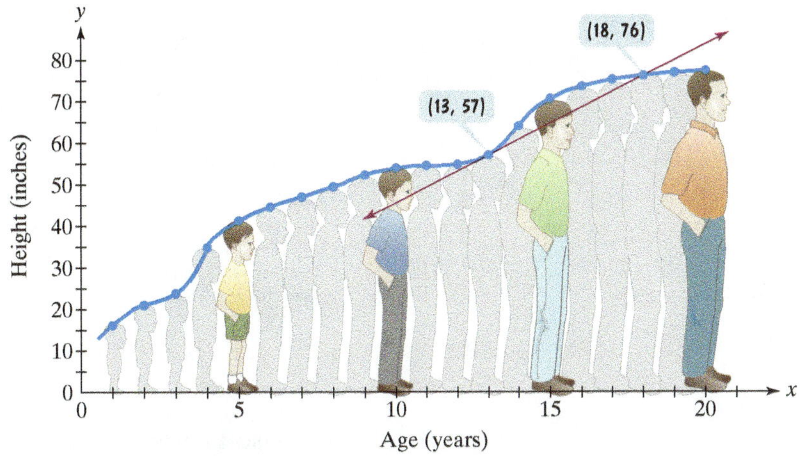

FIGURE 2.28 Height as a function of age

The man's average growth rate between ages 13 and 18 is the slope of the line containing (13, 57) and (18, 76):

$$m = \frac{\text{Change in } y}{\text{Change in } x} = \frac{76 - 57}{18 - 13} = \frac{19}{5} = 3\frac{4}{5}.$$

This man's average rate of change, or average growth rate, from age 13 to age 18 was $3\frac{4}{5}$, or 3.8, inches per year.

For any function, $y = f(x)$, the slope of the line between any two points is the **average change in *y* per unit change in *x*.**

EXAMPLE 9 Finding the Average Rate of Change

When a person receives a drug injected into a muscle, the concentration of the drug in the body, measured in milligrams per 100 milliliters, is a function of the time elapsed after the injection, measured in hours. **Figure 2.29** shows the graph of such a function, where x represents hours after the injection and $f(x)$ is the drug's concentration at time x. Find the average rate of change in the drug's concentration between 3 and 7 hours.

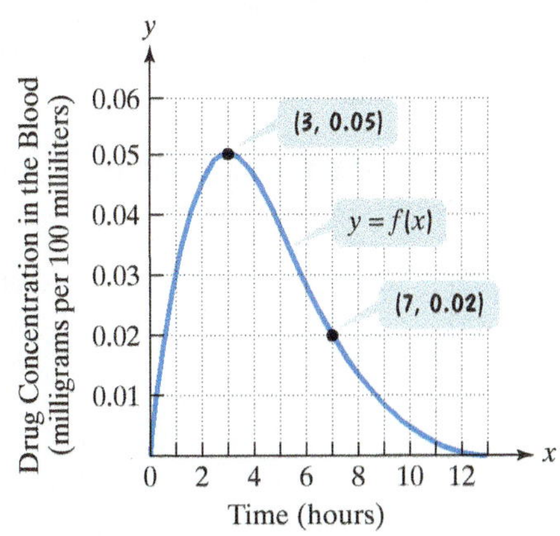

FIGURE 2.29 Concentration of a drug as a function of time

Solution At 3 hours, the drug's concentration is 0.05 and at 7 hours, the concentration is 0.02. The average rate of change in its concentration between 3 and 7 hours is the slope of the line connecting the points (3, 0.05) and (7, 0.02).

$$m = \frac{\text{Change in } y}{\text{Change in } x} = \frac{0.02 - 0.05}{7 - 3} = \frac{-0.03}{4} = -0.0075$$

The average rate of change is -0.0075. This means that the drug's concentration is decreasing at an average rate of 0.0075 milligram per 100 milliliters per hour.

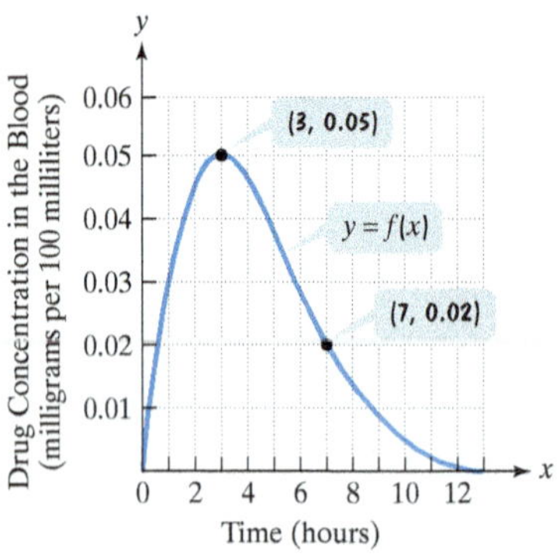

FIGURE 2.29 (repeated)

Study Tip

Units used to describe x and y tend to "pile up" when expressing the rate of change of y with respect to x. The unit used to express the rate of change of y with respect to x is

the unit used to describe y **per** the unit used to describe x.

In Figure 2.29, y, or drug concentration, is described in milligrams per 100 milliliters.

In Figure 2.29, x, or time, is described in hours.

In **Figure 2.29**, the rate of change is described in terms of milligrams per 100 milliliters per hour.

CHECK POINT 9 Use **Figure 2.29** to find the average rate of change in the drug's concentration between 1 hour and 3 hours.

Blitzer Bonus

How Calculus Studies Change

Take a rapid sequence of still photographs of a moving scene and project them onto a screen at thirty shots a second or faster. Our eyes see the results as continuous motion. The small difference between one frame and the next cannot be detected by the human visual system. The idea of calculus likewise regards continuous motion as made up of a sequence of still configurations. Calculus masters the mystery of movement by "freezing the frame" of a continuous changing process, instant by instant. For example, **Figure 2.30** shows a male's changing height over intervals of time. Over the period of time from P to D, his average rate of growth is his change in height—that is, his height at time D minus his height at time P—divided by the change in time from P to D. This is the slope of line PD.

The lines PD, PC, PB, and PA shown in **Figure 2.30** have slopes that show average growth rates for successively shorter periods of time. Calculus makes these time frames so small that they approach a single point—that is, a single instant in time. This point is shown as point P in **Figure 2.30**. The slope of the line that touches the graph at P gives the male's growth rate at one instant in time, P.

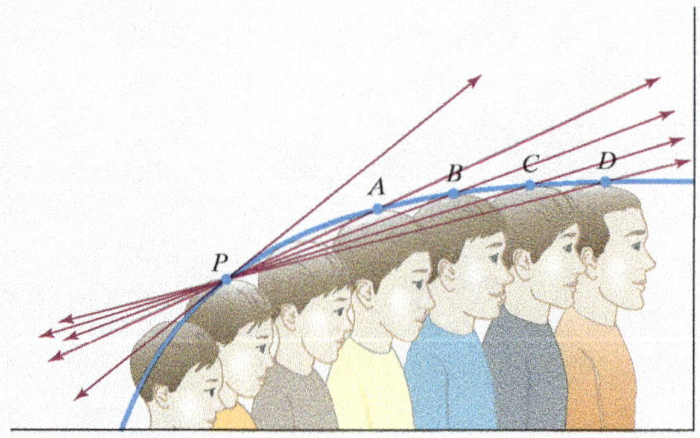

FIGURE 2.30 Analyzing continuous growth over intervals of time and at an instant in time

8 Use slope and y-intercept to model data.

Modeling Data with the Slope-Intercept Form of the Equation of a Line

Linear functions are useful for modeling data that fall on or near a line. For example, the bar graph in **Figure 2.31(a)** gives the average ticket price for a U.S. movie in the indicated year. The data are displayed as a set of four points in the scatter plot in **Figure 2.31(b)**.

Average Ticket Price for a Movie in the United States

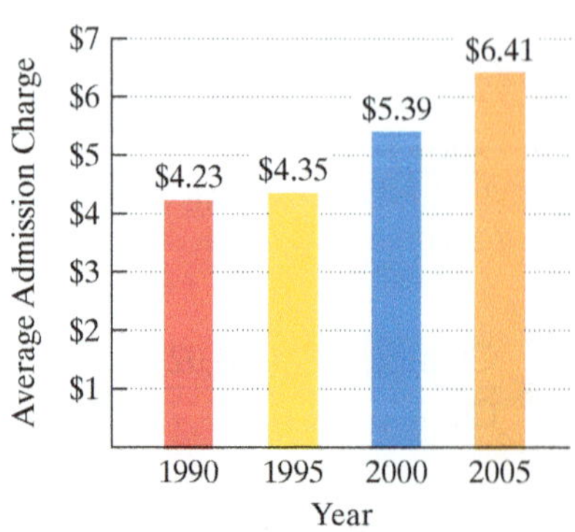

FIGURE 2.31(a)

Source: National Association of Theatre Owners

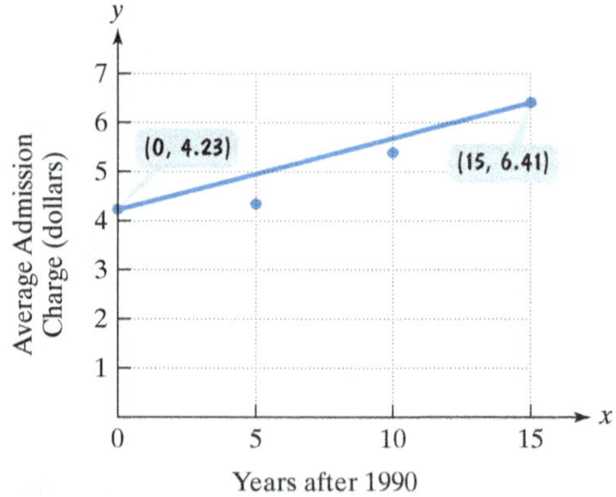

FIGURE 2.31(b)

Also shown on the scatter plot in **Figure 2.31(b)** is a line that passes through or near the four points. Example 10 illustrates how we can use the equation $y = mx + b$ to obtain a model for the data and make predictions about what you can expect to pay for a movie ticket in the future.

EXAMPLE 10 Modeling with the Slope-Intercept Form of the Equation

a. Use the scatter plot in **Figure 2.31(b)** to find a function in the form $T(x) = mx + b$ that models the average ticket price for a movie, $T(x)$, x years after 1990.

b. Use the model to predict the average ticket price in 2010.

Solution

a. We will use the line segment passing through the points $(0, 4.23)$ and $(15, 6.41)$ to obtain a model. We need values for m, the slope, and $(0, b)$, the y-intercept.

$$T(x) = mx + b$$

$m = \frac{\text{Change in } y}{\text{Change in } x}$

$m = \frac{6.41 - 4.23}{15 - 0} \approx 0.15$

The point $(0, 4.23)$ lies on the line segment, so the y-intercept is 4.23: $b = 4.23$.

The average ticket price for a movie, $T(x)$, x years after 1990 can be modeled by the linear function

$$T(x) = 0.15x + 4.23.$$

The slope, approximately 0.15, indicates an increase in ticket price of about \$0.15 per year from 1990 through 2005.

b. Now let's use this function to predict the average ticket price in 2010. Because 2010 is 20 years after 1990, substitute 20 for x in $T(x) = 0.15x + 4.23$ and evaluate the function at 20.

$$T(20) = 0.15(20) + 4.23 = 7.23$$

Our model predicts an average ticket price of \$7.23 in 2010.

✓ CHECK POINT 10 The table shows the median age of first marriage for U.S. women. (The median age is the age in the middle when all the ages of first-married women are arranged from youngest to oldest.) **Figure 2.32** shows a scatter plot based on the data, as well as a line that passes through or near the four points.

Women's Median Age of First Marriage

Year	Median Age
1990	23.9
1995	24.5
2000	25.1
2003	25.3

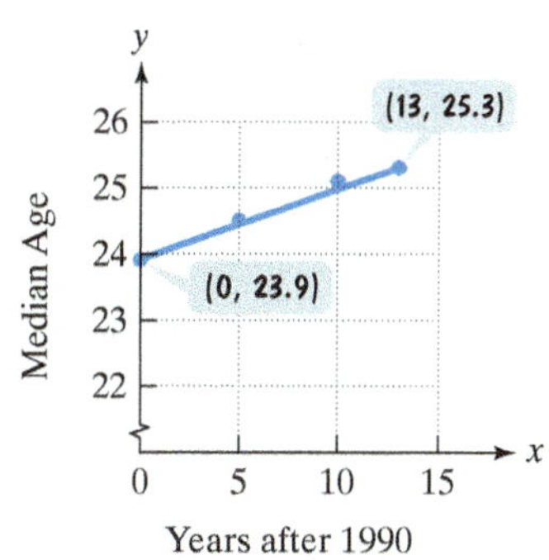

FIGURE 2.32

Source: U.S. Census Bureau

a. Use **Figure 2.32** on page 139 to find a function in the form $A(x) = mx + b$ that models the median age of first marriage for U.S. women, $A(x)$, x years after 1990.

b. Use the model to predict the median age of first marriage for U.S. women in 2030.

2.4 EXERCISE SET

MyMathLab | MathXL PRACTICE | WATCH | DOWNLOAD | READ | REVIEW

Practice Exercises

In Exercises 1–14, use intercepts and a checkpoint to graph each linear function.

1. $x + y = 4$
2. $x + y = 2$
3. $x + 3y = 6$
4. $2x + y = 4$
5. $6x - 2y = 12$
6. $6x - 9y = 18$
7. $3x - y = 6$
8. $x - 4y = 8$
9. $x - 3y = 9$
10. $2x - y = 5$
11. $2x = 3y + 6$
12. $3x = 5y - 15$
13. $6x - 3y = 15$
14. $8x - 2y = 12$

In Exercises 15–24, find the slope of the line passing through each pair of points or state that the slope is undefined. Then indicate whether the line through the points rises, falls, is horizontal, or is vertical.

15. $(2, 4)$ and $(3, 8)$
16. $(3, 1)$ and $(5, 4)$
17. $(-1, 4)$ and $(2, 5)$
18. $(-3, -2)$ and $(2, 5)$
19. $(2, 5)$ and $(-1, 5)$
20. $(-6, -3)$ and $(4, -3)$
21. $(-7, 1)$ and $(-4, -3)$
22. $(2, -1)$ and $(-6, 3)$
23. $\left(\frac{7}{2}, -2\right)$ and $\left(\frac{7}{2}, \frac{1}{4}\right)$
24. $\left(\frac{3}{2}, -6\right)$ and $\left(\frac{3}{2}, \frac{1}{6}\right)$

In Exercises 25–26, find the slope of each line.

25.

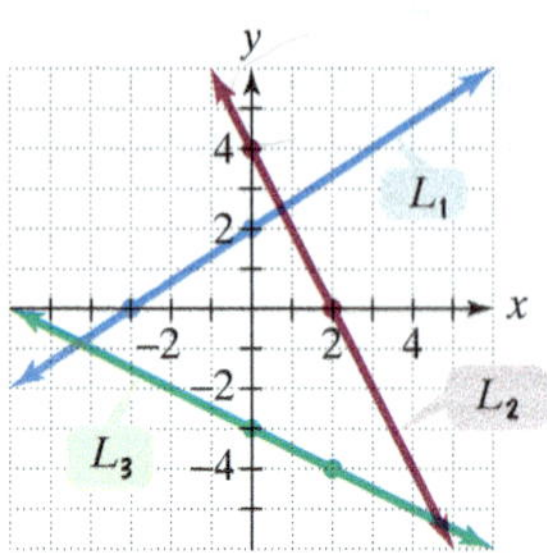

26.

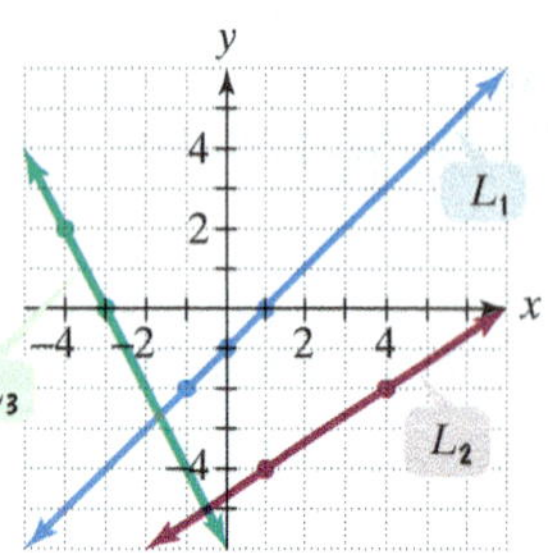

In Exercises 27–38, give the slope and y-intercept of each line whose equation is given. Then graph the linear function.

27. $y = 2x + 1$
28. $y = 3x + 2$
29. $y = -2x + 1$
30. $y = -3x + 2$
31. $f(x) = \frac{3}{4}x - 2$
32. $f(x) = \frac{3}{4}x - 3$
33. $f(x) = -\frac{3}{5}x + 7$
34. $f(x) = -\frac{2}{5}x + 6$
35. $y = -\frac{1}{2}x$
36. $y = -\frac{1}{3}x$
37. $y = -\frac{1}{2}$
38. $y = -\frac{1}{3}$

In Exercises 39–46,

a. *Rewrite the given equation in slope-intercept form by solving for y.*

b. *Give the slope and y-intercept.*

c. *Use the slope and y-intercept to graph the linear function.*

39. $2x + y = 0$
40. $3x + y = 0$
41. $5y = 4x$
42. $4y = 3x$
43. $3x + y = 2$
44. $2x + y = 4$
45. $5x + 3y = 15$
46. $7x + 2y = 14$

In Exercises 47–60, graph each equation in a rectangular coordinate system.

47. $y = 3$
48. $y = 5$
49. $f(x) = -2$
50. $f(x) = -4$
51. $3y = 18$
52. $5y = -30$
53. $f(x) = 2$
54. $f(x) = 1$
55. $x = 5$
56. $x = 4$
57. $3x = -12$
58. $4x = -12$
59. $x = 0$
60. $y = 0$

Practice PLUS

In Exercises 61–64, find the slope of the line passing through each pair of points or state that the slope is undefined. Assume that all variables represent positive real numbers. Then indicate whether the line through the points rises, falls, is horizontal, or is vertical.

61. $(0, a)$ and $(b, 0)$

62. $(-a, 0)$ and $(0, -b)$

63. (a, b) and $(a, b + c)$

64. $(a - b, c)$ and $(a, a + c)$

In Exercises 65–66, give the slope and y-intercept of each line whose equation is given. Assume that $B \neq 0$.

65. $Ax + By = C$

66. $Ax = By - C$

In Exercises 67–68, find the value of y if the line through the two given points is to have the indicated slope.

67. $(3, y)$ and $(1, 4)$, $m = -3$

68. $(-2, y)$ and $(4, -4)$, $m = \frac{1}{3}$

In Exercises 69–70, graph each linear function.

69. $3x - 4f(x) = 6$

70. $6x - 5f(x) = 20$

71. If one point on a line is $(3, -1)$ and the line's slope is -2, find the y-intercept.

72. If one point on a line is $(2, -6)$ and the line's slope is $-\frac{3}{2}$, find the y-intercept.

Use the figure to make the lists in Exercises 73–74.

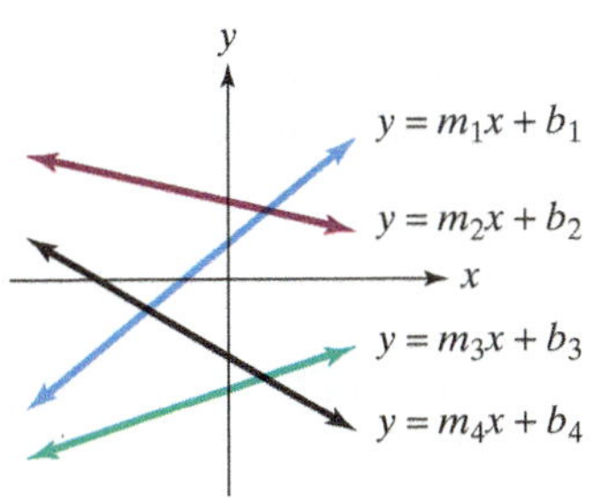

73. List the slopes m_1, m_2, m_3, and m_4 in order of decreasing size.

74. List the y-intercepts b_1, b_2, b_3, and b_4 in order of decreasing size.

Application Exercises

In Exercises 75–78, a linear function that models data is described. Find the slope of each model. Then describe what this means in terms of the rate of change of the dependent variable per unit change in the independent variable.

75. The linear function $f(x) = 0.01x + 57.7$ models the global average temperature of Earth, $f(x)$, in degrees Fahrenheit, x years after 1995.

76. The linear function $f(x) = 2x + 10$ models the amount, $f(x)$, in billions of dollars, that the drug industry spent on marketing information about drugs to doctors x years after 2000. (*Source:* IMS Health)

77. The linear function $f(x) = -0.52x + 24.7$ models the percentage of U.S. adults who smoked cigarettes, $f(x)$, x years after 1997. (*Source:* National Center for Health Statistics)

78. The linear function $f(x) = -0.28x + 1.7$ models the percentage of U.S. taxpayers who were audited by the IRS, $f(x)$, x years after 1996. (*Source:* IRS)

Divorce rates are typically higher for couples who marry in their teens. The graph shows the percentage of marriages ending in divorce by wife's age at marriage. Use the information shown to solve Exercises 79–80.

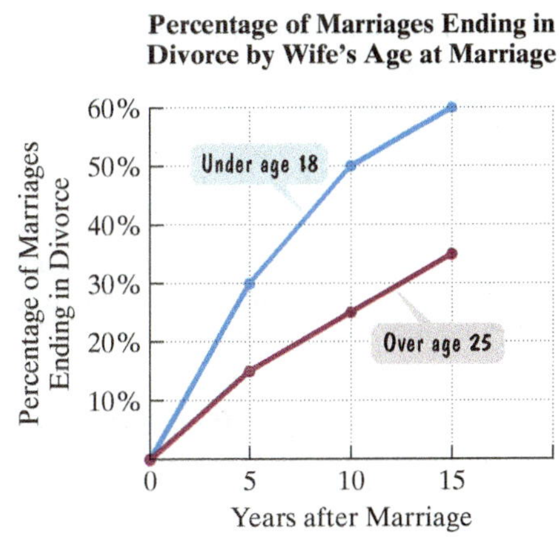

Source: B. E. Pruitt et al., *Human Sexuality*, Prentice Hall, 2007

79. **a.** What percentage of marriages in which the wife is under 18 when she marries end in divorce within the first five years?

b. What percentage of marriages in which the wife is under 18 when she marries end in divorce within the first ten years?

c. Find the average rate of change in the percentage of marriages ending in divorce between five and ten years of marriage in which the wife is under 18 when she marries.

80. **a.** What percentage of marriages in which the wife is over age 25 when she marries end in divorce within the first five years?

b. What percentage of marriages in which the wife is over age 25 when she marries end in divorce within the first ten years?

c. Find the average rate of change in the percentage of marriages ending in divorce between five and ten years of marriage in which the wife is over age 25 when she marries.

81. Shown, again, is the scatter plot that indicates a relationship between the percentage of adult females in a country who are literate and the mortality of children under five. Also shown is a line that passes through or near the points.

Literacy and Child Mortality

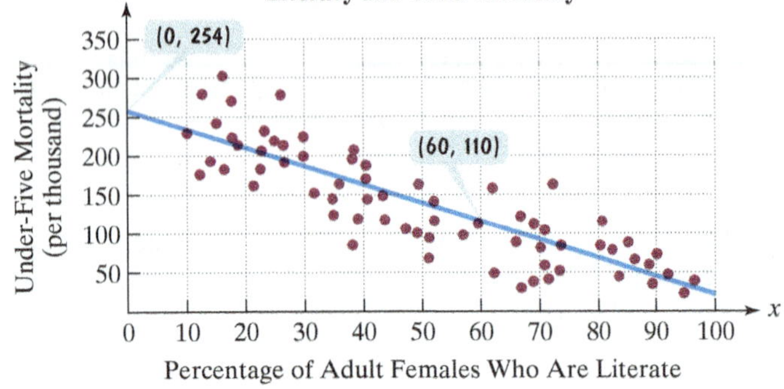

Source: United Nations

a. According to the graph, what is the y-intercept of the line? Describe what this represents in this situation.

b. Use the coordinates of the two points shown to compute the slope of the line. Describe what this means in terms of the rate of change.

c. Use the y-intercept from part (a) and the slope from part (b) to write a linear function that models child mortality, $f(x)$, per thousand, for children under five in a country where x% of adult women are literate.

d. Use the function from part (c) to predict the mortality rate of children under five in a country where 50% of adult females are literate.

82. The scatter plot shows the number of college students in the United States, in thousands, enrolled exclusively in online education from 2002 through 2007. Also shown is a line that passes through or near the six data points.

Number of United States College Students Enrolled Exclusively in Online Education

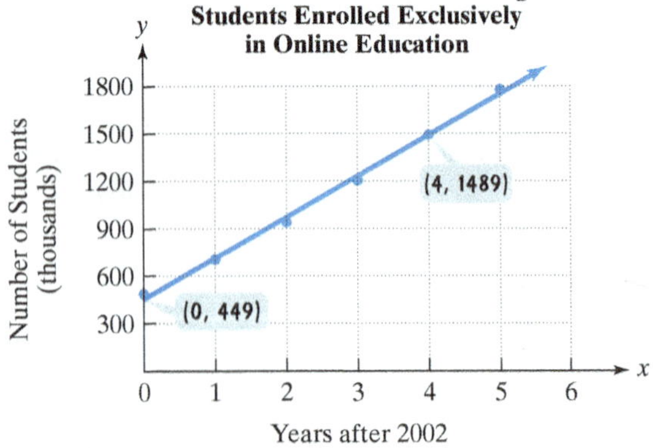

Source: U.S. Distance Learning Association

a. Use the coordinates of the two points shown to compute the slope of the line. Describe what this means in terms of the rate of change.

b. Use the y-intercept shown and the slope from part (a) to write a linear function that models the number of college students enrolled exclusively in online education, $L(x)$, in thousands, x years after 2002.

c. Use the function from part (b) to predict the number of college students who will be enrolled exclusively in online education in 2010.

The bar graph shows that as online news has grown, traditional news media have slipped.

Percentage of Americans Who Regularly Use Various News Media

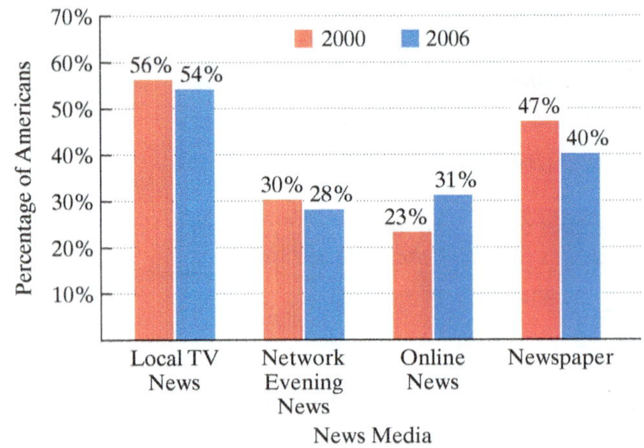

Source: Pew Research Center

In Exercises 83–84, find a linear function in slope-intercept form that models the given description. Each function should model the percentage of Americans, $P(x)$, who regularly used the news outlet x years after 2000.

83. In 2000, 47% of Americans regularly used newspapers for getting news and this has decreased at an average rate of approximately 1.2% per year since then.

84. In 2000, 23% of Americans regularly used online news for getting news and this has increased at an average rate of approximately 1.3% per year since then.

Writing in Mathematics

85. What is a scatter plot?

86. What is a regression line?

87. What is the standard form of the equation of a line?

88. What is an x-intercept of a graph?

89. What is a y-intercept of a graph?

90. If you are given the standard form of the equation of a line, explain how to find the x-intercept.

91. If you are given the standard form of the equation of a line, explain how to find the y-intercept.

92. What is the slope of a line?

93. Describe how to calculate the slope of a line passing through two points.

94. What does it mean if the slope of a line is zero?

95. What does it mean if the slope of a line is undefined?

96. Describe how to find the slope of a line whose equation is given.

97. Describe how to graph a line using the slope and y-intercept. Provide an original example with your description.

98. Describe the graph of $y = b$.

99. Describe the graph of $x = a$.

100. If the graph of a function is not a straight line, explain how to find the average rate of change between two points.

101. Take another look at the scatter plot in Exercise 81. Although there is a relationship between literacy and child mortality, we cannot conclude that increased literacy causes child mortality to decrease. Offer two or more possible explanations for the data in the scatter plot.

Technology Exercises

102. Use a graphing utility to verify any three of your hand-drawn graphs in Exercises 1–14. Solve the equation for y before entering it.

In Exercises 103–106, use a graphing utility to graph each linear function. Then use the TRACE *feature to trace along the line and find the coordinates of two points. Use these points to compute the line's slope. Check your result by using the coefficient of x in the line's equation.*

103. $y = 2x + 4$

104. $y = -3x + 6$

105. $f(x) = -\frac{1}{2}x - 5$

106. $f(x) = \frac{3}{4}x - 2$

Critical Thinking Exercises

Make Sense? *In Exercises 107–110, determine whether each statement "makes sense" or "does not make sense" and explain your reasoning.*

107. The graph of my linear function at first rose from left to right, reached a maximum point, and then fell from left to right.

108. A linear function that models tuition and fees at public four-year colleges from 2000 through 2006 has negative slope.

109. The function $S(x) = 49{,}100x + 1700$ models the average salary for a college professor, $S(x)$, x years after 2000.

110. The federal minimum wage was $5.15 per hour from 1997 through 2006, so $f(x) = 5.15$ models the minimum wage, $f(x)$, in dollars, for the domain $\{1997, 1998, 1999, \ldots, 2006\}$.

In Exercises 111–114, determine whether each statement is true or false. If the statement is false, make the necessary change(s) to produce a true statement.

111. A linear function with nonnegative slope has a graph that rises from left to right.

112. Every line in the rectangular coordinate system has an equation that can be expressed in slope-intercept form.

113. The graph of the linear function $5x + 6y = 30$ is a line passing through the point $(6, 0)$ with slope $-\frac{5}{6}$.

114. The graph of $x = 7$ in the rectangular coordinate system is the single point $(7, 0)$.

In Exercises 115–116, find the coefficients that must be placed in each shaded area so that the function's graph will be a line satisfying the specified conditions.

115. $\square x + \square y = 12$; x-intercept $= -2$; y-intercept $= 4$

116. $\square x + \square y = 12$; y-intercept $= -6$; slope $= \frac{1}{2}$

117. For the linear function

$$f(x) = mx + b,$$

a. Find $f(x_1 + x_2)$.

b. Find $f(x_1) + f(x_2)$.

c. Is $f(x_1 + x_2) = f(x_1) + f(x_2)$?

Review Exercises

118. Simplify: $\left(\frac{4x^2}{y^{-3}}\right)^2$. (Section 1.6, Example 9)

119. Multiply and write the answer in scientific notation:

$$(8 \times 10^{-7})(4 \times 10^3).$$

(Section 1.7, Example 4)

120. Simplify: $5 - [3(x - 4) - 6x]$. (Section 1.2, Example 14)

Preview Exercises

Exercises 121–123 will help you prepare for the material covered in the next section.

121. Write the equation $y - 5 = 7(x + 4)$ in slope-intercept form.

122. Write the equation $y + 3 = -\frac{7}{3}(x - 1)$ in slope-intercept form.

123. The equation of a line is $x + 4y - 8 = 0$.

a. Write the equation in slope-intercept form and determine the slope.

b. The product of the line's slope in part (a) and the slope of a second line is -1. What is the slope of the second line?

SECTION 2.5

The Point-Slope Form of the Equation of a Line

Objectives

1. Use the point-slope form to write equations of a line.
2. Model data with linear functions and make predictions.
3. Find slopes and equations of parallel and perpendicular lines.

If present trends continue, is it possible that our descendants could live to be 200 years of age? To answer this question, we need to develop a function that models life expectancy by birth year. In this section, you will learn to use another form of a line's equation to obtain functions that model data.

Point-Slope Form

We can use the slope of a line to obtain another useful form of the line's equation. Consider a nonvertical line that has slope m and contains the point (x_1, y_1). Now, let (x, y) represent any other point on the line, shown in **Figure 2.33**. Keep in mind that the point (x, y) is arbitrary and is not in one fixed position. By contrast, the point (x_1, y_1) is fixed.

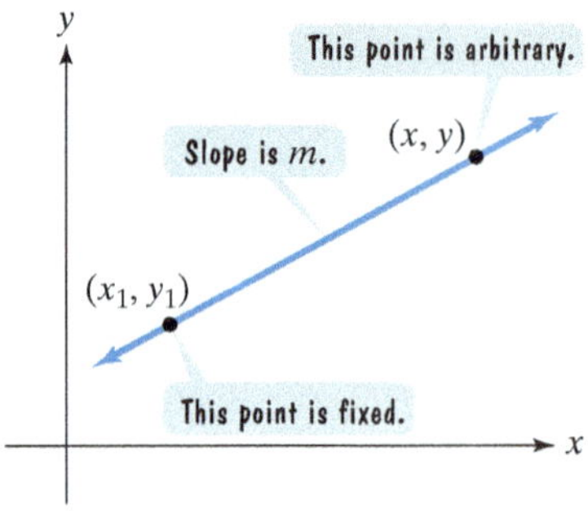

FIGURE 2.33 A line passing through (x_1, y_1) with slope m

Regardless of where the point (x, y) is located, the steepness of the line in **Figure 2.33** remains the same. Thus, the ratio for slope stays a constant m. This means that for all points (x, y) along the line

$$m = \frac{\text{Change in } y}{\text{Change in } x} = \frac{y - y_1}{x - x_1}.$$

We can clear the fraction by multiplying both sides by $x - x_1$, the least common denominator, where $x - x_1 \neq 0$.

$$m = \frac{y - y_1}{x - x_1} \quad \text{This is the slope of the line in Figure 2.33.}$$

$$m(x - x_1) = \frac{y - y_1}{x - x_1} \cdot (x - x_1) \quad \text{Multiply both sides by } x - x_1.$$

$$m(x - x_1) = y - y_1 \quad \text{Simplify: } \frac{y - y_1}{\cancel{x - x_1}} \cdot (\cancel{x - x_1}) = y - y_1.$$

Now, if we reverse the two sides, we obtain the **point-slope form** of the equation of a line.

Study Tip

When writing the point-slope form of a line's equation, you will never substitute numbers for x and y. You will substitute values for x_1, y_1, and m.

Point-Slope Form of the Equation of a Line

The **point-slope form of the equation** of a nonvertical line with slope m that passes through the point (x_1, y_1) is

$$y - y_1 = m(x - x_1).$$

For example, the point-slope form of the equation of the line passing through $(1, 5)$ with slope 2 ($m = 2$) is

$$y - 5 = 2(x - 1).$$

1 Use the point-slope form to write equations of a line.

Using the Point-Slope Form to Write a Line's Equation

If we know the slope of a line and a point not containing the y-intercept through which the line passes, the point-slope form is the equation that we should use. Once we have obtained this equation, it is customary to solve for y and write the equation in slope-intercept form. Examples 1 and 2 illustrate these ideas.

EXAMPLE 1 Writing the Point-Slope Form and the Slope-Intercept Form

Write the point-slope form and the slope-intercept form of the equation of the line with slope 7 that passes through the point $(-4, 5)$.

Solution We begin with the point-slope form of the equation of a line with $m = 7$, $x_1 = -4$, and $y_1 = 5$.

$$y - y_1 = m(x - x_1) \quad \text{This is the point-slope form of the equation.}$$

$$y - 5 = 7[x - (-4)] \quad \text{Substitute the given values.}$$

$$y - 5 = 7(x + 4) \quad \text{We now have the point-slope form of the equation of the given line.}$$

Now we solve this equation for y and write an equivalent equation in slope-intercept form $(y = mx + b)$.

We need to isolate y.

$$y - 5 = 7(x + 4) \quad \text{This is the point-slope form of the equation.}$$

$$y - 5 = 7x + 28 \quad \text{Use the distributive property.}$$

$$y = 7x + 33 \quad \text{Add 5 to both sides.}$$

The slope-intercept form of the line's equation is $y = 7x + 33$. Using function notation, the equation is $f(x) = 7x + 33$.

✓ **CHECK POINT 1** Write the point-slope form and the slope-intercept form of the equation of the line with slope -2 that passes through the point $(4, -3)$.

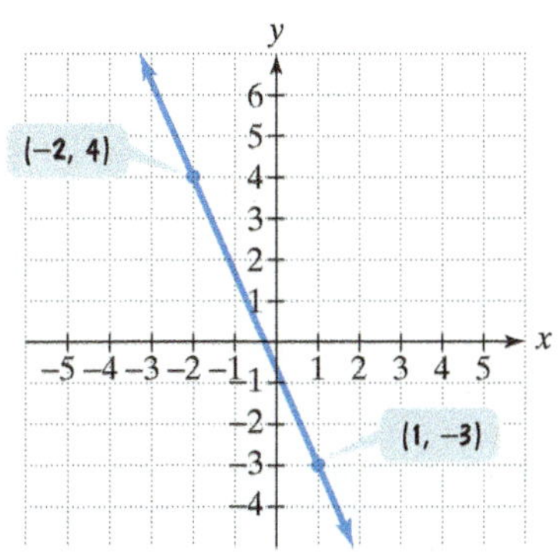

FIGURE 2.34 Line passing through $(1, -3)$ and $(-2, 4)$

EXAMPLE 2 Writing the Point-Slope Form and the Slope-Intercept Form

A line passes through the points $(1, -3)$ and $(-2, 4)$. (See **Figure 2.34**.) Find an equation of the line

a. in point-slope form. **b.** in slope-intercept form.

Solution

a. To use the point-slope form, we need to find the slope. The slope is the change in the y-coordinates divided by the corresponding change in the x-coordinates.

$$m = \frac{4 - (-3)}{-2 - 1} = \frac{7}{-3} = -\frac{7}{3} \quad \text{This is the definition of slope using } (1, -3) \text{ and } (-2, 4).$$

We can take either point on the line to be (x_1, y_1). Let's use $(x_1, y_1) = (1, -3)$. Now, we are ready to write the point-slope form of the equation.

$$y - y_1 = m(x - x_1) \quad \text{This is the point-slope form of the equation.}$$

$$y - (-3) = -\frac{7}{3}(x - 1) \quad \text{Substitute: } (x_1, y_1) = (1, -3) \text{ and } m = -\frac{7}{3}.$$

$$y + 3 = -\frac{7}{3}(x - 1) \quad \text{Simplify.}$$

This equation is the point-slope form of the equation of the line shown in **Figure 2.34**.

b. Now, we solve this equation for y and write an equivalent equation in slope-intercept form ($y = mx + b$).

We need to isolate y.

$$y + 3 = -\frac{7}{3}(x - 1)$$ This is the point-slope form of the equation.

$$y + 3 = -\frac{7}{3}x + \frac{7}{3}$$ Use the distributive property.

$$y = -\frac{7}{3}x - \frac{2}{3}$$ Subtract 3 from both sides: $\frac{7}{3} - 3 = \frac{7}{3} - \frac{9}{3} = -\frac{2}{3}$.

This equation is the slope-intercept form of the equation of the line shown in **Figure 2.34** on the previous page. Using function notation, the equation is $f(x) = -\frac{7}{3}x - \frac{2}{3}$.

Discover for Yourself

If you are given two points on a line, you can use either point for (x_1, y_1) when you write the point-slope form of its equation. Rework Example 2 using $(-2, 4)$ for (x_1, y_1). Once you solve for y, you should obtain the same slope-intercept form of the equation as the one shown in the last line of the solution to Example 2.

✓ **CHECK POINT 2** A line passes through the points $(6, -3)$ and $(2, 5)$. Find an equation of the line

a. in point-slope form.

b. in slope-intercept form.

Here is a summary of the various forms for equations of lines:

Equations of Lines

1. Standard form: $Ax + By = C$
2. Slope-intercept form: $y = mx + b$ or $f(x) = mx + b$
3. Horizontal line: $y = b$
4. Vertical line: $x = a$
5. Point-slope form: $y - y_1 = m(x - x_1)$

In Examples 1 and 2, we eventually wrote a line's equation in slope-intercept form, or in function notation. But where do we start our work?

Starting with $y = mx + b$	Starting with $y - y_1 = m(x - x_1)$
Begin with the slope-intercept form if you know • The slope of the line and the y-intercept. or • Two points on the line, one of which contains the y-intercept.	Begin with the point-slope form if you know • The slope of the line and a point on the line not containing the y-intercept or • Two points on the line, neither of which contains the y-intercept.

2 Model data with linear functions and make predictions.

Applications

We have seen that linear functions are useful for modeling data that fall on or near a line. For example, the bar graph in **Figure 2.35(a)** gives the life expectancy for American men and women born in the indicated year. The data for the men are displayed as a set of six points in the scatter plot in **Figure 2.35(b)**.

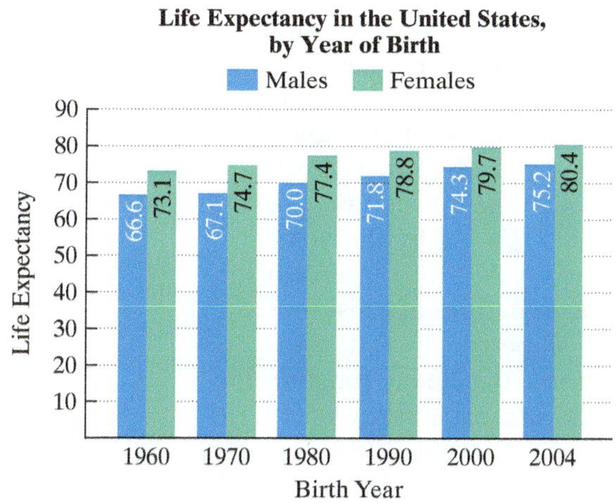

FIGURE 2.35(a) Data for men and women

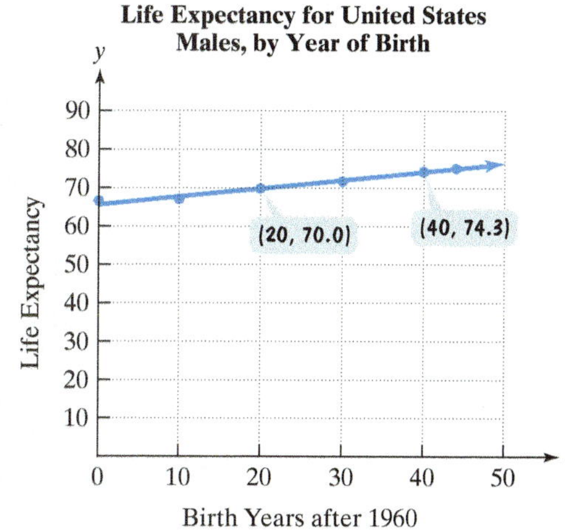

FIGURE 2.35(b) A scatter plot for the men's data

Source: National Center for Health Statistics

Also shown on the scatter plot in **Figure 2.35(b)** is a line that passes through or near the six points. By writing the equation of this line, we can obtain a model for life expectancy and make predictions about how long American men will live in the future.

EXAMPLE 3 Modeling Life Expectancy

Write the slope-intercept form of the equation of the line shown in **Figure 2.35(b)**. Use the equation to predict the life expectancy of an American man born in 2020.

Using Technology

You can use a graphing utility to obtain a model for a scatter plot in which the data points fall on or near a straight line. After entering the data in **Figure 2.35(b)**, a graphing utility displays a scatter plot of the data and the regression line, that is, the line that best fits the data.

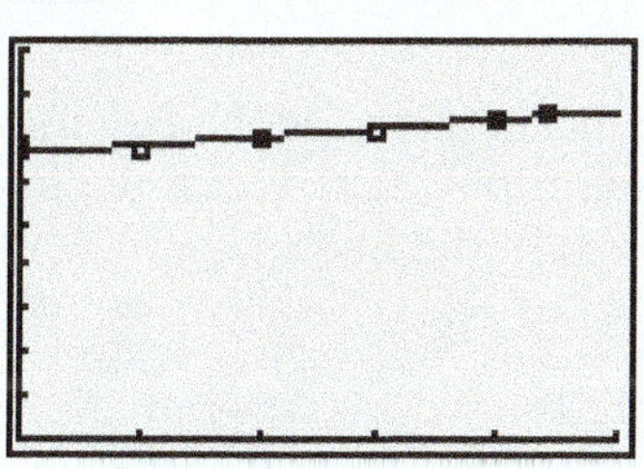

[0, 50, 10] by [0, 90, 10]

Also displayed is the regression line's equation.

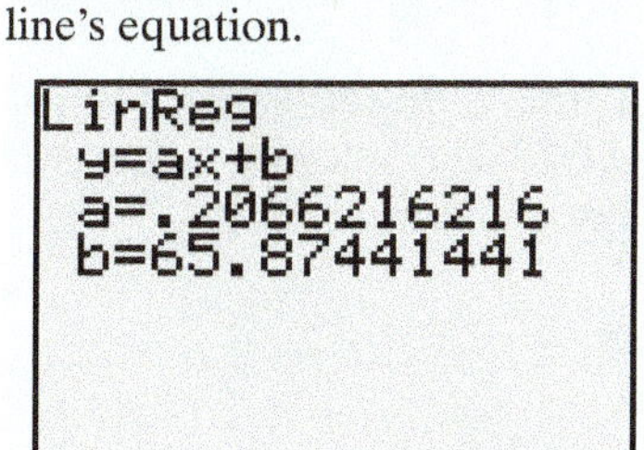

Solution The line in **Figure 2.35(b)** passes through (20, 70.0) and (40, 74.3). We start by finding its slope.

$$m = \frac{\text{Change in } y}{\text{Change in } x} = \frac{74.3 - 70.0}{40 - 20} = \frac{4.3}{20} = 0.215$$

The slope indicates that for each subsequent birth year, a man's life expectancy is increasing by 0.215 years.

Now we write the line's equation in slope-intercept form.

$$y - y_1 = m(x - x_1)$$ Begin with the point-slope form.

$$y - 70.0 = 0.215(x - 20)$$ Either ordered pair can be (x_1, y_1). Let $(x_1, y_1) = (20, 70.0)$. From above, $m = 0.215$.

$$y - 70.0 = 0.215x - 4.3$$ Apply the distributive property: $0.215(20) = 4.3$.

$$y = 0.215x + 65.7$$ Add 70 to both sides and solve for y.

A linear function that models life expectancy, $f(x)$, for American men born x years after 1960 is

$$f(x) = 0.215x + 65.7.$$

Now let's use this function to predict the life expectancy of an American man born in 2020. Because 2020 is 60 years after 1960, substitute 60 for x and evaluate the function at 60.

$$f(60) = 0.215(60) + 65.7 = 78.6$$

Our model predicts that American men born in 2020 will have a life expectancy of 78.6 years.

✓ **CHECK POINT 3** The data for the life expectancy for American women are displayed as a set of six points in the scatter plot in **Figure 2.36**. Also shown is a line that passes through or near the six points. Use the data points labeled by the voice balloons to write the slope-intercept form of the equation of this line. Round the slope to two decimal places. Then use the linear function to predict the life expectancy of an American woman born in 2020.

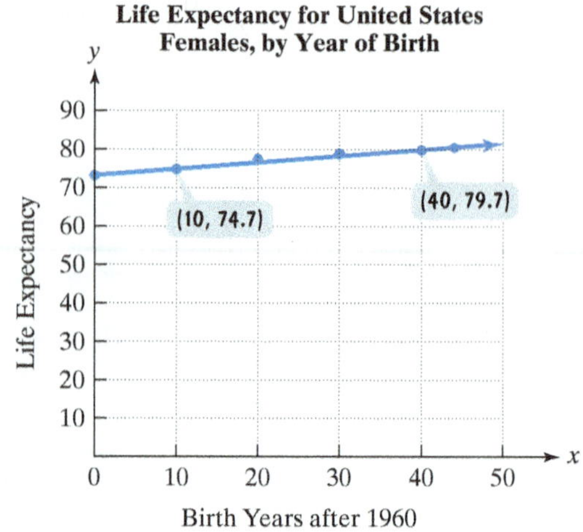

FIGURE 2.36

Parallel and Perpendicular Lines

3 Find slopes and equations of parallel and perpendicular lines.

Two nonintersecting lines that lie in the same plane are **parallel**. If two lines do not intersect, the ratio of the vertical change to the horizontal change is the same for each line. Because two parallel lines have the same "steepness," they must have the same slope.

Slope and Parallel Lines

1. If two nonvertical lines are parallel, then they have the same slope.
2. If two distinct nonvertical lines have the same slope, then they are parallel.
3. Two distinct vertical lines, both with undefined slopes, are parallel.

EXAMPLE 4 Writing Equations of a Line Parallel to a Given Line

Write an equation of the line passing through $(-3, 1)$ and parallel to the line whose equation is $y = 2x + 1$. Express the equation in point-slope form and slope-intercept form.

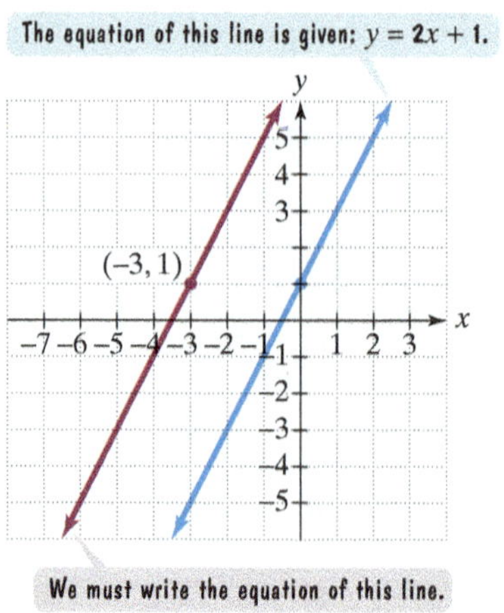

FIGURE 2.37

Solution The situation is illustrated in **Figure 2.37**. We are looking for the equation of the red line shown on the left. How do we obtain this equation? Notice that the line passes through the point $(-3, 1)$. Using the point-slope form of the line's equation, we have $x_1 = -3$ and $y_1 = 1$.

$$y - y_1 = m(x - x_1)$$

$y_1 = 1$ $\quad$ $x_1 = -3$

Now the only thing missing from the equation of the red line is m, the slope. Do we know anything about the slope of either line in **Figure 2.37**? The answer is yes; we know the slope of the blue line on the right, whose equation is given.

$$y = 2x + 1$$

The slope of the blue line on the right in Figure 2.37 is 2.

Parallel lines have the same slope. Because the slope of the blue line is 2, the slope of the red line, the line whose equation we must write, is also 2: $m = 2$. We now have values for x_1, y_1, and m for the red line.

$$y - y_1 = m(x - x_1)$$

$y_1 = 1$ $m = 2$ $x_1 = -3$

The point-slope form of the red line's equation is

$$y - 1 = 2[x - (-3)] \text{ or}$$
$$y - 1 = 2(x + 3).$$

Solving for y, we obtain the slope-intercept form of the equation.

$$y - 1 = 2x + 6$$ Apply the distributive property.

$$y = 2x + 7$$ Add 1 to both sides. This is the slope-intercept form, $y = mx + b$, of the equation. Using function notation, the equation is $f(x) = 2x + 7$.

CHECK POINT 4 Write an equation of the line passing through $(-2, 5)$ and parallel to the line whose equation is $y = 3x + 1$. Express the equation in point-slope form and slope-intercept form.

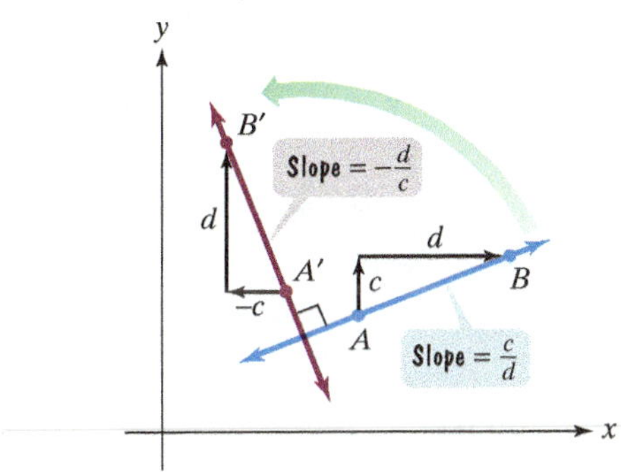

FIGURE 2.38
Slopes of perpendicular lines

Two lines that intersect at a right angle (90°) are said to be **perpendicular**, shown in **Figure 2.38**. The relationship between the slopes of perpendicular lines is not as obvious as the relationship between parallel lines. **Figure 2.38** shows line AB, with slope $\frac{c}{d}$. Rotate line AB through 90° counterclockwise to obtain line $A'B'$, perpendicular to line AB. The figure indicates that the rise and the run of the new line are reversed from the original line, but the run is now negative. This means that the slope of the new line is $-\frac{d}{c}$. Notice that the product of the slopes of the two perpendicular lines is -1:

$$\left(\frac{c}{d}\right)\left(-\frac{d}{c}\right) = -1.$$

This relationship holds for all nonvertical perpendicular lines and is summarized in the following box:

Slope and Perpendicular Lines

1. If two nonvertical lines are perpendicular, then the product of their slopes is -1.
2. If the product of the slopes of two lines is -1, then the lines are perpendicular.
3. A horizontal line having zero slope is perpendicular to a vertical line having undefined slope.

An equivalent way of stating this relationship is to say that **one line is perpendicular to another line if its slope is the *negative reciprocal* of the slope of the other line**. For example, if a line has slope 5, any line having slope $-\frac{1}{5}$ is perpendicular to it. Similarly, if a line has slope $-\frac{3}{4}$, any line having slope $\frac{4}{3}$ is perpendicular to it.

EXAMPLE 5 Writing Equations of a Line Perpendicular to a Given Line

a. Find the slope of any line that is perpendicular to the line whose equation is $x + 4y = 8$.

b. Write the equation of the line passing through $(3, -5)$ and perpendicular to the line whose equation is $x + 4y = 8$. Express the equation in point-slope form and slope-intercept form.

Solution

a. We begin by writing the equation of the given line, $x + 4y = 8$, in slope-intercept form. Solve for y.

$$x + 4y = 8 \quad \text{This is the given equation.}$$
$$4y = -x + 8 \quad \text{To isolate the y-term, subtract x from both sides.}$$
$$y = -\frac{1}{4}x + 2 \quad \text{Divide both sides by 4.}$$

Slope is $-\frac{1}{4}$.

The given line has slope $-\frac{1}{4}$. Any line perpendicular to this line has a slope that is the negative reciprocal of $-\frac{1}{4}$. Thus, the slope of any perpendicular line is 4.

b. Let's begin by writing the point-slope form of the perpendicular line's equation. Because the line passes through the point $(3, -5)$, we have $x_1 = 3$ and $y_1 = -5$. In part (a), we determined that the slope of any line perpendicular to $x + 4y = 8$ is 4, so the slope of this particular perpendicular line must also be 4: $m = 4$.

$$y - y_1 = m(x - x_1)$$

$y_1 = -5$ $m = 4$ $x_1 = 3$

The point-slope form of the perpendicular line's equation is

$$y - (-5) = 4(x - 3) \text{ or}$$
$$y + 5 = 4(x - 3).$$

How can we express this equation in slope-intercept form, $y = mx + b$? We need to solve for y.

$$y + 5 = 4(x - 3) \quad \text{This is the point-slope form of the line's equation.}$$
$$y + 5 = 4x - 12 \quad \text{Apply the distributive property.}$$
$$y = 4x - 17 \quad \text{Subtract 5 from both sides of the equation and solve for y.}$$

The point-slope form of the perpendicular line's equation is

$$y = 4x - 17 \quad \text{or} \quad f(x) = 4x - 17.$$

✓ CHECK POINT 5

a. Find the slope of any line that is perpendicular to the line whose equation is $x + 3y = 12$.

b. Write the equation of the line passing through $(-2, -6)$ and perpendicular to the line whose equation is $x + 3y = 12$. Express the equation in point-slope form and slope-intercept form.

2.5 EXERCISE SET

MyMathLab MathXL PRACTICE WATCH DOWNLOAD READ REVIEW

Practice Exercises

Write the point-slope form of the line's equation satisfying each of the conditions in Exercises 1–28. Then use the point-slope form of the equation to write the slope-intercept form of the equation in function notation.

1. Slope $= 3$, passing through $(2, 5)$
2. Slope $= 4$, passing through $(3, 1)$
3. Slope $= 5$, passing through $(-2, 6)$
4. Slope $= 8$, passing through $(-4, 1)$
5. Slope $= -4$, passing through $(-3, -2)$
6. Slope $= -6$, passing through $(-2, -4)$
7. Slope $= -5$, passing through $(-2, 0)$

8. Slope $= -4$, passing through $(0, -3)$

9. Slope $= -1$, passing through $\left(-2, -\frac{1}{2}\right)$

10. Slope $= -1$, passing through $\left(-\frac{1}{4}, -4\right)$

11. Slope $= \frac{1}{4}$, passing through the origin

12. Slope $= \frac{1}{5}$, passing through the origin

13. Slope $= -\frac{2}{3}$, passing through $(6, -4)$

14. Slope $= -\frac{2}{5}$, passing through $(15, -4)$

15. Passing through $(6, 3)$ and $(5, 2)$

16. Passing through $(1, 3)$ and $(2, 4)$

17. Passing through $(-2, 0)$ and $(0, 4)$

18. Passing through $(2, 0)$ and $(0, -1)$

19. Passing through $(-6, 13)$ and $(-2, 5)$

20. Passing through $(-3, 2)$ and $(2, -8)$

21. Passing through $(1, 9)$ and $(4, -2)$

22. Passing through $(4, -8)$ and $(8, -3)$

23. Passing through $(-2, -5)$ and $(3, -5)$

24. Passing through $(-1, -4)$ and $(3, -4)$

25. Passing through $(7, 8)$ with x-intercept is $(3, 0)$

26. Passing through $(-4, 5)$ with y-intercept is $(0, -3)$

27. x-intercept is $(2, 0)$ and y-intercept is $(0, -1)$

28. x-intercept is $(-2, 0)$ and y-intercept is $(0, 4)$

In Exercises 29–44, the equation of a line is given. Find the slope of a line that is **a.** *parallel to the line with the given equation; and* **b.** *perpendicular to the line with the given equation.*

29. $y = 5x$

30. $y = 3x$

31. $y = -7x$

32. $y = -9x$

33. $y = \dfrac{1}{2}x + 3$

34. $y = \dfrac{1}{4}x - 5$

35. $y = -\dfrac{2}{5}x - 1$

36. $y = -\dfrac{3}{7}x - 2$

37. $4x + y = 7$

38. $8x + y = 11$

39. $2x + 4y = 8$

40. $3x + 2y = 6$

41. $2x - 3y = 5$

42. $3x - 4y = -7$

43. $x = 6$

44. $y = 9$

In Exercises 45–48, write an equation for line L in point-slope form and slope-intercept form.

45.

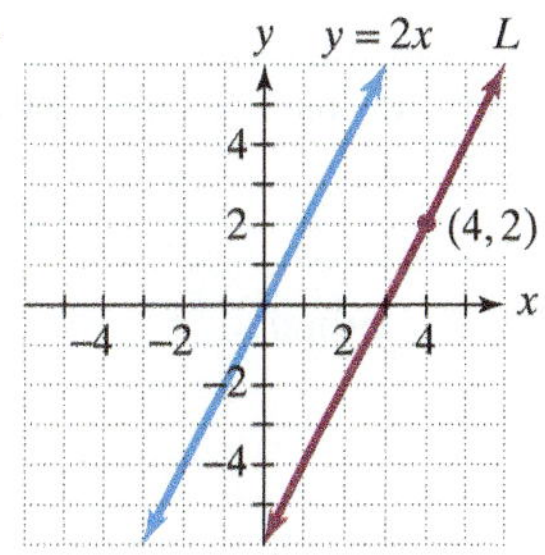

L is parallel to $y = 2x$.

46.

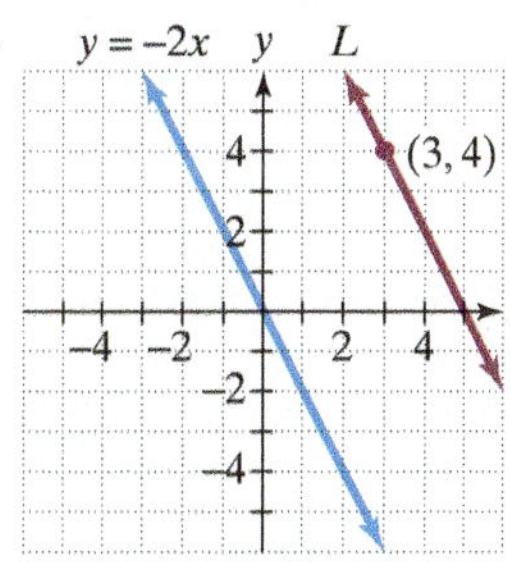

L is parallel to $y = -2x$.

47.

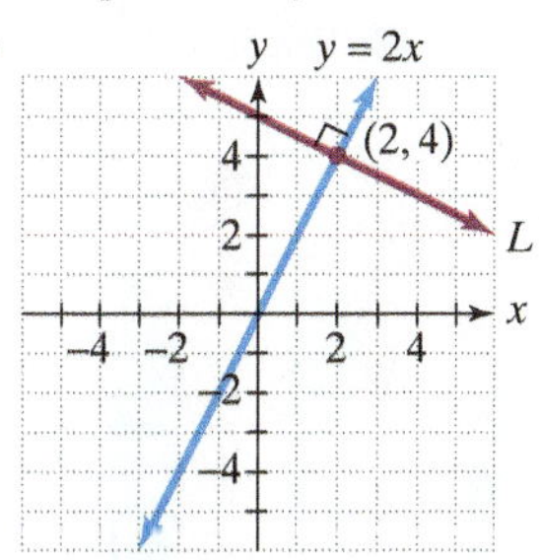

L is perpendicular to $y = 2x$.

48.

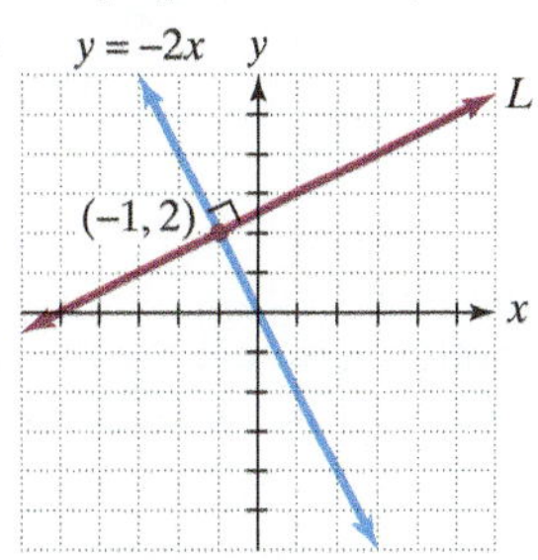

L is perpendicular to $y = -2x$.

In Exercises 49–56, use the given conditions to write an equation for each line in point-slope form and slope-intercept form.

49. Passing through $(-8, -10)$ and parallel to the line whose equation is $y = -4x + 3$

50. Passing through $(-2, -7)$ and parallel to the line whose equation is $y = -5x + 4$

51. Passing through $(2, -3)$ and perpendicular to the line whose equation is $y = \frac{1}{5}x + 6$

52. Passing through $(-4, 2)$ and perpendicular to the line whose equation is $y = \frac{1}{3}x + 7$

53. Passing through $(-2, 2)$ and parallel to the line whose equation is $2x - 3y = 7$

54. Passing through $(-1, 3)$ and parallel to the line whose equation is $3x - 2y = 5$

55. Passing through $(4, -7)$ and perpendicular to the line whose equation is $x - 2y = 3$

56. Passing through $(5, -9)$ and perpendicular to the line whose equation is $x + 7y = 12$

Practice PLUS

In Exercises 57–64, write the slope-intercept form of the equation of a function f whose graph satisfies the given conditions.

57. The graph of f passes through $(-1, 5)$ and is perpendicular to the line whose equation is $x = 6$.

58. The graph of f passes through $(-2, 6)$ and is perpendicular to the line whose equation is $x = -4$.

59. The graph of f passes through $(-6, 4)$ and is perpendicular to the line that has x-intercept $(2, 0)$ and y-intercept $(0, -4)$.

60. The graph of f passes through $(-5, 6)$ and is perpendicular to the line that has x-intercept $(3, 0)$ and y-intercept $(0, -9)$.

61. The graph of f is perpendicular to the line whose equation is $3x - 2y = 4$ and has the same y-intercept as this line.

62. The graph of f is perpendicular to the line whose equation is $4x - y = 6$ and has the same y-intercept as this line.

63. The graph of f is the graph of $g(x) = 4x - 3$ shifted down 2 units.

64. The graph of f is the graph of $g(x) = 2x - 5$ shifted up 3 units.

65. What is the slope of a line that is parallel to the line whose equation is $Ax + By = C, B \neq 0$?

66. What is the slope of a line that is perpendicular to the line whose equation is $Ax + By = C, A \neq 0$ and $B \neq 0$?

Application Exercises

Americans are getting married later in life, or not getting married at all. In 2006, nearly half of Americans ages 25 through 29 were unmarried. The bar graph shows the percentage of never-married men and women in this age group. The data are displayed as two sets of four points each, one scatter plot for the percentage of never-married American men and one for the percentage of never-married American women. Also shown for each scatter plot is a line that passes through or near the four points. Use these lines to solve Exercises 67–68.

Percentage of United States Population Never Married, Ages 25–29

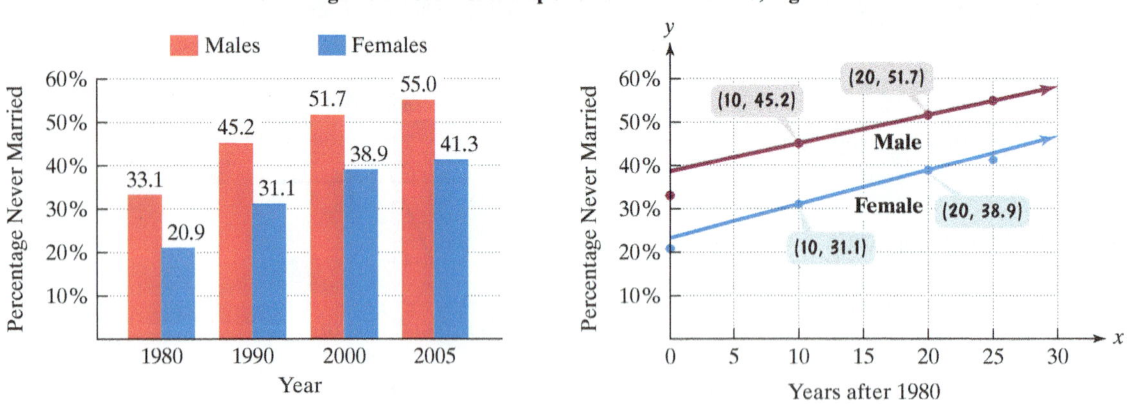

Source: U.S. Census Bureau

67. In this exercise, you will use the blue line for the women shown on the scatter plot to develop a model for the percentage of never-married American females ages 25–29.

a. Use the two points whose coordinates are shown by the voice balloons to find the point-slope form of the equation of the line that models the percentage of never-married American females ages 25–29, y, x years after 1980.

b. Write the equation from part (a) in slope-intercept form. Use function notation.

c. Use the linear function to predict the percentage of never-married American females, ages 25–29, in 2020.

68. In this exercise, you will use the red line for the men shown on the scatter plot on page 152 to develop a model for the percentage of never-married American males ages 25–29.

a. Use the two points whose coordinates are shown by the voice balloons to find the point-slope form of the equation of the line that models the percentage of never-married American males ages 25–29, y, x years after 1980.

b. Write the equation from part (a) in slope-intercept form. Use function notation.

c. Use the linear function to predict the percentage of never-married American males, ages 25–29, in 2015.

The amount of carbon dioxide in the atmosphere, measured in parts per million, has been increasing as a result of the burning of oil and coal. The buildup of gases and particles traps heat and raises the planet's temperature, a phenomenon called the greenhouse effect. In Exercises 69–70, you will develop linear models involving variables related to global warming.

69. The bar graph shows the average global temperature, in degrees Fahrenheit, for seven selected years.

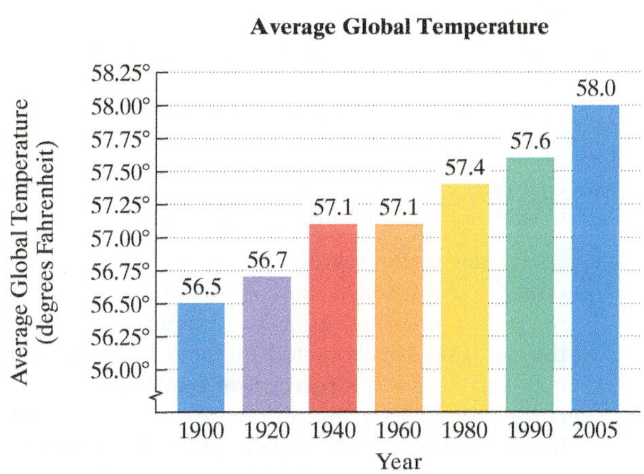

Source: National Oceanic and Atmospheric Administration

a. Let x represent the number of years after 1900 and let y represent the average global temperature. Create a scatter plot that displays the data as a set of seven points in a rectangular coordinate system.

b. Draw a line through the two points that show the average global temperatures for 1940 and 1990. Use the coordinates of these points to write the line's equation in point-slope form and slope-intercept form.

c. Use the slope-intercept form of the equation from part (b) to predict the average global temperature in 2050.

70. The pre-industrial concentration of atmospheric carbon dioxide was 280 parts per million. The bar graph at the top of the next column shows the average atmospheric concentration of carbon dioxide, in parts per million, for seven selected years.

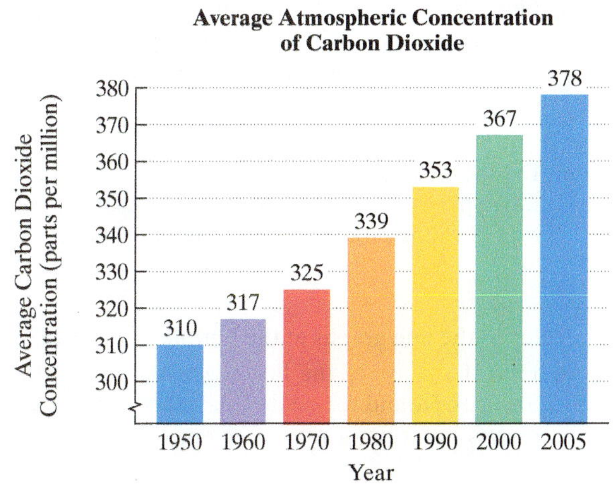

Source: National Oceanic and Atmospheric Administration

a. Let x represent the number of years after 1950 and let y represent the average atmospheric concentration of carbon dioxide. Create a scatter plot that displays the data as a set of seven points in a rectangular coordinate system.

b. Draw a line through the two points that show the average atmospheric concentration of carbon dioxide for 1960 and 2000. Use the coordinates of these points to write the line's equation in point-slope form and slope-intercept form.

c. Use the slope-intercept form of the equation from part (b) to predict the average atmospheric concentration of carbon dioxide in 2050.

In 2007, the U.S. government faced the prospect of paying out more and more in Social Security, Medicare, and Medicaid benefits. The line graphs show the projected costs of these entitlement programs, in billions of dollars, from 2007 through 2016. Use this information to solve Exercises 71–72.

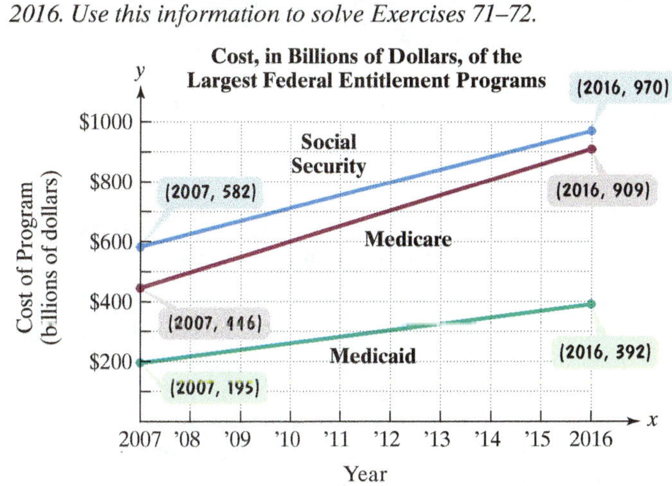

Source: Congressional Budget Office

71. a. Find the slope of the line segment representing Social Security. Round to one decimal place. Describe what this means in terms of rate of change.

b. Find the slope of the line segment representing Medicare. Round to one decimal place. Describe what this means in terms of rate of change.

c. Do the line segments for Social Security and Medicare lie on parallel lines? What does this mean in terms of the rate of change for these entitlement programs?

72. Refer to the line graphs at the bottom of page 153.

a. Find the slope of the line segment representing Social Security. Round to one decimal place. Describe what this means in terms of rate of change.

b. Find the slope of the line segment representing Medicaid. Round to one decimal place. Describe what this means in terms of rate of change.

c. Do the line segments for Social Security and Medicaid lie on parallel lines? What does this mean in terms of the rate of change for these entitlement programs?

73. Just as money doesn't buy happiness for individuals, the two don't necessarily go together for countries either. However, the scatter plot does show a relationship between a country's annual per capita income and the percentage of people in that country who call themselves "happy."

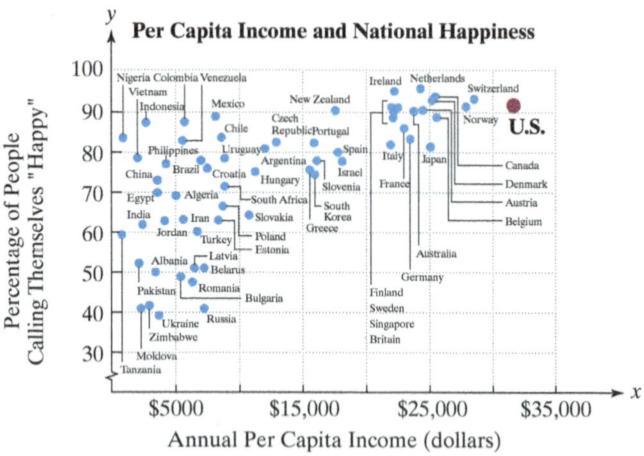

Source: Richard Layard, *Happiness: Lessons from a New Science*, Penguin, 2005

Draw a line that fits the data so that the spread of the data points around the line is as small as possible. Use the coordinates of two points along your line to write the slope-intercept form of its equation. Express the equation in function notation and use the linear function to make a prediction about national happiness based on per capita income.

Writing in Mathematics

74. Describe how to write the equation of a line if its slope and a point along the line are known.

75. Describe how to write the equation of a line if two points along the line are known.

76. If two lines are parallel, describe the relationship between their slopes.

77. If two lines are perpendicular, describe the relationship between their slopes.

78. If you know a point on a line and you know the equation of a line parallel to this line, explain how to write the line's equation.

79. In Example 3 on page 147, we developed a model that predicted American men born in 2020 will have a life expectancy of 78.6 years. Describe something that might occur that would make this prediction inaccurate.

Technology Exercises

80. The lines whose equations are $y = \frac{1}{3}x + 1$ and $y = -3x - 2$ are perpendicular because the product of their slopes, $\frac{1}{3}$ and -3, respectively, is -1.

a. Use a graphing utility to graph the equations in a $[-10, 10, 1]$ by $[-10, 10, 1]$ viewing rectangle. Do the lines appear to be perpendicular?

b. Now use the zoom square feature of your utility. Describe what happens to the graphs. Explain why this is so.

81. a. Use the statistical menu of your graphing utility to enter the seven data points shown in the scatter plot that you drew in Exercise 69(a).

b. Use the DRAW menu and the scatter plot capability to draw a scatter plot of the data points.

c. Select the linear regression option. Use your utility to obtain values for a and b for the equation of the regression line, $y = ax + b$. Compare this equation to the one that you obtained by hand in Exercise 69. You may also be given a **correlation coefficient**, r. Values of r close to 1 indicate that the points can be modeled by a linear function and the regression line has a positive slope. Values of r close to -1 indicate that the points can be modeled by a linear function and the regression line has a negative slope. Values of r close to 0 indicate no linear relationship between the variables. In this case, a linear model does not accurately describe the data.

d. Use the appropriate sequence (consult your manual) to graph the regression equation on top of the points in the scatter plot.

82. Repeat Exercise 81 using the seven data points shown in the scatter plot that you drew in Exercise 70(a).

Critical Thinking Exercises

Make Sense? *In Exercises 83–86, determine whether each statement "makes sense" or "does not make sense" and explain your reasoning.*

83. I can use any two points in a scatter plot to write the point-slope form of the equation of the line through those points. However, the other data points in the scatter plot might not fall on, or even near, this line.

84. I have linear functions that model changes for men and women over the same time period. The functions have the same slope, so their graphs are parallel lines, indicating that the rate of change for men is the same as the rate of change for women.

85. Some of the steel girders in this photo of the Eiffel Tower appear to be perpendicular. I can verify my observation by determining that their slopes are negative reciprocals.

86. When writing equations of lines, it's always easiest to begin by writing the point-slope form of the equation.

In Exercises 87–90, determine whether each statement is true or false. If the statement is false, make the necessary change(s) to produce a true statement.

87. The standard form of the equation of a line passing through $(-3, -1)$ and perpendicular to the line whose equation is $y = -\frac{2}{5}x - 4$ is $5x - 2y = -13$.

88. If I change the subtraction signs to addition signs in $y - 12 = 8(x - 2)$, the y-intercept of the corresponding graph will change from -4 to 4.

89. $y - 5 = 2(x - 1)$ is an equation of a line passing through $(4, 11)$.

90. The function $\{(-1, 4), (3, 6), (5, 7), (11, 10)\}$ can be described using $y - 7 = \frac{1}{2}(x - 5)$ with a domain of $\{-1, 3, 5, 11\}$.

91. Determine the value of B so that the line whose equation is $By = 8x - 1$ has slope -2.

92. Determine the value of A so that the line whose equation is $Ax + y = 2$ is perpendicular to the line containing the points $(1, -3)$ and $(-2, 4)$.

93. Consider a line whose x-intercept is $(-3, 0)$ and whose y-intercept is $(0, -6)$. Provide the missing coordinate for the following two points that lie on this line: $(-40, \ \ )$ and $(\ \ , -200)$.

94. Prove that the equation of a line passing through $(a, 0)$ and $(0, b)(a \neq 0, b \neq 0)$ can be written in the form $\frac{x}{a} + \frac{y}{b} = 1$. Why is this called the *intercept form* of a line?

Review Exercises

95. If $f(x) = 3x^2 - 8x + 5$, find $f(-2)$. (Section 2.1, Example 3)

96. If $f(x) = x^2 - 3x + 4$ and $g(x) = 2x - 5$, find $(fg)(-1)$. (Section 2.3, Example 4)

97. The sum of the angles of a triangle is 180°. Find the three angles of a triangle if one angle is 20° greater than the smallest angle and the third angle is twice the smallest angle. (Section 1.5, Example 1)

Preview Exercises

Exercises 98–100 will help you prepare for the material covered in the first section of the next chapter.

98. **a.** Does $(-5, -6)$ satisfy $2x - y = -4$?

b. Does $(-5, -6)$ satisfy $3x - 5y = 15$?

99. Graph $y = -x - 1$ and $4x - 3y = 24$ in the same rectangular coordinate system. At what point do the graphs intersect?

100. Solve: $7x - 2(-2x + 4) = 3$.

GROUP PROJECT

CHAPTER 2

In Example 3 on page 147, we used the data in **Figure 2.35** to develop a linear function that modeled life expectancy. For this group exercise, you might find it helpful to pattern your work after **Figure 2.35** and the solution to Example 3. Group members should begin by consulting an almanac, newspaper, magazine, or the Internet to find data that appear to lie approximately on or near a line. Working by hand or using a graphing utility, group members should construct scatter plots for the data that were collected. If working by hand, draw a line that approximately fits the data in each scatter plot and then write its equation as a function in slope-intercept form. If using a graphing utility, obtain the equation of each regression line. Then use each linear function's equation to make predictions about what might occur in the future. Are there circumstances that might affect the accuracy of the prediction? List some of these circumstances.

Chapter 2 Summary

Definitions and Concepts / Examples

Section 2.1 Introduction to Functions

A relation is any set of ordered pairs. The set of first components of the ordered pairs is the domain and the set of second components is the range. A function is a relation in which each member of the domain corresponds to exactly one member of the range. No two ordered pairs of a function can have the same first component and different second components.

Example: The domain of the relation $\{(1, 2), (3, 4), (3, 7)\}$ is $\{1, 3\}$. The range is $\{2, 4, 7\}$. The relation is not a function: 3, in the domain, corresponds to both 4 and 7 in the range.

If a function is defined by an equation, the notation $f(x)$, read "f of x" or "f at x," describes the value of the function at the number, or input, x.

Example: If $f(x) = 7x - 5$, then

$$\begin{aligned} f(a+2) &= 7(a+2) - 5 \\ &= 7a + 14 - 5 \\ &= 7a + 9. \end{aligned}$$

Section 2.2 Graphs of Functions

The graph of a function is the graph of its ordered pairs.

The Vertical Line Test for Functions

If any vertical line intersects a graph in more than one point, the graph does not define y as a function of x.

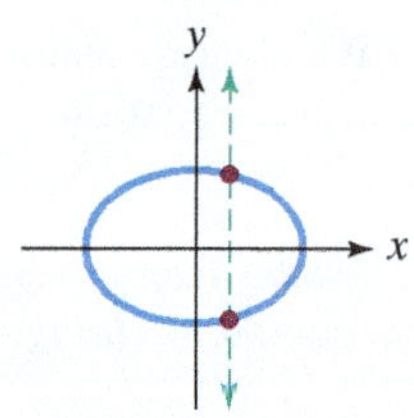

Not the graph of a function

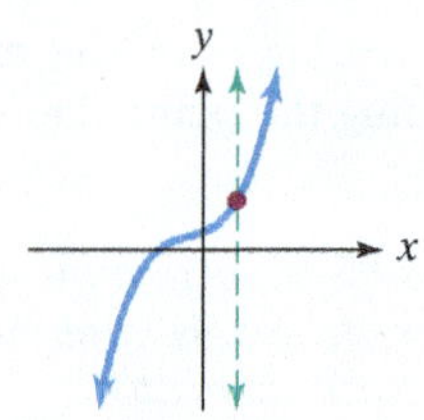

The graph of a function

At the left or right of a function's graph, you will often find closed dots, open dots, or arrows. A closed dot shows that the graph ends and the point belongs to the graph. An open dot shows that the graph ends and the point does not belong to the graph. An arrow indicates that the graph extends indefinitely.

The graph of a function can be used to determine the function's domain and its range. To find the domain, look for all the inputs on the x-axis that correspond to points on the graph. To find the range, look for all the outputs on the y-axis that correspond to points on the graph.

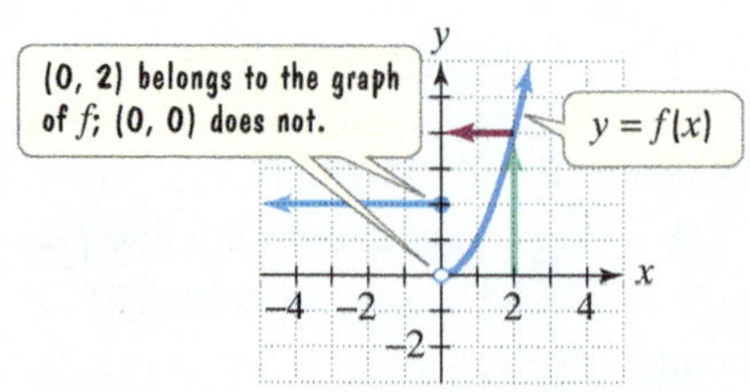

To find $f(2)$, locate 2 on the x-axis. The graph shows $f(2) = 4$.

Domain of $f = \{x \mid x \text{ is a real number}\}$

Range of $f = \{y \mid y > 0\}$

Section 2.3 The Algebra of Functions

A Function's Domain

If a function f does not model data or verbal conditions, its domain is the largest set of real numbers for which the value of $f(x)$ is a real number. Exclude from a function's domain real numbers that cause division by zero and real numbers that result in a square root of a negative number.

Example:

$$f(x) = 7x + 13$$

Domain of $f = \{x \mid x \text{ is a real number}\}$

$$g(x) = \frac{7x}{12 - x}$$

Domain of $g = \{x \mid x \text{ is a real number and } x \neq 12\}$

Definitions and Concepts	Examples

Section 2.3 The Algebra of Functions (continued)

The Algebra of Functions

Let f and g be two functions. The sum $f + g$, the difference $f - g$, the product fg, and the quotient $\frac{f}{g}$ are functions whose domains are the set of all real numbers common to the domains of f and g, defined as follows:

1. Sum: $(f + g)(x) = f(x) + g(x)$
2. Difference: $(f - g)(x) = f(x) - g(x)$
3. Product: $(fg)(x) = f(x) \cdot g(x)$
4. Quotient: $\left(\frac{f}{g}\right)(x) = \frac{f(x)}{g(x)}, g(x) \neq 0.$

Let $f(x) = x^2 + 2x$ and $g(x) = 4 - x$.

- $(f + g)(x) = (x^2 + 2x) + (4 - x) = x^2 + x + 4$
  $(f + g)(-2) = (-2)^2 + (-2) + 4 = 4 - 2 + 4 = 6$
- $(f - g)(x) = (x^2 + 2x) - (4 - x) = x^2 + 2x - 4 + x$
  $= x^2 + 3x - 4$
  $(f - g)(5) = 5^2 + 3 \cdot 5 - 4 = 25 + 15 - 4 = 36$
- $(fg)(1) = f(1) \cdot g(1) = (1^2 + 2 \cdot 1)(4 - 1)$
  $= 3(3) = 9$
- $\left(\frac{f}{g}\right)(x) = \frac{x^2 + 2x}{4 - x}, x \neq 4$
  $\left(\frac{f}{g}\right)(3) = \frac{3^2 + 2 \cdot 3}{4 - 3} = \frac{9 + 6}{1} = 15$

Section 2.4 Linear Functions and Slope

Data presented in a visual form as a set of points is called a scatter plot. A line that best fits the data points is called a regression line.
A function whose graph is a straight line is called a linear function. All linear functions can be written in the form $f(x) = mx + b$.

$f(x) = 3x + 10$ is a linear function.

$g(x) = 3x^2 + 10$ is not a linear function.

If a graph intersects the x-axis at $(a, 0)$, then a is an x-intercept. If a graph intersects the y-axis at $(0, b)$, then b is a y-intercept. The standard form of the equation of a line,

$$Ax + By = C,$$

can be graphed using intercepts and a checkpoint.

Graph using intercepts: $4x + 3y = 12$.

x-intercept: (Set $y = 0$.) $4x = 12$, $x = 3$ — Line passes through (3, 0).

y-intercept: (Set $x = 0$.) $3y = 12$, $y = 4$ — Line passes through (0, 4).

Checkpoint: Let $x = 2$.

$$4 \cdot 2 + 3y = 12$$
$$8 + 3y = 12$$
$$3y = 4$$
$$y = \frac{4}{3}$$

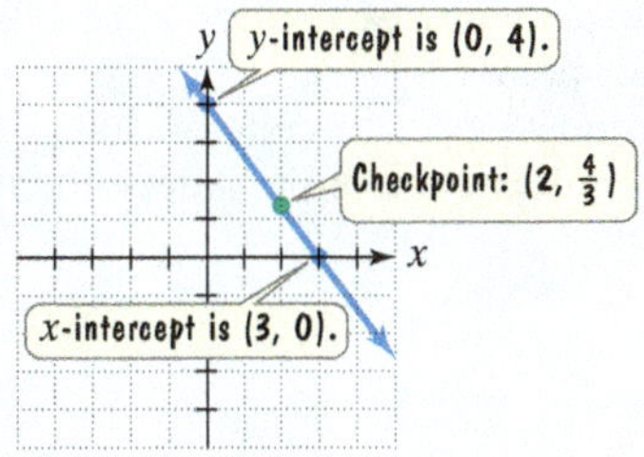

Definitions and Concepts / Examples

Section 2.4 Linear Functions and Slope (continued)

The slope, m, of the line through the points (x_1, y_1) and (x_2, y_2) is

$$m = \frac{y_2 - y_1}{x_2 - x_1}, \quad x_2 - x_1 \neq 0.$$

If the slope is positive, the line rises from left to right. If the slope is negative, the line falls from left to right. The slope of a horizontal line is 0. The slope of a vertical line is undefined.

For points $(-7, 2)$ and $(3, -4)$, the slope of the line through the points is

$$m = \frac{\text{Change in } y}{\text{Change in } x} = \frac{-4 - 2}{3 - (-7)} = \frac{-6}{10} = -\frac{3}{5}.$$

The slope is negative, so the line falls.

For points $(2, -5)$ and $(2, 16)$, the slope of the line through the points is:

$$m = \frac{\text{Change in } y}{\text{Change in } x} = \frac{16 - (-5)}{2 - 2} = \frac{21}{0}.$$

undefined

The slope is undefined, so the line is vertical.

The slope-intercept form of the equation of a nonvertical line with slope m and y-intercept b is

$$y = mx + b.$$

Using function notation, the equation is

$$f(x) = mx + b.$$

Graph: $f(x) = -\frac{3}{4}x + 1.$

Slope is $-\frac{3}{4}$. y-intercept is (0, 1).

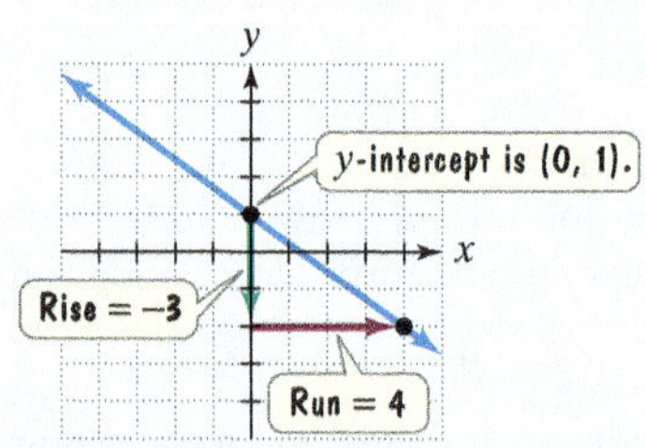

Horizontal and Vertical Lines

The graph of $y = b$, or $f(x) = b$, is a horizontal line. The y-intercept is $(0, b)$. The linear function $f(x) = b$ is called a constant function.
The graph of $x = a$ is a vertical line. The x-intercept is $(a, 0)$. A vertical line is not a linear function.

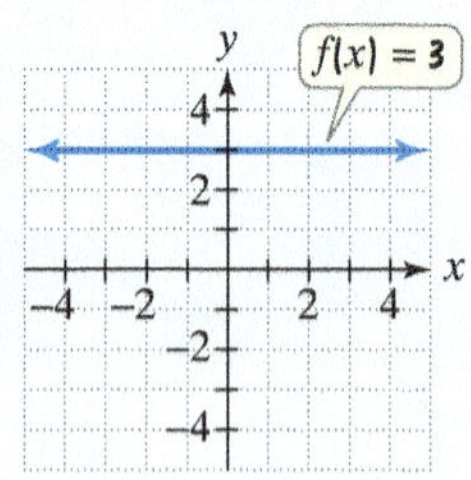

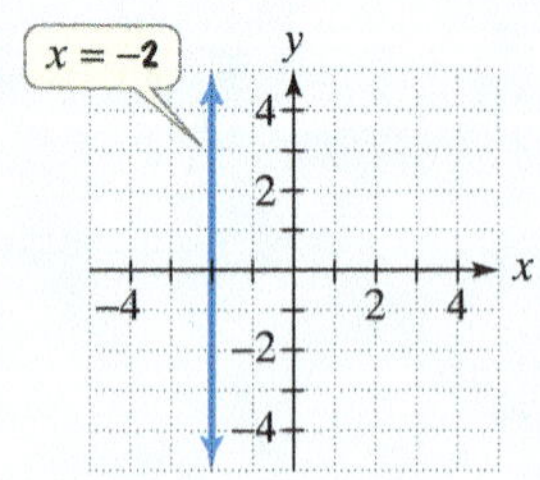

If the graph of a function is not a straight line, the average rate of change between any two points is the slope of the line containing the two points.
For a linear function, slope is the rate of change of the dependent variable per unit change of the independent variable.

The function

$$p(t) = -0.59t + 80.75$$

slope

models the percentage, $p(t)$, of Americans smoking cigarettes t years after 1900. The slope, -0.59, shows that the percentage of smokers is decreasing by 0.59% per year.

Definitions and Concepts

Examples

Section 2.5 The Point-Slope Form of the Equation of a Line

The point-slope form of the equation of a nonvertical line with slope m that passes through the point (x_1, y_1) is

$$y - y_1 = m(x - x_1).$$

Slope $= -4$, passing through $(-1, 5)$

$m = -4$ $x_1 = -1$ $y_1 = 5$

The point-slope form of the line's equation is

$$y - 5 = -4[x - (-1)].$$

Simplify:

$$y - 5 = -4(x + 1).$$

To write the point-slope form of the line passing through two points, begin by using the points to compute the slope, m. Use either given point as (x_1, y_1) and write the point-slope equation:

$$y - y_1 = m(x - x_1).$$

Solving this equation for y gives the slope-intercept form of the line's equation.

Write equations in point-slope form and in slope-intercept form of the line passing through $(4, 1)$ and $(3, -2)$.

$$m = \frac{-2 - 1}{3 - 4} = \frac{-3}{-1} = 3$$

Using $(4, 1)$ as (x_1, y_1), the point-slope form of the equation is

$$y - 1 = 3(x - 4).$$

Solve for y to obtain the slope-intercept form.

$$y - 1 = 3x - 12$$
$$y = 3x - 11$$

In function notation,

$$f(x) = 3x - 11.$$

Nonvertical parallel lines have the same slope. If the product of the slopes of two lines is -1, then the lines are perpendicular. One line is perpendicular to another line if its slope is the negative reciprocal of the slope of the other. A horizontal line having zero slope is perpendicular to a vertical line having undefined slope.

Write equations in point-slope form and in slope-intercept form of the line passing through $(2, -1)$

x_1 y_1

and perpendicular to $y = -\frac{1}{5}x + 6$.

slope

The slope, m, of the perpendicular line is 5, the negative reciprocal of $-\frac{1}{5}$.

$$y - (-1) = 5(x - 2)$$ Point-slope form of the equation

$$y + 1 = 5(x - 2)$$
$$y + 1 = 5x - 10$$
$$y = 5x - 11 \text{ or } f(x) = 5x - 11$$

Slope-intercept form of the equation

CHAPTER 2 REVIEW EXERCISES

2.1 *In Exercises 1–3, determine whether each relation is a function. Give the domain and range for each relation.*

1. $\{(3, 10), (4, 10), (5, 10)\}$

2. $\{(1, 12), (2, 100), (3, \pi), (4, -6)\}$

3. $\{(13, 14), (15, 16), (13, 17)\}$

In Exercises 4–5, find the indicated function values.

4. $f(x) = 7x - 5$
   a. $f(0)$ b. $f(3)$ c. $f(-10)$
   d. $f(2a)$ e. $f(a + 2)$
5. $g(x) = 3x^2 - 5x + 2$
   a. $g(0)$ b. $g(5)$ c. $g(-4)$
   d. $g(b)$ e. $g(4a)$

2.2 *In Exercises 6–7, graph the given functions, f and g, in the same rectangular coordinate system. Select integers for x, starting with −2 and ending with 2. Once you have obtained your graphs, describe how the graph of g is related to the graph of f.*

6. $f(x) = x^2, \quad g(x) = x^2 - 1$

7. $f(x) = |x|, \quad g(x) = |x| + 2$

In Exercises 8–13, use the vertical line test to identify graphs in which y is a function of x.

8.

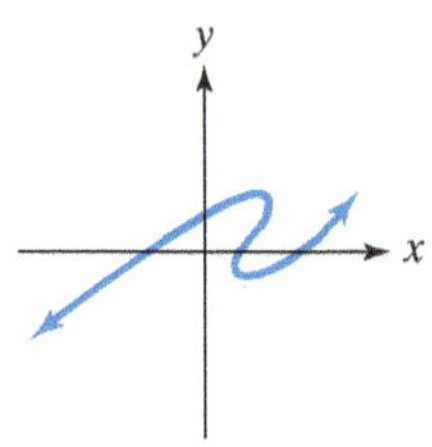

9.

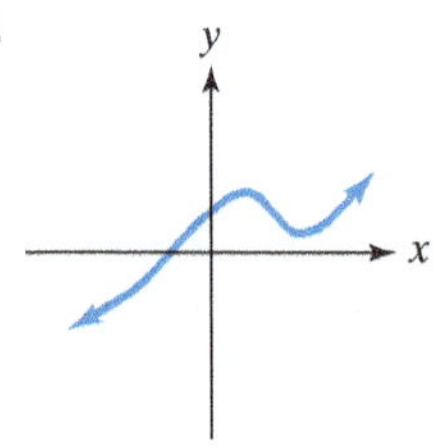

10.

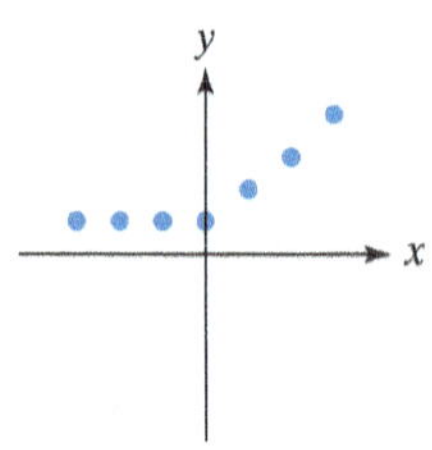

11.

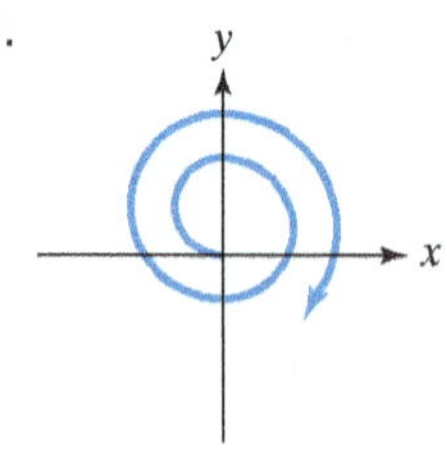

12.

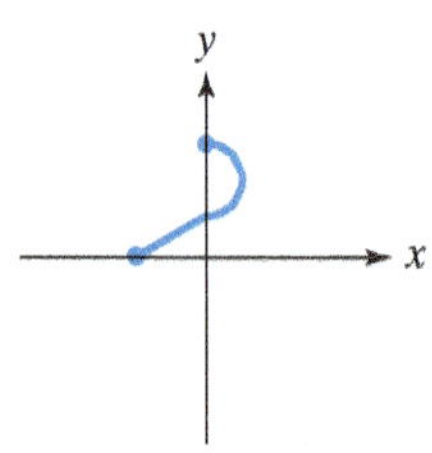

13.

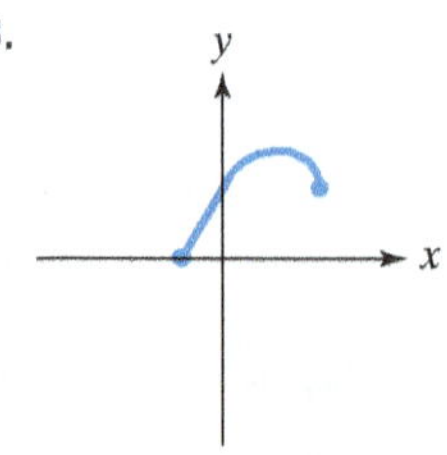

Use the graph of f to solve Exercises 14–18.

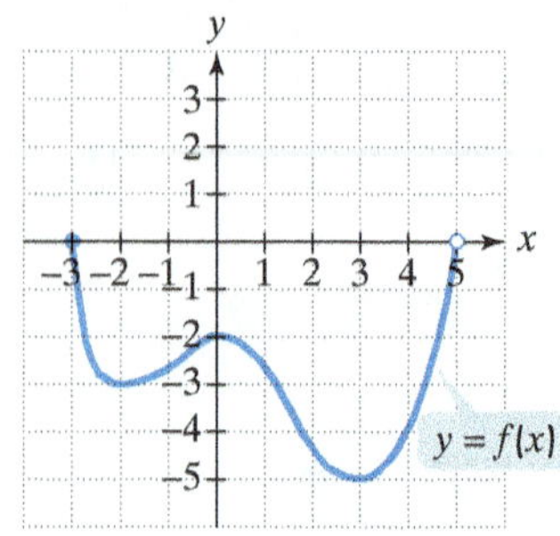

14. Find $f(-2)$. 15. Find $f(0)$.
16. For what value of x is $f(x) = -5$?
17. Find the domain of f.
18. Find the range of f.
19. The graph shows the height, in meters, of an eagle in terms of its time, in seconds, in flight.

   a. Use the graph to explain why the eagle's height is a function of its time in flight.
   b. Find $f(15)$. Describe what this means in practical terms.
   c. What is a reasonable estimate of the eagle's maximum height?
   d. For what values of x is $f(x) = 20$? Describe what this means in practical terms.
   e. Use the graph of the function to write a description of the eagle's flight.

2.3 *In Exercises 20–22, find the domain of each function.*

20. $f(x) = 7x - 3$

21. $g(x) = \dfrac{1}{x + 8}$

22. $f(x) = x + \dfrac{3x}{x - 5}$

In Exercises 23–24, find **a.** $(f + g)(x)$ *and* **b.** $(f + g)(3)$.

23. $f(x) = 4x - 5, g(x) = 2x + 1$

24. $f(x) = 5x^2 - x + 4, g(x) = x - 3$

In Exercises 25–26, for each pair of functions, f and g, determine the domain of f + g.

25. $f(x) = 3x + 4, g(x) = \frac{5}{4 - x}$

26. $f(x) = \frac{7x}{x + 6}, g(x) = \frac{4}{x + 1}$

In Exercises 27–34, let

$$f(x) = x^2 - 2x \quad \text{and} \quad g(x) = x - 5.$$

Find each of the following.

27. $(f + g)(x)$ and $(f + g)(-2)$

28. $f(3) + g(3)$

29. $(f - g)(x)$ and $(f - g)(1)$

30. $f(4) - g(4)$

31. $(fg)(-3)$

32. $\left(\frac{f}{g}\right)(x)$ and $\left(\frac{f}{g}\right)(4)$

33. The domain of $f - g$

34. The domain of $\frac{f}{g}$

2.4 *In Exercises 35–37, use intercepts and a checkpoint to graph each linear function.*

35. $x + 2y = 4$

36. $2x - 3y = 12$

37. $4x = 8 - 2y$

In Exercises 38–41, find the slope of the line passing through each pair of points or state that the slope is undefined. Then indicate whether the line through the points rises, falls, is horizontal, or is vertical.

38. $(5, 2)$ and $(2, -4)$

39. $(-2, 3)$ and $(7, -3)$

40. $(3, 2)$ and $(3, -1)$

41. $(-3, 4)$ and $(-1, 4)$

In Exercises 42–44, give the slope and y-intercept of each line whose equation is given. Then graph the linear function.

42. $y = 2x - 1$

43. $f(x) = -\frac{1}{2}x + 4$

44. $y = \frac{2}{3}x$

In Exercises 45–47, rewrite the equation in slope-intercept form. Give the slope and y-intercept.

45. $2x + y = 4$

46. $-3y = 5x$

47. $5x + 3y = 6$

In Exercises 48–52, graph each equation in a rectangular coordinate system.

48. $y = 2$

49. $7y = -21$

50. $f(x) = -4$

51. $x = 3$

52. $2x = -10$

53. The function $f(t) = -0.27t + 70.45$ models record time, $f(t)$, in seconds, for the women's 400-meter run t years after 1900. What is the slope of this model? Describe what this means in terms of rate of change.

54. The stated intent of the 1994 "don't ask, don't tell" policy was to reduce the number of discharges of gay men and lesbians from the military. The line graph shows the number of active-duty gay servicemembers discharged from the military for homosexuality under the policy.

Number of Active-Duty Gay Servicemembers Discharged from the Military for Homosexuality

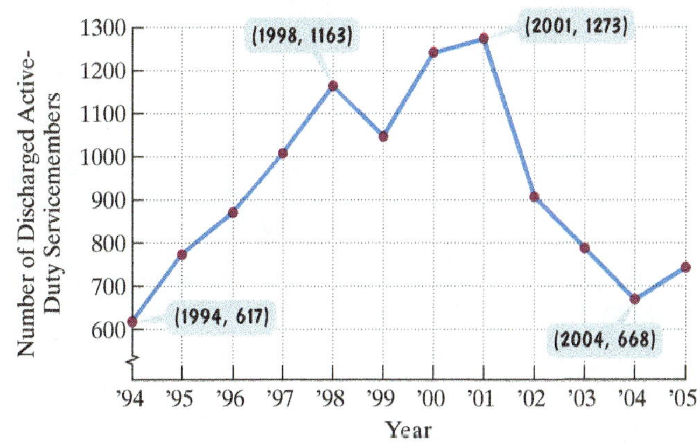

Source: General Accountability Office

a. Find the average rate of change, rounded to the nearest whole number, from 1994 through 1998. Describe what this means.

b. Find the average rate of change, rounded to the nearest whole number, from 2001 through 2004. Describe what this means.

55. The graph shows that a linear function describes the relationship between Fahrenheit temperature, F, and Celsius temperature, C.

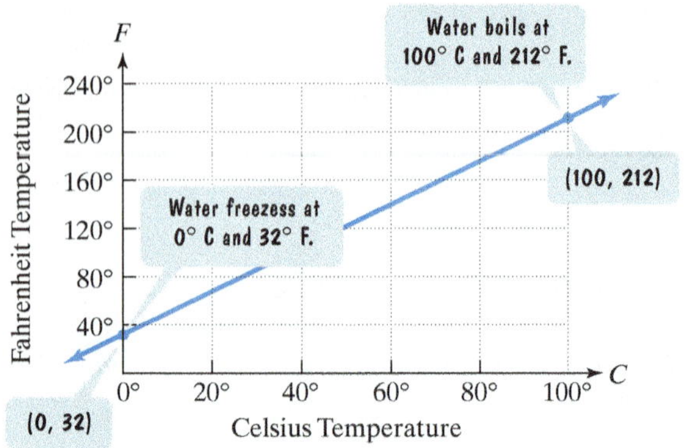

a. Use the points labeled by the voice balloons to find a function in the form $F = mC + b$ that expresses Fahrenheit temperature, F, in terms of Celsius temperature, C.

b. Use the function from part (a) to find the Fahrenheit temperature when the Celsius temperature is 30°.

2.5 *In Exercises 56–59, use the given conditions to write an equation for each line in point-slope form and in slope-intercept form.*

56. Pasing through $(-3, 2)$ with slope -6

57. Passing through $(1, 6)$ and $(-1, 2)$

58. Passing through $(4, -7)$ and parallel to the line whose equation is $3x + y - 9 = 0$

59. Passing through $(-2, 6)$ and perpendicular to the line whose equation is $y = \frac{1}{3}x + 4$

60. The bar graph shows the number of Americans, in millions, living below the poverty level from 2001 through 2005. The data are displayed as five points in a scatter plot. Also shown is a line that passes through or near the points.

Number of Americans in Poverty

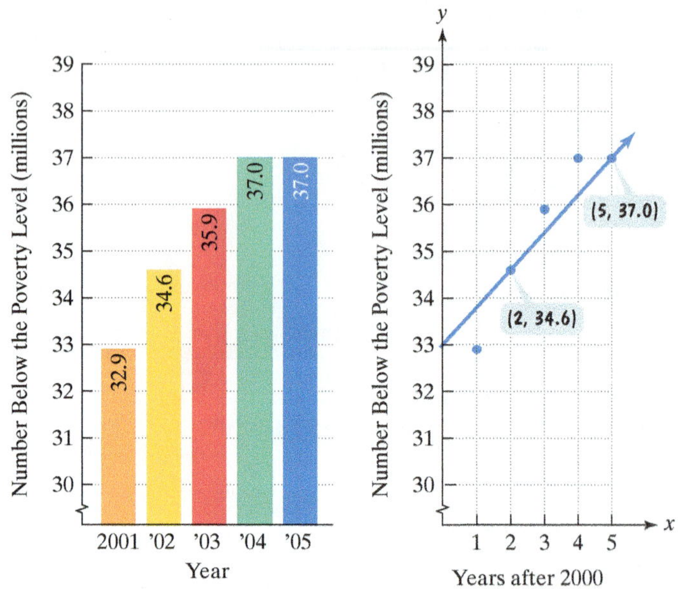

Source: U.S. Census Bureau

a. Use the two points whose coordinates are shown by the voice balloons to find the point-slope form of the equation of the line that models the number of Americans, y, in millions, living below the poverty level x years after 2000.

b. Write the equation from part (a) in slope-intercept form. Use function notation.

c. If trends shown from 2001 through 2005 continue, use the linear function to predict the number of Americans who will be living below the poverty level in 2010.

CHAPTER 2 TEST

Remember to use your Chapter Test Prep Video CD to see the worked-out solutions to the test questions you want to review.

In Exercises 1–2, determine whether each relation is a function. Give the domain and range for each relation.

1. $\{(1, 2), (3, 4), (5, 6), (6, 6)\}$

2. $\{(2, 1), (4, 3), (6, 5), (6, 6)\}$

3. If $f(x) = 3x - 2$, find $f(a + 4)$.

4. If $f(x) = 4x^2 - 3x + 6$, find $f(-2)$.

5. Graph $f(x) = x^2 - 1$ and $g(x) = x^2 + 1$ in the same rectangular coordinate system. Select integers for x, starting with -2 and ending with 2. Once you have obtained your graphs, describe how the graph of g is related to the graph of f.

In Exercises 6–7, identify the graph or graphs in which y is a function of x.

6.

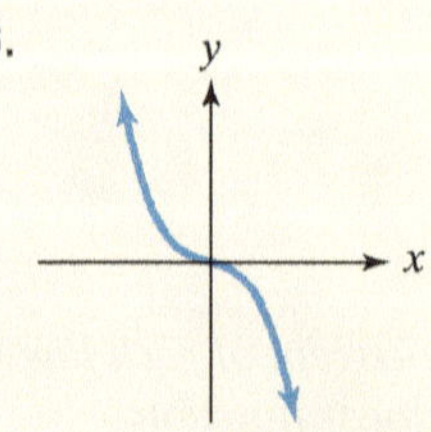

7.

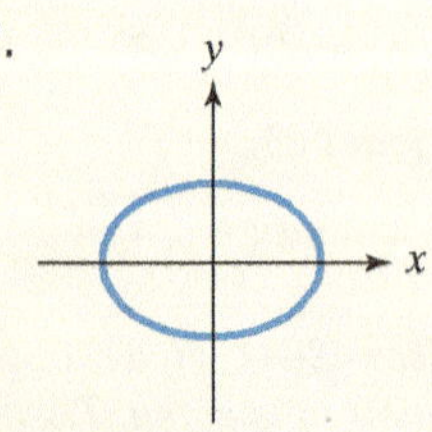

Use the graph of f to solve Exercises 8–11.

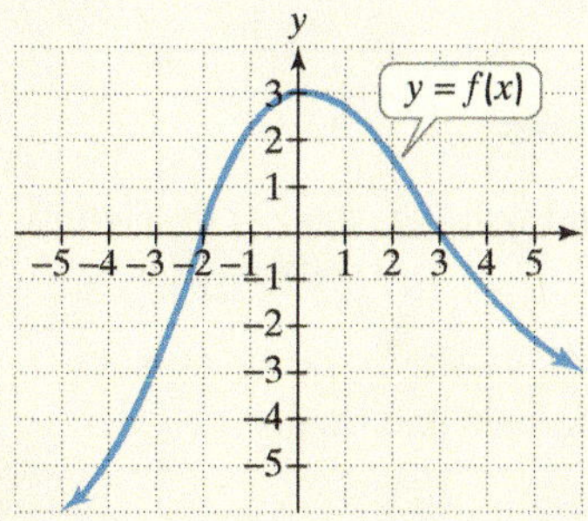

8. Find $f(6)$.
9. List two values of x for which $f(x) = 0$.
10. Find the domain of f.
11. Find the range of f.
12. Find the domain of $f(x) = \dfrac{6}{10 - x}$.

In Exercises 13–17, let

$$f(x) = x^2 + 4x \quad and \quad g(x) = x + 2.$$

Find each of the following.

13. $(f + g)(x)$ and $(f + g)(3)$
14. $(f - g)(x)$ and $(f - g)(-1)$
15. $(fg)(-5)$
16. $\left(\dfrac{f}{g}\right)(x)$ and $\left(\dfrac{f}{g}\right)(2)$
17. The domain of $\dfrac{f}{g}$

In Exercises 18–20, graph each linear function.

18. $4x - 3y = 12$
19. $f(x) = -\dfrac{1}{3}x + 2$
20. $f(x) = 4$

In Exercises 21–22, find the slope of the line passing through each pair of points or state that the slope is undefined. Then indicate whether the line through the points rises, falls, is horizontal, or is vertical.

21. $(5, 2)$ and $(1, 4)$
22. $(4, 5)$ and $(4, -5)$

The function $V(t) = 3.6t + 140$ models the number of Super Bowl viewers, $V(t)$, in millions, t years after 1995. Use the model to solve Exercises 23–24.

23. Find $V(10)$. Describe what this means in terms of the variables in the model.
24. What is the slope of this model? Describe what this means in terms of rate of change.

In Exercises 25–27, use the given conditions to write an equation for each line in point-slope form and slope-intercept form.

25. Passing through $(-1, -3)$ and $(4, 2)$
26. Passing through $(-2, 3)$ and perpendicular to the line whose equation is $y = -\frac{1}{2}x - 4$
27. Passing through $(6, -4)$ and parallel to the line whose equation is $x + 2y = 5$
28. The scatter plot shows the number of sentenced inmates in the United States per 100,000 residents from 2001 through 2005. Also shown is a line that passes through or near the data points.

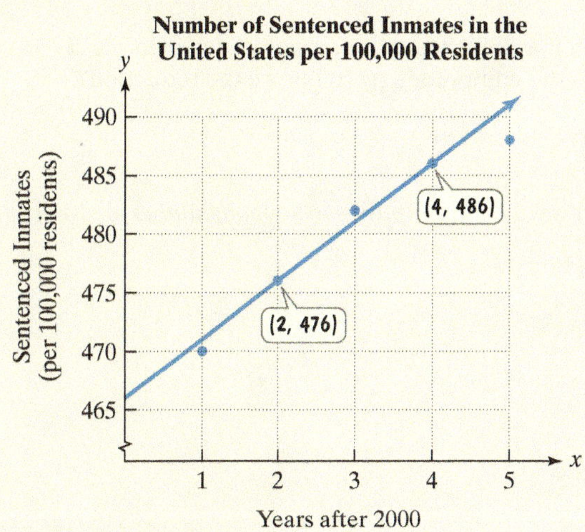

Source: U.S. Justice Department

a. Use the two points whose coordinates are shown by the voice balloons to find the point-slope form of the equation of the line that models the number of inmates per 100,000 residents, y, x years after 2000.

b. Write the equation from part (a) in slope-intercept form. Use function notation.

c. Use the linear function to predict the number of sentenced inmates in the United States per 100,000 residents in 2010.

CUMULATIVE REVIEW EXERCISES (CHAPTERS 1-2)

1. Use the roster method to list the elements in the set:

   $\{x \mid x \text{ is a whole number less than } 4\}$.

2. Determine whether the following statement is true or false:

   $\pi \notin \{x \mid x \text{ is an irrational number}\}$.

In Exercises 3–4, use the order of operations to simplify each expression.

3. $\dfrac{8 - 3^2 \div 9}{|-5| - [5 - (18 \div 6)]^2}$

4. $4 - (2 - 9)^0 + 3^2 \div 1 + 3$

5. Simplify: $3 - [2(x - 2) - 5x]$.

In Exercises 6–8, solve each equation. If the solution set is ∅ or ℝ, classify the equation as an inconsistent equation or an identity.

6. $2 + 3x - 4 = 2(x - 3)$

7. $4x + 12 - 8x = -6(x - 2) + 2x$

8. $\dfrac{x - 2}{4} = \dfrac{2x + 6}{3}$

9. After a 20% reduction, a computer sold for \$1800. What was the computer's price before the reduction?

10. Solve for t: $A = p + prt$.

In Exercises 11–12, simplify each exponential expression.

11. $(3x^4y^{-5})^{-2}$

12. $\left(\dfrac{3x^2y^{-4}}{x^{-3}y^2}\right)^2$

13. Multiply and write the answer in scientific notation:

    $(7 \times 10^{-8})(3 \times 10^2)$.

14. Is $\{(1, 5), (2, 5), (3, 5), (4, 5), (6, 5)\}$ a function? Give the relation's domain and range.

15. Graph $f(x) = |x| - 1$ and $g(x) = |x| + 2$ in the same rectangular coordinate system. Select integers for x, starting with -2 and ending with 2. Once you have obtained your graphs, describe how the graph of g is related to the graph of f.

16. Find the domain of $f(x) = \dfrac{1}{15 - x}$.

17. If $f(x) = 3x^2 - 4x + 2$ and $g(x) = x^2 - 5x - 3$, find $(f - g)(x)$ and $(f - g)(-1)$.

In Exercises 18-19, graph each linear function.

18. $f(x) = -2x + 4$.

19. $x - 2y = 6$

20. Write equations in point-slope form and slope-intercept form for the line passing through $(3, -5)$ and parallel to the line whose equation is $y = 4x + 7$.

CHAPTER 3

Systems of Linear Equations

Held in a different country every five years, with contributions from more than 120 countries, the World Expo is a showcase of cool gizmos, a multinational take on what the future might be. Who needs a human friend when you can connect with an adorable robot that recognizes you, engages in (meaningful?) conversation in any language, takes orders, sings, dances, and even does algebra? And what do the entrepreneurs who create these robots and other global visions want to do? Generate profit, of course. In this chapter, you'll learn how algebra models every business venture, from a kid selling lemonade to companies producing innovative "personal partner" robots.

Functions of business, including modeling business ventures with profit functions, are discussed in Section 3.2, pages 189–191.

SECTION 3.1

Systems of Linear Equations in Two Variables

Objectives

1. Determine whether an ordered pair is a solution of a system of linear equations.
2. Solve systems of linear equations by graphing.
3. Solve systems of linear equations by substitution.
4. Solve systems of linear equations by addition.
5. Select the most efficient method for solving a system of linear equations.
6. Identify systems that do not have exactly one ordered-pair solution.

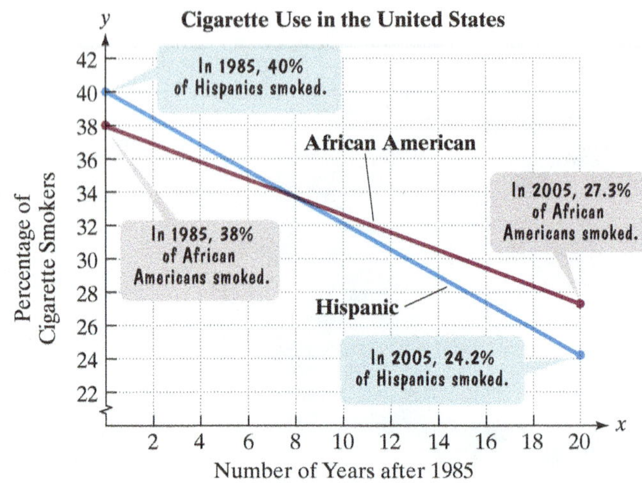

FIGURE 3.1

Source: Department of Health and Human Services

Although we still see celebrities smoking in movies, in music videos, and on television, there has been a remarkable decline in the percentage of cigarette smokers in the United States. The decline among African Americans and Hispanics, illustrated in **Figure 3.1**, can be analyzed using a pair of linear models in two variables.

In the first three sections of this chapter, you will learn to model your world with two equations in two variables and three equations in three variables. The methods you learn for solving these systems provide the foundation for solving complex problems involving thousands of equations containing thousands of variables. In this section's exercise set, you will apply these methods to analyze the decrease in cigarette use among whites, African Americans, and Hispanics.

1 Determine whether an ordered pair is a solution of a system of linear equations.

Systems of Linear Equations and Their Solutions

We have seen that all equations in the form $Ax + By = C$ are straight lines when graphed. Two such equations are called a **system of linear equations**, or a **linear system**. A **solution of a system of linear equations** is an ordered pair that satisfies both equations in the system. For example, (3, 4) satisfies the system

$$x + y = 7 \quad (3 + 4 \text{ is, indeed, } 7.)$$
$$x - y = -1. \quad (3 - 4 \text{ is, indeed, } -1.)$$

Thus, (3, 4) satisfies both equations and is a solution of the system. The solution can be described by saying that $x = 3$ and $y = 4$. The solution can also be described using set notation. The solution set of the system is $\{(3, 4)\}$—that is, the set consisting of the ordered pair (3, 4).

A system of linear equations can have exactly one solution, no solution, or infinitely many solutions. We begin with systems with exactly one solution.

EXAMPLE 1 Determining Whether Ordered Pairs Are Solutions of a System

Consider the system:

$$x + 2y = -7$$
$$2x - 3y = 0.$$

Determine if each ordered pair is a solution of the system:

a. $(-3, -2)$ b. $(1, -4)$.

Solution

a. We begin by determining whether $(-3, -2)$ is a solution. Because -3 is the x-coordinate and -2 is the y-coordinate of $(-3, -2)$, we replace x with -3 and y with -2.

$$\begin{aligned} x + 2y &= -7 \\ -3 + 2(-2) &\stackrel{?}{=} -7 \\ -3 + (-4) &\stackrel{?}{=} -7 \\ -7 &= -7, \quad \text{true} \end{aligned} \qquad \begin{aligned} 2x - 3y &= 0 \\ 2(-3) - 3(-2) &\stackrel{?}{=} 0 \\ -6 - (-6) &\stackrel{?}{=} 0 \\ -6 + 6 &\stackrel{?}{=} 0 \\ 0 &= 0, \quad \text{true} \end{aligned}$$

The pair $(-3, -2)$ satisfies both equations: It makes each equation true. Thus, the ordered pair is a solution, of the system.

b. To determine whether $(1, -4)$ is a solution, we replace x with 1 and y with -4.

$$\begin{aligned} x + 2y &= -7 \\ 1 + 2(-4) &\stackrel{?}{=} -7 \\ 1 + (-8) &\stackrel{?}{=} -7 \\ -7 &= -7, \quad \text{true} \end{aligned} \qquad \begin{aligned} 2x - 3y &= 0 \\ 2 \cdot 1 - 3(-4) &\stackrel{?}{=} 0 \\ 2 - (-12) &\stackrel{?}{=} 0 \\ 2 + 12 &\stackrel{?}{=} 0 \\ 14 &= 0, \quad \text{false} \end{aligned}$$

The pair $(1, -4)$ fails to satisfy *both* equations: It does not make both equations true. Thus, the ordered pair is not a solution of the system.

✓ CHECK POINT 1 Consider the system:

$$\begin{aligned} 2x + 5y &= -24 \\ 3x - 5y &= 14. \end{aligned}$$

Determine if each ordered pair is a solution of the system:

a. $(-7, -2)$ b. $(-2, -4)$.

2 Solve systems of linear equations by graphing.

Solving Linear Systems by Graphing

The solution of a system of two linear equations in two variables can be found by graphing both of the equations in the same rectangular coordinate system. For a system with one solution, **the coordinates of the point of intersection give the system's solution.**

Study Tip

When solving linear systems by graphing, neatly drawn graphs are essential for determining points of intersection.

- Use rectangular coordinate graph paper.
- Use a ruler or straightedge.
- Use a pencil with a sharp point.

Solving Systems of Two Linear Equations in Two Variables, x and y, by Graphing

1. Graph the first equation.
2. Graph the second equation on the same set of axes.
3. If the lines intersect at a point, determine the coordinates of this point of intersection. The ordered pair is the solution to the system.
4. Check the solution in both equations.

EXAMPLE 2 Solving a Linear System by Graphing

Solve by graphing:

$$y = -x - 1$$
$$4x - 3y = 24.$$

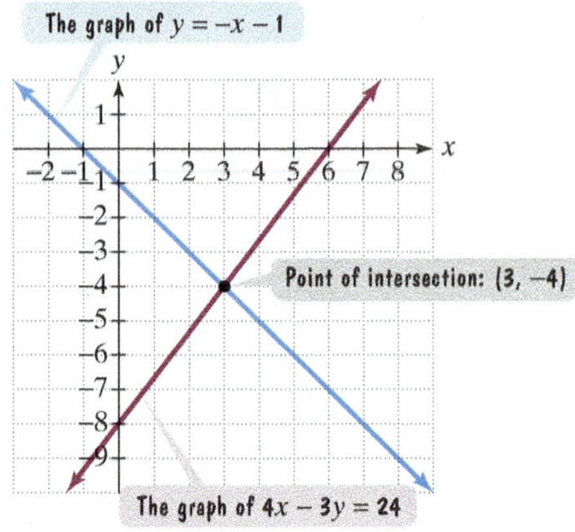

FIGURE 3.2

Solution

Step 1. Graph the first equation. We use the y-intercept and the slope to graph $y = -x - 1$.

$$y = -x - 1$$

The slope is −1. The y-intercept is (0, −1)

The graph of the linear function is shown as a blue line in **Figure 3.2.**

Step 2. Graph the second equation on the same axes. We use intercepts to graph $4x - 3y = 24$.

x-intercept (Set $y = 0$.)	y-intercept (Set $x = 0$.)
$4x - 3 \cdot 0 = 24$	$4 \cdot 0 - 3y = 24$
$4x = 24$	$-3y = 24$
$x = 6$	$y = -8$

The x-intercept is (6, 0). The y-intercept (0, −8). The graph of $4x - 3y = 24$ is shown as a red line in **Figure 3.2.**

Step 3. Determine the coordinates of the intersection point. This ordered pair is the system's solution. Using **Figure 3.2**, it appears that the lines intersect at (3, −4). The "apparent" solution of the system is (3, −4).

Step 4. Check the solution in both equations.

Check (3, −4) in $y = -x - 1$:	Check (3, −4) in $4x - 3y = 24$:
$-4 \stackrel{?}{=} -3 - 1$	$4(3) - 3(-4) \stackrel{?}{=} 24$
$-4 = -4$, true	$12 + 12 \stackrel{?}{=} 24$
	$24 = 24$, true

Because both equations are satisfied, (3, −4) is the solution and $\{(3, -4)\}$ is the solution set.

Using Technology

A graphing utility can be used to solve the system in Example 2. Solve each equation for y, graph the equations, and use the intersection feature. The utility displays the solution (3, −4) as $x = 3$, $y = -4$.

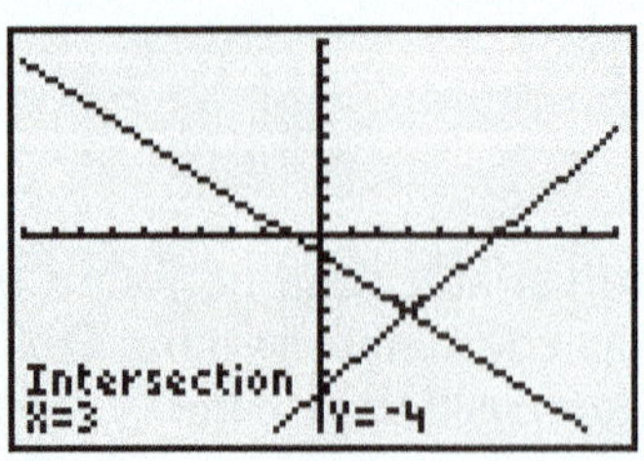

[−10, 10, 1] by [−10, 10, 1]

CHECK POINT 2 Solve by graphing:

$$y = -2x + 6$$
$$2x - y = -2.$$

3 Solve systems of linear equations by substitution.

Eliminating a Variable Using the Substitution Method

Finding the solution to a linear system by graphing equations may not be easy to do. For example, a solution of $\left(-\frac{2}{3}, \frac{157}{29}\right)$ would be difficult to "see" as an intersection point on a graph.

Let's consider a method that does not depend on finding a system's solution visually: the substitution method. This method involves converting the system to one equation in one variable by an appropriate substitution.

Study Tip

In step 1, you can choose which variable to isolate in which equation. If possible, solve for a variable whose coefficient is 1 or -1 to avoid working with fractions.

Solving Linear Systems by Substitution

1. Solve either of the equations for one variable in terms of the other. (If one of the equations is already in this form, you can skip this step.)
2. Substitute the expression found in step 1 into the other equation. This will result in an equation in one variable.
3. Solve the equation containing one variable.
4. Back-substitute the value found in step 3 into one of the original equations. Simplify and find the value of the remaining variable.
5. Check the proposed solution in both of the system's given equations.

EXAMPLE 3 Solving a System by Substitution

Solve by the substitution method:

$$y = -2x + 4$$
$$7x - 2y = 3.$$

Solution

Step 1. Solve either of the equations for one variable in terms of the other. This step has already been done for us. The first equation, $y = -2x + 4$, is solved for y in terms of x.

Step 2. Substitute the expression from step 1 into the other equation. We substitute the expression $-2x + 4$ for y in the second equation:

$$y = \boxed{-2x + 4} \qquad 7x - 2\boxed{y} = 3. \qquad \text{Substitute } -2x + 4 \text{ for } y.$$

This gives us an equation in one variable, namely

$$7x - 2(-2x + 4) = 3.$$

The variable y has been eliminated.

Step 3. Solve the resulting equation containing one variable.

$$7x - 2(-2x + 4) = 3 \quad \text{This is the equation containing one variable.}$$
$$7x + 4x - 8 = 3 \quad \text{Apply the distributive property.}$$
$$11x - 8 = 3 \quad \text{Combine like terms.}$$
$$11x = 11 \quad \text{Add 8 to both sides.}$$
$$x = 1 \quad \text{Divide both sides by 11.}$$

Step 4. Back-substitute the obtained value into one of the original equations. We now know that the x-coordinate of the solution is 1. To find the y-coordinate, we back-substitute the x-value into either original equation. We will use

$$y = -2x + 4.$$

Substitute 1 for x.

$$y = -2 \cdot 1 + 4 = -2 + 4 = 2$$

With $x = 1$ and $y = 2$, the proposed solution is $(1, 2)$.

Step 5. Check the proposed solution in both of the system's given equations. Replace x with 1 and y with 2.

$$\begin{aligned} y &= -2x + 4 \\ 2 &\stackrel{?}{=} -2 \cdot 1 + 4 \\ 2 &\stackrel{?}{=} -2 + 4 \\ 2 &= 2, \quad \text{true} \end{aligned} \qquad \begin{aligned} 7x - 2y &= 3 \\ 7(1) - 2(2) &\stackrel{?}{=} 3 \\ 7 - 4 &\stackrel{?}{=} 3 \\ 3 &= 3, \quad \text{true} \end{aligned}$$

The pair $(1, 2)$ satisfies both equations. The solution is $(1, 2)$ and the system's solution set is $\{(1, 2)\}$.

CHECK POINT 3 Solve by the substitution method:

$$\begin{aligned} y &= 3x - 7 \\ 5x - 2y &= 8. \end{aligned}$$

EXAMPLE 4 Solving a System by Substitution

Solve by the substitution method:

$$\begin{aligned} 5x + 2y &= 1 \\ x - 3y &= 7. \end{aligned}$$

Solution

Step 1. Solve either of the equations for one variable in terms of the other. We begin by isolating one of the variables in either of the equations. By solving for x in the second equation, which has a coefficient of 1, we can avoid fractions.

$$\begin{aligned} x - 3y &= 7 && \text{This is the second equation in the given system.} \\ x &= 3y + 7 && \text{Solve for } x \text{ by adding } 3y \text{ to both sides.} \end{aligned}$$

Step 2. Substitute the expression from step 1 into the other equation. We substitute $3y + 7$ for x in the first equation.

$$x = \boxed{3y + 7} \qquad 5\boxed{x} + 2y = 1$$

This gives us an equation in one variable, namely

$$5(3y + 7) + 2y = 1.$$

The variable x has been eliminated.

Step 3. Solve the resulting equation containing one variable.

$$\begin{aligned} 5(3y + 7) + 2y &= 1 && \text{This is the equation containing one variable.} \\ 15y + 35 + 2y &= 1 && \text{Apply the distributive property.} \\ 17y + 35 &= 1 && \text{Combine like terms.} \\ 17y &= -34 && \text{Subtract 35 from both sides.} \\ y &= -2 && \text{Divide both sides by 17.} \end{aligned}$$

Step 4. Back-substitute the obtained value into one of the original equations. We back-substitute -2 for y into one of the original equations to find x. Let's use both equations to show that we obtain the same value for x in either case.

Using the first equation:	Using the second equation:
$5x + 2y = 1$	$x - 3y = 7$
$5x + 2(-2) = 1$	$x - 3(-2) = 7$
$5x - 4 = 1$	$x + 6 = 7$
$5x = 5$	$x = 1$
$x = 1$	

With $x = 1$ and $y = -2$, the proposed solution is $(1, -2)$.

Study Tip

The equation from step 1, in which one variable is expressed in terms of the other, is equivalent to one of the original equations. It is often easiest to back-substitute an obtained value into this equation to find the value of the other variable. After obtaining both values, get into the habit of checking the ordered-pair solution in *both* equations of the system.

Step 5. Check. Take a moment to show that $(1, -2)$ satisfies both given equations. The solution is $(1, -2)$ and the solution set is $\{(1, -2)\}$.

CHECK POINT 4 Solve by the substitution method:

$$\begin{aligned} 3x + 2y &= 4 \\ 2x + y &= 1. \end{aligned}$$

4 Solve systems of linear equations by addition.

Eliminating a Variable Using the Addition Method

The substitution method is most useful if one of the given equations has an isolated variable. A third method for solving a linear system is the addition method. Like the substitution method, the addition method involves eliminating a variable and ultimately solving an equation containing only one variable. However, this time we eliminate a variable by adding the equations.

For example, consider the following system of linear equations:

$$\begin{aligned} 3x - 4y &= 11 \\ -3x + 2y &= -7. \end{aligned}$$

When we add these two equations, the x-terms are eliminated. This occurs because the coefficients of the x-terms, 3 and -3, are opposites (additive inverses) of each other:

$$\begin{aligned} 3x - 4y &= 11 \\ -3x + 2y &= -7 \\ \text{Add: } 0x - 2y &= 4 \quad \text{The sum is an equation in one variable.} \\ -2y &= 4 \\ y &= -2. \quad \text{Divide both sides by } -2 \text{ and solve for } y. \end{aligned}$$

Now we can back-substitute -2 for y into one of the original equations to find x. It does not matter which equation you use; you will obtain the same value for x in either case. If we use either equation, we can show that $x = 1$ and the solution $(1, -2)$ satisfies both equations in the system.

When we use the addition method, we want to obtain two equations whose sum is an equation containing only one variable. The key step is to **obtain, for one of the variables, coefficients that differ only in sign.** To do this, we may need to multiply one or both equations by some nonzero number so that the coefficients of one of the variables, x or y, become opposites. Then when the two equations are added, this variable will be eliminated.

EXAMPLE 5 Solving a System by the Addition Method

Solve by the addition method:

$$\begin{aligned} 3x + 4y &= -10 \\ 5x - 2y &= 18. \end{aligned}$$

Solution We must rewrite one or both equations in equivalent forms so that the coefficients of the same variable (either x or y) are opposites of each other. Consider the terms in y in each equation, that is, $4y$ and $-2y$. To eliminate y, we can multiply each term of the second equation by 2 and then add equations.

$$\begin{aligned} 3x + 4y = -10 &\xrightarrow{\text{No change}} & 3x + 4y &= -10 \\ 5x - 2y = 18 &\xrightarrow{\text{Multiply by 2.}} & 10x - 4y &= 36 \\ & & \text{Add: } 13x + 0y &= 26 \\ & & 13x &= 26 \\ & & x &= 2 \quad \text{Divide both sides by 13 and solve for } x. \end{aligned}$$

Study Tip

Although the addition method is also known as the elimination method, variables are eliminated when using both the substitution and addition methods. The name *addition method* specifically tells us that the elimination of a variable is accomplished by adding two equations.

Thus, $x = 2$. We back-substitute this value into either one of the given equations. We'll use the first one.

$$3x + 4y = -10 \quad \text{This is the first equation in the given system.}$$
$$3(2) + 4y = -10 \quad \text{Substitute 2 for x.}$$
$$6 + 4y = -10 \quad \text{Multiply.}$$
$$4y = -16 \quad \text{Subtract 6 from both sides.}$$
$$y = -4 \quad \text{Divide both sides by 4.}$$

We see that $x = 2$ and $y = -4$. The ordered pair $(2, -4)$ can be shown to satisfy both equations in the system. Consequently, the solution is $(2, -4)$ and the solution set is $\{(2, -4)\}$.

Solving Linear Systems by Addition

1. If necessary, rewrite both equations in the form $Ax + By = C$.
2. If necessary, multiply either equation or both equations by appropriate nonzero numbers so that the sum of the x-coefficients or the sum of the y-coefficients is 0.
3. Add the equations in step 2. The sum will be an equation in one variable.
4. Solve the equation in one variable.
5. Back-substitute the value obtained in step 4 into either of the given equations and solve for the other variable.
6. Check the solution in both of the original equations.

✓ CHECK POINT 5 Solve by the addition method:

$$4x - 7y = -16$$
$$2x + 5y = 9.$$

EXAMPLE 6 Solving a System by the Addition Method

Solve by the addition method:

$$7x = 5 - 2y$$
$$3y = 16 - 2x.$$

Solution

Step 1. Rewrite both equations in the form $Ax + By = C$. We first arrange the system so that variable terms appear on the left and constants appear on the right. We obtain

$$7x + 2y = 5 \quad \text{Add 2y to both sides of the first equation.}$$
$$2x + 3y = 16. \quad \text{Add 2x to both sides of the second equation.}$$

Step 2. If necessary, multiply either equation or both equations by appropriate numbers so that the sum of the x-coefficients or the sum of the y-coefficients is 0. We can eliminate x or y. Let's eliminate y by multiplying the first equation by 3 and the second equation by -2.

$$7x + 2y = 5 \xrightarrow{\text{Multiply by 3.}} 21x + 6y = 15$$
$$2x + 3y = 16 \xrightarrow{\text{Multiply by } -2.} -4x - 6y = -32$$

Step 3. Add the equations. Add: $17x + 0y = -17$

$$17x = -17$$

Step 4. Solve the equation in one variable. We solve $17x = -17$ by dividing both sides by 17.

$$\frac{17x}{17} = \frac{-17}{17} \quad \text{Divide both sides by 17.}$$

$$x = -1 \quad \text{Simplify.}$$

Step 5. Back-substitute and find the value of the other variable. We can back-substitute -1 for x into either one of the given equations. We'll use the second one.

$$3y = 16 - 2x \quad \text{This is the second equation in the given system.}$$

$$3y = 16 - 2(-1) \quad \text{Substitute } -1 \text{ for } x.$$

$$3y = 16 + 2 \quad \text{Multiply.}$$

$$3y = 18 \quad \text{Add.}$$

$$y = 6 \quad \text{Divide both sides by 3.}$$

We found that $x = -1$ and $y = 6$. The proposed solution is $(-1, 6)$.

Step 6. Check. Take a moment to show that $(-1, 6)$ satisfies both given equations. The solution is $(-1, 6)$ and the solution set is $\{(-1, 6)\}$.

✓ CHECK POINT 6 Solve by the addition method:

$$3x = 2 - 4y$$
$$5y = -1 - 2x.$$

Some linear systems have solutions that are not integers. If the value of one variable turns out to be a "messy" fraction, back-substitution might lead to cumbersome arithmetic. If this happens, you can return to the original system and use the addition method a second time to find the value of the other variable.

EXAMPLE 7 Solving a System by the Addition Method

Solve by the addition method:

$$\frac{x}{2} - 5y = 32$$
$$\frac{3x}{2} - 7y = 45.$$

Solution

Step 1. Rewrite both equations in the form $Ax + By = C$. Although each equation is already in this form, the coefficients of x are not integers. There is less chance for error if the coefficients for x and y in $Ax + By = C$ are integers. Consequently, we begin by clearing fractions. Multiply both sides of each equation by 2.

$$\frac{x}{2} - 5y = 32 \xrightarrow{\text{Multiply by 2.}} x - 10y = 64$$

$$\frac{3x}{2} - 7y = 45 \xrightarrow{\text{Multiply by 2.}} 3x - 14y = 90$$

Step 2. If necessary, multiply either equation or both equations by appropriate numbers so that the sum of the x-coefficients or the sum of the y-coefficients is 0. We will eliminate x. Multiply the first equation with integral coefficients by -3 and leave the second equation unchanged.

$$x - 10y = 64 \xrightarrow{\text{Multiply by } -3.} -3x + 30y = -192$$

$$3x - 14y = 90 \xrightarrow{\text{No change}} \underline{3x - 14y = 90}$$

Step 3. Add the equations.

$$\text{Add: } 0x + 16y = -102$$

$$16y = -102$$

Step 4. Solve the equation in one variable. We solve $16y = -102$ by dividing both sides by 16.

$$\frac{16y}{16} = \frac{-102}{16} \quad \text{Divide both sides by 16.}$$

$$y = -\frac{102}{16} = -\frac{51}{8} \quad \text{Simplify.}$$

Step 5. Back-substitute and find the value of the other variable. Back-substitution of $-\frac{51}{8}$ for y into either of the given equations results in cumbersome arithmetic. Instead, let's use the addition method on the system with integral coefficients from step 1 to find the value of x. Thus, we eliminate y by multiplying the first equation by -7 and the second equation by 5.

$$x - 10y = 64 \xrightarrow{\text{Multiply by } -7.} -7x + 70y = -448$$

$$3x - 14y = 90 \xrightarrow{\text{Multiply by } 5.} \underline{15x - 70y = 450}$$

$$\text{Add: } 8x = 2$$

$$x = \frac{2}{8} = \frac{1}{4}$$

We found that $x = \frac{1}{4}$ and $y = -\frac{51}{8}$. The proposed solution is $\left(\frac{1}{4}, -\frac{51}{8}\right)$.

Step 6. Check. For this system, a calculator is helpful in showing that $\left(\frac{1}{4}, -\frac{51}{8}\right)$ satisfies both of the original equations of the system. The solution is $\left(\frac{1}{4}, -\frac{51}{8}\right)$ and the solution set is $\left\{\left(\frac{1}{4}, -\frac{51}{8}\right)\right\}$.

CHECK POINT 7 Solve by the addition method:

$$\frac{3x}{2} - 2y = \frac{5}{2}$$

$$x - \frac{5y}{2} = -\frac{3}{2}.$$

5 Select the most efficient method for solving a system of linear equations.

Comparing the Three Solution Methods

The following chart compares the graphing, substitution, and addition methods for solving systems of linear equations in two variables. With increased practice, you will find it easier to select the best method for solving a particular linear system.

Comparing Solution Methods

Method	Advantages	Disadvantages
Graphing	You can see the solutions.	If the solutions do not involve integers or are too large to be seen on the graph, it's impossible to tell exactly what the solutions are.
Substitution	Gives exact solutions. Easy to use if a variable is on one side by itself.	Solutions cannot be seen. Introduces extensive work with fractions when no variable has a coefficient of 1 or -1.
Addition	Gives exact solutions. Easy to use if no variable has a coefficient of 1 or -1.	Solutions cannot be seen.

6 Identify systems that do not have exactly one ordered-pair solution.

Linear Systems Having No Solution or Infinitely Many Solutions

We have seen that a system of linear equations in two variables represents a pair of lines. The lines either intersect at one point, are parallel, or are identical. Thus, there are three possibilities for the number of solutions to a system of two linear equations.

The Number of Solutions to a System of Two Linear Equations

The number of solutions to a system of two linear equations in two variables is given by one of the following. (See **Figure 3.3.**)

Number of Solutions	What This Means Graphically
Exactly one ordered-pair solution	The two lines intersect at one point.
No solution	The two lines are parallel.
Infinitely many solutions	The two lines are identical.

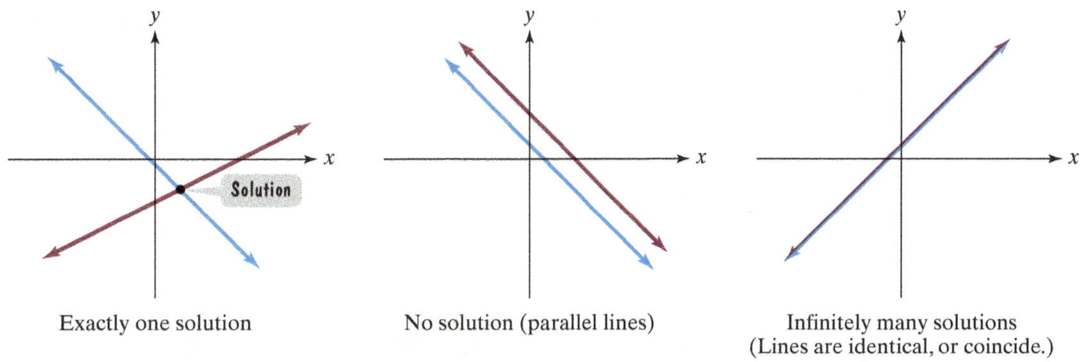

FIGURE 3.3 Possible graphs for a system of two linear equations in two variables

A linear system with no solution is called an **inconsistent system**. If you attempt to solve such a system by substitution or addition, you will eliminate both variables. A false statement, such as $0 = 6$, will be the result.

EXAMPLE 8 A System with No Solution

Solve the system:

$$\begin{aligned} 3x - 2y &= 6 \\ 6x - 4y &= 18. \end{aligned}$$

Solution Because no variable is isolated, we will use the addition method. To obtain coefficients of x that differ only in sign, we multiply the first equation by -2.

$$\begin{aligned} 3x - 2y &= 6 \xrightarrow{\text{Multiply by } -2.} & -6x + 4y &= -12 \\ 6x - 4y &= 18 \xrightarrow{\text{No change}} & 6x - 4y &= 18 \\ & \text{Add:} & 0 &= 6 \end{aligned}$$

There are no values of x and y for which $0 = 6$. No values of x and y satisfy $0x + 0y = 6$.

The false statement $0 = 6$ indicates that the system is inconsistent and has no solution. The solution set is the empty set, $\varnothing$.

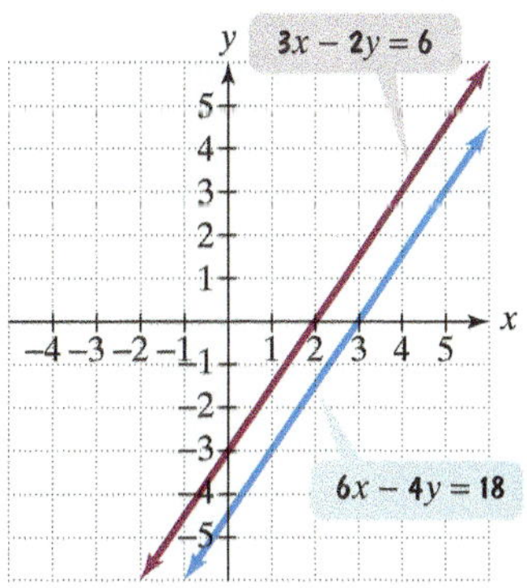

FIGURE 3.4 The graph of an inconsistent system

The lines corresponding to the two equations in Example 8 are shown in **Figure 3.4**. The lines are parallel and have no point of intersection.

Discover for Yourself

Show that the graphs of $3x - 2y = 6$ and $6x - 4y = 18$ must be parallel lines by solving each equation for y. What is the slope and the y-intercept for each line? What does this mean? If a linear system is inconsistent, what must be true about the slopes and the y-intercepts for the system's graphs?

✓ CHECK POINT 8 Solve the system:

$$5x - 2y = 4$$
$$-10x + 4y = 7$$

A linear system that has at least one solution is called a **consistent system.** Lines that intersect and lines that coincide both represent consistent systems. If the lines coincide, then the consistent system has infinitely many solutions, represented by every point on the coinciding lines.

The equations in a linear system with infinitely many solutions are called **dependent**. If you attempt to solve such a system by substitution or addition, you will eliminate both variables. However, a true statement, such as $10 = 10$, will be the result.

EXAMPLE 9 A System with Infinitely Many Solutions

Solve the system:

$$y = 3x - 2$$
$$15x - 5y = 10.$$

Solution Because the variable y is isolated in $y = 3x - 2$, the first equation, we can use the substitution method. We substitute the expression for y into the second equation.

$y = \boxed{3x - 2}$ $\qquad 15x - 5\boxed{y} = 10$ Substitute $3x - 2$ for y.

$15x - 5(3x - 2) = 10$ The substitution results in an equation in one variable.

$15x - 15x + 10 = 10$ Apply the distributive property.

$10 = 10$ Simplify.

This statement is true for all values of x and y.

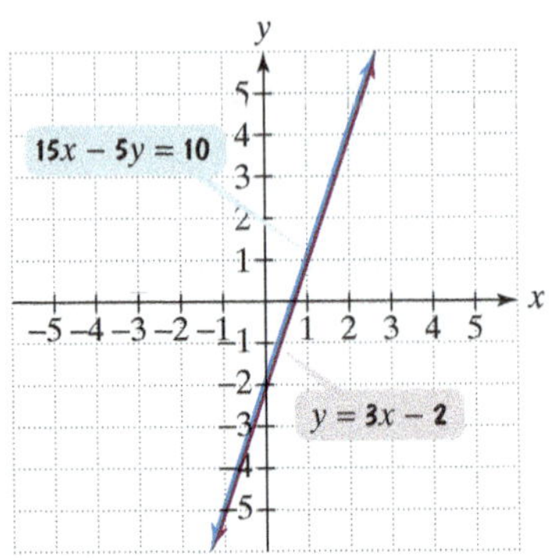

FIGURE 3.5 The graph of a system with infinitely many solutions

In our final step, both variables have been eliminated and the resulting statement, $10 = 10$, is true. This true statement indicates that the system has infinitely many solutions. The solution set consists of all points (x, y) lying on either of the coinciding lines, $y = 3x - 2$ or $15x - 5y = 10$, as shown in **Figure 3.5**.

We express the solution set for the system in one of two equivalent ways:

$$\{(x, y) \mid y = 3x - 2\} \quad \text{or} \quad \{(x, y) \mid 15x - 5y = 10\}.$$

The set of all ordered pairs (x, y) such that $y = 3x - 2$ / The set of all ordered pairs (x, y) such that $15x - 5y = 10$

Study Tip

Although the system in Example 9 has infinitely many solutions, this does not mean that any ordered pair of numbers you can form will be a solution. The ordered pair (x, y) must satisfy one of the system's equations, $y = 3x - 2$ or $15x - 5y = 10$, and there are infinitely many such ordered pairs. Because the graphs are coinciding lines, the ordered pairs that are solutions of one of the equations are also solutions of the other equation.

✓ CHECK POINT 9 Solve the system:

$$x = 4y - 8$$
$$5x - 20y = -40.$$

3.1 EXERCISE SET

MyMathLab

MathXL PRACTICE WATCH DOWNLOAD READ REVIEW

Practice Exercises

In Exercises 1–6, determine whether the given ordered pair is a solution of the system.

1. $(7, -5)$
$x - y = 12$
$x + y = 2$

2. $(-3, 1)$
$x - y = -4$
$2x + 10y = 4$

3. $(2, -1)$
$3x + 4y = 2$
$2x + 5y = 1$

4. $(4, 2)$
$2x - 5y = -2$
$3x + 4y = 18$

5. $(5, -3)$
$y = 2x - 13$
$4x + 9y = -7$

6. $(-3, -4)$
$y = 3x + 5$
$5x - 2y = -7$

In Exercises 7–24, solve each system by graphing. Identify systems with no solution and systems with infinitely many solutions, using set notation to express their solution sets.

7. $x + y = 4$
$x - y = 2$

8. $x + y = 6$
$x - y = -4$

9. $2x + y = 4$
$y = 4x + 1$

10. $x + 2y = 4$
$y = -2x - 1$

11. $3x - 2y = 6$
$x - 4y = -8$

12. $4x + y = 4$
$3x - y = 3$

13. $2x + 3y = 6$
$4x = -6y + 12$

14. $3x - 3y = 6$
$2x = 2y + 4$

15. $y = 2x - 2$
$y = -5x + 5$

16. $y = -x + 1$
$y = 3x + 5$

17. $3x - y = 4$
$6x - 2y = 4$

18. $2x - y = -4$
$4x - 2y = 6$

19. $2x + y = 4$
$4x + 3y = 10$

20. $4x - y = 9$
$x - 3y = 16$

21. $x - y = 2$
$y = 1$

22. $x + 2y = 1$
$x = 3$

23. $3x + y = 3$
$6x + 2y = 12$

24. $2x - 3y = 6$
$4x - 6y = 24$

In Exercises 25–42, solve each system by the substitution method. Identify inconsistent systems and systems with dependent equations, using set notation to express their solution sets.

25. $x + y = 6$
$y = 2x$

26. $x + y = 10$
$y = 4x$

27. $2x + 3y = 9$
$x = y + 2$

28. $3x - 4y = 18$
$y = 1 - 2x$

29. $y = -3x + 7$
$5x - 2y = 8$

30. $x = 3y + 8$
$2x - y = 6$

31. $4x + y = 5$
$2x - 3y = 13$

32. $x - 3y = 3$
$3x + 5y = -19$

33. $x - 2y = 4$
$2x - 4y = 5$

34. $x - 3y = 6$
$2x - 6y = 5$

35. $2x + 5y = -4$
$3x - y = 11$

36. $2x + 5y = 1$
$-x + 6y = 8$

37. $2(x - 1) - y = -3$
$y = 2x + 3$

38. $x + y - 1 = 2(y - x)$
$y = 3x - 1$

39. $\frac{x}{4} - \frac{y}{4} = -1$
$x + 4y = -9$

40. $\frac{x}{6} - \frac{y}{2} = \frac{1}{3}$
$x + 2y = -3$

41. $y = \frac{2}{5}x - 2$
$2x - 5y = 10$

42. $y = \frac{1}{3}x + 4$
$3y = x + 12$

In Exercises 43–58, solve each system by the addition method. Identify inconsistent systems and systems with dependent equations, using set notation to express their solution sets.

43. $x + y = 7$
$x - y = 3$

44. $2x + y = 3$
$x - y = 3$

45. $12x + 3y = 15$
$2x - 3y = 13$

46. $4x + 2y = 12$
$3x - 2y = 16$

47. $x + 3y = 2$
$4x + 5y = 1$

48. $x + 2y = -1$
$2x - y = 3$

49. $6x - y = -5$
$4x - 2y = 6$

50. $x - 2y = 5$
$5x - y = 2$

51. $3x - 5y = 11$
$2x - 6y = 2$

52. $4x - 3y = 12$
$3x - 4y = 2$

53. $2x - 5y = 13$
$5x + 3y = 17$

54. $4x + 5y = -9$
$6x - 3y = -3$

55. $2x + 6y = 8$
$3x + 9y = 12$

56. $x - 3y = -6$
$3x - 9y = 9$

57. $2x - 3y = 4$
$4x + 5y = 3$

58. $4x - 3y = 8$
$2x - 5y = -14$

In Exercises 59–82, solve each system by the method of your choice. Identify inconsistent systems and systems with dependent equations, using set notation to express solution sets.

59. $3x - 7y = 1$
$2x - 3y = -1$

60. $2x - 3y = 2$
$5x + 4y = 51$

61. $x = y + 4$
$3x + 7y = -18$

62. $y = 3x + 5$
$5x - 2y = -7$

63. $9x + \frac{4y}{3} = 5$
$4x - \frac{y}{3} = 5$

64. $\frac{x}{6} - \frac{y}{5} = -4$
$\frac{x}{4} - \frac{y}{6} = -2$

65. $\frac{1}{4}x - \frac{1}{9}y = \frac{2}{3}$
$\frac{1}{2}x - \frac{1}{3}y = 1$

66. $\frac{1}{16}x - \frac{3}{4}y = -1$
$\frac{3}{4}x + \frac{5}{2}y = 11$

67. $x = 3y - 1$
$2x - 6y = -2$

68. $x = 4y - 1$
$2x - 8y = -2$

69. $y = 2x + 1$
$y = 2x - 3$

70. $y = 2x + 4$
$y = 2x - 1$

71. $0.4x + 0.3y = 2.3$
$0.2x - 0.5y = 0.5$

72. $0.2x - y = -1.4$
$0.7x - 0.2y = -1.6$

73. $5x - 40 = 6y$
$2y = 8 - 3x$

74. $4x - 24 = 3y$
$9y = 3x - 1$

75. $3(x + y) = 6$
$3(x - y) = -36$

76. $4(x - y) = -12$
$4(x + y) = -20$

77. $3(x - 3) - 2y = 0$
$2(x - y) = -x - 3$

78. $5x + 2y = -5$
$4(x + y) = 6(2 - x)$

79. $x + 2y - 3 = 0$
$12 = 8y + 4x$

80. $2x - y - 5 = 0$
$10 = 4x - 2y$

81. $3x + 4y = 0$
$7x = 3y$

82. $5x + 8y = 20$
$4y = -5x$

Practice PLUS

In Exercises 83–84, solve each system by the method of your choice.

83. $\frac{x + 2}{2} - \frac{y + 4}{3} = 3$
$\frac{x + y}{5} = \frac{x - y}{2} - \frac{5}{2}$

84. $\frac{x - y}{3} = \frac{x + y}{2} - \frac{1}{2}$
$\frac{x + 2}{2} - 4 = \frac{y + 4}{3}$

In Exercises 85–86, solve each system for x and y, expressing either value in terms of a or b, if necessary. Assume that $a \neq 0$ and $b \neq 0$.

85. $5ax + 4y = 17$
$ax + 7y = 22$

86. $4ax + by = 3$
$6ax + 5by = 8$

87. For the linear function $f(x) = mx + b$, $f(-2) = 11$ and $f(3) = -9$. Find m and b.

88. For the linear function $f(x) = mx + b$, $f(-3) = 23$ and $f(2) = -7$. Find m and b.

Use the graphs of the linear functions to solve Exercises 89–90.

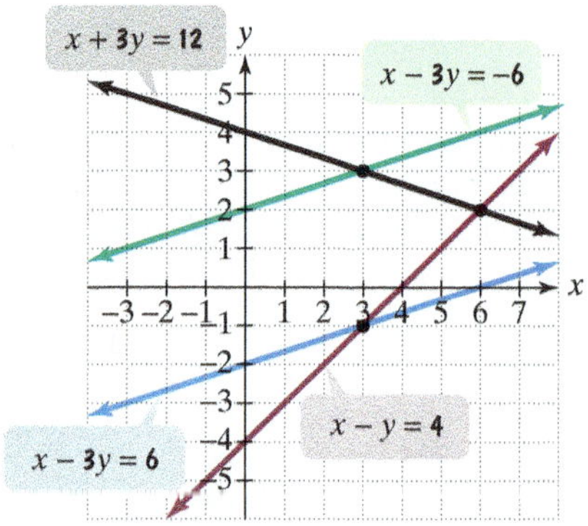

89. Write the linear system whose solution set is $\{(6, 2)\}$. Express each equation in the system in slope-intercept form.

90. Write the linear system whose solution set is $\varnothing$. Express each equation in the system in slope-intercept form.

Application Exercises

91. In 1915, the average U.S. household contained more than four people. In 2007, the average was 2.5. Large families are increasingly rare. The graphs below illustrate this trend.

Smaller United States Families

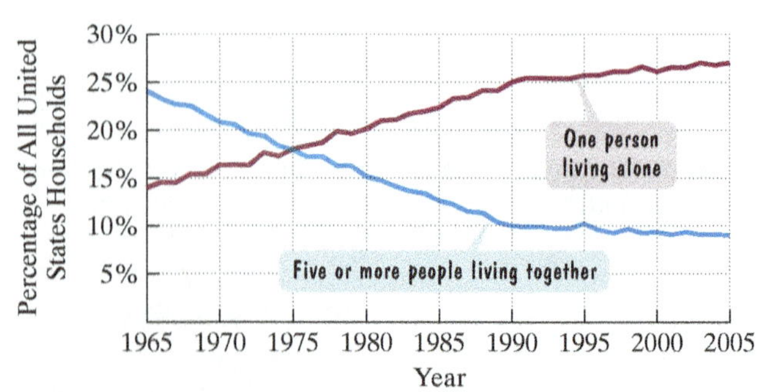

Source: U.S. Census Bureau

a. Use the graphs to estimate the point of intersection. In what year was the percentage of all U.S. households consisting of five or more people the same as the percentage of all U.S. households consisting of one person living alone? What percentage of all U.S. households did each of these groups comprise?

b. The function $0.6x + y = 24$ models the percentage, y, of all U.S. households consisting of five or more people x years after 1965. The function $y = 0.4x + 14$ models the percentage, y, of all U.S. households consisting of one person living alone x years after 1965. Use these models to determine when the percentage of all U.S. households consisting of five or more people was the same as the percentage of all U.S. households consisting of one person living alone. According to the models, what percentage of all U.S. households did each of these groups comprise?

c. How well do the models in part (b) describe the point of intersection of the graphs that you estimated in part (a)?

92. The graph shows that from 2000 through 2006, Americans unplugged land lines and switched to cellphones.

Number of Cellphone and Land-Line Customers in the United States

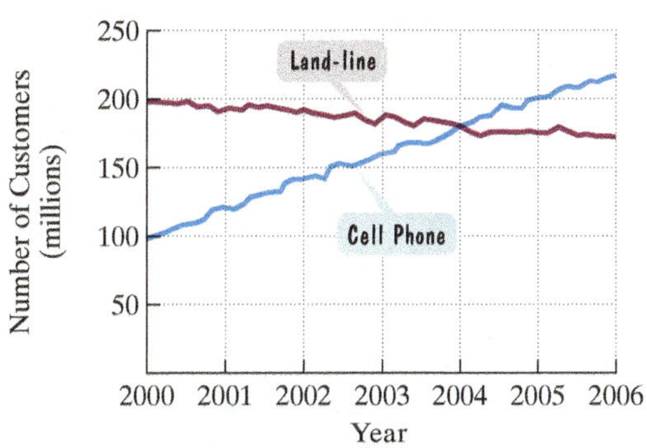

Source: Federal Communications Commission

a. Use the graphs to estimate the point of intersection. In what year was the number of cellphone and land-line customers the same? How many millions of customers were there for each?

b. The function $4.3x + y = 198$ models the number of land-line customers, in millions, x years after 2000. The function $y = 19.8x + 98$ models the number of cellphone customers, in millions, x years after 2000. Use these models to determine the year, rounded to the nearest year, when the number of cellphone and land-line customers was the same. According to the models, how many millions of customers, rounded to the nearest ten million, were there for each?

c. How well do the models in part (b) describe the point of intersection of the graphs that you estimated in part (a)?

93. Although Social Security is a problem, some projections indicate that there's a much bigger time bomb ticking in the federal budget, and that's Medicare. In 2000, the cost of Social Security was 5.48% of the gross domestic product, increasing by 0.04% of the GDP per year. In 2000, the cost of Medicare was 1.84% of the gross domestic product, increasing by 0.17% of the GDP per year. (*Source:* Congressional Budget Office)

a. Write a function that models the cost of Social Security as a percentage of the GDP x years after 2000.

b. Write a function that models the cost of Medicare as a percentage of the GDP x years after 2000.

c. In which year will the cost of Medicare and Social Security be the same? For that year, what will be the cost of each program as a percentage of the GDP? Which program will have the greater cost after that year?

94. The graph indicates that in 1984, there were 72 meals per person at take-out restaurants. For the period shown, this number increased by an average of 2.25 meals per person per year. In 1984, there were 94 meals per person at on-premise dining facilities and this number decreased by an average of 0.55 meal per person per year.

Eating Out in the United States: Average Number of Meals per Person

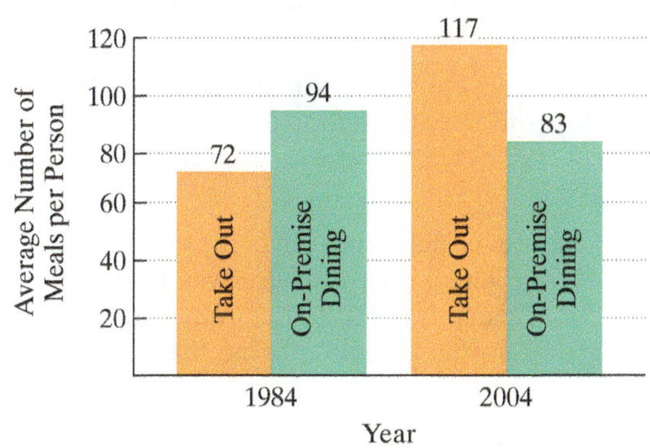

Source: The NPD Group

a. Write a function that models the average number of meals per person at take-out restaurants x years after 1984.

b. Write a function that models the average number of meals per person at on-premise dining facilities x years after 1984.

c. In which year, to the nearest whole year, was the average number of meals per person for take-out and on-premise restaurants the same? For that year, how many meals per person, to the nearest whole number, were there for each kind of restaurant? Which kind of restaurant had the greater number of meals per person after that year?

The bar graph shows the percentage of Americans who used cigarettes, by ethnicity, in 1985 and 2005. For each of the groups shown, cigarette use has been linearly decreasing. Use this information to solve Exercises 95–96.

Cigarette Use in the United States

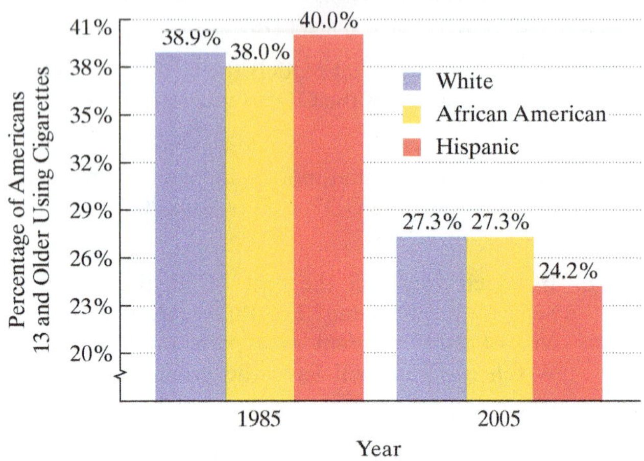

Source: Department of Health and Human Services

95. In this exercise, let x represent the number of years after 1985 and let y represent the percentage of Americans in one of the groups shown who used cigarettes.

a. Use the data points (0, 38) and (20, 27.3) to find the slope-intercept equation of the line that models the percentage of African Americans who used cigarettes, y, x years after 1985. Round the value of the slope m to two decimal places.

b. Use the data points (0, 40) and (20, 24.2) to find the slope-intercept equation of the line that models the percentage of Hispanics who used cigarettes, y, x years after 1985.

c. Use the models from parts (a) and (b) to find the year during which cigarette use was the same for African Americans and Hispanics. What percentage of each group used cigarettes during that year?

96. In this exercise, let x represent the number of years after 1985 and let y represent the percentage of Americans in one of the groups shown who used cigarettes.

a. Use the data points (0, 38.9) and (20, 27.3) to find the slope-intercept equation of the line that models the percentage of whites who used cigarettes, y, x years after 1985.

b. Use the data points (0, 40) and (20, 24.2) to find the slope-intercept equation of the line that models the percentage of Hispanics who used cigarettes, y, x years after 1985.

c. Use the models from parts (a) and (b) to find the year, to the nearest whole year, during which cigarette use was the same for whites and Hispanics. What percentage of each group, to the nearest percent, used cigarettes during that year?

An important application of systems of equations arises in connection with supply and demand. As the price of a product increases, the demand for that product decreases. However, at higher prices, suppliers are willing to produce greater quantities of the product. Exercises 97–98 involve supply and demand.

97. A chain of electronics stores sells hand-held color televisions. The weekly demand and supply models are given as follows:

Number sold per week — Demand model

$$N = -5p + 750$$

Price of television

Number supplied to the chain per week — Supply model

$$N = 2.5p.$$

a. How many hand-held color televisions can be sold and supplied at \$120 per television?

b. Find the price at which supply and demand are equal. At this price, how many televisions can be supplied and sold each week?

98. At a price of p dollars per ticket, the number of tickets to a rock concert that can be sold is given by the demand model $N = -25p + 7800$.
At a price of p dollars per ticket, the number of tickets that the concert's promoters are willing to make available is given by the supply model $N = 5p + 6000$.

a. How many tickets can be sold and supplied for \$50 per ticket?

b. Find the ticket price at which supply and demand are equal. At this price, how many tickets will be supplied and sold?

Writing in Mathematics

99. What is a system of linear equations? Provide an example with your description.

100. What is a solution of a system of linear equations?

101. Explain how to determine if an ordered pair is a solution of a system of linear equations.

102. Explain how to solve a system of linear equations by graphing.

103. Explain how to solve a system of equations using the substitution method. Use $y = 3 - 3x$ and $3x + 4y = 6$ to illustrate your explanation.

104. Explain how to solve a system of equations using the addition method. Use $5x + 8y = -1$ and $3x + y = 7$ to illustrate your explanation.

105. When is it easier to use the addition method rather than the substitution method to solve a system of equations?

106. When using the addition or substitution method, how can you tell if a system of linear equations has no solution? What is the relationship between the graphs of the two equations?

107. When using the addition or substitution method, how can you tell if a system of linear equations has infinitely many solutions? What is the relationship between the graphs of the two equations?

Technology Exercise

108. Verify your solutions to any five exercises from Exercises 7–24 by using a graphing utility to graph the two equations in the system in the same viewing rectangle. Then use the intersection feature to display the solution.

Critical Thinking Exercises

Make Sense? *In Exercises 109–112, determine whether each statement "makes sense" or "does not make sense" and explain your reasoning.*

109. Even if a linear system has a solution set involving fractions, such as $\{(\frac{8}{11}, \frac{43}{11})\}$, I can use graphs to determine if the solution set is reasonable.

110. If I add the equations on the right and solve the resulting equation for x, I will obtain the x-coordinate of the intersection point of the lines represented by the equations on the left.

$$\begin{aligned} 4x - 6y &= 1 \longrightarrow 20x - 30y = 5 \\ 3x + 5y &= -8 \longrightarrow 18x + 30y = -8 \end{aligned}$$

111. In the previous chapter, we developed models for life expectancy, y, for U.S. men and women born x years after 1960:

$$y = 0.22x + 65.7 \quad \text{Men}$$
$$y = 0.17x + 72.9. \quad \text{Women}$$

The system indicates that life expectancy for men is increasing at a faster rate than for women, so if these trends continue, life expectancies for men and women will be the same for some future birth year.

112. Here are two models that describe winning times for the Olympic 400-meter run, y, in seconds, x years after 1968:

$$y = -0.02433x + 44.43 \quad \text{Men}$$
$$y = -0.08883x + 50.86. \quad \text{Women}$$

The system indicates that winning times have been decreasing more rapidly for women than for men, so if these trends continue, there will be a year when winning times for men and women are the same.

In Exercises 113–116, determine whether each statement is true or false. If the statement is false, make the necessary change(s) to produce a true statement.

113. The addition method cannot be used to eliminate either variable in a system of two equations in two variables.

114. The solution set of the system

$$\begin{aligned} 5x - y &= 1 \\ 10x - 2y &= 2 \end{aligned}$$

is $\{(2, 9)\}$.

115. A system of linear equations can have a solution set consisting of precisely two ordered pairs.

116. The solution set of the system

$$\begin{aligned} y &= 4x - 3 \\ y &= 4x + 5 \end{aligned}$$

is the empty set.

117. Determine a and b so that $(2, 1)$ is a solution of this system:

$$\begin{aligned} ax - by &= 4 \\ bx + ay &= 7. \end{aligned}$$

118. Write a system of equations having $\{(-2, 7)\}$ as a solution set. (More than one system is possible.)

119. Solve the system for x and y in terms of a_1, b_1, c_1, a_2, b_2, and c_2:

$$a_1x + b_1y = c_1$$

$$a_2x + b_2y = c_2.$$

Review Exercises

120. Solve: $6x = 10 + 5(3x - 4)$. (Section 1.4, Example 3)

121. Simplify: $(4x^2y^4)^2(-2x^5y^0)^3$. (Section 1.6, Example 9)

122. If $f(x) = x^2 - 3x + 7$, find $f(-1)$. (Section 2.1, Example 3)

Preview Exercises

Exercises 123–125 will help you prepare for the material covered in the next section.

123. The formula $I = Pr$ is used to find the simple interest, I, earned for one year when the principal, P, is invested at an annual interest rate, r. Write an expression for the total interest earned on a principal of x dollars at a rate of 15% ($r = 0.15$) and a principal of y dollars at a rate of 7% ($r = 0.07$).

124. A chemist working on a flu vaccine needs to obtain 50 milliliters of a 30% sodium-iodine solution. How many milliliters of sodium-iodine are needed in the solution?

125. A company that manufactures running shoes sells them at \$80 per pair. Write an expression for the revenue that is generated by selling x pairs of shoes.

SECTION 3.2

Problem Solving and Business Applications Using Systems of Equations

Objectives

1. Solve problems using systems of equations.
2. Use functions to model revenue, cost, and profit, and perform a break-even analysis.

Driving through your neighborhood, you see kids selling lemonade. Would it surprise you to know that this activity can be analyzed using functions and systems of equations? By doing so, you will view profit and loss in the business world in a new way. In this section, we use systems of equations to solve problems and model business ventures.

1 Solve problems using systems of equations.

A Strategy for Solving Word Problems Using Systems of Equations

When we solved problems in Chapter 1, we let x represent a quantity that was unknown. Problems in this section involve two unknown quantities. We will let x and y represent these quantities. We then translate from the verbal conditions of the problem into a *system* of linear equations.

EXAMPLE 1 Solving a Problem Involving Energy Efficiency of Building Materials

A heat-loss survey by an electric company indicated that a wall of a house containing 40 square feet of glass and 60 square feet of plaster lost 1920 Btu (British thermal units) of heat. A second wall containing 10 square feet of glass and 100 square feet of plaster lost 1160 Btu of heat. Determine the heat lost per square foot for the glass and for the plaster.

Study Tip

There is great value in reasoning through a word problem. This value comes from the problem-solving skills that are attained and is often more important than the specific problem or its solution.

Solution

Step 1. Use variables to represent unknown quantities.

Let x = the heat lost per square foot for the glass.

Let y = the heat lost per square foot for the plaster.

Step 2. Write a system of equations that models the problem's conditions. The heat loss for each wall is the heat lost by the glass plus the heat lost by the plaster. One wall containing 40 square feet of glass and 60 square feet of plaster lost 1920 Btu of heat.

Heat lost by the glass + heat lost by the plaster = total heat lost.

(Number of ft²) · (heat lost per ft²) + (number of ft²) · (heat lost per ft²) = total heat lost.

$$40 \cdot x + 60 \cdot y = 1920$$

A second wall containing 10 square feet of glass and 100 square feet of plaster lost 1160 Btu of heat.

Heat lost by the glass + heat lost by the plaster = total heat lost.

$$\begin{pmatrix}\text{Number}\\ \text{of ft}^2\end{pmatrix} \cdot \begin{pmatrix}\text{heat lost}\\ \text{per ft}^2\end{pmatrix} + \begin{pmatrix}\text{number}\\ \text{of ft}^2\end{pmatrix} \cdot \begin{pmatrix}\text{heat lost}\\ \text{per ft}^2\end{pmatrix} = \text{total heat lost.}$$

$$10 \cdot x + 100 \cdot y = 1160$$

Step 3. Solve the system and answer the problem's question. The system

$$\begin{aligned} 40x + 60y &= 1920 \\ 10x + 100y &= 1160 \end{aligned}$$

can be solved by addition. We'll multiply the second equation by -4 and then add equations to eliminate x.

$$\begin{aligned} 40x + 60y = 1920 &\xrightarrow{\text{No change}} & 40x + 60y &= 1920 \\ 10x + 100y = 1160 &\xrightarrow{\text{Multiply by } -4.} & -40x - 400y &= -4640 \\ &\text{Add:} & -340y &= -2720 \\ && y &= \frac{-2720}{-340} = 8 \end{aligned}$$

Now we can find the value of x by back-substituting 8 for y in either of the system's equations.

$$10x + 100y = 1160 \quad \text{We'll use the second equation.}$$

$$10x + 100(8) = 1160 \quad \text{Back-substitute 8 for } y.$$

$$10x + 800 = 1160 \quad \text{Multiply.}$$

$$10x = 360 \quad \text{Subtract 800 from both sides.}$$

$$x = 36 \quad \text{Divide both sides by 10.}$$

We see that $x = 36$ and $y = 8$. Because x represents heat lost per square foot for the glass and y for the plaster, the glass lost 36 Btu of heat per square foot and the plaster lost 8 Btu per square foot.

Step 4. Check the proposed solution in the original wording of the problem. The problem states that the wall with 40 square feet of glass and 60 square feet of plaster lost 1920 Btu.

$$40(36) + 60(8) = 1440 + 480 = 1920 \text{ Btu of heat}$$

Proposed solution is 36 Btu per ft² for glass and 8 Btu per ft² for plaster.

Our proposed solution checks with the first statement. The problem also states that the wall with 10 square feet of glass and 100 square feet of plaster lost 1160 Btu.

$$10(36) + 100(8) = 360 + 800 = 1160 \text{ Btu of heat}$$

Our proposed solution also checks with the second statement.

✓ CHECK POINT 1 University of Arkansas researchers discovered that we underestimate the number of calories in restaurant meals. The next time you eat out, take the number of calories you think you ate and double it. The researchers concluded that this number should be a more accurate estimate. The actual number of calories in one portion of hamburger and fries and two portions of fettuccine Alfredo is 4240. The actual number of calories in two portions of hamburger and fries and one portion of fettuccine Alfredo is 3980. Find the actual number of calories in each of these dishes. (*Source*: *Consumer Reports*, January/February, 2007)

Test Your Calorie I.Q.

Hamburger and Fries
Average guess:
777 calories

Fettuccine Alfredo
Average guess:
704 calories

Next, we will solve problems involving investments, mixtures, and motion with systems of equations. We will continue using our four-step problem-solving strategy. We will also use tables to help organize the information in the problems.

Dual Investments with Simple Interest

Simple interest involves interest calculated only on the amount of money that we invest, called the **principal**. The formula $I = Pr$ is used to find the simple interest, I, earned for one year when the principal, P, is invested at an annual interest rate, r. Dual investment problems involve different amounts of money in two or more investments, each paying a different rate.

EXAMPLE 2 Solving a Dual Investment Problem

Your grandmother needs your help. She has \$50,000 to invest. Part of this money is to be invested in noninsured bonds paying 15% annual interest. The rest of this money is to be invested in a government-insured certificate of deposit paying 7% annual interest. She told you that she requires \$6000 per year in extra income from both of these investments. How much money should be placed in each investment?

Solution

Step 1. Use variables to represent unknown quantities.

Let x = the amount invested in the 15% noninsured bonds.

Let y = the amount invested in the 7% certificate of deposit.

Step 2. Write a system of equations that models the problem's conditions. Because Grandma has \$50,000 to invest,

The amount invested at 15%	plus	the amount invested at 7%	equals	\$50,000.
x	$+$	y	$=$	50,000

Furthermore, Grandma requires \$6000 in total interest. We can use a table to organize the information in the problem and obtain a second equation.

	Principal (amount invested)	×	Interest rate	=	Interest earned
15% Investment	x		0.15		$0.15x$
7% Investment	y		0.07		$0.07y$

The interest for the two investments combined must be $6000.

Interest from the 15% investment	plus	interest from the 7% investment	is	$6000.
$0.15x$	$+$	$0.07y$	$=$	6000

Step 3. Solve the system and answer the problem's question. The system

$$x + y = 50{,}000$$
$$0.15x + 0.07y = 6000$$

can be solved by substitution or addition. Substitution works well because both variables in the first equation have coefficients of 1. Addition also works well; if we multiply the first equation by -0.15 or -0.07, adding equations will eliminate a variable. We will use addition.

$$x + y = 50{,}000 \xrightarrow{\text{Multiply by } -0.07.} -0.07x - 0.07y = -3500$$
$$0.15x + 0.07y = 6000 \xrightarrow{\text{No change}} \underline{0.15x + 0.07y = 6000}$$
$$\text{Add: } 0.08x = 2500$$
$$x = \frac{2500}{0.08}$$
$$x = 31{,}250$$

Because x represents the amount that should be invested at 15%, Grandma should place $31,250 in 15% noninsured bonds. Now we can find y, the amount that she should place in the 7% certificate of deposit. We do so by back-substituting 31,250 for x in either of the system's equations.

$x + y = 50{,}000$ We'll use the first equation.

$31{,}250 + y = 50{,}000$ Back-substitute 31,250 for x.

$y = 18{,}750$ Subtract 31,250 from both sides.

Because $x = 31{,}250$ and $y = 18{,}750$, Grandma should invest $31,250 at 15% and $18,750 at 7%.

Step 4. Check the proposed answers in the original wording of the problem. Has Grandma invested $50,000?

$$\$31{,}250 + \$18{,}750 = \$50{,}000$$

Yes, all her money was placed in the dual investments. Can she count on $6000 interest? The interest earned on $31,250 at 15% is ($31,250)(0.15), or $4687.50. The interest earned on $18,750 at 7% is ($18,750)(0.07), or $1312.50. The total interest is $4687.50 + $1312.50, or $6000, exactly as it should be. You've made your grandmother happy. (Now if you would just visit her more often . . .)

✓ **CHECK POINT 2** You inherited $5000 with the stipulation that for the first year the money had to be invested in two funds paying 9% and 11% annual interest. How much did you invest at each rate if the total interest earned for the year was $487?

Problems Involving Mixtures

Chemists and pharmacists often have to change the concentration of solutions and other mixtures. In these situations, the amount of a particular ingredient in the solution or mixture is expressed as a percentage of the total.

EXAMPLE 3 Solving a Mixture Problem

A chemist working on a flu vaccine needs to mix a 10% sodium-iodine solution with a 60% sodium-iodine solution to obtain 50 milliliters of a 30% sodium-iodine solution. How many milliliters of the 10% solution and of the 60% solution should be mixed?

Solution

Step 1. Use variables to represent unknown quantities.

Let x = the number of milliliters of the 10% solution to be used in the mixture.

Let y = the number of milliliters of the 60% solution to be used in the mixture.

Step 2. Write a system of equations that models the problem's conditions. The situation is illustrated in **Figure 3.6**. The chemist needs 50 milliliters of a 30% sodium-iodine solution. We form a table that shows the amount of sodium-iodine in each of the three solutions.

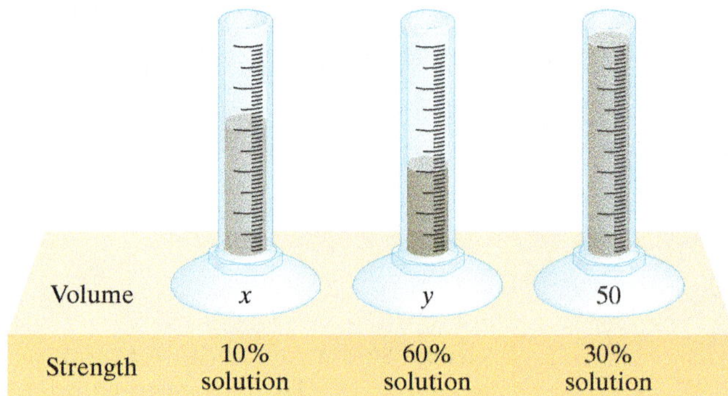

FIGURE 3.6

Solution	Number of Milliliters	×	Percent of Sodium-Iodine	=	Amount of Sodium-Iodine
10% Solution	x		$10\% = 0.1$		$0.1x$
60% Solution	y		$60\% = 0.6$		$0.6y$
30% Mixture	50		$30\% = 0.3$		$0.3(50) = 15$

The chemist needs to obtain a 50-milliliter mixture.

The number of milliliters used of the 10% solution	plus	the number of milliliters used of the 60% solution	must equal	50 milliliters.
x	$+$	y	$=$	50

The 50-milliliter mixture must be 30% sodium-iodine. The amount of sodium-iodine must be 30% of 50, or $(0.3)(50) = 15$ milliliters.

Amount of sodium-iodine in the 10% solution	plus	amount of sodium-iodine in the 60% solution	equals	amount of sodium-iodine in the mixture.
$0.1x$	$+$	$0.6y$	$=$	15

Step 3. Solve the system and answer the problem's question. The system

$$x + y = 50$$
$$0.1x + 0.6y = 15$$

can be solved by substitution or addition. Let's use substitution. The first equation can easily be solved for x or y. Solving for y, we obtain $y = 50 - x$.

$$y = \boxed{50 - x} \qquad 0.1x + 0.6\boxed{y} = 15$$

We substitute $50 - x$ for y in the second equation. This gives us an equation in one variable.

$$0.1x + 0.6(50 - x) = 15 \quad \text{This equation contains one variable, } x.$$
$$0.1x + 30 - 0.6x = 15 \quad \text{Apply the distributive property.}$$
$$-0.5x + 30 = 15 \quad \text{Combine like terms.}$$
$$-0.5x = -15 \quad \text{Subtract 30 from both sides.}$$
$$x = \frac{-15}{-0.5} = 30 \quad \text{Divide both sides by } -0.5.$$

Back-substituting 30 for x in either of the system's equations ($x + y = 50$ is easier to use) gives $y = 20$. Because x represents the number of milliliters of the 10% solution and y the number of milliliters of the 60% solution, the chemist should mix 30 milliliters of the 10% solution with 20 milliliters of the 60% solution.

Step 4. **Check the proposed solution in the original wording of the problem.** The problem states that the chemist needs 50 milliliters of a 30% sodium-iodine solution. The amount of sodium-iodine in this mixture is 0.3(50), or 15 milliliters. The amount of sodium-iodine in 30 milliliters of the 10% solution is 0.1(30), or 3 milliliters. The amount of sodium-iodine in 20 milliliters of the 60% solution is $0.6(20) = 12$ milliliters. The amount of sodium-iodine in the two solutions used in the mixture is 3 milliliters + 12 milliliters, or 15 milliliters, exactly as it should be.

✓ **CHECK POINT 3** A chemist needs to mix a 12% acid solution with a 20% acid solution to obtain 160 ounces of a 15% acid solution. How many ounces of each of the acid solutions must be used?

Study Tip

Problems involving dual investments and problems involving mixtures are both based on the same idea: The total amount times the rate gives the amount.

Dual Investment Problems: principal · rate = interest

(principal: total money amount; rate: rate of interest; interest: amount of interest)

Percents are expressed as decimals in these equations.

Mixture Problems: solution · concentration = ingredient

(solution: total solution amount; concentration: percent of ingredient, or rate of concentration; ingredient: amount of ingredient)

Our dual investment problem involved mixing two investments. Our mixture problem involved mixing two liquids. The equations in these problems are obtained from similar conditions:

Dual Investment Problems

Interest from investment 1 + interest from investment 2 = amount of interest from mixed investments.

Mixture Problems

Ingredient amount in solution 1 + ingredient amount in solution 2 = amount of ingredient in mixture.

Being aware of the similarities between dual investment and mixture problems should make you a better problem solver in a variety of situations that involve mixtures.

Problems Involving Motion

We have seen that the distance, d, covered by any moving body is the product of its average rate, r, and its time in motion, t:

$$d = rt. \quad \text{Distance equals rate times time.}$$

Wind and water current have the effect of increasing or decreasing a traveler's rate.

Study Tip

It is not always necessary to use x and y to represent a problem's variables. Select letters that help you remember what the variables represent. For example, in Example 4, you may prefer using p and w rather than x and y:

p = plane's average rate in still air

w = wind's average rate.

EXAMPLE 4 Solving a Motion Problem

When a small airplane flies with the wind, it can travel 450 miles in 3 hours. When the same airplane flies in the opposite direction against the wind, it takes 5 hours to fly the same distance. Find the average rate of the plane in still air and the average rate of the wind.

Solution

Step 1. Use variables to represent unknown quantities.

Let x = the average rate of the plane in still air.

Let y = the average rate of the wind.

Step 2. Write a system of equations that models the problem's conditions. As it travels with the wind, the plane's rate is increased. The net rate is its rate in still air, x, plus the rate of the wind, y, given by the expression $x + y$. As it travels against the wind, the plane's rate is decreased. The net rate is its rate in still air, x, minus the rate of the wind, y, given by the expression $x - y$. Here is a chart that summarizes the problem's information and includes the increased and decreased rates.

	Rate	×	Time	=	Distance
Trip with the Wind	$x + y$		3		$3(x + y)$
Trip against the Wind	$x - y$		5		$5(x - y)$

The problem states that the distance in each direction is 450 miles. We use this information to write our system of equations.

The distance of the trip with the wind | is | 450 miles.

$$3(x + y) = 450$$

The distance of the trip against the wind | is | 450 miles.

$$5(x - y) = 450$$

Step 3. Solve the system and answer the problem's question. We can simplify the system by dividing both sides of the equations by 3 and 5, respectively.

$$3(x + y) = 450 \xrightarrow{\text{Divide by 3.}} x + y = 150$$

$$5(x - y) = 450 \xrightarrow{\text{Divide by 5.}} x - y = 90$$

Solve the system on the right by the addition method.

$$\begin{aligned} x + y &= 150 \\ x - y &= 90 \\ \text{Add: } 2x \quad &= 240 \\ x &= 120 \quad \text{Divide both sides by 2.} \end{aligned}$$

Back-substituting 120 for x in either of the system's equations gives $y = 30$. Because $x = 120$ and $y = 30$, the average rate of the plane in still air is 120 miles per hour and the average rate of the wind is 30 miles per hour.

Step 4. Check the proposed solution in the original wording of the problem. The problem states that the distance in each direction is 450 miles. The average rate of the plane with the wind is $120 + 30 = 150$ miles per hour. In 3 hours, it travels $150 \cdot 3$, or 450 miles, which checks with the stated condition. Furthermore, the average rate of the plane against the wind is $120 - 30 = 90$ miles per hour. In 5 hours, it travels $90 \cdot 5 = 450$ miles, which is the stated distance.

CHECK POINT 4 With the current, a motorboat can travel 84 miles in 2 hours. Against the current, the same trip takes 3 hours. Find the average rate of the boat in still water and the average rate of the current.

---

2 Use functions to model revenue, cost, and profit, and perform a break-even analysis.

Functions of Business: Break-Even Analysis

Suppose that a company produces and sells x units of a product. Its *revenue* is the money generated by selling x units of the product. Its *cost* is the cost of producing x units of the product.

Revenue and Cost Functions

A company produces and sells x units of a product.

Revenue Function

$$R(x) = (\text{price per unit sold})x$$

Cost Function

$$C(x) = \text{fixed cost} + (\text{cost per unit produced})x$$

The point of intersection of the graphs of the revenue and cost functions is called the **break-even point**. The x-coordinate of the point reveals the number of units that a company must produce and sell so that money coming in, the revenue, is equal to money going out, the cost. The y-coordinate of the break-even point gives the amount of money coming in and going out. Example 5 illustrates the use of the substitution method in determining a company's break-even point.

EXAMPLE 5 Finding a Break-Even Point

Technology is now promising to bring light, fast, and beautiful wheelchairs to millions of disabled people. A company is planning to manufacture these radically different wheelchairs. Fixed cost will be \$500,000 and it will cost \$400 to produce each wheelchair. Each wheelchair will be sold for \$600.

a. Write the cost function, C, of producing x wheelchairs.

b. Write the revenue function, R, from the sale of x wheelchairs.

c. Determine the break-even point. Describe what this means.

Solution

a. The cost function is the sum of the fixed cost and variable cost.

Fixed cost of \$500,000 | plus | Variable cost: \$400 for each chair produced

$$C(x) = 500{,}000 + 400x$$

b. The revenue function is the money generated from the sale of x wheelchairs.

Revenue per chair, \$600, times | the number of chairs sold

$$R(x) = 600x$$

c. The break-even point occurs where the graphs of C and R intersect. Thus, we find this point by solving the system

$$\begin{aligned} C(x) &= 500{,}000 + 400x \\ R(x) &= 600x \end{aligned} \quad \text{or} \quad \begin{aligned} y &= 500{,}000 + 400x \\ y &= 600x. \end{aligned}$$

Using substitution, we can substitute $600x$ for y in the first equation.

$$600x = 500{,}000 + 400x$$ Substitute $600x$ for y in $y = 500{,}000 + 400x$.

$$200x = 500{,}000$$ Subtract $400x$ from both sides.

$$x = 2500$$ Divide both sides by 200.

Back-substituting 2500 for x in either of the system's equations (or functions), we obtain

$$R(2500) = 600(2500) = 1{,}500{,}000.$$

We used $R(x) = 600x$.

The break-even point is (2500, 1,500,000). This means that the company will break even if it produces and sells 2500 wheelchairs. At this level, the money coming in is equal to the money going out: \$1,500,000.

Figure 3.7 shows the graphs of the revenue and cost functions for the wheelchair business. Similar graphs and models apply no matter how small or large a business venture may be.

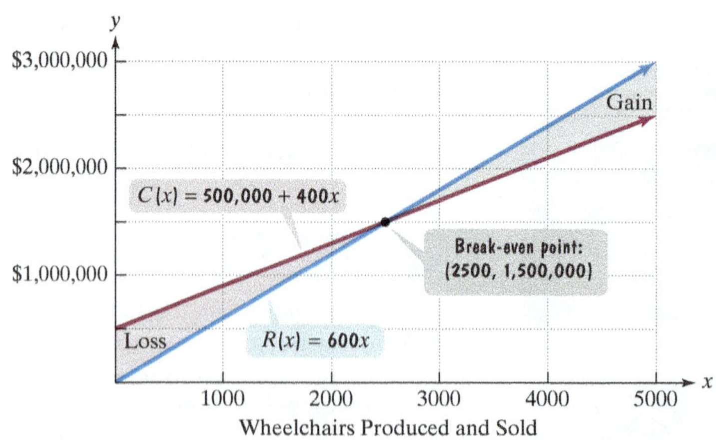

FIGURE 3.7

The intersection point confirms that the company breaks even by producing and selling 2500 wheelchairs. Can you see what happens for $x < 2500$? The red cost graph lies above the blue revenue graph. The cost is greater than the revenue and the business is losing money. Thus, if they sell fewer than 2500 wheelchairs, the result is a *loss*. By contrast, look at what happens for $x > 2500$. The blue revenue graph lies above the red cost graph. The revenue is greater than the cost and the business is making money. Thus, if they sell more than 2500 wheelchairs, the result is a *gain*.

✓ CHECK POINT 5 A company that manufactures running shoes has a fixed cost of \$300,000. Additionally, it costs \$30 to produce each pair of shoes. The shoes are sold at \$80 per pair.

a. Write the cost function, C, of producing x pairs of running shoes.

b. Write the revenue function, R, from the sale of x pairs of running shoes.

c. Determine the break-even point. Describe what this means.

What does every entrepreneur, from a kid selling lemonade to Donald Trump, want to do? Generate profit, of course. The *profit* made is the money taken in, or the revenue, minus the money spent, or the cost. This relationship between revenue and cost allows us to define the *profit function*, $P(x)$.

The Profit Function

The profit, $P(x)$, generated after producing and selling x units of a product is given by the **profit function**

$$P(x) = R(x) - C(x),$$

where R and C are the revenue and cost functions, respectively.

EXAMPLE 6 Writing a Profit Function

Use the revenue and cost functions for the wheelchair business in Example 5,

$$R(x) = 600x \quad \text{and} \quad C(x) = 500{,}000 + 400x,$$

to write the profit function for producing and selling x wheelchairs.

Solution The profit function is the difference between the revenue function and the cost function.

$$\begin{aligned} P(x) &= R(x) - C(x) && \text{This is the definition of the profit function.} \\ &= 600x - (500{,}000 + 400x) && \text{Substitute the given functions.} \\ &= 600x - 500{,}000 - 400x && \text{Distribute } -1 \text{ to each term in parentheses.} \\ &= 200x - 500{,}000 && \text{Simplify: } 600x - 400x = 200x. \end{aligned}$$

The profit function is $P(x) = 200x - 500{,}000$. ■

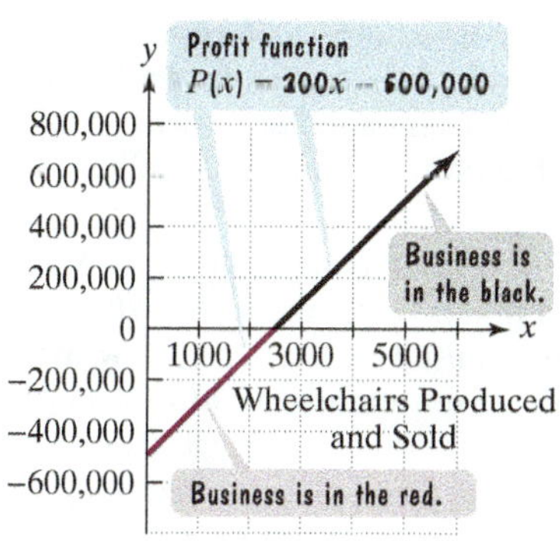

FIGURE 3.8

The graph of the profit function for the wheelchair business, $P(x) = 200x - 500{,}000$, is shown in **Figure 3.8**. The red portion lies below the x-axis and shows a loss when fewer than 2500 wheelchairs are sold. The business is "in the red." The black portion lies above the x-axis and shows a gain when more than 2500 wheelchairs are sold. The wheelchair business is "in the black."

✓ CHECK POINT 6 Use the revenue and cost functions that you obtained in Check Point 5 to write the profit function for producing and selling x pairs of running shoes. $P(x) = 50x - 300{,}000$

3.2 EXERCISE SET

MyMathLab — Math XL PRACTICE — WATCH — DOWNLOAD — READ — REVIEW

Practice Exercises

In Exercises 1–4, let x represent one number and let y represent the other number. Use the given conditions to write a system of equations. Solve the system and find the numbers.

1. The sum of two numbers is 7. If one number is subtracted from the other, the result is −1. Find the numbers.
2. The sum of two numbers is 2. If one number is subtracted from the other, the result is 8. Find the numbers.
3. Three times a first number decreased by a second number is 1. The first number increased by twice the second number is 12. Find the numbers.
4. The sum of three times a first number and twice a second number is 8. If the second number is subtracted from twice the first number, the result is 3. Find the numbers.

In Exercises 5–8, cost and revenue functions for producing and selling x units of a product are given. Cost and revenue are expressed in dollars.

a. *Find the number of units that must be produced and sold to break even. At this level, what is the dollar amount coming in and going out?*
b. *Write the profit function from producing and selling x units of the product.*

5. $C(x) = 25{,}500 + 15x$
   $R(x) = 32x$
6. $C(x) = 15{,}000 + 12x$
   $R(x) = 32x$
7. $C(x) = 105x + 70{,}000$
   $R(x) = 245x$
8. $C(x) = 1.2x + 1500$
   $R(x) = 1.7x$

Application Exercises

In Exercises 9–40, use the four-step strategy to solve each problem.

9. At some point, it's time to kick, or gently ease, kids off the parental gravy train. The circle graph shows the percentage of parents who think significant financial support should end at various milestones.

Percentage of Parents Ending a Child's Financial Support at Various Milestones

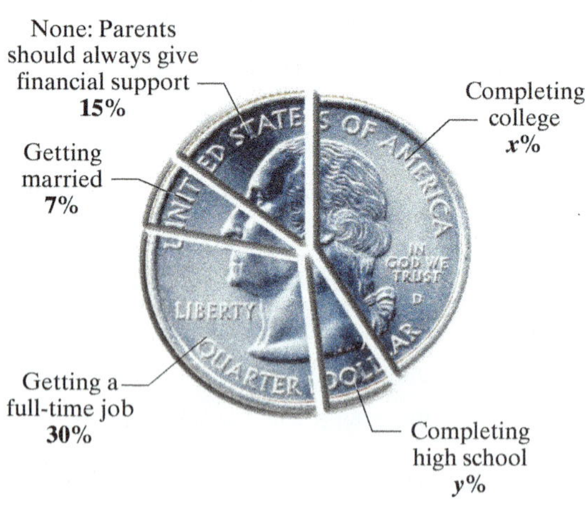

Source: Consumer Reports Money Adviser, July 2006

A total of 48% of parents would end financial support after completing education. The difference in the percentage who would end this support after completing college and after completing high school is 34%. Find the percentage of parents who would end financial support after a child completes college and the percentage who would end financial support after a child completes high school.

10. In 2007, there were approximately 730,000 homeless people in the United States. The circle graph shows the breakdown of the nation's homeless population.

The United States Homeless Population

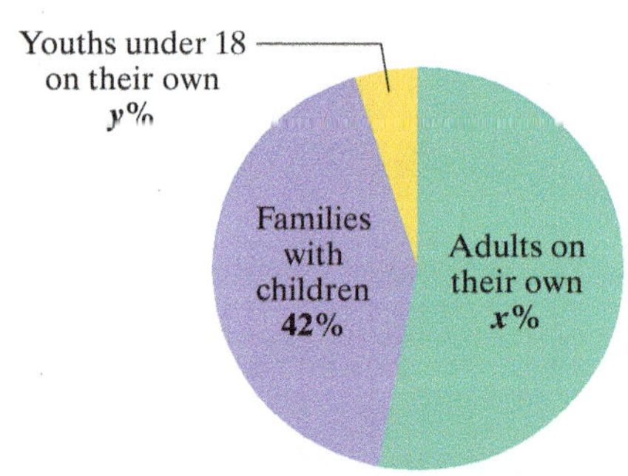

Source: U.S. Department of Housing and Urban Development

A total of 58% of the homeless consist of people on their own. The difference in the percentage of the population consisting of adults on their own and youths on their own is 48%. Find the percentage of the U.S. homeless population consisting of adults on their own and the percentage consisting of youths under 18 on their own.

11. One week a computer store sold a total of 36 computers and external hard drives. The revenue from these sales was $27,710. If computers sold for $1180 per unit and hard drives for $125 per unit, how many of each did the store sell?
12. There were 180 people at a civic club fundraiser. Members paid $4.50 per ticket and nonmembers paid $8.25 per ticket. If total receipts amounted to $1222.50, how many members and how many nonmembers attended the fundraiser?
13. You invested $7000 in two accounts paying 6% and 8% annual interest. If the total interest earned for the year was $520, how much was invested at each rate?

14. You invested $11,000 in stocks and bonds, paying 5% and 8% annual interest. If the total interest earned for the year was $730, how much was invested in stocks and how much was invested in bonds?

15. You invested money in two funds. Last year, the first fund paid a dividend of 9% and the second a dividend of 3%, and you received a total of $900. This year, the first fund paid a 10% dividend and the second only 1%, and you received a total of $860. How much money did you invest in each fund?

16. You invested money in two funds. Last year, the first fund paid a dividend of 8% and the second a dividend of 5%, and you received a total of $1330. This year, the first fund paid a 12% dividend and the second only 2%, and you received a total of $1500. How much money did you invest in each fund?

17. Things did not go quite as planned. You invested $20,000, part of it in a stock that paid 12% annual interest. However, the rest of the money suffered a 5% loss. If the total annual income from both investments was $1890, how much was invested at each rate?

18. Things did not go quite as planned. You invested $30,000, part of it in a stock that paid 14% annual interest. However, the rest of the money suffered a 6% loss. If the total annual income from both investments was $200, how much was invested at each rate?

19. A wine company needs to blend a California wine with a 5% alcohol content and a French wine with a 9% alcohol content to obtain 200 gallons of wine with a 7% alcohol content. How many gallons of each kind of wine must be used?

20. A jeweler needs to mix an alloy with a 16% gold content and an alloy with a 28% gold content to obtain 32 ounces of a new alloy with a 25% gold content. How many ounces of each of the original alloys must be used?

21. For thousands of years, gold has been considered one of Earth's most precious metals. One hundred percent pure gold is 24-karat gold, which is too soft to be made into jewelry. In the United States, most gold jewelry is 14-karat gold, approximately 58% gold. If 18-karat gold is 75% gold and 12-karat gold is 50% gold, how much of each should be used to make a 14-karat gold bracelet weighing 300 grams?

23. The manager of a candystand at a large multiplex cinema has a popular candy that sells for $1.60 per pound. The manager notices a different candy worth $2.10 per pound that is not selling well. The manager decides to form a mixture of both types of candy to help clear the inventory of the more expensive type. How many pounds of each kind of candy should be used to create a 75-pound mixture selling for $1.90 per pound?

24. A grocer needs to mix raisins at $2.00 per pound with granola at $3.25 per pound to obtain 10 pounds of a mixture that costs $2.50 per pound. How many pounds of raisins and how many pounds of granola must be used?

25. A coin purse contains a mixture of 15 coins in nickels and dimes. The coins have a total value of $1.10. Determine the number of nickels and the number of dimes in the purse.

26. A coin purse contains a mixture of 15 coins in dimes and quarters. The coins have a total value of $3.30. Determine the number of dimes and the number of quarters in the purse.

27. When a small plane flies with the wind, it can travel 800 miles in 5 hours. When the plane flies in the opposite direction, against the wind, it takes 8 hours to fly the same distance. Find the rate of the plane in still air and the rate of the wind.

28. When a plane flies with the wind, it can travel 4200 miles in 6 hours. When the plane flies in the opposite direction, against the wind, it takes 7 hours to fly the same distance. Find the rate of the plane in still air and the rate of the wind.

29. A boat's crew rowed 16 kilometers downstream, with the current, in 2 hours. The return trip upstream, against the current, covered the same distance, but took 4 hours. Find the crew's rowing rate in still water and the rate of the current.

30. A motorboat traveled 36 miles downstream, with the current, in 1.5 hours. The return trip upstream, against the current, covered the same distance, but took 2 hours. Find the boat's rate in still water and the rate of the current.

22. In the "Peanuts" cartoon shown, solve the problem that is sending Peppermint Patty into an agitated state. How much cream and how much milk, to the nearest hundredth of a gallon, must be mixed together to obtain 50 gallons of cream that contains 12.5% butterfat?

PEANUTS © United Feature Syndicate, Inc.

31. With the current, you can canoe 24 miles in 4 hours. Against the same current, you can canoe only $\frac{3}{4}$ of this distance in 6 hours. Find your rate in still water and the rate of the current.

32. With the current, you can row 24 miles in 3 hours. Against the same current, you can row only $\frac{2}{3}$ of this distance in 4 hours. Find your rowing rate in still water and the rate of the current.

33. A student has two test scores. The difference between the scores is 12 and the mean, or average, of the scores is 80. What are the two test scores?

34. A student has two test scores. The difference between the scores is 8 and the mean, or average, of the scores is 88. What are the two test scores?

In Exercises 35–36, an isosceles triangle containing two angles with equal measure is shown. The degree measure of each triangle's three interior angles and an exterior angle is represented with variables. Find the measure of the three interior angles.

35.

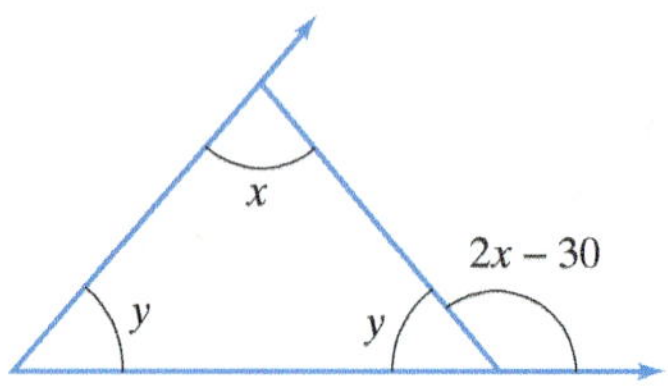

36.

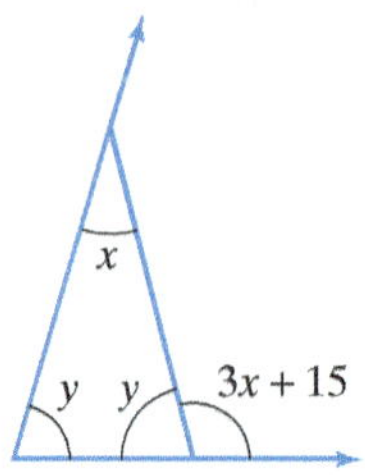

37. A rectangular lot whose perimeter is 220 feet is fenced along three sides. An expensive fencing along the lot's length costs \$20 per foot, and an inexpensive fencing along the two side widths costs only \$8 per foot. The total cost of the fencing along the three sides comes to \$2040. What are the lot's dimensions?

38. A rectangular lot whose perimeter is 260 feet is fenced along three sides. An expensive fencing along the lot's length costs \$16 per foot, and an inexpensive fencing along the two side widths costs only \$5 per foot. The total cost of the fencing along the three sides comes to \$1780. What are the lot's dimensions?

39. A new restaurant is to contain two-seat tables and four-seat tables. Fire codes limit the restaurant's maximum occupancy to 56 customers. If the owners have hired enough servers to handle 17 tables of customers, how many of each kind of table should they purchase?

40. A hotel has 200 rooms. Those with kitchen facilities rent for \$100 per night and those without kitchen facilities rent for \$80 per night. On a night when the hotel was completely occupied, revenues were \$17,000. How many of each type of room does the hotel have?

The figure shows the graphs of the cost and revenue functions for a company that manufactures and sells small radios. Use the information in the figure to solve Exercises 41–46.

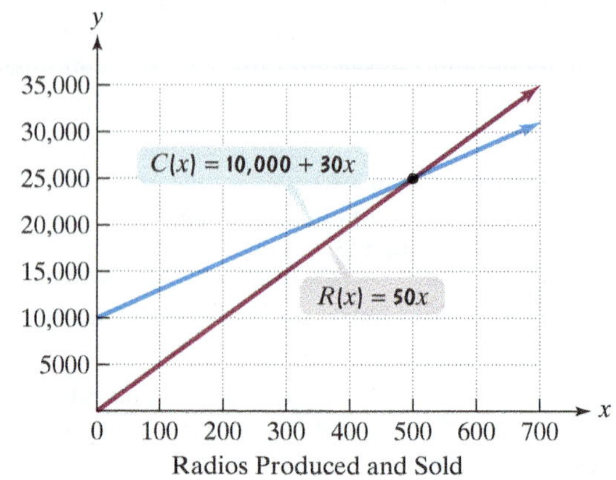

41. How many radios must be produced and sold for the company to break even?

42. More than how many radios must be produced and sold for the company to have a profit?

43. Use the formulas shown in the voice balloons to find $R(200) - C(200)$. Describe what this means for the company.

44. Use the formulas shown in the voice balloons to find $R(300) - C(300)$. Describe what this means for the company.

45. **a.** Use the formulas shown in the voice balloons to write the company's profit function, P, from producing and selling x radios.
b. Find the company's profit if 10,000 radios are produced and sold.

46. **a.** Use the formulas shown in the voice balloons to write the company's profit function, P, from producing and selling x radios.
b. Find the company's profit if 20,000 radios are produced and sold.

Exercises 47–50 describe a number of business ventures. For each exercise,

a. *Write the cost function, C.*
b. *Write the revenue function, R.*
c. *Determine the break-even point. Describe what this means.*

47. A company that manufactures small canoes has a fixed cost of \$18,000. It costs \$20 to produce each canoe. The selling price is \$80 per canoe. (In solving this exercise, let x represent the number of canoes produced and sold.)

48. A company that manufactures bicycles has a fixed cost of \$100,000. It costs \$100 to produce each bicycle. The selling price is \$300 per bike. (In solving this exercise, let x represent the number of bicycles produced and sold.)

49. You invest in a new play. The cost includes an overhead of \$30,000, plus production costs of \$2500 per performance. A sold-out performance brings in \$3125. (In solving this exercise, let x represent the number of sold-out performances.)

50. You invested \$30,000 and started a business writing greeting cards. Supplies cost 2¢ per card and you are selling each card for 50¢. (In solving this exercise, let x represent the number of cards produced and sold.)

Writing in Mathematics

51. Describe the conditions in a problem that enable it to be solved using a system of linear equations.

52. Write a word problem that can be solved by translating to a system of linear equations. Then solve the problem.

53. Describe a revenue function for a business venture.

54. Describe a cost function for a business venture. What are the two kinds of costs that are modeled by this function?

55. What is the profit function for a business venture and how is it determined?

56. Describe the break-even point for a business.

57. The law of supply and demand states that, in a free market economy, a commodity tends to be sold at its equilibrium price. At this price, the amount that the seller will supply is the same amount that the consumer will buy. Explain how graphs can be used to determine the equilibrium price.

58. Many students hate mixture problems and decide to ignore them, stating, "I'll just skip that one on the test." If you share this opinion, describe what you find particularly unappealing about this kind of problem.

Technology Exercises

In Exercises 59–60, graph the revenue and cost functions in the some viewing rectangle. Then use the intersection feature to determine the break-even point.

59. $R(x) = 50x, \quad C(x) = 20x + 180$

60. $R(x) = 92.5x, \quad C(x) = 52x + 1782$

61. Use the procedure in Exercises 59–60 to verify your work for any one of the break-even points that you found in Exercises 47–50.

Critical Thinking Exercises

Make Sense? *In Exercises 62–65, determine whether each statement "makes sense" or "does not make sense" and explain your reasoning.*

62. A system of linear equations can be used to model and compare the fees charged by two different taxicab companies.

63. I should mix 6 liters of a 50% acid solution with 4 liters of a 25% acid solution to obtain 10 liters of a 75% acid solution.

64. If I know the perimeter of this rectangle and triangle, each in the same unit of measure, I can use a system of linear equations to determine values for x and y.

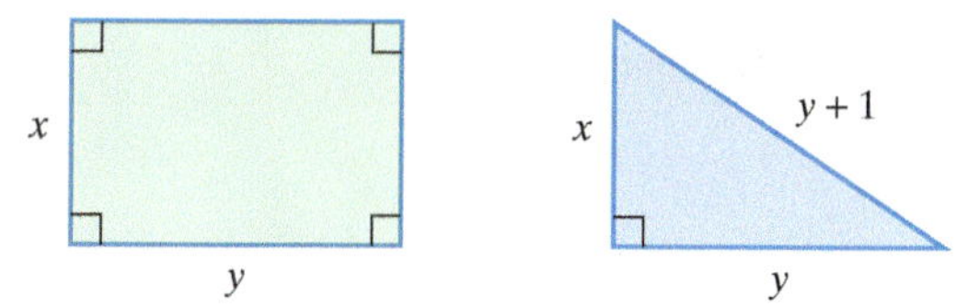

65. You told me that you flew against the wind from Miami to Seattle, 2800 miles, in 7 hours and, at the same time, your friend flew with the wind from Seattle to Miami in only 5.6 hours. You have not given me enough information to determine the average rate of the wind.

66. The radiator in your car contains 4 gallons of antifreeze and water. The mixture is 45% antifreeze. How much of this mixture should be drained and replaced with pure antifreeze in order to have a 60% antifreeze solution? Round to the nearest tenth of a gallon.

67. A marching band has 52 members, and there are 24 in the pom-pom squad. They wish to form several hexagons and squares like those diagrammed below. Can it be done with no people left over?

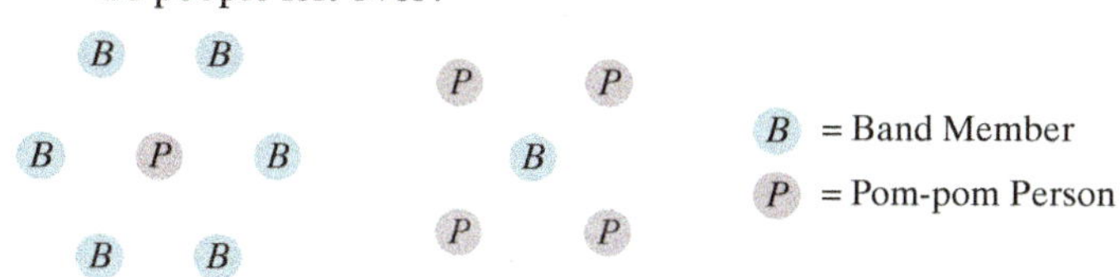

68. A boy has as many brothers as he has sisters. Each of his sisters has twice as many brothers as she has sisters. How many boys and girls are in this family?

69. When entering your test score into a computer, your professor accidently reversed the two digits. This error reduced your score by 36 points. Your professor told you that the sum of the digits of your actual score was 14, corrected the error, and agreed to give you extra credit if you could determine the actual score without looking back at the test. What was your actual test score? (*Hint*: Let t = the tens-place digit of your actual score and let u = the units-place digit of your actual score. Thus, $10t + u$ represents your actual test score.)

70. A dealer paid a total of \$67 for mangos and avocados. The mangos were sold at a profit of 20% on the dealer's cost, but the avocados started to spoil, resulting in a selling price of a 2% loss on the dealer's cost. The dealer made a profit of \$8.56 on the total transaction. How much did the dealer pay for the mangos and for the avocados?

Review Exercises

In Exercises 71–72, use the given conditions to write an equation for each line in point-slope form and slope-intercept form.

71. Passing through $(-2, 5)$ and $(-6, 13)$

(Section 2.5, Example 2)

72. Passing through $(-3, 0)$ and parallel to the line whose equation is $-x + y = 7$

(Section 2.5, Example 4)

73. Find the domain of $g(x) = \frac{x-2}{3-x}$.

(Section 2.3, Example 1)

Preview Exercises

Exercises 74–76 will help you prepare for the material covered in the next section.

74. If $x = 3$, $y = 2$, and $z = -3$, does the ordered triple (x, y, z) satisfy the equation $2x - y + 4z = -8$?

75. Consider the following equations:

$$5x - 2y - 4z = 3 \quad \text{Equation 1}$$
$$3x + 3y + 2z = -3. \quad \text{Equation 2}$$

Use these equations to eliminate z. Copy Equation 1 and multiply Equation 2 by 2. Then add the equations.

76. Write an equation involving a, b, and c based on the following description:

When the value of x in $y = ax^2 + bx + c$ is 4, the value of y is 1682.

GROUP PROJECT

CHAPTER 3

The group is going into business for the next year. Your first task is to determine the product that you plan to manufacture and sell. Choose something unique, but realistic, reflecting the abilities and interests of group members. How much will it cost to produce each unit of your product? What will be the selling price for each unit? What is a reasonable estimate of your fixed cost for the entire year? Be sure to include utilities, labor, materials, marketing, and anything else that is relevant to your business venture. Once you have determined the product and these three figures,

a. Write the cost function, C, of producing x units of your product.
b. Write the revenue function, R, from the sale of x units of your product.
c. Determine the break-even point. Does this seem realistic in terms of your product and its target market?
d. Graph the cost function and the revenue function in the same rectangular coordinate system. Label the figure just like **Figure 3.7** on page 190. Indicate the regions that show the loss and gain for your business venture.
e. Write the profit function, P, from producing and selling x units of your product. Graph the profit function, in the same style as **Figure 3.8** on page 191. Indicate where the business is in the red and where it is in the black.
f. MTV's *The Real World* has offered to pay the fixed cost for the business venture, videotaping all group interactions for its forthcoming *The Real World: Profit and Loss*. Group members need to determine whether or not to accept the offer.

Chapter 3 Summary

Definitions and Concepts	Examples

Section 3.1 Systems of Linear Equations in Two Variables

A system of linear equations in two variables, x and y, consists of two equations that can be written in the form $Ax + By = C$. The solution set is the set of all ordered pairs that satisfy both equations. Using the graphing method, a solution of a linear system is a point common to the graphs of both equations in the system.

Solve by graphing: $2x - y = 6$
$x + y = 6$.

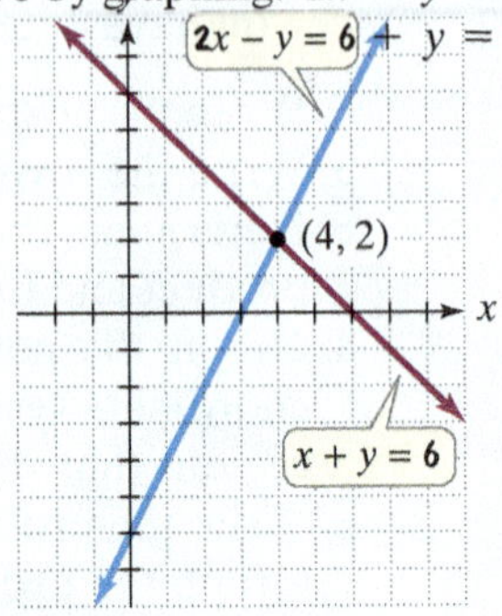

The intersection point gives the solution: (4, 2). The solution set is $\{(4, 2)\}$.

To solve a linear system by the substitution method,

1. Solve either equation for one variable in terms of the other.
2. Substitute the expression for that variable into the other equation.
3. Solve the equation in one variable.
4. Back-substitute the value of the variable into one of the original equations and find the value of the other variable.
5. Check the proposed solution in both equations.

Solve by the substitution method:

$$y = 2x - 3$$
$$4x - 3y = 5.$$

Substitute $2x - 3$ for y in the second equation.

$$\begin{aligned} 4x - 3(2x - 3) &= 5 \\ 4x - 6x + 9 &= 5 \\ -2x + 9 &= 5 \\ -2x &= -4 \\ x &= 2 \end{aligned}$$

Find y. Substitute 2 for x in $y = 2x - 3$.

$$y = 2(2) - 3 = 4 - 3 = 1$$

The ordered pair (2, 1) checks. The solution is (2, 1) and $\{(2, 1)\}$ is the solution set.

To solve a linear system by the addition method,

1. Write equations in $Ax + By = C$ form.
2. Multiply one or both equations by nonzero numbers so that coefficients of one variable are opposites.
3. Add equations.
4. Solve the resulting equation for the variable.
5. Back-substitute the value of the variable into either original equation and find the value of the remaining variable.
6. Check the proposed solution in both equations.

Solve by the addition method:

$$2x + y = 10$$
$$3x + 4y = 25.$$

Eliminate y. Multiply the first equation by -4.

$$\begin{aligned} -8x - 4y &= -40 \\ 3x + 4y &= 25 \\ \text{Add: } -5x \quad &= -15 \\ x &= 3 \end{aligned}$$

Find y. Back-substitute 3 for x. Use the first equation, $2x + y = 10$.

$$\begin{aligned} 2(3) + y &= 10 \\ 6 + y &= 10 \\ y &= 4 \end{aligned}$$

The ordered pair (3, 4) checks. The solution is (3, 4) and $\{(3, 4)\}$ is the solution set.

Definitions and Concepts	Examples

Section 3.1 Systems of Linear Equations in Two Variables (continued)

A linear system with at least one solution is a consistent system. A system that has no solution, with ∅ as its solution set, is an inconsistent system. A linear system with infinitely many solutions has dependent equations. Solving inconsistent systems by substitution or addition leads to a false statement, such as $0 = 3$. Solving systems with dependent equations leads to a true statement, such as $7 = 7$.

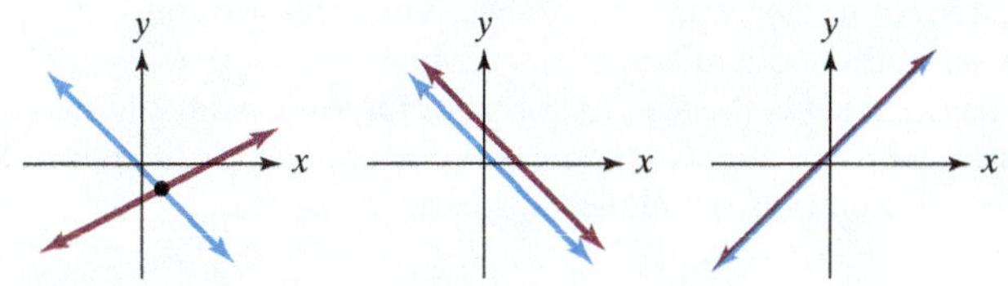

One solution: consistent — No solution: inconsistent — Infinitely many solutions: dependent and consistent

Section 3.2 Problem Solving and Business Applications Using Systems of Equations

A Problem-Solving Strategy

1. Use variables, usually x and y, to represent unknown quantities.
2. Write a system of equations describing the problem's conditions.
3. Solve the system and answer the problem's question.
4. Check proposed answers in the problem's wording.

You invested \$14,000 in two stocks paying 7% and 9% interest. Total year-end interest was \$1180. How much was invested at each rate?

Let x = amount invested at 7% and
y = amount invested at 9%.

(amount invested at 7%) (amount invested at 9%)

$$x + y = 14{,}000$$

(interest from 7% investment) (interest from 9% investment)

$$0.07x + 0.09y = 1180$$

Solving by substitution or addition, $x = 4000$ and $y = 10{,}000$. Thus, \$4000 was invested at 7% and \$10,000 at 9%.

Functions of Business

A company produces and sells x units of a product.

Revenue Function

$$R(x) = (\text{price per unit sold})x$$

Cost Function

$$C(x) = \text{fixed cost} + (\text{cost per unit produced})x$$

Profit Function

$$P(x) = R(x) - C(x)$$

The point of intersection of the graphs of R and C is the break even point. The x-coordinate of the point reveals the number of units that a company must produce and sell so that the money coming in, the revenue, is equal to the money going out, the cost. The y-coordinate gives the amount of money coming in and going out.

A company that manufactures lamps has a fixed cost of \$80,000 and it costs \$20 to produce each lamp. Lamps are sold for \$70.

a. Write the cost function.

$$C(x) = 80{,}000 + 20x$$

(Fixed cost) (Variable cost: \$20 per lamp.)

b. Write the revenue function.

$$R(x) = 70x$$

(Revenue per lamp, \$70, times number of lamps sold)

c. Find the break-even point.

Solve

$$y = 80{,}000 + 20x$$
$$y = 70x$$

by substitution. Solving

$$70x = 80{,}000 + 20x$$

yields $x = 1600$. Back-substituting, $y = 112{,}000$. The break-even point is (1600, 112,000): The company breaks even if it sells 1600 lamps. At this level, money coming in equals money going out: \$112,000.

Definitions and Concepts	Examples

Section 3.3 Systems of Linear Equations in Three Variables

A system of linear equations in three variables, x, y, and z, consists of three equations of the form $Ax + By + Cz = D$. The solution set is the set consisting of the ordered triple that satisfies all three equations. The solution represents the point of intersection of three planes in space.

Is $(2, -1, 3)$ a solution of

$$\begin{aligned} 3x + 5y - 2z &= -5 \\ 2x + 3y - z &= -2 \\ 2x + 4y + 6z &= 18? \end{aligned}$$

Replace x with 2, y with -1, and z with 3. Using the first equation, we obtain:

$$\begin{aligned} 3\cdot 2 + 5(-1) - 2(3) &\stackrel{?}{=} -5 \\ 6 - 5 - 6 &\stackrel{?}{=} -5 \\ -5 &= -5, \quad \text{true} \end{aligned}$$

The ordered triple $(2, -1, 3)$ satisfies the first equation. In a similar manner, it satisfies the other two equations and is a solution.

To solve a linear system in three variables by eliminating variables,

1. Reduce the system to two equations in two variables.
2. Solve the resulting system of two equations in two variables.
3. Use back-substitution in one of the equations in two variables to find the value of the second variable.
4. Back-substitute the values for two variables into one of the original equations to find the value of the third variable.
5. Check.

If all variables are eliminated and a false statement results, the system is inconsistent and has no solution. If a true statement results, the system contains dependent equations and has infinitely many solutions.

Solve:

$$\begin{aligned} 2x + 3y - 2z &= 0 \quad \text{Equation 1} \\ x + 2y - z &= 1 \quad \text{Equation 2} \\ 3x - y + z &= -15. \quad \text{Equation 3} \end{aligned}$$

Add Equations 2 and 3 to eliminate z.

$$4x + y = -14 \quad \text{Equation 4}$$

Eliminate z again. Multiply Equation 3 by 2 and add to Equation 1.

$$8x + y = -30 \quad \text{Equation 5}$$

Multiply Equation 4 by -1 and add to Equation 5.

$$\begin{aligned} -4x - y &= 14 \\ 8x + y &= -30 \\ \text{Add: } 4x \quad &= -16 \\ x &= -4 \end{aligned}$$

Substitute -4 for x in Equation 4.

$$\begin{aligned} 4(-4) + y &= -14 \\ y &= 2 \end{aligned}$$

Substitute -4 for x and 2 for y in Equation 3.

$$\begin{aligned} 3(-4) - 2 + z &= -15 \\ -14 + z &= -15 \\ z &= -1 \end{aligned}$$

Checking verifies that $(-4, 2, -1)$ is the solution and $\{(-4, 2, -1)\}$ is the solution set.

Definitions and Concepts	Examples

Section 3.3 Systems of Linear Equations in Three Variables (continued)

Curve Fitting

Curve fitting is determining a function whose graph contains given points. Three points that do not lie on a line determine the graph of a quadratic function

$$y = ax^2 + bx + c.$$

Use the three given points to create a system of three equations. Solve the system to find a, b, and c.

Find the quadratic function $y = ax^2 + bx + c$ whose graph passes through the points $(-1, 2)$, $(1, 8)$, and $(2, 14)$. Use $y = ax^2 + bx + c$.

When $x = -1, y = 2$: $2 = a(-1)^2 + b(-1) + c$

When $x = 1, y = 8$: $8 = a \cdot 1^2 + b \cdot 1 + c$

When $x = 2, y = 14$: $14 = a \cdot 2^2 + b \cdot 2 + c$

Solving,

$$\begin{aligned} a - b + c &= 2 \\ a + b + c &= 8 \\ 4a + 2b + c &= 14, \end{aligned}$$

$a = 1$, $b = 3$, and $c = 4$. The quadratic function, $y = ax^2 + bx + c$, is $y = x^2 + 3x + 4$.

Section 3.4 Matrix Solutions to Linear Systems

A matrix is a rectangular array of numbers. The augmented matrix of a linear system is obtained by writing the coefficients of each variable, a vertical bar, and the constants of the system.

$$\begin{aligned} x + 4y &= 9 \\ 3x + y &= 5 \end{aligned}$$

The augmented matrix is $\left[\begin{array}{cc|c} 1 & 4 & 9 \\ 3 & 1 & 5 \end{array}\right]$.

The following row operations produce matrices that represent systems with the same solution. Two matrices are row equivalent if one can be obtained from the other by a sequence of these row operations.

1. Interchange the elements in the ith and jth rows: $R_i \leftrightarrow R_j$.
2. Multiply each element in the ith row by k: kR_i.
3. Add k times the elements in row i to the corresponding elements in row j: $kR_i + R_j$.

Find the result of the row operation $-4R_1 + R_2$:

$$\left[\begin{array}{ccc|c} 1 & 0 & -2 & 5 \\ 4 & -1 & 2 & 6 \\ 3 & -7 & 9 & 10 \end{array}\right].$$

Add -4 times the elements in row 1 to the corresponding elements in row 2.

$$\left[\begin{array}{ccc|c} 1 & 0 & -2 & 5 \\ -4(1) + 4 & -4(0) + (-1) & -4(-2) + 2 & -4(5) + 6 \\ 3 & -7 & 9 & 10 \end{array}\right]$$

$$= \left[\begin{array}{ccc|c} 1 & 0 & -2 & 5 \\ 0 & -1 & 10 & -14 \\ 3 & -7 & 9 & 10 \end{array}\right]$$

Solving Linear Systems Using Matrices

1. Write the augmented matrix for the system.
2. Use matrix row operations to simplify the matrix to row-eschelon form, with 1s down the main diagonal from upper left to lower right, and 0s below the 1s.
3. Write the system of linear equations corresponding to the matrix from step 2 and use back-substitution to find the system's solution.

If you obtain a matrix with a row containing 0s to the left of the vertical bar and a nonzero number on the right, the system is inconsistent. If 0s appear across an entire row, the system contains dependent equations.

Solve using matrices:

$$\begin{aligned} 3x + y &= 5 \\ x + 4y &= 9. \end{aligned}$$

$$\left[\begin{array}{cc|c} 3 & 1 & 5 \\ 1 & 4 & 9 \end{array}\right] \xrightarrow{R_1 \leftrightarrow R_2} \left[\begin{array}{cc|c} 1 & 4 & 9 \\ 3 & 1 & 5 \end{array}\right]$$

$$\xrightarrow{-3R_1 + R_2} \left[\begin{array}{cc|c} 1 & 4 & 9 \\ 0 & -11 & -22 \end{array}\right] \xrightarrow{-\frac{1}{11}R_2} \left[\begin{array}{cc|c} 1 & 4 & 9 \\ 0 & 1 & 2 \end{array}\right]$$

$$\rightarrow \begin{aligned} x + 4y &= 9 \\ y &= 2. \end{aligned}$$

When $y = 2$, $x + 4 \cdot 2 = 9$, so $x = 1$. The solution is $(1, 2)$ and the solution set is $\{(1, 2)\}$.

Definitions and Concepts

Examples

Section 3.5 Determinants and Cramer's Rule

A square matrix has the same number of rows as columns. A determinant is a real number associated with a square matrix. The determinant is denoted by placing vertical bars about the array of numbers. The value of a second-order determinant is

$$\begin{vmatrix} a_1 & b_1 \\ a_2 & b_2 \end{vmatrix} = a_1b_2 - a_2b_1.$$

Evaluate:

$$\begin{vmatrix} 2 & -1 \\ 3 & 4 \end{vmatrix} = 2(4) - 3(-1) = 8 + 3 = 11.$$

Cramer's Rule for Two Linear Equations in Two Variables

If

$$a_1x + b_1y = c_1$$
$$a_2x + b_2y = c_2,$$

then

$$x = \frac{\begin{vmatrix} c_1 & b_1 \\ c_2 & b_2 \end{vmatrix}}{\begin{vmatrix} a_1 & b_1 \\ a_2 & b_2 \end{vmatrix}} = \frac{D_x}{D} \quad \text{and} \quad y = \frac{\begin{vmatrix} a_1 & c_1 \\ a_2 & c_2 \end{vmatrix}}{\begin{vmatrix} a_1 & b_1 \\ a_2 & b_2 \end{vmatrix}} = \frac{D_y}{D}, \quad D \neq 0.$$

If $D = 0$ and any numerator is not zero, the system is inconsistent and has no solution. If all determinants are 0, the system contains dependent equations and has infinitely many solutions.

Solve by Cramer's rule:

$$5x + 3y = 7$$
$$-x + 2y = 9.$$

$$D = \begin{vmatrix} 5 & 3 \\ -1 & 2 \end{vmatrix} = 5(2) - (-1)(3) = 10 + 3 = 13$$

$$D_x = \begin{vmatrix} 7 & 3 \\ 9 & 2 \end{vmatrix} = 7 \cdot 2 - 9 \cdot 3 = 14 - 27 = -13$$

$$D_y = \begin{vmatrix} 5 & 7 \\ -1 & 9 \end{vmatrix} = 5(9) - (-1)(7) = 45 + 7 = 52$$

$$x = \frac{D_x}{D} = \frac{-13}{13} = -1, \; y = \frac{D_y}{D} = \frac{52}{13} = 4$$

The solution is $(-1, 4)$ and the solution set is $\{(-1, 4)\}$.

The value of a third-order determinant is

$$\begin{vmatrix} a_1 & b_1 & c_1 \\ a_2 & b_2 & c_2 \\ a_3 & b_3 & c_3 \end{vmatrix}$$

$$= a_1\begin{vmatrix} b_2 & c_2 \\ b_3 & c_3 \end{vmatrix} - a_2\begin{vmatrix} b_1 & c_1 \\ b_3 & c_3 \end{vmatrix} + a_3\begin{vmatrix} b_1 & c_1 \\ b_2 & c_2 \end{vmatrix}.$$

Each second-order determinant is called a minor.

Evaluate:

$$\begin{vmatrix} 1 & -2 & 1 \\ 3 & 1 & -2 \\ 5 & 5 & 3 \end{vmatrix}$$

$$= 1\begin{vmatrix} 1 & -2 \\ 5 & 3 \end{vmatrix} - 3\begin{vmatrix} -2 & 1 \\ 5 & 3 \end{vmatrix} + 5\begin{vmatrix} -2 & 1 \\ 1 & -2 \end{vmatrix}$$

$$= 1(3 - (-10)) - 3(-6 - 5) + 5(4 - 1)$$

$$= 1(13) - 3(-11) + 5(3)$$

$$= 13 + 33 + 15 = 61.$$

Definitions and Concepts	Examples

Section 3.5 Determinants and Cramer's Rule (continued)

Cramer's Rule for Three Linear Equations in Three Variables

If

$$a_1x + b_1y + c_1z = d_1$$
$$a_2x + b_2y + c_2z = d_2$$
$$a_3x + b_3y + c_3z = d_3,$$

then

$$x = \frac{D_x}{D}, \quad y = \frac{D_y}{D}, \quad z = \frac{D_z}{D}.$$

$$D = \begin{vmatrix} a_1 & b_1 & c_1 \\ a_2 & b_2 & c_2 \\ a_3 & b_3 & c_3 \end{vmatrix} \neq 0, \quad D_x = \begin{vmatrix} d_1 & b_1 & c_1 \\ d_2 & b_2 & c_2 \\ d_3 & b_3 & c_3 \end{vmatrix}$$

$$D_y = \begin{vmatrix} a_1 & d_1 & c_1 \\ a_2 & d_2 & c_2 \\ a_3 & d_3 & c_3 \end{vmatrix}, \quad D_z = \begin{vmatrix} a_1 & b_1 & d_1 \\ a_2 & b_2 & d_2 \\ a_3 & b_3 & d_3 \end{vmatrix}$$

If $D = 0$ and any numerator is not zero, the system is inconsistent. If all determinants are 0, the system contains dependent equations.

Solve by Cramer's rule:

$$\begin{aligned} x - 2y + z &= 4 \\ 3x + y - 2z &= 3 \\ 5x + 5y + 3z &= -8. \end{aligned}$$

$$D = \begin{vmatrix} 1 & -2 & 1 \\ 3 & 1 & -2 \\ 5 & 5 & 3 \end{vmatrix} = 61$$

This evaluation is shown on the bottom of page 234.

$$D_x = \begin{vmatrix} 4 & -2 & 1 \\ 3 & 1 & -2 \\ -8 & 5 & 3 \end{vmatrix} = 61$$

$$D_y = \begin{vmatrix} 1 & 4 & 1 \\ 3 & 3 & -2 \\ 5 & -8 & 3 \end{vmatrix} = -122$$

$$D_z = \begin{vmatrix} 1 & -2 & 4 \\ 3 & 1 & 3 \\ 5 & 5 & -8 \end{vmatrix} = -61$$

$$x = \frac{D_x}{D} = \frac{61}{61} = 1, \quad y = \frac{D_y}{D} = \frac{-122}{61} = -2,$$

$$z = \frac{D_z}{D} = \frac{-61}{61} = -1.$$

The solution is $(1, -2, -1)$ and the solution set is $\{(1, -2, -1)\}$.

CHAPTER 3 REVIEW EXERCISES

3.1 *In Exercises 1–2, determine whether the given ordered pair is a solution of the system.*

1. $(4, 2)$
$$2x - 5y = -2$$
$$3x + 4y = 4$$

2. $(-5, 3)$
$$-x + 2y = 11$$
$$y = -\frac{x}{3} + \frac{4}{3}$$

In Exercises 3–6, solve each system by graphing. Identify systems with no solution and systems with infinitely many solutions, using set notation to express their solution sets.

3. $x + y = 5$
$3x - y = 3$

4. $3x - 2y = 6$
$6x - 4y = 12$

5. $y = \frac{3}{5}x - 3$
$2x - y = -4$

6. $y = -x + 4$
$3x + 3y = -6$

In Exercises 7–13, solve each system by the substitution method or the addition method. Identify systems with no solution and systems with infinitely many solutions, using set notation to express their solution sets.

7. $2x - y = 2$
$x + 2y = 11$

8. $y = -2x + 3$
$3x + 2y = -17$

9. $3x + 2y = -8$
$2x + 5y = 2$

10. $5x - 2y = 14$
$3x + 4y = 11$

11. $y = 4 - x$
$3x + 3y = 12$

12. $\frac{x}{8} + \frac{3y}{4} = \frac{19}{8}$
$-\frac{x}{2} + \frac{3y}{4} = \frac{1}{2}$

13. $x - 2y + 3 = 0$
$2x - 4y + 7 = 0$

3.2 *In Exercises 14–18, use the four-step strategy to solve each problem.*

14. An appliance store is having a sale on small TVs and stereos. One day a salesperson sells 3 of the TVs and 4 stereos for \$2530. The next day the salesperson sells 4 of the same TVs and 3 of the same stereos for \$2510. What are the prices of a TV and a stereo?

15. You invested \$9000 in two funds paying 4% and 7% annual interest. At the end of the year, the total interest from these investments was \$555. How much was invested at each rate?

16. A chemist needs to mix a solution that is 34% silver nitrate with one that is 4% silver nitrate to obtain 100 milliliters of a mixture that is 7% silver nitrate. How many milliliters of each of the solutions must be used?

17. When a plane flies with the wind, it can travel 2160 miles in 3 hours. When the plane flies in the opposite direction, against the wind, it takes 4 hours to fly the same distance. Find the rate of the plane in still air and the rate of the wind.

18. The perimeter of a rectangular table top is 34 feet. The difference between 4 times the length and 3 times the width is 33 feet. Find the dimensions.

The cost and revenue functions for producing and selling x units of a new graphing calculator are

$$C(x) = 22{,}500 + 40x \quad \text{and} \quad R(x) = 85x.$$

Use these functions to solve Exercises 19–21.

19. Find the loss or the gain from selling 400 graphing calculators.

20. Determine the break-even point. Describe what this means.

21. Write the profit function, P, from producing and selling x graphing calculators.

22. A company is planning to manufacture computer desks. The fixed cost will be \$60,000 and it will cost \$200 to produce each desk. Each desk will be sold for \$450.

a. Write the cost function, C, of producing x desks.

b. Write the revenue function, R, from the sale of x desks.

c. Determine the break-even point. Describe what this means.

3.3

23. Is $(-3, -2, 5)$ a solution of the system

$$\begin{aligned} x + y + z &= 0 \\ 2x - 3y + z &= 5 \\ 4x + 2y + 4z &= 3? \end{aligned}$$

Solve each system in Exercises 24–26 by eliminating variables using the addition method. If there is no solution or if there are infinitely many solutions and a system's equations are dependent, so state.

24. $$\begin{aligned} 2x - y + z &= 1 \\ 3x - 3y + 4z &= 5 \\ 4x - 2y + 3z &= 4 \end{aligned}$$

25. $$\begin{aligned} x + 2y - z &= 5 \\ 2x - y + 3z &= 0 \\ 2y + z &= 1 \end{aligned}$$

26. $$\begin{aligned} 3x - 4y + 4z &= 7 \\ x - y - 2z &= 2 \\ 2x - 3y + 6z &= 5 \end{aligned}$$

27. Find the quadratic function $y = ax^2 + bx + c$ whose graph passes through the points $(1, 4)$, $(3, 20)$, and $(-2, 25)$.

28. The bar graph shows the average debt in the United States, not including real estate mortgages, by age group. The difference between the average debt for the 30–39 age group and the 18–29 age group is \$8100. The difference between the average debt for the 40–49 age group and the 30–39 age group is \$3100. The combined average debt for these three age groups is \$44,200. Find the average debt for each of these age groups.

Average Debt by Age

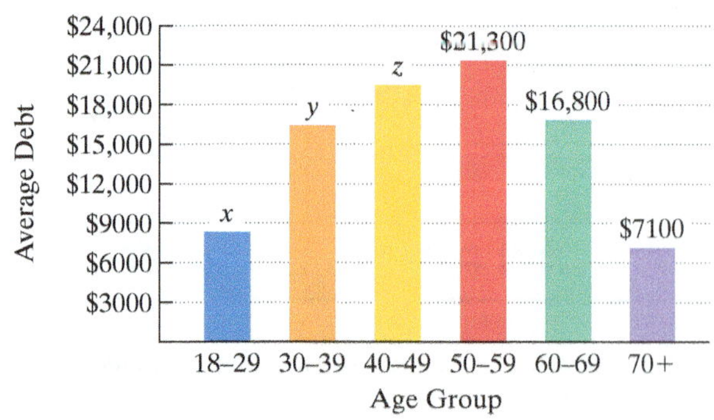

Source: Experian

3.4 *In Exercises 29–32, perform each matrix row operation and write the new matrix.*

29. $\left[\begin{array}{cc|c} 1 & -8 & 3 \\ 0 & 7 & -14 \end{array}\right] \frac{1}{7}R_2$

30. $\left[\begin{array}{cc|c} 1 & -3 & 1 \\ 2 & 1 & -5 \end{array}\right] -2R_1 + R_2$

31. $\left[\begin{array}{ccc|c} 2 & -2 & 1 & -1 \\ 1 & 2 & -1 & 2 \\ 6 & 4 & 3 & 5 \end{array}\right] \frac{1}{2}R_1$

32. $\left[\begin{array}{ccc|c} 1 & 2 & 2 & 2 \\ 0 & 1 & -1 & 2 \\ 0 & 5 & 4 & 1 \end{array}\right] -5R_2 + R_3$

In Exercises 33–36, solve each system using matrices. If there is no solution or if a system's equations are dependent, so state.

33. $x + 4y = 7$
$3x + 5y = 0$

34. $2x - 3y = 8$
$-6x + 9y = 4$

35. $x + 2y + 3z = -5$
$2x + y + z = 1$
$x + y - z = 8$

36. $x - 2y + z = 0$
$y - 3z = -1$
$2y + 5z = -2$

3.5 *In Exercises 37–40, evaluate each determinant.*

37. $\begin{vmatrix} 3 & 2 \\ -1 & 5 \end{vmatrix}$

38. $\begin{vmatrix} -2 & -3 \\ -4 & -8 \end{vmatrix}$

39. $\begin{vmatrix} 2 & 4 & -3 \\ 1 & -1 & 5 \\ -2 & 4 & 0 \end{vmatrix}$

40. $\begin{vmatrix} 4 & 7 & 0 \\ -5 & 6 & 0 \\ 3 & 2 & -4 \end{vmatrix}$

In Exercises 41–44, use Cramer's rule to solve each system. If there is no solution or if a system's equations are dependent, so state.

41. $x - 2y = 8$
$3x + 2y = -1$

42. $7x + 2y = 0$
$2x + y = -3$

43. $x + 2y + 2z = 5$
$2x + 4y + 7z = 19$
$-2x - 5y - 2z = 8$

44. $2x + y = -4$
$y - 2z = 0$
$3x - 2z = -11$

45. Use the quadratic function $y = ax^2 + bx + c$ to model the following data:

x (Age of a Driver)	y (Average Number of Automobile Accidents per Day in the United States)
20	400
40	150
60	400

Use Cramer's rule to determine values for a, b, and c. Then use the model to write a statement about the average number of automobile accidents in which 30-year-old drivers and 50-year-old drivers are involved daily.

CHAPTER 3 TEST

Remember to use your Chapter Test Prep Video CD to see the worked-out solutions to the test questions you want to review.

1. Solve by graphing

$$x + y = 6$$
$$4x - y = 4.$$

In Exercises 2–4, solve each system by the substitution method or the addition method. Identify systems with no solution and systems with infinitely many solutions, using set notation to express their solution sets.

2. $5x + 4y = 10$
$3x + 5y = -7$

3. $x = y + 4$
$3x + 7y = -18$

4. $4x = 2y + 6$
$y = 2x - 3$

In Exercises 5–8, solve each problem.

5. In a new development, 50 one- and two-bedroom condominiums were sold. Each one-bedroom condominium sold for \$120 thousand and each two-bedroom condominium sold for \$150 thousand. If sales totaled \$7050 thousand, how many of each type of unit was sold?

6. You invested \$9000 in two funds paying 6% and 7% annual interest. At the end of the year, the total interest from these investments was \$610. How much was invested at each rate?

7. You need to mix a 6% peroxide solution with a 9% peroxide solution to obtain 36 ounces of an 8% peroxide solution. How many ounces of each of the solutions must be used?

8. A paddleboat on the Mississippi River travels 48 miles downstream, with the current, in 3 hours. The return trip, against the current, takes the paddleboat 4 hours. Find the boat's rate in still water and the rate of the current.

Use this information to solve Exercises 9–11: A company is planning to produce and sell a new line of computers. The fixed cost will be \$360,000 and it will cost \$850 to produce each computer. Each computer will be sold for \$1150.

9. Write the cost function, C, of producing x computers.

10. Write the revenue function, R, from the sale of x computers.

11. Determine the break-even point. Describe what this means.

12. The cost and revenue functions for producing and selling x units of a toaster oven are

$$C(x) = 40x + 350{,}000 \quad \text{and} \quad R(x) = 125x.$$

Write the profit function, P, from producing and selling x toaster ovens.

13. Solve by eliminating variables using the addition method:

$$\begin{aligned} x + y + z &= 6 \\ 3x + 4y - 7z &= 1 \\ 2x - y + 3z &= 5. \end{aligned}$$

14. Perform the indicated matrix row operation and write the new matrix.

$$\left[\begin{array}{ccc|c} 1 & 0 & -4 & 5 \\ 6 & -1 & 2 & 10 \\ 2 & -1 & 4 & -3 \end{array}\right] -6R_1 + R_2$$

In Exercises 15–16, solve each system using matrices.

15. $$\begin{aligned} 2x + y &= 6 \\ 3x - 2y &= 16 \end{aligned}$$

16. $$\begin{aligned} x - 4y + 4z &= -1 \\ 2x - y + 5z &= 6 \\ -x + 3y - z &= 5 \end{aligned}$$

In Exercises 17–18, evaluate each determinant.

17. $\begin{vmatrix} -1 & -3 \\ 7 & 4 \end{vmatrix}$

18. $\begin{vmatrix} 3 & 4 & 0 \\ -1 & 0 & -3 \\ 4 & 2 & 5 \end{vmatrix}$

In Exercises 19–20, use Cramer's rule to solve each system.

19. $$\begin{aligned} 4x - 3y &= 14 \\ 3x - y &= 3 \end{aligned}$$

20. $$\begin{aligned} 2x + 3y + z &= 2 \\ 3x + 3y - z &= 0 \\ x - 2y - 3z &= 1 \end{aligned}$$

CUMULATIVE REVIEW EXERCISES (CHAPTERS 1–3)

1. Simplify: $\dfrac{6(8 - 10)^3 + (-2)}{(-5)^2(-2)}$.

2. Simplify: $7x - [5 - 2(4x - 1)]$.

In Exercises 3–5, solve each equation.

3. $5 - 2(3 - x) = 2(2x + 5) + 1$

4. $\dfrac{3x}{5} + 4 = \dfrac{x}{3}$

5. $3x - 4 = 2(3x + 2) - 3x$

6. For a summer sales job, you are choosing between two pay arrangements: a weekly salary of \$200 plus 5% commission on sales, or a straight 15% commission. For how many dollars of sales will the earnings be the same regardless of the pay arrangement?

7. Simplify: $\dfrac{-5x^6y^{-10}}{20x^{-2}y^{20}}$.

8. If $f(x) = -4x + 5$, find $f(a + 2)$.

9. Find the domain of $f(x) = \dfrac{4}{x + 3}$.

10. If $f(x) = 2x^2 - 5x + 2$ and $g(x) = x^2 - 2x + 3$, find $(f - g)(x)$ and $(f - g)(3)$.

In Exercises 11–12, graph each linear function.

11. $f(x) = -\frac{2}{3}x + 2$

12. $2x - y = 6$

In Exercises 13–14, use the given conditions to write an equation for each line in point-slope form and slope-intercept form.

13. Passing through $(2, 4)$ and $(4, -2)$

14. Passing through $(-1, 0)$ and parallel to the line whose equation is $3x + y = 6$

In Exercises 15–16, solve each system by eliminating variables using the addition method.

15. $$\begin{aligned} 3x + 12y &= 25 \\ 2x - 6y &= 12 \end{aligned}$$

16. $$\begin{aligned} x + 3y - z &= 5 \\ -x + 2y + 3z &= 13 \\ 2x - 5y - z &= -8 \end{aligned}$$

17. If two pads of paper and 19 pens are sold for \$5.40 and 7 of the same pads and 4 of the same pens sell for \$6.40, find the cost of one pad and one pen.

18. Evaluate:

$$\begin{vmatrix} 0 & 1 & -2 \\ -7 & 0 & -4 \\ 3 & 0 & 5 \end{vmatrix}.$$

19. Solve using matrices:

$$\begin{aligned} 2x + 3y - z &= -1 \\ x + 2y + 3z &= 2 \\ 3x + 5y - 2z &= -3. \end{aligned}$$

20. Solve using Cramer's rule (determinants):

$$\begin{aligned} 3x + 4y &= -1 \\ -2x + y &= 8. \end{aligned}$$

CHAPTER 4

Inequalities and Problem Solving

You are in Yosemite National Park in California, surrounded by evergreen forests, alpine meadows, and sheer walls of granite. The beauty of soaring cliffs, plunging waterfalls, gigantic trees, rugged canyons, mountains, and valleys is overwhelming. This is so different from where you live and attend college, a region in which grasslands predominate. What variables affect whether regions are forests, grasslands, or deserts, and what kinds of mathematical models are used to describe the incredibly diverse land that forms the surface of our planet?

The role that temperature and precipitation play in determining whether regions are forests, grasslands, or deserts can be modeled using inequalities in two variables. You will use these models and their graphs in Example 4 of Section 4.4.

SECTION 4.1

Objectives

1. Use interval notation.
2. Solve linear inequalities.
3. Recognize inequalities with no solution or all real numbers as solutions.
4. Solve applied problems using linear inequalities.

Solving Linear Inequalities

You can go online and obtain a list of telecommunication companies that provide residential long-distance phone service. The list contains the monthly fee, the monthly minimum, and the rate per minute for each service provider. You've chosen a plan that has a monthly fee of \$15 with a charge of \$0.08 per minute for all long-distance calls. Suppose you are limited by how much money you can spend for the month: You can spend at most \$35. If we let x represent the number of minutes of long-distance calls in a month, we can write an inequality that describes the given conditions:

The monthly fee of \$15	plus	the charge of \$0.08 per minute for x minutes	must be less than or equal to	\$35.
15	+	$0.08x$	$\le$	35.

Using the commutative property of addition, we can express this inequality as

$$0.08x + 15 \le 35.$$

Placing an inequality symbol between a linear expression $(mx + b)$ and a constant results in a *linear inequality in one variable*. In this section, we will study how to solve linear inequalities such as the one shown above. **Solving an inequality** is the process of finding the set of numbers that make the inequality a true statement. These numbers are called the **solutions** of the inequality and we say that they **satisfy** the inequality. The set of all solutions is called the **solution set** of the inequality. Set-builder notation and a new notation, called *interval notation*, are used to represent solution sets. We begin this section by looking at interval notation.

1 Use interval notation.

Interval Notation

Some sets of real numbers can be represented using **interval notation**. Suppose that a and b are two real numbers such that $a < b$.

Interval Notation	Graph
The **open interval** (a, b) represents the set of real numbers between, but not including, a and b. $(a, b) = \{x \mid a < x < b\}$ x is greater than a $(a < x)$ and x is less than b $(x < b)$.	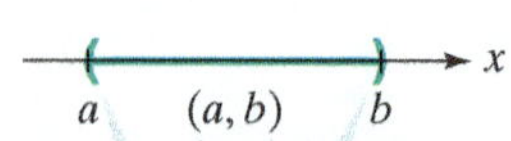 The parentheses in the graph and in interval notation indicate that a and b, the endpoints, are excluded from the interval.

(*continued*)

Interval Notation	Graph
The **closed interval** $[a, b]$ represents the set of real numbers between, and including, a and b. $[a, b] = \{x \mid a \leq x \leq b\}$ x is greater than or equal to a $(a \leq x)$ and x is less than or equal to b $(x \leq b)$.	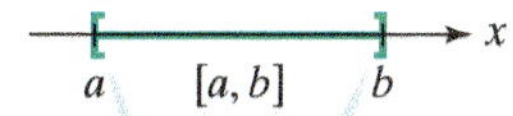 The square brackets in the graph and in interval notation indicate that a and b, the endpoints, are included in the interval.
The **infinite interval** (a, ∞) represents the set of real numbers that are greater than a. $(a, \infty) = \{x \mid x > a\}$ The infinity symbol does not represent a real number. It indicates that the interval extends indefinitely to the right.	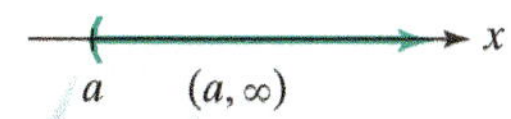 The parenthesis indicates that a is excluded from the interval.
The **infinite interval** $(-\infty, b]$ represents the set of real numbers that are less than or equal to b. $(-\infty, b] = \{x \mid x \leq b\}$ The negative infinity symbol indicates that the interval extends indefinitely to the left.	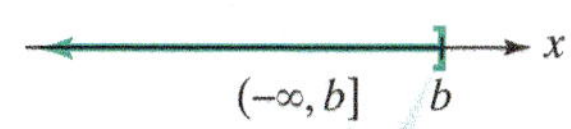 The square bracket indicates that b is included in the interval.

Parentheses and Brackets in Interval Notation

Parentheses indicate endpoints that are not included in an interval. Square brackets indicate endpoints that are included in an interval. Parentheses are always used with ∞ or $-\infty$.

Table 4.1 lists nine possible types of intervals used to describe sets of real numbers.

Table 4.1 Intervals on the Real Number Line

Let a and b be real numbers such that $a < b$.

Interval Notation	Set-Builder Notation	Graph
(a, b)	$\{x \mid a < x < b\}$	
$[a, b]$	$\{x \mid a \leq x \leq b\}$	
$[a, b)$	$\{x \mid a \leq x < b\}$	
$(a, b]$	$\{x \mid a < x \leq b\}$	
(a, ∞)	$\{x \mid x > a\}$	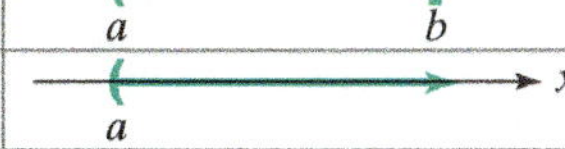
$[a, \infty)$	$\{x \mid x \geq a\}$	
$(-\infty, b)$	$\{x \mid x < b\}\{x \mid x < b\}$	
$(-\infty, b]$	$\{x \mid x \leq b\}$	
$(-\infty, \infty)$	$\{x \mid x \text{ is a real number}\}$ or $\mathbb{R}$ (set of all real numbers)	

EXAMPLE 1 Interpreting Interval Notation

Express each interval in set-builder notation and graph:

a. $(-1, 4]$ **b.** $[2.5, 4]$ **c.** $(-4, \infty)$.

Solution

a. $(-1, 4] = \{x | -1 < x \leq 4\}$

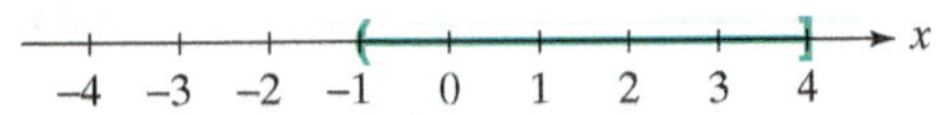

b. $[2.5, 4] = \{x | 2.5 \leq x \leq 4\}$

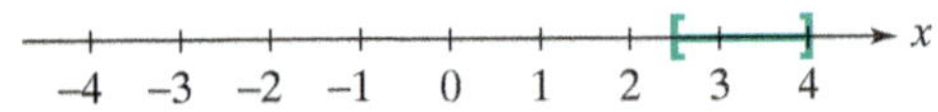

c. $(-4, \infty) = \{x | x > -4\}$

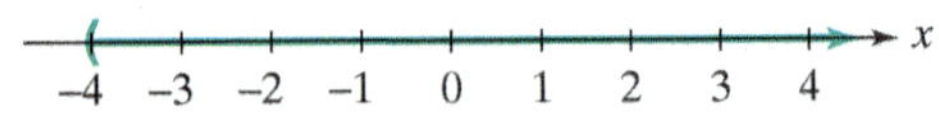

See graphing answer section.

✓ CHECK POINT 1 Express each interval in set-builder notation and graph:

a. $[-2, 5)$ **b.** $[1, 3.5]$ **c.** $(-\infty, -1)$.

---

2 Solve linear inequalities.

Solving Linear Inequalities in One Variable

We know that a linear equation in x can be expressed as $ax + b = 0$. A **linear inequality in x** can be written in one of the following forms: $ax + b < 0$, $ax + b \leq 0$, $ax + b > 0$, $ax + b \geq 0$. In each form, $a \neq 0$.

Back to our question that opened this section: How many minutes of long-distance calls can you make in a month if you can spend at most \$35? We answer the question by solving the linear inequality

$$0.08x + 15 \leq 35$$

for x. The solution procedure is nearly identical to that for solving the equation

$$0.08x + 15 = 35.$$

Our goal is to get x by itself on the left side. We do this by first subtracting 15 from both sides to isolate $0.08x$:

$$0.08x + 15 \leq 35 \quad \text{This is the given inequality.}$$
$$0.08x + 15 - 15 \leq 35 - 15 \quad \text{Subtract 15 from both sides.}$$
$$0.08x \leq 20. \quad \text{Simplify.}$$

Finally, we isolate x from $0.08x$ by dividing both sides of the inequality by 0.08:

$$\frac{0.08x}{0.08} \leq \frac{20}{0.08} \quad \text{Divide both sides by 0.08.}$$
$$x \leq 250. \quad \text{Simplify.}$$

With at most \$35 per month to spend, you can make no more than 250 minutes of long-distance calls each month.

We started with the inequality $0.08x + 15 \leq 35$ and obtained the inequality $x \leq 250$ in the final step. Both of these inequalities have the same solution set, namely $\{x | x \leq 250\}$. Inequalities such as these, with the same solution set, are said to be **equivalent**.

We isolated x from $0.08x$ by dividing both sides of $0.08x \leq 20$ by 0.08, a positive number. Let's see what happens if we divide both sides of an inequality by a negative number. Consider the inequality $10 < 14$. Divide 10 and 14 by -2:

$$\frac{10}{-2} = -5 \quad \text{and} \quad \frac{14}{-2} = -7.$$

Because −5 lies to the right of −7 on the number line, −5 is greater than −7:

$$-5 > -7.$$

Notice that the direction of the inequality symbol is reversed:

$$10 < 14$$
$$\downarrow$$
$$-5 > -7.$$

Dividing by −2 changes the direction of the inequality symbol.

In general, **when we multiply or divide both sides of an inequality by a negative number, the direction of the inequality symbol is reversed.** When we reverse the direction of the inequality symbol, we say that we change the *sense* of the inequality.

We can isolate a variable in a linear inequality the same way we can isolate a variable in a linear equation. The following properties are used to create equivalent inequalities:

Properties of Inequalities

Property	The Property in Words	Example
The Addition Property of Inequality If $a < b$, then $a + c < b + c$. If $a < b$, then $a - c < b - c$.	If the same quantity is added to or subtracted from both sides of an inequality, the resulting inequality is equivalent to the original one.	$2x + 3 < 7$ Subtract 3: $2x + 3 - 3 < 7 - 3$. Simplify: $2x < 4$.
The Positive Multiplication Property of Inequality If $a < b$ and c is positive, then $ac < bc$. If $a < b$ and c is positive, then $\frac{a}{c} < \frac{b}{c}$.	If we multiply or divide both sides of an inequality by the same positive quantity, the resulting inequality is equivalent to the original one.	$2x < 4$ Divide by 2: $\frac{2x}{2} < \frac{4}{2}$. Simplify: $x < 2$.
The Negative Multiplication Property of Inequality If $a < b$ and c is negative, then $ac > bc$. If $a < b$ and c is negative, then $\frac{a}{c} > \frac{b}{c}$.	If we multiply or divide both sides of an inequality by the same negative quantity and reverse the direction of the inequality symbol, the resulting inequality is equivalent to the original one.	$-4x < 20$ Divide by −4 and change the sense of the inequality: $\frac{-4x}{-4} > \frac{20}{-4}$. Simplify: $x > -5$.

If an inequality does not contain fractions, it can be solved using the following procedure. (In Example 4, we will see how to clear fractions.) Notice, again, how similar this procedure is to the procedure for solving a linear equation.

Solving a Linear Inequality

1. Simplify the algebraic expression on each side.
2. Use the addition property of inequality to collect all the variable terms on one side and all the constant terms on the other side.
3. Use the multiplication property of inequality to isolate the variable and solve. Change the sense of the inequality when multiplying or dividing both sides by a negative number.
4. Express the solution set in set-builder or interval notation and graph the solution set on a number line.

EXAMPLE 2 Solving a Linear Inequality

Solve and graph the solution set on a number line:

$$3x - 5 > -17.$$

Solution

Step 1. Simplify each side. Because each side is already simplified, we can skip this step.

Step 2. Collect variable terms on one side and constant terms on the other side. The variable term, $3x$, is already on the left side of $3x - 5 > -17$. We will collect constant terms on the right side by adding 5 to both sides.

$$3x - 5 > -17 \quad \text{This is the given inequality.}$$
$$3x - 5 + 5 > -17 + 5 \quad \text{Add 5 to both sides.}$$
$$3x > -12 \quad \text{Simplify.}$$

Step 3. Isolate the variable and solve. We isolate the variable, x, by dividing both sides by 3. Because we are dividing by a positive number, we do not reverse the direction of the inequality symbol.

$$\frac{3x}{3} > \frac{-12}{3} \quad \text{Divide both sides by 3.}$$
$$x > -4 \quad \text{Simplify.}$$

Step 4. Express the solution set in set-builder or interval notation and graph the set on a number line. The solution set consists of all real numbers that are greater than -4, expressed as $\{x|x > -4\}$ in set-builder notation. The interval notation for this solution set is $(-4, \infty)$. The graph of the solution set is shown as follows:

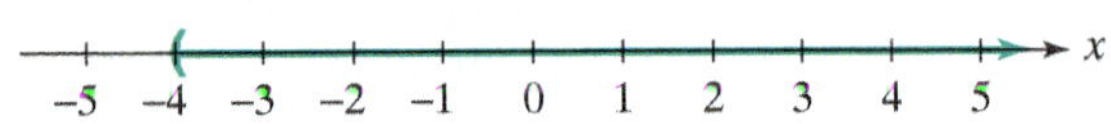

Discover for Yourself

As a partial check, select one number from the solution set of $3x - 5 > -17$, the inequality in Example 2. Substitute that number into the original inequality. Perform the resulting computations. You should obtain a true statement.

Is it possible to perform a partial check using a number that is not in the solution set? What should happen in this case? Try doing this.

CHECK POINT 2 Solve and graph the solution set on a number line: $4x - 3 > -23$.

EXAMPLE 3 Solving a Linear Inequality

Solve and graph the solution set on a number line:

$$-2x - 4 > x + 5.$$

Solution

Step 1. Simplify each side. Because each side is already simplified, we can skip this step.

Step 2. Collect variable terms on one side and constant terms on the other side. We will collect variable terms on the left and constant terms on the right.

$$-2x - 4 > x + 5 \quad \text{This is the given inequality.}$$
$$-2x - 4 - x > x + 5 - x \quad \text{Subtract } x \text{ from both sides.}$$
$$-3x - 4 > 5 \quad \text{Simplify.}$$
$$-3x - 4 + 4 > 5 + 4 \quad \text{Add 4 to both sides.}$$
$$-3x > 9 \quad \text{Simplify.}$$

Step 3. Isolate the variable and solve. We isolate the variable, x, by dividing both sides by -3. Because we are dividing by a negative number, we must reverse the direction of the inequality symbol.

$$\frac{-3x}{-3} < \frac{9}{-3} \quad \text{Divide both sides by } -3 \text{ and change the sense of the inequality.}$$
$$x < -3 \quad \text{Simplify.}$$

Study Tip

You can solve

$$-2x - 4 > x + 5$$

by isolating x on the right side. Add $2x$ to both sides.

$$-2x - 4 + 2x > x + 5 + 2x$$
$$-4 > 3x + 5$$

Now subtract 5 from both sides.

$$-4 - 5 > 3x + 5 - 5$$
$$-9 > 3x$$

Finally, divide both sides by 3.

$$\frac{-9}{3} > \frac{3x}{3}$$
$$-3 > x$$

This last inequality means the same thing as $x < -3$.

Step 4. Express the solution set in set-builder or interval notation and graph the set on a number line. The solution set consists of all real numbers that are less than -3, expressed in set-builder notation as $\{x|x < -3\}$. The interval notation for this solution set is $(-\infty, -3)$. The graph of the solution set is shown as follows:

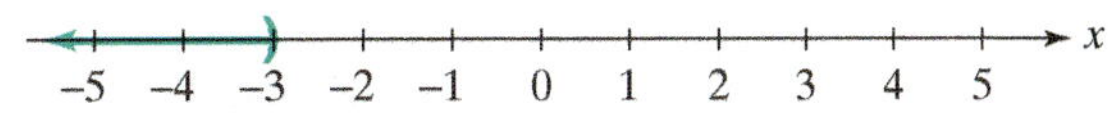

Using Technology

Numeric and Graphic Connections

You can use a graphing utility to check the solution set of a linear inequality. Enter each side of the inequality separately under y_1 and y_2. Then use the table or the graphs. To use the table, first locate the x-value for which the y-values are the same. Then scroll up or down to locate x values for which y_1 is greater than y_2 or for which y_1 is less than y_2. To use the graphs, locate the intersection point and then find the x-values for which the graph of y_1 lies above the graph of y_2 $(y_1 > y_2)$ or for which the graph of y_1 lies below the graph of y_2 $(y_1 < y_2)$.

Let's verify our work in Example 3 and show that $(-\infty, -3)$ is the solution set of

$$-2x - 4 > x + 5.$$

Enter $y_1 = -2x - 4$ in the $\boxed{y=}$ screen.

Enter $y_2 = x + 5$ in the $\boxed{y=}$ screen.

We are looking for values of x for which y_1 is greater than y_2.

Numeric Check

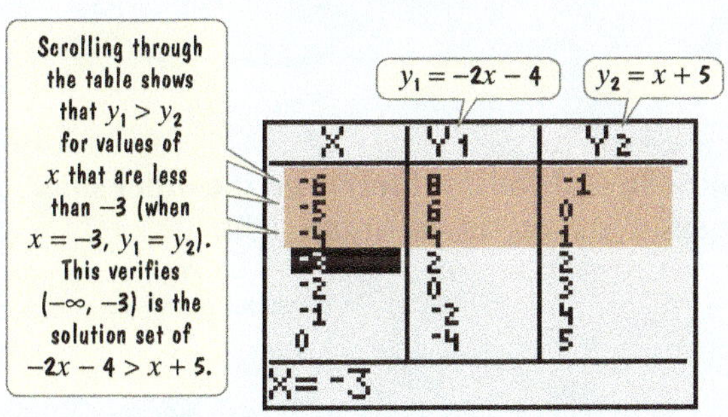

Graphic Check

Display the graphs for y_1 and y_2. Use the intersection feature. The solution set is the set of x-values for which the graph of y_1 lies above the graph of y_2.

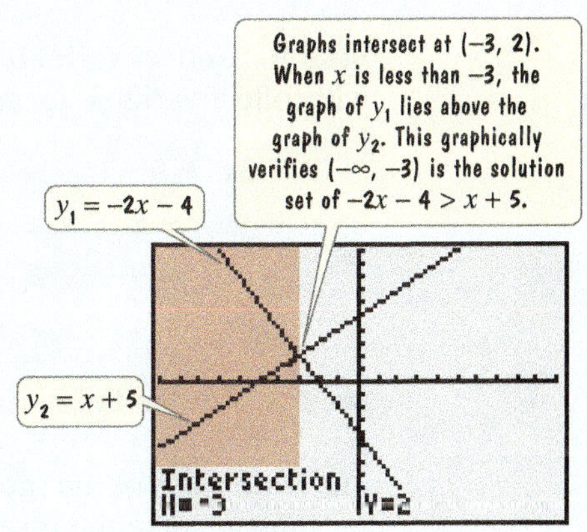

[−10, 10, 1] by [−10, 10, 1]

CHECK POINT 3 Solve and graph the solution set: $3x + 1 > 7x - 15$.

If an inequality contains fractions, begin by multiplying both sides by the least common denominator. This will clear the inequality of fractions.

EXAMPLE 4 Solving a Linear Inequality Containing Fractions

Solve and graph the solution set on a number line:

$$\frac{x+3}{4} \geq \frac{x-2}{3} + \frac{1}{4}.$$

Solution The denominators are 4, 3, and 4. The least common denominator is 12. We begin by multiplying both sides of the inequality by 12.

$$\frac{x+3}{4} \geq \frac{x-2}{3} + \frac{1}{4}$$ This is the given inequality.

$$12\left(\frac{x+3}{4}\right) \geq 12\left(\frac{x-2}{3} + \frac{1}{4}\right)$$ Multiply both sides by 12. Multiplying by a positive number preserves the sense of the inequality.

$$\frac{12}{1}\cdot\frac{x+3}{4} \geq \frac{12}{1}\cdot\frac{x-2}{3} + \frac{12}{1}\cdot\frac{1}{4}$$ Multiply each term by 12. Use the distributive property on the right side.

$$\frac{\overset{3}{\cancel{12}}}{1}\cdot\frac{x+3}{\underset{1}{\cancel{4}}} \geq \frac{\overset{4}{\cancel{12}}}{1}\cdot\frac{x-2}{\underset{1}{\cancel{3}}} + \frac{\overset{3}{\cancel{12}}}{1}\cdot\frac{1}{\underset{1}{\cancel{4}}}$$ Divide out common factors in each multiplication.

$$3(x+3) \geq 4(x-2) + 3$$ The fractions are now cleared.

Now that the fractions have been cleared, we follow the four steps that we used in the previous examples.

Step 1. Simplify each side.

$$3(x+3) \geq 4(x-2) + 3$$ This is the inequality with the fractions cleared.

$$3x + 9 \geq 4x - 8 + 3$$ Use the distributive property.

$$3x + 9 \geq 4x - 5$$ Simplify.

Step 2. Collect variable terms on one side and constant terms on the other side. We will collect variable terms on the left and constant terms on the right.

$$3x + 9 - 4x \geq 4x - 5 - 4x$$ Subtract 4x from both sides.

$$-x + 9 \geq -5$$ Simplify.

$$-x + 9 - 9 \geq -5 - 9$$ Subtract 9 from both sides.

$$-x \geq -14$$ Simplify.

Step 3. Isolate the variable and solve. To isolate x, we must eliminate the negative sign in front of the x. Because $-x$ means $-1x$, we can do this by multiplying (or dividing) both sides of the inequality by -1. We are multiplying by a negative number. Thus, we must reverse the direction of the inequality symbol.

$$(-1)(-x) \leq (-1)(-14)$$ Multiply both sides by -1 and change the sense of the inequality.

$$x \leq 14$$ Simplify.

Step 4. Express the solution set in set-builder or interval notation and graph the set on a number line. The solution set consists of all real numbers that are less than or equal to 14, expressed in set-builder notation as $\{x \mid x \leq 14\}$. The interval notation for this solution set is $(-\infty, 14]$. The graph of the solution set is shown as follows:

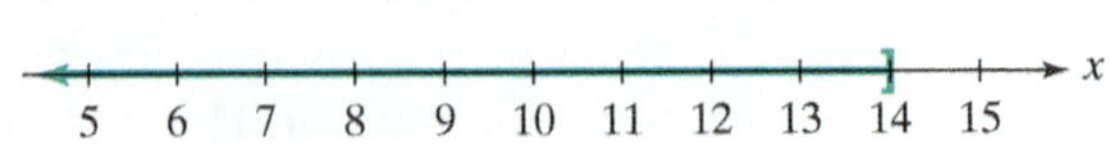

CHECK POINT 4 Solve and graph the solution set on a number line:

$$\frac{x-4}{2} \geq \frac{x-2}{3} + \frac{5}{6}.$$

3 Recognize inequalities with no solution or all real numbers as solutions.

Inequalities with Unusual Solution Sets

We have seen that some equations have no solution. This is also true for some inequalities. An example of such an inequality is

$$x > x + 1.$$

There is no number that is greater than itself plus 1. This inequality has no solution and its solution set is $\varnothing$, the empty set.

By contrast, some inequalities are true for all real numbers. An example of such an inequality is

$$x < x + 1.$$

Every real number is less than itself plus 1. The solution set is $\{x|x \text{ is a real number}\}$ or $\mathbb{R}$. In interval notation, the solution set is $(-\infty, \infty)$.

If you attempt to solve an inequality that has no solution, you will eliminate the variable and obtain a false statement, such as $0 > 1$. If you attempt to solve an inequality that is true for all real numbers, you will eliminate the variable and obtain a true statement, such as $0 < 1$.

Using Technology

Graphic Connections

The graphs of

$$y_1 = 2(x + 4) \text{ and } y_2 = 2x + 3$$

are parallel lines. The graph of y_1 is always above the graph of y_2. Every value of x satisfies the inequality $y_1 > y_2$. Thus, the solution set of the inequality

$$2(x + 4) > 2x + 3$$

is $(-\infty, \infty)$.

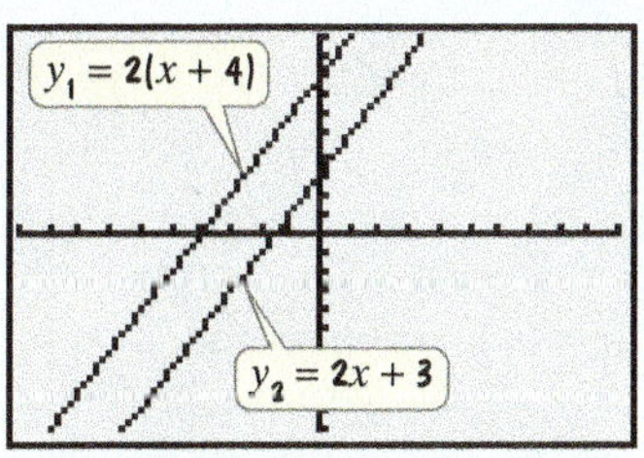

[−10, 10, 1] by [−10, 10, 1]

EXAMPLE 5 Solving Linear Inequalities

Solve each inequality:

a. $2(x + 4) > 2x + 3$ b. $x + 7 \leq x - 2$.

Solution

a.

$2(x + 4) > 2x + 3$	This is the given inequality.
$2x + 8 > 2x + 3$	Apply the distributive property.
$2x + 8 - 2x > 2x + 3 - 2x$	Subtract $2x$ from both sides.
$8 > 3$	Simplify. The statement $8 > 3$ is true.

The inequality $8 > 3$ is true for all values of x. Because this inequality is equivalent to the original inequality, the original inequality is true for all real numbers. The solution set is

$$\{x|x \text{ is a real number}\} \text{ or } \mathbb{R} \text{ or } (-\infty, \infty).$$

b.

$x + 7 \leq x - 2$	This is the given inequality.
$x + 7 - x \leq x - 2 - x$	Subtract x from both sides.
$7 \leq -2$	Simplify. The statement $7 \leq -2$ is false.

The inequality $7 \leq -2$ is false for all values of x. Because this inequality is equivalent to the original inequality, the original inequality has no solution. The solution set is $\varnothing$.

CHECK POINT 5 Solve each inequality:

a. $3(x + 1) > 3x + 2$

b. $x + 1 \leq x - 1$.

4 Solve applied problems using linear inequalities.

Applications

Commonly used English phrases such as "at least" and "at most" indicate inequalities. **Table 4.2** lists sentences containing these phrases and their algebraic translations into inequalities.

Table 4.2 English Sentences and Inequalities

English Sentence	Inequality
x is at least 5.	$x \geq 5$
x is at most 5.	$x \leq 5$
x is between 5 and 7.	$5 < x < 7$
x is no more than 5.	$x \leq 5$
x is no less than 5.	$x \geq 5$

Our next example shows how to use an inequality to select the better deal when considering two pricing options. We use our strategy for solving word problems, translating from the verbal conditions of the problem to a linear inequality.

EXAMPLE 6 Selecting the Better Deal

Acme Car rental agency charges $4 a day plus $0.15 per mile, whereas Interstate rental agency charges $20 a day and $0.05 per mile. How many miles must be driven to make the daily cost of an Acme rental a better deal than an Interstate rental?

Solution

Step 1. Let x represent one of the unknown quantities. We are looking for the number of miles that must be driven in a day to make Acme the better deal. Thus,

let x = the number of miles driven in a day.

Step 2. Represent other unknown quantities in terms of x. We are not asked to find another quantity, so we can skip this step.

Step 3. Write an inequality in x that models the conditions. Acme is a better deal than Interstate if the daily cost of Acme is less than the daily cost of Interstate.

The daily cost of Acme | is less than | the daily cost of Interstate.

4 dollars | plus | 15 cents times | the number of miles driven | | 20 dollars | plus | 5 cents times | the number of miles driven

$$4 + 0.15 \cdot x < 20 + 0.05 \cdot x$$

Step 4. Solve the inequality and answer the question.

$$4 + 0.15x < 20 + 0.05x$$ This is the inequality that models the verbal conditions.

$$4 + 0.15x - 0.05x < 20 + 0.05x - 0.05x$$ Subtract 0.05x from both sides.

$$4 + 0.1x < 20$$ Simplify.

$$4 + 0.1x - 4 < 20 - 4$$ Subtract 4 from both sides.

$$0.1x < 16$$ Simplify.

$$\frac{0.1x}{0.1} < \frac{16}{0.1}$$ Divide both sides by 0.1.

$$x < 160$$ Simplify.

Thus, driving fewer than 160 miles per day makes Acme the better deal.

Using Technology

Graphic Connections

The graphs of the daily cost models for the car rental agencies

$$y_1 = 4 + 0.15x$$
$$\text{and } y_2 = 20 + 0.05x$$

are shown in a [0, 300, 10] by [0, 40, 4] viewing rectangle. The graphs intersect at (160, 28). To the left of $x = 160$, the graph of Acme's daily cost lies below that of Interstate's daily cost. This shows that for fewer than 160 miles per day, Acme offers the better deal.

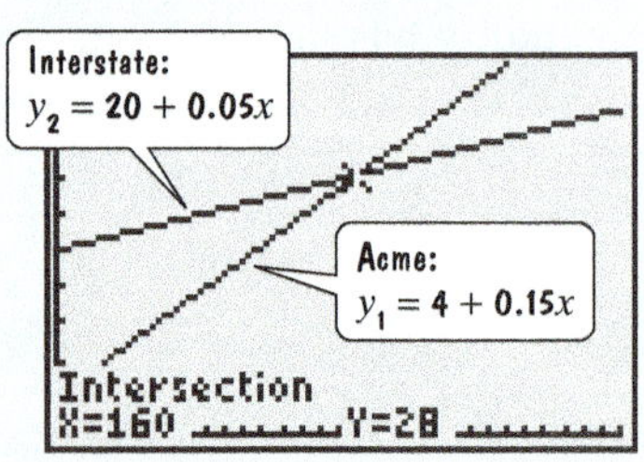

Step 5. Check the proposed solution in the original wording of the problem. One way to do this is to take a mileage less than 160 miles per day to see if Acme is the better deal. Suppose that 150 miles are driven in a day.

$$\text{Cost for Acme} = 4 + 0.15(150) = 26.50$$
$$\text{Cost for Interstate} = 20 + 0.05(150) = 27.50$$

Acme has a lower daily cost, making Acme the better deal.

CHECK POINT 6 A car can be rented from Basic Rental for $260 per week with no extra charge for mileage. Continental charges $80 per week plus 25 cents for each mile driven to rent the same car. How many miles must be driven in a week to make the rental cost for Basic Rental a better deal than Continental's?

4.1 EXERCISE SET

MyMathLab — MathXL PRACTICE — WATCH — DOWNLOAD — READ — REVIEW

Practice Exercises

In Exercises 1–14, express each interval in set-builder notation and graph the interval on a number line.

1. $(1, 6]$
2. $(-2, 4]$
3. $[-5, 2)$
4. $[-4, 3)$
5. $[-3, 1]$
6. $[-2, 5]$
7. $(2, \infty)$
8. $(3, \infty)$
9. $[-3, \infty)$
10. $[-5, \infty)$
11. $(-\infty, 3)$
12. $(-\infty, 2)$
13. $(-\infty, 5.5)$
14. $(-\infty, 3.5]$

In Exercises 15–46, solve each linear inequality. Other than ∅, graph the solution set on a number line.

15. $5x + 11 < 26$
16. $2x + 5 < 17$
17. $3x - 8 \geq 13$
18. $8x - 2 \geq 14$
19. $-9x \geq 36$
20. $-5x \leq 30$
21. $8x - 11 \leq 3x - 13$
22. $18x + 45 \leq 12x - 8$
23. $4(x + 1) + 2 \geq 3x + 6$
24. $8x + 3 > 3(2x + 1) + x + 5$
25. $2x - 11 < -3(x + 2)$
26. $-4(x + 2) > 3x + 20$
27. $1 - (x + 3) \geq 4 - 2x$
28. $5(3 - x) \leq 3x - 1$
29. $\frac{x}{4} - \frac{1}{2} \leq \frac{x}{2} + 1$
30. $\frac{3x}{10} + 1 \geq \frac{1}{5} - \frac{x}{10}$
31. $1 - \frac{x}{2} > 4$
32. $7 - \frac{4}{5}x < \frac{3}{5}$
33. $\frac{x - 4}{6} \geq \frac{x - 2}{9} + \frac{5}{18}$
34. $\frac{4x - 3}{6} + 2 \geq \frac{2x - 1}{12}$
35. $4(3x - 2) - 3x < 3(1 + 3x) - 7$
36. $3(x - 8) - 2(10 - x) < 5(x - 1)$
37. $8(x + 1) \leq 7(x + 5) + x$
38. $4(x - 1) \geq 3(x - 2) + x$
39. $3x < 3(x - 2)$
40. $5x < 5(x - 3)$
41. $7(x + 4) - 13 < 12 + 13(3 + x)$
42. $-3[7x - (2x - 3)] > -2(x + 1)$
43. $6 - \frac{2}{3}(3x - 12) \leq \frac{2}{5}(10x + 50)$
44. $\frac{2}{7}(7 - 21x) - 4 > 10 - \frac{3}{11}(11x - 11)$
45. $3[3(x + 5) + 8x + 7] + 5[3(x - 6) - 2(3x - 5)] < 2(4x + 3)$

46. $5[3(2 - 3x) - 2(5 - x)] - 6[5(x - 2) - 2(4x - 3)] < 3x + 19$

47. Let $f(x) = 3x + 2$ and $g(x) = 5x - 8$. Find all values of x for which $f(x) > g(x)$.

48. Let $f(x) = 2x - 9$ and $g(x) = 5x + 4$. Find all values of x for which $f(x) > g(x)$.

49. Let $f(x) = \frac{2}{5}(10x + 15)$ and $g(x) = \frac{1}{4}(8 - 12x)$. Find all values of x for which $g(x) \le f(x)$.

50. Let $f(x) = \frac{3}{5}(10x - 15) + 9$ and $g(x) = \frac{3}{8}(16 - 8x) - 7$. Find all values of x for which $g(x) \le f(x)$.

51. Let $f(x) = 1 - (x + 3) + 2x$. Find all values of x for which $f(x)$ is at least 4.

52. Let $f(x) = 2x - 11 + 3(x + 2)$. Find all values of x for which $f(x)$ is at most 0.

Practice PLUS

In Exercises 53–54, solve each linear inequality and graph the solution set on a number line.

53. $2(x + 3) > 6 - \{4[x - (3x - 4) - x] + 4\}$

54. $3(4x - 6) < 4 - \{5x - [6x - (4x - (3x + 2))]\}$

In Exercises 55–56, write an inequality with x isolated on the left side that is equivalent to the given inequality.

55. $ax + b > c$; Assume $a < 0$.

56. $\frac{ax + b}{c} > b$; Assume $a > 0$ and $c < 0$.

In Exercises 57–58, use the graphs of y_1 and y_2 to solve each inequality.

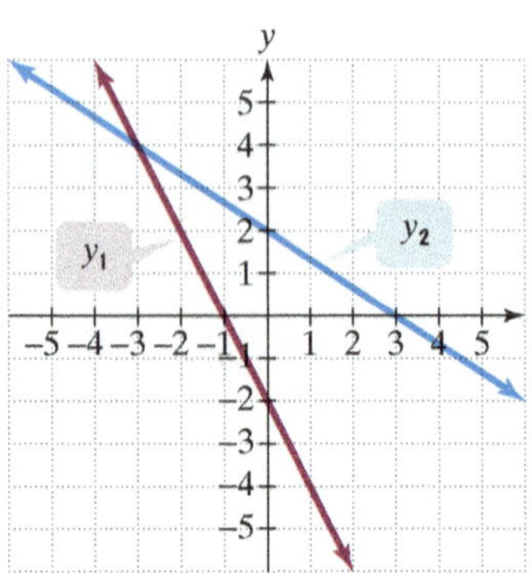

57. $y_1 \ge y_2$

58. $y_1 \le y_2$

In Exercises 59–60, use the table of values for the linear functions y_1 and y_2 to solve each inequality.

X	Y1	Y2
-1.8	9.4	7.4
-1.7	9.1	7.6
-1.6	8.8	7.8
-1.5	8.5	8
-1.4	8.2	8.2
-1.3	7.9	8.4
-1.2	7.6	8.6

X=-1.8

59. $y_1 < y_2$

60. $y_1 > y_2$

Application Exercises

The graphs show that the three components of love, namely passion, intimacy, and commitment, progress differently over time. Passion peaks early in a relationship and then declines. By contrast, intimacy and commitment build gradually. Use the graphs to solve Exercises 61–68.

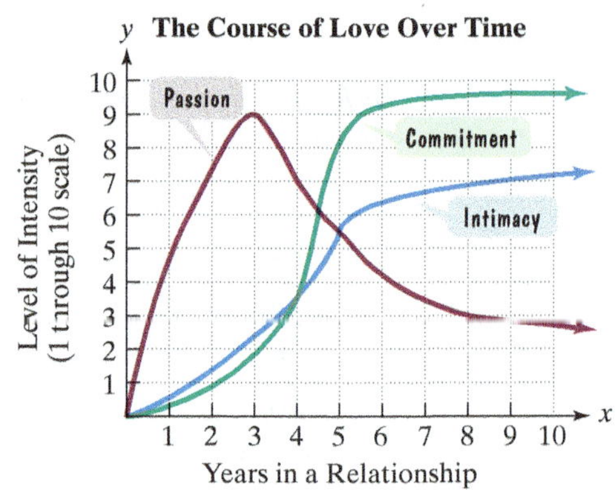

Source: R. J. Sternberg, A Triangular Theory of Love, *Psychological Review, 93, 119–135.*

61. Use interval notation to write an inequality that expresses for which years in a relationship intimacy is greater than commitment.

62. Use interval notation to write an inequality that expresses for which years in a relationship passion is greater than or equal to intimacy.

63. What is the relationship between passion and intimacy on the interval $[5, 7)$?

64. What is the relationship between intimacy and commitment on the interval $[4, 7)$?

65. What is the relationship between passion and commitment for $\{x | 6 < x < 8\}$?

66. What is the relationship between passion and commitment for $\{x | 7 < x < 9\}$?

67. What is the maximum level of intensity for passion? After how many years in a relationship does this occur?

68. After approximately how many years do levels of intensity for commitment exceed the maximum level of intensity for passion?

69. The percentage, P, of U.S. voters who use electronic voting systems, such as optical scans, in national elections can be modeled by the formula

$$P = 3.1x + 25.8,$$

where x is the number of years after 1994. In which years will more than 63% of U.S. voters use electronic systems?

70. The percentage, P, of U.S. voters who use punch cards or lever machines in national elections can be modeled by the formula

$$P = -2.5x + 63.1,$$

where x is the number of years after 1994. In which years will fewer than 38.1% of U.S. voters use punch cards or lever machines?

The Olympic 500-meter speed skating times have generally been decreasing over time. The formulas

$$W = -0.19t + 57 \quad \text{and} \quad M = -0.15t + 50$$

model the winning times, in seconds, for women, W, and men, M, t years after 1900. Use these models to solve Exercises 71–72.

71. Find values of t such that $W < M$. Describe what this means in terms of winning times.

72. Find values of t such that $W > M$. Describe what this means in terms of winning times.

In Exercises 73–80, use the strategy for solving word problems, translating from the verbal conditions of the problem to a linear inequality.

73. A truck can be rented from Basic Rental for \$50 a day plus \$0.20 per mile. Continental charges \$20 per day plus \$0.50 per mile to rent the same truck. How many miles must be driven in a day to make the rental cost for Basic Rental a better deal than Continental's?

74. You are choosing between two long-distance telephone plans. Plan A has a monthly fee of \$15 with a charge of \$0.08 per minute for all long-distance calls. Plan B has a monthly fee of \$3 with a charge of \$0.12 per minute for all long-distance calls. How many minutes of long-distance calls in a month make plan A the better deal?

75. A city commission has proposed two tax bills. The first bill requires that a homeowner pay \$1800 plus 3% of the assessed home value in taxes. The second bill requires taxes of \$200 plus 8% of the assessed home value. What price range of home assessment would make the first bill a better deal for the homeowner?

76. A local bank charges \$8 per month plus 5¢ per check. The credit union charges \$2 per month plus 8¢ per check. How many checks should be written each month to make the credit union a better deal?

77. A company manufactures and sells blank audiocassette tapes. The weekly fixed cost is \$10,000 and it costs \$0.40 to produce each tape. The selling price is \$2.00 per tape. How many tapes must be produced and sold each week for the company to have a profit?

78. A company manufactures and sells personalized stationery. The weekly fixed cost is \$3000 and it costs \$3.00 to produce each package of stationery. The selling price is \$5.50 per package. How many packages of stationery must be produced and sold each week for the company to have a profit?

79. An elevator at a construction site has a maximum capacity of 3000 pounds. If the elevator operator weighs 200 pounds and each cement bag weighs 70 pounds, how many bags of cement can be safely lifted on the elevator in one trip?

80. An elevator at a construction site has a maximum capacity of 2500 pounds. If the elevator operator weighs 160 pounds and each cement bag weighs 60 pounds, how many bags of cement can be safely lifted on the elevator in one trip?

Writing in Mathematics

81. When graphing the solutions of an inequality, what does a parenthesis signify? What does a bracket signify?

82. When solving an inequality, when is it necessary to change the sense of the inequality? Give an example.

83. Describe ways in which solving a linear inequality is similar to solving a linear equation.

84. Describe ways in which solving a linear inequality is different from solving a linear equation.

85. When solving a linear inequality, describe what happens if the solution set is $(-\infty, \infty)$.

86. When solving a linear inequality, describe what happens if the solution set is $\varnothing$.

87. What is the slope of each model in Exercises 69–70? What does this mean in terms of the percentage of U.S. voters using electronic voting systems and more traditional methods, such as punch cards or lever machines? What explanations can you offer for these changes in vote-counting systems?

Technology Exercises

In Exercises 88–89, solve each inequality using a graphing utility Graph each side separately. Then determine the values of x for which the graph on the left side lies above the graph on the right side.

88. $-3(x - 6) > 2x - 2$

89. $-2(x + 4) > 6x + 16$

90. Use a graphing utility's TABLE feature to verify your work in Exercises 88–89.

Use the same technique employed in Exercises 88–89 to solve each inequality in Exercises 91–92. In each case, what conclusion can you draw? What happens if you try solving the inequalities algebraically?

91. $12x - 10 > 2(x - 4) + 10x$

92. $2x + 3 > 3(2x - 4) - 4x$

93. A bank offers two checking account plans. Plan A has a base service charge of $4.00 per month plus 10¢ per check. Plan B charges a base service charge of $2.00 per month plus 15¢ per check.
   a. Write models for the total monthly costs for each plan if x checks are written.
   b. Use a graphing utility to graph the models in the same [0, 50, 1] by [0, 10, 1] viewing rectangle.
   c. Use the graphs (and the intersection feature) to determine for what number of checks per month plan A will be better than plan B.
   d. Verify the result of part (c) algebraically by solving an inequality.

Critical Thinking Exercises

Make Sense? *In Exercises 94–97, determine whether each statement "makes sense" or "does not make sense" and explain your reasoning.*

94. I began the solution of $5 - 3(x + 2) > 10x$ by simplifying the left side, obtaining $2x + 4 > 10x$.

95. I have trouble remembering when to reverse the direction of an inequality symbol, so I avoid this difficulty by collecting variable terms on an appropriate side.

96. If you tell me that three times a number is less than two times that number, it's obvious that no number statisfies this condition, and there is no need for me to write and solve an inequality.

97. Whenever I solve a linear inequality in which the coefficients of the variable on each side are the same, the solution set is $\varnothing$ or $(-\infty, \infty)$.

In Exercises 98–101, determine whether each statement is true or false. If the statement is false, make the necessary change(s) to produce a true statement.

98. The inequality $3x > 6$ is equivalent to $2 > x$.

99. The smallest real number in the solution set of $2x > 6$ is 4.

100. If x is at least 7, then $x > 7$.

101. The inequality $-3x > 6$ is equivalent to $-2 > x$.

$$ax + 4 \leq -12$$

102. Find a so that the solution set of is $[8, \infty)$.

103. What's wrong with this argument? Suppose x and y represent two real numbers, where $x > y$.

$2 > 1$	This is a true statement.
$2(y - x) > 1(y - x)$	Multiply both sides by $y - x$.
$2y - 2x > y - x$	Use the distributive property.
$y - 2x > -x$	Subtract y from both sides.
$y > x$	Add $2x$ to both sides.

The final inequality, $y > x$, is impossible because we were initially given $x > y$.

Review Exercises

104. If $f(x) = x^2 - 2x + 5$, find $f(-4)$. (Section 2.1, Example 3)

105. Solve the system:

$$\begin{aligned} 2x - y - z &= -3 \\ 3x - 2y - 2z &= -5 \\ -x + y + 2z &= 4. \end{aligned}$$

(Section 3.3, Example 2)

106. Simplify: $\left(\dfrac{2x^4y^{-2}}{4xy^3}\right)^3$. (Section 1.6, Example 9)

Preview Exercises

Exercises 107–109 will help you prepare for the material covered in the next section.

107. Consider the sets $A = \{1, 2, 3, 4\}$ and $B = \{3, 4, 5, 6, 7\}$.
   a. Write the set consisting of elements common to both set A and set B.
   b. Write the set consisting of elements that are members of set A or of set B or of both sets.

108. a. Solve: $x - 3 < 5$.
   b. Solve: $2x + 4 < 14$.
   c. Give an example of a number that satisfies the inequality in part (a) and the inequality in part (b).
   d. Give an example of a number that satisfies the inequality in part (a), but not the inequality in part (b).

109. a. Solve: $2x - 6 \geq -4$.
   b. Solve: $5x + 2 \geq 17$.
   c. Give an example of a number that satisfies the inequality in part (a) and the inequality in part (b).
   d. Give an example of a number that satisfies the inequality in part (a), but not the inequality in part (b).

SECTION 4.2

Compound Inequalities

Objectives

1. Find the intersection of two sets.
2. Solve compound inequalities involving *and*.
3. Find the union of two sets.
4. Solve compound inequalities involving *or*.

Sixty-six percent of U.S. adults are overweight or obese (30 or more pounds over a healthy weight). In this section's exercise set, you'll use *compound inequalities* to analyze data that show we are becoming a weightier nation.

A **compound inequality** is formed by joining two inequalities with the word *and* or the word *or*.

Examples of Compound Inequalities

- $x - 3 < 5$ and $2x + 4 < 14$
- $3x - 5 \leq 13$ or $5x + 2 > -3$

Compound inequalities illustrate the importance of the words *and* and *or* in mathematics, as well as in everyday English.

1 Find the intersection of two sets.

Compound Inequalities Involving *And*

If A and B are sets, we can form a new set consisting of all elements that are in both A and B. This set is called the *intersection* of the two sets.

Definition of the Intersection of Sets

The **intersection** of sets A and B, written $A \cap B$, is the set of elements common to both set A **and** set B. This definition can be expressed in set-builder notation as follows:

$$A \cap B = \{x | x \in A \text{ AND } x \in B\}.$$

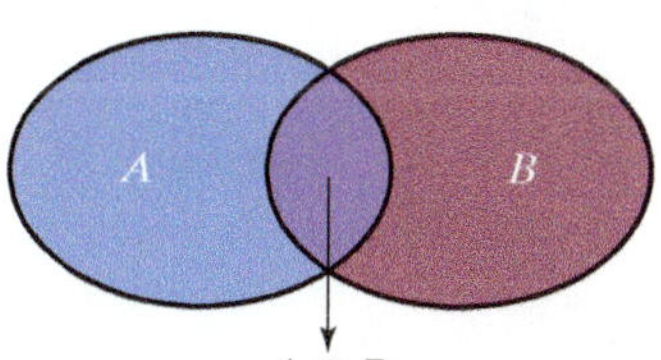

FIGURE 4.1 Picturing the intersection of two sets

Figure 4.1 shows a useful way of picturing the intersection of sets A and B. The figure indicates that $A \cap B$ contains those elements that belong to both A and B at the same time.

EXAMPLE 1 Finding the Intersection of Two Sets

Find the intersection: $\{7, 8, 9, 10, 11\} \cap \{6, 8, 10, 12\}$.

Solution The elements common to $\{7, 8, 9, 10, 11\}$ and $\{6, 8, 10, 12\}$ are 8 and 10. Thus,

$$\{7, 8, 9, 10, 11\} \cap \{6, 8, 10, 12\} = \{8, 10\}.$$

CHECK POINT 1 Find the intersection: $\{3, 4, 5, 6, 7\} \cap \{3, 7, 8, 9\}$.

2 Solve compound inequalities involving *and*.

A number is a **solution of a compound inequality formed by the word *and*** if it is a solution of both inequalities. For example, the solution set of the compound inequality

$$x \le 6 \quad \text{and} \quad x \ge 2$$

is the set of values of x that satisfy both $x \le 6$ and $x \ge 2$. Thus, the solution set is the intersection of the solution sets of the two inequalities.

What are the numbers that satisfy both $x \le 6$ and $x \ge 2$? These numbers are easier to see if we graph the solution set to each inequality on a number line. These graphs are shown in **Figure 4.2**. The intersection is shown in the third graph.

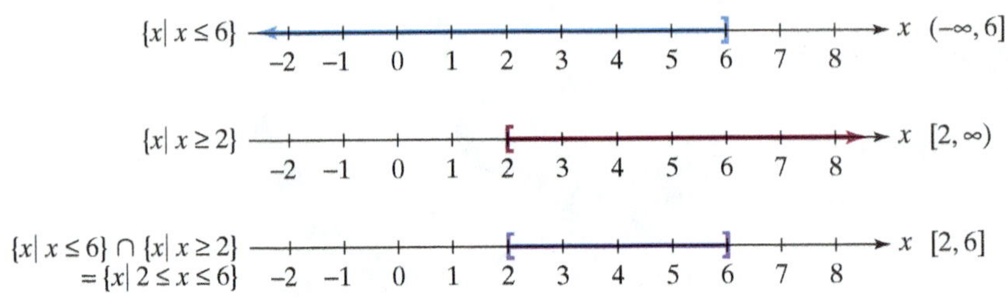

FIGURE 4.2 Numbers satisfying both $x \le 6$ and $x \ge 2$

The numbers common to both sets are those that are less than or equal to 6 and greater than or equal to 2. This set is $\{x|2 \le x \le 6\}$, or, in interval notation, $[2, 6]$.

Here is a procedure for finding the solution set of a compound inequality containing the word *and*.

Solving Compound Inequalities Involving *AND*

1. Solve each inequality separately.
2. Graph the solution set to each inequality on a number line and take the intersection of these solution sets. This intersection appears as the portion of the number line that the two graphs have in common.

EXAMPLE 2 Solving a Compound Inequality with *And*

Solve: $x - 3 < 5$ and $2x + 4 < 14$.

Solution

Step 1. Solve each inequality separately.

$$\begin{aligned} x - 3 &< 5 \quad \text{and} \quad & 2x + 4 &< 14 \\ x &< 8 & 2x &< 10 \\ & & x &< 5 \end{aligned}$$

Step 2. Take the intersection of the solution sets of the two inequalities. We graph the solution sets of $x < 8$ and $x < 5$. The intersection is shown in the third graph.

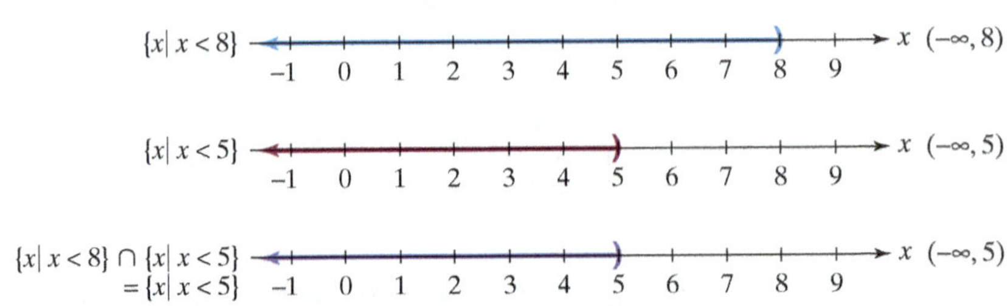

The numbers common to both sets are those that are less than 5. The solution set is $\{x|x < 5\}$, or, in interval notation, $(-\infty, 5)$. Take a moment to check that any number in $(-\infty, 5)$ satisfies both of the original inequalities.

✓ **CHECK POINT 2** Solve: $x + 2 < 5$ and $2x - 4 < -2$.

EXAMPLE 3 Solving a Compound Inequality with *And*

Solve: $2x - 7 > 3$ and $5x - 4 < 6$.

Solution

Step 1. Solve each inequality separately.

$$\begin{aligned} 2x - 7 &> 3 \quad \text{and} \quad & 5x - 4 &< 6 \\ 2x &> 10 & 5x &< 10 \\ x &> 5 & x &< 2 \end{aligned}$$

Step 2. Take the intersection of the solution sets of the two inequalities. We graph the solution sets of $x > 5$ and $x < 2$. We use these graphs to find their intersection.

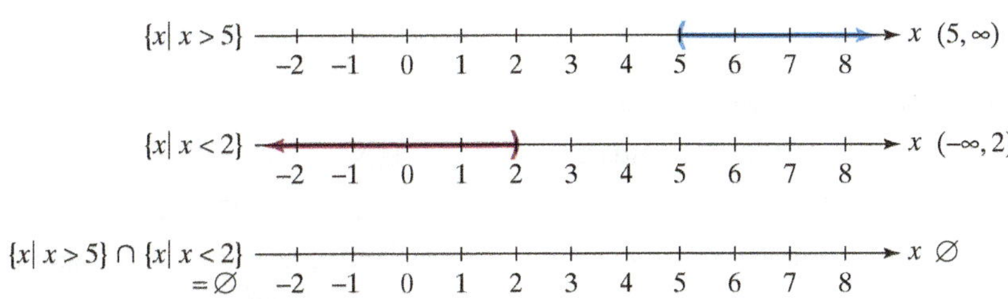

There is no number that is both greater than 5 and at the same time less than 2. Thus, the solution set is the empty set, $\varnothing$.

✓ **CHECK POINT 3** Solve: $4x - 5 > 7$ and $5x - 2 < 3$.

If $a < b$, the compound inequality

$$a < x \text{ and } x < b$$

can be written in the shorter form

$$a < x < b.$$

For example, the compound inequality

$$-3 < 2x + 1 \text{ and } 2x + 1 < 3$$

can be abbreviated

$$-3 < 2x + 1 < 3.$$

The word *and* does not appear when the inequality is written in the shorter form, although it is implied. The shorter form enables us to solve both inequalities at once. By performing the same operations on all three parts of the inequality, our goal is to **isolate x in the middle.**

EXAMPLE 4 Solving a Compound Inequality

Solve and graph the solution set:

$$-3 < 2x + 1 \leq 3.$$

Solution We would like to isolate x in the middle. We can do this by first subtracting 1 from all three parts of the compound inequality. Then we isolate x from $2x$ by dividing all three parts of the inequality by 2.

$$-3 < 2x + 1 \leq 3 \quad \text{This is the given inequality.}$$

$$-3 - 1 < 2x + 1 - 1 \leq 3 - 1 \quad \text{Subtract 1 from all three parts.}$$

$$-4 < 2x \leq 2 \quad \text{Simplify.}$$

$$\frac{-4}{2} < \frac{2x}{2} \leq \frac{2}{2} \quad \text{Divide each part by 2.}$$

$$-2 < x \leq 1 \quad \text{Simplify.}$$

The solution set consists of all real numbers greater than −2 and less than or equal to 1, represented by $\{x|-2 < x \leq 1\}$ in set-builder notation and $(-2, 1]$ in interval notation. The graph is shown as follows:

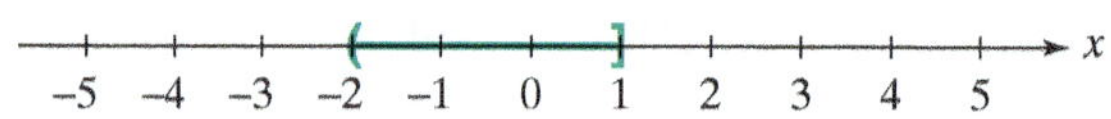

Using Technology

Numeric and Graphic Connections

Let's verify our work in Example 4 and show that $(-2, 1]$ is the solution set of $-3 < 2x + 1 \leq 3$.

Numeric Check

To check numerically, enter $y_1 = 2x + 1$.

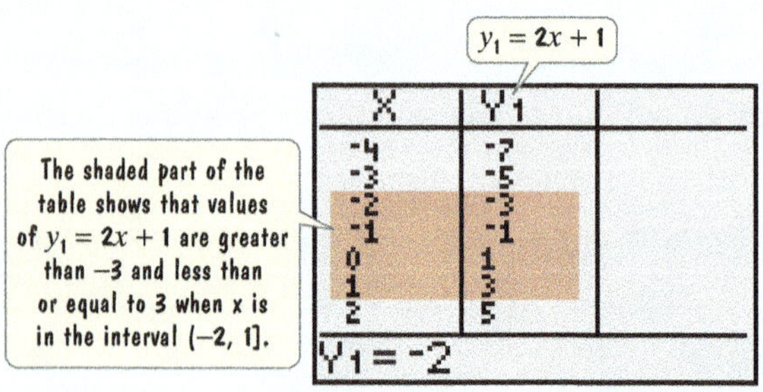

Graphic Check

To check graphically, graph each part of

$$-3 < 2x + 1 \leq 3.$$

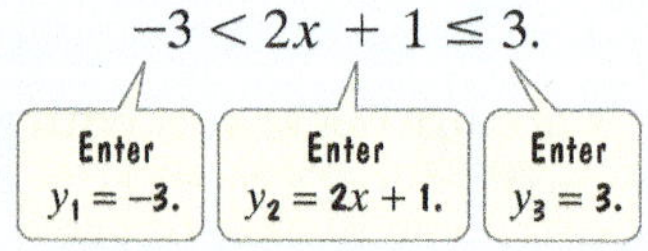

The figure shows that the graph of $y_2 = 2x + 1$ lies above the graph of $y_1 = -3$ and on or below the graph of $y_3 = 3$ when x is in the interval $(-2, 1]$.

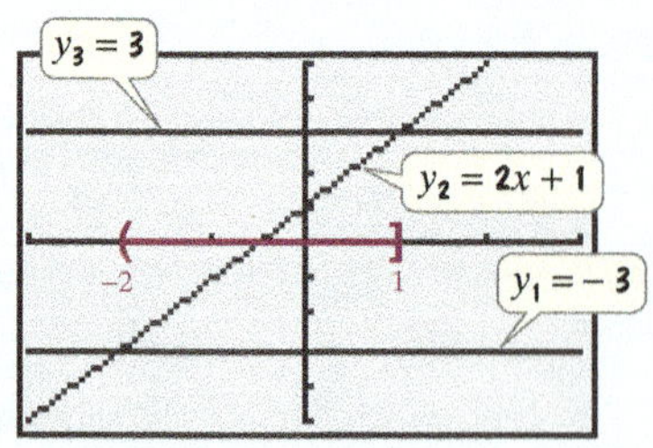

[−3, 3, 1] by [−5, 5, 1]

✓ **CHECK POINT 4** Solve and graph the solution set: $1 \leq 2x + 3 < 11$.

---

3 Find the union of two sets.

Compound Inequalities Involving *Or*

Another set that we can form from sets A and B consists of elements that are in A or B or in both sets. This set is called the *union* of the two sets.

Definition of the Union of Sets

The **union** of sets A and B, written $A \cup B$, is the set of elements that are members of set A **or** of set B or of both sets. This definition can be expressed in set-builder notation as follows:

$$A \cup B = \{x | x \in A \text{ OR } x \in B\}.$$

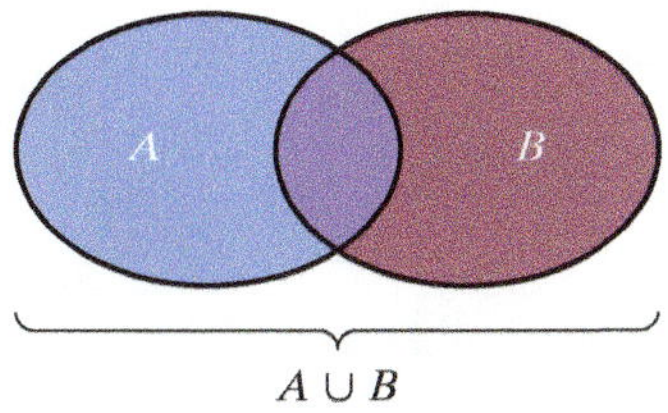

FIGURE 4.3 Picturing the union of two sets

Figure 4.3 shows a useful way of picturing the union of sets A and B. The figure indicates that $A \cup B$ is formed by joining the sets together.

We can find the union of set A and set B by listing the elements of set A. Then, we include any elements of set B that have not already been listed. Enclose all elements that are listed with braces. This shows that the union of two sets is also a set.

EXAMPLE 5 Finding the Union of Two Sets

Find the union: $\{7, 8, 9, 10, 11\} \cup \{6, 8, 10, 12\}$.

Solution To find $\{7, 8, 9, 10, 11\} \cup \{6, 8, 10, 12\}$, start by listing all the elements from the first set, namely 7, 8, 9, 10, and 11. Now list all the elements from the second set that are not in the first set, namely 6 and 12. The union is the set consisting of all these elements. Thus,

$$\{7, 8, 9, 10, 11\} \cup \{6, 8, 10, 12\} = \{6, 7, 8, 9, 10, 11, 12\}.$$

Although 8 and 10 appear in both sets, do not list 8 and 10 twice.

CHECK POINT 5 Find the union: $\{3, 4, 5, 6, 7\} \cup \{3, 7, 8, 9\}$.

Study Tip

The words *union* and *intersection* are helpful in distinguishing these two operations. Union, as in a marriage union, suggests joining things, or uniting them. Intersection, as in the intersection of two crossing streets, brings to mind the area common to both, suggesting things that overlap.

4 Solve compound inequalities involving *or*.

A number is a **solution of a compound inequality formed by the word *or*** if it is a solution of either inequality. Thus, the solution set of a compound inequality formed by the word *or* is the union of the solution sets of the two inequalities.

Solving Compound Inequalities Involving *OR*

1. Solve each inequality separately.
2. Graph the solution set to each inequality on a number line and take the union of these solution sets. This union appears as the portion of the number line representing the total collection of numbers in the two graphs.

EXAMPLE 6 Solving a Compound Inequality with *Or*

Solve: $2x - 3 < 7$ or $35 - 4x \leq 3$.

Solution

Step 1. Solve each inequality separately.

$$\begin{aligned} 2x - 3 &< 7 \quad \text{or} \quad & 35 - 4x &\leq 3 \\ 2x &< 10 & -4x &\leq -32 \\ x &< 5 & x &\geq 8 \end{aligned}$$

Step 2. Take the union of the solution sets of the two inequalities. We graph the solution sets of $x < 5$ and $x \geq 8$. We use these graphs to find their union.

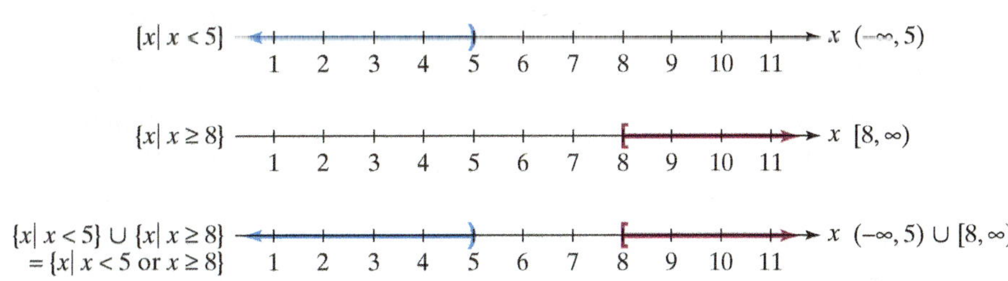

The solution set consists of all numbers that are less than 5 or greater than or equal to 8. The solution set is $\{x | x < 5 \text{ or } x \geq 8\}$, or, in interval notation, $(-\infty, 5) \cup [8, \infty)$. There is no shortcut way to express this union when interval notation is used.

CHECK POINT 6 Solve: $3x - 5 \le -2$ or $10 - 2x < 4$.

EXAMPLE 7 Solving a Compound Inequality with *Or*

Solve: $3x - 5 \le 13$ or $5x + 2 > -3$.

Solution

Step 1. Solve each inequality separately.

$$\begin{aligned} 3x - 5 &\le 13 \quad \text{or} \quad & 5x + 2 &> -3 \\ 3x &\le 18 & 5x &> -5 \\ x &\le 6 & x &> -1 \end{aligned}$$

Step 2. Take the union of the solution sets of the two inequalities. We graph the solution sets of $x \le 6$ and $x > -1$. We use these graphs to find their union.

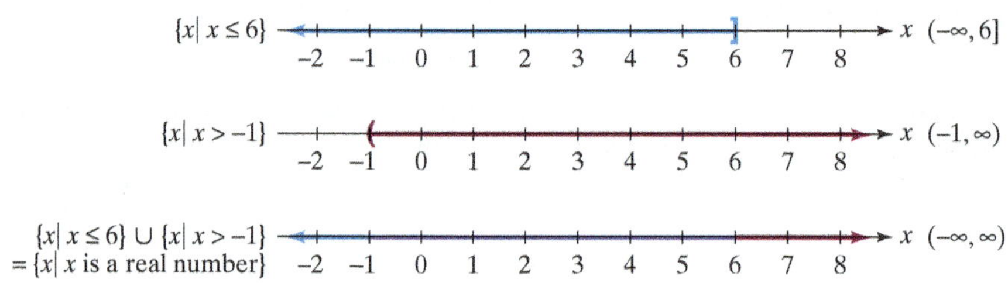

Because all real numbers are either less than or equal to 6 or greater than -1 or both, the union of the two sets fills the entire number line. Thus, the solution set is $\{x|x \text{ is a real number}\}$, or $\mathbb{R}$. The solution set in interval notation is $(-\infty, \infty)$. Any real number that you select will satisfy at least one of the original inequalities.

CHECK POINT 7 Solve: $2x + 5 \ge 3$ or $2x + 3 < 3$.

4.2 EXERCISE SET

MyMathLab — MathXL PRACTICE, WATCH, DOWNLOAD, READ, REVIEW

Practice Exercises

In Exercises 1–6, find the intersection of the sets.

1. $\{1, 2, 3, 4\} \cap \{2, 4, 5\}$
2. $\{1, 3, 7\} \cap \{2, 3, 8\}$
3. $\{1, 3, 5, 7\} \cap \{2, 4, 6, 8, 10\}$
4. $\{0, 1, 3, 5\} \cap \{-5, -3, -1\}$
5. $\{a, b, c, d\} \cap \varnothing$
6. $\{w, y, z\} \cap \varnothing$

In Exercises 7–24, solve each compound inequality. Use graphs to show the solution set to each of the two given inequalities, as well as a third graph that shows the solution set of the compound inequality. Except for the empty set, express the solution set in both set-builder and interval notations.

7. $x > 3$ and $x > 6$
8. $x > 2$ and $x > 4$
9. $x \le 5$ and $x \le 1$
10. $x \le 6$ and $x \le 2$

11. $x < 2$ and $x \geq -1$
12. $x < 3$ and $x \geq -1$
13. $x > 2$ and $x < -1$
14. $x > 3$ and $x < -1$
15. $5x < -20$ and $3x > -18$
16. $3x \leq 15$ and $2x > -6$
17. $x - 4 \leq 2$ and $3x + 1 > -8$
18. $3x + 2 > -4$ and $2x - 1 < 5$
19. $2x > 5x - 15$ and $7x > 2x + 10$
20. $6 - 5x > 1 - 3x$ and $4x - 3 > x - 9$
21. $4(1 - x) < -6$ and $\frac{x - 7}{5} \leq -2$
22. $5(x - 2) > 15$ and $\frac{x - 6}{4} \leq -2$
23. $x - 1 \leq 7x - 1$ and $4x - 7 < 3 - x$
24. $2x + 1 > 4x - 3$ and $x - 1 \geq 3x + 5$

In Exercises 25–32, solve each inequality and graph the solution set on a number line. Express the solution set in both set-builder and interval notations.

25. $6 < x + 3 < 8$
26. $7 < x + 5 < 11$
27. $-3 \leq x - 2 < 1$
28. $-6 < x - 4 \leq 1$
29. $-11 < 2x - 1 \leq -5$
30. $3 \leq 4x - 3 < 19$
31. $-3 \leq \frac{2x}{3} - 5 < -1$
32. $-6 \leq \frac{x}{2} - 4 < -3$

In Exercises 33–38, find the union of the sets.

33. $\{1, 2, 3, 4\} \cup \{2, 4, 5\}$
34. $\{1, 3, 7, 8\} \cup \{2, 3, 8\}$
35. $\{1, 3, 5, 7\} \cup \{2, 4, 6, 8, 10\}$
36. $\{0, 1, 3, 5\} \cup \{2, 4, 6\}$
37. $\{a, e, i, o, u\} \cup \varnothing$
38. $\{e, m, p, t, y\} \cup \varnothing$

In Exercises 39–54, solve each compound inequality. Use graphs to show the solution set to each of the two given inequalities, as well as a third graph that shows the solution set of the compound inequality. Express the solution set in both set-builder and interval notations.

39. $x > 3$ or $x > 6$
40. $x > 2$ or $x > 4$
41. $x \leq 5$ or $x \leq 1$
42. $x \leq 6$ or $x \leq 2$
43. $x < 2$ or $x \geq -1$
44. $x < 3$ or $x \geq -1$
45. $x \geq 2$ or $x < -1$
46. $x \geq 3$ or $x < -1$
47. $3x > 12$ or $2x < -6$
48. $3x < 3$ or $2x > 10$
49. $3x + 2 \leq 5$ or $5x - 7 \geq 8$
50. $2x - 5 \leq -11$ or $5x + 1 \geq 6$
51. $4x + 3 < -1$ or $2x - 3 \geq -11$
52. $2x + 1 < 15$ or $3x - 4 \geq -1$
53. $-2x + 5 > 7$ or $-3x + 10 > 2x$
54. $16 - 3x \geq -8$ or $13 - x > 4x + 3$
55. Let $f(x) = 2x + 3$ and $g(x) = 3x - 1$. Find all values of x for which $f(x) \geq 5$ and $g(x) > 11$.
56. Let $f(x) = 4x + 5$ and $g(x) = 3x - 4$. Find all values of x for which $f(x) \geq 5$ and $g(x) \leq 2$.
57. Let $f(x) = 3x - 1$ and $g(x) = 4 - x$. Find all values of x for which $f(x) < -1$ or $g(x) < -2$.
58. Let $f(x) = 2x - 5$ and $g(x) = 3 - x$. Find all values of x for which $f(x) \geq 3$ or $g(x) < 0$.

Practice PLUS

In Exercises 59–60, write an inequality with x isolated in the middle that is equivalent to the given inequality. Assume $a > 0$, $b > 0$, and $c > 0$.

59. $-c < ax - b < c$
60. $-2 < \frac{ax - b}{c} < 2$

In Exercises 61–62, use the graphs of y_1, y_2, and y_3 to solve each compound inequality.

61. $-3 \leq 2x - 1 \leq 5$

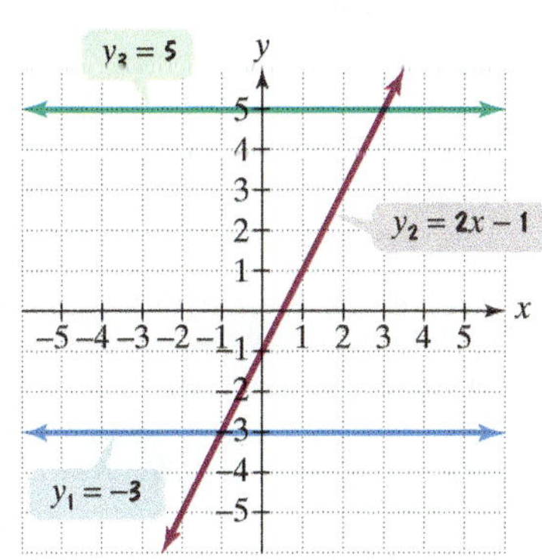

62. $x - 2 < 2x - 1 < x + 2$

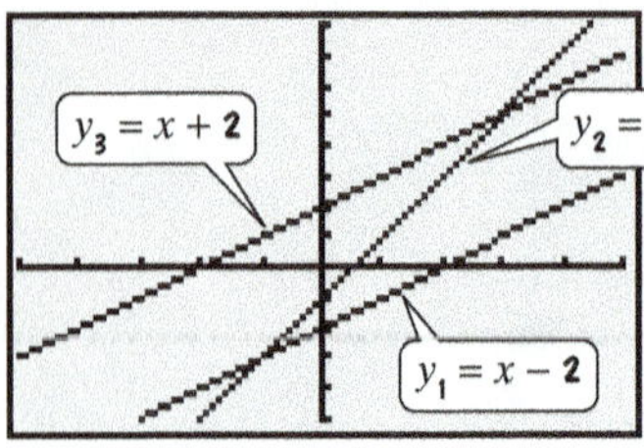

[−5, 5, 1] by [−5, 8, 1]

63. Solve $x - 2 < 2x - 1 < x + 2$, the inequality in Exercise 62, using algebraic methods. (*Hint:* Rewrite the inequality as $2x - 1 > x - 2$ and $2x - 1 < x + 2$.)

64. Use the hint given in Exercise 63 to solve $x \le 3x - 10 \le 2x$.

In Exercises 65–66, use the table to solve each inequality.

65. $-2 \le 5x + 3 < 13$

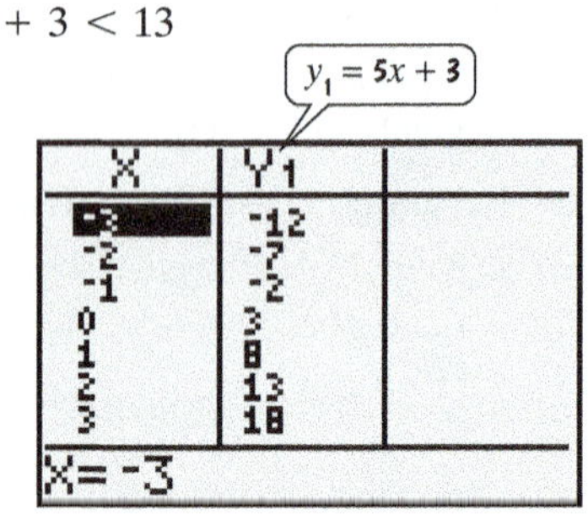

66. $-3 < 2x - 5 \le 3$

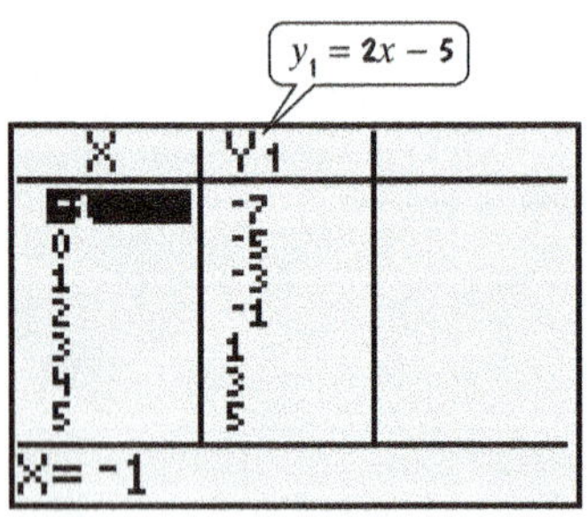

In Exercises 67–68, use the roster method to find the set of negative integers that are solutions of each inequality.

67. $5 - 4x \ge 1$ and $3 - 7x < 31$

68. $-5 < 3x + 4 \le 16$

Application Exercises

We are becoming weightier adults. The average weight of U.S. women, ages 20–74, has jumped 24 pounds over four decades, while average height has increased from 5-foot-3 to 5-foot-4. For U.S. men, ages 20–74, the average weight gain has been 25 pounds, while average height has increased from 5-foot-8 to 5-foot-9. The bar graph at the top of the next column shows the average weight of U.S. women and men, ages 20–74, for five selected years over four decades.

Average Weight of United States Women and Men for Five Selected Years

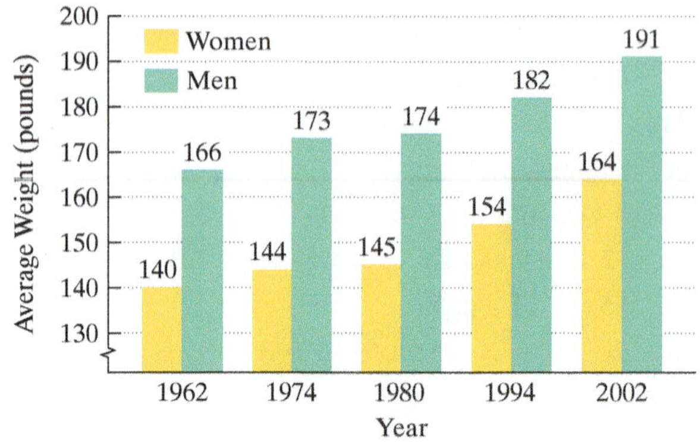

Source: Centers for Disease Control and Prevention

In Exercises 69–78, use the years shown in the graph to find each set.

69. $\{x|x$ is a year for which women's weight $\ge 144\}$ $\cap$ $\{x|x$ is a year for which men's weight $\le 182\}$

70. $\{x|x$ is a year for which women's weight $\ge 144\}$ $\cap$ $\{x|x$ is a year for which men's weight $< 182\}$

71. $\{x|x$ is a year for which women's weight $\ge 144\}$ $\cup$ $\{x|x$ is a year for which men's weight $\le 182\}$

72. $\{x|x$ is a year for which women's weight $\ge 144\}$ $\cup$ $\{x|x$ is a year for which men's weight $< 182\}$

73. $\{x|x$ is a year for which women's weight $< 144\}$ $\cup$ $\{x|x$ is a year for which men's weight $> 182\}$

74. $\{x|x$ is a year for which women's weight $< 144\}$ $\cup$ $\{x|x$ is a year for which men's weight $\ge 182\}$

75. $\{x|x$ is a year for which women's weight $< 144\}$ $\cap$ $\{x|x$ is a year for which men's weight $> 182\}$

76. $\{x|x$ is a year for which women's weight $< 144\}$ $\cap$ $\{x|x$ is a year for which men's weight $\ge 182\}$

77. $\{x|x$ is a year for which $166 \le$ men's weight $< 174\}$

78. $\{x|x$ is a year for which $145 <$ women's weight $\le 164\}$

79. A basic cellular phone plan costs \$20 per month for 60 calling minutes. Additional time costs \$0.40 per minute. The formula

$$C = 20 + 0.40(x - 60)$$

gives the monthly cost for this plan, C, for x calling minutes, where $x > 60$. How many calling minutes are possible for a monthly cost of at least \$28 and at most \$40?

80. The formula for converting Fahrenheit temperature, F, to Celsius temperature, C, is

$$C = \frac{5}{9}(F - 32).$$

If Celsius temperature ranges from 15° to 35°, inclusive, what is the range for the Fahrenheit temperature? Use interval notation to express this range.

81. On the first of four exams, your grades are 70, 75, 87, and 92. There is still one more exam, and you are hoping to earn a B in the course. This will occur if the average of your five exam grades is greater than or equal to 80 and less than 90. What range of grades on the fifth exam will result in earning a B? Use interval notation to express this range.

82. On the first of four exams, your grades are 82, 75, 80, and 90. There is still a final exam, and it counts as two grades. You are hoping to earn a B in the course: This will occur if the average of your six exam grades is greater than or equal to 80 and less than 90. What range of grades on the final exam will result in earning a B? Use interval notation to express this range.

83. The toll to a bridge is \$3.00. A three-month pass costs \$7.50 and reduces the toll to \$0.50. A six-month pass costs \$30 and permits crossing the bridge for no additional fee. How many crossing per three-month period does it take for the three-month pass to be the best deal?

84. Parts for an automobile repair cost \$175. The mechanic charges \$34 per hour. If you receive an estimate for at least \$226 and at most \$294 for fixing the car, what is the time interval that the mechanic will be working on the job?

Writing in Mathematics

85. Describe what is meant by the intersection of two sets. Give an example.
86. Explain how to solve a compound inequality involving *and*.
87. Why is $1 < 2x + 3 < 9$ a compound inequality? What are the two inequalities and what is the word that joins them?
88. Explain how to solve $1 < 2x + 3 < 9$.
89. Describe what is meant by the union of two sets. Give an example.
90. Explain how to solve a compound inequality involving *or*.

Technology Exercises

In Exercises 91–94, solve each inequality using a graphing utility. Graph each of the three parts of the inequality separately in the same viewing rectangle. The solution set consists of all values of x for which the graph of the linear function in the middle lies between the graphs of the constant functions on the left and the right.

91. $1 < x + 3 < 9$
92. $-1 < \frac{x + 4}{2} < 3$
93. $1 \le 4x - 7 \le 3$
94. $2 \le 4 - x \le 7$

95. Use a graphing utility's TABLE feature to verify your work in Exercises 91–94.

Critical Thinking Exercises

Make Sense? *In Exercises 96–99, determine whether each statement "makes sense" or "does not make sense" and explain your reasoning.*

96. I've noticed that when solving some compound inequalities with *or*, there is no way to express the solution set using a single interval, but this does not happen with *and* compound inequalities.
97. Compound inequalities with *and* have solutions that satisfy both inequalities, whereas compound inequalities with *or* have solutions that satisfy at least one of the inequalities.
98. I'm considering the compound inequality $x < 8$ and $x > a$, and I'm certain that there are no values of a that make the solution set $\varnothing$.
99. I'm considering the compound inequality $x < 8$ and $x > a$, and I'm certain that there are no values of a that make the solution set $(-\infty, \infty)$.

In Exercises 100–103, determine whether each statement is true or false. If the statement is false, make the necessary change(s) to produce a true statement.

100. $(-\infty, -1] \cap [-4, \infty) = [-4, -1]$
101. $(-\infty, 3) \cup (-\infty, -2) = (-\infty, -2)$
102. The union of two sets can never give the same result as the intersection of those same two sets.
103. The solution set of the compound inequality $x < a$ and $x > a$ is the set of all real numbers excluding a.
104. Solve and express the solution set in interval notation: $-7 \le 8 - 3x \le 20$ and $-7 < 6x - 1 < 41$.

The graphs of $f(x) = \sqrt{4 - x}$ and $g(x) = \sqrt{x + 1}$ are shown in a $[-3, 10, 1]$ by $[-2, 5, 1]$ viewing rectangle.

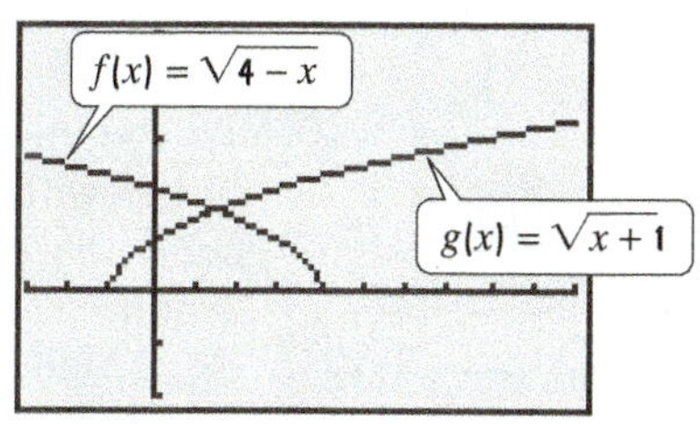

In Exercises 105–108, use the graphs and interval notation to express the domain of the given function.

105. The domain of f
106. The domain of g
107. The domain of $f + g$
108. The domain of $\frac{f}{g}$

109. At the end of the day, the change machine at a laundrette contained at least \$3.20 and at most \$5.45 in nickels, dimes, and quarters. There were 3 fewer dimes than twice the number of nickels and 2 more quarters than twice the number of nickels. What was the least possible number and the greatest possible number of nickels?

Review Exercises

110. If $f(x) = x^2 - 3x + 4$ and $g(x) = 2x - 5$, find $(g - f)(x)$ and $(g - f)(-1)$. (Section 2.3, Example 4)

111. Use function notation to write the equation of the line passing through (4, 2) and perpendicular to the line whose equation is $4x - 2y = 8$. (Section 2.5, Example 5)

112. Simplify: $4 - [2(x - 4) - 5]$. (Section 1.2, Example 14)

Preview Exercises

Exercises 113–115 will help you prepare for the material covered in the next section.

113. Find all values of x satisfying $1 - 4x = 3$ or $1 - 4x = -3$.

114. Find all values of x satisfying $3x - 1 = x + 5$ or $3x - 1 = -(x + 5)$.

115. a. Substitute -5 for x and determine whether -5 satisfies $|2x + 3| \geq 5$.

b. Does 0 satisfy $|2x + 3| \geq 5$?

SECTION 4.3

Equations and Inequalities Involving Absolute Value

Objectives

1. Solve absolute value equations.
2. Use boundary points to solve absolute value inequalities.
3. Use equivalent compound inequalities to solve absolute value inequalities.
4. Recognize absolute value inequalities with no solution or all real numbers as solutions.
5. Solve problems using absolute value inequalities.

*M*A*S*H was set in the early 1950s during the Korean War. By the final episode, the show had lasted four times as long as the Korean War.*

At the end of the twentieth century, there were 94 million households in the United States with television sets. The television program viewed by the greatest percentage of such households in that century was the final episode of $M*A*S*H$. Over 50 million American households watched this program.

Numerical information, such as the number of households watching a television program, is often given with a margin of error. Inequalities involving absolute value are used to describe errors in polling, as well as errors of measurement in manufacturing, engineering, science, and other fields. In this section, you will learn to solve equations and inequalities containing absolute value. With these skills, you will be able to analyze the percentage of households that watched the final episode of $M*A*S*H$.

1 Solve absolute value equations.

Equations Involving Absolute Value

We have seen that the absolute value of a, denoted $|a|$, is the distance from 0 to a on a number line. Now consider **absolute value equations**, such as

$$|x| = 2.$$

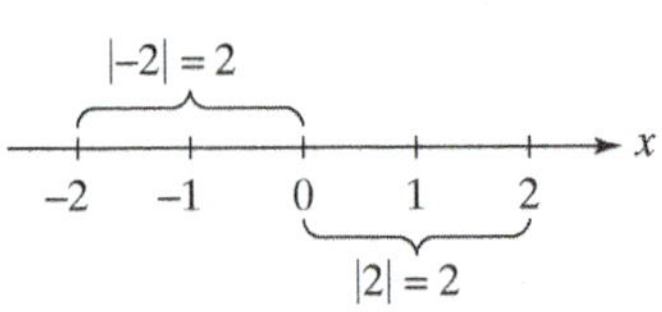

FIGURE 4.4 If $|x| = 2$, then $x = 2$ or $x = -2$.

This means that we must determine real numbers whose distance from the origin on a number line is 2. **Figure 4.4** shows that there are two numbers such that $|x| = 2$, namely, 2 and -2. We write $x = 2$ or $x = -2$. This observation can be generalized as follows:

Rewriting an Absolute Value Equation Without Absolute Value Bars

If c is a positive real number and X represents any algebraic expression, then $|X| = c$ is equivalent to $X = c$ or $X = -c$.

Using Technology

Graphic Connections

You can use a graphing utility to verify the solution set of an absolute value equation. Consider, for example,

$$|2x - 3| = 11.$$

Graph $y_1 = |2x - 3|$ and $y_2 = 11$. The graphs are shown in a $[-10, 10, 1]$ by $[-1, 15, 1]$ viewing rectangle. The x-coordinates of the intersection points are -4 and 7, verifying that $\{-4, 7\}$ is the solution set.

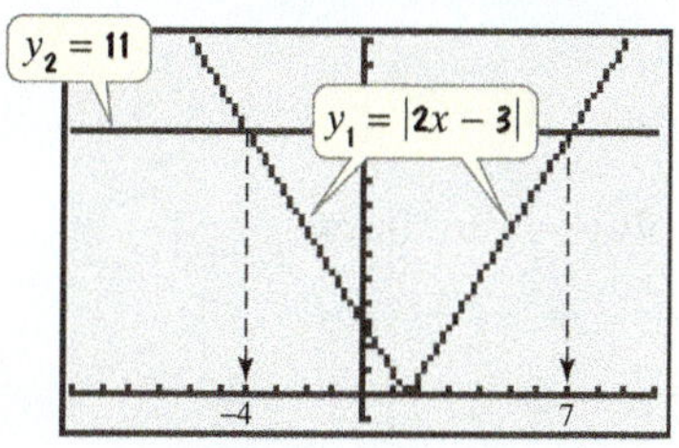

[-10, 10, 1] by [-1, 15, 1]

EXAMPLE 1 Solving an Equation Involving Absolute Value

Solve: $|2x - 3| = 11$.

Solution

$$|2x - 3| = 11 \quad \text{This is the given equation.}$$

$$2x - 3 = 11 \quad \text{or} \quad 2x - 3 = -11 \quad \text{Rewrite the equation without absolute value bars.}$$

$$2x = 14 \qquad 2x = -8 \quad \text{Add 3 to both sides of each equation.}$$

$$x = 7 \qquad x = -4 \quad \text{Divide both sides of each equation by 2.}$$

Check 7: | **Check −4:**

$$|2x - 3| = 11 \qquad |2x - 3| = 11 \quad \text{This is the original equation.}$$

$$|2(7) - 3| \stackrel{?}{=} 11 \qquad |2(-4) - 3| \stackrel{?}{=} 11 \quad \text{Substitute the proposed solutions.}$$

$$|14 - 3| \stackrel{?}{=} 11 \qquad |-8 - 3| \stackrel{?}{=} 11 \quad \text{Perform operations inside the absolute value bars.}$$

$$|11| \stackrel{?}{=} 11 \qquad |-11| \stackrel{?}{=} 11$$

$$11 = 11, \text{ true} \qquad 11 = 11, \text{ true} \quad \text{These true statements indicate that 7 and } -4 \text{ are solutions.}$$

The solutions are -4 and 7. We can also say that the solution set is $\{-4, 7\}$.

✓ CHECK POINT 1 Solve: $|2x - 1| = 5$.

EXAMPLE 2 Solving an Equation Involving Absolute Value

Solve: $5|1 - 4x| - 15 = 0$.

Solution

$$5|1 - 4x| - 15 = 0 \quad \text{This is the given equation.}$$

We need to isolate $|1 - 4x|$, the absolute value expression.

$$5|1 - 4x| = 15 \quad \text{Add 15 to both sides.}$$

$$|1 - 4x| = 3 \quad \text{Divide both sides by 5.}$$

$$1 - 4x = 3 \quad \text{or} \quad 1 - 4x = -3 \quad \text{Rewrite } |X| = c \text{ as } X = c \text{ or } X = -c.$$

$$-4x = 2 \qquad -4x = -4 \quad \text{Subtract 1 from both sides of each equation.}$$

$$x = -\tfrac{1}{2} \qquad x = 1 \quad \text{Divide both sides of each equation by } -4.$$

Take a moment to check $-\frac{1}{2}$ and 1, the proposed solutions, in the original equation, $5|1 - 4x| - 15 = 0$. In each case, you should obtain the true statement $0 = 0$. The solutions are $-\frac{1}{2}$ and 1, and the solution set is $\left\{-\frac{1}{2}, 1\right\}$.

✓ CHECK POINT 2 Solve: $2|1 - 3x| - 28 = 0$.

The absolute value of a number is never negative. Thus, if X is an algebraic expression and c is a negative number, then $|X| = c$ has no solution. For example, the equation $|3x - 6| = -2$ has no solution because $|3x - 6|$ cannot be negative. The solution set is $\varnothing$, the empty set.

The absolute value of 0 is 0. Thus, if X is an algebraic expression and $|X| = 0$, the solution is found by solving $X = 0$. For example, the solution of $|x - 2| = 0$ is obtained by solving $x - 2 = 0$. The solution is 2 and the solution set is $\{2\}$.

Some equations have two absolute value expressions, such as

$$|3x - 1| = |x + 5|.$$

These absolute value expressions are equal when the expressions inside the absolute value bars are equal to or opposites of each other.

Rewriting an Absolute Value Equation with Two Absolute Values Without Absolute Value Bars

If $|X_1| = |X_2|$, then $X_1 = X_2$ or $X_1 = -X_2$.

EXAMPLE 3 Solving an Absolute Value Equation with Two Absolute Values

Solve: $|3x - 1| = |x + 5|$.

Solution We rewrite the equation without absolute value bars.

$|X_1| = |X_2|$ means $X_1 = X_2$ or $X_1 = -X_2$

$|3x - 1| = |x + 5|$ means $3x - 1 = x + 5$ or $3x - 1 = -(x + 5)$.

We now solve the two equations that do not contain absolute value bars.

$$\begin{aligned} 3x - 1 &= x + 5 \quad \text{or} & 3x - 1 &= -(x + 5) \\ 2x - 1 &= 5 & 3x - 1 &= -x - 5 \\ 2x &= 6 & 4x - 1 &= -5 \\ x &= 3 & 4x &= -4 \\ & & x &= -1 \end{aligned}$$

Take a moment to complete the solution process by checking the two proposed solutions in the original equation. The solutions are -1 and 3, and the solution set is $\{-1, 3\}$.

CHECK POINT 3 Solve: $|2x - 7| = |x + 3|$.

2 Use boundary points to solve absolute value inequalities.

Solving Inequalities Involving Absolute Value Using Boundary Points

Graphs can help us visualize the solution sets of equations and inequalities involving absolute value. Let's first consider the equation

$$|x| = 2.$$

Figure 4.5 shows the graphs of $f(x) = |x|$ and $g(x) = 2$. The x-coordinates of the intersection points illustrate that $\{-2, 2\}$ is the solution set of $|x| = 2$.

Now let's see what the graphs of $f(x) = |x|$ and $g(x) = 2$ can tell us about the solution sets of the following inequalities involving absolute value:

$$|x| < 2 \qquad \text{and} \qquad |x| > 2.$$

Look for values of x where the graph of $f(x) = |x|$ lies *below* the graph of $g(x) = 2$.

Look for values of x where the graph of $f(x) = |x|$ lies *above* the graph of $g(x) = 2$.

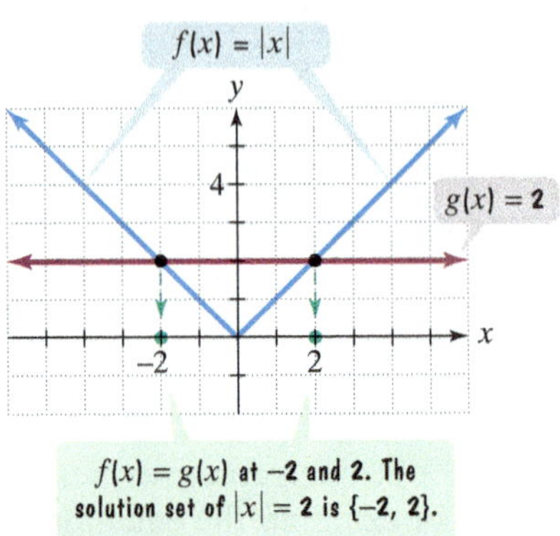

FIGURE 4.5

Figure 4.6(a) illustrates the solution set of $|x| < 2$. **Figure 4.6(b)** illustrates the solution set of $|x| > 2$.

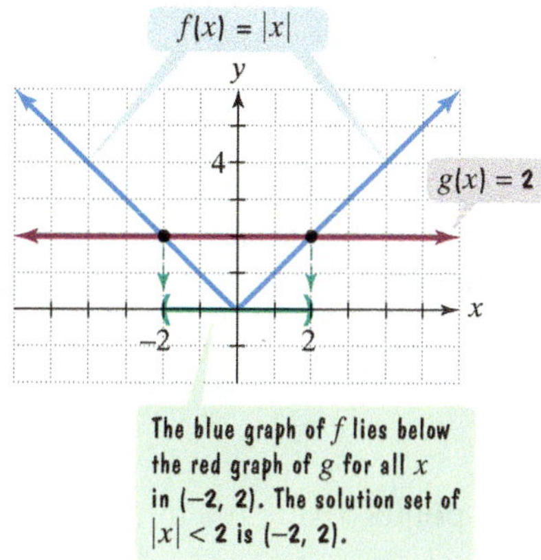

FIGURE 4.6(a)

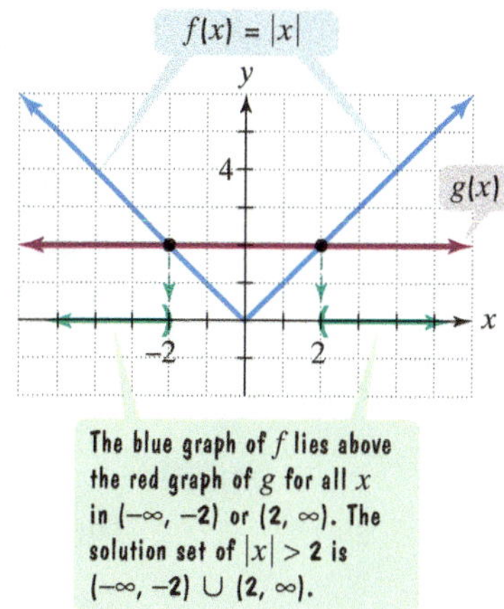

FIGURE 4.6(b)

Can you see how -2 and 2, the solutions of $|x| = 2$, serve as boundary points that divide the x-axis into intervals? On each interval, the blue graph of f is either below the red graph of g or above the red graph of g.

Boundary points play a fundamental role in solving absolute value inequalities in the form $|X| < c$ or $|X| > c$, where X is an algebraic expression and c is a positive number. The boundary points are found by solving the equation $|X| = c$.

Using Boundary Points to Solve Absolute Value Inequalities

If X is an algebraic expression and c is a positive number, the inequalities $|X| < c$ and $|X| > c$ can be solved by the following procedure.

1. Solve the equation $|X| = c$. The solutions are the **boundary points**.
2. Locate these boundary points on a number line, thereby dividing the number line into intervals.
3. Choose one representative number, called a **test value**, within each interval and substitute that number into the given inequality.
   a. If a true statement results, then all numbers, x, in the interval satisfy the given inequality.
   b. If a false statement results, then no numbers, x, in the interval satisfy the given inequality.
4. Write the solution set, selecting the interval or intervals that satisfy the given inequality.

This procedure is valid if $<$ is replaced by $\leq$, or $>$ is replaced by $\geq$. However, if the inequality involves $\leq$ or $\geq$, include the boundary points (the solutions of $|X| = c$) in the solution set.

Study Tip

Each test value must be chosen from the *interior* of an interval. Test values should not be endpoints of intervals.

EXAMPLE 4 Solving an Absolute Value Inequality Using Boundary Points

Solve and graph the solution set on a number line:

$$|x - 4| < 3.$$

Solution

Step 1. Solve the equation $|X| = c$. We find the boundary points for $|x - 4| < 3$ by solving $|x - 4| = 3$.

$$|x - 4| = 3 \qquad \text{This is the equation needed to find the boundary points.}$$

$$x - 4 = 3 \quad \text{or} \quad x - 4 = -3 \qquad \text{Rewrite } |X| = c \text{ as } X = c \text{ or } X = -c.$$

$$x = 7 \qquad\qquad x = 1 \qquad \text{Add 4 to both sides of each equation and solve for } x.$$

The boundary points are 1 and 7.

Step 2. Locate the boundary points on a number line and separate the line into intervals. The number line with the boundary points is shown as follows:

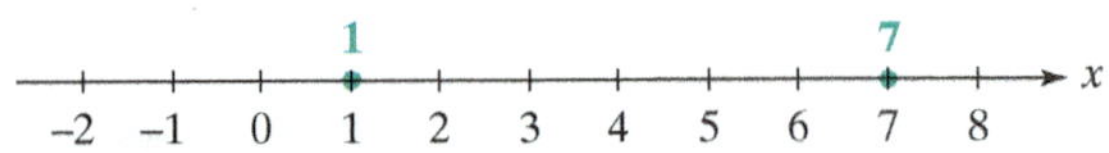

The boundary points divide the number line into three intervals:

$$(-\infty, 1) \qquad (1, 7) \qquad (7, \infty).$$

Step 3. Choose one test value within each interval and substitute that value into the given inequality.

Interval	Test Value	Substitute into $|x - 4| < 3$	Conclusion
$(-\infty, 1)$	0	$|0 - 4| \stackrel{?}{<} 3$ $|-4| \stackrel{?}{<} 3$ $4 < 3$, false	$(-\infty, 1)$ does not belong to the solution set.
$(1, 7)$	2	$|2 - 4| \stackrel{?}{<} 3$ $|-2| \stackrel{?}{<} 3$ $2 < 3$, true	$(1, 7)$ belongs to the solution set.
$(7, \infty)$	8	$|8 - 4| \stackrel{?}{<} 3$ $|-4| \stackrel{?}{<} 3$ $4 < 3$, false	$(7, \infty)$ does not belong to the solution set.

Step 4. Write the solution set, selecting the interval or intervals that satisfy the given inequality. Based on our work in step 3, we see that the solution set of the given inequality, $|x - 4| < 3$, is $(1, 7)$. The solution set can be expressed in set-builder notation as $\{x | 1 < x < 7\}$. The graph of the solution set is shown as follows:

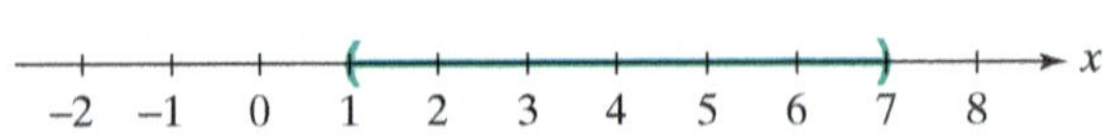

We can use the rectangular coordinate system to visualize the solution set of

$$|x - 4| < 3.$$

Figure 4.7 shows the graphs of $f(x) = |x - 4|$ and $g(x) = 3$. The solution set of $|x - 4| < 3$ consists of all values of x for which the blue graph of f lies below the red graph of g. These x-values make up the interval $(1, 7)$, which is the solution set.

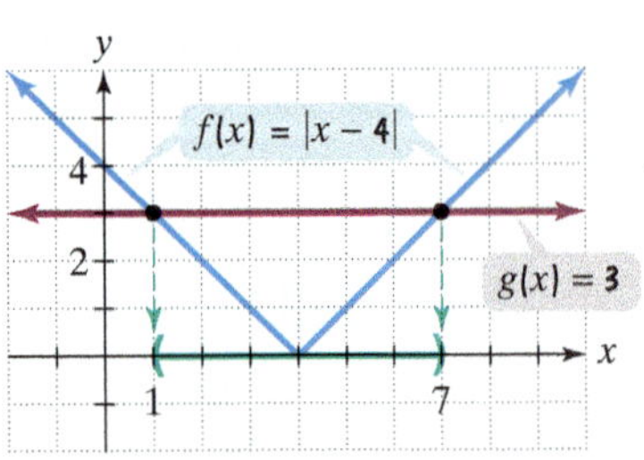

FIGURE 4.7 The solution set of $|x - 4| < 3$ is $(1, 7)$.

☑ **CHECK POINT 4** Solve and graph the solution set on a number line: $|x - 2| < 5$.

EXAMPLE 5 Solving an Absolute Value Inequality Using Boundary Points

Solve and graph the solution set on a number line:

$$|2x + 3| \geq 5.$$

Solution

Step 1. Solve the equation $|X| = c$. We find the boundary points by solving $|2x + 3| = 5$.

$|2x + 3| = 5$		This is the equation needed to find the boundary points.
$2x + 3 = 5$ or	$2x + 3 = -5$	Rewrite $|X| = c$ as $X = c$ or $X = -c$.
$2x = 2$	$2x = -8$	Subtract 3 from both sides of each equation.
$x = 1$	$x = -4$	Divide both sides of each equation by 2 and solve for x.

The boundary points are -4 and 1.

Step 2. Locate the boundary points on a number line and separate the line into intervals. The number line with the boundary points is shown as follows:

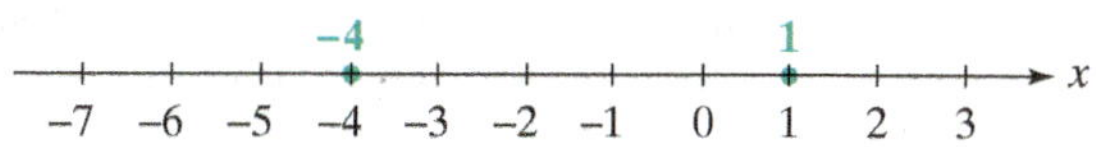

The boundary points divide the number line into three intervals:

$$(-\infty, -4) \qquad (-4, 1) \qquad (1, \infty).$$

Step 3. Choose one test value within each interval and substitute that value into the given inequality.

Interval	Test Value	Substitute into $|2x + 3| \geq 5$	Conclusion
$(-\infty, -4)$	-5	$|2(-5) + 3| \stackrel{?}{\geq} 5$ $|-10 + 3| \stackrel{?}{\geq} 5$ $|-7| \stackrel{?}{\geq} 5$ $7 \geq 5$, true	$(-\infty, -4)$ belongs to the solution set.
$(-4, 1)$	0	$|2 \cdot 0 + 3| \stackrel{?}{\geq} 5$ $|0 + 3| \stackrel{?}{\geq} 5$ $|3| \stackrel{?}{\geq} 5$ $3 \geq 5$, false	$(-4, 1)$ does not belong to the solution set.
$(1, \infty)$	2	$|2 \cdot 2 + 3| \stackrel{?}{\geq} 5$ $|4 + 3| \stackrel{?}{\geq} 5$ $|7| \stackrel{?}{\geq} 5$ $7 \geq 5$, true	$(1, \infty)$ belongs to the solution set.

Step 4. Write the solution set, selecting the interval or intervals that satisfy the given inequality. Based on our work in step 3, we see that all x in $(-\infty, -4)$ or $(1, \infty)$ belong to the solution set of $|2x + 3| \geq 5$. However, because the inequality

involves ≥ (greater than or *equal to*), we must also include the solutions of $|2x + 3| = 5$, namely the boundary points -4 and 1, in the solution set. Thus, the solution set of $|2x + 3| \geq 5$ is

$$(-\infty, -4] \cup [1, \infty)$$
$$\text{or} \quad \{x | x \leq -4 \quad \text{or} \quad x \geq 1\}.$$

The graph of the solution set on a number line is shown as follows:

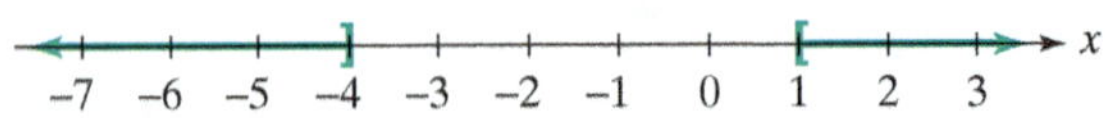

CHECK POINT 5 Solve and graph the solution set on a number line: $|2x - 5| \geq 3$.

3 Use equivalent compound inequalities to solve absolute value inequalities.

Solving Inequalities Involving Absolute Value Using Equivalent Compound Inequalities

In **Figure 4.6** on page 265, we used graphs of $f(x) = |x|$ and $g(x) = 2$ to visualize the solution sets of two absolute value inequalities:

- The solution set of $|x| < 2$ is $(-2, 2)$ or $\{x | -2 < x < 2\}$.
- The solution set of $|x| > 2$ is $(-\infty, -2) \cup (2, \infty)$ or $\{x | x < -2 \text{ or } x > 2\}$.

We can verify these results by interpreting $|x|$ as the distance from 0 to x on a number line.

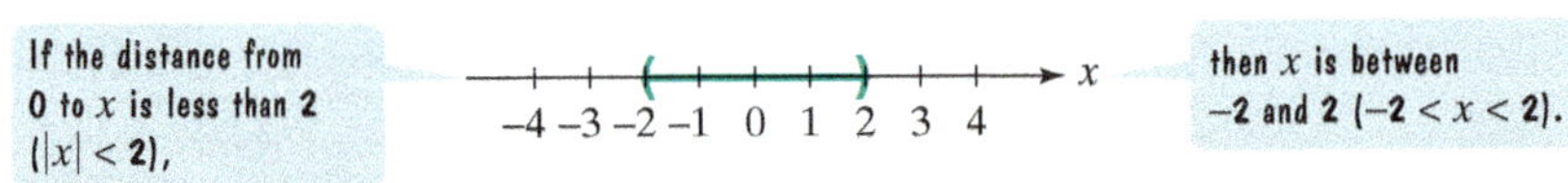

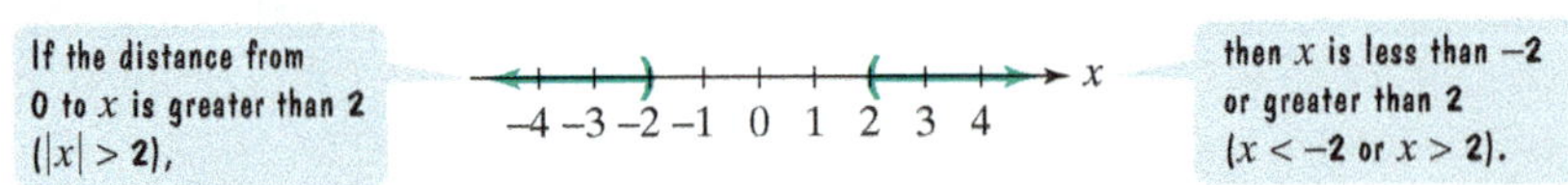

Generalizing from these observations gives us a second method for solving inequalities with absolute value. This method involves rewriting the given inequality without absolute value bars.

Using Equivalent Compound Inequalities to Solve Absolute Value Inequalities

If X is an algebraic expression and c is a positive number,

1. The solutions of $|X| < c$ are the numbers that satisfy $-c < X < c$.
2. The solutions of $|X| > c$ are the numbers that satisfy $X < -c$ or $X > c$.

These rules are valid if $<$ is replaced by $\leq$ and $>$ is replaced by $\geq$.

Let's rework Examples 4 and 5 by converting to equivalent compound inequalities. Although this method is faster than using boundary points, it's easy to make an error when rewriting absolute value inequalities without absolute value bars. Ask your instructor if one method is preferred over another.

EXAMPLE 6 Solving Absolute Value Inequalities Using Equivalent Compound Inequalities

Solve each inequality:

a. $|x - 4| < 3$ b. $|2x + 3| \geq 5$.

Solution

a. We rewrite the inequality without absolute value bars.

$|X| < c$ means $-c < X < c$.

$$|x - 4| < 3 \quad \text{means} \quad -3 < x - 4 < 3.$$

We solve the compound inequality by adding 4 to all three parts.

$$-3 < x - 4 < 3$$
$$-3 + 4 < x - 4 + 4 < 3 + 4$$
$$1 < x < 7$$

The solution set is all real numbers greater than 1 and less than 7, denoted by $\{x|1 < x < 7\}$ or $(1, 7)$.

b. We rewrite the inequality without absolute value bars.

$|X| \geq c$ means $X \leq -c$ or $X \geq c$.

$$|2x + 3| \geq 5 \quad \text{means} \quad 2x + 3 \leq -5 \quad \text{or} \quad 2x + 3 \geq 5.$$

We solve this compound inequality by solving each of these inequalities separately. Then we take the union of their solution sets.

$2x + 3 \leq -5$ or $2x + 3 \geq 5$ These are the inequalities without absolute value bars.

$2x \leq -8$ $2x \geq 2$ Subtract 3 from both sides.

$x \leq -4$ $x \geq 1$ Divide both sides by 2.

The solution set consists of all numbers that are less than or equal to -4 or greater than or equal to 1. The solution set is $\{x|x \leq -4 \text{ or } x \geq 1\}$, or, in interval notation, $(-\infty, -4] \cup [1, \infty)$.

Study Tip

If X is a linear expression, the graph of the solution set for $|X| > c$ will be divided into two intervals whose union cannot be represented as a single interval. The graph of the solution set for $|X| < c$ will be a single interval. Avoid the common error of rewriting $|X| > c$ as $-c < X > c$.

✓ CHECK POINT 6 Solve each inequality using equivalent compound inequalities. (These are the inequalities that you solved in Check Points 4 and 5 using boundary points.)

a. $|x - 2| < 5$ b. $|2x - 5| \geq 3$

EXAMPLE 7 Solving an Absolute Value Inequality Using an Equivalent Compound Inequality

Solve and graph the solution set on a number line: $-2|3x + 5| + 7 \geq -13$.

Study Tip

If you use the boundary point method, you'll still need to first isolate the absolute value expression. You can then use boundary points to solve

$$|3x + 5| \leq 10.$$

Solution

$$-2|3x + 5| + 7 \geq -13$$ This is the given inequality.

We need to isolate $|3x + 5|$, the absolute value expression.

$$-2|3x + 5| + 7 - 7 \geq -13 - 7$$ Subtract 7 from both sides.

$$-2|3x + 5| \geq -20$$ Simplify.

$$\frac{-2|3x + 5|}{-2} \leq \frac{-20}{-2}$$ Divide both sides by -2 and change the sense of the inequality.

$$|3x + 5| \leq 10$$ Simplify.

$$-10 \leq 3x + 5 \leq 10$$ Rewrite without absolute value bars: $|X| \leq c$ means $-c \leq X \leq c$.

Now we need to isolate x in the middle.

$$-10 - 5 \leq 3x + 5 - 5 \leq 10 - 5$$ Subtract 5 from all three parts.

$$-15 \leq 3x \leq 5$$ Simplify.

$$\frac{-15}{3} \leq \frac{3x}{3} \leq \frac{5}{3}$$ Divide each part by 3.

$$-5 \leq x \leq \frac{5}{3}$$ Simplify.

The solution set is $\{x | -5 \leq x \leq \frac{5}{3}\}$ in set-builder notation and $[-5, \frac{5}{3}]$ in interval notation. The graph is shown as follows:

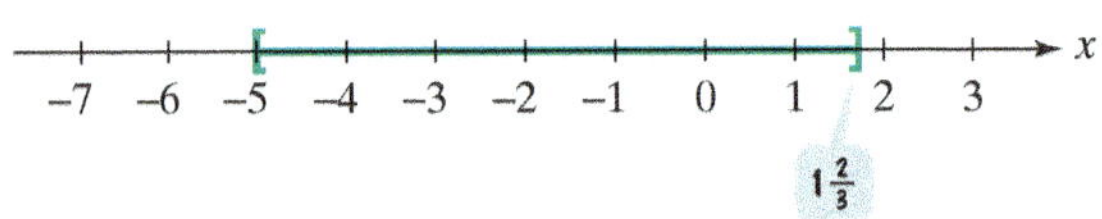

CHECK POINT 7 Solve and graph the solution set on a number line: $-3|5x - 2| + 20 \geq -19$.

---

4 Recognize absolute value inequalities with no solution or all real numbers as solutions.

Absolute Value Inequalities with Unusual Solution Sets

We have been working with $|X| < c$ and $|X| > c$, where c is a positive number. Now let's see what happens to these inequalities if c is a negative number. Consider, for example, $|x| < -2$. Because $|x|$ always has a value that is greater than or equal to 0, there is no number whose absolute value is less than -2. The inequality $|x| < -2$ has no solution. The solution set is $\varnothing$.

Now consider the inequality $|x| > -2$. Because $|x|$ is never negative, all numbers have an absolute value greater than -2. All real numbers satisfy the inequality $|x| > -2$. The solution set is $(-\infty, \infty)$.

Absolute Value Inequalities with Unusual Solution Sets

If X is an algebraic expression and c is a negative number,

1. The inequality $|X| < c$ has no solution.
2. The inequality $|X| > c$ is true for all real numbers for which X is defined.

5 Solve problems using absolute value inequalities.

Applications

When you were between the ages of 6 and 14, how would you have responded to this question:

What is bad about being a kid?

Table 4.3 What Is Bad about Being a Kid?

Kids Say	
Getting bossed around	17%
School, homework	15%
Can't do everything I want	11%
Chores	9%
Being grounded	9%

Source: Penn, Schoen, and Berland using 1172 interviews with children ages 6 to 14 from May 14 to June 1, 1999, Margin of error: ±2.9 %

Note the margin of error.

In a random sample of 1172 children ages 6 to 14, 17% of the children responded, "Getting bossed around." The problem is that this is a single random sample. Do 17% of kids in the entire population of children ages 6 to 14 think that getting bossed around is a bad thing?

If you look at the results of a poll like the one in **Table 4.3**, you will observe that a **margin of error** is reported. The margin of error is ±2.9%. This means that the actual percentage of children who feel getting bossed around is a bad thing is at most 2.9% greater than or less than 17%. If x represents the percentage of children in the population who think that getting bossed around is a bad thing, then the poll's margin of error can be expressed as an absolute value inequality:

$$|x - 17| \le 2.9.$$

EXAMPLE 8 Analyzing a Poll's Margin of Error

The inequality

$$|x - 9| \le 2.9$$

describes the percentage of children in the population who think that being grounded is a bad thing about being a kid. (See **Table 4.3**.) Solve the inequality and interpret the solution.

Solution We can solve the inequality using boundary points or an equivalent compound inequality. Let's use an equivalent compound inequality and rewrite without absolute value bars.

$|X| \le c$ means $-c \le X \le c$.

$$|x - 9| \le 2.9 \quad \text{means} \quad -2.9 \le x - 9 \le 2.9.$$

We solve the compound inequality by adding 9 to all three parts.

$$-2.9 \le x - 9 \le 2.9$$
$$-2.9 + 9 \le x - 9 + 9 \le 2.9 + 9$$
$$6.1 \le x \le 11.9$$

The percentage of children in the population who think that being grounded is a bad thing is somewhere between a low of 6.1% and a high of 11.9%. Notice that these percents are 2.9% above and below the given 9%, and that 2.9% is the poll's margin of error.

✓ **CHECK POINT 8** Solve the inequality:

$$|x - 11| \le 2.9.$$

Interpret the solution in terms of the information in **Table 4.3**.

4.3 EXERCISE SET

MyMathLab MathXL PRACTICE WATCH DOWNLOAD READ REVIEW

Practice Exercises

In Exercises 1–38, find the solution set for each equation.

1. $|x| = 8$
2. $|x| = 6$
3. $|x - 2| = 7$
4. $|x + 1| = 5$
5. $|2x - 1| = 7$
6. $|2x - 3| = 11$
7. $\left|\frac{4x - 2}{3}\right| = 2$
8. $\left|\frac{3x - 1}{5}\right| = 1$
9. $|x| = -8$
10. $|x| = -6$
11. $|x + 3| = 0$
12. $|x + 2| = 0$
13. $2|y + 6| = 10$
14. $3|y + 5| = 12$
15. $3|2x - 1| = 21$
16. $2|3x - 2| = 14$
17. $|6y - 2| + 4 = 32$
18. $|3y - 1| + 10 = 25$
19. $7|5x| + 2 = 16$
20. $7|3x| + 2 = 16$
21. $|x + 1| + 5 = 3$
22. $|x + 1| + 6 = 2$
23. $|4y + 1| + 10 = 4$
24. $|3y - 2| + 8 = 1$
25. $|2x - 1| + 3 = 3$
26. $|3x - 2| + 4 = 4$
27. $|5x - 8| = |3x + 2|$
28. $|4x - 9| = |2x + 1|$
29. $|2x - 4| = |x - 1|$
30. $|6x| = |3x - 9|$
31. $|2x - 5| = |2x + 5|$
32. $|3x - 5| = |3x + 5|$
33. $|x - 3| = |5 - x|$
34. $|x - 3| = |6 - x|$
35. $|2y - 6| = |10 - 2y|$
36. $|4y + 3| = |4y + 5|$
37. $\left|\frac{2x}{3} - 2\right| = \left|\frac{x}{3} + 3\right|$
38. $\left|\frac{x}{2} - 2\right| = \left|x - \frac{1}{2}\right|$

In Exercises 39–74, solve and graph the solution set on a number line. Use either boundary points or an equivalent compound inequality, or the method specified by your instructor.

39. $|x| < 3$
40. $|x| < 5$
41. $|x - 2| < 1$
42. $|x - 1| < 5$
43. $|x + 2| \le 1$
44. $|x + 1| \le 5$
45. $|2x - 6| < 8$
46. $|3x + 5| < 17$
47. $|x| > 3$
48. $|x| > 5$
49. $|x + 3| > 1$
50. $|x - 2| > 5$
51. $|x - 4| \ge 2$
52. $|x - 3| \ge 4$
53. $|3x - 8| > 7$
54. $|5x - 2| > 13$
55. $|2(x - 1) + 4| \le 8$
56. $|3(x - 1) + 2| \le 20$
57. $\left|\frac{2x + 6}{3}\right| < 2$
58. $\left|\frac{3x - 3}{4}\right| < 6$
59. $\left|\frac{2x + 2}{4}\right| \ge 2$
60. $\left|\frac{3x - 3}{9}\right| \ge 1$
61. $\left|3 - \frac{2x}{3}\right| > 5$
62. $\left|3 - \frac{3x}{4}\right| > 9$
63. $|x - 2| < -1$
64. $|x - 3| < -2$
65. $|x + 6| > -10$
66. $|x + 4| > -12$
67. $|x + 2| + 9 \le 16$
68. $|x - 2| + 4 \le 5$
69. $2|2x - 3| + 10 > 12$
70. $3|2x - 1| + 2 > 8$
71. $-4|1 - x| < -16$
72. $-2|5 - x| < -6$
73. $3 \le |2x - 1|$

74. $9 \le |4x + 7|$

75. Let $f(x) = |5 - 4x|$. Find all values of x for which $f(x) = 11$.

76. Let $f(x) = |2 - 3x|$. Find all values of x for which $f(x) = 13$.

77. Let $f(x) = |3 - x|$ and $g(x) = |3x + 11|$. Find all values of x for which $f(x) = g(x)$.

78. Let $f(x) = |3x + 1|$ and $g(x) = |6x - 2|$. Find all values of x for which $f(x) = g(x)$.

79. Let $g(x) = |-1 + 3(x + 1)|$. Find all values of x for which $g(x) \le 5$.

80. Let $g(x) = |-3 + 4(x + 1)|$. Find all values of x for which $g(x) \le 3$.

81. Let $h(x) = |2x - 3| + 1$. Find all values of x for which $h(x) > 6$.

82. Let $h(x) = |2x - 4| - 6$. Find all values of x for which $h(x) > 18$.

Practice PLUS

83. When 3 times a number is subtracted from 4, the absolute value of the difference is at least 5. Use interval notation to express the set of all real numbers that satisfy this condition.

84. When 4 times a number is subtracted from 5, the absolute value of the difference is at most 13. Use interval notation to express the set of all real numbers that satisfy this condition.

In Exercises 85–86, solve each inequality. Assume that $a > 0$ and $c > 0$. Use set-builder notation to express each solution set.

85. $|ax + b| < c$

86. $|ax + b| \ge c$

In Exercises 87–88, use the graph of $f(x) = |4 - x|$ to solve each equation or inequality.

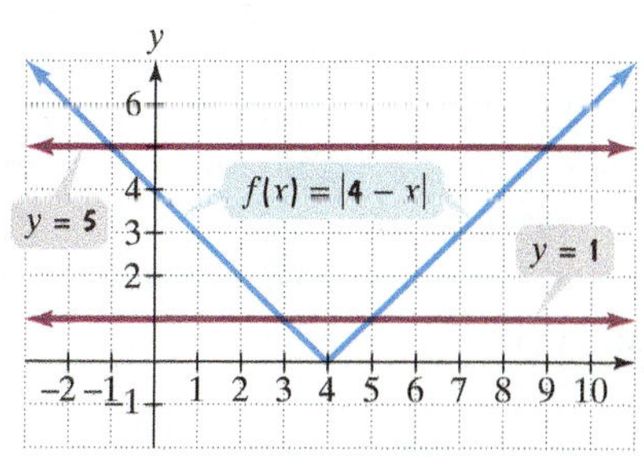

87. $|4 - x| = 1$

88. $|4 - x| < 5$

In Exercises 89–90, use the table to solve each inequality.

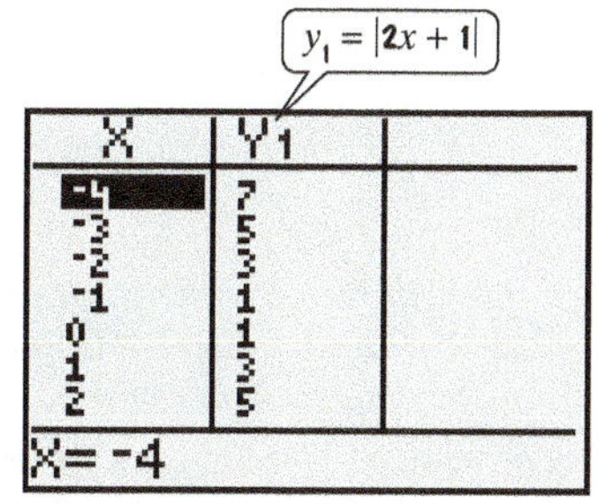

89. $|2x + 1| \le 3$

90. $|2x + 1| \ge 3$

Application Exercises

The three television programs viewed by the greatest percentage of U.S. households in the twentieth century are shown in the table. The data are from a random survey of 4000 TV households by Nielsen Media Research. In Exercises 91–92, let x represent the actual viewing percentage in the U.S. population.

TV Programs with the Greatest U.S. Audience Viewing Percentage of the Twentieth Century

Program	Viewing Percentage in Survey
1 *M*A*S*H* Feb. 28, 1983	60.2%
2 *Dallas* Nov. 21, 1980	53.3%
3 *Roots Part 8* Jan. 30, 1977	51.1%

Source: Nielsen Media Research

91. Solve the inequality: $|x - 60.2| \le 1.6$. Interpret the solution in terms of the information in the table. What is the margin of error?

92. Solve the inequality: $|x - 51.1| \le 1.6$. Interpret the solution in terms of the information in the table. What is the margin of error?

93. The inequality $|T - 57| \le 7$ describes the range of monthly average temperature, T, in degrees Fahrenheit, for San Francisco, California. Solve the inequality and interpret the solution.

94. The inequality $|T - 50| \le 22$ describes the range of monthly average temperature, T, in degrees Fahrenheit, for Albany, New York. Solve the inequality and interpret the solution.

The specifications for machine parts are given with tolerance limits that describe a range of measurements for which the part is acceptable. In Exercises 95–96, x represents the length of a machine part, in centimeters. The tolerance limit is 0.01 centimeter.

95. Solve: $|x - 8.6| \le 0.01$. If the length of the machine part is supposed to be 8.6 centimeters, interpret the solution.

96. Solve: $|x - 9.4| \le 0.01$. If the length of the machine part is supposed to be 9.4 centimeters, interpret the solution.

97. If a coin is tossed 100 times, we would expect approximately 50 of the outcomes to be heads. It can be demonstrated that a coin is unfair if h, the number of outcomes that result in heads, satisfies $\left|\frac{h - 50}{5}\right| \ge 1.645$. Describe the number of outcomes that result in heads that determine an unfair coin that is tossed 100 times.

Writing in Mathematics

98. Explain how to solve an equation containing one absolute value expression.

99. Explain why the procedure that you described in Exercise 98 does not apply to the equation $|x - 5| = -3$. What is the solution set of this equation?

100. Describe how to solve an absolute value equation with two absolute values.

101. Describe one method for solving an absolute value inequality.

102. Explain why the procedure that you described in Exercise 101 does not apply to the inequality $|x - 5| < -3$. What is the solution set of this inequality?

103. Explain why the procedure that you described in Exercise 101 does not apply to the inequality $|x - 5| > -3$. What is the solution set of this inequality?

104. The final episode of $M * A * S * H$ was viewed by more than 58% of U.S. television households. Is it likely that a popular television series in the twenty-first century will achieve a 58% market share? Explain your answer.

Technology Exercises

In Exercises 105–107, solve each equation using a graphing utility. Graph each side separately in the same viewing rectangle. The solutions are the x-coordinates of the intersection points.

105. $|x + 1| = 5$

106. $|3(x + 4)| = 12$

107. $|2x - 3| = |9 - 4x|$

In Exercises 108–110, solve each inequality using a graphing utility. Graph each side separately in the same viewing rectangle. The solution set consists of all values of x for which the graph of the left side lies below the graph of the right side.

108. $|2x + 3| < 5$

109. $\left|\frac{2x - 1}{3}\right| < \frac{5}{3}$

110. $|x + 4| < -1$

In Exercises 111–113, solve each inequality using a graphing utility. Graph each side separately in the same viewing rectangle. The solution set consists of all values of x for which the graph of the left side lies above the graph of the right side.

111. $|2x - 1| > 7$

112. $|0.1x - 0.4| + 0.4 > 0.6$

113. $|x + 4| > -1$

114. Use a graphing utility to verify the solution sets for any five equations or inequalities that you solved by hand in Exercises 1–74.

Critical Thinking Exercises

Make Sense? *In Exercises 115–118, determine whether each statement "makes sense" or "does not make sense" and explain your reasoning.*

115. I have problems setting up the appropriate compound inequality for an absolute value inequality, so I use the boundary point method.

116. I noticed that the graph of $f(x) = |x - 3|$ lies below the graph of $g(x) = 4$ in $(-1, 7)$, so the solution set of $|x - 3| > 4$ must be $(-\infty, -1) \cup (7, \infty)$.

117. Because the absolute value of any expression is never less than a negative number, I can immediately conclude that the inequality $|2x - 5| - 9 < -4$ has no solution.

118. I'll win the contest if I can complete the crossword puzzle in 20 minutes plus or minus 5 minutes, so my winning time, x, is modeled by $|x - 20| \le 5$.

In Exercises 119–122, determine whether each statement is true or false. If the statement is false, make the necessary change(s) to produce a true statement.

119. All absolute value equations have two solutions.

120. The equation $|x| = -6$ is equivalent to $x = 6$ or $x = -6$.

121. Values of -5 and 5 satisfy $|x| = 5$, $|x| \le 5$, and $|x| \ge -5$.

122. The absolute value of any linear expression is greater than 0 for all real numbers except the number for which the expression is equal to 0.

123. Write an absolute value inequality for which the interval shown is the solution.

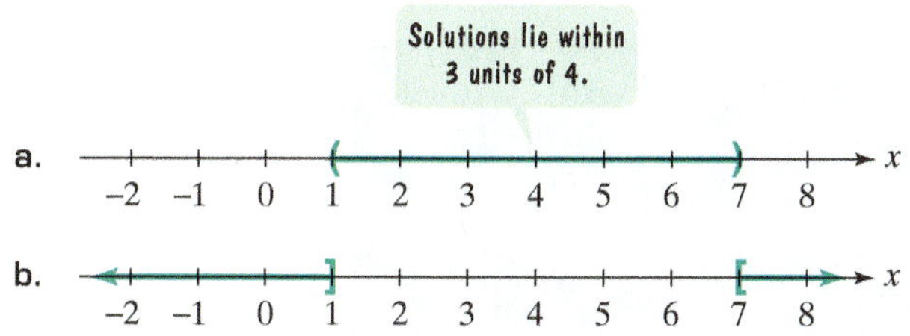

124. The percentage, p, of defective products manufactured by a company is given by $|p - 0.3\%| \le 0.2\%$. If 100,000 products are manufactured and the company offers a \$5 refund for each defective product, describe the company's cost for refunds.

125. Solve: $|2x + 5| = 3x + 4$.

Review and Preview Exercises

Exercises 126–128 will enable you to review graphing linear functions. In addition, they will help you prepare for the material covered in the next section. In each exercise, graph the linear function.

126. $3x - 5y = 15$ (Section 2.4, Example 1)

127. $f(x) = -\frac{2}{3}x$ (Section 2.4, Example 5)

128. $f(x) = -2$ (Section 2.4, Example 6)

MID-CHAPTER CHECK POINT Section 4.1–Section 4.3

What You Know: We learned to solve linear inequalities, expressing solution sets in set-builder and interval notations. We know that it is necessary to change the sense of an inequality when multiplying or dividing both sides by a negative number. We solved compound inequalities with *and* by finding the intersection of solution sets and with *or* by finding the union of solution sets. Finally, we solved equations and inequalities involving absolute value by carefully rewriting the given equation or inequality without absolute value bars, or, in the case of inequalities, using boundary points. For positive values of c, we wrote $|X| = c$ as $X = c$ or $X = -c$. Without using boundary points, we wrote $|X| < c$ as $-c < X < c$, and we wrote $|X| > c$ as $X < -c$ or $X > c$.

In Exercises 1–18, solve each inequality or equation.

1. $4 - 3x \ge 12 - x$
2. $5 \le 2x - 1 < 9$
3. $|4x - 7| = 5$
4. $-10 - 3(2x + 1) > 8x + 1$
5. $2x + 7 < -11$ or $-3x - 2 < 13$
6. $|3x - 2| \le 4$
7. $|x + 5| = |5x - 8|$
8. $5 - 2x \ge 9$ and $5x + 3 > -17$
9. $3x - 2 > -8$ or $2x + 1 < 9$
10. $\frac{x}{2} + 3 \le \frac{x}{3} + \frac{5}{2}$
11. $\frac{2}{3}(6x - 9) + 4 > 5x + 1$
12. $|5x + 3| > 2$
13. $7 - \left|\frac{x}{2} + 2\right| \le 4$
14. $5(x - 2) - 3(x + 4) \ge 2x - 20$
15. $\frac{x + 3}{4} < \frac{1}{3}$
16. $5x + 1 \ge 4x - 2$ and $2x - 3 > 5$
17. $3 - |2x - 5| = -6$
18. $3 + |2x - 5| = -6$

In Exercises 19–22, solve each problem.

19. A car rental agency rents a certain car for \$40 per day with unlimited mileage or \$24 per day plus \$0.20 per mile. How far can a customer drive this car per day for the \$24 option to cost no more than the unlimited mileage option?
20. To receive a B in a course, you must have an average of at least 80% but less than 90% on five exams. Your grades on the first four exams were 95%, 79%, 91%, and 86%. What range of grades on the fifth exam will result in a B for the course?
21. A retiree requires an annual income of at least \$9000 from an investment paying 7.5% annual interest. How much should the retiree invest to achieve the desired return?
22. A company that manufactures compact discs has fixed monthly overhead costs of \$60,000. Each disc costs \$0.18 to produce and sells for \$0.30. How many discs should be produced and sold each month for the company to have a profit of at least \$30,000?

SECTION 4.4

Linear Inequalities in Two Variables

Objectives

1. Graph a linear inequality in two variables.
2. Use mathematical models involving linear inequalities.
3. Graph a system of linear inequalities.

This book was written in Point Reyes National Seashore, 40 miles north of San Francisco. The park consists of 75,000 acres with miles of pristine surf-washed beaches, forested ridges, and bays bordered by white cliffs.

Like your author, many people are kept inspired and energized surrounded by nature's unspoiled beauty. In this section, you will see how systems of inequalities model whether a region's natural beauty manifests itself in forests, grasslands, or deserts.

Linear Inequalities in Two Variables and Their Solutions

We have seen that equations in the form $Ax + By = C$ are straight lines when graphed. If we change the symbol $=$ to $>$, $<$, $\geq$, or $\leq$, we obtain a **linear inequality in two variables**. Some examples of linear inequalities in two variables are $x + y > 2$, $3x - 5y \leq 15$, and $2x - y < 4$.

A **solution of an inequality in two variables**, x and y, is an ordered pair of real numbers with the following property: When the x-coordinate is substituted for x and the y-coordinate is substituted for y in the inequality, we obtain a true statement. For example, (3, 2) is a solution of the inequality $x + y > 1$. When 3 is substituted for x and 2 is substituted for y, we obtain the true statement $3 + 2 > 1$, or $5 > 1$. Because there are infinitely many pairs of numbers that have a sum greater than 1, the inequality $x + y > 1$ has infinitely many solutions. Each ordered-pair solution is said to **satisfy** the inequality. Thus, (3, 2) satisfies the inequality $x + y > 1$.

1 Graph a linear inequality in two variables.

The Graph of a Linear Inequality in Two Variables

We know that the graph of an equation in two variables is the set of all points whose coordinates satisfy the equation. Similarly, the **graph of an inequality in two variables** is the set of all points whose coordinates satisfy the inequality.

Let's use **Figure 4.8** to get an idea of what the graph of a linear inequality in two variables looks like. Part of the figure shows the graph of the linear equation $x + y = 2$. The line divides the points in the rectangular coordinate system into three sets. First, there is the set of points along the line, satisfying $x + y = 2$. Next, there is the set of points in the green region above the line. Points in the green region satisfy the linear inequality $x + y > 2$. Finally, there is the set of points in the purple region below the line. Points in the purple region satisfy the linear inequality $x + y < 2$.

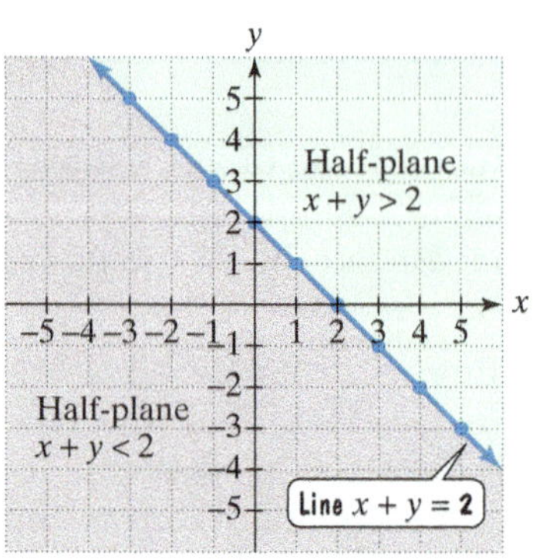

FIGURE 4.8

A **half-plane** is the set of all the points on one side of a line. In **Figure 4.8**, the green region is a half-plane. The purple region is also a half-plane. A half-plane is the graph of a linear inequality that involves $>$ or $<$. The graph of a linear inequality that involves $\geq$ or $\leq$ is a half-plane and a line. A solid line is used to show that a line is part of a graph. A dashed line is used to show that a line is not part of a graph.

Graphing a Linear Inequality in Two Variables

1. Replace the inequality symbol with an equal sign and graph the corresponding linear equation. Draw a solid line if the original inequality contains a $\leq$ or $\geq$ symbol. Draw a dashed line if the original inequality contains a $<$ or $>$ symbol.
2. Choose a test point from one of the half-planes. (Do not choose a point on the line.) Substitute the coordinates of the test point into the inequality.
3. If a true statement results, shade the half-plane containing this test point. If a false statement results, shade the half-plane not containing this test point.

EXAMPLE 1 Graphing a Linear Inequality in Two Variables

Graph: $2x - 3y \geq 6$.

Solution

Step 1. Replace the inequality symbol by = and graph the linear equation. We need to graph $2x - 3y = 6$. We can use intercepts to graph this line.

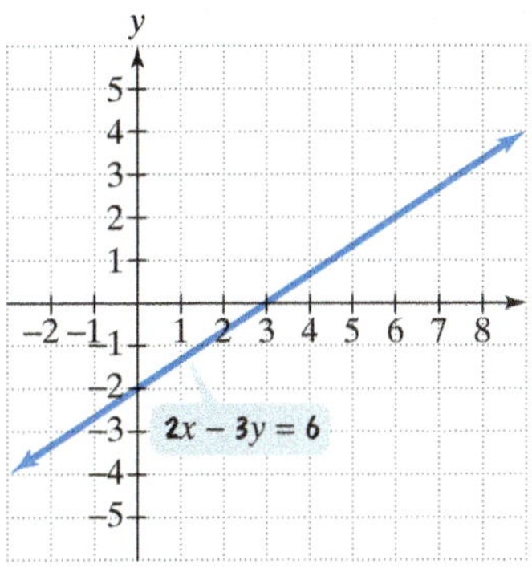

FIGURE 4.9 Preparing to graph $2x - 3y \geq 6$

We set $y = 0$ to find the x-intercept:	We set $x = 0$ to find the y-intercept:
$2x - 3y = 6$	$2x - 3y = 6$
$2x - 3 \cdot 0 = 6$	$2 \cdot 0 - 3y = 6$
$2x = 6$	$-3y = 6$
$x = 3.$	$y = -2.$

The x-intercept is (3, 0). The y-intercept is (0, −2). Using the intercepts, the line is shown in **Figure 4.9** as a solid line. This is because the inequality $2x - 3y \geq 6$ contains a $\geq$ symbol, in which equality is included.

Step 2. Choose a test point from one of the half-planes and not from the line. Substitute its coordinates into the inequality. The line $2x - 3y = 6$ divides the plane into three parts—the line itself and two half-planes. The points in one half-plane satisfy $2x - 3y > 6$. The points in the other half-plane satisfy $2x - 3y < 6$. We need to find which half-plane belongs to the solution of $2x - 3y \geq 6$. To do so, we test a point from either half-plane. The origin, (0, 0), is the easiest point to test.

$$2x - 3y \geq 6 \quad \text{This is the given inequality.}$$

$$2 \cdot 0 - 3 \cdot 0 \overset{?}{\geq} 6 \quad \text{Test (0, 0) by substituting 0 for } x \text{ and 0 for } y.$$

$$0 - 0 \overset{?}{\geq} 6 \quad \text{Multiply.}$$

$$0 \geq 6 \quad \text{This statement is false.}$$

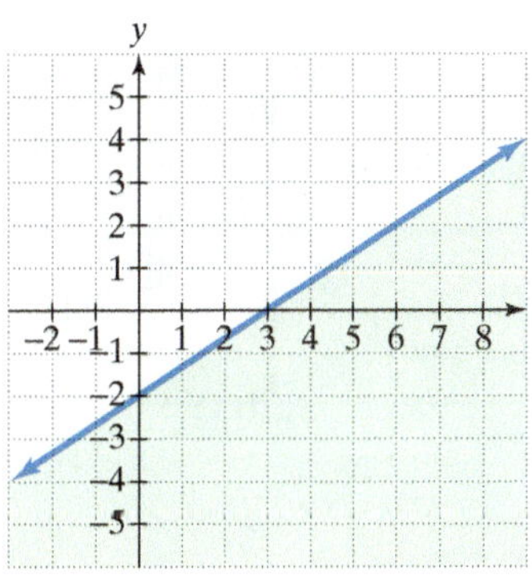

FIGURE 4.10 The graph of $2x - 3y \geq 6$

Step 3. If a false statement results, shade the half-plane not containing the test point. Because 0 is not greater than or equal to 6, the test point, (0, 0), is not part of the solution set. Thus, the half-plane below the solid line $2x - 3y = 6$ is part of the solution set. The solution set is the line and the half-plane that does not contain the point (0, 0), indicated by shading this half-plane. The graph is shown using green shading and a blue line in **Figure 4.10**.

✓ **CHECK POINT 1** Graph: $4x - 2y \geq 8$.

When graphing a linear inequality, choose a test point that lies in one of the half-planes and *not on the line dividing the half-planes*. The test point (0, 0) is convenient because it is easy to calculate when 0 is substituted for each variable. However, if (0, 0) lies on the dividing line and not in a half-plane, a different test point must be selected.

EXAMPLE 2 Graphing a Linear Inequality in Two Variables

Graph: $y > -\frac{2}{3}x.$

Solution

Step 1. Replace the inequality symbol by = and graph the linear equation. Because we are interested in graphing $y > -\frac{2}{3}x$, we begin by graphing $y = -\frac{2}{3}x$. We can use the slope and the y-intercept to graph this linear function.

$$y = -\frac{2}{3}x + 0$$

Slope $= \frac{-2}{3} = \frac{\text{rise}}{\text{run}}$ y-intercept is (0, 0)

The y-intercept is 0, so the line passes through (0, 0). Using the y-intercept and the slope, the line is shown in **Figure 4.11** as a dashed line. This is because the inequality $y > -\frac{2}{3}x$ contains a > symbol, in which equality is not included.

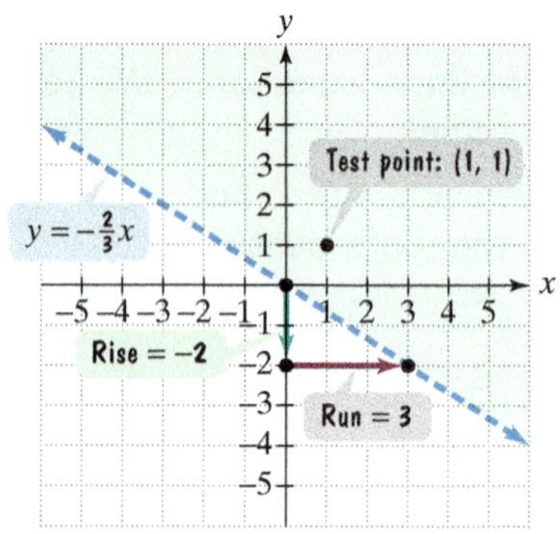

FIGURE 4.11 The graph of $y > -\frac{2}{3}x$

Step 2. Choose a test point from one of the half-planes and not from the line. Substitute its coordinates into the inequality. We cannot use (0, 0) as a test point because it lies on the line and not in a half-plane. Let's use (1, 1), which lies in the half-plane above the line.

$y > -\frac{2}{3}x$ This is the given inequality.

$1 \overset{?}{>} -\frac{2}{3} \cdot 1$ Test (1, 1) by substituting 1 for x and 1 for y.

$1 > -\frac{2}{3}$ This statement is true.

Step 3. If a true statement results, shade the half-plane containing the test point. Because 1 is greater than $-\frac{2}{3}$, the test point (1, 1) is part of the solution set. All the points on the same side of the line $y = -\frac{2}{3}x$ as the point (1, 1) are members of the solution set. The solution set is the half-plane that contains the point (1, 1), indicated by shading this half-plane. The graph is shown using green shading and a dashed blue line in **Figure 4.11**.

Using Technology

Most graphing utilities can graph inequalities in two variables with the SHADE feature. The procedure varies by model, so consult your manual. For most graphing utilities, you must first solve for y if it is not already isolated. The figure shows the graph of $y > -\frac{2}{3}x$. Most displays do not distinguish between dashed and solid boundary lines.

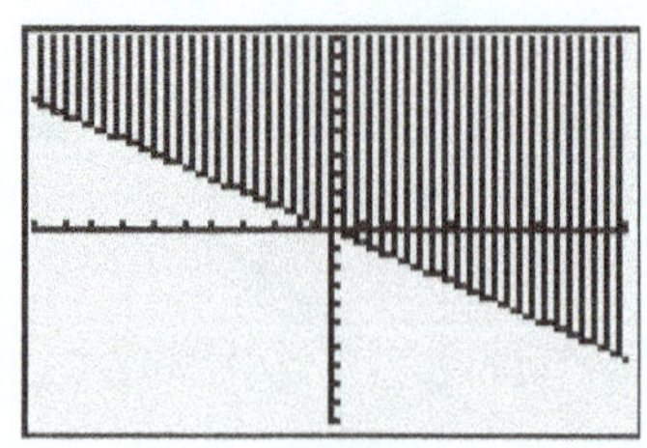

✓ CHECK POINT 2 Graph: $y > -\frac{3}{4}x.$

Graphing Linear Inequalities without Using Test Points

You can graph inequalities in the form $y > mx + b$ or $y < mx + b$ without using test points. The inequality symbol indicates which half-plane to shade.

- If $y > mx + b$, shade the half-plane above the line $y = mx + b$.
- If $y < mx + b$, shade the half-plane below the line $y = mx + b$.

Observe how this is illustrated in **Figure 4.11**. The graph of $y > -\frac{2}{3}x$ is the half-plane above the line $y = -\frac{2}{3}x$.

Study Tip

Continue using test points to graph inequalities in the form $Ax + By > C$ or $Ax + By < C$. The graph of $Ax + By > C$ can lie above or below the line given by $Ax + By = C$, depending on the value of B. The same comment applies to the graph of $Ax + By < C$.

It is also not necessary to use test points when graphing inequalities involving half-planes on one side of a vertical or a horizontal line.

For the Vertical Line $x = a$:	For the Horizontal Line $y = b$:
• If $x > a$, shade the half-plane to the right of $x = a$.	• If $y > b$, shade the half-plane above $y = b$.
• If $x < a$, shade the half-plane to the left of $x = a$.	• If $y < b$, shade the half-plane below $y = b$.

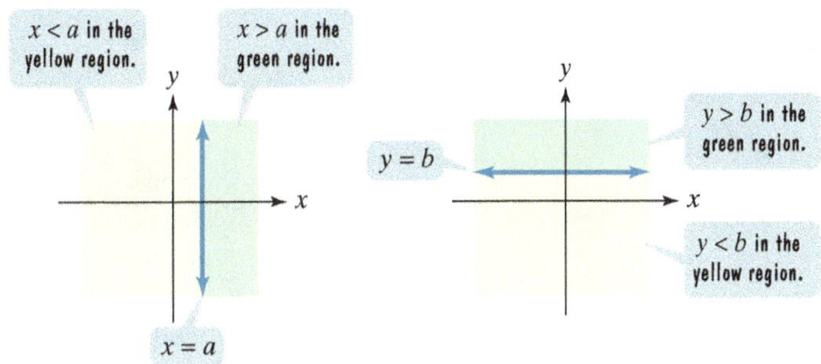

EXAMPLE 3 Graphing Inequalities without Using Test Points

Graph each inequality in a rectangular coordinate system:

a. $y \leq -3$ b. $x > 2$.

Solution

a. $y \leq -3$

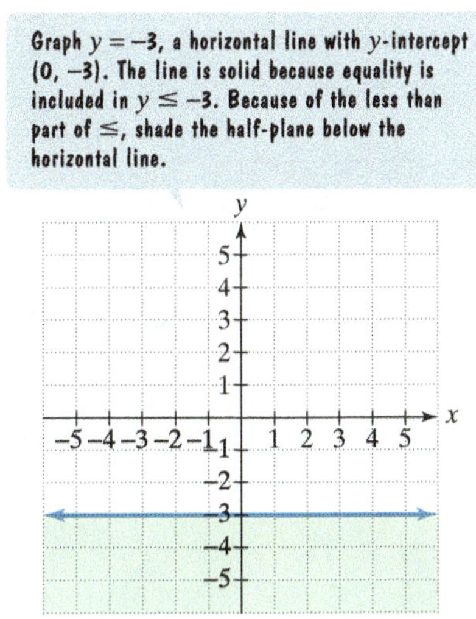

b. $x > 2$

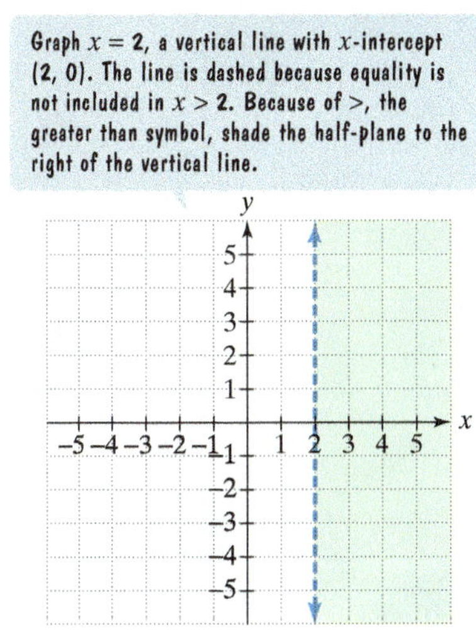

✓ CHECK POINT 3 Graph each inequality in a rectangular coordinate system:

a. $y > 1$ b. $x \leq -2$.

2 Use mathematical models involving linear inequalities.

Modeling with Systems of Linear Inequalities

Just as two or more linear equations make up a system of linear equations, two or more linear inequalities make up a **system of linear inequalities**. A **solution of a system of linear inequalities** in two variables is an ordered pair that satisfies each inequality in the system.

EXAMPLE 4 Forests, Grasslands, Deserts, and Systems of Inequalities

Temperature and precipitation affect whether or not trees and forests can grow. At certain levels of precipitation and temperature, only grasslands and deserts will exist. **Figure 4.12** shows three kinds of regions—deserts, grasslands, and forests—that result from various ranges of temperature, T, and precipitation, P.

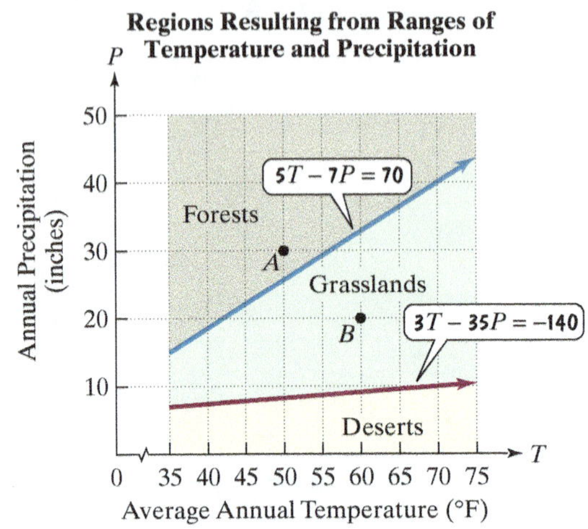

FIGURE 4.12

Source: A. Miller and J. Thompson, *Elements of Meteorology*

Systems of inequalities can be used to model where forests, grasslands, and deserts occur. Because these regions occur when the average annual temperature, T, is 35°F or greater, each system contains the inequality $T \geq 35$.

Forests occur if	Grasslands occur if	Deserts occur if
$T \geq 35$ $5T - 7P < 70.$	$T \geq 35$ $5T - 7P \geq 70$ $3T - 35P \leq -140.$	$T \geq 35$ $3T - 35P > -140.$

Show that point A in **Figure 4.12** is a solution of the system of inequalities that models where forests occur.

Solution Point A has coordinates (50, 30). This means that if a region has an average annual temperature of 50°F and an average annual precipitation of 30 inches, a forest occurs. We can show that (50, 30) satisfies the system of inequalities for forests by substituting 50 for T and 30 for P in each inequality in the system.

$$T \geq 35$$
$$50 \geq 35, \quad \text{true}$$

$$5T - 7P < 70$$
$$5 \cdot 50 - 7 \cdot 30 \stackrel{?}{<} 70$$
$$250 - 210 \stackrel{?}{<} 70$$
$$40 < 70, \quad \text{true}$$

The coordinates (50, 30) make each inequality true. Thus, (50, 30) satisfies the system for forests.

✓ **CHECK POINT 4** Show that point B in **Figure 4.12** is a solution of the system of inequalities that models where grasslands occur.

3 Graph a system of linear inequalities.

Graphing Systems of Linear Inequalities

The **solution set of a system of linear inequalities in two variables** is the set of all ordered pairs that satisfy each inequality in the system. Thus, to graph a system of inequalities in two variables, begin by graphing each individual inequality in the same rectangular coordinate system. Then find the region, if there is one, that is common to every graph in the system. This region of intersection gives a picture of the system's solution set.

EXAMPLE 5 Graphing a System of Linear Inequalities

Graph the solution set of the system:

$$x - y < 1$$
$$2x + 3y \geq 12.$$

Solution Replacing each inequality symbol with an equal sign indicates that we need to graph $x - y = 1$ and $2x + 3y = 12$. We can use intercepts to graph these lines.

$\mathbf{x - y = 1}$		$\mathbf{2x + 3y = 12}$
x-intercept: $x - 0 = 1$, $x = 1$. The line passes through $(1, 0)$.	Set $y = 0$ in each equation.	x-intercept: $2x + 3 \cdot 0 = 12$, $2x = 12$, $x = 6$. The line passes through $(6, 0)$.
y-intercept: $0 - y = 1$, $-y = 1$, $y = -1$. The line passes through $(0, -1)$	Set $x = 0$ in each equation.	y-intercept: $2 \cdot 0 + 3y = 12$, $3y = 12$, $y = 4$. The line passes through $(0, 4)$.

Now we are ready to graph the solution set of the system of linear inequalities.

Graph $x - y < 1$. The blue line, $x - y = 1$, is dashed: Equality is not included in $x - y < 1$. Because $(0, 0)$ makes the inequality true $(0 - 0 < 1$, or $0 < 1$, is true), shade the half-plane containing $(0, 0)$ in yellow.

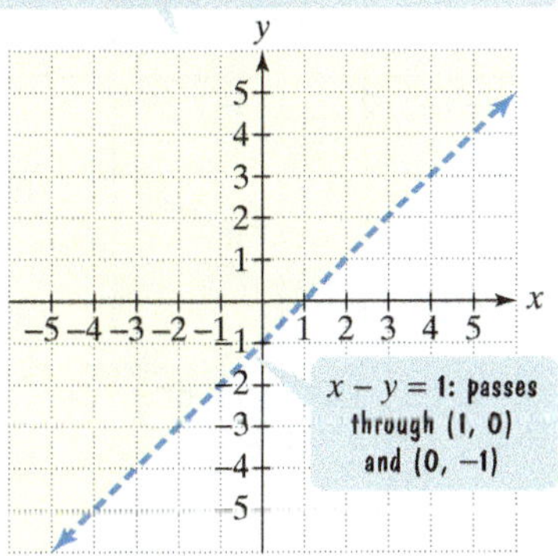

The graph of $x - y < 1$

Add the graph of $2x + 3y \geq 12$. The red line, $2x + 3y = 12$, is solid: Equality is included in $2x + 3y \geq 12$. Because $(0, 0)$ makes the inequality false $(2 \cdot 0 + 3 \cdot 0 \geq 12$, or $0 \geq 12$, is false), shade the half-plane not containing $(0, 0)$ using green vertical shading.

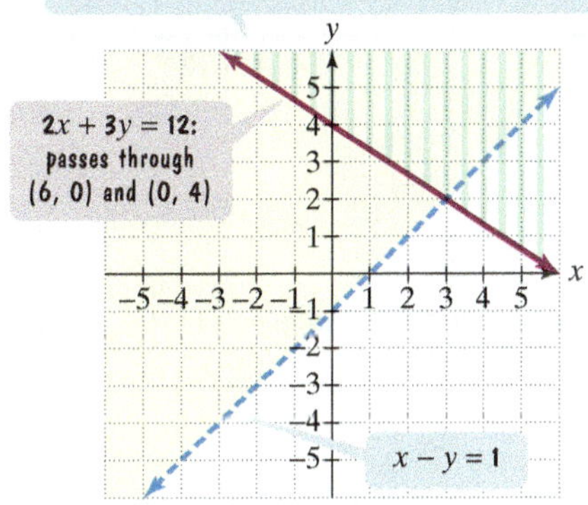

Adding the graph of $2x + 3y \geq 12$

The solution set of the system is graphed as the intersection (the overlap) of the two half-planes. This is the region in which the yellow shading and the green vertical shading overlap.

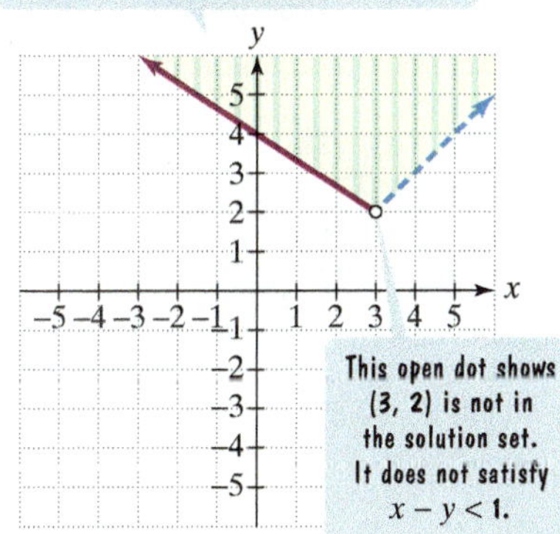

The graph of $x - y < 1$ and $2x + 3y \geq 12$

CHECK POINT 5 Graph the solution set of the system:

$$x - 3y < 6$$
$$2x + 3y \geq -6.$$

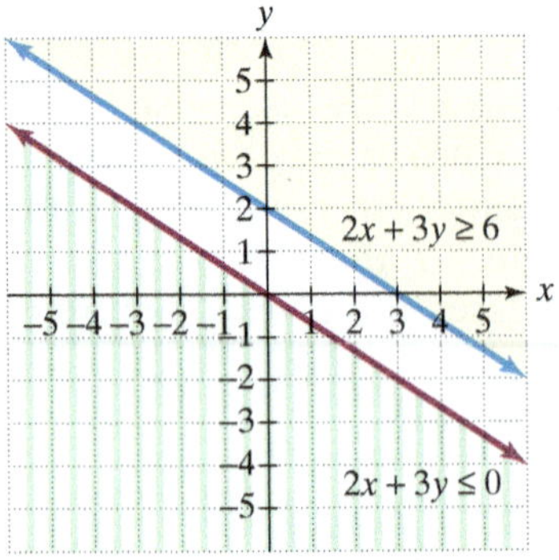

FIGURE 4.13 A system of inequalities with no solution

A system of inequalities has no solution if there are no points in the rectangular coordinate system that simultaneously satisfy each inequality in the system. For example, the system

$$2x + 3y \geq 6$$
$$2x + 3y \leq 0,$$

whose separate graphs are shown in **Figure 4.13**, has no overlapping region. Thus, the system has no solution. The solution set is ∅, the empty set.

EXAMPLE 6 Graphing a System of Inequalities

Graph the solution set of the system:

$$x - y < 2$$
$$-2 \leq x < 4$$
$$y < 3.$$

Solution We begin by graphing $x - y < 2$, the first given inequality. The line $x - y = 2$ has an x-intercept of 2 and a y-intercept of -2. The test point $(0, 0)$ makes the inequality $x - y < 2$ true. The graph of $x - y < 2$ is shown in **Figure 4.14**.

Now, let's consider the second given inequality, $-2 \leq x < 4$. Replacing the inequality symbols by =, we obtain $x = -2$ and $x = 4$, graphed as red vertical lines in **Figure 4.15**. The line of $x = 4$ is not included. Because x is between -2 and 4, we shade the region between the vertical lines. We must intersect this region with the yellow region in **Figure 4.14**. The resulting region is shown in yellow and green vertical shading in **Figure 4.15**.

Finally, let's consider the third given inequality, $y < 3$. Replacing the inequality symbol by =, we obtain $y = 3$, which graphs as a horizontal line. Because of the less than symbol in $y < 3$, the graph consists of the half-plane below the line $y = 3$. We must intersect this half-plane with the region in **Figure 4.15**. The resulting region is shown in yellow and green vertical shading in **Figure 4.16**. This region represents the graph of the solution set of the given system.

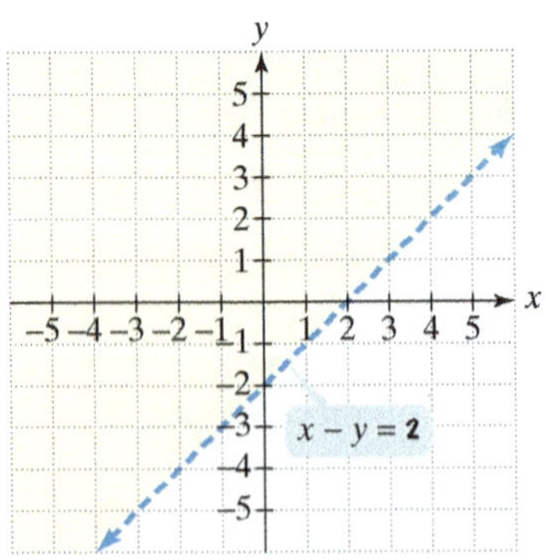

FIGURE 4.14 The graph of $x - y < 2$

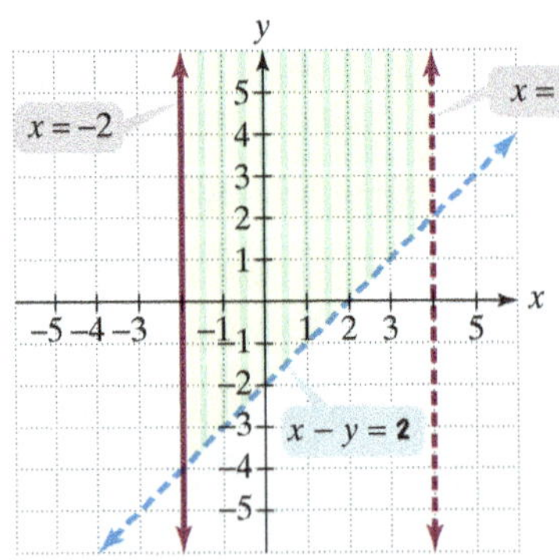

FIGURE 4.15 The graph of $x - y < 2$ and $-2 \leq x < 4$

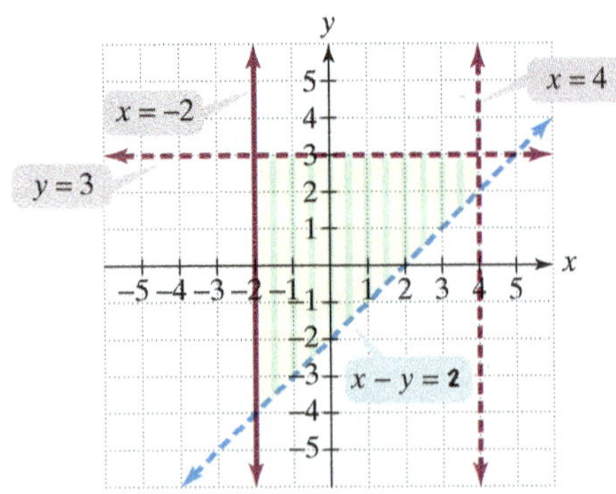

FIGURE 4.16 The graph of $x - y < 2$ and $-2 \leq x < 4$ and $y < 3$

CHECK POINT 6 Graph the solution set of the system:

$$x + y < 2$$
$$-2 \leq x < 1$$
$$y > -3.$$

4.4 EXERCISE SET

Practice Exercises

In Exercises 1–22, graph each inequality.

1. $x + y \geq 3$

2. $x + y \geq 2$

3. $x - y < 5$

4. $x - y < 6$

5. $x + 2y > 4$

6. $2x + y > 6$

7. $3x - y \leq 6$

8. $x - 3y \leq 6$

9. $\frac{x}{2} + \frac{y}{3} < 1$

10. $\frac{x}{4} + \frac{y}{2} < 1$

11. $y > \frac{1}{3}x$

12. $y > \frac{1}{4}x$

13. $y \leq 3x + 2$

14. $y \leq 2x - 1$

15. $y < -\frac{1}{4}x$

16. $y < -\frac{1}{3}x$

17. $x \leq 2$

18. $x \leq -4$

19. $y > -4$

20. $y > -2$

21. $y \geq 0$

22. $x \leq 0$

In Exercises 23–46, graph the solution set of each system of inequalities or indicate that the system has no solution.

23. $3x + 6y \leq 6$
$2x + y \leq 8$

24. $x - y \geq 4$
$x + y \leq 6$

25. $2x - 5y \leq 10$
$3x - 2y > 6$

26. $2x - y \leq 4$
$3x + 2y > -6$

27. $y > 2x - 3$
$y < -x + 6$

28. $y < -2x + 4$
$y < x - 4$

29. $x + 2y \leq 4$
$y \geq x - 3$

30. $x + y \leq 4$
$y \geq 2x - 4$

31. $x \leq 2$
$y \geq -1$

32. $x \leq 3$
$y \leq -1$

33. $-2 \leq x < 5$

34. $-2 < y \leq 5$

35. $x - y \leq 1$
$x \geq 2$

36. $4x - 5y \geq -20$
$x \geq -3$

37. $x + y > 4$
$x + y < -1$

38. $x + y > 3$
$x + y < -2$

39. $x + y > 4$
$x + y > -1$

40. $x + y > 3$
$x + y > 2$

41. $x - y \leq 2$
$x \geq -2$
$y \leq 3$

42. $3x + y \leq 6$
$x \geq -2$
$y \leq 4$

43. $x \geq 0$
$y \geq 0$
$2x + 5y \leq 10$
$3x + 4y \leq 12$

44. $x \geq 0$
$y \geq 0$
$2x + y \leq 4$
$2x - 3y \leq 6$

45. $3x + y \leq 6$
$2x - y \leq -1$
$x \geq -2$
$y \leq 4$

46. $2x + y \leq 6$
$x + y \geq 2$
$1 \leq x \leq 2$
$y \leq 3$

Practice PLUS

In Exercises 47–48, write each sentence as a linear inequality in two variables. Then graph the inequality.

47. The y-variable is at least 4 more than the product of -2 and the x-variable.

48. The y-variable is at least 2 more than the product of -3 and the x-variable.

In Exercises 49–50, write the given sentences as a system of linear inequalities in two variables. Then graph the system.

49. The sum of the x-variable and the y-variable is at most 4. The y-variable added to the product of 3 and the x-variable does not exceed 6.

50. The sum of the x-variable and the y-variable is at most 3. The y-variable added to the product of 4 and the x-variable does not exceed 6.

In Exercises 51–52, rewrite each inequality in the system without absolute value bars. Then graph the rewritten system in rectangular coordinates.

51. $|x| \leq 2$
$|y| \leq 3$

52. $|x| \leq 1$
$|y| \leq 2$

*The graphs of solution sets of systems of inequalities involve finding the intersection of the solution sets of two or more inequalities. By contrast, in Exercises 53–54 you will be graphing the **union** of the solution sets of two inequalities.*

53. Graph the union of $y > \frac{3}{2}x - 2$ and $y < 4$.

54. Graph the union of $x - y \geq -1$ and $5x - 2y \leq 10$.

Without graphing, in Exercises 55–58, determine if each system has no solution or infinitely many solutions.

55. $3x + y < 9$
$3x + y > 9$

56. $6x - y \leq 24$
$6x - y > 24$

57. $3x + y \leq 9$
$3x + y \geq 9$

58. $6x - y \leq 24$
$6x - y \geq 24$

Application Exercises

Maximum heart rate, H, in beats per minute is a function of age, a, modeled by the formula

$$H = 220 - a,$$

where $10 \leq a \leq 70$. The bar graph shows the target heart rate ranges for four types of exercise goals in terms of maximum heart rate.

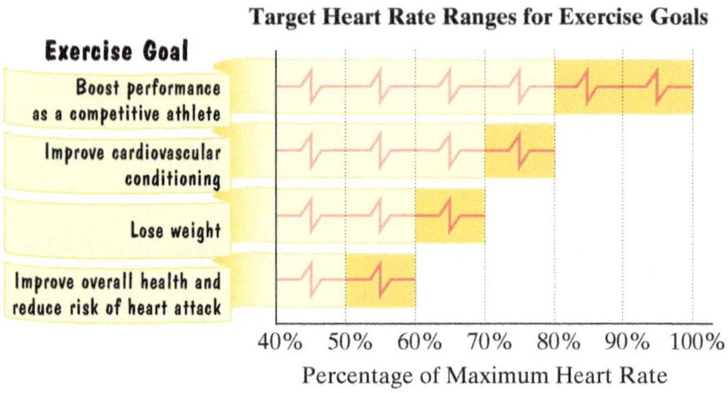

Source: Vitality

In Exercises 59–62, systems of inequalities will be used to model three of the target heart rate ranges shown in the bar graph. We begin with the target heart rate range for cardiovascular conditioning, modeled by the following system of inequalities:

$10 \leq a \leq 70$ — Heart rate ranges apply to ages 10 through 70, inclusive.

$H \geq 0.7(220 - a)$ — Target heart rate range is greater than or equal to 70% of maximum heart rate

$H \leq 0.8(220 - a).$ — and less than or equal to 80% of maximum heart rate.

The graph of this system is shown in the figure. Use the graph to solve Exercises 59–60.

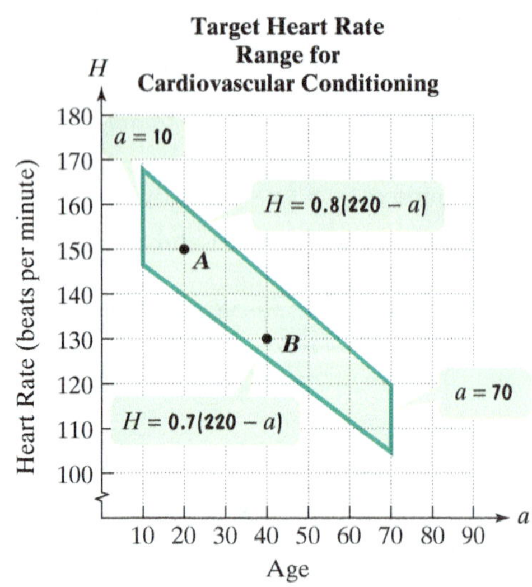

59. a. What are the coordinates of point A and what does this mean in terms of age and heart rate?

b. Show that point A is a solution of the system of inequalities.

60. a. What are the coordinates of point B and what does this mean in terms of age and heart rate?

b. Show that point B is a solution of the system of inequalities.

61. Write a system of inequalities that models the target heart rate range for the goal of losing weight.

62. Write a system of inequalities that models the target heart rate range for improving overall health.

63. On your next vacation, you will divide lodging between large resorts and small inns. Let x represent the number of nights spent in large resorts. Let y represent the number of nights spent in small inns.

a. Write a system of inequalities that models the following conditions:

> You want to stay at least 5 nights. At least one night should be spent at a large resort. Large resorts average \$200 per night and small inns average \$100 per night. Your budget permits no more than \$700 for lodging.

b. Graph the solution set of the system of inequalities in part (a).

c. Based on your graph in part (b), how many nights could you spend at a large resort and still stay within your budget?

64. a. An elevator can hold no more than 2000 pounds. If children average 80 pounds and adults average 160 pounds, write a system of inequalities that models when the elevator holding x children and y adults is overloaded.

b. Graph the solution set of the system of inequalities in part (a).

Writing in Mathematics

65. What is a linear inequality in two variables? Provide an example with your description.

66. How do you determine if an ordered pair is a solution of an inequality in two variables, x and y?

67. What is a half-plane?

68. What does a solid line mean in the graph of an inequality?

69. What does a dashed line mean in the graph of an inequality?

70. Explain how to graph $x - 2y < 4$.

71. What is a system of linear inequalities?

72. What is a solution of a system of linear inequalities?

73. Explain how to graph the solution set of a system of inequalities.

74. What does it mean if a system of linear inequalities has no solution?

Technology Exercises

Graphing utilities can be used to shade regions in the rectangular coordinate system, thereby graphing an inequality in two variables. Read the section of the user's manual for your graphing utility that describes how to shade a region. Then use your graphing utility to graph the inequalities in Exercises 75–78.

75. $y \leq 4x + 4$

76. $y \geq \frac{2}{3}x - 2$

77. $2x + y \leq 6$

78. $3x - 2y \geq 6$

79. Does your graphing utility have any limitations in terms of graphing inequalities? If so, what are they?

80. Use a graphing utility with a SHADE feature to verify any five of the graphs that you drew by hand in Exercises 1–22.

81. Use a graphing utility with a SHADE feature to verify any five of the graphs that you drew by hand for the systems in Exercises 23–46.

Critical Thinking Exercises

Make Sense? *In Exercises 82–85, determine whether each statement "makes sense' or "does not make sense" and explain your reasoning.*

82. When graphing a linear inequality, I should always use (0, 0) as a test point because it's easy to perform the calculations when 0 is substituted for each variable.

83. If you want me to graph $x < 3$, you need to tell me whether to use a number line or a rectangular coordinate system.

84. When graphing $3x - 4y < 12$, it's not necessary for me to graph the linear equation $3x - 4y = 12$ because the inequality contains a $<$ symbol, in which equality is not included.

85. Linear inequalities can model situations in which I'm interested in purchasing two items at different costs, I can spend no more than a specified amount on both items, and I want to know how many of each item I can purchase.

In Exercises 86–89, determine whether each statement is true or false. If the statement is false, make the necessary change(s) to produce a true statement.

86. The graph of $3x - 5y < 10$ consists of a dashed line and a shaded half-plane below the line.

87. The graph of $y \geq -x + 1$ consists of a solid line that rises from left to right and a shaded half-plane above the line.

88. The ordered pair $(-2, 40)$ satisfies the following system:

$$y \geq 9x + 11$$
$$13x + y > 14.$$

89. For the graph of $y < x - 3$, the points $(0, -3)$ and $(8, 5)$ lie on the graph of the corresponding linear equation, but neither point is a solution of the inequality.

90. Write a linear inequality in two variables whose graph is shown.

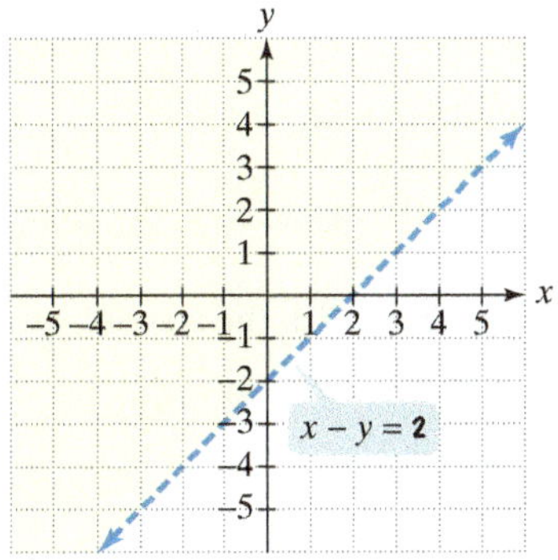

In Exercises 91–92, write a system of inequalities for each graph.

91.

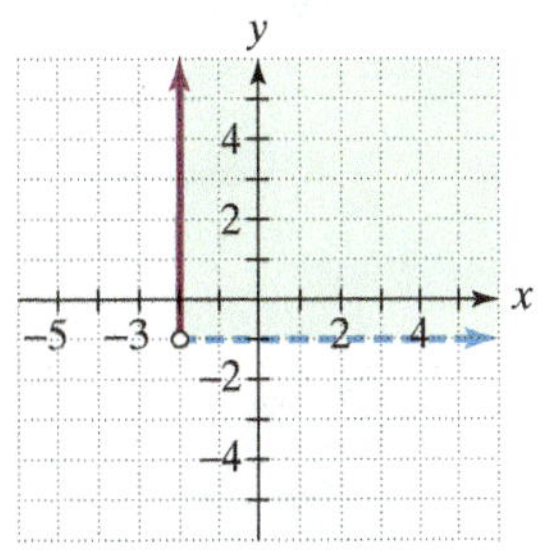

92.

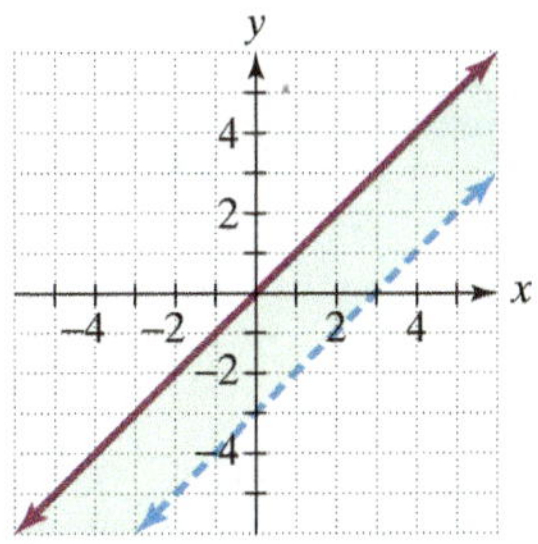

93. Write a linear inequality in two variables satisfying the following conditions: The points $(-3, -8)$ and $(4, 6)$ lie on the graph of the corresponding linear equation and each point is a solution of the inequality. The point $(1, 1)$ is also a solution.

94. Write a system of inequalities whose solution set includes every point in the rectangular coordinate system.

95. Sketch the graph of the solution set for the following system of inequalities:

$$y \geq nx + b \ (n < 0, b > 0)$$
$$y \leq mx + b \ (m > 0, b > 0).$$

Review Exercises

96. Solve using matrices:

$$3x - y = 8$$
$$x - 5y = -2.$$

(Section 3.4, Example 2)

97. Solve by graphing:

$$y = 3x - 2$$
$$y = -2x + 8.$$

(Section 3.1, Example 2)

98. Evaluate:

$$\begin{vmatrix} 8 & 2 & -1 \\ 3 & 0 & 5 \\ 6 & -3 & 4 \end{vmatrix}.$$

(Section 3.5, Example 3)

Preview Exercises

Exercises 99–101 will help you prepare for the material covered in the next section.

99. **a.** Graph the solution set of the system:

$$\begin{aligned} x + y &\geq 6 \\ x &\leq 8 \\ y &\leq 5. \end{aligned}$$

b. List the points that form the corners of the graphed region in part (a).

c. Evaluate $3x + 2y$ at each of the points obtained in part (b).

100. **a.** Graph the solution set of the system:

$$\begin{aligned} x &\geq 0 \\ y &\geq 0 \\ 3x - 2y &\leq 6 \\ y &\leq -x + 7. \end{aligned}$$

b. List the points that form the corners of the graphed region in part (a).

c. Evaluate $2x + 5y$ at each of the points obtained in part (b).

101. Bottled water and medical supplies are to be shipped to survivors of an earthquake by plane. The bottled water weighs 20 pounds per container and medical kits weigh 10 pounds per kit. Each plane can carry no more than 80,000 pounds. If x represents the number of bottles of water to be shipped per plane and y represents the number of medical kits per plane, write an inequality that models each plane's 80,000 pound weight restriction.

SECTION 4.5

Linear Programming

Objectives

1 Write an objective function modeling a quantity that must be maximized or minimized.

2 Use inequalities to model limitations in a situation.

3 Use linear programming to solve problems.

West Berlin children at Tempelhof airport watch fleets of U.S. airplanes bringing in supplies to circumvent the Soviet blockade. The airlift began June 28, 1948 and continued for 15 months.

The Berlin Airlift (1948–1949) was an operation by the United States and Great Britain in response to military action by the former Soviet Union: Soviet troops closed all roads and rail lines between West Germany and Berlin, cutting off supply routes to the city. The Allies used a mathematical technique developed during World War II to maximize the quantities of supplies transported. During the 15-month airlift, 278,228 flights provided basic necessities to blockaded Berlin, saving one of the world's great cities.

In this section, we will look at an important application of systems of linear inequalities. Such systems arise in **linear programming**, a method for solving problems in which a particular quantity that must be maximized or minimized is limited by other factors. Linear programming is one of the most widely used tools in management science. It helps businesses allocate resources to manufacture products in a way that will maxmize profit. Linear programming accounts for more than 50% and perhaps as much as 90% of all computing time used for management decisions in business. The Allies used linear programming to save Berlin.

1 Write an objective function modeling a quantity that must be maximized or minimized.

Objective Functions in Linear Programming

Many problems involve quantities that must be maximized or minimized. Businesses are interested in maximizing profit. A relief operation in which bottled water and medical kits are shipped to earthquake survivors needs to maximize the number of survivors helped by this shipment. An **objective function** is an algebraic expression in two or more variables describing a quantity that must be maximized or minimized.

EXAMPLE 1 Writing an Objective Function

Bottled water and medical supplies are to be shipped to survivors of an earthquake by plane. Each container of bottled water will serve 10 people and each medical kit will aid 6 people. If x represents the number of bottles of water to be shipped and y represents the number of medical kits, write the objective function that models the number of people that can be helped.

Solution Because each bottle of water serves 10 people and each medical kit aids 6 people, we have

The number of people helped	is	10 times the number of bottles of water	plus	6 times the number of medical kits.
	$=$	$10x$	$+$	$6y.$

Using z to represent the number of people helped, the objective function is

$$z = 10x + 6y.$$

Unlike the functions that we have seen so far, the objective function is an equation in three variables. For a value of x and a value of y, there is one and only one value of z. Thus, z is a function of x and y.

CHECK POINT 1 A company manufactures bookshelves and desks for computers. Let x represent the number of bookshelves manufactured daily and y the number of desks manufactured daily. The company's profits are \$25 per bookshelf and \$55 per desk. Write the objective function that models the company's total daily profit, z, from x bookshelves and y desks. (Check Points 2 through 4 are related to this situation, so keep track of your answers.)

2 Use inequalities to model limitations in a situation.

Constraints in Linear Programming

Ideally, the number of earthquake survivors helped in Example 1 should increase without restriction so that every survivor receives water and medical supplies. However, the planes that ship these supplies are subject to weight and volume restrictions. In linear programming problems, such restrictions are called **constraints**. Each constraint is expressed as a linear inequality. The list of constraints forms a system of linear inequalities.

EXAMPLE 2 Writing a Constraint

Each plane can carry no more than 80,000 pounds. The bottled water weighs 20 pounds per container and each medical kit weighs 10 pounds. Let x represent the number of bottles of water to be shipped and y the number of medical kits. Write an inequality that models this constraint.

Solution Because each plane can carry no more than 80,000 pounds, we have

The total weight of the water bottles	plus	the total weight of the medical kits	must be less than or equal to	80,000 pounds.
$20x$	$+$	$10y$	$\leq$	80,000.
Each bottle weighs 20 pounds.		Each kit weighs 10 pounds.		

The plane's weight constraint is modeled by the inequality

$$20x + 10y \leq 80{,}000.$$

✓ **CHECK POINT 2** To maintain high quality, the company in Check Point 1 should not manufacture more than a total of 80 bookshelves and desks per day. Write an inequality that models this constraint.

In addition to a weight constraint on its cargo, each plane has a limited amount of space in which to carry supplies. Example 3 demonstrates how to express this constraint.

EXAMPLE 3 Writing a Constraint

Each plane can carry a total volume of supplies that does not exceed 6000 cubic feet. Each water bottle is 1 cubic foot and each medical kit also has a volume of 1 cubic foot. With x still representing the number of water bottles and y the number of medical kits, write an inequality that models this second constraint.

Solution Because each plane can carry a volume of supplies that does not exceed 6000 cubic feet, we have

The total volume of the water	plus	the total volume of the medical kits	must be less than or equal to	6000 cubic feet.
$1x$	$+$	$1y$	$\leq$	$6000.$

$1x$: Each bottle is 1 cubic foot.

$1y$: Each kit is 1 cubic foot.

The plane's volume constraint is modeled by the inequality $x + y \leq 6000.$

In summary, here's what we have described so far in this aid-to-earthquake-survivors situation:

$$z = 10x + 6y$$

This is the objective function modeling the number of people helped with x bottles of water and y medical kits.

$$20x + 10y \leq 80{,}000$$
$$x + y \leq 6000.$$

These are the constraints based on each plane's weight and volume limitations.

✓ **CHECK POINT 3** To meet customer demand, the company in Check Point 1 must manufacture between 30 and 80 bookshelves per day, inclusive. Furthermore, the company must manufacture at least 10 and no more than 30 desks per day. Write an inequality that models each of these sentences. Then summarize what you have described about this company by writing the objective function for its profits and the three constraints.

3 Use linear programming to solve problems.

Solving Problems with Linear Programming

The problem in the earthquake situation described previously is to maximize the number of survivors who can be helped, subject to each plane's weight and volume constraints. The process of solving this problem is called *linear programming*, based on a theorem that was proven during World War II.

Solving a Linear Programming Problem

Let $z = ax + by$ be an objective function that depends on x and y. Furthermore, z is subject to a number of linear constraints on x and y. If a maximum or minimum value of z exists, it can be determined as follows:

1. Graph the system of inequalities representing the constraints.
2. Find the value of the objective function at each corner, or **vertex**, of the graphed region. The maximum and minimum of the objective function occur at one or more of the corner points.

EXAMPLE 4 Solving a Linear Programming Problem

Determine how many bottles of water and how many medical kits should be sent on each plane to maximize the number of earthquake survivors who can be helped.

Solution We must maximize $z = 10x + 6y$ subject to the following constraints:

$$20x + 10y \le 80{,}000$$
$$x + y \le 6000.$$

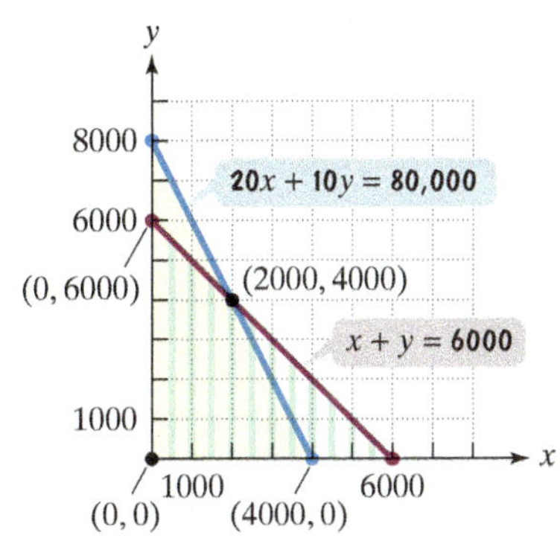

FIGURE 4.17 The region in quadrant I representing the constraints $20x + 10y \le 80{,}000$ and $x + y \le 6000$

Step 1. Graph the system of inequalities representing the constraints. Because x (the number of bottles of water per plane) and y (the number of medical kits per plane) must be nonnegative, we need to graph the system of inequalities in quadrant I and its boundary only.

To graph the inequality $20x + 10y \le 80{,}000$, we graph the equation $20x + 10y = 80{,}000$ as a solid blue line (**Figure 4.17**). Setting $y = 0$, the x-intercept is $(4000, 0)$ and setting $x = 0$, the y-intercept is $(0, 8000)$. Using $(0, 0)$ as a test point, the inequality is satisfied, so we shade below the blue line, as shown in yellow in **Figure 4.17**.

Now we graph $x + y \le 6000$ by first graphing $x + y = 6000$ as a solid red line. Setting $y = 0$, the x-intercept is $(6000, 0)$. Setting $x = 0$, the y-intercept is $(0, 6000)$. Using $(0, 0)$ as a test point, the inequality is satisfied, so we shade below the red line, as shown using green vertical shading in **Figure 4.17**.

We use the addition method to find where the lines $20x + 10y = 80{,}000$ and $x + y = 6000$ intersect.

$$\begin{array}{rcl} 20x + 10y = 80{,}000 & \xrightarrow{\text{No change}} & 20x + 10y = 80{,}000 \\ x + y = 6000 & \xrightarrow{\text{Multiply by } -10.} & -10x - 10y = -60{,}000 \\ & \text{Add:} & 10x = 20{,}000 \\ & & x = 2000 \end{array}$$

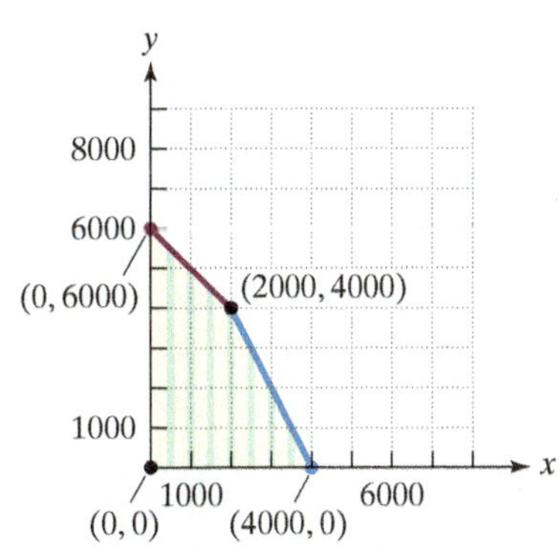

FIGURE 4.18

Back-substituting 2000 for x in $x + y = 6000$, we find $y = 4000$, so the intersection point is $(2000, 4000)$.

The system of inequalities representing the constraints is shown by the region in which the yellow shading and the green vertical shading overlap in **Figure 4.17**. The graph of the system of inequalities is shown again in **Figure 4.18**. The red and blue line segments are included in the graph.

Step 2. Find the value of the objective function at each corner of the graphed region. The maximum and minimum of the objective function occur at one or more of the corner points. We must evaluate the objective function, $z = 10x + 6y$, at the four corners, or vertices, of the region in **Figure 4.18**.

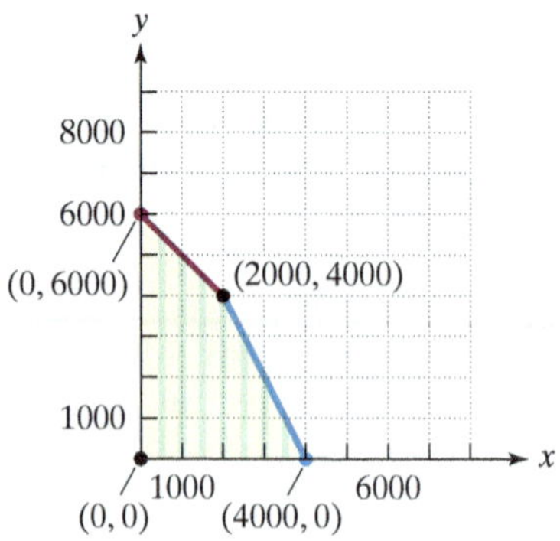

FIGURE 4.18 (repeated)

Corner (x, y)	Objective Function $z = 10x + 6y$
$(0, 0)$	$z = 10(0) + 6(0) = 0$
$(4000, 0)$	$z = 10(4000) + 6(0) = 40{,}000$
$(2000, 4000)$	$z = 10(2000) + 6(4000) = 44{,}000$ ← maximum
$(0, 6000)$	$z = 10(0) + 6(6000) = 36{,}000$

Thus, the maximum value of z is 44,000 and this occurs when $x = 2000$ and $y = 4000$. In practical terms, this means that the maximum number of earthquake survivors who can be helped with each plane shipment is 44,000. This can be accomplished by sending 2000 water bottles and 4000 medical kits per plane.

✓ CHECK POINT 4 For the company in Check Points 1–3, how many bookshelves and how many desks should be manufactured per day to obtain maximum profit? What is the maximum daily profit?

EXAMPLE 5 Solving a Linear Programming Problem

Find the maximum value of the objective function

$$z = 2x + y$$

subject to the following constraints:

$$x \geq 0, y \geq 0$$
$$x + 2y \leq 5$$
$$x - y \leq 2.$$

Solution We begin by graphing the region in quadrant I ($x \geq 0, y \geq 0$) formed by the constraints. The graph is shown in **Figure 4.19**.

Now we evaluate the objective function at the four vertices of this region.

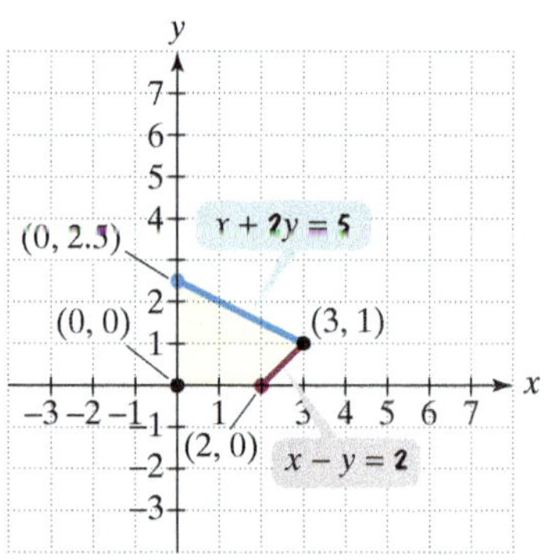

FIGURE 4.19 The graph of $x + 2y \leq 5$ and $x - y \leq 2$ in quadrant I

Objective function: $z = 2x + y$

At $(0, 0)$: $z = 2 \cdot 0 + 0 = 0$
At $(2, 0)$: $z = 2 \cdot 2 + 0 = 4$
At $(3, 1)$: $z = 2 \cdot 3 + 1 = 7$ — Maximum value of z
At $(0, 2.5)$: $z = 2 \cdot 0 + 2.5 = 2.5$

Thus, the maximum value of z is 7, and this occurs when $x = 3$ and $y = 1$.

We can see why the objective function in Example 5 has a maximum value that occurs at a vertex by solving the equation for y.

$$z = 2x + y \qquad \text{This is the objective function of Example 5.}$$
$$y = -2x + z \qquad \text{Solve for } y\text{. Recall that the slope-intercept form of the equation of a line is } y = mx + b.$$

Slope = −2; y-intercept = z (0, 2)

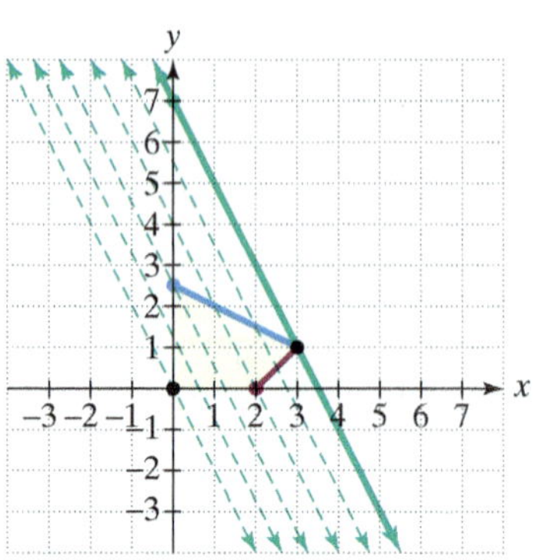

FIGURE 4.20 The line with slope −2 with the greatest y-intercept that intersects the shaded region passes through one of the vertices of the region.

In this form, z represents the y-intercept of the objective function. The equation describes infinitely many parallel lines (one for each value of z), each with slope −2. The process in linear programming involves finding the maximum z-value for all lines that intersect the region determined by the constraints. Of all the lines whose slope is −2, we're looking for the one with the greatest y-intercept that intersects the given region. As we see in **Figure 4.20**, such a line will pass through one (or possibly more) of the vertices of the region.

CHECK POINT 5 Find the maximum value of the objective function $z = 3x + 5y$ subject to the constraints $x \geq 0, y \geq 0, x + y \geq 1, x + y \leq 6$.

4.5 EXERCISE SET

Practice Exercises

In Exercises 1–4, find the value of the objective function at each corner of the graphed region. What is the maximum value of the objective function? What is the minimum value of the objective function?

1. Objective Function
$z = 5x + 6y$

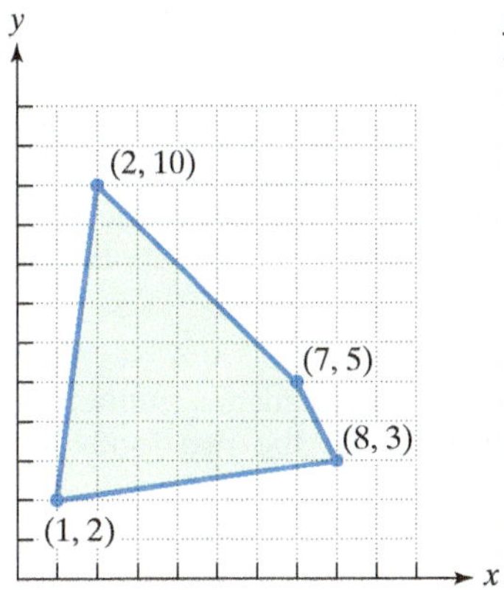

2. Objective Function
$z = 3x + 2y$

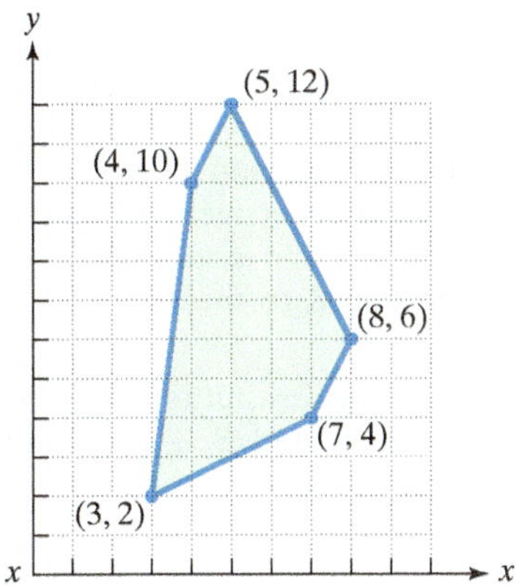

3. Objective Function
$z = 40x + 50y$

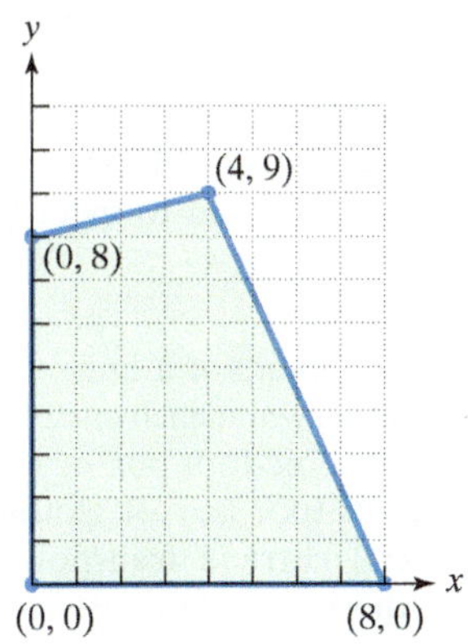

4. Objective Function
$z = 30x + 45y$

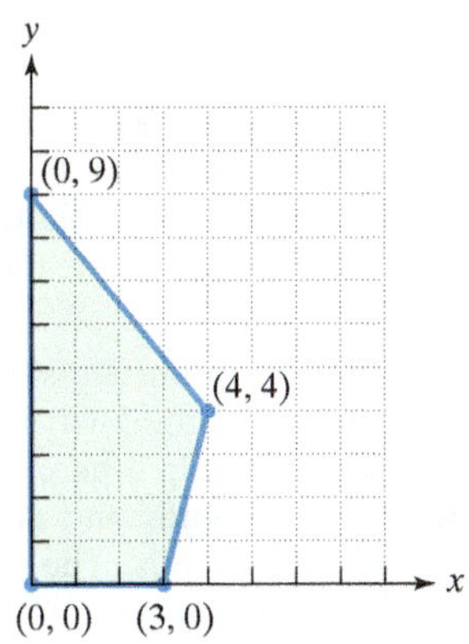

In Exercises 5–14, an objective function and a system of linear inequalities representing constraints are given.

a. *Graph the system of inequalities representing the constraints.*

b. *Find the value of the objective function at each corner of the graphed region.*

c. *Use the values in part (b) to determine the maximum value of the objective function and the values of x and y for which the maximum occurs.*

5. Objective Function $z = 3x + 2y$
Constraints $x \geq 0, y \geq 0$
$2x + y \leq 8$
$x + y \geq 4$

6. Objective Function $z = 2x + 3y$
Constraints $x \geq 0, y \geq 0$
$2x + y \leq 8$
$2x + 3y \leq 12$

7. Objective Function $z = 4x + y$
Constraints $x \geq 0, y \geq 0$
$2x + 3y \leq 12$
$x + y \geq 3$

8. Objective Function $z = x + 6y$
Constraints $x \geq 0, y \geq 0$
$2x + y \leq 10$
$x - 2y \geq -10$

9. Objective Function $z = 3x - 2y$
Constraints $1 \leq x \leq 5$
$y \geq 2$
$x - y \geq -3$

10. Objective Function $z = 5x - 2y$
Constraints $0 \leq x \leq 5$
$0 \leq y \leq 3$
$x + y \geq 2$

11. Objective Function $z = 4x + 2y$
Constraints $x \geq 0, y \geq 0$
$2x + 3y \leq 12$
$3x + 2y \leq 12$
$x + y \geq 2$

12. Objective Function $z = 2x + 4y$
Constraints $x \geq 0, y \geq 0$
$x + 3y \geq 6$
$x + y \geq 3$
$x + y \leq 9$

13. Objective Function $z = 10x + 12y$
Constraints $x \geq 0, y \geq 0$
$2x + y \leq 10$
$2x + 3y \leq 18$

14. Objective Function $z = 5x + 6y$
Constraints
$$x \geq 0, y \geq 0$$
$$2x + y \geq 10$$
$$x + 2y \geq 10$$
$$x + y \leq 10$$

Application Exercises

15. A television manufacturer makes rear-projection and plasma televisions. The profit per unit is $125 for the rear-projection televisions and $200 for the plasma televisions.
   a. Let x = the number of rear-projection televisions manufactured in a month and y = the number of plasma televisions manufactured in a month. Write the objective function that models the total monthly profit.
   b. The manufacturer is bound by the following constraints:
      - Equipment in the factory allows for making at most 450 rear-projection televisions in one month.
      - Equipment in the factory allows for making at most 200 plasma televisions in one month.
      - The cost to the manufacturer per unit is $600 for the rear-projection televisions and $900 for the plasma televisions. Total monthly costs cannot exceed $360,000.

      Write a system of three inequalities that models these constraints.
   c. Graph the system of inequalities in part (b). Use only the first quadrant and its boundary, because x and y must both be nonnegative.
   d. Evaluate the objective function for total monthly profit at each of the five vertices of the graphed region. [The vertices should occur at (0, 0), (0, 200), (300, 200), (450, 100), and (450, 0).]
   e. Complete the missing portions of this statement: The television manufacturer will make the greatest profit by manufacturing ____ rear-projection televisions each month and ____ plasma televisions each month. The maximum monthly profit is $______.

16. a. A student earns $10 per hour for tutoring and $7 per hour as a teacher's aid. Let x = the number of hours each week spent tutoring and y = the number of hours each week spent as a teacher's aid. Write the objective function that models total weekly earnings.
   b. The student is bound by the following constraints:
      - To have enough time for studies, the student can work no more than 20 hours a week.
      - The tutoring center requires that each tutor spend at least three hours a week tutoring.
      - The tutoring center requires that each tutor spend no more than eight hours a week tutoring.

      Write a system of three inequalities that models these constraints.
   c. Graph the system of inequalities in part (b). Use only the first quadrant and its boundary, because x and y are nonnegative.
   d. Evaluate the objective function for total weekly earnings at each of the four vertices of the graphed region. [The vertices should occur at (3, 0), (8, 0), (3, 17), and (8, 12).]
   e. Complete the missing portions of this statement: The student can earn the maximum amount per week by tutoring for __ hours per week and working as a teacher's aid for ___ hours per week. The maximum amount that the student can earn each week is $____.

Use the two steps for solving a linear programming problem, given in the box on page 289, to solve the problems in Exercises 17–23.

17. A manufacturer produces two models of mountain bicycles. The times (in hours) required for assembling and painting each model are given in the following table:

	Model A	Model B
Assembling	5	4
Painting	2	3

The maximum total weekly hours available in the assembly department and the paint department are 200 hours and 108 hours, respectively. The profits per unit are $25 for model A and $15 for model B. How many of each type should be produced to maximize profit?

18. A large institution is preparing lunch menus containing foods A and B. The specifications for the two foods are given in the following table:

Food	Units of Fat per Ounce	Units of Carbohydrates per Ounce	Units of Protein per Ounce
A	1	2	1
B	1	1	1

Each lunch must provide at least 6 units of fat per serving, no more than 7 units of protein, and at least 10 units of carbohydrates. The institution can purchase food A for $0.12 per ounce and food B for $0.08 per ounce. How many ounces of each food should a serving contain to meet the dietary requirement at the least cost?

19. Food and clothing are shipped to survivors of a hurricane. Each carton of food will feed 12 people, while each carton of clothing will help 5 people. Each 20-cubic-foot box of food weighs 50 pounds and each 10-cubic-foot box of clothing weighs 20 pounds. The commercial carriers transporting food and clothing are bound by the following constraints:
   - The total weight per carrier cannot exceed 19,000 pounds.
   - The total volume must be less than 8000 cubic feet.

   How many cartons of food and how many cartons of clothing should be sent with each plane shipment to maximize the number of people who can be helped?

20. On June 24, 1948, the former Soviet Union blocked all land and water routes through East Germany to Berlin. A gigantic airlift was organized using American and British planes to bring food, clothing, and other supplies to the more than 2 million people in West Berlin. The cargo capacity was 30,000 cubic feet for an American plane and 20,000

cubic feet for a British plane. To break the Soviet blockade, the Western Allies had to maximize cargo capacity, but were subject to the following restrictions:

- No more than 44 planes could be used.
- The larger American planes required 16 personnel per flight, double that of the requirement for the British planes. The total number of personnel available could not exceed 512.
- The cost of an American flight was \$9000 and the cost of a British flight was \$5000. Total weekly costs could not exceed \$300,000.

Find the number of American planes and the number of British planes that were used to maximize cargo capacity.

21. A theater is presenting a program on drinking and driving for students and their parents. The proceeds will be donated to a local alcohol information center. Admission is \$2.00 for parents and \$1.00 for students. However, the situation has two constraints: The theater can hold no more than 150 people and every two parents must bring at least one student. How many parents and students should attend to raise the maximum amount of money?

22. You are about to take a test that contains computation problems worth 6 points each and word problems worth 10 points each. You can do a computation problem in 2 minutes and a word problem in 4 minutes. You have 40 minutes to take the test and may answer no more than 12 problems. Assuming you answer all the problems attempted correctly, how many of each type of problem must you do to maximize your score? What is the maximum score?

23. In 1978, a ruling by the Civil Aeronautics Board allowed Federal Express to purchase larger aircraft. Federal Express's options included 20 Boeing 727s that United Airlines was retiring and/or the French-built Dassault Fanjet Falcon 20. To aid in their decision, executives at Federal Express analyzed the following data:

	Boeing 727	Falcon 20
Direct Operating Cost	\$1400 per hour	\$500 per hour
Payload	42,000 pounds	6000 pounds

Federal Express was faced with the following constraints:

- Hourly operating cost was limited to \$35,000.
- Total payload had to be at least 672,000 pounds.
- Only twenty 727s were available.

Given the constraints, how many of each kind of aircraft should Federal Express have purchased to maximize the number of aircraft?

Writing in Mathematics

24. What kinds of problems are solved using the linear programming method?

25. What is an objective function in a linear programming problem?

26. What is a constraint in a linear programming problem? How is a constraint represented?

27. In your own words, describe how to solve a linear programming problem.

28. Describe a situation in your life in which you would like to maximize something, but are limited by at least two constraints. Can linear programming be used in this situation? Explain your answer.

Critical Thinking Exercises

Make Sense? *In Exercises 29–32, determine whether each statement "makes sense" or "does not make sense" and explain your reasoning.*

29. In order to solve a linear programming problem, I use the graph representing the constraints and the graph of the objective function.

30. I use the coordinates of each vertex from my graph representing the constraints to find the values that maximize or minimize an objective function.

31. I need to be able to graph systems of linear inequalities in order to solve linear programming problems.

32. An important application of linear programming for businesses involves maximizing profit.

33. Suppose that you inherit \$10,000. The will states how you must invest the money. Some (or all) of the money must be invested in stocks and bonds. The requirements are that at least \$3000 be invested in bonds, with expected returns of \$0.08 per dollar, and at least \$2000 be invested in stocks, with expected returns of \$0.12 per dollar. Because the stocks are medium risk, the final stipulation requires that the investment in bonds should never be less than the investment in stocks. How should the money be invested so as to maximize your expected returns?

34. Consider the objective function $z = Ax + By$ ($A > 0$ and $B > 0$) subject to the following constraints: $2x + 3y \leq 9$, $x - y \leq 2$, $x \geq 0$, and $y \geq 0$. Prove that the objective function will have the same maximum value at the vertices $(3, 1)$ and $(0, 3)$ if $A = \frac{2}{3}B$.

Review Exercises

35. Simplify: $(2x^4y^3)(3xy^4)^3$. (Section 1.6, Example 9)

36. Solve for L: $3P = \dfrac{2L - W}{4}$. (Section 1.5, Example 8)

37. If $f(x) = x^3 + 2x^2 - 5x + 4$, find $f(-1)$. (Section 2.1, Example 3)

Preview Exercises

Exercises 38–40 will help you prepare for the material covered in the first section of the next chapter.

In Exercises 38–39, simplify each algebraic expression.

38. $(-9x^3 + 7x^2 - 5x + 3) + (13x^3 + 2x^2 - 8x - 6)$

39. $(7x^3 - 8x^2 + 9x - 6) - (2x^3 - 6x^2 - 3x + 9)$

40. The figures show the graphs of two functions.

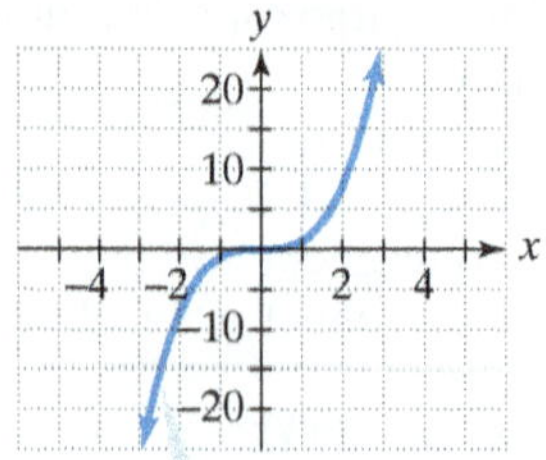

$f(x) = x^3$

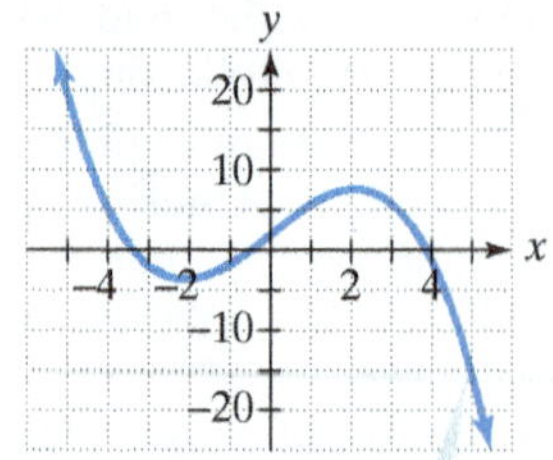

$g(x) = -0.3x^3 + 4x + 2$

a. Which function, f or g, has a graph that rises to the left and falls to the right?

b. Which function, f or g, has a graph that falls to the left and rises to the right?

GROUP PROJECT

CHAPTER 4

Each group member should research one situation that provides two different pricing options. These can involve areas such as public transportation options (with or without coupon books) or long-distance telephone plans or anything of interest. Be sure to bring in all the details for each option. At the group meeting, select the two pricing situations that are most interesting and relevant. Using each situation, write a word problem about selecting the better of the two options. The word problem should be one that can be solved using a linear inequality. The group should turn in the two problems and their solutions.

Chapter 4 Summary

Definitions and Concepts	Examples

Section 4.1 Solving Linear Inequalities

A linear inequality in one variable can be written in the form $ax + b < 0$, $ax + b \le 0$, $ax + b > 0$, or $ax + b \ge 0$. The set of all numbers that make the inequality a true statement is its solution set. Graphs of solution sets are shown on a number line by shading all points representing numbers that are solutions. Parentheses indicate endpoints that are not solutions. Square brackets indicate endpoints that are solutions.

- $(-2, 1] = \{x | -2 < x \le 1\}$

- $[-2, \infty) = \{x | x \ge -2\}$

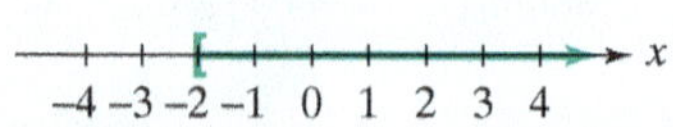

Solving a Linear Inequality

1. Simplify each side.
2. Collect variable terms on one side and constant terms on the other side.
3. Isolate the variable and solve.

If an inequality is multiplied or divided by a negative number, the direction of the inequality symbol must be reversed.

Solve: $2(x + 3) - 5x \le 15$.

$$2x + 6 - 5x \le 15$$
$$-3x + 6 \le 15$$
$$-3x \le 9$$
$$\frac{-3x}{-3} \ge \frac{9}{-3}$$
$$x \ge -3$$

Solution set: $\{x | x \ge -3\}$ or $[-3\ \infty)$

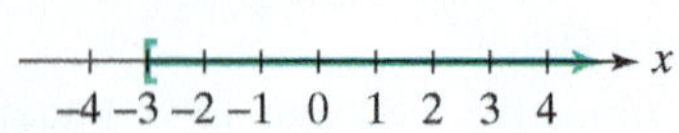

Definitions and Concepts	Examples

Section 4.2 Compound Inequalities

Intersection ($\cap$) and Union ($\cup$)

$A \cap B$ is the set of elements common to both set A and set B.
$A \cup B$ is the set of elements that are members of set A or set B or of both sets.

$\{1, 3, 5, 7\} \cap \{5, 7, 9, 11\} = \{5, 7\}$
$\{1, 3, 5, 7\} \cup \{5, 7, 9, 11\} = \{1, 3, 5, 7, 9, 11\}$

A compound inequality is formed by joining two inequalities with the word *and* or *or*.

When the connecting word is *and*, graph each inequality separately and take the intersection of their solution sets.

Solve: $x + 1 > 3$ and $x + 4 \le 8$.

$x > 2$ and $x \le 4$

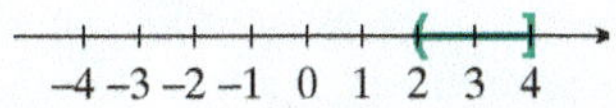

Solution set: $\{x | 2 < x \le 4\}$ or $(2, 4]$

The compound inequality $a < x < b$ means $a < x$ and $x < b$. Solve by isolating the variable in the middle.

Solve: $-1 < \dfrac{2x + 1}{3} \le 2$.

$-3 < 2x + 1 \le 6$ Multiply by 3.

$-4 < 2x \le 5$ Subtract 1.

$-2 < x \le \dfrac{5}{2}$ Divide by 2.

Solution set: $\left\{x \middle| -2 < x \le \dfrac{5}{2}\right\}$ or $\left(-2, \dfrac{5}{2}\right]$

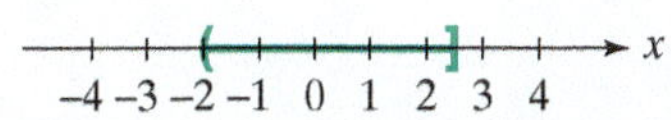

When the connecting word in a compound inequality is *or*, graph each inequality separately and take the union of their solution sets.

Solve: $x - 2 > -3$ or $2x \le -6$.

$x > -1$ or $x \le -3$

Solution set: $\{x | x \le -3 \text{ or } x > -1\}$ or $(-\infty, -3] \cup (-1, \infty)$

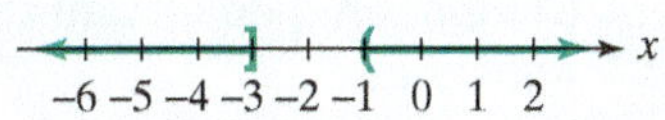

Section 4.3 Equations and Inequalities Involving Absolute Value

Absolute Value Equations

1. If $c > 0$, then $|X| = c$ means $X = c$ or $X = -c$.
2. If $c < 0$, then $|X| = c$ has no solution.
3. If $c = 0$, then $|X| = 0$ means $X = 0$.

Solve: $|2x - 7| = 3$.

$2x - 7 = 3$ or $2x - 7 = -3$

$2x = 10$ $\quad$ $2x = 4$

$x = 5$ $\quad$ $x = 2$

The solution set is $\{2, 5\}$.

Absolute Value Equations with Two Absolute Value Bars

If $|X_1| = |X_2|$, then $X_1 = X_2$ or $X_1 = -X_2$.

Solve: $|x - 6| = |2x + 1|$.

$x - 6 = 2x + 1$ or $x - 6 = -(2x + 1)$

$-x - 6 = 1$ $\quad$ $x - 6 = -2x - 1$

$-x = 7$ $\quad$ $3x - 6 = -1$

$x = -7$ $\quad$ $3x = 5$

$x = \dfrac{5}{3}$

The solutions are -7 and $\dfrac{5}{3}$, and the solution set is $\left\{-7, \dfrac{5}{3}\right\}$.

Definitions and Concepts	Examples

Section 4.3 Equations and Inequalities Involving Absolute Value (continued)

Using Boundary Points to Solve Absolute Value Inequalities

If c is a positive number, solve $|X| < c$ and $|X| > c$ using the following procedure.

1. Solve $|X| = c$. The solutions are the boundary points.
2. Locate these boundary points on a number line, thereby dividing the number line into intervals.
3. Choose a test value within each interval and substitute that number into the given inequality. If a true statement results, all numbers in the interval satisfy the inequality.
4. Write the solution set, selecting the interval(s) that satisfy the given inequality.

This procedure is valid if $<$ is replaced by $\le$ or if $>$ is replaced by $\ge$. In these cases, include the boundary points in the solution set.

Solve: $|3x + 6| > 12$.

First solve $|3x + 6| = 12$.

$$3x + 6 = 12 \quad \text{or} \quad 3x + 6 = -12$$
$$3x = 6 \qquad\qquad 3x = -18$$
$$x = 2 \qquad\qquad x = -6$$

Boundary points are -6 and 2.

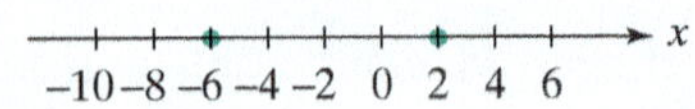

Intervals are $(-\infty, -6)$ $(-6, 2)$ $(2, \infty)$

Test value: −7 Test value: 0 Test value: 3

Test each value in $|3x + 6| > 12$.

Test -7: $|3(-7) + 6| \overset{?}{>} 12$

$15 > 12$, true

$(-\infty, -6)$ is in the solution set.

Test 0: $|3 \cdot 0 + 6| \overset{?}{>} 12$

$6 > 12$, false

$(-6, 2)$ is not in the solution set.

Test 3: $|3 \cdot 3 + 6| \overset{?}{>} 12$

$15 > 12$, true

$(2, \infty)$ is in the solution set.

The solution set is $(-\infty, -6) \cup (2, \infty)$ or $\{x|x < -6 \text{ or } x > 2\}$.

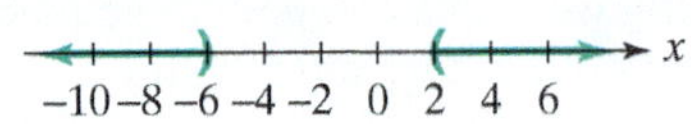

Using Equivalent Compound Inequalities to Solve Absolute Value Inequalities

If c is a positive number,

1. The solutions of $|X| < c$ are the numbers that satisfy $-c < X < c$.
2. The solutions of $|X| > c$ are the numbers that satisfy $X < -c$ or $X > c$.

In each case, the absolute value inequality is rewritten as an equivalent compound inequality without absolute value bars.

Solve: $|x - 4| < 3$.

$$-3 < x - 4 < 3$$
$$1 < x < 7 \quad \text{Add 4.}$$

The solution set is $\{x|1 < x < 7\}$ or $(1, 7)$.

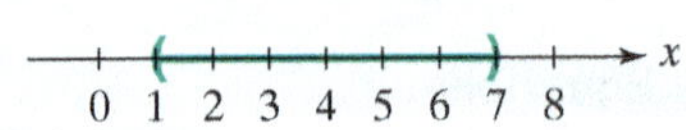

Solve: $\left|\frac{x}{3} - 1\right| \ge 2$.

$$\frac{x}{3} - 1 \le -2 \quad \text{or} \quad \frac{x}{3} - 1 \ge 2.$$
$$x - 3 \le -6 \quad \text{or} \quad x - 3 \ge 6 \quad \text{Multiply by 3.}$$
$$x \le -3 \quad \text{or} \quad x \ge 9 \quad \text{Add 3.}$$

The solution set is $\{x|x \le -3 \text{ or } x \ge 9\}$ or $(-\infty, -3] \cup [9, \infty)$.

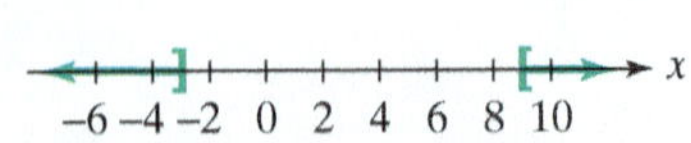

Definitions and Concepts	Examples

Section 4.3 Equations and Inequalities Involving Absolute Value (continued)

Absolute Value Inequalities with Unusual Solution Sets

If c is a negative number,

1. $|X| < c$ has no solution.
2. $|X| > c$ is true for all real numbers for which X is defined.

- $|x - 4| < -3$ has no solution. The solution set is $\varnothing$.
- $|3x + 6| > -12$ is true for all real numbers. The solution set is $(-\infty, \infty)$.

Section 4.4 Linear Inequalities in Two Variables

If the equal sign in $Ax + By = C$ is replaced with an inequality symbol, the result is a linear inequality in two variables. Its graph is the set of all points whose coordinates satisfy the inequality. To obtain the graph,

1. Replace the inequality symbol with an equal sign and graph the boundary line. Use a solid line for $\leq$ or $\geq$ and a dashed line for $<$ or $>$.
2. Choose a test point not on the line and substitute its coordinates into the inequality.
3. If a true statement results, shade the half-plane containing the test point. If a false statement results, shade the half-plane not containing the test point.

Graph: $x - 2y \leq 4$.

1. Graph $x - 2y = 4$. Use a solid line because the inequality symbol is $\leq$.
2. Test $(0, 0)$.

$$x - 2y \leq 4$$
$$0 - 2 \cdot 0 \overset{?}{\leq} 4$$
$$0 \leq 4, \quad \text{true}$$

3. The inequality is true. Shade the half-plane containing $(0, 0)$.

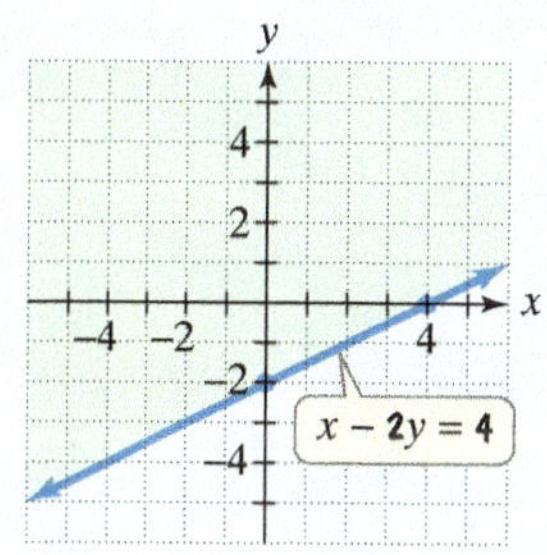

Two or more linear inequalities make up a system of linear inequalities. A solution is an ordered pair satisfying all inequalities in the system. To graph a system of inequalities, graph each inequality in the system. The overlapping region, if there is one, represents the solutions of the system. If there is no overlapping region, the system has no solution.

Graph the solutions of the system:

$$y \leq -2x$$
$$x - y \geq 3.$$

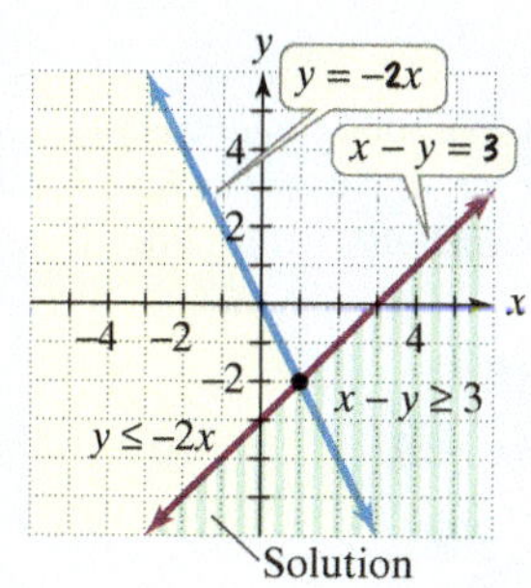

Definitions and Concepts	Examples

Section 4.5 Linear Programming

Linear programming is a method for solving problems in which a particular quantity that must be maximized or minimized is limited. An objective function is an algebraic expression in three variables modeling a quantity that must be maximized or minimized. Constraints are restrictions, expressed as linear inequalities.

Solving a Linear Programming Problem

1. Graph the system of inequalities representing the constraints.
2. Find the value of the objective function at each corner, or vertex, of the graphed region. The maximum and minimum of the objective function occur at one or more vertices.

Find the maximum value of the objective function $z = 3x + 2y$ subject to the following constraints: $x \geq 0, y \geq 0, 2x + 3y \leq 18, 2x + y \leq 10$.

1. Graph the system of inequalities representing the constraints.

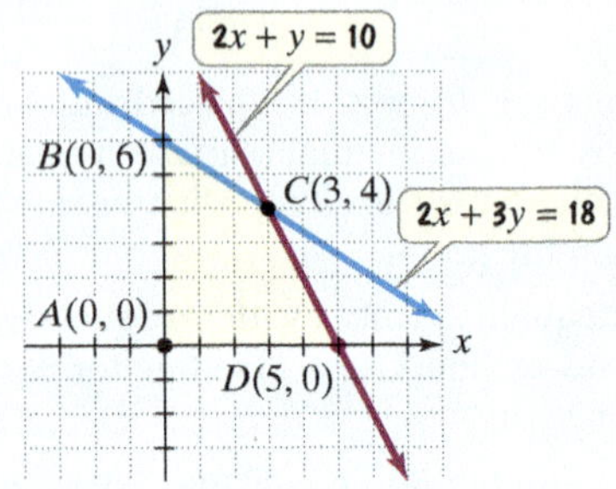

2. Evaluate the objective function at each vertex.

Vertex	$z = 3x + 2y$
$A(0,0)$	$z = 3(0) + 2(0) = 0$
$B(0,6)$	$z = 3(0) + 2(6) = 12$
$C(3,4)$	$z = 3(3) + 2(4) = 17$
$D(5,0)$	$z = 3(5) + 2(0) = 15$

The maximum value of the objective function is 17.

CHAPTER 4 REVIEW EXERCISES

4.1 *In Exercises 1–3, express each interval in set-builder notation and graph the interval on a number line.*

1. $(-2, 3]$
2. $[-1.5, 2]$
3. $(-1, \infty)$

In Exercises 4–9, solve each linear inequality. Other than ∅, graph the solution set on a number line. Express the solution set in both set-builder and interval notations.

4. $-6x + 3 \leq 15$
5. $6x - 9 \geq -4x - 3$
6. $\frac{x}{3} - \frac{3}{4} - 1 > \frac{x}{2}$
7. $6x + 5 > -2(x - 3) - 25$
8. $3(2x - 1) - 2(x - 4) \geq 7 + 2(3 + 4x)$
9. $2x + 7 \leq 5x - 6 - 3x$
10. A person can choose between two charges on a checking account. The first method involves a fixed cost of \$11 per month plus 6¢ for each check written. The second method involves a fixed cost of \$4 per month plus 20¢ for each check written. How many checks should be written to make the first method a better deal?
11. A salesperson earns \$500 per month plus a commission of 20% of sales. Describe the sales needed to receive a total income that exceeds \$3200 per month.

4.2 *In Exercises 12–15, let $A = \{a, b, c\}$, $B = \{a, c, d, e\}$, and $C = \{a, d, f, g\}$. Find the indicated set.*

12. $A \cap B$
13. $A \cap C$
14. $A \cup B$
15. $A \cup C$

In Exercises 16–26, solve each compound inequality. Except for the empty set, express the solution set in both set-builder and interval notations. Graph the solution set on a number line.

16. $x \le 3$ and $x < 6$
17. $x \le 3$ or $x < 6$
18. $-2x < -12$ and $x - 3 < 5$
19. $5x + 3 \le 18$ and $2x - 7 \le -5$
20. $2x - 5 > -1$ and $3x < 3$
21. $2x - 5 > -1$ or $3x < 3$
22. $x + 1 \le -3$ or $-4x + 3 < -5$
23. $5x - 2 \le -22$ or $-3x - 2 > 4$
24. $5x + 4 \ge -11$ or $1 - 4x \ge 9$
25. $-3 < x + 2 \le 4$
26. $-1 \le 4x + 2 \le 6$
27. To receive a B in a course, you must have an average of at least 80% but less than 90% on five exams. Your grades on the first four exams were 95%, 79%, 91%, and 86%. What range of grades on the fifth exam will result in a B for the course? Use interval notation to express this range.

4.3

In Exercises 28–31, find the solution set for each equation.

28. $|2x + 1| = 7$
29. $|3x + 2| = -5$
30. $2|x - 3| - 7 = 10$
31. $|4x - 3| = |7x + 9|$

In Exercises 32–36, solve and graph the solution set on a number line. Except for the empty set, express the solution set in both set-builder and interval notations.

32. $|2x + 3| \le 15$
33. $\left|\frac{2x + 6}{3}\right| > 2$
34. $|2x + 5| - 7 < -6$
35. $-4|x + 2| + 5 \le -7$
36. $|2x - 3| + 4 \le -10$
37. Approximately 90% of the population sleeps h hours daily, where h is modeled by the inequality $|h - 6.5| \le 1$. Write a sentence describing the range for the number of hours that most people sleep. Do *not* use the phrase "absolute value" in your description.

4.4

In Exercises 38–43, graph each inequality in a rectangular coordinate system.

38. $3x - 4y > 12$
39. $x - 3y \le 6$
40. $y \le -\frac{1}{2}x + 2$
41. $y > \frac{3}{5}x$
42. $x \le 2$
43. $y > -3$

In Exercises 44–52, graph the solution set of each system of inequalities or indicate that the system has no solution.

44. $2x - y \le 4$, $x + y \ge 5$
45. $y < -x + 4$, $y > x - 4$
46. $-3 \le x < 5$
47. $-2 < y \le 6$
48. $x \ge 3$, $y \le 0$
49. $2x - y > -4$, $x \ge 0$
50. $x + y \le 6$, $y \ge 2x - 3$
51. $3x + 2y \ge 4$, $x - y \le 3$, $x \ge 0, y \ge 0$
52. $2x - y > 2$, $2x - y < -2$

4.5

53. Find the value of the objective function $z = 2x + 3y$ at each corner of the graphed region shown. What is the maximum value of the objective function? What is the minimum value of the objective function?

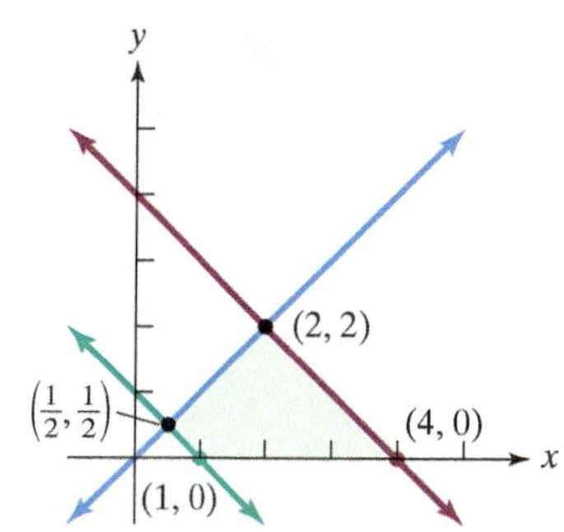

In Exercises 54–56, graph the region determined by the constraints. Then find the maximum value of the given objective function, subject to the constraints.

54. Objective Function $z = 2x + 3y$
Constraints $x \ge 0, y \ge 0$; $x + y \le 8$; $3x + 2y \ge 6$
55. Objective Function $z = x + 4y$
Constraints $0 \le x \le 5, 0 \le y \le 7$; $x + y \ge 3$
56. Objective Function $z = 5x + 6y$
Constraints $x \ge 0, y \ge 0$; $y \le x$; $2x + y \le 12$; $2x + 3y \ge 6$
57. A paper manufacturing company converts wood pulp to writing paper and newsprint. The profit on a unit of writing paper is \$500 and the profit on a unit of newsprint is \$350.
    a. Let x represent the number of units of writing paper produced daily. Let y represent the number of units of newsprint produced daily. Write the objective function that models total daily profit.

b. The manufacturer is bound by the following constraints:

- Equipment in the factory allows for making at most 200 units of paper (writing paper and newsprint) in a day.
- Regular customers require at least 10 units of writing paper and at least 80 units of newsprint daily.

Write a system of inequalities that models these constraints.

c. Graph the inequalities in part (b). Use only the first quadrant, because x and y must both be positive. (*Suggestion*: Let each unit along the x- and y-axes represent 20.)

d. Evaluate the objective function at each of the three vertices of the graphed region.

e. Complete the missing portions of this statement: The company will make the greatest profit by producing ____ units of writing paper and ___ units of newsprint each day. The maximum daily profit is $______.

58. A manufacturer of lightweight tents makes two models whose specifications are given in the following table.

	Cutting Time per Tent	Assembly Time per Tent
Model A	0.9 hour	0.8 hour
Model B	1.8 hours	1.2 hours

Each month, the manufacturer has no more than 864 hours of labor available in the cutting department and at most 672 hours in the assembly division. The profits come to $25 per tent for model A and $40 per tent for model B. How many of each should be manufactured monthly to maximize the profit?

CHAPTER 4 TEST

Remember to use your Chapter Test Prep Video CD to see the worked-out solutions to the test questions you want to review.

In Exercises 1–2, express each interval in set-builder notation and graph the interval on a number line.

1. $[-3, 2)$

2. $(-\infty, -1]$

In Exercises 3–4, solve and graph the solution set on a number line. Express the solution set in both set-builder and interval notations.

3. $3(x + 4) \geq 5x - 12$

4. $\frac{x}{6} + \frac{1}{8} \leq \frac{x}{2} - \frac{3}{4}$

5. You are choosing between two telephone plans for local calls. Plan A charges $25 per month for unlimited calls. Plan B has a monthly fee of $13 with a charge of $0.06 per local call. How many local telephone calls in a month make plan A the better deal?

6. Find the intersection: $\{2, 4, 6, 8, 10\} \cap \{4, 6, 12, 14\}$.

7. Find the union: $\{2, 4, 6, 8, 10\} \cup \{4, 6, 12, 14\}$.

In Exercises 8–12, solve each compound inequality. Except for the empty set, express the solution set in both set-builder and interval notations. Graph the solution set on a number line.

8. $2x + 4 < 2$ and $x - 3 > -5$

9. $x + 6 \geq 4$ and $2x + 3 \geq -2$

10. $2x - 3 < 5$ or $3x - 6 \leq 4$

11. $x + 3 \leq -1$ or $-4x + 3 < -5$

12. $-3 \leq \frac{2x + 5}{3} < 6$

In Exercises 13–14, find the solution set for each equation.

13. $|5x + 3| = 7$

14. $|6x + 1| = |4x + 15|$

In Exercises 15–16, solve and graph the solution set on a number line. Express the solution set in both set-builder and interval notations.

15. $|2x - 1| < 7$

16. $|2x - 3| \geq 5$

17. The inequality $|b - 98.6| > 8$ describes a person's body temperature, b, in degrees Fahrenheit, when hyperthermia (extremely high body temperature) or hypothermia (extremely low body temperature) occurs. Solve the inequality and interpret the solution.

In Exercises 18–20, graph each inequality in a rectangular coordinate system.

18. $3x - 2y < 6$ **19.** $y \geq \frac{1}{2}x - 1$ **20.** $y \leq -1$

In Exercises 21–23, graph the solution set of each system of inequalities.

21. $x + y \geq 2$
$x - y \geq 4$

22. $3x + y \leq 9$
$2x + 3y \geq 6$
$x \geq 0, y \geq 0$

23. $-2 < x \leq 4$

24. Find the maximum value of the objective function $z = 3x + 5y$ subject to the following constraints: $x \geq 0$, $y \geq 0$, $x + y \leq 6$, $x \geq 2$.

25. A manufacturer makes two types of jet skis, regular and deluxe. The profit on a regular jet ski is \$200 and the profit on the deluxe model is \$250. To meet customer demand, the company must manufacture at least 50 regular jet skis per week and at least 75 deluxe models. To maintain high quality, the total number of both models of jet skis manufactured by the company should not exceed 150 per week. How many jet skis of each type should be manufactured per week to obtain maximum profit? What is the maximum weekly profit?

CUMULATIVE REVIEW EXERCISES (CHAPTERS 1–4)

In Exercises 1–2, solve each equation.

1. $5(x + 1) + 2 = x - 3(2x + 1)$

2. $\dfrac{2(x + 6)}{3} = 1 + \dfrac{4x - 7}{3}$

3. Simplify: $\dfrac{-10x^2y^4}{15x^7y^{-3}}$.

4. If $f(x) = x^2 - 3x + 4$, find $f(-3)$ and $f(2a)$.

5. If $f(x) = 3x^2 - 4x + 1$ and $g(x) = x^2 - 5x - 1$, find $(f - g)(x)$ and $(f - g)(2)$.

6. Use function notation to write the equation of the line passing through $(2, 3)$ and perpendicular to the line whose equation is $y = 2x - 3$.

In Exercises 7–10, graph each equation or inequality in a rectangular coordinate system.

7. $f(x) = 2x + 1$

8. $y > 2x$

9. $2x - y \geq 6$

10. $f(x) = -1$

11. Solve the system:

$$\begin{aligned} 3x - y + z &= -15 \\ x + 2y - z &= 1 \\ 2x + 3y - 2z &= 0 \end{aligned}$$

12. Solve using matrices:

$$\begin{aligned} 2x - y &= -4 \\ x + 3y &= 5. \end{aligned}$$

13. Evaluate: $\begin{vmatrix} 4 & 3 \\ -1 & -5 \end{vmatrix}$.

14. A motel with 60 rooms charges \$90 per night for rooms with kitchen facilities and \$80 per night for rooms without kitchen facilities. When all rooms are occupied, the nightly revenue is \$5260. How many rooms of each kind are there?

15. Which of the following are functions?

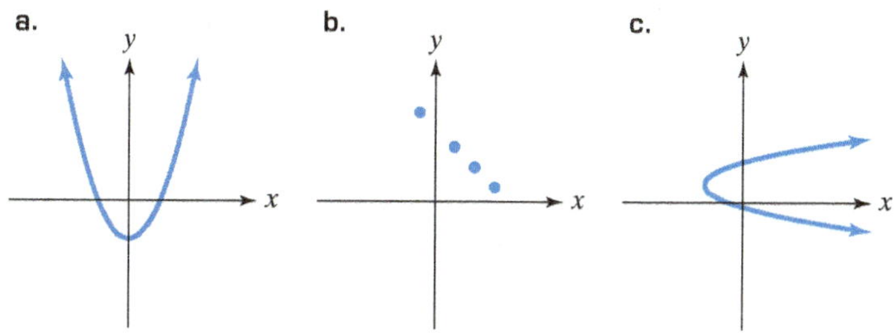

In Exercises 16–20, solve and graph the solution set on a number line. Express the solution set in both set-builder and interval notations.

16. $\dfrac{x}{4} - \dfrac{3}{4} - 1 \leq \dfrac{x}{2}$

17. $2x + 5 \leq 11$ and $-3x > 18$

18. $x - 4 \geq 1$ or $-3x + 1 \geq -5 - x$

19. $|2x + 3| \leq 17$

20. $|3x - 8| > 7$

CHAPTER 5

Polynomials, Polynomial Functions, and Factoring

New carry-on restrictions are wreaking havoc at the airport. You were made paranoid by news reports of lost, damaged, delayed, and pilfered checked luggage, so you overpacked a carry-on bag. The good news is that everything in your bag is permitted under the new restrictions. The bad news is that your bag is too large to carry on. The airline informed you that the sum of the length, width, and depth of a piece of carry-on luggage cannot exceed 40 inches. This bit of unexpected bad news, propelled by the airport's overall chaos, is enough to throw you into a real hissy fit. However, here's something that might help you get a grip on your emotions, if not on your luggage: your carry-on debacle has been modeled by a function, with an accompanying graph no less, for your reading pleasure on the plane. Have a good flight!

A function that models this scenario, $V(x) = -2x^3 + 10x^2 + 300x$, is developed, with the promised graph, as Exercise 119 in Exercise Set 5.2. The algebraic expression on the right side contains variables to powers that are whole numbers and is an example of a polynomial. Much of what we do in algebra involves polynomials and polynomial functions.

SECTION 5.1

Introduction to Polynomials and Polynomial Functions

Objectives

1. Use the vocabulary of polynomials.
2. Evaluate polynomial functions.
3. Determine end behavior.
4. Add polynomials.
5. Subtract polynomials.

In 1980, U.S. doctors diagnosed 41 cases of a rare form of cancer, Kaposi's sarcoma, that involved skin lesions, pneumonia, and severe immunological deficiencies. All cases involved gay men ranging in age from 26 to 51. By the end of 2003, approximately 930,000 Americans, straight and gay, male and female, old and young, were infected with the HIV virus.

Modeling AIDS-related data and making predictions about the epidemic's havoc is serious business. **Figure 5.1** shows the number of AIDS cases diagnosed in the United States from 1983 through 2003.

AIDS Cases Diagnosed in the United States, 1983–2003

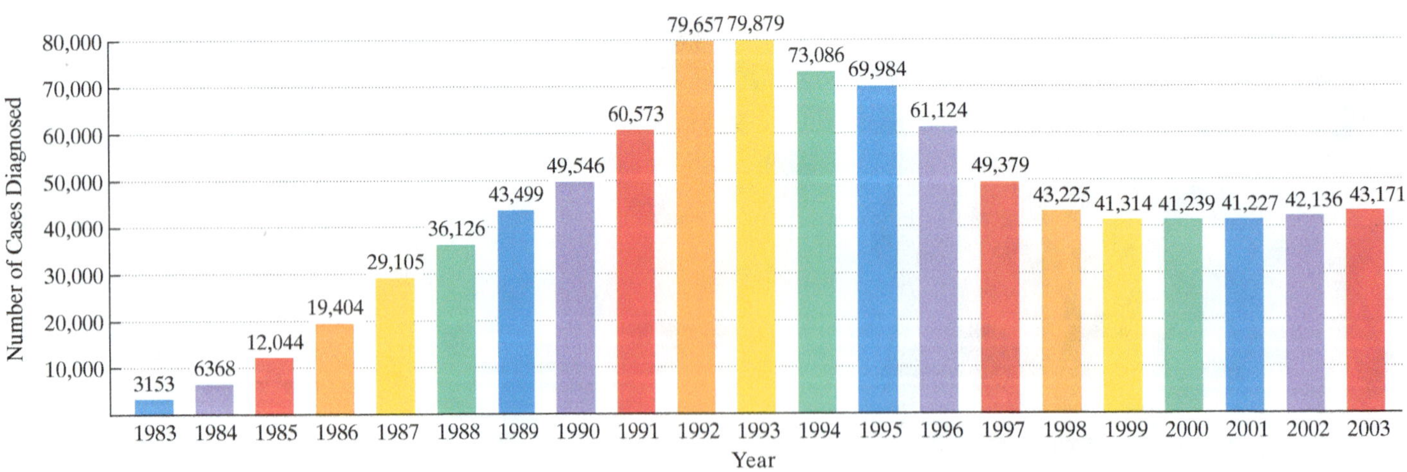

FIGURE 5.1

Source: Department of Health and Human Services

Changing circumstances and unforeseen events can result in models for AIDS-related data that are not particularly useful over long periods of time. For example, the function

$$f(x) = -49x^3 + 806x^2 + 3776x + 2503$$

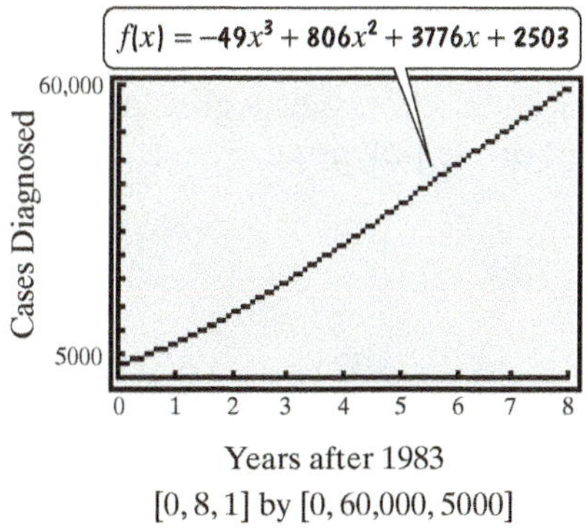

FIGURE 5.2 The graph of a function modeling the number of AIDS cases from 1983 through 1991

models the number of AIDS cases diagnosed in the United States x years after 1983. The model was obtained using a portion of the data shown in **Figure 5.1**, namely cases diagnosed from 1983 through 1991, inclusive. **Figure 5.2** shows the graph of f from 1983 through 1991.

The voice balloon in **Figure 5.2** displays the algebraic expression used to define f. This algebraic expression is an example of a *polynomial*. A **polynomial** is a single term or the sum of two or more terms containing variables with whole-number exponents. Functions containing polynomials are used in such diverse areas as science, business, medicine, psychology, and sociology. In this section, we present basic ideas about polynomials and polynomial functions. We will use our knowledge of combining like terms to find sums and differences of polynomials.

How We Describe Polynomials

Consider the polynomial

$$-49x^3 + 806x^2 + 3776x + 2503.$$

1 Use the vocabulary of polynomials.

This polynomial contains four terms. It is customary to write the terms in the order of descending powers of the variable. This is the **standard form** of a polynomial.

Some polynomials contain only one variable. Each term of such a polynomial in x is of the form ax^n. If $a \neq 0$, the **degree** of ax^n is n. For example, the degree of the term $-49x^3$ is 3.

The Degree of ax^n

If $a \neq 0$ and n is a whole number, the degree of ax^n is n. The degree of a nonzero constant is 0. The constant 0 has no defined degree.

Study Tip

We can express 0 in many ways, including $0x$, $0x^2$, and $0x^3$. It is impossible to assign a single exponent on the variable. This is why 0 has no defined degree.

Here is the polynomial modeling AIDS cases and the degree of each of its four terms:

$$-49x^3 + 806x^2 + 3776x + 2503$$

degree 3 | degree 2 | degree 1 | degree of nonzero constant: 0

Notice that the exponent on x for the term $3776x$ is understood to be 1: $3776x^1$. For this reason, the degree of $3776x$ is 1. You can think of 2503 as $2503x^0$; thus, its degree is 0.

A polynomial is simplified when it contains no grouping symbols and no like terms. A simplified polynomial that has exactly one term is called a **monomial**. A **binomial** is a simplified polynomial that has two terms. A **trinomial** is a simplified polynomial with three terms. Simplified polynomials with four or more terms have no special names.

Some polynomials contain two or more variables. Here is an example of a polynomial in two variables, x and y:

$$7x^2y^3 - 17x^4y^2 + xy - 6y^2 + 9.$$

A polynomial in two variables, x and y, contains the sum of one or more monomials of the form ax^ny^m. The constant a is the **coefficient**. The exponents, n and m, represent whole numbers. The **degree of the term** ax^ny^m is the sum of the exponents of the variables, $n + m$.

The **degree of a polynomial** is the greatest degree of any term of the polynomial. If there is precisely one term of the greatest degree, it is called the **leading term**. Its coefficient is called the **leading coefficient**.

EXAMPLE 1 Using the Vocabulary of Polynomials

Determine the coefficient of each term, the degree of each term, the degree of the polynomial, the leading term, and the leading coefficient of the polynomial

$$7x^2y^3 - 17x^4y^2 + xy - 6y^2 + 9.$$

Solution

Term	Coefficient	Degree (Sum of Exponents on the Variables)
$7x^2y^3$	7	$2 + 3 = 5$
$-17x^4y^2$	-17	$4 + 2 = 6$
xy	1	$1 + 1 = 2$
$-6y^2$	-6	$0 + 2 = 2$
9	9	$0 + 0 = 0$

Think of xy as $1x^1y^1$.

Think of $-6y^2$ as $-6x^0y^2$.

Think of 9 as $9x^0y^0$.

The degree of the polynomial is the greatest degree of any term of the polynomial, which is 6. The leading term is the term of the greatest degree, which is $-17x^4y^2$. Its coefficient, -17, is the leading coefficient.

CHECK POINT 1 Determine the coefficient of each term, the degree of each term, the degree of the polynomial, the leading term, and the leading coefficient of the polynomial

$$8x^4y^5 - 7x^3y^2 - x^2y - 5x + 11.$$

If a polynomial contains three or more variables, the degree of a term is the sum of the exponents of all the variables. Here is an example of a polynomial in three variables, x, y, and z:

The coefficients are $\frac{1}{4}$, -2, 6 and 5.

$$\frac{1}{4}xy^2z^4 - 2xyz + 6x^2 + 5.$$

Degree: $1 + 2 + 4 = 7$ | Degree: $1 + 1 + 1 = 3$ | Degree: 2 | Degree: 0

The degree of this polynomial is the greatest degree of any term of the polynomial, which is 7.

2 Evaluate polynomial functions.

Polynomial Functions

The expression $4x^3 - 5x^2 + 3$ is a polynomial. If we write

$$f(x) = 4x^3 - 5x^2 + 3,$$

then we have a **polynomial function**. In a polynomial function, the expression that defines the function is a polynomial. How do we evaluate a polynomial function? Use substitution, just as we did to evaluate other functions in Chapter 2.

EXAMPLE 2 Evaluating a Polynomial Function

The polynomial function

$$f(x) = -49x^3 + 806x^2 + 3776x + 2503$$

models the number of AIDS cases diagnosed in the United States, $f(x)$, x years after 1983, where $0 \leq x \leq 8$. Find $f(6)$ and describe what this means in practical terms.

Solution To find $f(6)$, or f of 6, we replace each occurrence of x in the function's formula with 6.

$f(x) = -49x^3 + 806x^2 + 3776x + 2503$ This is the given function.

$f(6) = -49(6)^3 + 806(6)^2 + 3776(6) + 2503$ Replace each occurrence of x with 6.

$= -49(216) + 806(36) + 3776(6) + 2503$ Evaluate exponential expressions.

$= -10{,}584 + 29{,}016 + 22{,}656 + 2503$ Multiply.

$= 43{,}591$ Add.

Thus, $f(6) = 43{,}591$. According to the model, this means that 6 years after 1983, in 1989, there were 43,591 AIDS cases diagnosed in the United States. (The actual number, shown in **Figure 5.1**, is 43,499.)

Using Technology

Once each occurrence of x in $f(x) = -49x^3 + 806x^2 + 3776x + 2503$ is replaced with 6, the resulting computation can be performed using a scientific calculator or a graphing calculator.

$$-49(6)^3 + 806(6)^2 + 3776(6) + 2503$$

Many Scientific Calculators

49 [+/−] [×] 6 [y^x] 3 [+] 806 [×] 6 [y^x] 2 [+] 3776 [×] 6 [+] 2503 [=]

Many Graphing Calculators

[(−)] 49 [×] 6 [^] 3 [+] 806 [×] 6 [^] 2 [+] 3776 [×] 6 [+] 2503 [ENTER]

The display should be 43591. This number can also be obtained by using a graphing utility's feature that evaluates a function or by using its table feature.

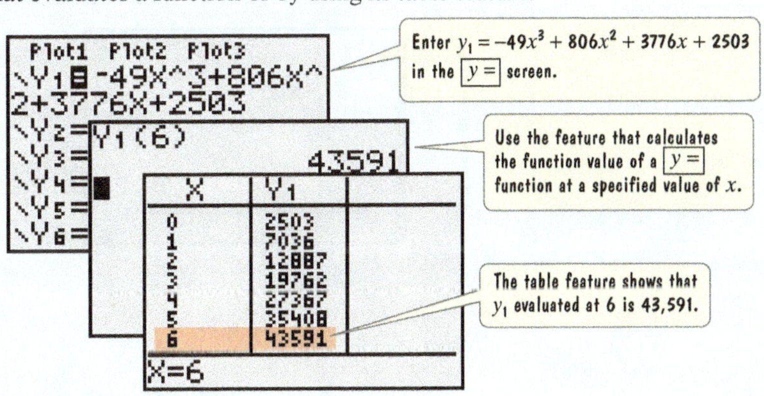

CHECK POINT 2 For the polynomial function

$$f(x) = 4x^3 - 3x^2 - 5x + 6,$$

find $f(2)$.

Smooth, Continuous Graphs

Polynomial functions of degree 2 or higher have graphs that are *smooth* and *continuous*. By **smooth**, we mean that the graph contains only rounded curves with no sharp corners. By **continuous**, we mean that the graph has no breaks and can be drawn without lifting your pencil from the rectangular coordinate system. These ideas are illustrated in **Figure 5.3**.

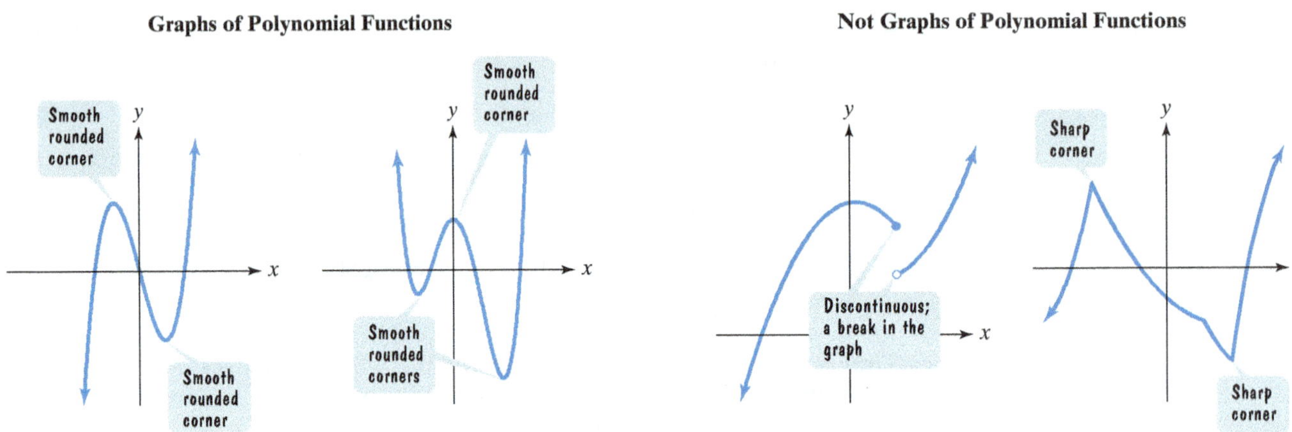

FIGURE 5.3 Recognizing graphs of polynomial functions

3 Determine end behavior.

End Behavior of Polynomial Functions

Figure 5.4 shows the graph of the function

$$f(x) = -49x^3 + 806x^2 + 3776x + 2503,$$

which models U.S. AIDS cases from 1983 through 1991. Look what happens to the graph when we extend the year up through 2005. By year 21 (2004), the values of y are negative and the function no longer models AIDS cases. We've added an arrow to the graph at the far right to emphasize that it continues to decrease without bound. This far-right *end behavior* of the graph is one reason that this function is inappropriate for modeling AIDS cases into the future.

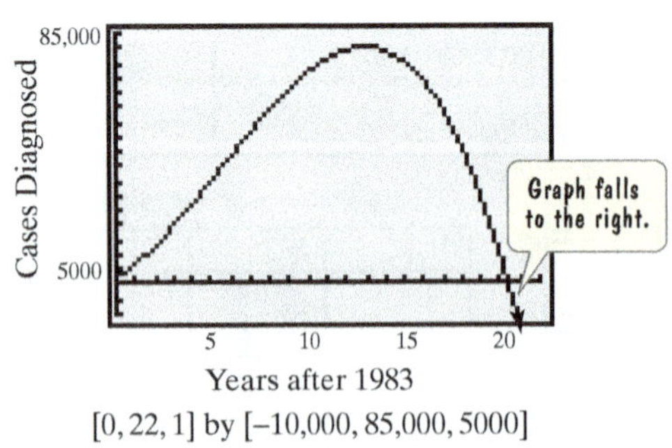

FIGURE 5.4 By extending the viewing rectangle, we see that y is eventually negative and the function no longer models the number of AIDS cases.

The behavior of the graph of a function to the far left or the far right is called its **end behavior**. Although the graph of a polynomial function may have intervals where it increases and intervals where it decreases, the graph will eventually rise or fall without bound as it moves far to the left or far to the right.

How can you determine whether the graph of a polynomial function goes up or down at each end? **The end behavior depends upon the leading term.** In particular, the sign of the leading coefficient and the degree of the polynomial reveal the graph's end behavior. With regard to end behavior, only the leading term—that is, the term of the greatest degree—counts, as summarized by the **Leading Coefficient Test**.

The Leading Coefficient Test

As x increases or decreases without bound, the graph of a polynomial function eventually rises or falls. In particular,

1. For odd-degree polynomials:

If the leading coefficient is positive, the graph falls to the left and rises to the right. (↙, ↗)

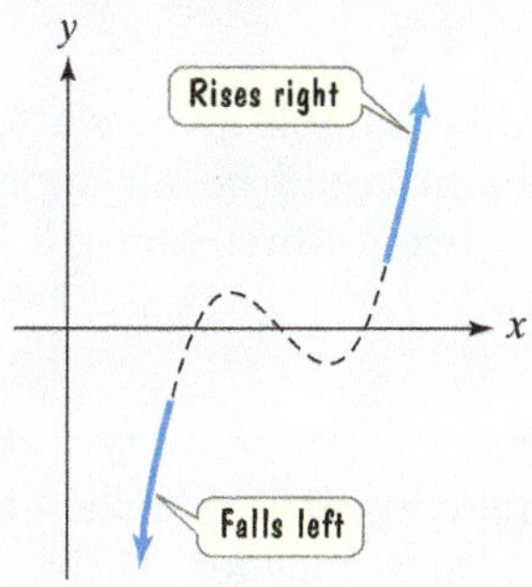

If the leading coefficient is negative, the graph rises to the left and falls to the right. (↖, ↘)

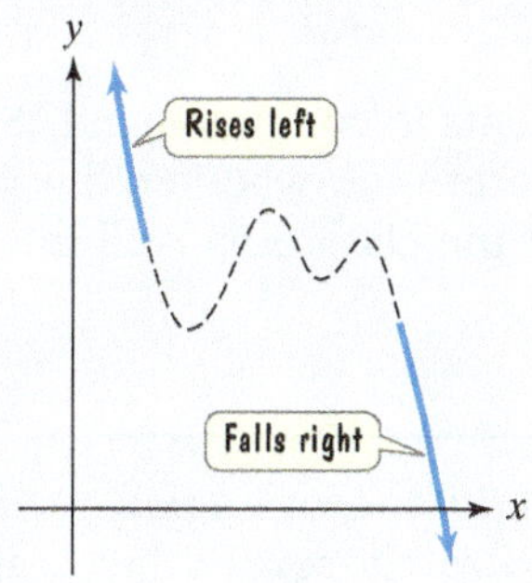

2. For even-degree polynomials:

If the leading coefficient is positive, the graph rises to the left and rises to the right. (↖, ↗)

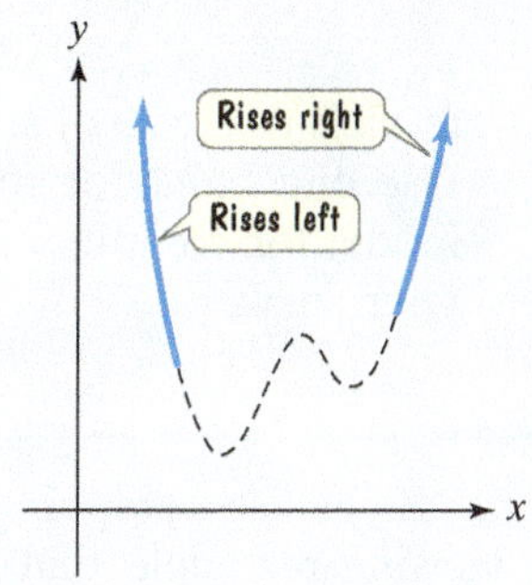

If the leading coefficient is negative, the graph falls to the left and falls to the right. (↙, ↘)

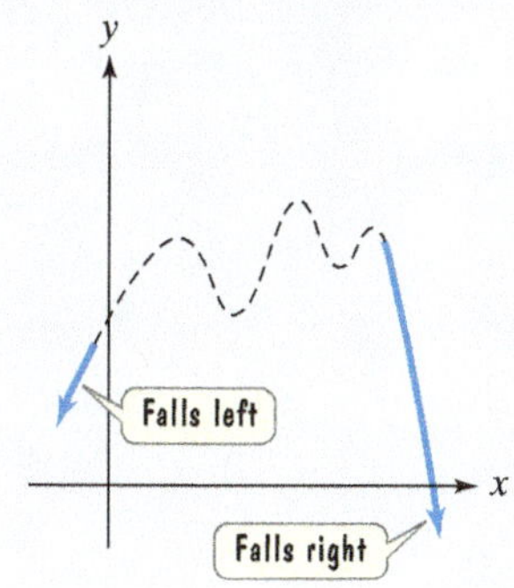

Study Tip

Odd-degree polynomial functions have graphs with opposite behavior at each end. Even-degree polynomial functions have graphs with the same behavior at each end.

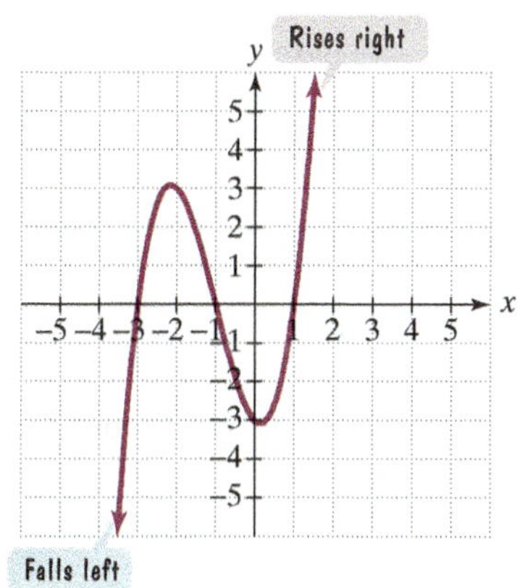

FIGURE 5.5 The graph of $f(x) = x^3 + 3x^2 - x - 3$

EXAMPLE 3 Using the Leading Coefficient Test

Use the Leading Coefficient Test to determine the end behavior of the graph of

$$f(x) = x^3 + 3x^2 - x - 3.$$

Solution We begin by identifying the sign of the leading coefficient and the degree of the polynomial.

$$f(x) = x^3 + 3x^2 - x - 3$$

The leading coefficient, 1, is positive. The degree of the polynomial, 3, is odd.

The degree of the function f is 3, which is odd. Odd-degree polynomial functions have graphs with opposite behavior at each end. The leading coefficient, 1, is positive. Thus, the graph falls to the left and rises to the right (↙, ↗). The graph of f is shown in **Figure 5.5**.

✓ CHECK POINT 3 Use the Leading Coefficient Test to determine the end behavior of the graph of $f(x) = x^4 - 4x^2$.

EXAMPLE 4 Using the Leading Coefficient Test

Use end behavior to explain why

$$f(x) = -49x^3 + 806x^2 + 3776x + 2503$$

is only an appropriate model for AIDS cases for a limited time period.

Solution We begin by identifying the sign of the leading coefficient and the degree of the polynomial.

$$f(x) = -49x^3 + 806x^2 + 3776x + 2503$$

The leading coefficient, −49, is negative. The degree of the polynomial, 3, is odd.

The degree of f in $f(x) = -49x^3 + 806x^2 + 3776x + 2053$ is 3, which is odd. Odd-degree polynomial functions have graphs with opposite behavior at each end. The leading coefficient, -49, is negative. Thus, the graph rises to the left and falls to the right (↖, ↘). The fact that the graph falls to the right indicates that at some point the number of AIDS cases will be negative, an impossibility. If a function has a graph that decreases without bound over time, it will not be capable of modeling nonnegative phenomena over long time periods. Model breakdown will eventually occur.

✓ CHECK POINT 4 The polynomial function

$$f(x) = -0.27x^3 + 9.2x^2 - 102.9x + 400$$

models the ratio of students to computers in U.S. public schools x years after 1980. Use end behavior to determine whether this function could be an appropriate model for computers in the classroom well into the twenty-first century. Explain your answer.

If you use a graphing utility to graph a polynomial function, it is important to select a viewing rectangle that accurately reveals the graph's end behavior. If the viewing rectangle, or window, is too small, it may not accurately show a complete graph with the appropriate end behavior.

EXAMPLE 5 Using the Leading Coefficient Test

The graph of $f(x) = -x^4 + 8x^3 + 4x^2 + 2$ was obtained with a graphing utility using a $[-8, 8, 1]$ by $[-10, 10, 1]$ viewing rectangle. The graph is shown in **Figure 5.6**. Is this a complete graph that shows the end behavior of the function?

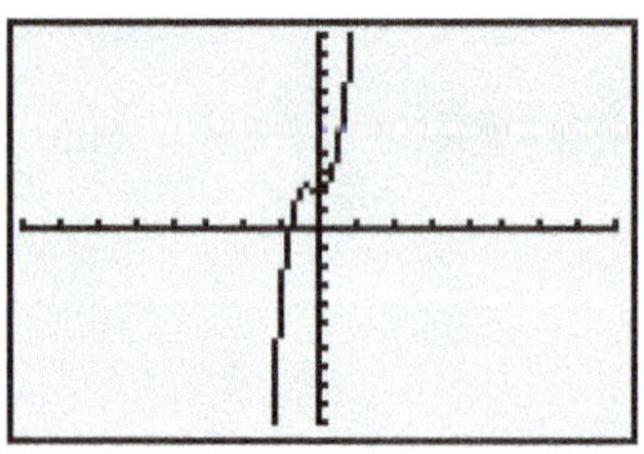

$[-8, 8, 1]$ by $[-10, 10, 1]$

FIGURE 5.6

Solution We begin by identifying the sign of the leading coefficient and the degree of the polynomial.

$$f(x) = -x^4 + 8x^3 + 4x^2 + 2$$

The leading coefficient, -1, is negative.

The degree of the polynomial, 4, is even.

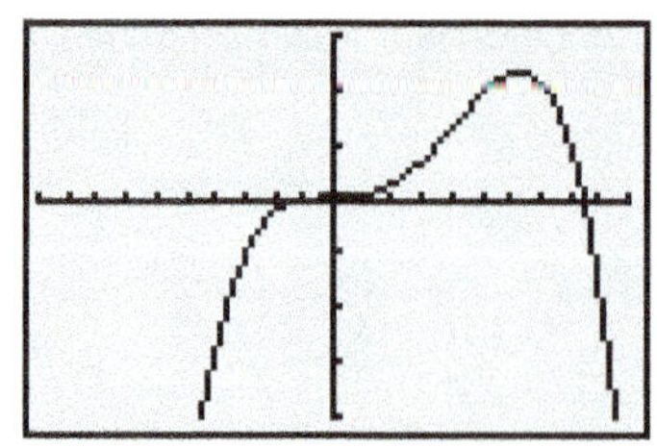

$[-10, 10, 1]$ by $[-1000, 750, 250]$

FIGURE 5.7

The degree of f is 4, which is even. Even-degree polynomial functions have graphs with the same behavior at each end. The leading coefficient, -1, is negative. Thus, the graph should fall to the left and fall to the right (↙, ↘). The graph in **Figure 5.6** is falling to the left, but it is not falling to the right. Therefore, the graph is not complete enough to show end behavior. A more complete graph of the function is shown in a larger viewing rectangle in **Figure 5.7**.

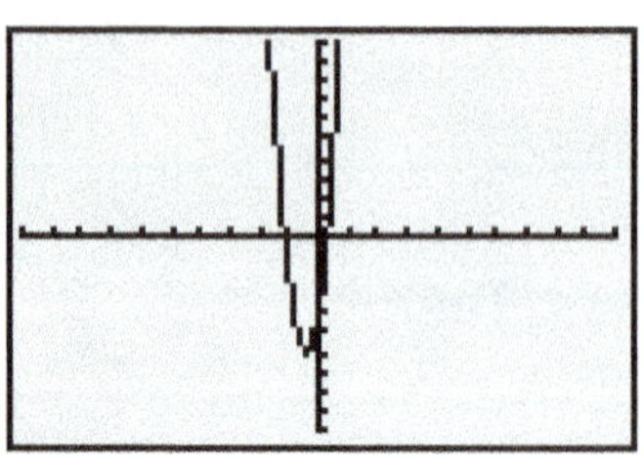

FIGURE 5.8

✓ CHECK POINT 5 The graph of $f(x) = x^3 + 13x^2 + 10x - 4$ is shown in a standard viewing rectangle in **Figure 5.8**. Use the Leading Coefficient Test to determine whether this is a complete graph that shows the end behavior of the function. Explain your answer.

---

4 Add polynomials.

Adding Polynomials

Polynomials are added by combining like terms. Here are two examples that illustrate the use of the distributive property in adding monomials and combining like terms:

$$-9x^3 + 13x^3 = (-9 + 13)x^3 = 4x^3$$

Add coefficients and keep the same variable factor(s).

$$-7x^3y^2 + 4x^3y^2 = (-7 + 4)x^3y^2 = -3x^3y^2.$$

EXAMPLE 6 Adding Polynomials

Add: $(-6x^3 + 5x^2 + 4) + (2x^3 + 7x^2 - 10)$.

Solution

$$(-6x^3 + 5x^2 + 4) + (2x^3 + 7x^2 - 10)$$

$$= -6x^3 + 5x^2 + 4 + 2x^3 + 7x^2 - 10$$ Remove the parentheses. Like terms are shown in the same color.

$$= \underbrace{-6x^3 + 2x^3} + \underbrace{5x^2 + 7x^2} + \underbrace{4 - 10}$$ Rearrange the terms so that like terms are adjacent.

$$= -4x^3 \quad + 12x^2 \quad - 6$$ Combine like terms.

$$= -4x^3 + 12x^2 - 6$$ This is the same sum as above, written more concisely.

CHECK POINT 6 Add: $(-7x^3 + 4x^2 + 3) + (4x^3 + 6x^2 - 13)$.

EXAMPLE 7 Adding Polynomials

Add: $(5x^3y - 4x^2y - 7y) + (2x^3y + 6x^2y - 4y - 5)$.

Solution

$$(5x^3y - 4x^2y - 7y) + (2x^3y + 6x^2y - 4y - 5)$$ The given problem involves adding polynomials in two variables.

$$= 5x^3y - 4x^2y - 7y + 2x^3y + 6x^2y - 4y - 5$$ Remove the parentheses. Like terms are shown in the same color.

$$= \underbrace{5x^3y + 2x^3y} - \underbrace{4x^2y + 6x^2y} - \underbrace{7y - 4y} \underbrace{- 5}$$ Rearrange the terms so that like terms are adjacent.

$$= 7x^3y \quad + 2x^2y \quad - 11y \quad - 5$$ Combine like terms.

Polynomials can be added by arranging like terms in columns. Then combine like terms, column by column. Here's the solution to Example 7 using columns and a vertical format:

$$\begin{array}{r} 5x^3y - 4x^2y - 7y \phantom{{}-5} \\ 2x^3y + 6x^2y - 4y - 5 \\ \hline 7x^3y + 2x^2y - 11y - 5 \end{array}$$

CHECK POINT 7 Add: $(7xy^3 - 5xy^2 - 3y) + (2xy^3 + 8xy^2 - 12y - 9)$.

5 Subtract polynomials.

Subtracting Polynomials

We subtract real numbers by adding the opposite, or additive inverse, of the number being subtracted. For example,

$$8 - 3 = 8 + (-3) = 5.$$

Similarly, we subtract one polynomial from another by adding the opposite of the polynomial being subtracted.

Subtracting Polynomials

To subtract two polynomials, change the sign of every term of the second polynomial. Add this result to the first polynomial.

Study Tip

You can also subtract polynomials using a vertical format. Here's the solution to Example 8 using a vertical format. Notice that you still distribute the negative sign, thereby adding the opposite.

$$\begin{array}{r} 7x^3 - 8x^2 + 9x - 6 \\ -(2x^3 - 6x^2 - 3x + 9) \\ \hline 7x^3 - 8x^2 + 9x - 6 \\ + -2x^3 + 6x^2 + 3x - 9 \\ \hline 5x^3 - 2x^2 + 12x - 15 \end{array}$$

EXAMPLE 8 Subtracting Polynomials

Subtract: $(7x^3 - 8x^2 + 9x - 6) - (2x^3 - 6x^2 - 3x + 9)$.

Solution

$(7x^3 - 8x^2 + 9x - 6) - (2x^3 - 6x^2 - 3x + 9)$

$= (7x^3 - 8x^2 + 9x - 6) + (-2x^3 + 6x^2 + 3x - 9)$ Change the sign of each term of the second polynomial and add the two polynomials. Like terms are shown in the same color.

$= 7x^3 - 2x^3 - 8x^2 + 6x^2 + 9x + 3x - 6 - 9$ Rearrange terms.

$= 5x^3 - 2x^2 + 12x - 15$ Combine like terms.

CHECK POINT 8 Subtract: $(14x^3 - 5x^2 + x - 9) - (4x^3 - 3x^2 - 7x + 1)$.

Study Tip

Be careful of the order in Example 9. For example, subtracting 2 from 5 is equivalent to 5 − 2. In general, subtracting B from A means $A - B$. The order of the resulting algebraic expression is not the same as the order in English.

EXAMPLE 9 Subtracting Polynomials

Subtract $-2x^5y^2 - 3x^3y + 7$ from $3x^5y^2 - 4x^3y - 3$.

Solution

$(3x^5y^2 - 4x^3y - 3) - (-2x^5y^2 - 3x^3y + 7)$

$= 3x^5y^2 - 4x^3y - 3 + 2x^5y^2 + 3x^3y - 7$ Change subtraction to addition and change the sign of every term of the second polynomial. Like terms are shown in the same color.

$= 3x^5y^2 + 2x^5y^2 - 4x^3y + 3x^3y - 3 - 7$ Rearrange terms.

$= 5x^5y^2 - x^3y - 10$ Combine like terms.

CHECK POINT 9 Subtract $-7x^2y^5 - 4xy^3 + 2$ from $6x^2y^5 - 2xy^3 - 8$.

5.1 EXERCISE SET

MyMathLab MathXL PRACTICE WATCH DOWNLOAD READ REVIEW

Practice Exercises

In Exercises 1–10, determine the coefficient of each term, the degree of each term, the degree of the polynomial, the leading term, and the leading coefficient of the polynomial.

1. $-x^4 + x^2$

2. $x^3 - 4x^2$

3. $5x^3 + 7x^2 - x + 9$

4. $11x^3 - 6x^2 + x + 3$

5. $3x^2 - 7x^4 - x + 6$

6. $2x^2 - 9x^4 - x + 5$

7. $x^3y^2 - 5x^2y^7 + 6y^2 - 3$

8. $12x^4y - 5x^3y^7 - x^2 + 4$

9. $x^5 + 3x^2y^4 + 7xy + 9x - 2$

10. $3x^6 + 4x^4y^4 - x^3y + 4x^2 - 5$

In Exercises 11–20, let

$$f(x) = x^2 - 5x + 6 \quad \text{and} \quad g(x) = 2x^3 - x^2 + 4x - 1.$$

Find the indicated function values.

11. $f(3)$ 12. $f(4)$ 13. $f(-1)$
14. $f(-2)$ 15. $g(3)$ 16. $g(2)$
17. $g(-2)$ 18. $g(-3)$ 19. $g(0)$
20. $f(0)$

In Exercises 21–24, identify which graphs are not those of polynomial functions.

21.

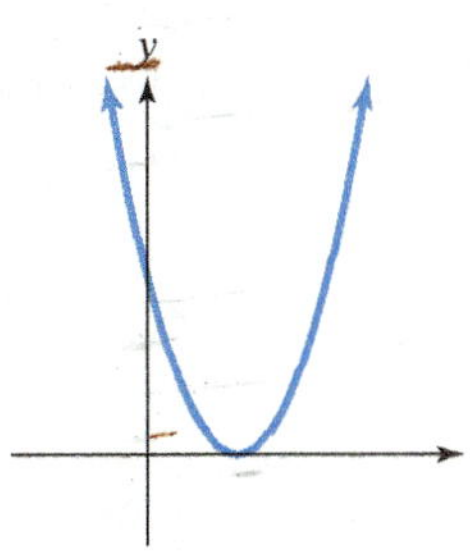

22.

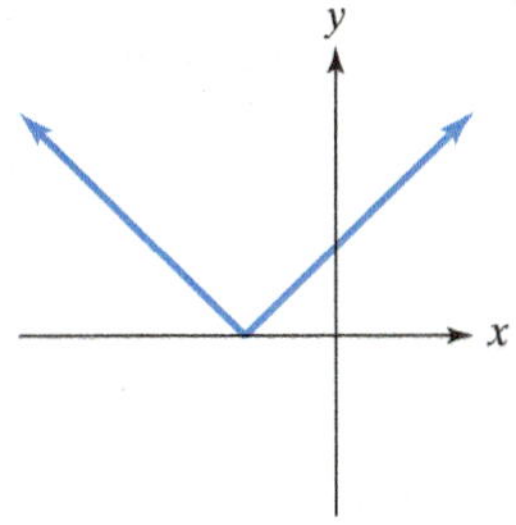

23.

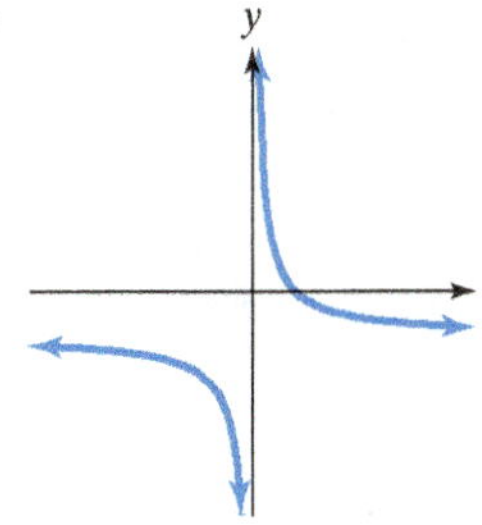

24.

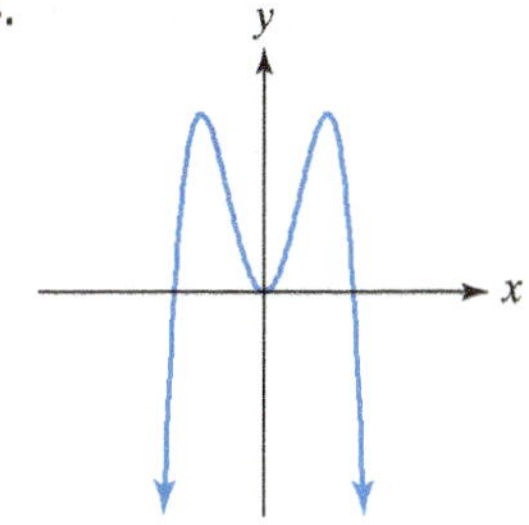

In Exercises 25–28, use the Leading Coefficient Test to determine the end behavior of the graph of the given polynomial function. Then use this end behavior to match the polynomial function with its graph. [The graphs are labeled (a) through (d).]

25. $f(x) = -x^4 + x^2$

26. $f(x) = x^3 - 4x^2$

27. $f(x) = x^2 - 6x + 9$

28. $f(x) = -x^3 - x^2 + 5x - 3$

a.

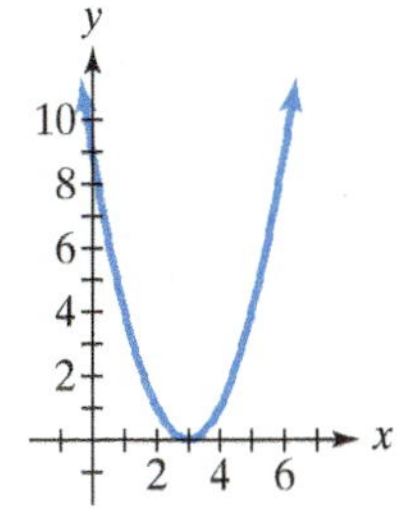

b.

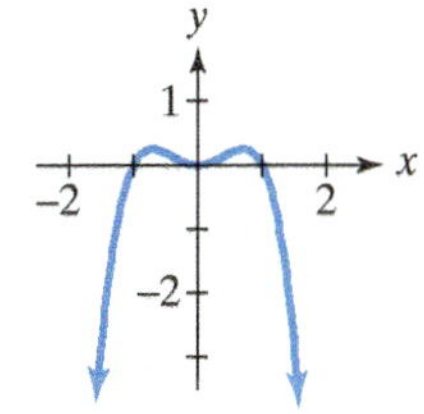

c.

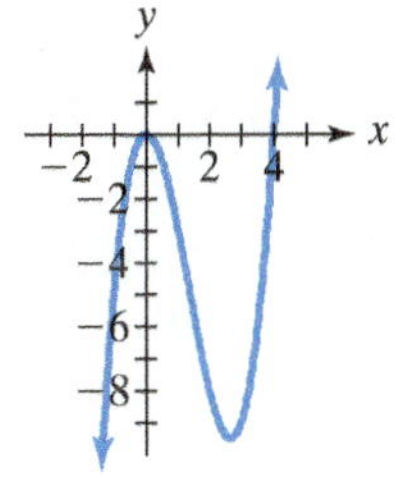

d.

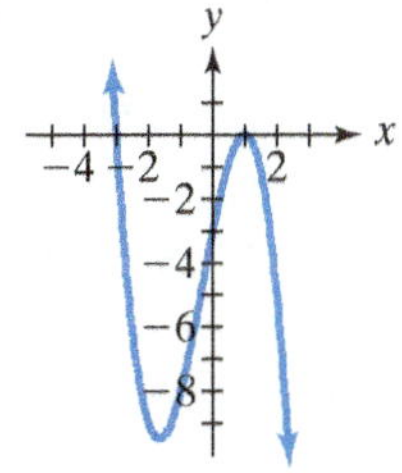

In Exercises 29–40, add the polynomials. Assume that all variable exponents represent whole numbers.

29. $(-6x^3 + 5x^2 - 8x + 9) + (17x^3 + 2x^2 - 4x - 13)$

30. $(-7x^3 + 6x^2 - 11x + 13) + (19x^3 - 11x^2 + 7x - 17)$

31. $\left(\frac{2}{5}x^4 + \frac{2}{3}x^3 + \frac{5}{8}x^2 + 7\right) + \left(-\frac{4}{5}x^4 + \frac{1}{3}x^3 - \frac{1}{4}x^2 - 7\right)$

32. $\left(\frac{1}{5}x^4 + \frac{1}{3}x^3 + \frac{3}{8}x^2 + 6\right) + \left(-\frac{3}{5}x^4 + \frac{2}{3}x^3 - \frac{1}{2}x^2 - 6\right)$

33. $(7x^2y - 5xy) + (2x^2y - xy)$

34. $(-4x^2y + xy) + (7x^2y + 8xy)$

35. $(5x^2y + 9xy + 12) + (-3x^2y + 6xy + 3)$

36. $(8x^2y + 12xy + 14) + (-2x^2y + 7xy + 4)$

37. $(9x^4y^2 - 6x^2y^2 + 3xy) + (-18x^4y^2 - 5x^2y - xy)$

38. $(10x^4y^2 - 3x^2y^2 + 2xy) + (-16x^4y^2 - 4x^2y - xy)$

39. $(x^{2n} + 5x^n - 8) + (4x^{2n} - 7x^n + 2)$

40. $(6y^{2n} + y^n + 5) + (3y^{2n} - 4y^n - 15)$

In Exercises 41–50, subtract the polynomials. Assume that all variable exponents represent whole numbers.

41. $(17x^3 - 5x^2 + 4x - 3) - (5x^3 - 9x^2 - 8x + 11)$

42. $(18x^3 - 2x^2 - 7x + 8) - (9x^3 - 6x^2 - 5x + 7)$

43. $(13y^5 + 9y^4 - 5y^2 + 3y + 6) - (-9y^5 - 7y^3 + 8y^2 + 11)$

44. $(12y^5 + 7y^4 - 3y^2 + 6y + 7) - (-10y^5 - 8y^3 + 3y^2 + 14)$

45. $(x^3 + 7xy - 5y^2) - (6x^3 - xy + 4y^2)$

46. $(x^4 - 7xy - 5y^3) - (6x^4 - 3xy + 4y^3)$

47. $(3x^4y^2 + 5x^3y - 3y) - (2x^4y^2 - 3x^3y - 4y + 6x)$

48. $(5x^4y^2 + 6x^3y - 7y) - (3x^4y^2 - 5x^3y - 6y + 8x)$

49. $(7y^{2n} + y^n - 4) - (6y^{2n} - y^n - 1)$

50. $(8x^{2n} + x^n - 4) - (9x^{2n} - x^n - 2)$

51. Subtract $-5a^2b^4 - 8ab^2 - ab$ from $3a^2b^4 - 5ab^2 + 7ab$.

52. Subtract $-7a^2b^4 - 8ab^2 - ab$ from $13a^2b^4 - 17ab^2 + ab$.

53. Subtract $-4x^3 - x^2y + xy^2 + 3y^3$ from $x^3 + 2x^2y - y^3$.

54. Subtract $-6x^3 + x^2y - xy^2 + 2y^3$ from $x^3 + 2xy^2 - y^3$.

Practice PLUS

55. Add $6x^4 - 5x^3 + 2x$ to the difference between $4x^3 + 3x^2 - 1$ and $x^4 - 2x^2 + 7x - 3$.

56. Add $5x^4 - 2x^3 + 7x$ to the difference between $2x^3 + 5x^2 - 3$ and $-x^4 - x^2 - x - 1$.

57. Subtract $9x^2y^2 - 3x^2 - 5$ from the sum of $-6x^2y^2 - x^2 - 1$ and $5x^2y^2 + 2x^2 - 1$.

58. Subtract $6x^2y^3 - 2x^2 - 7$ from the sum of $-5x^2y^3 + 3x^2 - 4$ and $4x^2y^3 - 2x^2 - 6$.

In Exercises 59–64, let

$$f(x) = -3x^3 - 2x^2 - x + 4$$
$$g(x) = x^3 - x^2 - 5x - 4$$
$$h(x) = -2x^3 + 5x^2 - 4x + 1.$$

Find the indicated function, function value, or polynomial.

59. $(f - g)(x)$ and $(f - g)(-1)$

60. $(g - h)(x)$ and $(g - h)(-1)$

61. $(f + g - h)(x)$ and $(f + g - h)(-2)$

62. $(g + h - f)(x)$ and $(g + h - f)(-2)$

63. $2f(x) - 3g(x)$

64. $-2g(x) - 3h(x)$

Application Exercises

The polynomial function

$$f(x) = -1844x^2 + 54{,}923x + 111{,}568$$

models the cumulative number of deaths from AIDS in the United States, $f(x)$, x years after 1990. Use this function to solve Exercises 65–66.

65. Find and interpret $f(10)$.

66. Find and interpret $f(8)$.

The graph of the polynomial function in Exercises 65–66 is shown in the figure. Use the graph to solve Exercises 67–68.

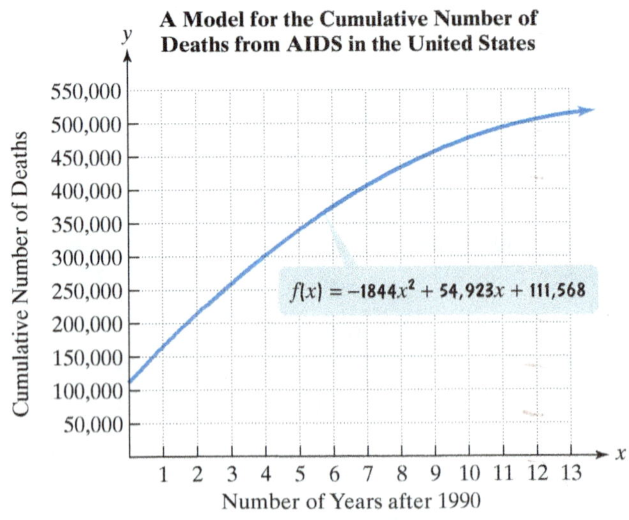

67. Identify your answer from Exercise 65 as a point on the graph.

68. Identify your answer from Exercise 66 as a point on the graph.

The bar graph shows the actual data for the cumulative number of deaths from AIDS in the United States from 1990 through 2003.

Cumulative Number of Deaths from AIDS in the United States

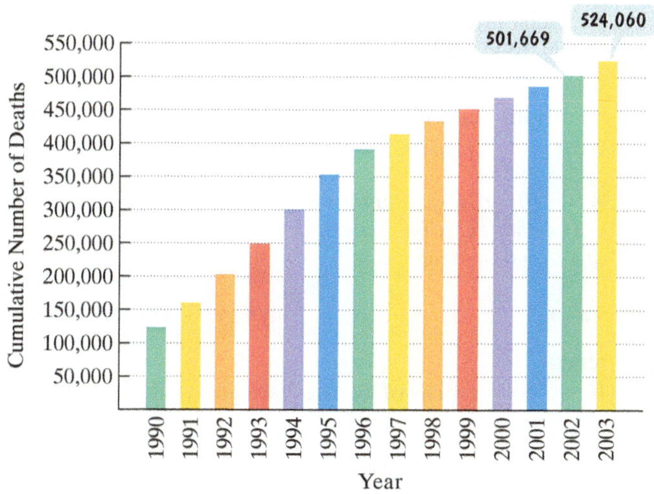

Source: Centers for Disease Control

The data in the bar graph can be modeled by the following second- and third-degree polynomial functions:

Cumulative number of AIDS deaths x years after 1990

$$f(x) = -1844x^2 + 54{,}923x + 111{,}568$$
$$g(x) = 11x^3 - 2066x^2 + 56{,}036x + 110{,}590$$

Use these functions to solve Exercises 69–71.

69. Use both functions to find the cumulative number of AIDS deaths in 2003. Which function provides a better description for the actual number shown in the bar graph?

70. Use both functions to find the cumulative number of AIDS deaths in 2002. Which function provides a better description for the actual number shown in the bar graph?

71. Use the Leading Coefficient Test to determine the end behavior to the right for the graph of f. Will this function be useful in modeling the cumulative number of AIDS deaths over an extended period of time? Explain your answer.

72. The common cold is caused by a rhinovirus. After x days of invasion by the viral particles, the number of particles in our bodies, $f(x)$, in billions, can be modeled by the polynomial function

$$f(x) = -0.75x^4 + 3x^3 + 5.$$

Use the Leading Coefficient Test to determine the graph's end behavior to the right. What does this mean about the number of viral particles in our bodies over time?

73. The polynomial function

$$f(x) = -0.87x^3 + 0.35x^2 + 81.62x + 7684.94$$

models the number of thefts, $f(x)$, in thousands, in the United States x years after 1987. Will this function be useful in modeling the number of thefts over an extended period of time? Explain your answer.

74. A herd of 100 elk is introduced to a small island. The number of elk, $f(x)$, after x years is modeled by the polynomial function

$$f(x) = -x^4 + 21x^2 + 100.$$

Use the Leading Coefficient Test to determine the graph's end behavior to the right. What does this mean about what will eventually happen to the elk population?

Writing in Mathematics

75. What is a polynomial?

76. Explain how to determine the degree of each term of a polynomial.

77. Explain how to determine the degree of a polynomial.

78. Explain how to determine the leading coefficient of a polynomial.

79. What is a polynomial function?

80. What do we mean when we describe the graph of a polynomial function as smooth and continuous?

81. What is meant by the end behavior of a polynomial function?

82. Explain how to use the Leading Coefficient Test to determine the end behavior of a polynomial function.

83. Why is a polynomial function of degree 3 with a negative leading coefficient not appropriate for modeling nonnegative real-world phenomena over a long period of time?

84. Explain how to add polynomials.

85. Explain how to subtract polynomials.

86. In a favorable habitat and without natural predators, a population of reindeer is introduced to an island preserve. The reindeer population t years after their introduction is modeled by the polynomial function

$$f(t) = -0.125t^5 + 3.125t^4 + 4000.$$

Discuss the growth and decline of the reindeer population. Describe the factors that might contribute to this population model.

Technology Exercises

Write a polynomial function that imitates the end behavior of each graph in Exercises 87–90. The dashed portions of the graphs indicate that you should focus only on imitating the left and right end behavior of the graph. You can be flexible about what occurs between the left and right ends. Then use your graphing utility to graph the polynomial function and verify that you imitated the end behavior shown in the given graph.

87.

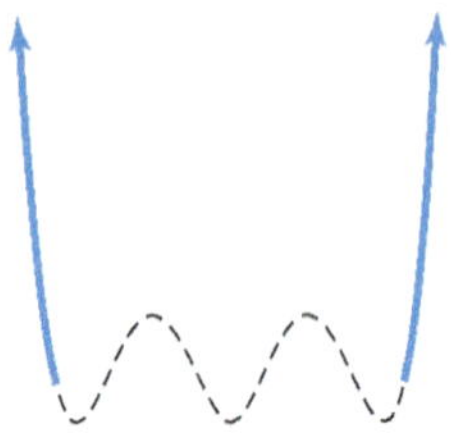

88.

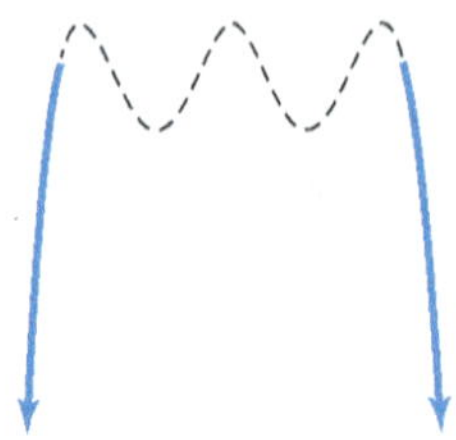

89.

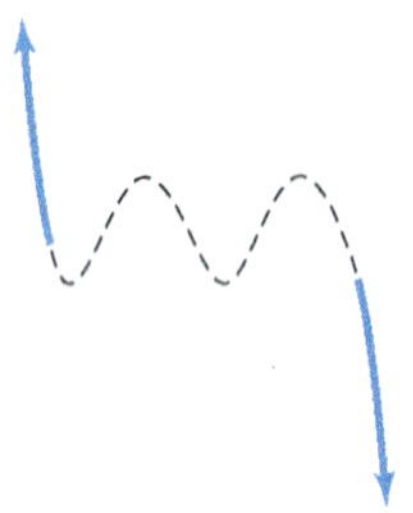

90.

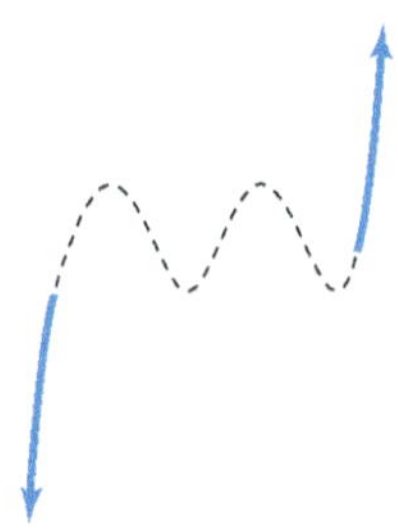

In Exercises 91–94, use a graphing utility with a viewing rectangle large enough to show end behavior to graph each polynomial function.

91. $f(x) = x^3 + 13x^2 + 10x - 4$

92. $f(x) = -2x^3 + 6x^2 + 3x - 1$

93. $f(x) = -x^4 + 8x^3 + 4x^2 + 2$

94. $f(x) = -x^5 + 5x^4 - 6x^3 + 2x + 20$

In Exercises 95–96, use a graphing utility to graph f and g in the same viewing rectangle. Then use the ZOOM OUT *feature to show that f and g have identical end behavior.*

95. $f(x) = x^3 - 6x + 1, \quad g(x) = x^3$

96. $f(x) = -x^4 + 2x^3 - 6x, \quad g(x) = -x^4$

Critical Thinking Exercises

Make Sense? *In Exercises 97–100, determine whether each statement "makes sense" or "does not make sense" and explain your reasoning.*

97. Many English words have prefixes with meanings similar to those used to describe polynomials, such as *monologue*, *binocular*, and *tricuspid*.

98. I can determine a polynomial's leading coefficient by inspecting the coefficient of the first term.

99. When I'm trying to determine end behavior, it's the coefficient of the first term of a polynomial function written in standard form that I should inspect.

100. When I rearrange the terms of a polynomial, it's important that I move the sign in front of a term with that term.

In Exercises 101–104, determine whether each statement is true or false. If the statement is false, make the necessary change(s) to produce a true statement.

101. $4x^3 + 7x^2 - 5x + \frac{2}{x}$ is a polynomial containing four terms.

102. If two polynomials of degree 2 are added, the sum must be a polynomial of degree 2.

103. $(x^2 - 7x) - (x^2 - 4x) = -11x$ for all values of x.

104. All terms of a polynomial are monomials.

In Exercises 105–106, perform the indicated operations. Assume that exponents represent whole numbers.

105. $(x^{2n} - 3x^n + 5) + (4x^{2n} - 3x^n - 4) - (2x^{2n} - 5x^n - 3)$

106. $(y^{3n} - 7y^{2n} + 3) - (-3y^{3n} - 2y^{2n} - 1) + (6y^{3n} - y^{2n} + 1)$

107. From what polynomial must $4x^2 + 2x - 3$ be subtracted to obtain $5x^2 - 5x + 8$?

Review Exercises

108. Solve: $9(x - 1) = 1 + 3(x - 2)$. (Section 1.4, Example 3)

109. Graph: $2x - 3y < -6$. (Section 4.4, Example 1)

110. Write the point-slope form and slope-intercept form of equations of a line passing through the point $(-2, 5)$ and parallel to the line whose equation is $3x - y = 9$. (Section 2.5, Example 4)

Preview Exercises

Exercises 111–113 will help you prepare for the material covered in the next section.

111. Multiply: $(2x^3y^2)(5x^4y^7)$.

112. Use the distributive property to multiply: $2x^4(8x^4 + 3x)$.

113. Simplify and express the polynomial in standard form:

$$3x(x^2 + 4x + 5) + 7(x^2 + 4x + 5).$$

SECTION 5.2

Multiplication of Polynomials

Objectives

1. Multiply monomials.
2. Multiply a monomial and a polynomial.
3. Multiply polynomials when neither is a monomial.
4. Use FOIL in polynomial multiplication.
5. Square binomials.
6. Multiply the sum and difference of two terms.
7. Find the product of functions.
8. Use polynomial multiplication to evaluate functions.

Old Dog... New Chicks

Can that be Axl, your author's yellow lab, sharing a special moment with a baby chick? And if it is (it is), what possible relevance can this have to multiplying polynomials? An answer is promised before you reach the exercise set. For now, let's begin by reviewing how to multiply monomials, a skill that you will apply in every polynomial multiplication problem.

Multiplying Monomials

1 Multiply monomials.

To multiply monomials, begin by multiplying the coefficients. Then multiply the variables. Use the product rule for exponents to multiply the variables: Retain the variable and add the exponents.

EXAMPLE 1 Multiplying Monomials

Multiply:

a. $(5x^3y^4)(-6x^7y^8)$ **b.** $(4x^3y^2z^5)(2x^5y^2z^4)$.

Solution

a.

$$(5x^3y^4)(-6x^7y^8) = 5(-6)x^3 \cdot x^7 \cdot y^4 \cdot y^8$$ Rearrange factors. This step is usually done mentally.

$$= -30x^{3+7}y^{4+8}$$ Multiply coefficients and add exponents.

$$= -30x^{10}y^{12}$$ Simplify.

b.

$$(4x^3y^2z^5)(2x^5y^2z^4) = 4 \cdot 2 \cdot x^3 \cdot x^5 \cdot y^2 \cdot y^2 \cdot z^5 \cdot z^4$$ Rearrange factors.

$$= 8x^{3+5}y^{2+2}z^{5+4}$$ Multiply coefficients and add exponents.

$$= 8x^8y^4z^9$$ Simplify.

CHECK POINT 1 Multiply:

a. $(6x^5y^7)(-3x^2y^4)$ **b.** $(10x^4y^3z^6)(3x^6y^3z^2)$.

2 Multiply a monomial and a polynomial.

Multiplying a Monomial and a Polynomial That Is Not a Monomial

We use the distributive property to multiply a monomial and a polynomial that is not a monomial. For example,

$$3x^2(2x^3 + 5x) = 3x^2 \cdot 2x^3 + 3x^2 \cdot 5x = 3 \cdot 2x^{2+3} + 3 \cdot 5x^{2+1} = 6x^5 + 15x^3.$$

Monomial Binomial Multiply coefficients and add exponents.

To multiply a monomial and a polynomial, multiply each term of the polynomial by the monomial. Once the monomial factor is distributed, we multiply the resulting monomials using the procedure shown in Example 1.

EXAMPLE 2 Multiplying a Monomial and a Trinomial

Multiply:

a. $4x^3(6x^5 - 2x^2 + 3)$ b. $5x^3y^4(2x^7y - 6x^4y^3 - 3)$.

Solution

a. $4x^3(6x^5 - 2x^2 + 3) = 4x^3 \cdot 6x^5 - 4x^3 \cdot 2x^2 + 4x^3 \cdot 3$ Use the distributive property.

$= 24x^8 - 8x^5 + 12x^3$ Multiply coefficients and add exponents.

b. $5x^3y^4(2x^7y - 6x^4y^3 - 3)$

$= 5x^3y^4 \cdot 2x^7y - 5x^3y^4 \cdot 6x^4y^3 - 5x^3y^4 \cdot 3$ Use the distributive property.

$= 10x^{10}y^5 - 30x^7y^7 - 15x^3y^4$ Multiply coefficients and add exponents.

CHECK POINT 2 Multiply:

a. $6x^4(2x^5 - 3x^2 + 4)$

b. $2x^4y^3(5xy^6 - 4x^3y^4 - 5)$.

3 Multiply polynomials when neither is a monomial.

Multiplying Polynomials when Neither Is a Monomial

How do we multiply two polynomials if neither is a monomial? For example, consider

$$(3x + 7)(x^2 + 4x + 5).$$

Binomial Trinomial

One way to perform this multiplication is to distribute $3x$ throughout the trinomial

$$3x(x^2 + 4x + 5)$$

and 7 throughout the trinomial

$$7(x^2 + 4x + 5).$$

Then combine the like terms that result.

Multiplying Polynomials when Neither Is a Monomial

Multiply each term of one polynomial by each term of the other polynomial. Then combine like terms.

EXAMPLE 3 Multiplying a Binomial and a Trinomial

Multiply: $(3x + 7)(x^2 + 4x + 5)$.

Solution

$(3x + 7)(x^2 + 4x + 5)$

$= 3x(x^2 + 4x + 5) + 7(x^2 + 4x + 5)$ — Multiply the trinomial by each term of the binomial.

$= 3x \cdot x^2 + 3x \cdot 4x + 3x \cdot 5 + 7x^2 + 7 \cdot 4x + 7 \cdot 5$ — Use the distributive property.

$= 3x^3 + 12x^2 + 15x + 7x^2 + 28x + 35$ — Multiply monomials: Multiply coefficients and add exponents.

$= 3x^3 + 19x^2 + 43x + 35$ — Combine like terms: $12x^2 + 7x^2 = 19x^2$ and $15x + 28x = 43x$.

CHECK POINT 3 Multiply: $(3x + 2)(2x^2 - 2x + 1)$.

Another method for solving Example 3 is to use a vertical format similar to that used for multiplying whole numbers.

$$\begin{array}{rrrrl} & x^2 & + 4x & + 5 & \\ & & 3x & + 7 & \\ \hline & 7x^2 & + 28x & + 35 & \leftarrow 7(x^2+4x+5) \\ 3x^3 & + 12x^2 & + 15x & & \leftarrow 3x(x^2+4x+5) \\ \hline 3x^3 & + 19x^2 & + 43x & + 35 & \text{Combine like terms.} \end{array}$$

Write like terms in the same column.

EXAMPLE 4 Multiplying a Binomial and a Trinomial

Multiply: $(2x^2y + 3y)(5x^4y - 4x^2y + y)$.

Solution

$(2x^2y + 3y)(5x^4y - 4x^2y + y)$

$= 2x^2y(5x^4y - 4x^2y + y) + 3y(5x^4y - 4x^2y + y)$ — Multiply the trinomial by each term of the binomial.

$= 2x^2y \cdot 5x^4y - 2x^2y \cdot 4x^2y + 2x^2y \cdot y + 3y \cdot 5x^4y - 3y \cdot 4x^2y + 3y \cdot y$ — Use the distributive property.

$= 10x^6y^2 - 8x^4y^2 + 2x^2y^2 + 15x^4y^2 - 12x^2y^2 + 3y^2$ — Multiply coefficients and add exponents.

$= 10x^6y^2 + 7x^4y^2 - 10x^2y^2 + 3y^2$ — Combine like terms: $-8x^4y^2 + 15x^4y^2 = 7x^4y^2$, $2x^2y^2 - 12x^2y^2 = -10x^2y^2$.

CHECK POINT 4 Multiply: $(4xy^2 + 2y)(3xy^4 - 2xy^2 + y)$.

4 Use FOIL in polynomial multiplication.

The Product of Two Binomials: FOIL

Frequently we need to find the product of two binomials. One way to perform this multiplication is to distribute each term in the first binomial throughout the second binomial. For example, we can find the product of the binomials $7x + 2$ and $4x + 5$ as follows:

$$\begin{aligned}(7x + 2)(4x + 5) &= 7x(4x + 5) + 2(4x + 5)\\ &= 7x(4x) + 7x(5) + 2(4x) + 2(5)\\ &= 28x^2 + 35x + 8x + 10.\end{aligned}$$

Distribute $7x$ over $4x + 5$.

Distribute 2 over $4x + 5$.

We'll combine these like terms later. For now, our interest is in how to obtain *each* of these four terms.

We can also find the product of $7x + 2$ and $4x + 5$ using a method called FOIL, which is based on our work shown above. Any two binomials can be quickly multiplied using the FOIL method, in which **F** represents the product of the **first** terms in each binomial, **O** represents the product of the **outside** terms, **I** represents the product of the two **inside** terms, and **L** represents the product of the **last**, or second, terms in each binomial. For example, we can use the FOIL method to find the product of the binomials $7x + 2$ and $4x + 5$ as follows:

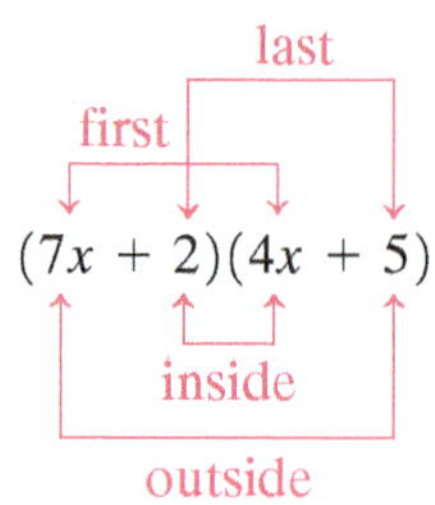

$$(7x + 2)(4x + 5) = 28x^2 + 35x + 8x + 10$$

F: Product of First terms; O: Product of Outside terms; I: Product of Inside terms; L: Product of Last terms

$$= 28x^2 + 43x + 10 \quad \text{Combine like terms.}$$

In general, here's how to use the FOIL method to find the product of $ax + b$ and $cx + d$:

Using the FOIL Method to Multiply Binomials

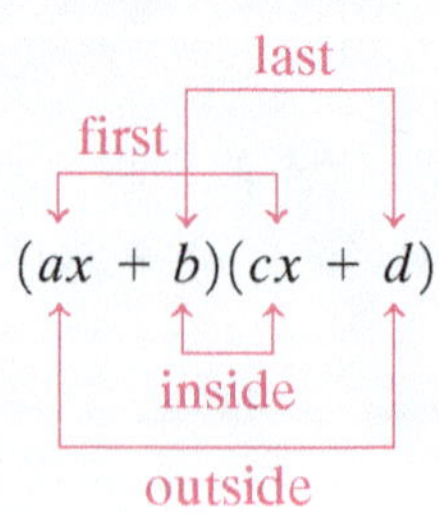

$$(ax + b)(cx + d) = ax \cdot cx + ax \cdot d + b \cdot cx + b \cdot d$$

F: Product of First terms; O: Product of Outside terms; I: Product of Inside terms; L: Product of Last terms

EXAMPLE 5 Using the FOIL Method

Multiply:

a. $(x + 3)(x + 2)$ **b.** $(3x + 5y)(x - 2y)$ **c.** $(5x^3 - 6)(4x^3 - x)$.

Solution

a. first, last, inside, outside — F O I L

$$(x + 3)(x + 2) = x \cdot x + x \cdot 2 + 3 \cdot x + 3 \cdot 2$$
$$= x^2 + 2x + 3x + 6$$
$$= x^2 + 5x + 6 \quad \text{Combine like terms.}$$

b. first, last, inside, outside — F O I L

$$(3x + 5y)(x - 2y) = 3x \cdot x + 3x(-2y) + 5y \cdot x + 5y(-2y)$$
$$= 3x^2 - 6xy + 5xy - 10y^2$$
$$= 3x^2 - xy - 10y^2 \quad \text{Combine like terms.}$$

c. first, last, inside, outside — F O I L

$$(5x^3 - 6)(4x^3 - x) = 5x^3 \cdot 4x^3 + 5x^3(-x) + (-6)(4x^3) + (-6)(-x)$$
$$= 20x^6 - 5x^4 - 24x^3 + 6x$$

There are no like terms to combine.

✓ CHECK POINT 5 Multiply:

a. $(x + 5)(x + 3)$ b. $(7x + 4y)(2x - y)$ c. $(4x^3 - 5)(x^3 - 3x)$.

5 Square binomials.

The Square of a Binomial

Let us find $(A + B)^2$, the square of a binomial sum. To do so, we begin with the FOIL method and look for a general rule.

F O I L

$$(A + B)^2 = (A + B)(A + B) = A \cdot A + A \cdot B + A \cdot B + B \cdot B$$
$$= A^2 + 2AB + B^2$$

This result implies the following rule, which is often called a **special-product formula**:

The Square of a Binomial Sum

$$(A + B)^2 = A^2 + 2AB + B^2$$

The square of a binomial sum is first term squared plus 2 times the product of the terms plus last term squared.

Study Tip

Caution! The square of a sum is *not* the sum of the squares.

$$(A + B)^2 \neq A^2 + B^2$$

The middle term $2AB$ is missing.

$$(x + 5)^2 \neq x^2 + 25$$

Incorrect

Show that $(x + 5)^2$ and $x^2 + 25$ are not equal by substituting 3 for x in each expression and simplifying.

EXAMPLE 6 Finding the Square of a Binomial Sum

Multiply:

a. $(x + 5)^2$ b. $(3x + 2y)^2$.

Solution Use the special-product formula shown.

$$(A + B)^2 = A^2 + 2AB + B^2$$

	(First Term)^2	+	2 · Product of the Terms	+	(Last Term)^2	= Product
a. $(x + 5)^2 =$	x^2	+	$2 \cdot x \cdot 5$	+	5^2	$= x^2 + 10x + 25$
b. $(3x + 2y)^2 =$	$(3x)^2$	+	$2 \cdot 3x \cdot 2y$	+	$(2y)^2$	$= 9x^2 + 12xy + 4y^2$

✓ CHECK POINT 6 Multiply:

a. $(x + 8)^2$ b. $(4x + 5y)^2$.

The formula for the square of a binomial sum can be interpreted geometrically by analyzing the areas in **Figure 5.9**.

Area of the large square: $(A + B)^2$

Sum of the areas of the four smaller rectangles inside the large square: $A^2 + AB + AB + B^2$
$= A^2 + 2AB + B^2$

Conclusion:
$(A + B)^2 = A^2 + 2AB + B^2$

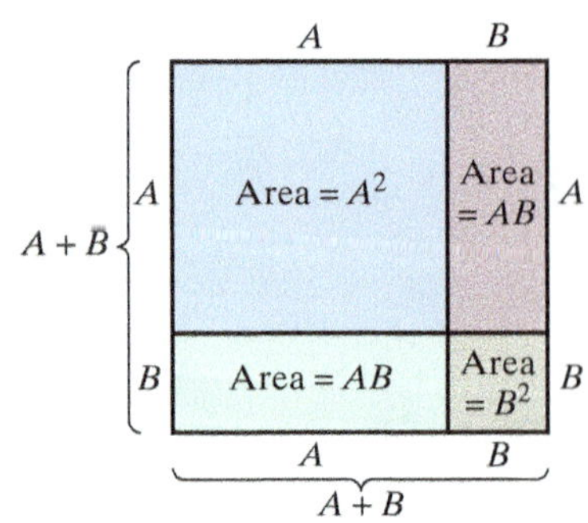

FIGURE 5.9

A similar pattern occurs for $(A - B)^2$, the square of a binomial difference. Using the FOIL method on $(A - B)^2$, we obtain the following rule:

The Square of a Binomial Difference

$$(A - B)^2 = A^2 - 2AB + B^2$$

The square of a binomial difference is first term squared minus 2 times the product of the terms plus last term squared.

EXAMPLE 7 Finding the Square of a Binomial Difference

Multiply:

a. $(x - 8)^2$ b. $\left(\frac{1}{2}x - 4y^3\right)^2$.

Solution Use the special-product formula shown.

$$(A - B)^2 = A^2 - 2AB + B^2$$

	(First Term)^2	−	2 · Product of the Terms	+	(Last Term)^2	= Product
a. $(x-8)^2 =$	x^2	−	$2 \cdot x \cdot 8$	+	8^2	$= x^2 - 16x + 64$
b. $\left(\frac{1}{2}x - 4y^3\right)^2 =$	$\left(\frac{1}{2}x\right)^2$	−	$2 \cdot \frac{1}{2}x \cdot 4y^3$	+	$(4y^3)^2$	$= \frac{1}{4}x^2 - 4xy^3 + 16y^6$

CHECK POINT 7 Multiply:

a. $(x-5)^2$ b. $(2x - 6y^4)^2$.

6 Multiply the sum and difference of two terms.

Multiplying the Sum and Difference of Two Terms

We can use the FOIL method to multiply $A + B$ and $A - B$ as follows:

$$(A+B)(A-B) = \underset{F}{A^2} - \underset{O}{AB} + \underset{I}{AB} - \underset{L}{B^2} = A^2 - B^2.$$

Notice that the outside and inside products have a sum of 0 and the terms cancel. The FOIL multiplication provides us with a quick rule for multiplying the sum and difference of two terms, which is another example of a special-product formula.

The Product of the Sum and Difference of Two Terms

$$(A+B)(A-B) = A^2 - B^2$$

The product of the sum and the difference of the same two terms is the square of the first term minus the square of the second term.

EXAMPLE 8 **Finding the Product of the Sum and Difference of Two Terms**

Multiply:

a. $(x+8)(x-8)$ b. $(9x+5y)(9x-5y)$ c. $(6a^2b - 3b)(6a^2b + 3b)$.

Solution Use the special-product formula shown.

	$(A+B)(A-B)$	=	A^2 First term squared	−	B^2 Second term squared	=	Product
a.	$(x+8)(x-8)$	=	x^2	−	8^2	=	$x^2 - 64$
b.	$(9x+5y)(9x-5y)$	=	$(9x)^2$	−	$(5y)^2$	=	$81x^2 - 25y^2$
c.	$(6a^2b-3b)(6a^2b+3b)$	=	$(6a^2b)^2$	−	$(3b)^2$	=	$36a^4b^2 - 9b^2$

✓ **CHECK POINT 8** Multiply:

a. $(x + 3)(x - 3)$
b. $(5x + 7y)(5x - 7y)$
c. $(5ab^2 - 4a)(5ab^2 + 4a)$.

Special products can sometimes be used to find the products of certain trinomials, as illustrated in Example 9.

EXAMPLE 9 Using the Special Products

Multiply:

a. $(7x + 5 + 4y)(7x + 5 - 4y)$ b. $(3x + y + 1)^2$.

Solution

a. By grouping the first two terms within each set of parentheses, we can find the product using the form for the sum and difference of two terms.

$$(A + B) \cdot (A - B) = A^2 - B^2$$

$$[(7x + 5) + 4y] \cdot [(7x + 5) - 4y] = (7x + 5)^2 - (4y)^2$$
$$= (7x)^2 + 2 \cdot 7x \cdot 5 + 5^2 - (4y)^2$$
$$= 49x^2 + 70x + 25 - 16y^2$$

b. We can group the terms so that the formula for the square of a binomial can be applied.

Discover for Yourself

Group $(3x + y + 1)^2$ as $[3x + (y + 1)]^2$. Verify that you get the same product as we obtained in Example 9(b).

$$(A + B)^2 = A^2 + 2 \cdot A \cdot B + B^2$$

$$[(3x + y) + 1]^2 = (3x + y)^2 + 2 \cdot (3x + y) \cdot 1 + 1^2$$
$$= 9x^2 + 6xy + y^2 + 6x + 2y + 1$$

✓ **CHECK POINT 9** Multiply:

a. $(3x + 2 + 5y)(3x + 2 - 5y)$
b. $(2x + y + 3)^2$.

7 Find the product of functions.

Multiplication of Polynomial Functions

In Chapter 2, we developed an algebra of functions, defining the product of functions f and g as follows:

$$(fg)(x) = f(x) \cdot g(x).$$

Now that we know how to multiply polynomials, we can find the product of functions.

EXAMPLE 10 Using the Algebra of Functions

Let $f(x) = x - 5$ and $g(x) = x - 2$. Find:

a. $(fg)(x)$ b. $(fg)(1)$.

5.2 EXERCISE SET

MyMathLab — Math XL PRACTICE | WATCH | DOWNLOAD | READ | REVIEW

Practice Exercises

Throughout the practice exercises, assume that any variable exponents represent whole numbers.

In Exercises 1–8, multiply the monomials.

1. $(3x^2)(5x^4)$
2. $(4x^2)(6x^4)$
3. $(3x^2y^4)(5xy^7)$
4. $(6x^4y^2)(3x^7y)$
5. $(-3xy^2z^5)(2xy^7z^4)$
6. $(11x^2yz^4)(-3xy^5z^6)$
7. $(-8x^{2n}y^{n-5})\left(-\frac{1}{4}x^ny^3\right)$
8. $(-9x^{3n}y^{n-3})\left(-\frac{1}{3}x^ny^2\right)$

In Exercises 9–22, multiply the monomial and the polynomial.

9. $4x^2(3x + 2)$
10. $5x^2(6x + 7)$
11. $2y(y^2 - 5y)$
12. $3y(y^2 - 4y)$
13. $5x^3(2x^5 - 4x^2 + 9)$
14. $6x^3(3x^5 - 5x^2 + 7)$
15. $4xy(7x + 3y)$
16. $5xy(8x + 3y)$
17. $3ab^2(6a^2b^3 + 5ab)$
18. $5ab^2(10a^2b^3 + 7ab)$
19. $-4x^2y(3x^4y^2 - 7xy^3 + 6)$
20. $-3x^2y(10x^2y^4 - 2xy^3 + 7)$
21. $-4x^n\left(3x^{2n} - 5x^n + \frac{1}{2}x\right)$
22. $-10x^n\left(4x^{2n} - 3x^n + \frac{1}{5}x\right)$

In Exercises 23–34, find each product using either a horizontal or a vertical format.

23. $(x - 3)(x^2 + 2x + 5)$
24. $(x + 4)(x^2 - 5x + 8)$
25. $(x - 1)(x^2 + x + 1)$
26. $(x - 2)(x^2 + 2x + 4)$
27. $(a - b)(a^2 + ab + b^2)$
28. $(a + b)(a^2 - ab + b^2)$
29. $(x^2 + 2x - 1)(x^2 + 3x - 4)$
30. $(x^2 - 2x + 3)(x^2 + x + 1)$
31. $(x - y)(x^2 - 3xy + y^2)$
32. $(x - y)(x^2 - 4xy + y^2)$
33. $(xy + 2)(x^2y^2 - 2xy + 4)$
34. $(xy + 3)(x^2y^2 - 2xy + 5)$

In Exercises 35–54, use the FOIL method to multiply the binomials.

35. $(x + 4)(x + 7)$
36. $(x + 5)(x + 8)$
37. $(y + 5)(y - 6)$
38. $(y + 5)(y - 8)$
39. $(5x + 3)(2x + 1)$
40. $(4x + 3)(5x + 1)$
41. $(3y - 4)(2y - 1)$
42. $(5y - 2)(3y - 1)$
43. $(3x - 2)(5x - 4)$
44. $(2x - 3)(4x - 5)$
45. $(x - 3y)(2x + 7y)$
46. $(3x - y)(2x + 5y)$
47. $(7xy + 1)(2xy - 3)$
48. $(3xy - 1)(5xy + 2)$
49. $(x - 4)(x^2 - 5)$
50. $(x - 5)(x^2 - 3)$
51. $(8x^3 + 3)(x^2 - 5)$
52. $(7x^3 + 5)(x^2 - 2)$
53. $(3x^n - y^n)(x^n + 2y^n)$
54. $(5x^n - y^n)(x^n + 4y^n)$

In Exercises 55–68, multiply using one of the rules for the square of a binomial.

55. $(x + 3)^2$
56. $(x + 4)^2$
57. $(y - 5)^2$
58. $(y - 6)^2$
59. $(2x + y)^2$
60. $(4x + y)^2$
61. $(5x - 3y)^2$
62. $(3x - 4y)^2$
63. $(2x^2 + 3y)^2$
64. $(4x^2 + 5y)^2$
65. $(4xy^2 - xy)^2$
66. $(5xy^2 - xy)^2$
67. $(a^n + 4b^n)^2$
68. $(3a^n - b^n)^2$

In Exercises 69–82, multiply using the rule for the product of the sum and difference of two terms.

69. $(x + 4)(x - 4)$
70. $(x + 5)(x - 5)$
71. $(5x + 3)(5x - 3)$
72. $(3x + 2)(3x - 2)$

73. $(4x + 7y)(4x - 7y)$
74. $(8x + 7y)(8x - 7y)$
75. $(y^3 + 2)(y^3 - 2)$
76. $(y^3 + 3)(y^3 - 3)$
77. $(1 - y^5)(1 + y^5)$
78. $(2 - y^5)(2 + y^5)$
79. $(7xy^2 - 10y)(7xy^2 + 10y)$
80. $(3xy^2 - 4y)(3xy^2 + 4y)$
81. $(5a^n - 7)(5a^n + 7)$
82. $(10b^n - 3)(10b^n + 3)$

In Exercises 83–94, find each product.

83. $[(2x + 3) + 4y][(2x + 3) - 4y]$
84. $[(3x + 2) + 5y][(3x + 2) - 5y]$
85. $(x + y + 3)(x + y - 3)$
86. $(x + y + 4)(x + y - 4)$
87. $(5x + 7y - 2)(5x + 7y + 2)$
88. $(7x + 5y - 2)(7x + 5y + 2)$
89. $[5y + (2x + 3)][5y - (2x + 3)]$
90. $[8y + (3x + 2)][8y - (3x + 2)]$
91. $(x + y + 1)^2$
92. $(x + y + 2)^2$
93. $(x + 1)(x - 1)(x^2 + 1)$
94. $(x + 2)(x - 2)(x^2 + 4)$
95. Let $f(x) = x - 2$ and $g(x) = x + 6$. Find each of the following.
   a. $(fg)(x)$
   b. $(fg)(-1)$
   c. $(fg)(0)$
96. Let $f(x) = x - 4$ and $g(x) = x + 10$. Find each of the following.
   a. $(fg)(x)$
   b. $(fg)(-1)$
   c. $(fg)(0)$
97. Let $f(x) = x - 3$ and $g(x) = x^2 + 3x + 9$. Find each of the following.
   a. $(fg)(x)$
   b. $(fg)(-2)$
   c. $(fg)(0)$
98. Let $f(x) = x + 3$ and $g(x) = x^2 - 3x + 9$. Find each of the following.
   a. $(fg)(x)$
   b. $(fg)(-2)$
   c. $(fg)(0)$

In Exercises 99–102, find each of the following and simplify:
a. $f(a + 2)$ b. $f(a + h) - f(a)$.

99. $f(x) = x^2 - 3x + 7$
100. $f(x) = x^2 - 4x + 9$
101. $f(x) = 3x^2 + 2x - 1$
102. $f(x) = 4x^2 + 5x - 1$

Practice PLUS

In Exercises 103–112, perform the indicated operation or operations.

103. $(3x + 4y)^2 - (3x - 4y)^2$
104. $(5x + 2y)^2 - (5x - 2y)^2$
105. $(5x - 7)(3x - 2) - (4x - 5)(6x - 1)$
106. $(3x + 5)(2x - 9) - (7x - 2)(x - 1)$
107. $(2x + 5)(2x - 5)(4x^2 + 25)$
108. $(3x + 4)(3x - 4)(9x^2 + 16)$
109. $(x - 1)^3$
110. $(x - 2)^3$
111. $\dfrac{(2x - 7)^5}{(2x - 7)^3}$
112. $\dfrac{(5x - 3)^6}{(5x - 3)^4}$

Application Exercises

In Exercises 113–114, find the area of the large rectangle in two ways:

a. *Find the sum of the areas of the four smaller rectangles.*
b. *Multiply the length and the width of the large rectangle using the FOIL method. Compare this product with your answer to part (a).*

113.
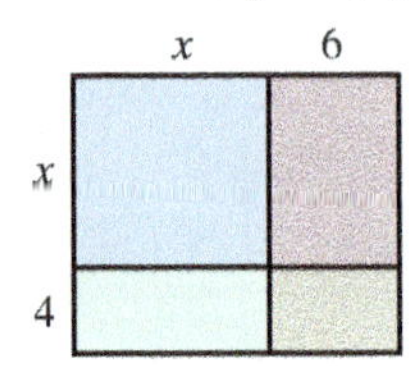

114.
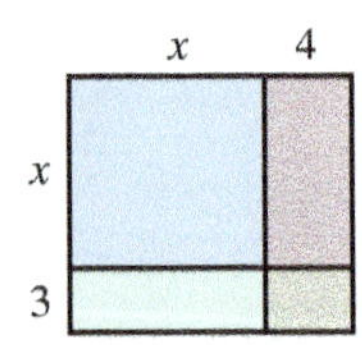

In Exercises 115–116, express each polynomial in standard form—that is, in descending powers of x.

a. *Write a polynomial that represents the area of the large rectangle.*
b. *Write a polynomial that represents the area of the small, unshaded rectangle.*
c. *Write a polynomial that represents the area of the shaded region.*

115.
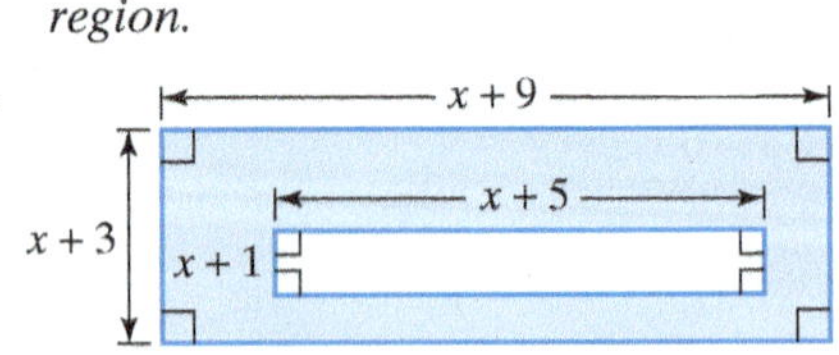

116.
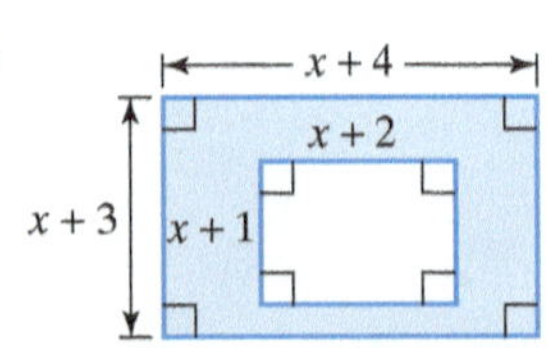

In Exercises 117–118, express each polynomial in standard form.

a. *Write a polynomial that represents the area of the rectangular base of the open box.*

b. *Write a polynomial that represents the volume of the open box.*

117.

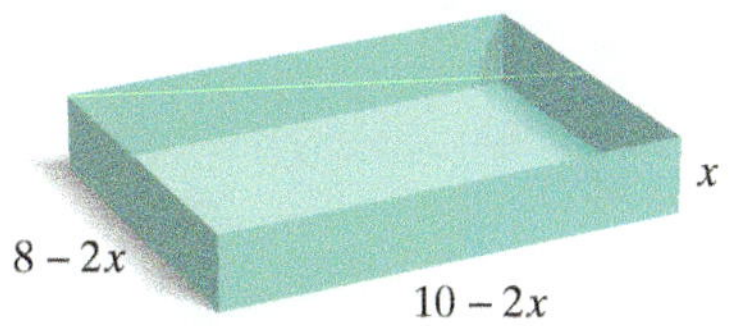

118.

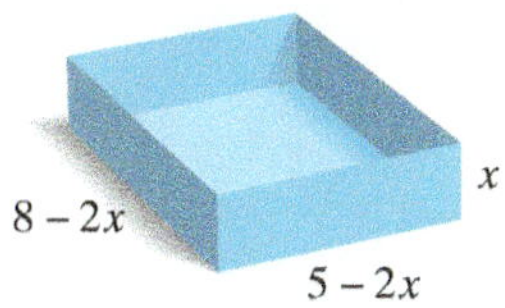

119. A popular model of carry-on luggage has a length that is 10 inches greater than its depth. Airline regulations require that the sum of the length, width, and depth cannot exceed 40 inches. These conditions, with the assumption that this sum *is* 40 inches, can be modeled by a function that gives the volume of the luggage, V, in cubic inches, in terms of its depth, x, in inches.

Volume = depth · length · width: 40 − (depth + length)

$$V(x) = x \cdot (x+10) \cdot [40-(x+x+10)]$$

$$V(x) = x(x+10)(30-2x)$$

a. Perform the multiplications in the formula for $V(x)$ and express the formula in standard form.

b. Use the function's formula from part (a) and the Leading Coefficient Test to determine the end behavior of its graph.

c. Does the end behavior to the right make this function useful in modeling the volume of carry-on luggage as its depth continues to increase?

d. Use the formula from part (a) to find $V(10)$. Describe what this means in practical terms.

e. The graph of the function modeling the volume of carry-on luggage is shown below. Identify your answer from part (d) as a point on the graph.

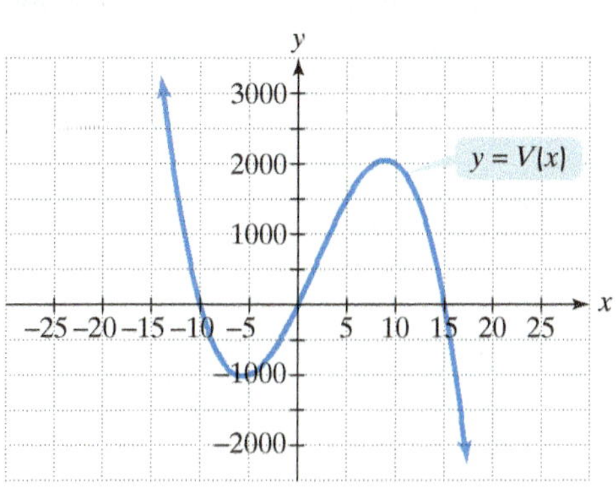

f. Use the graph to describe a realistic domain, x, for the volume function, where x represents the depth of the carry-on luggage. Use set-builder notation or interval notation to express this realistic domain.

120. Before working this exercise, be sure that you have read the Blitzer Bonus on page 326. The table shows the four combinations of color genes that a YY yellow lab and a BY black lab can pass to their offspring.

	B	Y
Y	BY	YY
Y	BY	YY

a. How many combinations result in a yellow lab with two recessive yellow genes? What is the probability of a yellow lab?

b. How many combinations result in a black lab with a recessive yellow gene? What is the probability of a black lab?

c. Find the product of Y and $\frac{1}{2}\text{B} + \frac{1}{2}\text{Y}$. How does this product model the probabilities that you determined in parts (a) and (b)?

Writing in Mathematics

121. Explain how to multiply monomials. Give an example.

122. Explain how to multiply a monomial and a polynomial that is not a monomial. Give an example.

123. Explain how to multiply a binomial and a trinomial.

124. What is the FOIL method and when is it used? Give an example of the method.

125. Explain how to square a binomial sum. Give an example.

126. Explain how to square a binomial difference. Give an example.

127. Explain how to find the product of the sum and difference of two terms. Give an example with your explanation.

128. How can the graph of function fg be obtained from the graphs of functions f and g?

129. Explain how to find $f(a+h) - f(a)$ for a given function f.

Technology Exercises

In Exercises 130–133, use a graphing utility to graph the functions y_1 and y_2. Select a viewing rectangle that is large enough to show the end behavior of y_2. What can you conclude? Verify your conclusions using polynomial multiplication.

130. $y_1 = (x-2)^2$
$y_2 = x^2 - 4x + 4$

131. $y_1 = (x-4)(x^2 - 3x + 2)$
$y_2 = x^3 - 7x^2 + 14x - 8$

132. $y_1 = (x - 1)(x^2 + x + 1)$
$y_2 = x^3 - 1$

133. $y_1 = (x + 1.5)(x - 1.5)$
$y_2 = x^2 - 2.25$

134. Graph $f(x) = x + 4$, $g(x) = x - 2$, and the product function, fg, in a $[-6, 6, 1]$ by $[-10, 10, 1]$ viewing rectangle. Trace along the curves and show that $(fg)(1) = f(1) \cdot g(1)$.

Critical Thinking Exercises

Make Sense? *In Exercises 135–138, determine whether each statement "makes sense" or "does not make sense" and explain your reasoning.*

135. Knowing the difference between factors and terms is important: In $(3x^2y)^2$, I can distribute the exponent 2 on each factor, but in $(3x^2 + y)^2$, I cannot do the same thing on each term.

136. I used the FOIL method to find the product of $x + 5$ and $x^2 + 2x + 1$.

137. Instead of using the formula for the square of a binomial sum, I prefer to write the binomial sum twice and then apply the FOIL method.

138. Special-product formulas have patterns that make their multiplications quicker than using the FOIL method.

In Exercises 139–142, determine whether each statement is true or false. If the statement is false, make the necessary change(s) to produce a true statement.

139. If f is a polynomial function, then
$f(a + h) - f(a) = f(a) + f(h) - f(a) = f(h)$.

140. $(x - 5)^2 = x^2 - 5x + 25$

141. $(x + 1)^2 = x^2 + 1$

142. Suppose a square garden has an area represented by $9x^2$ square feet. If one side is made 7 feet longer and the other side is made 2 feet shorter, then the trinomial that represents the area of the larger garden is $9x^2 + 15x - 14$ square feet.

143. Express the area of the plane figure shown as a polynomial in standard form.

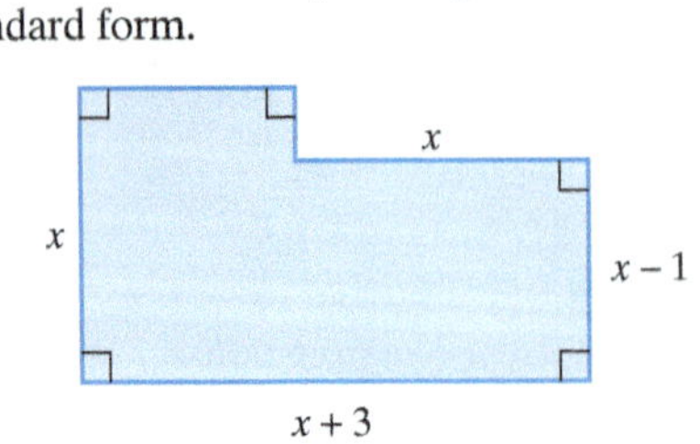

In Exercises 144–145, represent the volume of each figure as a polynomial in standard form.

144.

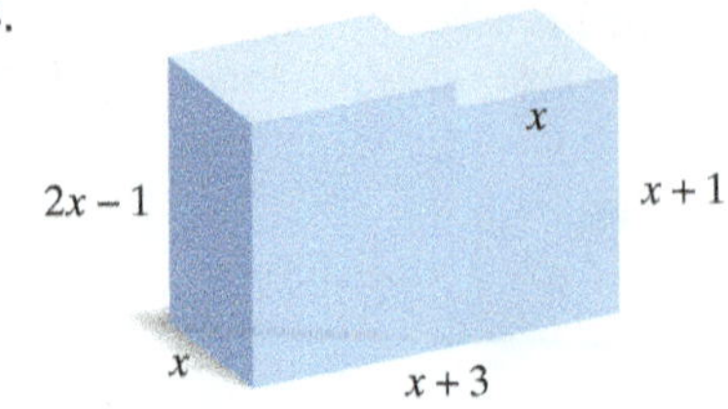

145.

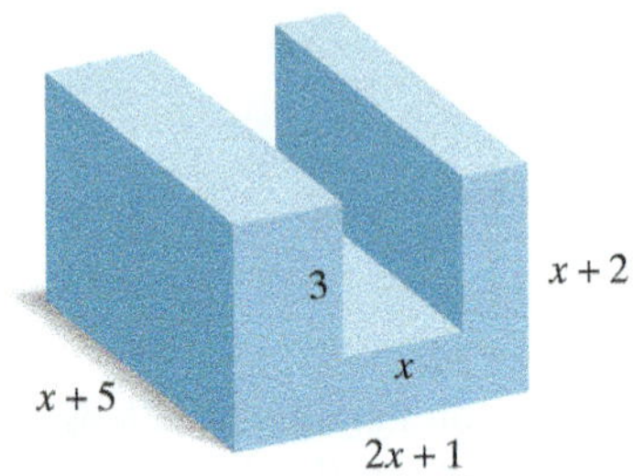

146. Simplify: $(y^n + 2)(y^n - 2) - (y^n - 3)^2$.

147. The product of two consecutive odd integers is 22 less than the square of the greater integer. Find the integers.

Review Exercises

148. Solve: $|3x + 4| \geq 10$. (Section 4.3, Example 5 or Example 6(b))

149. Solve: $2 - 6x \leq 20$. (Section 4.1, Example 2)

150. Write in scientific notation: 8,034,000,000. (Section 1.7, Example 2)

Preview Exercises

Exercises 151–153 will help you prepare for the material covered in the next section.

151. Replace each boxed question mark with a polynomial that results in the given product.

a. $3x^3 \cdot \boxed{?} = 9x^5$

b. $2x^3y^2 \cdot \boxed{?} = 12x^5y^4$

In Exercises 152–153, a polynomial is given in factored form. Use multipication to find the product of the factors.

152. $(x - 5)(x^2 + 3)$

153. $(x + 4)(3x - 2y)$

SECTION 5.3

Greatest Common Factors and Factoring By Grouping

Objectives

1. Factor out the greatest common factor of a polynomial.
2. Factor out a common factor with a negative coefficient.
3. Factor by grouping.

The inability to understand numbers and their meanings is called innumeracy *by mathematics professor John Allen Paulos. Paulos has written a book about mathematical illiteracy. Entitled* Innumeracy, *the book seeks to explain why so many people are numerically inept and to show how the problem can be corrected.*

Jasper Johns *0 Through 9*, 1961. Oil on canvas, 137 × 105 cm. The Saatchi Collection, Courtesy of the Leo Castelli Gallery. © Jasper Johns/Licensed by VAGA, New York, NY.

Did you know that one of the most common ways that you are given numerical information is with percents? Unfortunately, many people are innumerate when it comes to this topic. For example, a computer whose price has been reduced by 40% and then another 40% is not selling at 20% of its original price.

To cure this bout of innumeracy (see Exercise 81 in Exercise Set 5.3), we turn to a process that reverses polynomial multiplication. For example, we can multiply polynomials and show that

$$7x(3x + 4) = 21x^2 + 28x.$$

We can also reverse this process and express the resulting polynomial as

$$21x^2 + 28x = 7x(3x + 4).$$

Factoring a polynomial consisting of the sum of monomials means finding an equivalent expression that is a product.

Factoring $21x^2 + 28x$

Sum of monomials | Equivalent expression that is a product

$$21x^2 + 28x = 7x(3x + 4)$$

The factors of $21x^2 + 28x$ are $7x$ and $3x + 4$.

In this chapter, we will be factoring over the set of integers, meaning that the coefficients in the factors are integers. Polynomials that cannot be factored using integer coefficients are called **prime polynomials over the set of integers**.

1 Factor out the greatest common factor of a polynomial.

Factoring Out the Greatest Common Factor

In any factoring problem, the first step is to look for the *greatest common factor*. The **greatest common factor**, abbreviated GCF, is an expression with the greatest coefficient and of the highest degree that divides each term of the polynomial. Can you see that $7x$ is the greatest common factor of $21x^2 + 28x$? 7 is the greatest integer that divides both 21 and 28. Furthermore, x is the greatest power of x that divides x^2 and x.

The variable part of the greatest common factor always contains the *smallest* power of a variable that appears in all terms of the polynomial. For example, consider the polynomial

$$21x^2 + 28x.$$

x^1, or x, is the variable raised to the smallest exponent.

We see that x is the variable part of the greatest common factor, $7x$.

When factoring a monomial from a polynomial, determine the greatest common factor of all terms in the polynomial. Sometimes there may not be a GCF other than 1. When a GCF other than 1 exists, we use the following procedure:

Factoring a Monomial From a Polynomial

1. Determine the greatest common factor of all terms in the polynomial.
2. Express each term as the product of the GCF and its other factor.
3. Use the distributive property to factor out the GCF.

EXAMPLE 1 Factoring Out the Greatest Common Factor

Factor: $21x^2 + 28x$.

Solution The GCF of the two terms of the polynomial is $7x$.

$$21x^2 + 28x$$

$= 7x(3x) + 7x(4)$ Express each term as the product of the GCF and its other factor.

$= 7x(3x + 4)$ Factor out the GCF.

We can check this factorization by multiplying $7x$ and $3x + 4$, obtaining the original polynomial as the answer.

CHECK POINT 1 Factor: $20x^2 + 30x$.

EXAMPLE 2 Factoring Out the Greatest Common Factor

Factor:

a. $9x^5 + 15x^3$ b. $16x^2y^3 - 24x^3y^4$ c. $12x^5y^4 - 4x^4y^3 + 2x^3y^2$.

Solution

a. First, determine the greatest common factor.

3 is the greatest integer that divides 9 and 15.

$$9x^5 + 15x^3$$

x^3 is the variable raised to the smallest exponent.

The GCF of the two terms of the polynomial is $3x^3$.

$$9x^5 + 15x^3$$
$$= 3x^3 \cdot 3x^2 + 3x^3 \cdot 5$$ Express each term as the product of the GCF and its other factor.
$$= 3x^3(3x^2 + 5)$$ Factor out the GCF.

b. Begin by determining the greatest common factor.

8 is the greatest integer that divides 16 and 24.

$$16x^2y^3 - 24x^3y^4$$

The variables raised to the smallest exponents are x^2 and y^3.

The GCF of the two terms of the polynomial is $8x^2y^3$.

$$16x^2y^3 - 24x^3y^4$$
$$= 8x^2y^3 \cdot 2 - 8x^2y^3 \cdot 3xy$$ Express each term as the product of the GCF and its other factor.
$$= 8x^2y^3(2 - 3xy)$$ Factor out the GCF.

c. First, determine the greatest common factor of the three terms.

2 is the greatest integer that divides 12, 4, and 2.

$$12x^5y^4 - 4x^4y^3 + 2x^3y^2$$

The variables raised to the smallest exponents are x^3 and y^2.

The GCF is $2x^3y^2$.

$$12x^5y^4 - 4x^4y^3 + 2x^3y^2$$
$$= 2x^3y^2 \cdot 6x^2y^2 - 2x^3y^2 \cdot 2xy + 2x^3y^2 \cdot 1$$ Express each term as the product of the GCF and its other factor.

You can obtain the factors shown in black by dividing each term of the given polynomial by $2x^3y^2$, the GCF.

$$\frac{12x^5y^4}{2x^3y^2} = 6x^2y^2 \qquad \frac{4x^4y^3}{2x^3y^2} = 2xy \qquad \frac{2x^3y^2}{2x^3y^2} = 1$$

$$= 2x^3y^2(6x^2y^2 - 2xy + 1)$$ Factor out the GCF.

Because factoring reverses the process of multiplication, all factorizations can be checked by multiplying. Take a few minutes to check each of the three factorizations in Example 2. Use the distributive property to multiply the factors. This should give the original polynomial.

✓ CHECK POINT 2 Factor:

a. $9x^4 + 21x^2$

b. $15x^3y^2 - 25x^4y^3$

c. $16x^4y^5 - 8x^3y^4 + 4x^2y^3$.

---

2 Factor out a common factor with a negative coefficient.

When the leading coefficient of a polynomial is negative, it is often desirable to factor out a common factor with a negative coefficient. The common factor is the GCF preceded by a negative sign.

EXAMPLE 3 Using a Common Factor with a Negative Coefficient

Factor: $-3x^3 + 12x^2 - 15x$.

Solution The GCF is $3x$. Because the leading coefficient, -3, is negative, we factor out a common factor with a negative coefficient. We will factor out the opposite of the GCF, or $-3x$.

$$-3x^3 + 12x^2 - 15x$$

$$= -3x(x^2) - 3x(-4x) - 3x(5)$$ Express each term as the product of the common factor and its other factor.

$$= -3x(x^2 - 4x + 5)$$ Factor out the opposite of the GCF.

CHECK POINT 3 Factor out a common factor with a negative coefficient: $-2x^3 + 10x^2 - 6x$.

Factoring by Grouping

Up to now, we have factored a monomial from a polynomial. By contrast, in our next example, the greatest common factor of the polynomial is a binomial.

EXAMPLE 4 Factoring Out the Greatest Common Binomial Factor

Factor:

a. $2(x - 7) + 9a(x - 7)$ b. $5y(a - b) - (a - b)$.

Solution Let's identify the common binomial factor in each part of the problem.

$2(x - 7) + 9a(x - 7)$ — The GCF, a binomial, is $x - 7$.

$5y(a - b) - (a - b)$ — The GCF, a binomial, is $a - b$.

We factor out each common binomial factor as follows.

a. $2(x - 7) + 9a(x - 7)$

$= (x - 7)2 + (x - 7)9a$ This step, usually omitted, shows each term as the product of the GCF and its other factor, in that order.

$= (x - 7)(2 + 9a)$ Factor out the GCF.

b. $5y(a - b) - (a - b)$

$= 5y(a - b) - 1(a - b)$ Write $-(a - b$ as $-1(a - b)$ to aid in the factoring.

$= (a - b)(5y - 1)$ Factor out the GCF.

CHECK POINT 4 Factor:

a. $3(x - 4) + 7a(x - 4)$

b. $7x(a + b) - (a + b)$.

3 Factor by grouping.

Some polynomials have only a greatest common factor of 1. However, by a suitable grouping of the terms, it still may be possible to factor. This process, called **factoring by grouping**, is illustrated in Example 5.

EXAMPLE 5 Factoring by Grouping

Factor: $x^3 - 5x^2 + 3x - 15$.

Solution There is no factor other than 1 common to all four terms. However, we can group terms that have a common factor:

$$\boxed{x^3 - 5x^2} + \boxed{3x - 15}$$

Common factor is x^2. Common factor is 3.

Discover for Yourself

In Example 5, group the terms as follows:

$$x^2 + 3x - 5x^2 - 15.$$

Factor out the common factor from each group and complete the factoring process. Describe what happens. What can you conclude?

We now factor the given polynomial as follows:

$$x^3 - 5x^2 + 3x - 15$$
$$= (x^3 - 5x^2) + (3x - 15) \quad \text{Group terms with common factors.}$$
$$= x^2(x - 5) + 3(x - 5) \quad \text{Factor out the greatest common factor from the grouped terms. The remaining two terms have } x - 5 \text{ as a common binomial factor.}$$
$$= (x - 5)(x^2 + 3). \quad \text{Factor out the GCF.}$$

Thus, $x^3 - 5x^2 + 3x - 15 = (x - 5)(x^2 + 3)$. Check the factorization by multiplying the right side of the equation using the FOIL method. Because the factorization is correct, you should obtain the original polynomial.

✓ CHECK POINT 5 Factor: $x^3 - 4x^2 + 5x - 20$.

Factoring by Grouping

1. Group terms that have a common monomial factor. There will usually be two groups. Sometimes the terms must be rearranged.
2. Factor out the common monomial factor from each group.
3. Factor out the remaining common binomial factor (if one exists).

EXAMPLE 6 Factoring by Grouping

Factor: $3x^2 + 12x - 2xy - 8y$.

Solution There is no factor other than 1 common to all four terms. However, we can group terms that have a common factor:

$$\boxed{3x^2 + 12x} + \boxed{-2xy - 8y}.$$

Common factor is $3x$: $3x^2 + 12x = 3x(x + 4)$.

Use $-2y$, rather than $2y$, as the common factor: $-2xy - 8y = -2y(x + 4)$. In this way, the common binomial factor, $x + 4$, appears.

The voice balloons illustrate that it is sometimes necessary to use a factor with a negative coefficient to obtain a common binomial factor for the two groupings. We now factor the given polynomial as follows:

$$3x^2 + 12x - 2xy - 8y$$
$$= (3x^2 + 12x) + (-2xy - 8y) \quad \text{Group terms with common factors.}$$
$$= 3x(x + 4) - 2y(x + 4) \quad \text{Factor out the common factors from the grouped terms.}$$
$$= (x + 4)(3x - 2y). \quad \text{Factor out the GCF.}$$

Thus, $3x^2 + 12x - 2xy - 8y = (x + 4)(3x - 2y)$. Using the commutative property of multiplication, the factorization can also be expressed as $(3x - 2y)(x + 4)$. Verify the factorization by showing that, regardless of the order, FOIL multiplication gives the original polynomial.

CHECK POINT 6 Factor: $4x^2 + 20x - 3xy - 15y$.

Factoring by grouping sometimes requires that the terms be rearranged before the groupings are made. For example, consider the polynomial

$$3x^2 - 8y + 12x - 2xy.$$

The first two terms have no common factor other than 1. We must rearrange the terms and try a different grouping. Example 6 showed one such rearrangement of two groupings.

5.3 EXERCISE SET

MyMathLab MathXL PRACTICE WATCH DOWNLOAD READ REVIEW

Practice Exercises

Throughout the practice exercises, assume that any variable exponents represent whole numbers.

In Exercises 1–22, factor the greatest common factor from each polynomial.

1. $10x^2 + 4x$
2. $12x^2 + 9x$
3. $y^2 - 4y$
4. $y^2 - 7y$
5. $x^3 + 5x^2$
6. $x^3 + 7x^2$
7. $12x^4 - 8x^2$
8. $20x^4 - 8x^2$
9. $32x^4 + 2x^3 + 8x^2$
10. $9x^4 + 18x^3 + 6x^2$
11. $4x^2y^3 + 6xy$
12. $6x^3y^2 + 9xy$
13. $30x^2y^3 - 10xy^2$
14. $27x^2y^3 - 18xy^2$
15. $12xy - 6xz + 4xw$
16. $14xy - 10xz + 8xw$
17. $15x^3y^6 - 9x^4y^4 + 12x^2y^5$
18. $15x^4y^6 - 3x^3y^5 + 12x^4y^4$
19. $25x^3y^6z^2 - 15x^4y^4z^4 + 25x^2y^5z^3$
20. $49x^4y^3z^5 - 70x^3y^5z^4 + 35x^4y^4z^3$
21. $15x^{2n} - 25x^n$
22. $12x^{3n} - 9x^{2n}$

In Exercises 23–34, factor out the negative of the greatest common factor.

23. $-4x + 12$
24. $-5x + 20$
25. $-8x - 48$
26. $-7x - 63$
27. $-2x^2 + 6x - 14$
28. $-2x^2 + 8x - 12$
29. $-5y^2 + 40x$
30. $-9y^2 + 45x$
31. $-4x^3 + 32x^2 - 20x$
32. $-5x^3 + 50x^2 - 10x$
33. $-x^2 - 7x + 5$
34. $-x^2 - 8x + 8$

In Exercises 35–44, factor the greatest common binomial factor from each polynomial.

35. $4(x + 3) + a(x + 3)$
36. $5(x + 4) + a(x + 4)$
37. $x(y - 6) - 7(y - 6)$
38. $x(y - 9) - 5(y - 9)$
39. $3x(x + y) - (x + y)$
40. $7x(x + y) - (x + y)$
41. $4x^2(3x - 1) + 3x - 1$
42. $6x^2(5x - 1) + 5x - 1$
43. $(x + 2)(x + 3) + (x - 1)(x + 3)$
44. $(x + 4)(x + 5) + (x - 1)(x + 5)$

In Exercises 45–68, factor by grouping.

45. $x^2 + 3x + 5x + 15$
46. $x^2 + 2x + 4x + 8$
47. $x^2 + 7x - 4x - 28$
48. $x^2 + 3x - 5x - 15$
49. $x^3 - 3x^2 + 4x - 12$
50. $x^3 - 2x^2 + 5x - 10$
51. $xy - 6x + 2y - 12$

52. $xy - 5x + 9y - 45$
53. $xy + x - 7y - 7$
54. $xy + x - 5y - 5$
55. $10x^2 - 12xy + 35xy - 42y^2$
56. $3x^2 - 6xy + 5xy - 10y^2$
57. $4x^3 - x^2 - 12x + 3$
58. $3x^3 - 2x^2 - 6x + 4$
59. $x^2 - ax - bx + ab$
60. $x^2 + ax - bx - ab$
61. $x^3 - 12 - 3x^2 + 4x$
62. $2x^3 - 10 + 4x^2 - 5x$
63. $ay - by + bx - ax$
64. $cx - dx + dy - cy$
65. $ay^2 + 2by^2 - 3ax - 6bx$
66. $3a^2x + 6a^2y - 2bx - 4by$
67. $x^ny^n + 3x^n + y^n + 3$
68. $x^ny^n - x^n + 2y^n - 2$

Practice PLUS

In Exercises 69–78, factor each polynomial.

69. $ab - c - ac + b$
70. $ab - 3c - ac + 3b$
71. $x^3 - 5 + 4x^3y - 20y$
72. $x^3 - 2 + 3x^3y - 6y$
73. $2y^7(3x - 1)^5 - 7y^6(3x - 1)^4$
74. $3y^9(3x - 2)^7 - 5y^8(3x - 2)^6$
75. $ax^2 + 5ax - 2a + bx^2 + 5bx - 2b$
76. $ax^2 + 3ax - 11a + bx^2 + 3bx - 11b$
77. $ax + ay + az - bx - by - bz + cx + cy + cz$
78. $ax^2 + ay^2 - az^2 + bx^2 + by^2 - bz^2 + cx^2 + cy^2 - cz^2$

Application Exercises

79. A ball is thrown straight upward. The function

$$f(t) = -16t^2 + 40t$$

describes the ball's height above the ground, $f(t)$, in feet, t seconds after it is thrown.

a. Find and interpret $f(2)$.

b. Find and interpret $f(2.5)$.

c. Factor the polynomial $-16t^2 + 40t$ and write the function in factored form.

d. Use the factored form of the function to find $f(2)$ and $f(2.5)$. Do you get the same answers as you did in parts (a) and (b)? If so, does this prove that your factorization is correct? Explain.

80. An explosion causes debris to rise vertically. The function

$$f(t) = -16t^2 + 72t$$

describes the height of the debris above the ground, $f(t)$, in feet, t seconds after the explosion.

a. Find and interpret $f(2)$.

b. Find and interpret $f(4.5)$.

c. Factor the polynomial $-16t^2 + 72t$ and write the function in factored form.

d. Use the factored form of the function to find $f(2)$ and $f(4.5)$. Do you get the same answers as you did in parts (a) and (b)? If so, does this prove that your factorization is correct? Explain.

81. Your computer store is having an incredible sale. The price on one model is reduced by 40%. Then the sale price is reduced by another 40%. If x is the computer's original price, the sale price can be represented by

$$(x - 0.4x) - 0.4(x - 0.4x).$$

a. Factor out $(x - 0.4x)$ from each term. Then simplify the resulting expression.

b. Use the simplified expression from part (a) to answer these questions. With a 40% reduction followed by a 40% reduction, is the computer selling at 20% of its original price? If not, at what percentage of the original price is it selling?

82. Your local electronics store is having an end-of-the-year sale. The price on a plasma television had been reduced by 30%. Now the sale price is reduced by another 30%. If x is the television's original price, the sale price can be represented by

$$(x - 0.3x) - 0.3(x - 0.3x).$$

a. Factor out $(x - 0.3x)$ from each term. Then simplify the resulting expression.

b. Use the simplified expression from part (a) to answer these questions. With a 30% reduction followed by a 30% reduction, is the television selling at 40% of its original price? If not, at what percentage of the original price is it selling?

Exercises 83–84 involve compound interest. **Compound interest** *is interest computed on your original savings as well as on any accumulated interest.*

83. After 2 years, the balance, A, in an account with principal P and interest rate r compounded annually is given by the formula

$$A = P + Pr + (P + Pr)r.$$

Use factoring by grouping to express the formula as $A = P(1 + r)^2$.

84. After 3 years, the balance, A, in an account with principal P and interest rate r compounded annually is given by the formula

$$A = P(1 + r)^2 + P(1 + r)^2 r.$$

Use factoring by grouping to express the formula as $A = P(1 + r)^3$.

85. The area of the skating rink with semicircular ends shown is $A = \pi r^2 + 2rl$. Express the area, A, in factored form.

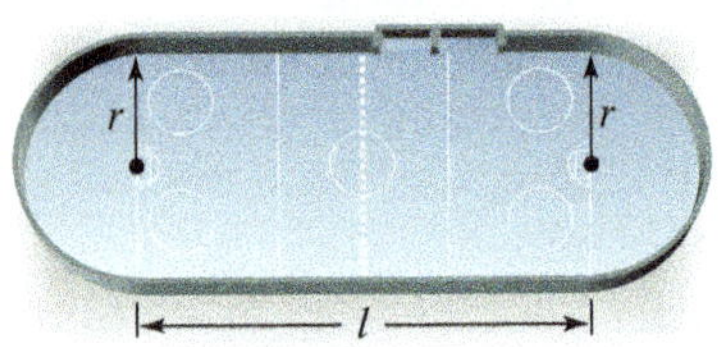

86. The amount of sheet metal needed to manufacture a cylindrical tin can, that is, its surface area, S, is $S = 2\pi r^2 + 2\pi rh$. Express the surface area, S, in factored form.

Writing in Mathematics

87. What is factoring?

88. If a polynomial has a greatest common factor other than 1, explain how to find its GCF.

89. Using an example, explain how to factor out the greatest common factor of a polynomial.

90. Suppose that a polynomial contains four terms and can be factored by grouping. Explain how to obtain the factorization.

91. Use two different groupings to factor

$$ac - ad + bd - bc$$

in two ways. Then explain why the two factorizations are the same.

92. Write a sentence that uses the word *factor* as a noun. Then write a sentence that uses the word *factor* as a verb.

Technology Exercises

In Exercises 93–96, use a graphing utility to graph the function on each side of the equation in the same viewing rectangle. Use end behavior to show a complete picture of the polynomial function on the left side. Do the graphs coincide? If so, this means that the polynomial on the left side has been factored correctly. If not, factor the polynomial correctly and then use your graphing utility to verify the factorization.

93. $x^2 - 4x = x(x - 4)$

94. $x^2 - 2x + 5x - 10 = (x - 2)(x - 5)$

95. $x^2 + 2x + x + 2 = x(x + 2) + 1$

96. $x^3 - 3x^2 + 4x - 12 = (x^2 + 4)(x - 3)$

Critical Thinking Exercises

Make Sense? *In Exercises 97–100, determine whether each statement "makes sense" or "does not make sense" and explain your reasoning.*

97. After I've factored a polynomial, my answer cannot always be checked by multiplication.

98. The word *greatest* in greatest common factor is helpful because it tells me to look for the greatest power of a variable appearing in all terms.

99. Although $20x^3$ appears in both $20x^3 + 8x^2$ and $20x^3 + 10x$, I'll need to factor $20x^3$ in different ways to obtain each polynomial's factorization.

100. You grouped the polynomial's terms using different groupings than I did, yet we both obtained the same factorization.

In Exercises 101–104, determine whether each statement is true or false. If the statement is false, make the necessary change(s) to produce a true statement.

101. Because the GCF of $9x^3 + 6x^2 + 3x$ is $3x$, it is not necessary to write the 1 when $3x$ is factored from the last term.

102. Some polynomials with four terms, such as $x^3 + x^2 + 4x - 4$, cannot be factored by grouping.

103. The polynomial $28x^3 - 7x^2 + 36x - 9$ can be factored by grouping terms as follows:

$$(28x^3 + 36x) + (-7x^2 - 9).$$

104. $x^2 - 2$ is a factor of $2 - 50x - x^2 + 25x^3$.

In Exercises 105–107, factor each polynomial. Assume that all variable exponents represent whole numbers.

105. $x^{4n} + x^{2n} + x^{3n}$

106. $3x^{3m}y^m - 6x^{2m}y^{2m}$

107. $8y^{2n+4} + 16y^{2n+3} - 12y^{2n}$

In Exercises 108–109, write a polynomial that fits the given description. Do not use a polynomial that appeared in this section or in the exercise set.

108. The polynomial has three terms and can be factored using a greatest common factor that has both a negative coefficient and a variable.

109. The polynomial has four terms and can be factored by grouping.

Review Exercises

110. Solve by Cramer's rule:

$$3x - 2y = 8$$
$$2x - 5y = 10.$$

(Section 3.5, Example 2)

111. Determine whether each relation is a function.

a. $\{(0, 5), (3, -5), (5, 5), (7, -5)\}$

b. $\{(1, 2), (3, 4), (5, 5), (5, 6)\}$ (Section 2.1, Example 2)

112. The length of a rectangle is 2 feet greater than twice its width. If the rectangle's perimeter is 22 feet, find the length and width. (Section 1.5, Example 5)

Preview Exercises

Exercises 113–115 will help you prepare for the material covered in the next section. In each exercise, replace the boxed question mark with an integer that results in the given product. Some trial and error may be necessary.

113. $(x + 3)(x + \boxed{?}) = x^2 + 7x + 12$

114. $(x - \boxed{?})(x - 12) = x^2 - 14x + 24$

115. $(x + 3y)(x - \boxed{?}y) = x^2 - 4xy - 21y^2$

SECTION 5.4

Factoring Trinomials

Objectives

1. Factor a trinomial whose leading coefficient is 1.
2. Factor using a substitution.
3. Factor a trinomial whose leading coefficient is not 1.
4. Factor trinomials by grouping.

A great deal of trial and error is involved in finding your way out of this maze. Trial and error play an important role in problem solving and can be helpful in leading to correct solutions. In this section, you will use trial and error to factor trinomials following a problem-solving process that is not very different from learning to traverse the maze.

1 Factor a trinomial whose leading coefficient is 1.

Factoring a Trinomial Whose Leading Coefficient Is 1

In Section 5.2, we used the FOIL method to multiply two binomials. The product was often a trinomial. The following are some examples:

Factored Form		F O I L		Trinomial Form
$(x + 3)(x + 4)$	=	$x^2 + 4x + 3x + 12$	=	$x^2 + 7x + 12$
$(x - 3)(x - 4)$	=	$x^2 - 4x - 3x + 12$	=	$x^2 - 7x + 12$
$(x + 3)(x - 5)$	=	$x^2 - 5x + 3x - 15$	=	$x^2 - 2x - 15$

Observe that each trinomial is of the form $x^2 + bx + c$, where the coefficient of the squared term is 1. Our goal in the first part of this section is to start with the trinomial form and, assuming that it is factorable, return to the factored form.

The first FOIL multiplication shown in our list indicates that

$$(x + 3)(x + 4) = x^2 + 7x + 12.$$

Let's reverse the sides of this equation:

$$x^2 + 7x + 12 = (x + 3)(x + 4).$$

We can use $x^2 + 7x + 12 = (x + 3)(x + 4)$ to make several important observations about the factors on the right side.

$$x^2 + 7x + 12 = (x + 3)(x + 4)$$

The first term of each factor is x. The product of the First terms is $x \cdot x = x^2$.

$$x^2 + 7x + 12 = (x + 3)(x + 4)$$

3 and 4 are factors of 12. The product of the Last terms is $3 \cdot 4 = 12$.

$$x^2 + 7x + 12 = (x + 3)(x + 4)$$

I: $3x$

O: $4x$

The sum of the Outside and Inside products is $4x + 3x = 7x$.

These observations provide us with a procedure for factoring $x^2 + bx + c$.

A Strategy for Factoring $x^2 + bx + c$

1. Enter x as the first term of each factor.

$$(x \qquad)(x \qquad) = x^2 + bx + c$$

2. List pairs of factors of the constant c.
3. Try various combinations of these factors. Select the combination in which the sum of the Outside and Inside products is equal to bx.

$$(x + \square)(x + \square) = x^2 + bx + c$$

I

O

Sum of O + I

4. Check your work by multiplying the factors using the FOIL method. You should obtain the original trinomial.

If none of the possible combinations yield an Outside product and an Inside product whose sum is equal to bx, the trinomial cannot be factored using integers and is called **prime** over the set of integers.

EXAMPLE 1 Factoring a Trinomial Whose Leading Coefficient Is 1

Factor: $x^2 + 5x + 6$.

Solution

Step 1. Enter x as the first term of each factor.

$$x^2 + 5x + 6 = (x \qquad)(x \qquad)$$

Step 2. List pairs of factors of the constant, 6.

Factors of 6	6, 1	3, 2	−6, −1	−3, −2

Step 3. Try various combinations of these factors. The correct factorization of $x^2 + 5x + 6$ is the one in which the sum of the Outside and Inside products is equal to $5x$. At the top of the next page is a list of the possible factorizations.

Possible Factorizations of $x^2 + 5x + 6$	Sum of Outside and Inside Products (Should Equal $5x$)
$(x + 6)(x + 1)$	$x + 6x = 7x$
$(x + 3)(x + 2)$	$2x + 3x = 5x$
$(x - 6)(x - 1)$	$-x - 6x = -7x$
$(x - 3)(x - 2)$	$-2x - 3x = -5x$

This is the required middle term.

Thus,

$$x^2 + 5x + 6 = (x + 3)(x + 2).$$

Check this result by multiplying the right side using the FOIL method. You should obtain the original trinomial. Because of the commutative property, the factorization can also be expressed as

$$x^2 + 5x + 6 = (x + 2)(x + 3).$$

In factoring a trinomial of the form $x^2 + bx + c$, you can speed things up by listing the factors of c and then finding their sums. We are interested in a sum of b. For example, in factoring $x^2 + 5x + 6$, we are interested in the factors of 6 whose sum is 5.

Factors of 6	6, 1	3, 2	−6, −1	−3, −2
Sum of Factors	7	5	−7	−5

This is the desired sum.

Thus, $x^2 + 5x + 6 = (x + 3)(x + 2)$.

Using Technology

Numeric and Graphic Connections

If a polynomial contains one variable, a graphing utility can be used to check its factorization. For example, the factorization in Example 1 can be checked graphically or numerically.

Given polynomial — The factorization we are checking

$$x^2 + 5x + 6 = (x + 3)(x + 2)$$

Enter $y_1 = x^2 + 5x + 6$ in the [y=] screen. — Enter $y_2 = (x + 3)(x + 2)$ in the [y=] screen.

Numeric Check

Use the [TABLE] feature.

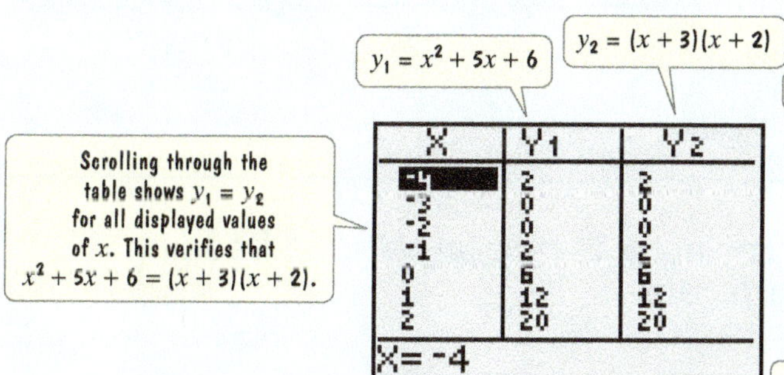

Graphic Check

Use the [GRAPH] feature to display graphs for y_1 and y_2.

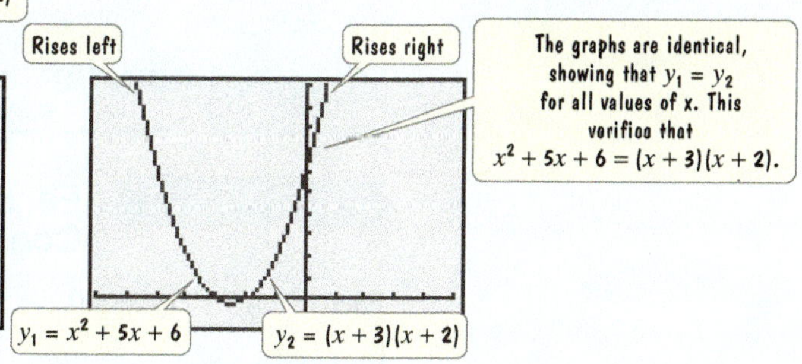

[−7, 5, 1] by [−1, 10, 1]

Notice that the graph of the quadratic function is shaped like a bowl. The graph of the even-degree quadratic function exhibits the same behavior at each end, rising to the left and rising to the right (↖, ↗).

CHECK POINT 1 Factor: $x^2 + 6x + 8$.

EXAMPLE 2 Factoring a Trinomial Whose Leading Coefficient Is 1

Factor: $x^2 - 14x + 24$.

Solution

Step 1. Enter x as the first term of each factor.

$$x^2 - 14x + 24 = (x \quad)(x \quad)$$

To find the second term of each factor, we must find two integers whose product is 24 and whose sum is -14.

Step 2. List pairs of factors of the constant, 24. Because the desired sum, -14, is negative, we will list only the negative pairs of factors of 24.

Negative Factors of 24	−24, −1	−12, −2	−8, −3	−6, −4

Step 3. Try various combinations of these factors. We are interested in the factors whose sum is -14.

Negative Factors of 24	−24, −1	−12, −2	−8, −3	−6, −4
Sum of Factors	−25	−14	−11	−10

This is the desired sum.

Thus, $x^2 - 14x + 24 = (x - 12)(x - 2)$.

Study Tip

To factor $x^2 + bx + c$ when c is positive, find two numbers with the same sign as the middle term.

$$x^2 + 5x + 6 = (x + 3)(x + 2)$$

Same signs

$$x^2 - 14x + 24 = (x - 12)(x - 2)$$

Same signs

CHECK POINT 2 Factor: $x^2 - 9x + 20$.

EXAMPLE 3 Factoring a Trinomial Whose Leading Coefficient Is 1

Factor: $y^2 + 7y - 60$.

Solution

Step 1. Enter y as the first term of each factor.

$$y^2 + 7y - 60 = (y \quad)(y \quad)$$

To find the second term of each factor, we must find two integers whose product is -60 and whose sum is 7.

Steps 2 and 3. List pairs of factors of the constant, −60, and try various combinations of these factors. Because the desired sum, 7, is positive, the positive factor of −60 must be farther from 0 than the negative factor is. Thus, we will only list pairs of factors of −60 in which the positive factor has the larger absolute value.

Some Factors of −60	60, −1	30, −2	20, −3	15, −4	12, −5	10, −6
Sum of Factors	59	28	17	11	7	4

This is the desired sum.

Thus, $y^2 + 7y - 60 = (y + 12)(y - 5)$.

Study Tip

To factor $x^2 + bx + c$ when c is negative, find two numbers with opposite signs whose sum is the coefficient of the middle term.

$$y^2 + 7y - 60 = (y + 12)(y - 5)$$

Negative — Opposite signs

CHECK POINT 3 Factor: $y^2 + 19y - 66$.

EXAMPLE 4 Factoring a Trinomial in Two Variables

Factor: $x^2 - 4xy - 21y^2$.

Solution

Step 1. Enter x as the first term of each factor. Because the last term of the trinomial contains y^2, the second term of each factor must contain y.

$$x^2 - 4xy - 21y^2 = (x \quad ?y)(x \quad ?y)$$

The question marks indicate that we are looking for the coefficients of y in each factor. To find these coefficients, we must find two integers whose product is −21 and whose sum is −4.

Steps 2 and 3. List pairs of factors of the coefficient of the last term, −21, and try various combinations of these factors. We are interested in the factors whose sum is −4.

Factors of −21	1, −21	3, −7	−1, 21	−3, 7
Sum of Factors	−20	−4	20	4

This is the desired sum.

Thus, $x^2 - 4xy - 21y^2 = (x + 3y)(x - 7y)$ or $(x - 7y)(x + 3y)$.

Step 4. Verify the factorization using the FOIL method.

$$(x + 3y)(x - 7y) = x^2 - 7xy + 3xy - 21y^2 = x^2 - 4xy - 21y^2$$

Because the product of the factors is the original polynomial, the factorization is correct.

CHECK POINT 4 Factor: $x^2 - 5xy + 6y^2$.

Can every trinomial be factored? The answer is no. For example, consider

$$x^2 + x - 5 = (x \quad)(x \quad).$$

To find the second term of each factor, we must find two integers whose product is -5 and whose sum is 1. Because no such integers exist, $x^2 + x - 5$ cannot be factored. This trinomial is prime.

To factor some polynomials, more than one technique must be used. **Always begin by trying to factor out the greatest common factor.** A polynomial is **factored completely** when it is written as the product of prime polynomials.

Study Tip

The technology box on page 341 shows how to use a graphing utility to check a polynomial's factorization graphically or numerically. The graphing utility verifies that the factorization is equivalent to the original polynomial. However, a graphing utility cannot verify that the factorization is complete.

EXAMPLE 5 Factoring Completely

Factor: $8x^3 - 40x^2 - 48x$.

Solution The GCF of the three terms of the polynomial is $8x$. We begin by factoring out $8x$. Then we factor the remaining trinomial.

$$8x^3 - 40x^2 - 48x$$

$$= 8x(x^2 - 5x - 6) \quad \text{Factor out the GCF.}$$

$$= 8x(x \quad)(x \quad) \quad \text{Begin factoring } x^2 - 5x - 6. \text{ Find two integers whose product is } -6 \text{ and whose sum is } -5.$$

$$= 8x(x - 6)(x + 1) \quad \text{The integers are } -6 \text{ and } 1.$$

Thus,

$$8x^3 - 40x^2 - 48x = 8x(x - 6)(x + 1).$$

Be sure to include the GCF in the factorization.

You can check this factorization by multiplying the binomials using the FOIL method. Then use the distributive property and multiply each term in this product by $8x$. Try doing this now. Because the factorization is correct, you should obtain the original polynomial.

CHECK POINT 5 Factor: $3x^3 - 15x^2 - 42x$.

Some trinomials, such as $-x^2 + 5x + 6$, have a leading coefficient of -1. Because it is easier to factor a trinomial with a positive leading coefficient, begin by factoring out -1. For example,

$$-x^2 + 5x + 6 = -1(x^2 - 5x - 6) = -(x - 6)(x + 1).$$

2 Factor using a substitution.

In some trinomials, the highest power is greater than 2, and the exponent in one of the terms is half that of the other term. By letting u equal the variable to the smaller power, the trinomial can be written in a form that makes its possible factorization more obvious. Here are some examples:

Given Trinomial	Substitution	New Trinomial
$x^6 - 8x^3 + 15$ or $(x^3)^2 - 8x^3 + 15$	$u = x^3$	$u^2 - 8u + 15$
$x^4 - 8x^2 - 9$ or $(x^2)^2 - 8x^2 - 9$	$u = x^2$	$u^2 - 8u - 9$

In each case, we factor the given trinomial by working with the new trinomial on the right. If a factorization is found, we replace all occurrences of u in the factorization with the substitution shown in the middle column.

EXAMPLE 6 Factoring by Substitution

Factor: $x^6 - 8x^3 + 15$.

Solution Notice that the exponent on x^3 is half that of the exponent on x^6. We will let u equal the variable to the power that is half of 6. Thus, let $u = x^3$.

$$(x^3)^2 - 8x^3 + 15 \quad \text{This is the given polynomial, with } x^6 \text{ written as } (x^3)^2.$$
$$= u^2 - 8u + 15 \quad \text{Let } u = x^3. \text{ Rewrite the trinomial in terms of } u.$$
$$= (u - 5)(u - 3) \quad \text{Factor.}$$
$$= (x^3 - 5)(x^3 - 3) \quad \text{Now substitute } x^3 \text{ for } u.$$

Thus, the given trinomial can be factored as

$$x^6 - 8x^3 + 15 = (x^3 - 5)(x^3 - 3).$$

Check this result using FOIL multiplication on the right.

CHECK POINT 6 Factor: $x^6 - 7x^3 + 10$.

3 Factor a trinomial whose leading coefficient is not 1.

Factoring a Trinomial Whose Leading Coefficient Is Not 1

How do we factor a trinomial such as $5x^2 - 14x + 8$? Notice that the leading coefficient is 5. We must find two binomials whose product is $5x^2 - 14x + 8$. The product of the First terms must be $5x^2$:

$$(5x \quad)(x \quad).$$

From this point on, the factoring strategy is exactly the same as the one we use to factor a trinomial whose leading coefficient is 1.

A Strategy for Factoring $ax^2 + bx + c$

Assume, for the moment, that there is no greatest common factor.

1. Find two **F**irst terms whose product is ax^2:

$$(\square x + \quad)(\square x + \quad) = ax^2 + bx + c.$$

2. Find two **L**ast terms whose product is c:

$$(\square x + \square)(\square x + \square) = ax^2 + bx + c.$$

3. By trial and error, perform steps 1 and 2 until the sum of the **O**utside product and **I**nside product is bx:

$$(\square x + \square)(\square x + \square) = ax^2 + bx + c.$$

I

O

Sum of O + I

If no such combinations exist, the polynomial is prime.

Study Tip

The *error* part of the factoring strategy plays an important role in the process. If you do not get the correct factorization the first time, this is not a bad thing. This error is often helpful in leading you to the correct factorization.

EXAMPLE 7 Factoring a Trinomial Whose Leading Coefficient Is Not 1

Factor: $5x^2 - 14x + 8$.

Solution

Step 1. Find two First terms whose product is $5x^2$.

$$5x^2 - 14x + 8 = (5x \quad)(x \quad)$$

Step 2. Find two Last terms whose product is 8. The number 8 has pairs of factors that are either both positive or both negative. Because the middle term, $-14x$, is negative, both factors must be negative. The negative factorizations of 8 are $-1(-8)$ and $-2(-4)$.

Step 3. Try various combinations of these factors. The correct factorization of $5x^2 - 14x + 8$ is the one in which the sum of the Outside and Inside products is equal to $-14x$. Here is a list of the possible factorizations:

Study Tip

With practice, you will find that you don't have to list every possible factorization of the trinomial. As you practice factoring, you will be able to narrow down the list of possible factorizations to just a few. When it comes to factoring, practice makes perfect.

Possible Factorizations of $5x^2 - 14x + 8$	Sum of Outside and Inside Products (Should Equal $-14x$)
$(5x - 1)(x - 8)$	$-40x - x = -41x$
$(5x - 8)(x - 1)$	$-5x - 8x = -13x$
$(5x - 2)(x - 4)$	$-20x - 2x = -22x$
$(5x - 4)(x - 2)$	$-10x - 4x = -14x$

This is the required middle term.

Thus,

$$5x^2 - 14x + 8 = (5x - 4)(x - 2).$$

Show that this factorization is correct by multiplying the factors using the FOIL method. You should obtain the original trinomial.

✓ **CHECK POINT 7** Factor: $3x^2 - 20x + 28$.

EXAMPLE 8 Factoring a Trinomial Whose Leading Coefficient Is Not 1

Factor: $8x^6 - 10x^5 - 3x^4$.

Solution The GCF of the three terms of the polynomial is x^4. We begin by factoring out x^4.

$$8x^6 - 10x^5 - 3x^4 = x^4(8x^2 - 10x - 3)$$

Now we factor the remaining trinomial, $8x^2 - 10x - 3$.

Step 1. Find two First terms whose product is $8x^2$.

$$8x^2 - 10x - 3 \stackrel{?}{=} (8x \quad)(x \quad)$$

$$8x^2 - 10x - 3 \stackrel{?}{=} (4x \quad)(2x \quad)$$

Step 2. Find two Last terms whose product is -3. The possible factorizations are $1(-3)$ and $-1(3)$.

Step 3. Try various combinations of these factors. The correct factorization of $8x^2 - 10x - 3$ is the one in which the sum of the Outside and Inside products is equal to $-10x$. At the top of the next page is a list of the possible factorizations.

Study Tip

Here are some suggestions for reducing the list of possible factorizations for $ax^2 + bx + c$.

1. If b is relatively small, avoid the larger factors of a.
2. If c is positive, the signs in both binomial factors must match the sign of b.
3. If the trinomial has no common factor, no binomial factor can have a common factor.
4. Reversing the signs in the binomial factors reverses the sign of bx, the middle term.

Possible Factorizations of $8x^2 - 10x - 3$	Sum of Outside and Inside Products (Should Equal $-10x$)
$(8x + 1)(x - 3)$	$-24x + x = -23x$
$(8x - 3)(x + 1)$	$8x - 3x = 5x$
$(8x - 1)(x + 3)$	$24x - x = 23x$
$(8x + 3)(x - 1)$	$-8x + 3x = -5x$
$(4x + 1)(2x - 3)$	$-12x + 2x = -10x$
$(4x - 3)(2x + 1)$	$4x - 6x = -2x$
$(4x - 1)(2x + 3)$	$12x - 2x = 10x$
$(4x + 3)(2x - 1)$	$-4x + 6x = 2x$

This is the required middle term.

The factorization of $8x^2 - 10x - 3$ is $(4x + 1)(2x - 3)$. Now we include the GCF in the complete factorization of the given polynomial. Thus,

$$8x^6 - 10x^5 - 3x^4 = x^4(8x^2 - 10x - 3) = x^4(4x + 1)(2x - 3).$$

This is the complete factorization with the GCF, x^4, included.

CHECK POINT 8 Factor: $6x^6 + 19x^5 - 7x^4$.

We have seen that not every trinomial can be factored. For example, consider

$$6x^2 + 14x + 7 = (6x + \square)(x + \square)$$
$$6x^2 + 14x + 7 = (3x + \square)(2x + \square).$$

The possible factors for the last term are 1 and 7. However, regardless of how these factors are placed in the boxes shown, the sum of the Outside and Inside products is not equal to $14x$. Thus, the trinomial $6x^2 + 14x + 7$ cannot be factored and is prime.

EXAMPLE 9 Factoring a Trinomial in Two Variables

Factor: $3x^2 - 13xy + 4y^2$.

Solution

Step 1. Find two First terms whose product is $3x^2$.

$$3x^2 - 13xy + 4y^2 = (3x \quad ?y)(x \quad ?y)$$

The question marks indicate that we are looking for the coefficients of y in each factor.

Steps 2 and 3. List pairs of factors of the coefficient of the last term, 4, and try various combinations of these factors. The correct factorization is the one in which the sum of the Outside and Inside products is equal to $-13xy$. Because of the negative coefficient, -13, we will consider only the negative pairs of factors of 4. The possible factorizations are $-1(-4)$ and $-2(-2)$.

Possible Factorizations of $3x^2 - 13xy + 4y^2$	Sum of Outside and Inside Products (Should Equal $-13xy$)
$(3x - y)(x - 4y)$	$-12xy - xy = -13xy$
$(3x - 4y)(x - y)$	$-3xy - 4xy = -7xy$
$(3x - 2y)(x - 2y)$	$-6xy - 2xy = -8xy$

This is the required middle term.

Thus,

$$3x^2 - 13xy + 4y^2 = (3x - y)(x - 4y).$$

CHECK POINT 9 Factor: $2x^2 - 7xy + 3y^2$.

EXAMPLE 10 Factoring by Substitution

Factor: $6y^4 + 13y^2 + 6$.

Solution Notice that the exponent on y^2 is half that of the exponent on y^4. We will let u equal the variable to the smaller power. Thus, let $u = y^2$.

$$6(y^2)^2 + 13y^2 + 6$$ This is the given polynomial, with y^4 written as $(y^2)^2$.

$$= 6u^2 + 13u + 6$$ Let $u = y^2$. Rewrite the trinomial in terms of u.

$$= (3u + 2)(2u + 3)$$ Factor the trinomial.

$$= (3y^2 + 2)(2y^2 + 3)$$ Now substitute y^2 for u.

Therefore, $6y^4 + 13y^2 + 6 = (3y^2 + 2)(2y^2 + 3)$. Check using FOIL multiplication.

CHECK POINT 10 Factor: $3y^4 + 10y^2 - 8$.

4 Factor trinomials by grouping.

Factoring Trinomials by Grouping

A second method for factoring $ax^2 + bx + c$, $a \neq 1$, is called the **grouping method**. This method involves both trial and error, as well as grouping. The trial and error in factoring $ax^2 + bx + c$ depends upon finding two numbers, p and q, for which $p + q = b$. Then we factor $ax^2 + px + qx + c$ using grouping.

Let's see how this works by looking at a particular factorization:

$$15x^2 - 7x - 2 = (3x - 2)(5x + 1).$$

If we multiply using FOIL on the right, we obtain

$$(3x - 2)(5x + 1) = 15x^2 + 3x - 10x - 2.$$

In this case, the desired numbers, p and q, are $p = 3$ and $q = -10$. Compare these numbers to ac and b in the given polynomial.

$a = 15$ $b = -7$ $c = -2$

$$15x^2 - 7x - 2$$

$ac = 15(-2) = -30$

Can you see that p and q, 3 and -10, are factors of ac, or -30? Furthermore, p and q have a sum of b, namely -7. By expressing the middle term, $-7x$, in terms of p and q, we can factor by grouping as follows:

$$15x^2 - 7x - 2$$

$$= 15x^2 + (3x - 10x) - 2$$ Rewrite $-7x$ as $3x - 10x$.

$$= (15x^2 + 3x) + (-10x - 2)$$ Group terms.

$$= 3x(5x + 1) - 2(5x + 1)$$ Factor from each group.

$$= (5x + 1)(3x - 2).$$ Factor out $5x + 1$, the common binomial factor.

Factoring $ax^2 + bx + c$ Using Grouping ($a \neq 1$)

1. Multiply the leading coefficient, a, and the constant, c.
2. Find the factors of ac whose sum is b.
3. Rewrite the middle term, bx, as a sum or difference using the factors from step 2.
4. Factor by grouping.

EXAMPLE 11 Factoring a Trinomial by Grouping

Factor by grouping: $12x^2 - 5x - 2$.

Solution The trinomial is of the form $ax^2 + bx + c$.

$$12x^2 - 5x - 2$$

$a = 12$ $b = -5$ $c = -2$

Step 1. Multiply the leading coefficient, a, and the constant, c. Using $a = 12$ and $c = -2$,

$$ac = 12(-2) = -24.$$

Step 2. Find the factors of ac whose sum is b. We want the factors of -24 whose sum is b, or -5. The factors of -24 whose sum is -5 are -8 and 3.

Step 3. Rewrite the middle term, $-5x$, as a sum or difference using the factors from step 2, -8 and 3.

$$12x^2 - 5x - 2 = 12x^2 - 8x + 3x - 2$$

Step 4. Factor by grouping.

$= (12x^2 - 8x) + (3x - 2)$ Group terms.

$= 4x(3x - 2) + 1(3x - 2)$ Factor from each group.

$= (3x - 2)(4x + 1)$ Factor out $3x - 2$, the common binomial factor.

Thus,

$$12x^2 - 5x - 2 = (3x - 2)(4x + 1).$$

Discover for Yourself

In step 2, we found that the desired numbers were -8 and 3. We wrote $-5x$ as $-8x + 3x$. What happens if we write $-5x$ as $3x - 8x$? Use factoring by grouping on

$$12x^2 - 5x - 2$$
$$= 12x^2 + 3x - 8x - 12.$$

Describe what happens.

CHECK POINT 11 Factor by grouping: $8x^2 - 22x + 5$.

5.4 EXERCISE SET

MyMathLab MathXL PRACTICE WATCH DOWNLOAD READ REVIEW

Practice Exercises

In Exercises 1–30, factor each trinomial, or state that the trinomial is prime. Check each factorization using FOIL multiplication.

1. $x^2 + 5x + 6$
2. $x^2 + 10x + 9$
3. $x^2 + 8x + 12$
4. $x^2 + 8x + 15$
5. $x^2 + 9x + 20$
6. $x^2 + 11x + 24$
7. $y^2 + 10y + 16$
8. $y^2 + 9y + 18$
9. $x^2 - 8x + 15$
10. $x^2 - 5x + 6$
11. $y^2 - 12y + 20$
12. $y^2 - 25y + 24$
13. $a^2 + 5a - 14$
14. $a^2 + a - 12$
15. $x^2 + x - 30$

16. $x^2 + 14x - 32$
17. $x^2 - 3x - 28$
18. $x^2 - 4x - 21$
19. $y^2 - 5y - 36$
20. $y^2 - 3y - 40$
21. $x^2 - x + 7$
22. $x^2 + 3x + 8$
23. $x^2 - 9xy + 14y^2$
24. $x^2 - 8xy + 15y^2$
25. $x^2 - xy - 30y^2$
26. $x^2 - 3xy - 18y^2$
27. $x^2 + xy + y^2$
28. $x^2 - xy + y^2$
29. $a^2 - 18ab + 80b^2$
30. $a^2 - 18ab + 45b^2$

In Exercises 31–38, factor completely.

31. $3x^2 + 3x - 18$
32. $4x^2 - 4x - 8$
33. $2x^3 - 14x^2 + 24x$
34. $2x^3 + 6x^2 + 4x$
35. $3y^3 - 15y^2 + 18y$
36. $4y^3 + 12y^2 - 72y$
37. $2x^4 - 26x^3 - 96x^2$
38. $3x^4 + 54x^3 + 135x^2$

In Exercises 39–44, factor by introducing an appropriate substitution.

39. $x^6 - x^3 - 6$
40. $x^6 + x^3 - 6$
41. $x^4 - 5x^2 - 6$
42. $x^4 - 4x^2 - 5$
43. $(x + 1)^2 + 6(x + 1) + 5$ (Let $u = x + 1$.)
44. $(x + 1)^2 + 8(x + 1) + 7$ (Let $u = x + 1$.)

In Exercises 45–68, use the method of your choice to factor each trinomial, or state that the trinomial is prime. Check each factorization using FOIL multiplication.

45. $3x^2 + 8x + 5$
46. $2x^2 + 9x + 7$
47. $5x^2 + 56x + 11$
48. $5x^2 - 16x + 3$
49. $3y^2 + 22y - 16$
50. $5y^2 + 33y - 14$
51. $4y^2 + 9y + 2$
52. $8y^2 + 10y + 3$
53. $10x^2 + 19x + 6$
54. $6x^2 + 19x + 15$
55. $8x^2 - 18x + 9$
56. $4x^2 - 27x + 18$
57. $6y^2 - 23y + 15$
58. $16y^2 - 6y - 27$
59. $6y^2 + 14y + 3$
60. $4y^2 + 22y - 5$
61. $3x^2 + 4xy + y^2$
62. $2x^2 + 3xy + y^2$
63. $6x^2 - 7xy - 5y^2$
64. $6x^2 - 5xy - 6y^2$
65. $15x^2 - 31xy + 10y^2$
66. $15x^2 + 11xy - 14y^2$
67. $3a^2 - ab - 14b^2$
68. $15a^2 - ab - 6b^2$

In Exercises 69–82, factor completely.

69. $15x^3 - 25x^2 + 10x$
70. $10x^3 + 24x^2 + 14x$
71. $24x^4 + 10x^3 - 4x^2$
72. $15x^4 - 39x^3 + 18x^2$
73. $15y^5 - 2y^4 - y^3$
74. $10y^5 - 17y^4 + 3y^3$
75. $24x^2 + 3xy - 27y^2$
76. $12x^2 + 10xy - 8y^2$
77. $6a^2b - 2ab - 60b$
78. $8a^2b + 34ab - 84b$
79. $12x^2y - 34xy^2 + 14y^3$
80. $12x^2y - 46xy^2 + 14y^3$
81. $13x^3y^3 + 39x^3y^2 - 52x^3y$
82. $4x^3y^5 + 24x^2y^5 - 64xy^5$

In Exercises 83–92, factor by introducing an appropriate substitution.

83. $2x^4 - x^2 - 3$
84. $5x^4 + 2x^2 - 3$
85. $2x^6 + 11x^3 + 15$
86. $2x^6 + 13x^3 + 15$
87. $2y^{10} + 7y^5 + 3$
88. $5y^{10} + 29y^5 - 42$
89. $5(x + 1)^2 + 12(x + 1) + 7$ (Let $u = x + 1$.)
90. $3(x + 1)^2 - 5(x + 1) + 2$ (Let $u = x + 1$.)
91. $2(x - 3)^2 - 5(x - 3) - 7$
92. $3(x - 2)^2 - 5(x - 2) - 2$

Practice PLUS

In Exercises 93–100, factor completely.

93. $x^2 - 0.5x + 0.06$
94. $x^2 + 0.3x - 0.04$
95. $x^2 - \frac{3}{49} + \frac{2}{7}x$
96. $x^2 - \frac{6}{25} + \frac{1}{5}x$

97. $acx^2 - bcx + adx - bd$

98. $acx^2 - bcx - adx + bd$

99. $-4x^5y^2 + 7x^4y^3 - 3x^3y^4$

100. $-5x^4y^3 + 7x^3y^4 - 2x^2y^5$

101. If $(fg)(x) = 3x^2 - 22x + 39$, find f and g.

102. If $(fg)(x) = 4x^2 - x - 5$, find f and g.

In Exercises 103–104, a large rectangle formed by a number of smaller rectangles is shown. Factor the sum of the areas of the smaller rectangles to determine the dimensions of the large rectangle.

103.

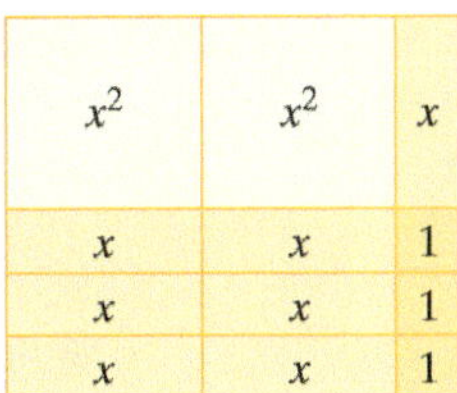

104.

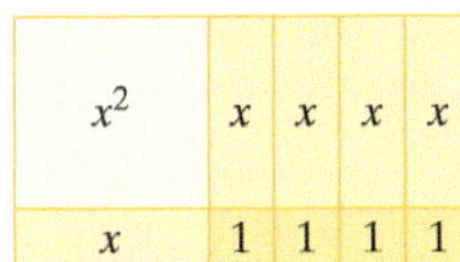

Application Exercises

105. A diver jumps directly upward from a board that is 32 feet high. The function

$$f(t) = -16t^2 + 16t + 32$$

describes the driver's height above the water, $f(t)$, in feet, after t seconds.

a. Find and interpret $f(1)$.

b. Find and interpret $f(2)$.

c. Factor the expression for $f(t)$ and write the function in completely factored form.

d. Use the factored form of the function to find $f(1)$ and $f(2)$.

106. The function $V(x) = 3x^3 - 2x^2 - 8x$ describes the volume, $V(x)$, in cubic inches, of the box shown whose height is x inches.

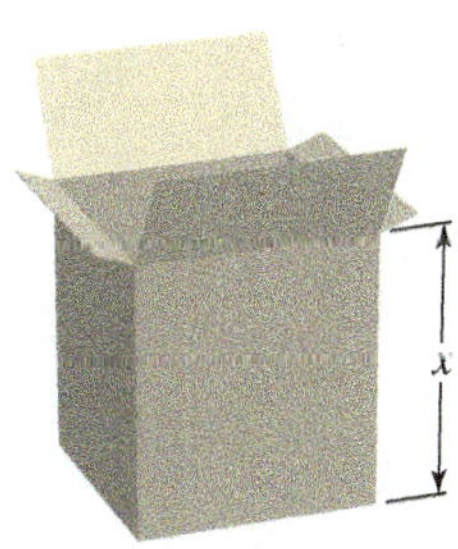

a. Find and interpret $V(4)$.

b. Factor the expression for $V(x)$ and write the function in completely factored form.

c. Use the factored form of the function to find $V(4)$ and $V(5)$.

107. Find the area of the large rectangle shown below in two ways.

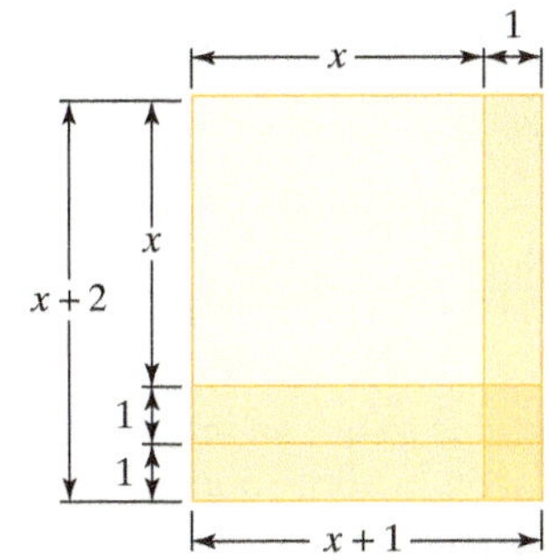

a. Find the sum of the areas of the six smaller rectangles and squares.

b. Express the area of the large rectangle as the product of its length and width.

c. Explain how the figure serves as a geometric model for the factorization of the sum that you wrote in part (a).

108. If x represents a positive integer, factor $x^3 + 3x^2 + 2x$ to show that the trinomial represents the product of three consecutive integers.

Writing in Mathematics

109. Explain how to factor $x^2 + 8x + 15$.

110. Give two helpful suggestions for factoring $x^2 - 5x + 6$.

111. In factoring $x^2 + bx + c$, describe how the last term in each binomial factor is related to b and to c.

112. Describe the first thing that you should try doing when factoring a polynomial.

113. What does it mean to factor completely?

114. Explain how to factor $x^6 - 7x^3 + 10$ by substitution.

115. Is it possible to factor $x^6 - 7x^3 + 10$ without using substitution? How might this be done?

116. Explain how to factor $2x^2 - x - 1$.

Technology Exercises

In Exercises 117–120, use a graphing utility to graph the function on each side of the equation in the same viewing rectangle. Use end behavior to show a complete picture of the polynomial function on the left side. Do the graphs coincide? If so, this means that the polynomial on the left side has been factored correctly. If not, factor the polynomial correctly and then use your graphing utility to verify the factorization.

117. $x^2 + 7x + 12 = (x + 4)(x + 3)$

118. $x^2 - 7x + 6 = (x - 2)(x - 3)$

119. $6x^3 + 5x^2 - 4x = x(3x + 4)(2x - 1)$

120. $x^4 - x^2 - 20 = (x^2 + 5)(x^2 - 4)$

121. Use the TABLE feature of a graphing utility to verify any two of your factorizations in Exercises 39–44.

Critical Thinking Exercises

Make Sense? *In Exercises 122–125, determine whether each statement "makes sense" or "does not make sense" and explain your reasoning.*

122. Although $(x + 2)(x - 5)$ is the same as $(x - 5)(x + 2)$, the factorization $(2 - x)(2 + x)$ is not the same as $-(x - 2)(x + 2)$.

123. I'm often able to use an incorrect factorization to lead me to the correct factorization.

124. My graphing calculator showed the same graph for $y_1 = 20x^3 - 70x^2 + 60x$ and $y_2 = 10x(2x^2 - 7x + 6)$, so I can conclude that the complete factorization of $20x^3 - 70x^2 + 60x$ is $10x(2x^2 - 7x + 6)$.

125. First factoring out the greatest common factor makes it easier for me to determine how to factor the remaining factor, assuming that it is not prime.

In Exercises 126–129, determine whether each statement is true or false. If the statement is false, make the necessary change(s) to produce a true statement.

126. Once a GCF is factored from $6y^6 - 19y^5 + 10y^4$, the remaining trinomial factor is prime.

127. One factor of $8y^2 - 51y + 18$ is $8y - 3$.

128. We can immediately tell that $6x^2 - 11xy - 10y^2$ is prime because 11 is a prime number and the polynomial contains two variables.

129. A factor of $12x^2 - 19xy + 5y^2$ is $4x - y$.

In Exercises 130–131, find all integers b so that the trinomial can be factored.

130. $4x^2 + bx - 1$

131. $3x^2 + bx + 5$

In Exercises 132–137, factor each polynomial. Assume that all variable exponents represent whole numbers.

132. $9x^{2n} + x^n - 8$

133. $4x^{2n} - 9x^n + 5$

134. $a^{2n+2} - a^{n+2} - 6a^2$

135. $b^{2n+2} + 3b^{n+2} - 10b^2$

136. $3c^{n+2} - 10c^{n+1} + 3c^n$

137. $2d^{n+2} - 5d^{n+1} + 3d^n$

Review Exercises

138. Solve: $-2x \leq 6$ and $-2x + 3 < -7$. (Section 4.2, Example 2)

139. Solve the system:

$$\begin{aligned} 2x - y - 2z &= -1 \\ x - 2y - z &= 1 \\ x + y + z &= 4. \end{aligned}$$

(Section 3.3, Example 2)

140. Factor: $4x^3 + 8x^2 - 5x - 10$. (Section 5.3, Example 5)

Preview Exercises

Exercises 141–143 will help you prepare for the material covered in the next section. In each exercise, factor the polynomial. (You'll soon be learning techniques that will shorten the factoring process.)

141. $x^2 + 14x + 49$

142. $x^2 - 8x + 16$

143. $x^2 - 25$ (or $x^2 + 0x - 25$)

MID-CHAPTER CHECK POINT Section 5.1–Section 5.4

What You Know: We learned the vocabulary of polynomials and observed the smooth, continuous graphs of polynomial functions. We used the Leading Coefficient Test to describe the end behavior of these graphs. We learned to add, subtract, and multiply polynomials. We used a number of fast methods for finding products of polynomials, including the FOIL method for multiplying binomials, special-product formulas for squaring binomials $[(A + B)^2 = A^2 + 2AB + B^2; (A - B)^2 = A^2 - 2AB + B^2]$, and a special-product formula for the product of the sum and difference of two terms $[(A + B)(A - B) = A^2 - B^2]$. We learned to factor out a polynomial's greatest common factor and to use grouping to factor polynomials with more than three terms. We factored polynomials with three terms, beginning with trinomials with leading coefficient 1 and moving on to $ax^2 + bx + c$, with $a \neq 1$. We saw that the factoring process should begin by looking for a GCF and, if there is one, factoring it out first.

In Exercises 1–18, perform the indicated operations.

1. $(-8x^3 + 6x^2 - x + 5) - (-7x^3 + 2x^2 - 7x - 12)$

2. $(6x^2yz^4)\left(-\frac{1}{3}x^5y^2z\right)$

3. $5x^2y\left(6x^3y^2 - 7xy - \frac{2}{5}\right)$

4. $(3x - 5)(x^2 + 3x - 8)$

5. $(x^2 - 2x + 1)(2x^2 + 3x - 4)$

6. $(x^2 - 2x + 1) - (2x^2 + 3x - 4)$

7. $(6x^3y - 11x^2y - 4y) + (-11x^3y + 5x^2y - y - 6) - (-x^3y + 2y - 1)$

8. $(2x + 5)(4x - 1)$

9. $(2xy - 3)(5xy + 2)$

10. $(3x - 2y)(3x + 2y)$

11. $(3xy + 1)(2x^2 - 3y)$

12. $(7x^3y + 5x)(7x^3y - 5x)$

13. $3(x + h)^2 - 2(x + h) + 5 - (3x^2 - 2x + 5)$

14. $(x^2 - 3)^2$

15. $(x^2 - 3)(x^3 + 5x + 2)$

16. $(2x + 5y)^2$

17. $(x + 6 + 3y)(x + 6 - 3y)$

18. $(x + y + 5)^2$

In Exercises 19–30, factor completely, or state that the polynomial is prime.

19. $x^2 - 5x - 24$

20. $15xy + 5x + 6y + 2$

21. $5x^2 + 8x - 4$

22. $35x^2 + 10x - 50$

23. $9x^2 - 9x - 18$

24. $10x^3y^2 - 20x^2y^2 + 35x^2y$

25. $18x^2 + 21x + 5$

26. $12x^2 - 9xy - 16x + 12y$

27. $9x^2 - 15x + 4$

28. $3x^6 + 11x^3 + 10$

29. $25x^3 + 25x^2 - 14x$

30. $2x^4 - 6x - x^3y + 3y$

SECTION 5.5

Factoring Special Forms

Objectives

1. Factor the difference of two squares.
2. Factor perfect square trinomials.
3. Use grouping to obtain the difference of two squares.
4. Factor the sum or difference of two cubes.

Bees use honeycombs to store honey and house larvae. They construct honey storage cells from wax. Each cell has the shape of a six-sided figure whose sides are all the same length and whose angles all have the same measure, called a regular hexagon. The cells fit together perfectly, preventing dirt or predators from entering. Squares or equilateral triangles would fit equally well, but regular hexagons provide the largest storage space for the amount of wax used.

In this section, we develop factoring techniques by reversing the formulas for special products discussed in Section 5.2. Like the construction of honeycombs, these factorizations can be visualized by perfectly fitting together "cells" of squares and rectangles to form larger rectangles.

1 Factor the difference of two squares.

Factoring the Difference of Two Squares

A method for factoring the difference of two squares is obtained by reversing the special product for the sum and difference of two terms.

The Difference of Two Squares

If A and B are real numbers, variables, or algebraic expressions, then

$$A^2 - B^2 = (A + B)(A - B).$$

In words: The difference of the squares of two terms factors as the product of a sum and a difference of those terms.

EXAMPLE 1 Factoring the Difference of Two Squares

Factor:

a. $9x^2 - 100$ b. $36y^6 - 49x^4$.

Solution We must express each term as the square of some monomial. Then we use the formula for factoring $A^2 - B^2$.

a. $9x^2 - 100 = (3x)^2 - 10^2 = (3x + 10)(3x - 10)$

$A^2 - B^2 = (A + B)(A - B)$

b. $36y^6 - 49x^4 = (6y^3)^2 - (7x^2)^2 = (6y^3 + 7x^2)(6y^3 - 7x^2)$

In order to apply the factoring formula for $A^2 - B^2$, each term must be the square of an integer or a polynomial.

- A number that is the square of an integer is called a **perfect square**. For example, 100 is a perfect square because $100 = 10^2$.
- Any exponential expression involving a perfect-square coefficient and variables to even powers is a perfect square. For example, $100y^6$ is a perfect square because $100y^6 = (10y^3)^2$.

Study Tip

It's helpful to recognize perfect squares. Here are 16 perfect squares, each printed in boldface.

1 $= 1^2$	**25** $= 5^2$	**81** $= 9^2$	**169** $= 13^2$
4 $= 2^2$	**36** $= 6^2$	**100** $= 10^2$	**196** $= 14^2$
9 $= 3^2$	**49** $= 7^2$	**121** $= 11^2$	**225** $= 15^2$
16 $= 4^2$	**64** $= 8^2$	**144** $= 12^2$	**256** $= 16^2$

CHECK POINT 1 Factor:

a. $16x^2 - 25$ b. $100y^6 - 9x^4$.

Be careful when determining whether or not to apply the factoring formula for the difference of two squares.

Prime Over the Integers

- $x^2 - 5$ — 5 is not a perfect square.
- $x^7 - 25$ — 7 is an odd power. x^7 is not the square of any integer power of x.

Factorable

- $1 - x^6y^4$ — Even powers. Perfect square: $1 = 1^2$. Perfect square: $x^6y^4 = (x^3y^2)^2$

When factoring, always check first for common factors. If there are common factors, factor out the GCF and then factor the resulting polynomial.

EXAMPLE 2 Factoring Out the GCF and Then Factoring the Difference of Two Squares

Factor: $3y - 3x^6y^5$.

Solution The GCF of the two terms of the polynomial is $3y$. We begin by factoring out $3y$.

$$3y - 3x^6y^5 = 3y(1 - x^6y^4) = 3y[1^2 - (x^3y^2)^2] = 3y(1 + x^3y^2)(1 - x^3y^2)$$

Factor out the GCF. $A^2 - B^2 = (A + B)(A - B)$

CHECK POINT 2 Factor: $6y - 6x^2y^7$.

We have seen that a polynomial is factored completely when it is written as the product of prime polynomials. To be sure that you have factored completely, check to see whether any factors with more than one term in the factored polynomial can be factored further. If so, continue factoring.

EXAMPLE 3 A Repeated Factorization

Factor completely: $81x^4 - 16$.

Solution

$81x^4 - 16 = (9x^2)^2 - 4^2$ — Express as the difference of two squares.

$= (9x^2 + 4)(9x^2 - 4)$ — The factors are the sum and difference of the expressions being squared.

$= (9x^2 + 4)[(3x)^2 - 2^2]$ — The factor $9x^2 - 4$ is the difference of two squares and can be factored.

$= (9x^2 + 4)(3x + 2)(3x - 2)$ — The factors of $9x^2 - 4$ are the sum and difference of the expressions being squared.

Study Tip

Factoring $81x^4 - 16$ as

$$(9x^2 + 4)(9x^2 - 4)$$

is not a complete factorization. The second factor, $9x^2 - 4$, is itself a difference of two squares and can be factored.

Are you tempted to further factor $9x^2 + 4$, the sum of two squares, in Example 3? Resist the temptation! **The sum of two squares, $A^2 + B^2$, with no common factor other than 1 is a prime polynomial.**

CHECK POINT 3 Factor completely: $16x^4 - 81$.

In our next example, we begin with factoring by grouping. We can then factor further using the difference of two squares.

EXAMPLE 4 Factoring Completely

Factor completely: $x^3 + 5x^2 - 9x - 45$.

Solution

$x^3 + 5x^2 - 9x - 45$

$= (x^3 + 5x^2) + (-9x - 45)$ — Group terms with common factors.

$= x^2(x + 5) - 9(x + 5)$ — Factor out the common factor from each group.

$= (x + 5)(x^2 - 9)$ — Factor out $x + 5$, the common binomial factor, from both terms.

$= (x + 5)(x + 3)(x - 3)$ — Factor $x^2 - 3^2$, the difference of two squares.

CHECK POINT 4 Factor completely: $x^3 + 7x^2 - 4x - 28$.

In Examples 1–4, we used the formula for factoring the difference of two squares. Although we obtained the formula by reversing the special product for the sum and difference of two terms, it can also be obtained geometrically.

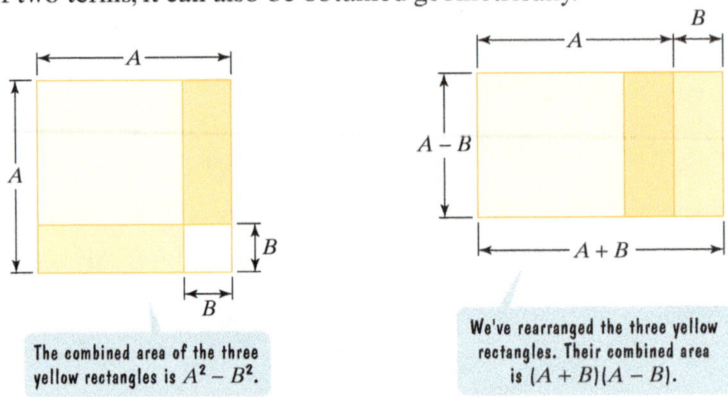

Because the three yellow rectangles make up the same combined area in both figures,

$$A^2 - B^2 = (A + B)(A - B).$$

2 Factor perfect square trinomials.

Factoring Perfect Square Trinomials

Our next factoring technique is obtained by reversing the special products for squaring binomials. The trinomials that are factored using this technique are called **perfect square trinomials**.

Factoring Perfect Square Trinomials

Let A and B be real numbers, variables, or algebraic expressions.

1. $A^2 + 2AB + B^2 = (A + B)^2$ (Same sign)
2. $A^2 - 2AB + B^2 = (A - B)^2$ (Same sign)

The two items in the box show that perfect square trinomials, $A^2 + 2AB + B^2$ and $A^2 - 2AB + B^2$, come in two forms: one in which the coefficient of the middle term is positive and one in which the coefficient of the middle term is negative. Here's how to recognize a perfect square trinomial:

1. The first and last terms are squares of monomials or integers.
2. The middle term is twice the product of the expressions being squared in the first and last terms.

EXAMPLE 5 Factoring Perfect Square Trinomials

Factor:

a. $x^2 + 14x + 49$ b. $4x^2 + 12xy + 9y^2$ c. $9y^4 - 12y^2 + 4$.

Solution

a. $x^2 + 14x + 49 = x^2 + 2 \cdot x \cdot 7 + 7^2 = (x + 7)^2$ The middle term has a positive sign.

$A^2 + 2AB + B^2 = (A + B)^2$

b. We suspect that $4x^2 + 12xy + 9y^2$ is a perfect square trinomial because $4x^2 = (2x)^2$ and $9y^2 = (3y)^2$. The middle term can be expressed as twice the product of $2x$ and $3y$.

$$4x^2 + 12xy + 9y^2 = (2x)^2 + 2 \cdot 2x \cdot 3y + (3y)^2 = (2x + 3y)^2$$

$$A^2 + 2AB + B^2 = (A + B)^2$$

c. $9y^4 - 12y^2 + 4 = (3y^2)^2 - 2 \cdot 3y^2 \cdot 2 + 2^2 = (3y^2 - 2)^2$

$$A^2 - 2AB + B^2 = (A - B)^2$$

The middle term has a negative sign.

CHECK POINT 5 Factor:

a. $x^2 + 6x + 9$

b. $16x^2 + 40xy + 25y^2$

c. $4y^4 - 20y^2 + 25$.

---

3 Use grouping to obtain the difference of two squares.

Using Special Forms When Factoring by Grouping

If a polynomial contains four terms, try factoring by grouping. In the next example, we group the terms to obtain the difference of two squares. One of the squares is a perfect square trinomial.

EXAMPLE 6 Using Grouping to Obtain the Difference of Two Squares

Factor: $x^2 - 8x + 16 - y^2$.

Solution

$x^2 - 8x + 16 - y^2$

$= (x^2 - 8x + 16) - y^2$ — Group as a perfect square trinomial minus y^2 to obtain a difference of two squares.

$= (x - 4)^2 - y^2$ — Factor the perfect square trinomial.

$= (x - 4 + y)(x - 4 - y)$ — Factor the difference of two squares. The factors are the sum and difference of the expressions being squared.

CHECK POINT 6 Factor: $x^2 + 10x + 25 - y^2$.

EXAMPLE 7 Using Grouping to Obtain the Difference of Two Squares

Factor: $a^2 - b^2 + 10b - 25$.

Solution Grouping into two groups of two terms does not result in a common binomial factor. Let's look for a perfect square trinomial. Can you see that the perfect square trinomial is the expression being subtracted from a^2?

$a^2 - b^2 + 10b - 25$

$= a^2 - (b^2 - 10b + 25)$ — Factor out −1 and group as a^2 − (perfect square trinomial) to obtain a difference of two squares.

$= a^2 - (b - 5)^2$ — Factor the perfect square trinomial.

$= [a + (b - 5)][a - (b - 5)]$ — Factor the difference of squares. The factors are the sum and difference of the expressions being squared.

$= (a + b - 5)(a - b + 5)$ — Simplify.

✓ **CHECK POINT 7** Factor: $a^2 - b^2 + 4b - 4$.

4 Factor the sum or difference of two cubes.

Factoring the Sum or Difference of Two Cubes

Here are two multiplications that lead to factoring formulas for the sum of two cubes and the difference of two cubes:

$$(A + B)(A^2 - AB + B^2) = A(A^2 - AB + B^2) + B(A^2 - AB + B^2)$$
$$= A^3 - A^2B + AB^2 + A^2B - AB^2 + B^3$$
$$= A^3 + B^3$$

The product results in the sum of two cubes.

Combine like terms: $-A^2B + A^2B = 0$ and $AB^2 - AB^2 = 0$.

and

$$(A - B)(A^2 + AB + B^2) = A(A^2 + AB + B^2) - B(A^2 + AB + B^2)$$
$$= A^3 + A^2B + AB^2 - A^2B - AB^2 - B^3$$
$$= A^3 - B^3.$$

The product results in the difference of two cubes.

Combine like terms: $A^2B - A^2B = 0$ and $AB^2 - AB^2 = 0$.

By reversing the two sides of these equations, we obtain formulas that allow us to factor a sum or difference of two cubes. These formulas should be memorized.

Factoring the Sum or Difference of Two Cubes

1. Factoring the Sum of Two Cubes

$$A^3 + B^3 = (A + B)(A^2 - AB + B^2)$$

Same signs Opposite signs

2. Factoring the Difference of Two Cubes

$$A^3 - B^3 = (A - B)(A^2 + AB + B^2)$$

Same signs Opposite signs

EXAMPLE 8 Factoring the Sum of Two Cubes

Factor:

a. $x^3 + 125$ b. $x^6 + 64y^3$.

Solution We must express each term as the cube of some monomial. Then we use the formula for factoring $A^3 + B^3$.

a. $x^3 + 125 = x^3 + 5^3 = (x + 5)(x^2 - x \cdot 5 + 5^2) = (x + 5)(x^2 - 5x + 25)$

$A^3 + B^3 = (A + B)(A^2 - AB + B^2)$

b. $x^6 + 64y^3 = (x^2)^3 + (4y)^3 = (x^2 + 4y)[(x^2)^2 - x^2 \cdot 4y + (4y)^2]$

$A^3 + B^3 = (A + B)(A^2 - AB + B^2)$

$= (x^2 + 4y)(x^4 - 4x^2y + 16y^2)$

Study Tip

When factoring the sum or difference of cubes, it is helpful to recognize the following cubes:

$$1 = 1^3$$
$$8 = 2^3$$
$$27 = 3^3$$
$$64 = 4^3$$
$$125 = 5^3$$
$$216 = 6^3$$
$$1000 = 10^3.$$

✓ CHECK POINT 8 Factor:

a. $x^3 + 27$

b. $x^6 + 1000y^3$.

EXAMPLE 9 Factoring the Difference of Two Cubes

Factor:

a. $x^3 - 216$ b. $8 - 125x^3y^3$.

Solution We must express each term as the cube of some monomial. Then we use the formula for factoring $A^3 - B^3$.

a. $x^3 - 216 = x^3 - 6^3 = (x - 6)(x^2 + x \cdot 6 + 6^2) = (x - 6)(x^2 + 6x + 36)$

$A^3 - B^3 = (A - B)(A^2 + AB + B^2)$

b. $8 - 125x^3y^3 = 2^3 - (5xy)^3 = (2 - 5xy)[2^2 + 2 \cdot 5xy + (5xy)^2]$

$A^3 - B^3 = (A - B)(A^2 + AB + B^2)$

$= (2 - 5xy)(4 + 10xy + 25x^2y^2)$

✓ CHECK POINT 9 Factor:

a. $x^3 - 8$

b. $1 - 27x^3y^3$.

5.5 EXERCISE SET

MyMathLab — MathXL PRACTICE, WATCH, DOWNLOAD, READ, REVIEW

Practice Exercises

In Exercises 1–22, factor each difference of two squares. Assume that any variable exponents represent whole numbers.

1. $x^2 - 4$
2. $x^2 - 16$
3. $9x^2 - 25$
4. $4x^2 - 9$
5. $9 - 25y^2$
6. $16 - 49y^2$
7. $36x^2 - 49y^2$
8. $64x^2 - 25y^2$
9. $x^2y^2 - 1$
10. $x^2y^2 - 100$
11. $9x^4 - 25y^6$
12. $25x^4 - 9y^6$
13. $x^{14} - y^4$
14. $x^4 - y^{10}$
15. $(x - 3)^2 - y^2$
16. $(x - 6)^2 - y^2$
17. $a^2 - (b - 2)^2$
18. $a^2 - (b - 3)^2$
19. $x^{2n} - 25$
20. $x^{2n} - 36$
21. $1 - a^{2n}$
22. $4 - b^{2n}$

In Exercises 23–48, factor completely, or state that the polynomial is prime.

23. $2x^3 - 8x$
24. $2x^3 - 72x$
25. $50 - 2y^2$
26. $72 - 2y^2$
27. $8x^2 - 8y^2$
28. $6x^2 - 6y^2$
29. $2x^3y - 18xy$

30. $2x^3y - 32xy$
31. $a^3b^2 - 49ac^2$
32. $4a^3c^2 - 16ax^2y^2$
33. $5y - 5x^2y^7$
34. $2y - 2x^6y^3$
35. $8x^2 + 8y^2$
36. $6x^2 + 6y^2$
37. $x^2 + 25y^2$
38. $x^2 + 36y^2$
39. $x^4 - 16$
40. $x^4 - 1$
41. $81x^4 - 1$
42. $1 - 81x^4$
43. $2x^5 - 2xy^4$
44. $3x^5 - 3xy^4$
45. $x^3 + 3x^2 - 4x - 12$
46. $x^3 + 3x^2 - 9x - 27$
47. $x^3 - 7x^2 - x + 7$
48. $x^3 - 6x^2 - x + 6$

In Exercises 49–64, factor any perfect square trinomials, or state that the polynomial is prime.

49. $x^2 + 4x + 4$
50. $x^2 + 2x + 1$
51. $x^2 - 10x + 25$
52. $x^2 - 14x + 49$
53. $x^4 - 4x^2 + 4$
54. $x^4 - 6x^2 + 9$
55. $9y^2 + 6y + 1$
56. $4y^2 + 4y + 1$
57. $64y^2 - 16y + 1$
58. $25y^2 - 10y + 1$
59. $x^2 - 12xy + 36y^2$
60. $x^2 + 16xy + 64y^2$
61. $x^2 - 8xy + 64y^2$
62. $x^2 - 9xy + 81y^2$
63. $9x^2 + 48xy + 64y^2$
64. $16x^2 - 40xy + 25y^2$

In Exercises 65–74, factor by grouping to obtain the difference of two squares.

65. $x^2 - 6x + 9 - y^2$
66. $x^2 - 12x + 36 - y^2$
67. $x^2 + 20x + 100 - x^4$
68. $x^2 + 16x + 64 - x^4$
69. $9x^2 - 30x + 25 - 36y^2$
70. $25x^2 - 20x + 4 - 81y^2$
71. $x^4 - x^2 - 2x - 1$
72. $x^4 - x^2 - 6x - 9$
73. $z^2 - x^2 + 4xy - 4y^2$
74. $z^2 - x^2 + 10xy - 25y^2$

In Exercises 75–94, factor using the formula for the sum or difference of two cubes.

75. $x^3 + 64$
76. $x^3 + 1$
77. $x^3 - 27$
78. $x^3 - 1000$
79. $8y^3 + 1$
80. $27y^3 + 1$
81. $125x^3 - 8$
82. $27x^3 - 8$
83. $x^3y^3 + 27$
84. $x^3y^3 + 64$
85. $64x - x^4$
86. $216x - x^4$
87. $x^6 + 27y^3$
88. $x^6 + 8y^3$
89. $125x^6 - 64y^6$
90. $125x^6 - y^6$
91. $x^9 + 1$
92. $x^9 - 1$
93. $(x - y)^3 - y^3$
94. $x^3 + (x + y)^3$

Practice PLUS

In Exercises 95–104, factor completely.

95. $0.04x^2 + 0.12x + 0.09$
96. $0.09x^2 - 0.12x + 0.04$
97. $8x^4 - \frac{x}{8}$
98. $27x^4 + \frac{x}{27}$
99. $x^6 - 9x^3 + 8$
100. $x^6 + 9x^3 + 8$
101. $x^8 - 15x^4 - 16$
102. $x^8 + 15x^4 - 16$
103. $x^5 - x^3 - 8x^2 + 8$
104. $x^5 - x^3 + 27x^2 - 27$

105. The figure shows four yellow rectangles that fit together to form a large square.

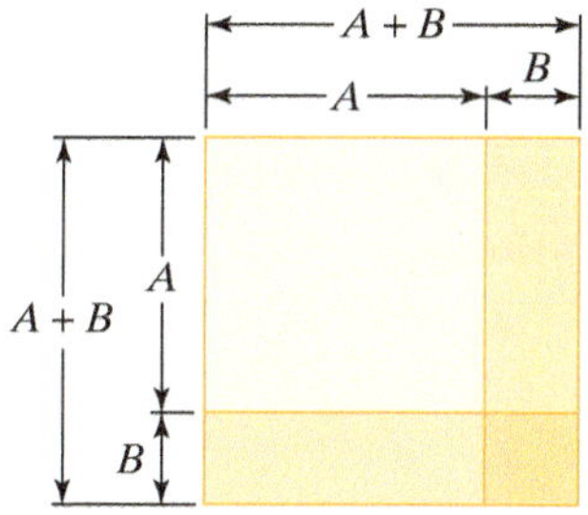

a. Express the area of the large square in terms of one of its sides, $A + B$.
b. Write an expression for the area of each of the four rectangles that form the large square.
c. Use the sum of the areas from part (b) to write a second expression for the area of the large square.
d. Set the expression from part (c) equal to the expression from part (a). What factoring technique have you established?

Application Exercises

In Exercises 106–109, find the formula for the area of the shaded region and express it in factored form.

106.

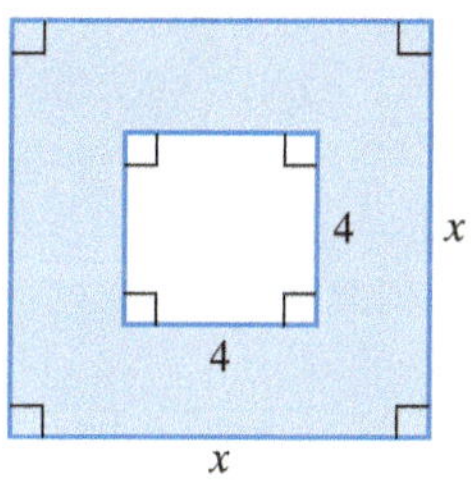

107.

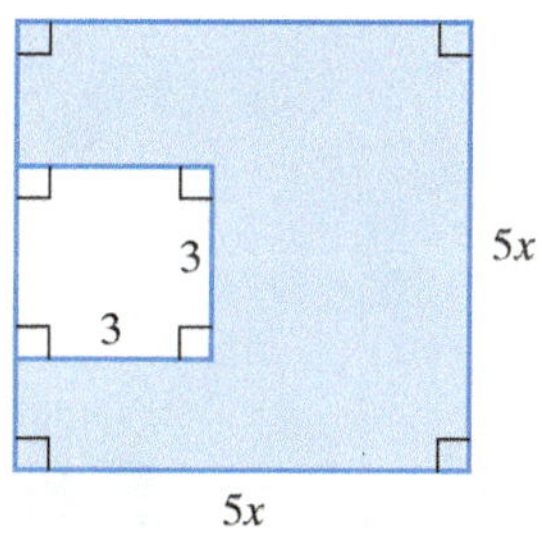

108.

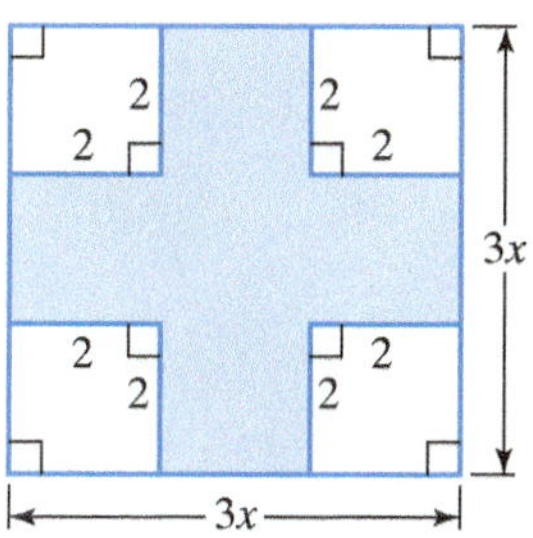

109.

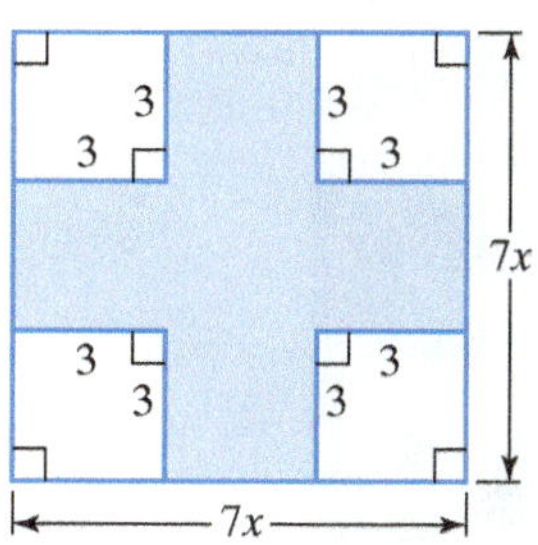

In Exercises 110–111, find the formula for the volume of the region outside the smaller rectangular solid and inside the larger rectangular solid. Then express the volume in factored form.

110.

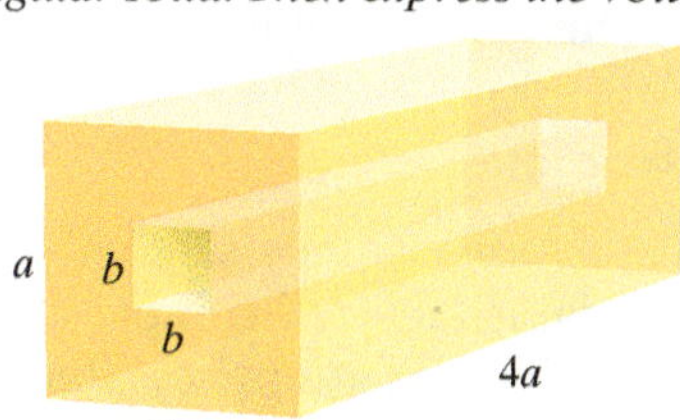

111.

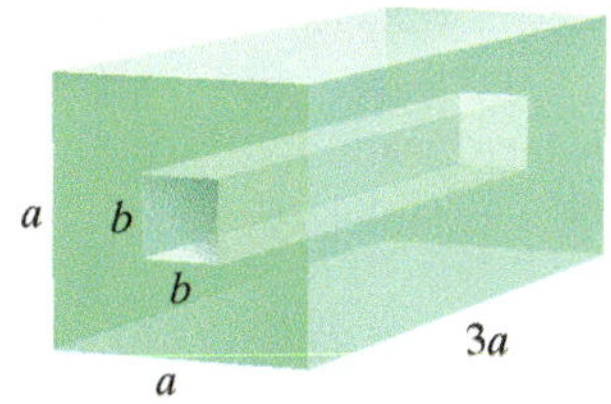

Writing in Mathematics

112. Explain how to factor the difference of two squares. Provide an example with your explanation.

113. What is a perfect square trinomial and how is it factored?

114. Explain how to factor $x^2 - y^2 + 8x - 16$. Should the expression be grouped into two groups of two terms? If not, why not, and what sort of grouping should be used?

115. Explain how to factor $x^3 + 1$.

Technology Exercises

In Exercises 116–123, use a graphing utility to graph the function on each side of the equation in the same viewing rectangle. Use end behavior to show a complete picture of the polynomial function on the left side. Do the graphs coincide? If so, this means that the polynomial on the left side has been factored correctly. If not, factor the polynomial correctly and then use your graphing utility to verify the factorization.

116. $9x^2 - 4 = (3x + 2)(3x - 2)$

117. $x^2 + 4x + 4 = (x + 4)^2$

118. $9x^2 + 12x + 4 = (3x + 2)^2$

119. $25 - (x^2 + 4x + 4) = (x + 7)(x - 3)$

120. $(2x + 3)^2 - 9 = 4x(x + 3)$

121. $(x - 3)^2 + 8(x - 3) + 16 = (x - 1)^2$

122. $x^3 - 1 = (x - 1)(x^2 - x + 1)$

123. $(x + 1)^3 + 1 = (x + 1)(x^2 + x + 1)$

124. Use the TABLE feature of a graphing utility to verify any two of your factorizations in Exercises 67–68 or 77–78.

Critical Thinking Exercises

Make Sense? *In Exercises 125–128, determine whether each statement "makes sense" or "does not make sense" and explain your reasoning.*

125. Although I can factor the difference of squares and perfect square trinomials using trial-and-error associated with FOIL, recognizing these special forms shortens the process.

126. Although $x^3 + 2x^2 - 5x - 6$ can be factored as $(x + 1)(x + 3)(x - 2)$, I have not yet learned techniques to obtain this factorization.

127. I factored $4x^2 - 100$ completely and obtained $(2x + 10)(2x - 10)$.

128. You told me that the area of a square is represented by $9x^2 + 12x + 4$ square inches, so I factored and concluded that the length of one side must be $3x + 2$ inches.

In Exercises 129–132, determine whether each statement is true or false. If the statement is false, make the necessary change(s) to produce a true statement.

129. $9x^2 + 15x + 25 = (3x + 5)^2$

130. $x^3 - 27 = (x - 3)(x^2 + 6x + 9)$

131. $x^3 - 64 = (x - 4)^3$

132. $4x^2 - 121 = (2x - 11)^2$

In Exercises 133–136, factor each polynomial completely. Assume that any variable exponents represent whole numbers.

133. $y^3 + x + x^3 + y$

134. $36x^{2n} - y^{2n}$

135. $x^{3n} + y^{12n}$

136. $4x^{2n} + 20x^n y^m + 25y^{2m}$

137. Factor $x^6 - y^6$ first as the difference of squares and then as the difference of cubes. From these two factorizations, determine a factorization for $x^4 + x^2y^2 + y^4$.

In Exercises 138–139, find all integers k so that the trinomial is a perfect square trinomial.

138. $kx^2 + 8xy + y^2$

139. $64x^2 - 16x + k$

Review Exercises

140. Solve: $2x + 2 \geq 12$ and $\frac{2x - 1}{3} \leq 7$. (Section 4.2, Example 2)

141. Solve using matrices:

$$\begin{aligned} 3x - 2y &= -8 \\ x + 6y &= 4. \end{aligned}$$

(Section 3.4, Example 2)

142. Factor: $3x^2 + 21x - xy - 7y$. (Section 5.3, Example 6)

Preview Exercises

Exercises 143–145 will help you prepare for the material covered in the next section. In each exercise, factor completely.

143. $2x^3 + 8x^2 + 8x$

144. $5x^3 - 40x^2y + 35xy^2$

145. $9b^2x + 9b^2y - 16x - 16y$

SECTION 5.6

A General Factoring Strategy

Objectives

1 Use the appropriate method for factoring a polynomial.

2 Use a general strategy for factoring polynomials.

Successful problem solving involves understanding the problem, devising a plan for solving it, and then carrying out the plan. In this section, you will learn a step-by-step strategy that provides a plan and direction for solving factoring problems.

1 Use the appropriate method for factoring a polynomial.

A Strategy for Factoring Polynomials

It is important to practice factoring a wide variety of polynomials so that you can quickly select the appropriate technique. The polynomial is factored completely when all its polynomial factors, except possibly for monomial factors, are prime. Because of the commutative property, the order of the factors does not matter.

Here is a general strategy for factoring polynomials:

2 Use a general strategy for factoring polynomials.

A Strategy for Factoring a Polynomial

1. If there is a common factor, factor out the GCF or factor out a common factor with a negative coefficient.
2. Determine the number of terms in the polynomial and try factoring as follows:
   a. If there are two terms, can the binomial be factored by using one of the following special forms?

      Difference of two squares: $A^2 - B^2 = (A + B)(A - B)$

      Sum of two cubes: $A^3 + B^3 = (A + B)(A^2 - AB + B^2)$

      Difference of two cubes: $A^3 - B^3 = (A - B)(A^2 + AB + B^2)$

   b. If there are three terms, is the trinomial a perfect square trinomial? If so, factor by using one of the following special forms:

      $$A^2 + 2AB + B^2 = (A + B)^2$$
      $$A^2 - 2AB + B^2 = (A - B)^2.$$

      If the trinomial is not a perfect square trinomial, try factoring by trial and error or grouping.

   c. If there are four or more terms, try factoring by grouping.
3. Check to see if any factors with more than one term in the factored polynomial can be factored further. If so, factor completely.

Remember to check the factored form by multiplying or by using the TABLE or GRAPH feature of a graphing utility.

The following examples and those in the exercise set are similar to the previous factoring problems. However, these factorizations are not all of the same type. They are intentionally mixed to promote the development of a general factoring strategy.

Using Technology

Graphic Connections

The polynomial functions $y_1 = 2x^3 + 8x^2 + 8x$ and $y_2 = 2x(x + 2)^2$ have identical graphs. This verifies that

$$2x^3 + 8x^2 + 8x = 2x(x + 2)^2.$$

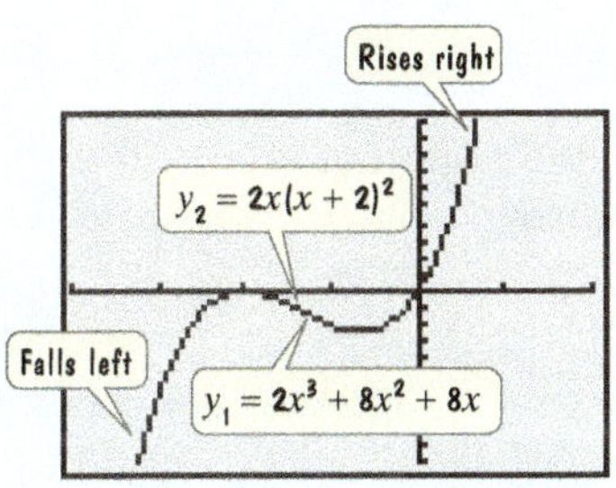

[−4, 2, 1] by [−10, 10, 1]

The degree of y_1 is 3, which is odd. Odd-degree polynomial functions have graphs with opposite behavior at each end. The leading coefficient, 2, is positive. The graph should fall to the left and rise to the right (↙, ↗). The viewing rectangle used is complete enough to show this end behavior.

EXAMPLE 1 Factoring a Polynomial

Factor: $2x^3 + 8x^2 + 8x$.

Solution

Step 1. If there is a common factor, factor out the GCF. Because $2x$ is common to all terms, we factor it out.

$$2x^3 + 8x^2 + 8x = 2x(x^2 + 4x + 4) \quad \text{Factor out the GCF.}$$

Step 2. Determine the number of terms and factor accordingly. The factor $x^2 + 4x + 4$ has three terms and is a perfect square trinomial. We factor using $A^2 + 2AB + B^2 = (A + B)^2$.

$$\begin{aligned} 2x^3 + 8x^2 + 8x &= 2x(x^2 + 4x + 4) \\ &= 2x(\underbrace{x^2}_{A^2} + \underbrace{2 \cdot x \cdot 2}_{2AB} + \underbrace{2^2}_{B^2}) \\ &= 2x(x + 2)^2 \qquad A^2 + 2AB + B^2 = (A + B)^2 \end{aligned}$$

Step 3. Check to see if factors can be factored further. In this problem, they cannot. Thus,

$$2x^3 + 8x^2 + 8x = 2x(x + 2)^2.$$

✓ **CHECK POINT 1** Factor: $3x^3 - 30x^2 + 75x$.

EXAMPLE 2 Factoring a Polynomial

Factor: $4x^2y - 16xy - 20y$.

Solution

Step 1. If there is a common factor, factor out the GCF. Because $4y$ is common to all terms, we factor it out.

$$4x^2y - 16xy - 20y = 4y(x^2 - 4x - 5) \quad \text{Factor out the GCF.}$$

Step 2. Determine the number of terms and factor accordingly. The factor $x^2 - 4x - 5$ has three terms, but it is not a perfect square trinomial. We factor it using trial and error.

$$4x^2y - 16xy - 20y = 4y(x^2 - 4x - 5) = 4y(x + 1)(x - 5)$$

Step 3. Check to see if factors can be factored further. In this case, they cannot, so we have factored completely.

✓ CHECK POINT 2 Factor: $3x^2y - 12xy - 36y$.

EXAMPLE 3 Factoring a Polynomial

Factor: $9b^2x - 16y - 16x + 9b^2y$.

Solution

Step 1. If there is a common factor, factor out the GCF. Other than 1 or -1, there is no common factor.

Step 2. Determine the number of terms and factor accordingly. There are four terms. We try factoring by grouping. Notice that the first and last terms have a common factor of $9b^2$ and the two middle terms have a common factor of -16. Thus, we begin by rearranging the terms.

$$9b^2x - 16y - 16x + 9b^2y$$
$$= (9b^2x + 9b^2y) + (-16x - 16y) \quad \text{Rearrange terms and group terms with common factors.}$$
$$= 9b^2(x + y) - 16(x + y) \quad \text{Factor from each group.}$$
$$= (x + y)(9b^2 - 16) \quad \text{Factor out the common binomial factor, } x + y.$$

Step 3. Check to see if factors can be factored further. We note that $9b^2 - 16$ is the difference of two squares, $(3b)^2 - 4^2$, so we continue factoring.

$$9b^2x - 16y - 16x + 9b^2y$$
$$= (x + y)[(3b)^2 - 4^2] \quad \text{Express } 9b^2 - 16 \text{ as the difference of squares.}$$
$$= (x + y)(3b + 4)(3b - 4) \quad \text{The factors of } 9b^2 - 16 \text{ are the sum and difference of the expressions being squared.}$$

✓ CHECK POINT 3 Factor: $16a^2x - 25y - 25x + 16a^2y$.

EXAMPLE 4 Factoring a Polynomial

Factor: $x^2 - 25a^2 + 8x + 16$.

Solution

Step 1. If there is a common factor, factor out the GCF. Other than 1 or -1, there is no common factor.

Step 2. Determine the number of terms and factor accordingly. There are four terms. We try factoring by grouping. Grouping into two groups of two terms does not result in a common binomial factor. Let's try grouping as a difference of squares.

$$x^2 - 25a^2 + 8x + 16$$

$$= (x^2 + 8x + 16) - 25a^2$$ Rearrange terms and group as a perfect square trinomial minus $25a^2$ to obtain a difference of squares.

$$= (x + 4)^2 - (5a)^2$$ Factor the perfect square trinomial.

$$= (x + 4 + 5a)(x + 4 - 5a)$$ Factor the difference of squares. The factors are the sum and difference of the expressions being squared.

Step 3. Check to see if factors can be factored further. In this case, they cannot, so we have factored completely.

CHECK POINT 4 Factor: $x^2 - 36a^2 + 20x + 100$.

EXAMPLE 5 Factoring a Polynomial

Factor: $3x^{10} + 3x$.

Solution

Step 1. If there is a common factor, factor out the GCF. Because $3x$ is common to both terms, we factor it out.

$$3x^{10} + 3x = 3x(x^9 + 1)$$ Factor out the GCF.

Step 2. Determine the number of terms and factor accordingly. The factor $x^9 + 1$ has two terms. This binomial can be expressed as $(x^3)^3 + 1^3$, so it can be factored as the sum of two cubes.

$$3x^{10} + 3x = 3x(x^9 + 1)$$

$$= 3x[(x^3)^3 + 1^3] = 3x(x^3 + 1)[(x^3)^2 - x^3 \cdot 1 + 1^2]$$

$$A^3 + B^3 = (A + B)(A^2 - AB + B^2)$$

$$= 3x(x^3 + 1)(x^6 - x^3 + 1)$$ Simplify.

Step 3. Check to see if factors can be factored further. We note that $x^3 + 1$ is the sum of two cubes, $x^3 + 1^3$, so we continue factoring.

$$3x^{10} + 3x$$

$$= 3x(x^3 + 1)(x^6 - x^3 + 1)$$ This is our factorization in the previous step.

$$A^3 + B^3 = (A + B)(A^2 - AB + B^2)$$

$$= 3x(x + 1)(x^2 - x + 1)(x^6 - x^3 + 1)$$ Factor completely by factoring $x^3 + 1^3$, the sum of cubes.

CHECK POINT 5 Factor: $x^{10} + 512x$. *Hint*: $512 = 8^3$.

5.6 EXERCISE SET

MyMathLab | MathXL PRACTICE | WATCH | DOWNLOAD | READ | REVIEW

Practice Exercises

In Exercises 1–68, factor completely, or state that the polynomial is prime.

1. $x^3 - 16x$
2. $x^3 - x$
3. $3x^2 + 18x + 27$
4. $8x^2 + 40x + 50$
5. $81x^3 - 3$
6. $24x^3 - 3$
7. $x^2y - 16y + 32 - 2x^2$
8. $12x^2y - 27y - 4x^2 + 9$
9. $4a^2b - 2ab - 30b$
10. $32y^2 - 48y + 18$
11. $ay^2 - 4a - 4y^2 + 16$
12. $ax^2 - 16a - 2x^2 + 32$
13. $11x^5 - 11xy^2$
14. $4x^9 - 400x$
15. $4x^5 - 64x$
16. $7x^5 - 7x$
17. $x^3 - 4x^2 - 9x + 36$
18. $x^3 - 5x^2 - 4x + 20$
19. $2x^5 + 54x^2$
20. $3x^5 + 24x^2$
21. $3x^4y - 48y^5$
22. $32x^4y - 2y^5$
23. $12x^3 + 36x^2y + 27xy^2$
24. $18x^3 + 48x^2y + 32xy^2$
25. $x^2 - 12x + 36 - 49y^2$
26. $x^2 - 10x + 25 - 36y^2$
27. $4x^2 + 25y^2$
28. $16x^2 + 49y^2$
29. $12x^3y - 12xy^3$
30. $9x^2y^2 - 36y^2$
31. $6bx^2 + 6by^2$
32. $6x^2 - 66$
33. $x^4 - xy^3 + x^3y - y^4$
34. $x^3 - xy^2 + x^2y - y^3$
35. $x^2 - 4a^2 + 12x + 36$
36. $x^2 - 49a^2 + 14x + 49$
37. $5x^3 + x^6 - 14$
38. $6x^3 + x^6 - 16$
39. $4x - 14 + 2x^3 - 7x^2$
40. $3x^3 + 8x + 9x^2 + 24$
41. $54x^3 - 16y^3$
42. $54x^3 - 250y^3$
43. $x^2 + 10x - y^2 + 25$
44. $x^2 + 6x - y^2 + 9$
45. $x^8 - y^8$
46. $x^8 - 1$
47. $x^3y - 16xy^3$
48. $x^3y - 100xy^3$
49. $x + 8x^4$
50. $x + 27x^4$
51. $16y^2 - 4y - 2$
52. $32y^2 + 4y - 6$
53. $14y^3 + 7y^2 - 10y$
54. $5y^3 - 45y^2 + 70y$
55. $27x^2 + 36xy + 12y^2$
56. $125x^2 + 50xy + 5y^2$
57. $12x^3 + 3xy^2$
58. $3x^4 + 27x^2$
59. $x^6y^6 - x^3y^3$
60. $x^3 - 2x^2 - x + 2$
61. $(x + 5)(x - 3) + (x + 5)(x - 7)$
62. $(x + 4)(x - 9) + (x + 4)(2x - 3)$
63. $a^2(x - y) + 4(y - x)$
64. $b^2(x - 3) + c^2(3 - x)$
65. $(c + d)^4 - 1$
66. $(c + d)^4 - 16$
67. $p^3 - pq^2 + p^2q - q^3$
68. $p^3 - pq^2 - p^2q + q^3$

Practice PLUS

In Exercises 69–80, factor completely.

69. $x^4 - 5x^2y^2 + 4y^4$
70. $x^4 - 10x^2y^2 + 9y^4$
71. $(x + y)^2 + 6(x + y) + 9$
72. $(x - y)^2 - 8(x - y) + 16$
73. $(x - y)^4 - 4(x - y)^2$
74. $(x + y)^4 - 100(x + y)^2$
75. $2x^2 - 7xy^2 + 3y^4$
76. $3x^2 + 5xy^2 + 2y^4$
77. $x^3 - y^3 - x + y$
78. $x^3 + y^3 + x^2 - y^2$
79. $x^6y^3 + x^3 - 8x^3y^3 - 8$
80. $x^6y^3 - x^3 + x^3y^3 - 1$

Application Exercises

In Exercises 81–86,

a. *Write an expression for the area of the shaded region.*

b. *Write the expression in factored form.*

81.

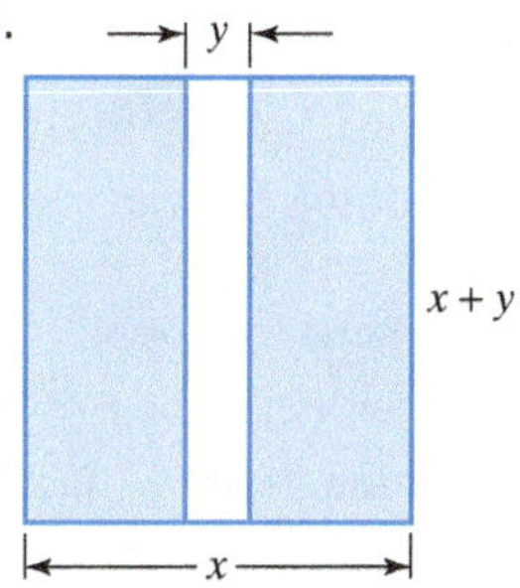

82.

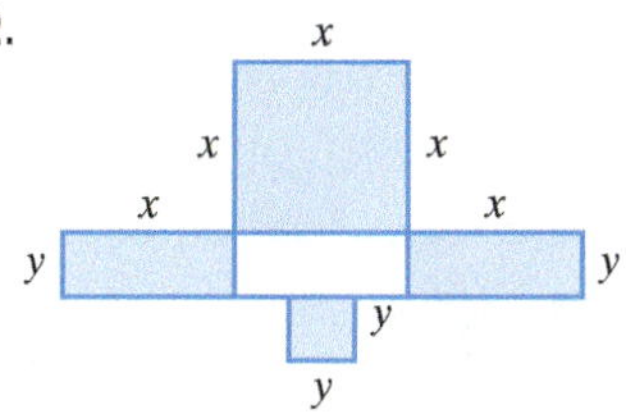

83.

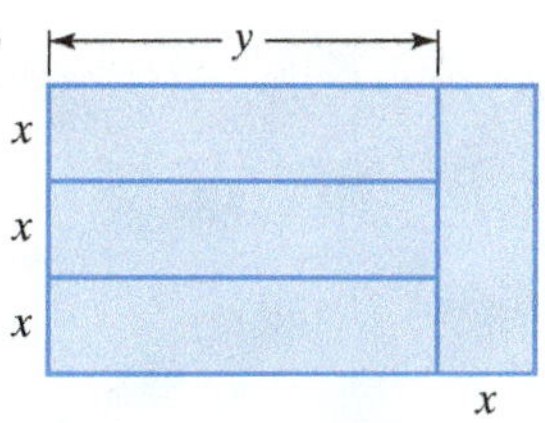

84.

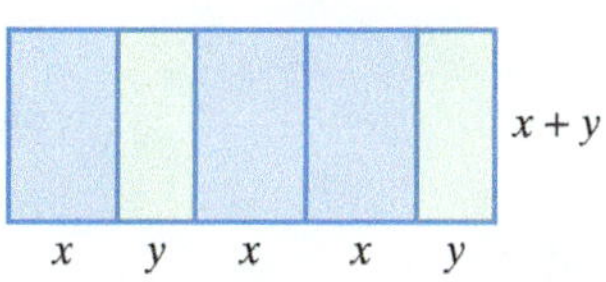

85.

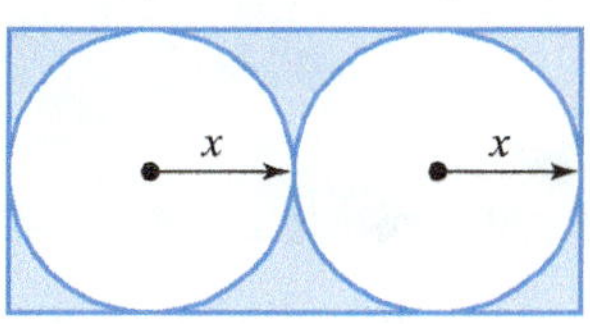

86.

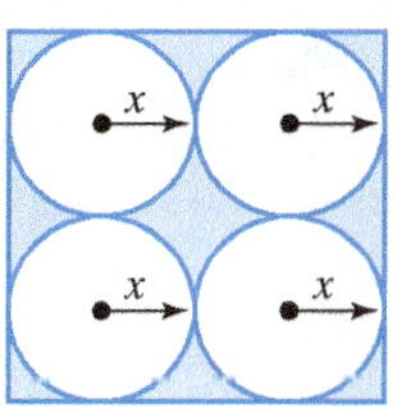

Writing in Mathematics

87. Describe a strategy that can be used to factor polynomials.

88. Describe some of the difficulties in factoring polynomials. What suggestions can you offer to overcome these difficulties?

Technology Exercises

In Exercises 89–92, use a graphing utility to graph the function on each side of the equation in the same viewing rectangle. Use end behavior to show a complete picture of the polynomial on the left side. Do the graphs coincide? If so, the factorization is correct. If not, factor correctly and then use your graphing utility to verify the factorization.

89. $4x^2 - 12x + 9 = (4x - 3)^2$

90. $2x^3 + 10x^2 - 2x - 10 = 2(x + 5)(x^2 + 1)$

91. $x^4 - 16 = (x^2 + 4)(x + 2)(x - 2)$

92. $x^3 + 1 = (x + 1)^3$

93. Use the TABLE feature of a graphing utility to verify any two of your complete factorizations in Exercises 15–20.

Critical Thinking Exercises

Make Sense? *In Exercises 94–97, determine whether each statement "makes sense" or "does not make sense" and explain your reasoning.*

94. It takes a great deal of practice to get good at factoring a wide variety of polynomials.

95. Multiplying polynomials is relatively mechanical, but factoring often requires a great deal of thought.

96. The factorable trinomial $4x^2 + 8x + 3$ and the prime trinomial $4x^2 + 8x + 1$ are in the form $ax^2 + bx + c$, but $b^2 - 4ac$ is a perfect square only in the case of the factorable trinomial.

97. You told me that the volume of a rectangular solid is represented by $5x^3 + 30x^2 + 40x$ cubic inches, so I factored completely and concluded that the dimensions are $5x$ inches, $x + 2$ inches, and $x + 5$ inches.

In Exercises 98–101, determine whether each statement is true or false. If the statement is false, make the necessary change(s) to produce a true statement.

98. $x^4 - 16$ is factored completely as $(x^2 + 4)(x^2 - 4)$.

99. The trinomial $x^2 - 4x - 4$ is a prime polynomial.

100. $x^2 + 36 = (x + 6)^2$

101. $x^3 - 64 = (x + 4)(x^2 + 4x - 16)$

In Exercises 102–104, factor completely. Assume that variable exponents represent whole numbers.

102. $x^{2n+3} - 10x^{n+3} + 25x^3$

103. $3x^{n+2} - 13x^{n+1} + 4x^n$

104. $x^{4n+1} - xy^{4n}$

105. In certain circumstances, the sum of two perfect squares can be factored by adding and subtracting the same perfect square. For example,

$$x^4 + 4 = x^4 + 4x^2 + 4 - 4x^2. \quad \text{Add and subtract } 4x^2.$$

Use this first step to factor $x^4 + 4$.

106. Express $x^3 + x + 2x^4 + 4x^2 + 2$ as the product of two polynomials of degree 2.

Review Exercises

107. Solve: $\frac{3x - 1}{5} + \frac{x + 2}{2} = -\frac{3}{10}$.

(Section 1.4, Example 4)

108. Simplify: $(4x^3y^{-1})^2(2x^{-3}y)^{-1}$.

(Section 1.6, Example 9)

109. Evaluate: $\begin{vmatrix} 0 & -3 & 2 \\ 1 & 5 & 3 \\ -2 & 1 & 4 \end{vmatrix}$.

(Section 3.5, Example 3)

Preview Exercises

Exercises 110–112 will help you prepare for the material covered in the next section.

110. Evaluate $(2x + 3)(x - 4)$ in your head for $x = 4$.

111. Evaluate $-16(t - 6)(t + 4)$ in your head for $t = 6$.

112. Express as an equivalent equation with a factored trinomial on the left side and zero on the right side:

$$x^2 + (x + 7)^2 = (x + 8)^2.$$

SECTION 5.7

Polynomial Equations and Their Applications

Objectives

1. Solve quadratic equations by factoring.
2. Solve higher-degree polynomial equations by factoring.
3. Solve problems using polynomial equations.

Motion and change are the very essence of life. Moving air brushes against our faces; rain falls on our heads; birds fly past us; plants spring from the earth, grow, and then die; and rocks thrown upward reach a maximum height before falling to the ground. In this section, you will use quadratic functions and factoring strategies to model and visualize motion. Analyzing the where and when of moving objects involves equations in which the highest exponent on the variable is 2, called *quadratic equations*.

The Standard Form of a Quadratic Equation

We begin by defining a quadratic equation.

Definition of a Quadratic Equation

A **quadratic equation** in x is an equation that can be written in the **standard form**

$$ax^2 + bx + c = 0,$$

where a, b, and c are real numbers, with $a \neq 0$. A quadratic equation in x is also called a **second-degree polynomial equation** in x.

Here is an example of a quadratic equation in standard form:

$$x^2 - 12x + 27 = 0.$$

$a = 1$ $b = -12$ $c = 27$

1 Solve quadratic equations by factoring.

Solving Quadratic Equations by Factoring

We can factor the left side of the quadratic equation $x^2 - 12x + 27 = 0$. We obtain $(x - 3)(x - 9) = 0$. If a quadratic equation has zero on one side and a factored expression on the other side, it can be solved using the **zero-product principle**.

The Zero-Product Principle

If the product of two algebraic expressions is zero, then at least one of the factors is equal to zero.

If $AB = 0$, then $A = 0$ or $B = 0$.

For example, consider the equation $(x - 3)(x - 9) = 0$. According to the zero-product principle, this product can be zero only if at least one of the factors is zero. We set each individual factor equal to zero and solve the resulting equations for x.

$$(x - 3)(x - 9) = 0$$
$$x - 3 = 0 \quad \text{or} \quad x - 9 = 0$$
$$x = 3 \qquad\qquad x = 9$$

The solutions of the original quadratic equation, $x^2 - 12x + 27 = 0$, are 3 and 9. The solution set is $\{3, 9\}$.

Solving a Quadratic Equation by Factoring

1. If necessary, rewrite the equation in the standard form $ax^2 + bx + c = 0$, moving all terms to one side, thereby obtaining zero on the other side.
2. Factor completely.
3. Apply the zero-product principle, setting each factor containing a variable equal to zero.
4. Solve the equations in step 3.
5. Check the solutions in the original equation.

EXAMPLE 1 Solving a Quadratic Equation by Factoring

Solve: $2x^2 - 5x = 12$.

Solution

Step 1. Move all terms to one side and obtain zero on the other side. Subtract 12 from both sides and write the equation in standard form.

$$2x^2 - 5x - 12 = 12 - 12$$
$$2x^2 - 5x - 12 = 0$$

Step 2. Factor.

$$(2x + 3)(x - 4) = 0$$

Steps 3 and 4. Set each factor equal to zero and solve the resulting equations.

$$2x + 3 = 0 \quad \text{or} \quad x - 4 = 0$$
$$2x = -3 \qquad\qquad x = 4$$
$$x = -\frac{3}{2}$$

Step 5. Check the solutions in the original equation.

Check $-\frac{3}{2}$:

$$2x^2 - 5x = 12$$
$$2\left(-\frac{3}{2}\right)^2 - 5\left(-\frac{3}{2}\right) \stackrel{?}{=} 12$$
$$2\left(\frac{9}{4}\right) - 5\left(-\frac{3}{2}\right) \stackrel{?}{=} 12$$
$$\frac{9}{2} + \frac{15}{2} \stackrel{?}{=} 12$$
$$\frac{24}{2} \stackrel{?}{=} 12$$
$$12 = 12, \quad \text{true}$$

Check 4:

$$2x^2 - 5x = 12$$
$$2(4)^2 - 5(4) \stackrel{?}{=} 12$$
$$2(16) - 5(4) \stackrel{?}{=} 12$$
$$32 - 20 \stackrel{?}{=} 12$$
$$12 = 12, \quad \text{true}$$

The solutions are $-\frac{3}{2}$ and 4, and the solution set is $\left\{-\frac{3}{2}, 4\right\}$.

CHECK POINT 1 Solve: $2x^2 - 9x = 5$.

Study Tip

Do not confuse factoring a polynomial with solving a quadratic equation by factoring.

Factoring a Polynomial

Factor: $2x^2 - 5x - 12$.

This is not an equation. There is no equal sign.

Solution: $(2x + 3)(x - 4)$

Stop! Avoid the common error of setting each factor equal to zero.

Solving a Quadratic Equation

Solve: $2x^2 - 5x - 12 = 0$.

This is an equation. There is an equal sign.

Solution: $(2x + 3)(x - 4) = 0$

$2x + 3 = 0$ or $x - 4 = 0$

$x = -\frac{3}{2}$ $\quad$ $x = 4$

The solution set is $\left\{-\frac{3}{2}, 4\right\}$.

There is an important relationship between a quadratic equation in standard form, such as

$$2x^2 - 5x - 12 = 0$$

and a quadratic function, such as

$$y = 2x^2 - 5x - 12.$$

The solutions of $ax^2 + bx + c = 0$ correspond to the x-intercepts of the graph of the quadratic function $y = ax^2 + bx + c$. For example, you can visualize the solutions of $2x^2 - 5x - 12 = 0$ by looking at the x-intercepts of the graph of the quadratic function $y = 2x^2 - 5x - 12$. The graph, shaped like a bowl, is shown in **Figure 5.10**. The solutions of the equation $2x^2 - 5x - 12 = 0$, $-\frac{3}{2}$ and 4, appear as the graph's x-intercepts.

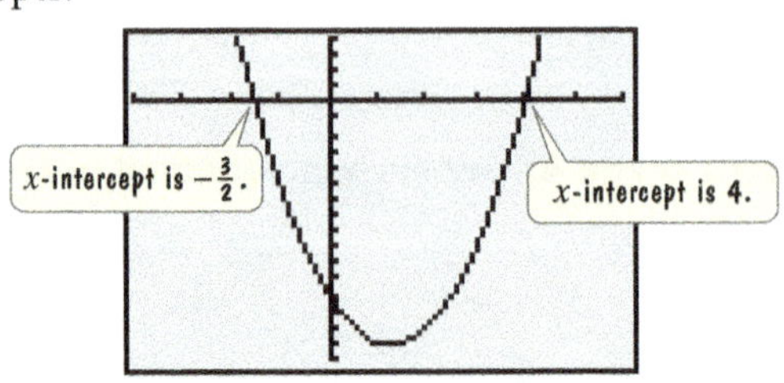

[−4, 6, 1] by [−16, 4, 1]

FIGURE 5.10

EXAMPLE 2 Solving Quadratic Equations by Factoring

Solve:

a. $5x^2 = 20x$ b. $x^2 + 4 = 8x - 12$ c. $(x - 7)(x + 5) = -20$.

Solution

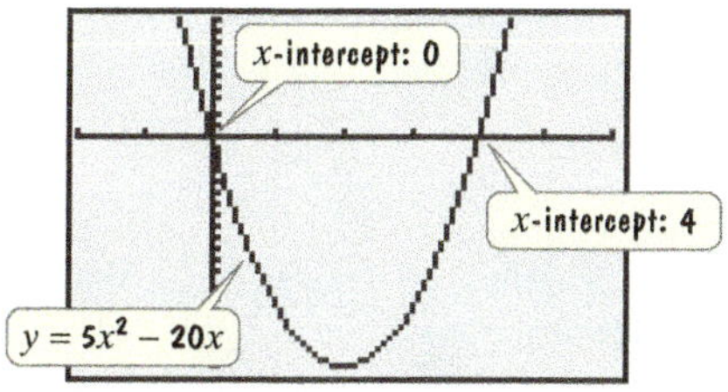

[−2, 6, 1] by [−20, 10, 1]

FIGURE 5.11 The solution set of $5x^2 = 20x$, or $5x^2 - 20x = 0$, is $\{0, 4\}$.

a.

$5x^2 = 20x$	This is the given equation.
$5x^2 - 20x = 0$	Subtract 20x from both sides and write the equation in standard form.
$5x(x - 4) = 0$	Factor.
$5x = 0$ or $x - 4 = 0$	Set each factor equal to 0.
$x = 0$ $x = 4$	Solve the resulting equations.

Check by substituting 0 and 4 into the given equation. The graph of $y = 5x^2 - 20x$, obtained with a graphing utility, is shown in **Figure 5.11**. The x-intercepts are 0 and 4. This verifies that the solutions are 0 and 4, and the solution set is $\{0, 4\}$.

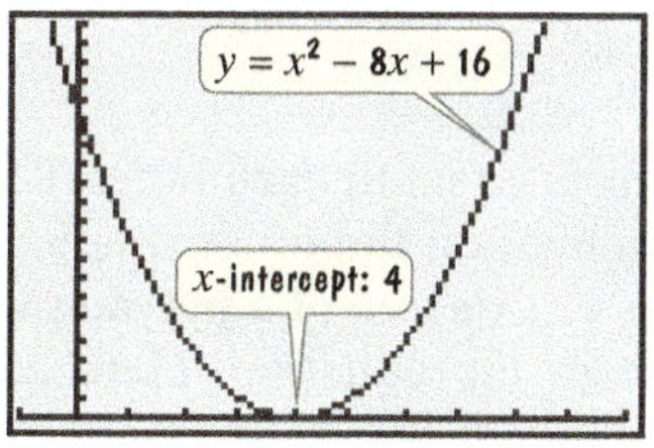

[−1, 10, 1] by [0, 20, 1]

FIGURE 5.12 The solution set of $x^2 + 4 = 8x - 12$, or $x^2 - 8x + 16 = 0$, is $\{4\}$.

b.

$x^2 + 4 = 8x - 12$	This is the given equation.
$x^2 - 8x + 16 = 0$	Write the equation in standard form by subtracting 8x and adding 12 on both sides.
$(x - 4)(x - 4) = 0$	Factor.
$x - 4 = 0$ or $x - 4 = 0$	Set each factor equal to 0.
$x = 4$ $x = 4$	Solve the resulting equations.

Notice that there is only one solution (or, if you prefer, a repeated solution.) The trinomial $x^2 - 8x + 16$ is a perfect square trinomial that could have been factored as $(x - 4)^2$. The graph of $y = x^2 - 8x + 16$, obtained with a graphing utility, is shown in **Figure 5.12**. The graph has only one x-intercept at 4. This verifies that the equation's solution is 4 and the solution set is $\{4\}$.

c. Be careful! Although the left side of $(x - 7)(x + 5) = -20$ is factored, we cannot use the zero-product principle. Why not? The right side of the equation is not 0. So we begin by multiplying the factors on the left side of the equation. Then we add 20 to both sides to obtain 0 on the right side.

$(x - 7)(x + 5) = -20$	This is the given equation.
$x^2 - 2x - 35 = -20$	Use the FOIL method to multiply on the left side.
$x^2 - 2x - 15 = 0$	Add 20 to both sides.
$(x + 3)(x - 5) = 0$	Factor.
$x + 3 = 0$ or $x - 5 = 0$	Set each factor equal to 0.
$x = -3$ $x = 5$	Solve the resulting equations.

Check by substituting -3 and 5 into the given equation. The graph of $y = x^2 - 2x - 15$, obtained with a graphing utility, is shown in **Figure 5.13**. The x-intercepts are -3 and 5. This verifies that the solutions are -3 and 5, and the solution set is $\{-3, 5\}$.

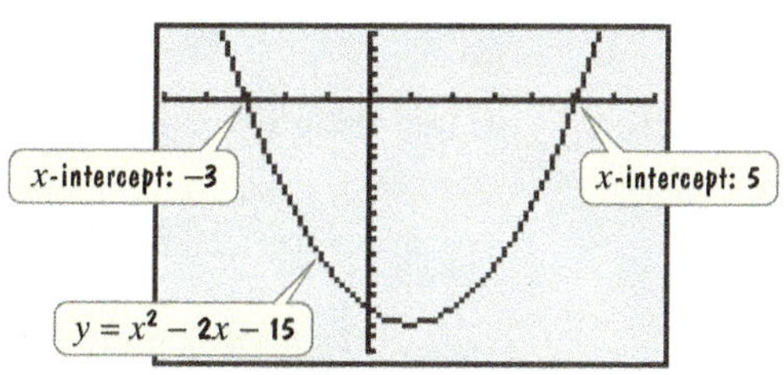

[−5, 7, 1] by [−18, 5, 1]

FIGURE 5.13 The solution set of $(x - 7)(x + 5) = -20$, or $x^2 - 2x - 15 = 0$, is $\{-3, 5\}$.

Study Tip

Avoid the following errors:

$$5x^2 = 20x$$
$$\frac{5x^2}{x} = \frac{20x}{x}$$
$$5x = 20$$
$$x = 4$$

(crossed out)

Never divide both sides of an equation by x. Division by zero is undefined and x may be zero. Indeed, the solutions for this equation (Example 2a) are 0 and 4. Dividing both sides by x does not permit us to find both solutions.

$$(x - 7)(x + 5) = -20$$
$$x - 7 = -20 \quad \text{or} \quad x + 5 = -20$$
$$x = -13 \quad \text{or} \quad x = -25$$

(crossed out)

The zero-product principle cannot be used because the right side of the equation is not equal to 0.

CHECK POINT 2 Solve:

a. $3x^2 = 2x$

b. $x^2 + 7 = 10x - 18$

c. $(x - 2)(x + 3) = 6.$

2 Solve higher-degree polynomial equations by factoring.

Polynomial Equations

A **polynomial equation** is the result of setting two polynomials equal to each other. The equation is in **standard form** if one side is 0 and the polynomial on the other side is in standard form, that is, in descending powers of the variable. The **degree of a polynomial equation** is the same as the highest degree of any term in the equation. Here are examples of three polynomial equations:

$$3x + 5 = 14 \qquad 2x^2 + 7x = 4 \qquad x^3 + x^2 = 4x + 4.$$

This equation is of degree 1 because 1 is the highest degree.

This equation is of degree 2 because 2 is the highest degree.

This equation is of degree 3 because 3 is the highest degree.

Notice that a polynomial equation of degree 1 is a linear equation. A polynomial equation of degree 2 is a quadratic equation.

Some polynomial equations of degree 3 or higher can be solved by moving all terms to one side, thereby obtaining 0 on the other side. Once the equation is in standard form, factor and then set each factor equal to 0.

EXAMPLE 3 Solving a Polynomial Equation by Factoring

Solve by factoring: $x^3 + x^2 = 4x + 4.$

Solution

Step 1. Move all terms to one side and obtain zero on the other side. Subtract $4x$ and subtract 4 from both sides.

$$x^3 + x^2 - 4x - 4 = 4x + 4 - 4x - 4$$
$$x^3 + x^2 - 4x - 4 = 0$$

Step 2. Factor. Use factoring by grouping. Group terms that have a common factor.

$$\boxed{x^3 + x^2} + \boxed{-4x - 4} = 0$$

Common factor is x^2.

Common factor is -4.

Using Technology

Numeric Connections

A graphing utility's TABLE feature can be used to numerically verify that $\{-2, -1, 2\}$ is the solution set of

$$x^3 + x^2 = 4x + 4$$

Enter $y_1 = x^3 + x^2$. Enter $y_2 = 4x + 4$.

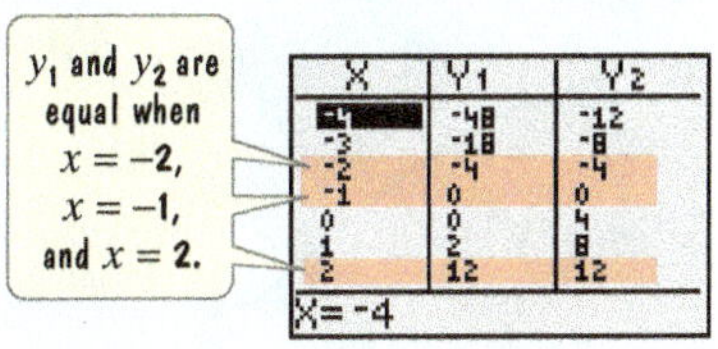

$$x^2(x + 1) - 4(x + 1) = 0 \quad \text{Factor } x^2 \text{ from the first two terms and } -4 \text{ from the last two terms.}$$

$$(x + 1)(x^2 - 4) = 0 \quad \text{Factor out the common binomial, } x + 1, \text{ from each term.}$$

$$(x + 1)(x + 2)(x - 2) = 0 \quad \text{Factor completely by factoring } x^2 - 4 \text{ as the difference of two squares.}$$

Steps 3 and 4. Set each factor equal to zero and solve the resulting equations.

$$x + 1 = 0 \quad \text{or} \quad x + 2 = 0 \quad \text{or} \quad x - 2 = 0$$
$$x = -1 \qquad\qquad x = -2 \qquad\qquad x = 2$$

Step 5. Check the solutions in the original equation. Check the three solutions, -1, -2, and 2, by substituting them into the original equation. Can you verify that the solutions are -1, -2, and 2, and the solution set is $\{-2, -1, 2\}$?

Using Technology

Graphic Connections

You can use a graphing utility to check the solutions to $x^3 + x^2 - 4x - 4 = 0$. Graph $y = x^3 + x^2 - 4x - 4$, as shown on the right. Is the graph complete? Because the degree, 3, is odd, and the leading coefficient, 1, is positive, it should fall to the left and rise to the right (↙, ↗). The graph shows this end behavior and is therefore complete. The x-intercepts are $-2, -1$, and 2, corresponding to the equation's solutions.

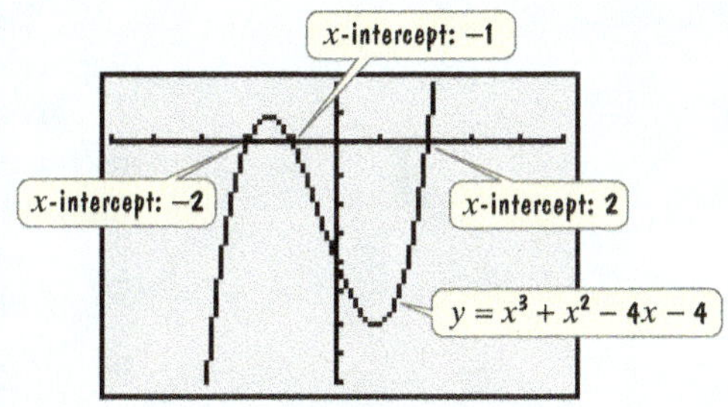

[−5, 5, 1] by [−8, 2, 1]

Discover for Yourself

Suggest a method involving intersecting graphs that can be used with a graphing utility to verify that $\{-2, -1, 2\}$ is the solution set of

$$x^3 + x^2 = 4x + 4.$$

Apply this method to verify the solution set.

CHECK POINT 3 Solve by factoring: $2x^3 + 3x^2 = 8x + 12$.

---

3 Solve problems using polynomial equations.

Applications of Polynomial Equations

Solving polynomial equations by factoring can be used to answer questions about variables contained in mathematical models.

EXAMPLE 4 Modeling Motion

You throw a ball straight up from a rooftop 384 feet high with an initial speed of 32 feet per second. The function

$$s(t) = -16t^2 + 32t + 384$$

describes the ball's height above the ground, $s(t)$, in feet, t seconds after you throw it. The ball misses the rooftop on its way down and eventually strikes the ground. How long will it take for the ball to hit the ground?

Solution The ball hits the ground when $s(t)$, its height above the ground, is 0 feet. Thus, we substitute 0 for $s(t)$ in the given function and solve for t.

$$s(t) = -16t^2 + 32t + 384 \quad \text{This is the function that models the ball's height.}$$

$$0 = -16t^2 + 32t + 384 \quad \text{Substitute 0 for } s(t).$$

$$0 = -16(t^2 - 2t - 24) \quad \text{Factor out } -16.$$

$$0 = -16(t - 6)(t + 4) \quad \text{Factor } t^2 - 2t - 24\text{, the trinomial.}$$

Do not set the constant, −16, equal to zero: $-16 \neq 0$.

$$t - 6 = 0 \quad \text{or} \quad t + 4 = 0 \quad \text{Set each variable factor equal to 0.}$$
$$t = 6 \qquad\qquad t = -4 \quad \text{Solve for } t.$$

Because we begin describing the ball's height at $t = 0$, we discard the solution $t = -4$. The ball hits the ground after 6 seconds.

Figure 5.14 shows the graph of the function $s(t) = -16t^2 + 32t + 384$. The horizontal axis is labeled t, for the ball's time in motion. The vertical axis is labeled $s(t)$, for the ball's height above the ground at time t. Because time and height are both positive, the function is graphed in quadrant I only.

The graph visually shows what we discovered algebraically: The ball hits the ground after 6 seconds. The graph also reveals that the ball reaches its maximum height, 400 feet, after 1 second. Then the ball begins to fall.

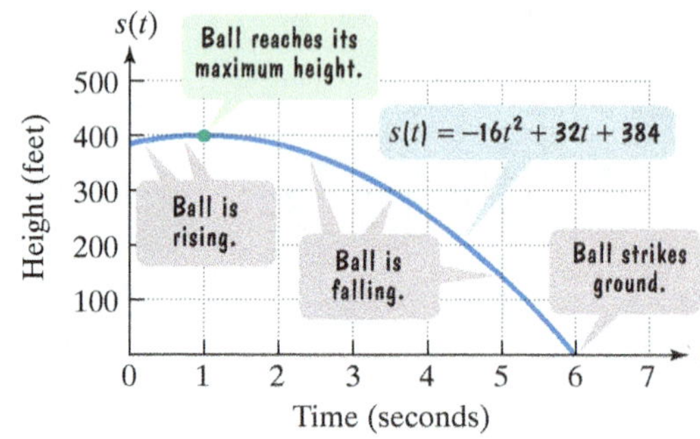

FIGURE 5.14

✓ **CHECK POINT 4** Use the function $s(t) = -16t^2 + 32t + 384$ to determine when the ball's height is 336 feet. Identify your meaningful solution as a point on the graph in **Figure 5.14**.

In our next example, we use our five-step strategy for solving word problems.

EXAMPLE 5 Solving a Problem Involving Landscape Design

A rectangular garden measures 80 feet by 60 feet. A large path of uniform width is to be added along both shorter sides and one longer side of the garden. The landscape designer doing the work wants to double the garden's area with the addition of this path. How wide should the path be?

Solution

Step 1. Let x represent one of the unknown quantities. We will let

$$x = \text{the width of the path.}$$

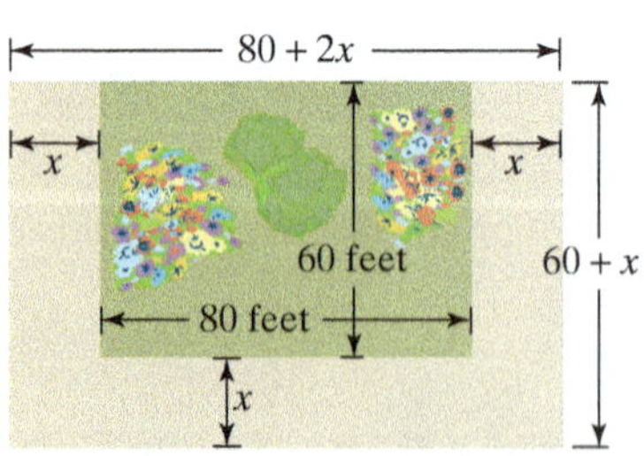

FIGURE 5.15 The garden's area is to be doubled by adding the path.

The situation is illustrated in **Figure 5.15**. The figure shows the original 80-by-60 foot rectangular garden and the path of width x added along both shorter sides and one longer side.

Step 2. Represent other unknown quantities in terms of x. Because the path is added along both shorter sides and one longer side, **Figure 5.15** shows that

$$80 + 2x = \text{the length of the new, expanded rectangle}$$
$$60 + x = \text{the width of the new, expanded rectangle.}$$

Step 3. Write an equation that models the conditions. The area of the rectangle must be doubled by the addition of the path.

The area, or length times width, of the new, expanded rectangle | must be | twice that of | the area of the garden.

$$(80 + 2x)(60 + x) = 2 \cdot 80 \cdot 60$$

Step 4. Solve the equation and answer the question.

$(80 + 2x)(60 + x) = 2 \cdot 80 \cdot 60$	This is the equation that models the problem's conditions.
$4800 + 200x + 2x^2 = 9600$	Multiply. Use FOIL on the left side.
$2x^2 + 200x - 4800 = 0$	Subtract 9600 from both sides and write the equation in standard form.
$2(x^2 + 100x - 2400) = 0$	Factor out 2, the GCF.
$2(x - 20)(x + 120) = 0$	Factor the trinomial.
$x - 20 = 0$ or $x + 120 = 0$	Set each variable factor equal to 0.
$x = 20$ or $x = -120$	Solve for x.

The path cannot have a negative width. Because −120 is geometrically impossible, we use $x = 20$. The width of the path should be 20 feet.

Step 5. Check the proposed solution in the original wording of the problem. Has the landscape architect doubled the garden's area with the 20-foot-wide path? The area of the garden is 80 feet times 60 feet, or 4800 square feet. Because $80 + 2x$ and $60 + x$ represent the length and width of the expanded rectangle,

$80 + 2x = 80 + 2 \cdot 20 = 120$ feet is the expanded rectangle's length.

$60 + x = 60 + 20 = 80$ feet is the expanded rectangle's width.

The area of the expanded rectangle is 120 feet times 80 feet, or 9600 square feet. This is double the area of the garden, 4800 square feet, as specified by the problem's conditions.

✓ CHECK POINT 5 A rectangular garden measures 16 feet by 12 feet. A path of uniform width is to be added so as to surround the entire garden. The landscape artist doing the work wants the garden and path to cover an area of 320 square feet. How wide should the path be?

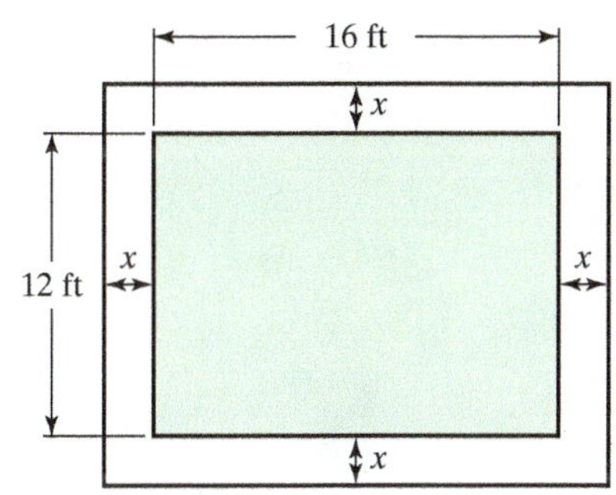

The solution to our next problem relies on knowing the **Pythagorean Theorem**. The theorem relates the lengths of the three sides of a **right triangle**, a triangle with one angle measuring 90°. The side opposite the 90° angle is called the **hypotenuse**. The other sides are called **legs**. The legs form the two sides of the right angle.

The Pythagorean Theorem

The sum of the squares of the lengths of the legs of a right triangle equals the square of the length of the hypotenuse.

If the legs have lengths a and b, and the hypotenuse has length c, then

$$a^2 + b^2 = c^2.$$

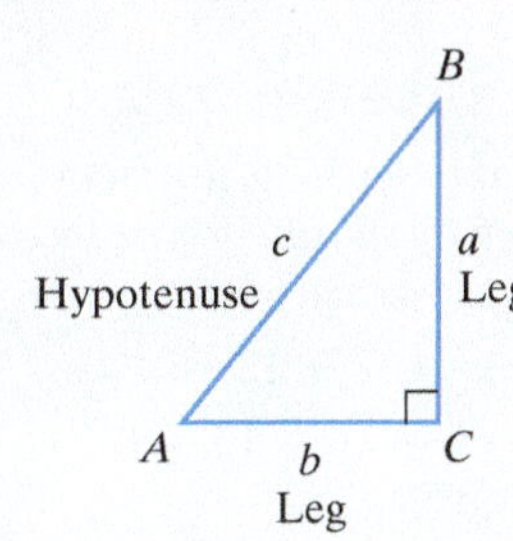

EXAMPLE 6 Using the Pythagorean Theorem to Obtain a Polynomial Equation

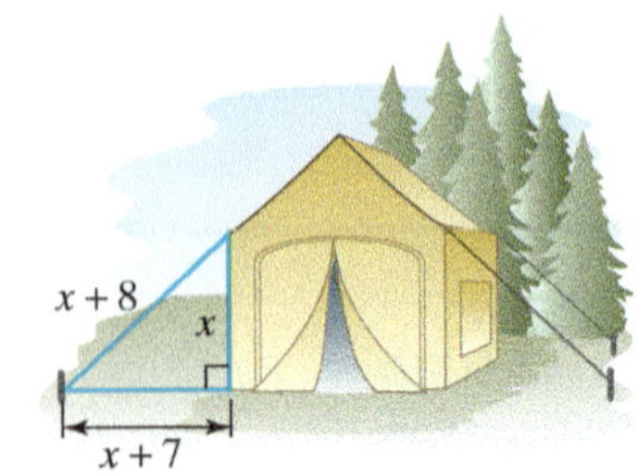

FIGURE 5.16

Figure 5.16 shows a tent with wires attached to help stabilize it. The length of each wire is 8 feet greater than the distance from the ground to where it is attached to the tent. The distance from the base of the tent to where the wire is anchored exceeds this height by 7 feet. Find the length of each wire used to stabilize the tent.

Solution **Figure 5.16** shows a right triangle. The length of the legs are x and $x + 7$. The length of the hypotenuse, $x + 8$, represents the length of the wire. We use the Pythagorean Theorem to find this length.

$$\text{leg}^2 + \text{leg}^2 = \text{hypotenuse}^2$$

$$x^2 + (x + 7)^2 = (x + 8)^2$$ This is the equation arising from the Pythagorean Theorem.

$$x^2 + x^2 + 14x + 49 = x^2 + 16x + 64$$ Square $x + 7$ and $x + 8$.

$$2x^2 + 14x + 49 = x^2 + 16x + 64$$ Combine like terms: $x^2 + x^2 = 2x^2$.

$$x^2 - 2x - 15 = 0$$ Subtract $x^2 + 16x + 64$ from both sides and write the quadratic equation in standard form.

$$(x - 5)(x + 3) = 0$$ Factor the trinomial.

$$x - 5 = 0 \quad \text{or} \quad x + 3 = 0$$ Set each factor equal to 0.

$$x = 5 \qquad\qquad x = -3$$ Solve for x.

Because x represents the distance from the ground to where the wire is attached, x cannot be negative. Thus, we only use $x = 5$. **Figure 5.16** shows that the length of the wire is $x + 8$ feet. The length of the wire is $5 + 8$ feet, or 13 feet.

We can check to see that the lengths of the three sides of the right triangle, x, $x + 7$, and $x + 8$, satisfy the Pythagorean Theorem when $x = 5$. The lengths are 5 feet, 12 feet, and 13 feet.

$$\text{leg}^2 + \text{leg}^2 = \text{hypotenuse}^2$$

$$5^2 + 12^2 \stackrel{?}{=} 13^2$$

$$25 + 144 \stackrel{?}{=} 169$$

$$169 = 169,$$ true

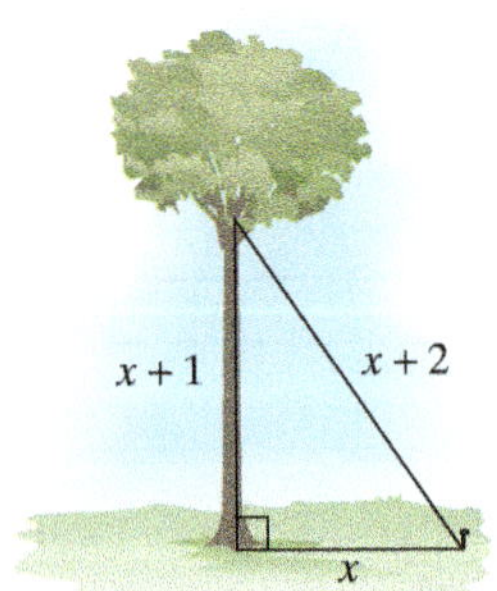

FIGURE 5.17

CHECK POINT 6 A guy wire is attached to a tree to help it grow straight. The situation is illustrated in **Figure 5.17**. The length of the wire is 2 feet greater than the distance from the base of the tree to the stake. Find the length of the wire.

5.7 EXERCISE SET

MyMathLab — MathXL PRACTICE — WATCH — DOWNLOAD — READ — REVIEW

Practice Exercises

In Exercises 1–36, use factoring to solve each quadratic equation. Check by substitution or by using a graphing utility and identifying x-intercepts.

1. $x^2 + x - 12 = 0$
2. $x^2 - 2x - 15 = 0$
3. $x^2 + 6x = 7$
4. $x^2 - 4x = 45$
5. $3x^2 + 10x - 8 = 0$
6. $2x^2 - 5x - 3 = 0$
7. $5x^2 = 8x - 3$
8. $7x^2 = 30x - 8$
9. $3x^2 = 2 - 5x$
10. $5x^2 = 2 + 3x$
11. $x^2 = 8x$

12. $x^2 = 4x$

13. $3x^2 = 5x$

14. $2x^2 = 5x$

15. $x^2 + 4x + 4 = 0$

16. $x^2 + 6x + 9 = 0$

17. $x^2 = 14x - 49$

18. $x^2 = 12x - 36$

19. $9x^2 = 30x - 25$

20. $4x^2 = 12x - 9$

21. $x^2 - 25 = 0$

22. $x^2 - 49 = 0$

23. $9x^2 = 100$

24. $4x^2 = 25$

25. $x(x - 3) = 18$

26. $x(x - 4) = 21$

27. $(x - 3)(x + 8) = -30$

28. $(x - 1)(x + 4) = 14$

29. $x(x + 8) = 16(x - 1)$

30. $x(x + 9) = 4(2x + 5)$

31. $(x + 1)^2 - 5(x + 2) = 3x + 7$

32. $(x + 1)^2 = 2(x + 5)$

33. $x(8x + 1) = 3x^2 - 2x + 2$

34. $2x(x + 3) = -5x - 15$

35. $\frac{x^2}{18} + \frac{x}{2} + 1 = 0$

36. $\frac{x^2}{4} - \frac{5x}{2} + 6 = 0$

In Exercises 37–46, use factoring to solve each polynomial equation. Check by substitution or by using a graphing utility and identifying x-intercepts.

37. $x^3 + 4x^2 - 25x - 100 = 0$

38. $x^3 - 2x^2 - x + 2 = 0$

39. $x^3 - x^2 = 25x - 25$

40. $x^3 + 2x^2 = 16x + 32$

41. $3x^4 - 48x^2 = 0$

42. $5x^4 - 20x^2 = 0$

43. $x^4 - 4x^3 + 4x^2 = 0$

44. $x^4 - 6x^3 + 9x^2 = 0$

45. $2x^3 + 16x^2 + 30x = 0$

46. $3x^3 - 9x^2 - 30x = 0$

In Exercises 47–50, determine the x-intercepts of the graph of each quadratic function. Then match the function with its graph, labeled (a)–(d). Each graph is shown in a $[-10, 10, 1]$ *by* $[-10, 10, 1]$ *viewing rectangle.*

47. $y = x^2 - 6x + 8$

48. $y = x^2 - 2x - 8$

49. $y = x^2 + 6x + 8$

50. $y = x^2 + 2x - 8$

a.

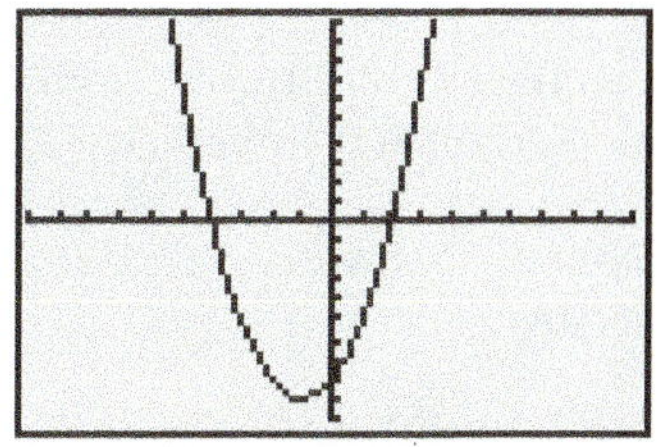

b.

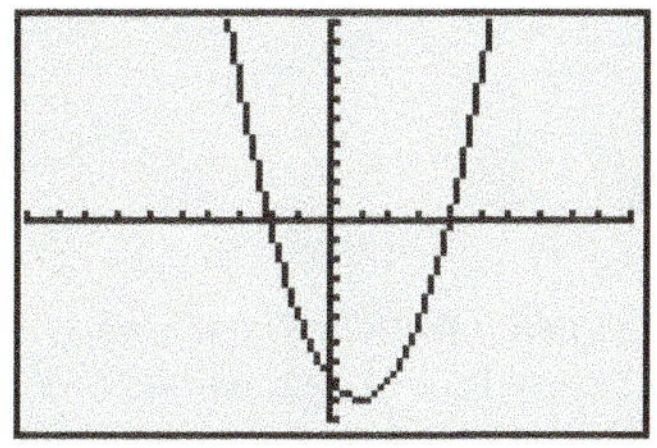

c.

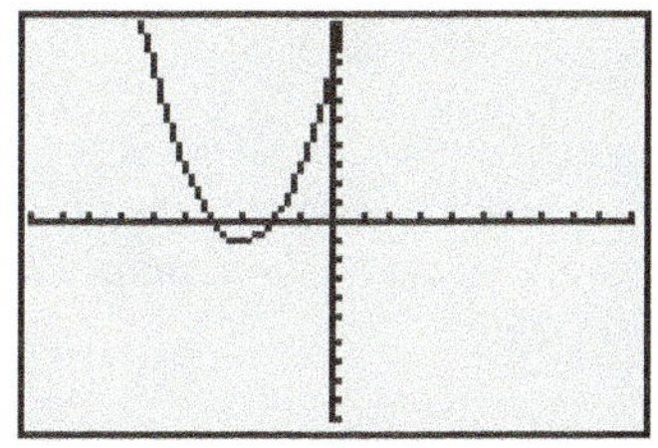

d.

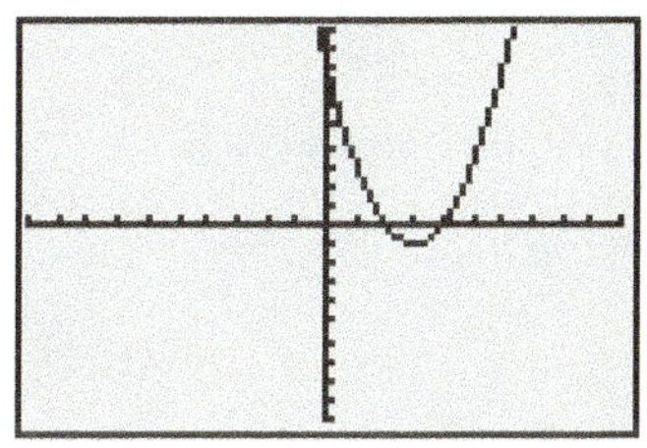

Practice PLUS

In Exercises 51–54, solve each polynomial equation.

51. $x(x + 1)^3 - 42(x + 1)^2 = 0$

52. $x(x - 2)^3 - 35(x - 2)^2 = 0$

53. $-4x[x(3x - 2) - 8](25x^2 - 40x + 16) = 0$

54. $-7x[x(2x - 5) - 12](9x^2 + 30x + 25) = 0$

In Exercises 55–58, find all values of c satisfying the given conditions.

55. $f(x) = x^2 - 4x - 27$ and $f(c) = 5$.

56. $f(x) = 5x^2 - 11x + 6$ and $f(c) = 4$.

57. $f(x) = 2x^3 + x^2 - 8x + 2$ and $f(c) = 6$.

58. $f(x) = x^3 + 4x^2 - x + 6$ and $f(c) = 10$.

In Exercises 59–62, find all numbers satisfying the given conditions.

59. The product of the number decreased by 1 and increased by 4 is 24.

60. The product of the number decreased by 6 and increased by 2 is 20.

61. If 5 is subtracted from 3 times the number, the result is the square of 1 less than the number.

62. If the square of the number is subtracted from 61, the result is the square of 1 more than the number.

In Exercises 63–64, list all numbers that must be excluded from the domain of the given function.

63. $f(x) = \dfrac{3}{x^2 + 4x - 45}$

64. $f(x) = \dfrac{7}{x^2 - 3x - 28}$

Application Exercises

A gymnast dismounts the uneven parallel bars at a height of 8 feet with an initial upward velocity of 8 feet per second. The function

$$s(t) = -16t^2 + 8t + 8$$

describes the height of the gymnast's feet above the ground, s(t), in feet, t seconds after dismounting. The graph of the function is shown, with unlabeled tick marks along the horizontal axis. Use the function to solve Exercises 65–66.

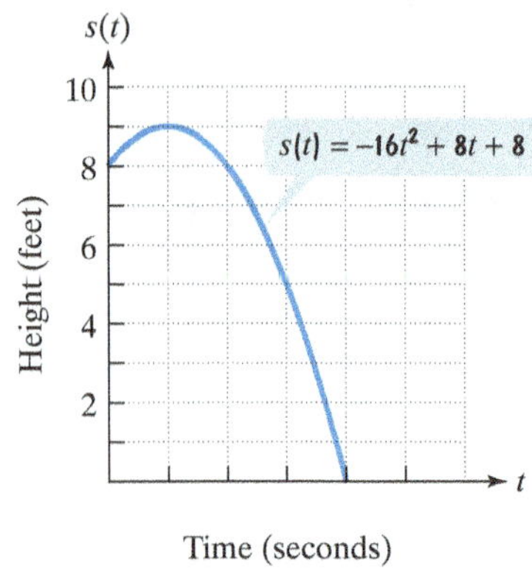

65. How long will it take the gymnast to reach the ground? Use this information to provide a number on each tick mark along the horizontal axis in the figure shown.

66. When will the gymnast be 8 feet above the ground? Identify the solution(s) as one or more points on the graph.

In a round-robin chess tournament, each player is paired with every other player once. The function

$$f(x) = \frac{x^2 - x}{2}$$

models the number of chess games, f(x), that must be played in a round-robin tournament with x chess players. Use this function to solve Exercises 67–68.

67. In a round-robin chess tournament, 21 games were played. How many players were entered in the tournament?

68. In a round-robin chess tournament, 36 games were played. How many players were entered in the tournament?

The graph of the quadratic function in Exercises 67–68 is shown. Use the graph to solve Exercises 69–70.

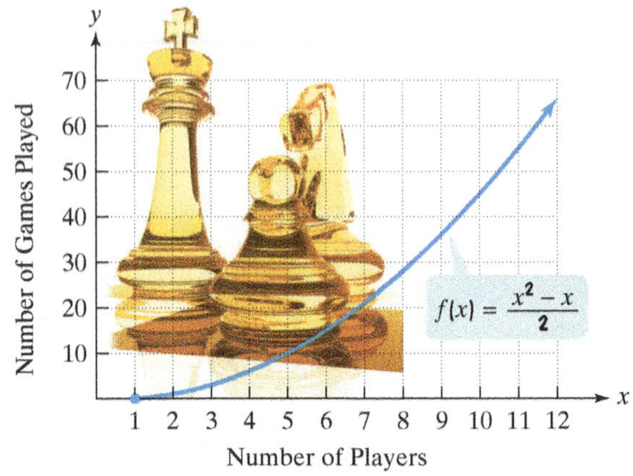

69. Identify your solution to Exercise 67 as a point on the graph.

70. Identify your solution to Exercise 68 as a point on the graph.

71. The length of a rectangular sign is 3 feet longer than the width. If the sign's area is 54 square feet, find its length and width.

72. A rectangular parking lot has a length that is 3 yards greater than the width. The area of the parking lot is 180 square yards. Find the length and the width.

73. Each side of a square is lengthened by 3 inches. The area of this new, larger square is 64 square inches. Find the length of a side of the original square.

74. Each side of a square is lengthened by 2 inches. The area of this new, larger square is 36 square inches. Find the length of a side of the original square.

75. A pool measuring 10 meters by 20 meters is surrounded by a path of uniform width, as shown in the figure. If the area of the pool and the path combined is 600 square meters, what is the width of the path?

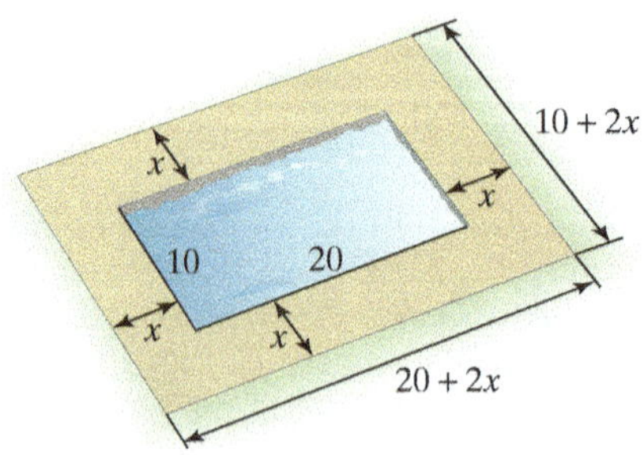

76. A vacant rectangular lot is being turned into a community vegetable garden measuring 15 meters by 12 meters. A path of uniform width is to surround the garden. If the area of the lot is 378 square meters, find the width of the path surrounding the garden.

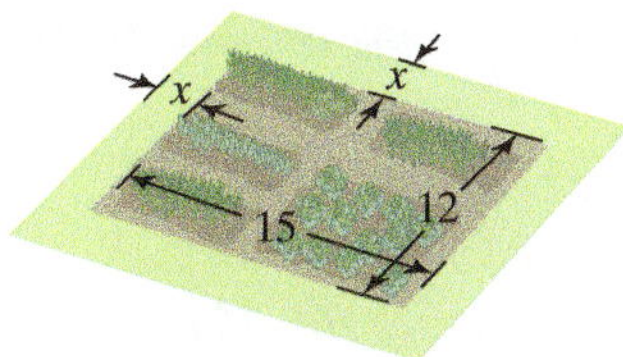

77. As part of a landscaping project, you put in a flower bed measuring 10 feet by 12 feet. You plan to surround the bed with a uniform border of low-growing plants.

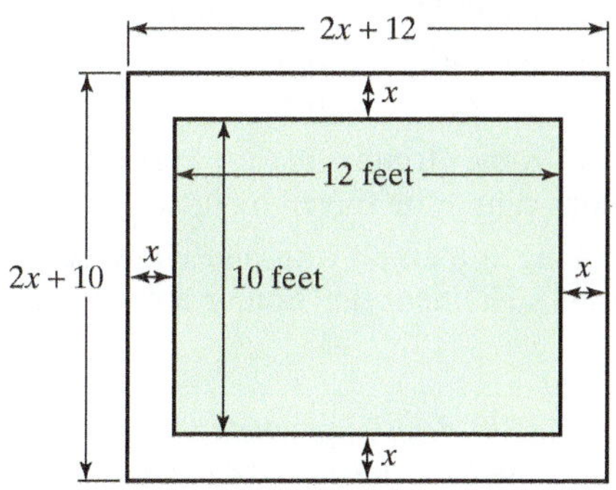

a. Write a polynomial that describes the area of the uniform border that surrounds your flower bed. *Hint:* The area of the border is the area of the large rectangle shown in the figure minus the area of the flower bed.

b. The low growing plants surrounding the flower bed require 1 square foot each when mature. If you have 168 of these plants, how wide a strip around the flower bed should you prepare for the border?

78. As part of a landscaping project, you put in a flower bed measuring 20 feet by 30 feet. To finish off the project, you are putting in a uniform border of pine bark around the outside of the rectangular garden. You have enough pine bark to cover 336 square feet. How wide should the border be?

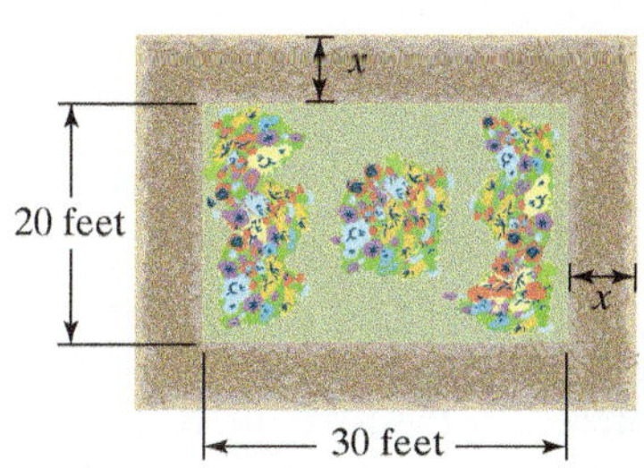

79. A machine produces open boxes using square sheets of metal. The figure illustrates that the machine cuts equal-sized squares measuring 2 inches on a side from the corners and then shapes the metal into an open box by turning up the sides. If each box must have a volume of 200 cubic inches, find the length and width of the open box.

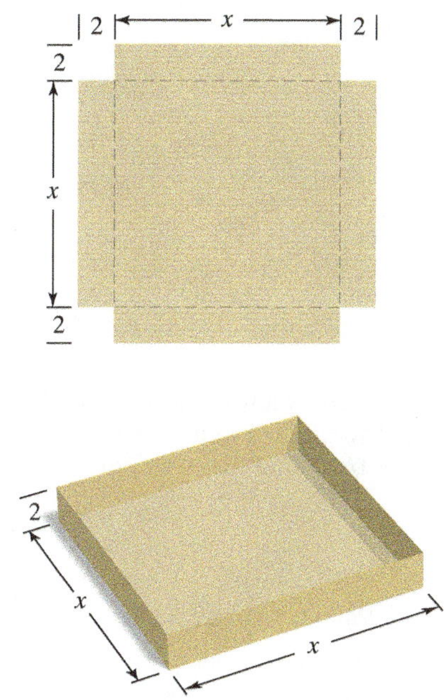

80. A machine produces open boxes using square sheets of metal. The machine cuts equal-sized squares measuring 3 inches on a side from the corners and then shapes the metal into an open box by turning up the sides. If each box must have a volume of 75 cubic inches, find the length and width of the open box.

81. The rectangular floor of a closet is divided into two right triangles by drawing a diagonal, as shown in the figure. One leg of the right triangle is 2 feet more than twice the other leg. The hypotenuse is 13 feet. Determine the closet's length and width.

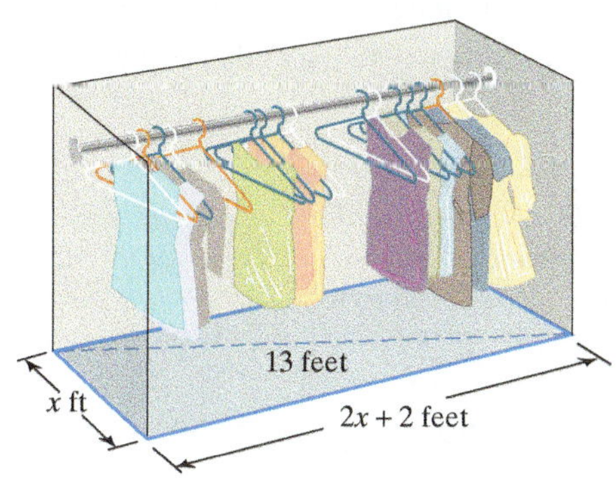

82. A piece of wire measuring 20 feet is attached to a telephone pole as a guy wire. The distance along the ground from the bottom of the pole to the end of the wire is 4 feet greater than the height where the wire is attached to the pole. How far up the pole does the guy wire reach?

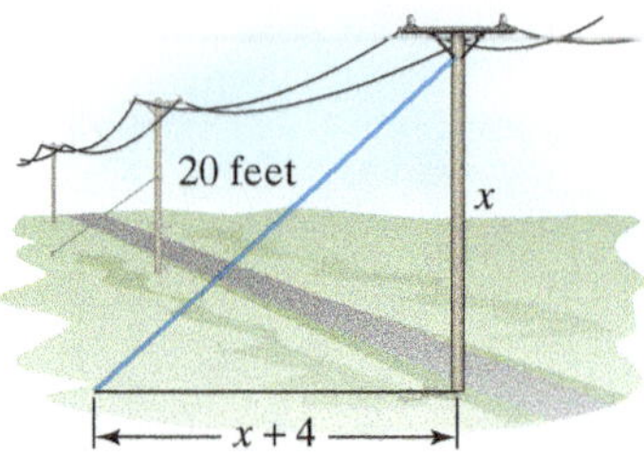

83. A tree is supported by a wire anchored in the ground 15 feet from its base. The wire is 4 feet longer than the height that it reaches on the tree. Find the length of the wire.

84. A tree is supported by a wire anchored in the ground 5 feet from its base. The wire is 1 foot longer than the height that it reaches on the tree. Find the length of the wire.

Writing in Mathematics

85. What is a quadratic equation?

86. What is the zero-product principle?

87. Explain how to solve $x^2 - x = 6$.

88. Describe the relationship between the solutions of a quadratic equation and the graph of the corresponding quadratic function.

89. What is a polynomial equation? When is it in standard form?

90. What is the degree of a polynomial equation? What are polynomial equations of degree 1 and degree 2, respectively, called?

91. Explain how to solve $x^3 + x^2 = x + 1$.

92. If something is thrown straight up, or possibly dropped, describe a situation in which it is important to know how long it will take the object to hit the ground or possibly the water.

93. A toy rocket is launched vertically upward. Using a quadratic equation, we find that the rocket will reach a height of 220 feet at 2.5 seconds and again at 5.5 seconds. How can this be?

94. Describe a situation in which a landscape designer might use polynomials and polynomial equations.

95. In your own words, state the Pythagorean Theorem.

Technology Exercises

In Exercises 96–99, use a graphing utility with a viewing rectangle large enough to show end behavior to graph each polynomial function. Then use the x-intercepts for the graph to solve the polynomial equation. Check by substitution.

96. Use the graph of $y = x^2 + 3x - 4$ to solve $x^2 + 3x - 4 = 0$.

97. Use the graph of $y = x^3 + 3x^2 - x - 3$ to solve $x^3 + 3x^2 - x - 3 = 0$.

98. Use the graph of $y = 2x^3 - 3x^2 - 11x + 6$ to solve $2x^3 - 3x^2 - 11x + 6 = 0$.

99. Use the graph of $y = -x^4 + 4x^3 - 4x^2$ to solve $-x^4 + 4x^3 - 4x^2 = 0$.

100. Use the TABLE feature of a graphing utility to verify the solution sets for any two equations in Exercises 31–32 or 39–40.

Critical Thinking Exercises

Make Sense? *In Exercises 101–104, determine whether each statement "makes sense" or "does not make sense" and explain your reasoning.*

101. I'm working with a quadratic function that describes the length of time a ball has been thrown into the air and its height above the ground, and I find the function's graph more meaningful than its equation.

102. I set the quadratic equation $2x^2 - 5x = 12$ equal to zero and obtained $2x^2 - 5x = 0$.

103. Because some trinomials are prime, some quadratic equations cannot be solved by factoring.

104. I'm looking at a graph with one x-intercept, so it must be the graph of a linear function or a vertical line.

In Exercises 105–108, determine whether each statement is true or false. If the statement is false, make the necessary change(s) to produce a true statement.

105. Quadratic equations solved by factoring always have two different solutions.

106. If $4x(x^2 + 49) = 0$, then

$$4x = 0 \quad \text{or} \quad x^2 + 49 = 0$$
$$x = 0 \quad \text{or} \quad x = 7 \quad \text{or} \quad x = -7.$$

107. If -4 is a solution of $7y^2 + (2k - 5)y - 20 = 0$, then k must equal 14.

108. Some quadratic equations have more than two solutions.

109. Write a quadratic equation in standard form whose solutions are -3 and 7.

110. Solve: $|x^2 + 2x - 36| = 12$.

Review Exercises

111. Solve: $|3x - 2| = 8$. (Section 4.3, Example 1)

112. Simplify: $3(5 - 7)^2 + \sqrt{16} + 12 \div (-3)$. (Section 1.2, Example 7)

113. You invested \$3000 in two accounts paying 5% and 8% annual interest. If the total interest earned for the year is \$189, how much was invested at each rate? (Section 3.2, Example 2)

Preview Exercises

Exercises 114–116 will help you prepare for the material covered in the first section of the next chapter.

114. If $f(x) = \frac{120x}{100 - x}$, find $f(20)$.

115. Find the domain of $f(x) = \frac{4}{x - 2}$.

116. Factor the numerator and the denominator. Then simplify by dividing out the common factor in the numerator and the denominator.

$$\frac{x^2 - 7x - 18}{2x^2 + 3x - 2}$$

GROUP PROJECT

CHAPTER 5

Divide the group in half. Without looking at any factoring problems in the book, each group should use polynomial multiplication to create five factoring problems. Make sure that some of your problems require at least two factoring strategies. Next, exchange problems with the other half of the group. Work to factor the five problems. After completing the factorizations, evaluate the factoring problems that you were given. Are they too easy? Too difficult? Can the polynomials really be factored? Share your responses with the half of the group that wrote the problems. Finally, grade each other's work in factoring the polynomials. Each factoring problem is worth 20 points. You may award partial credit. If you take off points, explain why points are deducted and how you decided to take off a particular number of points for the error(s) that you found.

Chapter 5 Summary

Definitions and Concepts	Examples

Section 5.1 Introduction to Polynomials and Polynomial Functions

Definitions and Concepts

A polynomial is a single term or the sum of two or more terms containing variables with whole-number exponents. A monomial is a polynomial with exactly one term; a binomial has exactly two terms; a trinomial has exactly three terms. If $a \neq 0$, the degree of ax^n is n and the degree of ax^ny^m is $n + m$. The degree of a nonzero constant is 0. The constant 0 has no defined degree. The degree of a polynomial is the greatest degree of any term. The leading term is the term of greatest degree. Its coefficient is called the leading coefficient.

Examples

$$7x^3y - 4x^5y^4 - 2x^4y$$

Degree is $3 + 1 = 4$. Degree is $5 + 4 = 9$. Degree is $4 + 1 = 5$.

The degree of the polynomial is 9. The leading term is $-4x^5y^4$. The leading coefficient is -4. This polynomial is a trinomial.

Definitions and Concepts

In a polynomial function, the expression that defines the function is a polynomial. Polynomial functions have graphs that are smooth and continuous. The behavior of the graph of a polynomial function to the far left or the far right is called its end behavior.

The Leading Coefficient Test

1. Odd degree polynomial functions have graphs with opposite behavior at each end. If the leading coefficient is positive, the graph falls to the left and rises to the right. If the leading coefficient is negative, the graph rises to the left and falls to the right.
2. Even-degree polynomial functions have graphs with the same behavior at each end. If the leading coefficient is positive, the graph rises to the left and rises to the right. If the leading coefficient is negative, the graph falls to the left and falls to the right.

Examples

Describe the end behavior of the graph of each polynomial function:

- $f(x) = -2x^3 + 3x^2 + 11x - 6$

  The degree, 3, is odd. The leading coefficient, -2, is negative. The graph rises to the left and falls to the right.

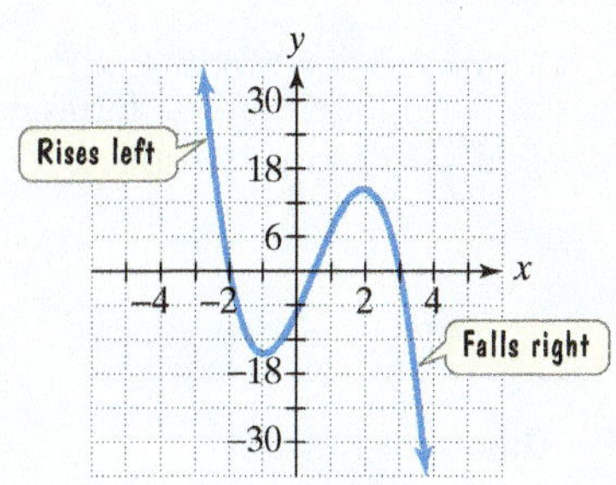

Definitions and Concepts	Examples
Section 5.1 Introduction to Polynomials and Polynomial Functions (continued)	
	• $f(x) = x^2 - 4$ The degree, 2, is even. The leading coefficient, 1, is positive. The graph rises to the left and rises to the right. 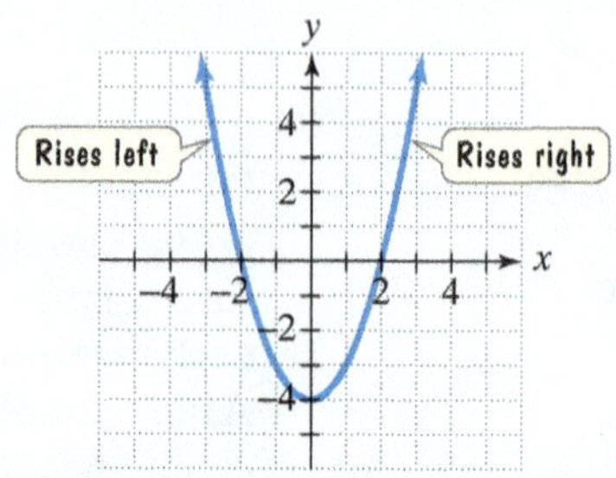
To add polynomials, add like terms.	$(6x^3y + 5x^2y - 7y) + (-9x^3y + x^2y + 6y)$ $= (6x^3y - 9x^3y) + (5x^2y + x^2y) + (-7y + 6y)$ $= -3x^3y + 6x^2y - y$
To subtract two polynomials, change the sign of every term of the second polynomial. Add this result to the first polynomial.	$(5y^3 - 9y^2 - 4) - (3y^3 - 12y^2 - 5)$ $= (5y^3 - 9y^2 - 4) + (-3y^3 + 12y^2 + 5)$ $= (5y^3 - 3y^3) + (-9y^2 + 12y^2) + (-4 + 5)$ $= 2y^3 + 3y^2 + 1$
Section 5.2 Multiplication of Polynomials	
To multiply monomials, multiply coefficients and add exponents.	$(-2x^2y^4)(-3x^3y)$ $= (-2)(-3)x^{2+3}y^{4+1} = 6x^5y^5$
To multiply a monomial and a polynomial, multiply each term of the polynomial by the monomial.	$7x^2y(4x^3y^5 - 2xy - 1)$ $= 7x^2y \cdot 4x^3y^5 - 7x^2y \cdot 2xy - 7x^2y \cdot 1$ $= 28x^5y^6 - 14x^3y^2 - 7x^2y$
To multiply polynomials if neither is a monomial, multiply each term of one by each term of the other.	$(x^3 + 2x)(5x^2 - 3x + 4)$ $= x^3(5x^2 - 3x + 4) + 2x(5x^2 - 3x + 4)$ $= 5x^5 - 3x^4 + 4x^3 + 10x^3 - 6x^2 + 8x$ $= 5x^5 - 3x^4 + 14x^3 - 6x^2 + 8x$
The FOIL method may be used when multiplying two binomials: First terms multiplied. Outside terms multiplied. Inside terms multiplied. Last terms multiplied.	F O I L $(5x - 3y)(2x + y) = 5x \cdot 2x + 5x \cdot y + (-3y) \cdot 2x + (-3y) \cdot y$ $= 10x^2 + 5xy - 6xy - 3y^2$ $= 10x^2 - xy - 3y^2$
The Square of a Binomial Sum $(A + B)^2 = A^2 + 2AB + B^2$	$(x^2 + 6)^2 = (x^2)^2 + 2 \cdot x^2 \cdot 6 + 6^2$ $= x^4 + 12x^2 + 36$

Definitions and Concepts	Examples

Section 5.2 Multiplication of Polynomials (continued)

The Square of a Binomial Difference

$$(A - B)^2 = A^2 - 2AB + B^2$$

$$(4x - 5)^2 = (4x)^2 - 2\cdot 4x \cdot 5 + 5^2$$
$$= 16x^2 - 40x + 25$$

The Product of the Sum and Difference of Two Terms

$$(A + B)(A - B) = A^2 - B^2$$

- $(3x + 7y)(3x - 7y) = (3x)^2 - (7y)^2$
  $= 9x^2 - 49y^2$
- $[(x + 2) - 4y][(x + 2) + 4y]$
  $= (x + 2)^2 - (4y)^2 = x^2 + 4x + 4 - 16y^2$

Section 5.3 Greatest Common Factors and Factoring by Grouping

Factoring a polynomial consisting of the sum of monomials means finding an equivalent expression that is a product. Polynomials that cannot be factored using integer coefficients are called prime polynomials over the integers. The greatest common factor, GCF, is an expression that divides every term of the polynomial. The GCF is the product of the largest common numerical factor and the variable of lowest degree common to every term of the polynomial. To factor a monomial from a polynomial, express each term as the product of the GCF and its other factor. Then use the distributive property to factor out the GCF. When the leading coefficient of a polynomial is negative, it is often desirable to factor out a common factor with a negative coefficient.

Factor: $4x^4y^2 - 12x^2y^3 + 20xy^2$.
(GCF is $4xy^2$)

$$= 4xy^2 \cdot x^3 - 4xy^2 \cdot 3xy + 4xy^2 \cdot 5$$
$$= 4xy^2(x^3 - 3xy + 5)$$

Factor: $-25x^3 + 10x^2 - 15x$.
(Use $-5x$ as a common factor.)

$$= -5x(5x^2) - 5x(-2x) - 5x(3)$$
$$= -5x(5x^2 - 2x + 3)$$

To factor by grouping, factor out the GCF from each group. Then factor out the remaining factor.

$$xy + 7x - 2y - 14$$
$$= x(y + 7) - 2(y + 7)$$
$$= (y + 7)(x - 2)$$

Section 5.4 Factoring Trinomials

To factor a trinomial of the form $x^2 + bx + c$, find two numbers whose product is c and whose sum is b. The factorization is

$$(x + \text{one number})(x + \text{other number}).$$

Factor: $x^2 + 9x + 20$.
Find two numbers whose product is 20 and whose sum is 9. The numbers are 4 and 5.

$$x^2 + 9x + 20 = (x + 4)(x + 5)$$

In some trinomials, the highest power is greater than 2, and the exponent in one of the terms is half that of the other term. Factor by introducing a substitution. Let u equal the variable to the smaller power.

Factor: $x^6 - 7x^3 + 12$.

$$= (x^3)^2 - 7x^3 + 12$$
$$= u^2 - 7u + 12 \quad \text{Let } u = x^3.$$
$$= (u - 4)(u - 3)$$
$$= (x^3 - 4)(x^3 - 3)$$

To factor $ax^2 + bx + c$ by trial and error, try various combinations of factors of ax^2 and c until a middle term of bx is obtained for the sum of the outside and inside products.

Factor: $2x^2 + 7x - 15$.

Factors of $2x^2$: $2x$, x

Factors of -15: 1 and -15, -1 and 15, 3 and -5, -3 and 5

$$(2x - 3)(x + 5)$$

Sum of outside and inside products should equal $7x$.

$$10x - 3x = 7x$$

Thus, $2x^2 + 7x - 15 = (2x - 3)(x + 5)$.

Definitions and Concepts	Examples
Section 5.4 Factoring Trinomials (continued)	
To factor $ax^2 + bx + c$ by grouping, find the factors of ac whose sum is b. Write bx as a sum or difference using these factors. Then factor by grouping.	Factor: $2x^2 + 7x - 15$. Find the factors of $2(-15)$, or -30, whose sum is 7. They are 10 and -3. $2x^2 + 7x - 15$ $= 2x^2 + 10x - 3x - 15$ $= 2x(x + 5) - 3(x + 5) = (x + 5)(2x - 3)$
Section 5.5 Factoring Special Forms	
The Difference of Two Squares $A^2 - B^2 = (A + B)(A - B)$	$16x^2 - 9y^2$ $= (4x)^2 - (3y)^2 = (4x + 3y)(4x - 3y)$
Perfect Square Trinomials $A^2 + 2AB + B^2 = (A + B)^2$ $A^2 - 2AB + B^2 = (A - B)^2$	• $x^2 + 20x + 100 = x^2 + 2 \cdot x \cdot 10 + 10^2 = (x + 10)^2$ • $9x^2 - 30x + 25 = (3x)^2 - 2 \cdot 3x \cdot 5 + 5^2 = (3x - 5)^2$
Sum or Difference of Cubes $A^3 + B^3 = (A + B)(A^2 - AB + B^2)$ $A^3 - B^3 = (A - B)(A^2 + AB + B^2)$	$125x^3 - 8 = (5x)^3 - 2^3$ $= (5x - 2)[(5x)^2 + 5x \cdot 2 + 2^2]$ $= (5x - 2)(25x^2 + 10x + 4)$
When using factoring by grouping, terms can sometimes be grouped to obtain the difference of two squares. One of the squares is a perfect square trinomial.	$\underbrace{x^2 + 18x + 81} - 25y^2$ $= (x + 9)^2 - (5y)^2$ $= (x + 9 + 5y)(x + 9 - 5y)$
Section 5.6 A General Factoring Strategy	
A Factoring Strategy 1. Factor out the GCF or a common factor with a negative coefficient. 2. a. If two terms, try $A^2 - B^2 = (A + B)(A - B)$ $A^3 + B^3 = (A + B)(A^2 - AB + B^2)$ $A^3 - B^3 = (A - B)(A^2 + AB + B^2)$. b. If three terms, try $A^2 + 2AB + B^2 = (A + B)^2$ $A^2 - 2AB + B^2 = (A - B)^2$. If not a perfect square trinomial, try trial and error or grouping. c. If four terms, try factoring by grouping. 3. See if any factors can be factored further.	Factor: $3x^4 + 12x^3 - 3x^2 - 12x$. The GCF is $3x$. $3x^4 + 12x^3 - 3x^2 - 12x$ $= 3x(x^3 + 4x^2 - x - 4)$ Four terms: Try grouping. $= 3x[x^2(x + 4) - 1(x + 4)]$ $= 3x(x + 4)(x^2 - 1)$ This can be factored further. $= 3x(x + 4)(x + 1)(x - 1)$

Definitions and Concepts	Examples

Section 5.7 Polynomial Equations and Their Applications

A quadratic equation in x can be written in the standard form

$$ax^2 + bx + c, a \neq 0.$$

A polynomial equation is the result of setting two polynomials equal to each other. The equation is in standard form if one side is 0 and the polynomial on the other side is in standard form, that is in descending powers of the variable. In standard form, its degree is the highest degree of any term in the equation. A polynomial equation of degree 1 is a linear equation and of degree 2 a quadratic equation. Some polynomial equations can be solved by writing the equation in standard form, factoring, and then using the zero-product principle: If a product is 0, then at least one of the factors is equal to 0.

Solve: $5x^2 + 7x = 6$.

$$5x^2 + 7x - 6 = 0$$
$$(5x - 3)(x + 2) = 0$$
$$5x - 3 = 0 \quad \text{or} \quad x + 2 = 0$$
$$5x = 3 \qquad\qquad x = -2$$
$$x = \frac{3}{5}$$

The solutions are -2 and $\frac{3}{5}$, and the solution set is $\left\{-2, \frac{3}{5}\right\}$. (The solutions are the x-intercepts of the graph of $y = 5x^2 + 7x - 6$.)

CHAPTER 5 REVIEW EXERCISES

5.1 *In Exercises 1–2, determine the coefficient of each term, the degree of each term, the degree of the polynomial, the leading term, and the leading coefficient of the polynomial.*

1. $-5x^3 + 7x^2 - x + 2$

2. $8x^4y^2 - 7xy^6 - x^3y$

3. If $f(x) = x^3 - 4x^2 + 3x - 1$, find $f(-2)$.

4. The bar graph shows the number of Americans living with AIDS from 1999 through 2003. The data can be modeled by the function

$$f(x) = 220x^3 - 986x^2 + 24{,}104x + 311{,}243,$$

where $f(x)$ is the number of Americans living with AIDS x years after 1999.

Number of People in the United States Living with AIDS

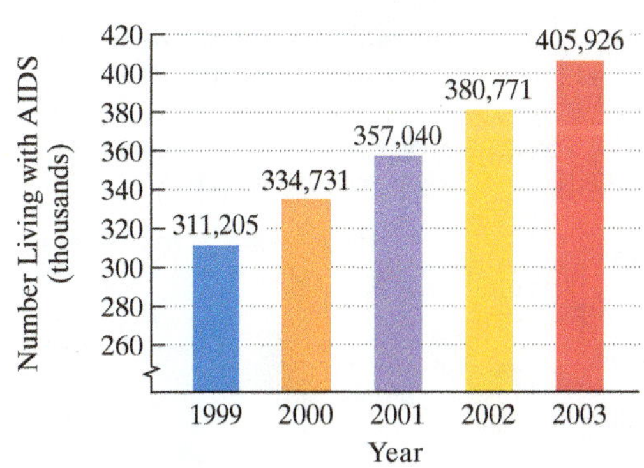

Source: Centers for Disease Control

a. Find and interpret $f(3)$.

b. Does your answer in part (a) underestimate or overestimate the actual number shown by the graph? By how much?

In Exercises 5–8, use the Leading Coefficient Test to determine the end behavior of the graph of the given polynomial function. Then use this end behavior to match the polynomial function with its graph. [The graphs are labeled (a) through (d).]

5. $f(x) = -x^3 + x^2 + 2x$
6. $f(x) = x^6 - 6x^4 + 9x^2$
7. $f(x) = x^5 - 5x^3 + 4x$
8. $f(x) = -x^4 + 1$

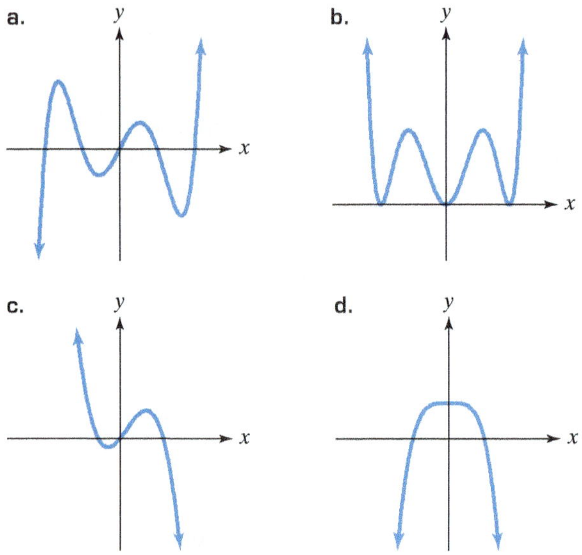

9. Tech firms might be rebounding from the dot-com bust, but enrollment in college computer programs keeps falling. The bar graph shows the number of newly-declared computer science and computer engineering majors for the fall term in U.S. and Canadian colleges from 1999 through 2003. The data can be modeled by the function

$$f(x) = -365x^4 + 2728x^3 - 7106x^2 + 7372x + 20{,}787,$$

where $f(x)$ is the number of newly-declared computer majors for the fall term x years after 1999.

Number of Newly-Declared Computer Majors in U.S. and Canadian Colleges

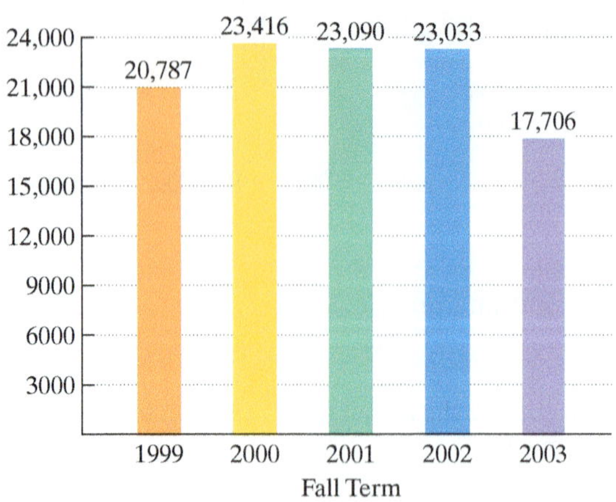

Source: Computing Research Association

a. Find $f(0)$ and $f(1)$. What do you observe when you compare these values with the appropriate numbers shown by the bar graph?

b. Use end behavior to explain why the function that models the data is valid only for a limited period of time.

In Exercises 10–11, add the polynomials.

10. $(-8x^3 + 5x^2 - 7x + 4) + (9x^3 - 11x^2 + 6x - 13)$
11. $(7x^3y - 13x^2y - 6y) + (5x^3y + 11x^2y - 8y - 17)$

In Exercises 12–13, subtract the polynomials.

12. $(7x^3 - 6x^2 + 5x - 11) - (-8x^3 + 4x^2 - 6x - 3)$
13. $(4x^3y^2 - 7x^3y - 4) - (6x^3y^2 - 3x^3y + 4)$
14. Subtract $-2x^3 - x^2y + xy^2 + 7y^3$ from $x^3 + 4x^2y - y^3$.

5.2 *In Exercises 15–27, multiply the polynomials.*

15. $(4x^2yz^5)(-3x^4yz^2)$
16. $6x^3\left(\frac{1}{3}x^5 - 4x^2 - 2\right)$
17. $7xy^2(3x^4y^2 - 5xy - 1)$
18. $(2x + 5)(3x^2 + 7x - 4)$
19. $(x^2 + x - 1)(x^2 + 3x + 2)$
20. $(4x - 1)(3x - 5)$
21. $(3xy - 2)(5xy + 4)$
22. $(3x + 7y)^2$
23. $(x^2 - 5y)^2$
24. $(2x + 7y)(2x - 7y)$
25. $(3xy^2 - 4x)(3xy^2 + 4x)$
26. $[(x + 3) + 5y][(x + 3) - 5y]$
27. $(x + y + 4)^2$
28. Let $f(x) = x - 3$ and $g(x) = 2x + 5$. Find $(fg)(x)$ and $(fg)(-4)$.
29. Let $f(x) = x^2 - 7x + 2$. Find each of the following and simplify:

a. $f(a - 1)$

b. $f(a + h) - f(a)$.

5.3 *In Exercises 30–33, factor the greatest common factor from each polynomial.*

30. $16x^3 + 24x^2$
31. $2x - 36x^2$
32. $21x^2y^2 - 14xy^2 + 7xy$
33. $18x^3y^2 - 27x^2y$

In Exercises 34–35, factor out a common factor with a negative coefficient.

34. $-12x^2 + 8x - 48$
35. $-x^2 - 11x + 14$

In Exercises 36–38, factor by grouping.

36. $x^3 - x^2 - 2x + 2$

37. $xy - 3x - 5y + 15$

38. $5ax - 15ay + 2bx - 6by$

5.4 *In Exercises 39–47, factor each trinomial completely, or state that the trinomial is prime.*

39. $x^2 + 8x + 15$

40. $x^2 + 16x - 80$

41. $x^2 + 16xy - 17y^2$

42. $3x^3 - 36x^2 + 33x$

43. $3x^2 + 22x + 7$

44. $6x^2 - 13x + 6$

45. $5x^2 - 6xy - 8y^2$

46. $6x^3 + 5x^2 - 4x$

47. $2x^2 + 11x + 15$

In Exercises 48–51, factor by introducing an appropriate substitution.

48. $x^6 + x^3 - 30$

49. $x^4 - 10x^2 - 39$

50. $(x + 5)^2 + 10(x + 5) + 24$

51. $5x^6 + 17x^3 + 6$

5.5 *In Exercises 52–55, factor each difference of two squares.*

52. $4x^2 - 25$

53. $1 - 81x^2y^2$

54. $x^8 - y^6$

55. $(x - 1)^2 - y^2$

In Exercises 56–60, factor any perfect square trinomials, or state that the polynomial is prime.

56. $x^2 + 16x + 64$

57. $9x^2 - 6x + 1$

58. $25x^2 + 20xy + 4y^2$

59. $49x^2 + 7x + 1$

60. $25x^2 - 40xy + 16y^2$

In Exercises 61–62, factor by grouping to obtain the difference of two squares.

61. $x^2 + 18x + 81 - y^2$

62. $z^2 - 25x^2 + 10x - 1$

In Exercises 63–65, factor using the formula for the sum or difference of two cubes.

63. $64x^3 + 27$

64. $125x^3 - 8$

65. $x^3y^3 + 1$

5.6 *In Exercises 66–90, factor completely, or state that the polynomial is prime.*

66. $15x^2 + 3x$

67. $12x^4 - 3x^2$

68. $20x^4 - 24x^3 + 28x^2 - 12x$

69. $x^3 - 15x^2 + 26x$

70. $-2y^4 + 24y^3 - 54y^2$

71. $9x^2 - 30x + 25$

72. $5x^2 - 45$

73. $2x^3 - x^2 - 18x + 9$

74. $6x^2 - 23xy + 7y^2$

75. $2y^3 + 12y^2 + 18y$

76. $x^2 + 6x + 9 - 4a^2$

77. $8x^3 - 27$

78. $x^5 - x$

79. $x^4 - 6x^2 + 9$

80. $x^2 + xy + y^2$

81. $4a^3 + 32$

82. $x^4 - 81$

83. $ax + 3bx - ay - 3by$

84. $27x^3 - 125y^3$

85. $10x^3y + 22x^2y - 24xy$

86. $6x^6 + 13x^3 - 5$

87. $2x + 10 + x^2y + 5xy$

88. $y^3 + 2y^2 - 25y - 50$

89. $a^8 - 1$

90. $9(x - 4) + y^2(4 - x)$

In Exercises 91–92,

a. *Write an expression for the area of the shaded region.*

b. *Write the expression in factored form.*

91.

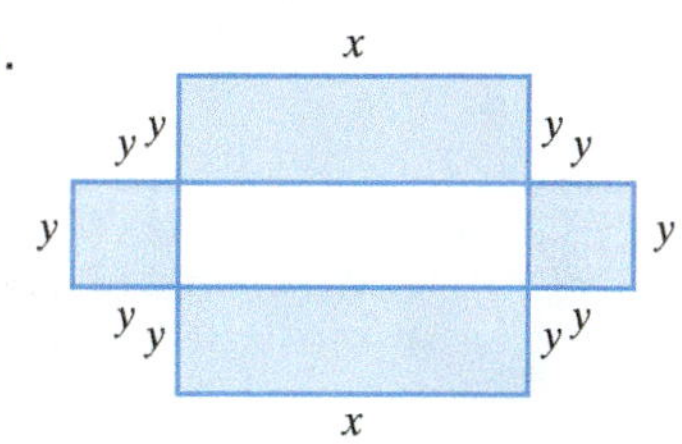

92.

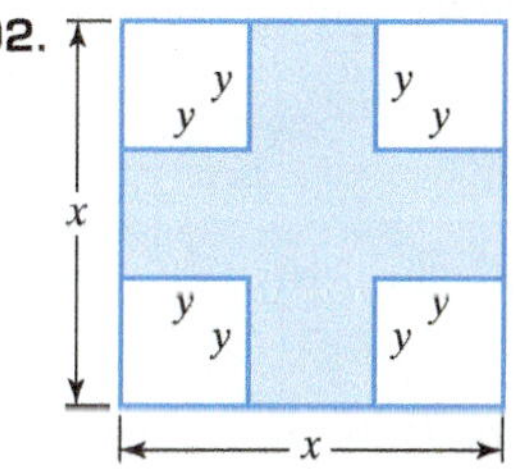

5.7 *In Exercises 93–97, use factoring to solve each polynomial equation.*

93. $x^2 + 6x + 5 = 0$

94. $3x^2 = 22x - 7$

95. $(x + 3)(x - 2) = 50$

96. $3x^2 = 12x$

97. $x^3 + 5x^2 = 9x + 45$

98. A model rocket is launched from the top of a cliff 144 feet above sea level. The function

$$s(t) = -16t^2 + 128t + 144$$

describes the rocket's height above the water, $s(t)$, in feet, t seconds after it is launched. The rocket misses the edge of the cliff on its way down and eventually lands in the ocean. How long will it take for the rocket to hit the water?

99. How much distance do you need to bring your car to a complete stop? A function used by those who study automobile safety is

$$d(x) = \frac{x^2}{20} + x,$$

where $d(x)$ is the stopping distance, in feet, for a car traveling at x miles per hour.

a. If it takes you 40 feet to come to a complete stop, how fast was your car traveling?

b. The graph of the quadratic function that models stopping distance is shown. Identify your solution from part (a) as a point on the graph.

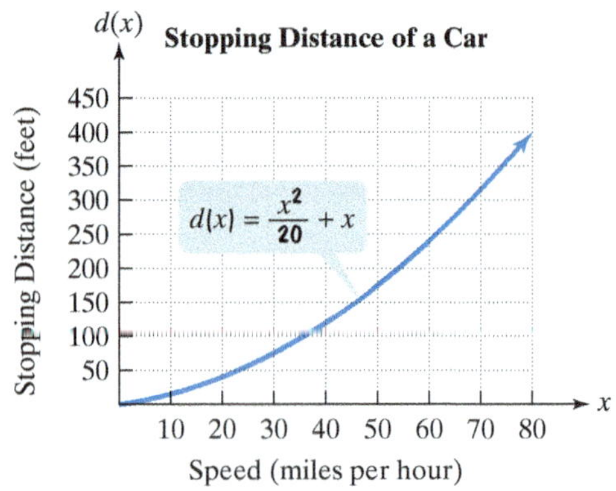

c. Describe the trend shown by the graph.

100. The length of a rectangular sign is 3 feet longer than the width. If the sign has space for 54 square feet of advertising, find its length and its width.

101. A painting measuring 10 inches by 16 inches is surrounded by a frame of uniform width. If the combined area of the painting and frame is 280 square inches, determine the width of the frame.

102. A lot is in the shape of a right triangle. The longer leg of the triangle is 20 yards longer than twice the length of the shorter leg. The hypotenuse is 30 yards longer than twice the length of the shorter leg. What are the lengths of the three sides?

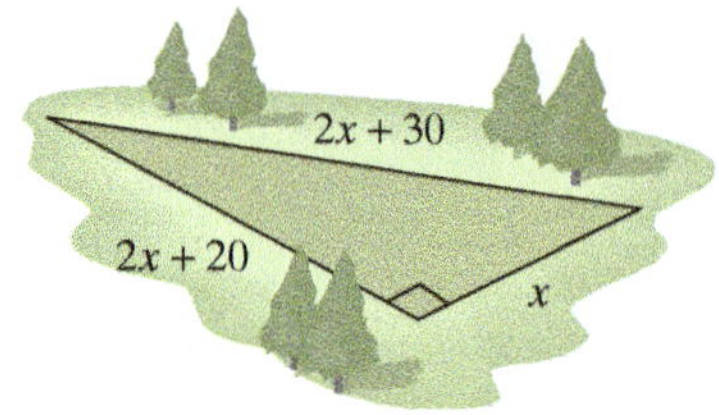

CHAPTER 5 TEST

Remember to use your Chapter Test Prep Video CD to see the worked-out solutions to the test questions you want to review.

In Exercises 1–2, give the degree and the leading coefficient of the polynomial.

1. $7x - 5 + x^2 - 6x^3$

2. $4xy^3 + 7x^4y^5 - 3xy^4$

3. If $f(x) = 3x^3 + 5x^2 - x + 6$, find $f(0)$ and $f(-2)$.

In Exercises 4–5, use the Leading Coefficient Test to describe the end behavior of the graph of the polynomial function.

4. $f(x) = -16x^2 + 160x$

5. $f(x) = 4x^3 + 12x^2 - x - 3$

In Exercises 6–13, perform the indicated operations.

6. $(4x^3y - 19x^2y - 7y) + (3x^3y + x^2y + 6y - 9)$

7. $(6x^2 - 7x - 9) - (-5x^2 + 6x - 3)$

8. $(-7x^3y)(-5x^4y^2)$

9. $(x - y)(x^2 - 3xy - y^2)$

10. $(7x - 9y)(3x + y)$

11. $(2x - 5y)(2x + 5y)$

12. $(4y - 7)^2$

13. $[(x + 2) + 3y][(x + 2) - 3y]$

14. Let $f(x) = x + 2$ and $g(x) = 3x - 5$. Find $(fg)(x)$ and $(fg)(-5)$.

15. Let $f(x) = x^2 - 5x + 3$. Find $f(a + h) - f(a)$ and simplify.

In Exercises 16–33, factor completely, or state that the polynomial is prime.

16. $14x^3 - 15x^2$

17. $81y^2 - 25$

18. $x^3 + 3x^2 - 25x - 75$

19. $25x^2 - 30x + 9$

20. $x^2 + 10x + 25 - 9y^2$

21. $x^4 + 1$

22. $y^2 - 16y - 36$

23. $14x^2 + 41x + 15$

24. $5x^3 - 5$

25. $12x^2 - 3y^2$

26. $12x^2 - 34x + 10$

27. $3x^4 - 3$

28. $x^8 - y^8$

29. $12x^2y^4 + 8x^3y^2 - 36x^2y$

30. $x^6 - 12x^3 - 28$

31. $x^4 - 2x^2 - 24$

32. $12x^2y - 27xy + 6y$

33. $y^4 - 3y^3 + 2y^2 - 6y$

In Exercises 34–37, solve each polynomial equation.

34. $3x^2 = 5x + 2$

35. $(5x + 4)(x - 1) = 2$

36. $15x^2 - 5x = 0$

37. $x^3 - 4x^2 - x + 4 = 0$

38. A baseball is thrown straight up from a rooftop 448 feet high. The function

$$s(t) = -16t^2 + 48t + 448$$

describes the ball's height above the ground, $s(t)$, in feet, t seconds after it is thrown. How long will it take for the ball to hit the ground?

39. An architect is allowed 15 square yards of floor space to add a small bedroom to a house. Because of the room's design in relationship to the existing structure, the width of the rectangular floor must be 7 yards less than two times the length. Find the length and width of the rectangular floor that the architect is permitted.

40. Find the lengths of the three sides of the right triangle in the figure shown.

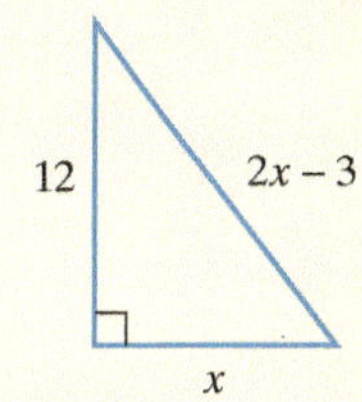

CUMULATIVE REVIEW EXERCISES (CHAPTERS 1–5)

In Exercises 1–7, solve each equation, inequality, or system of equations.

1. $8(x + 2) - 3(2 - x) = 4(2x + 6) - 2$

2. $2x + 4y = -6$
$x = 2y - 5$

3. $2x - y + 3z = 0$
$2y + z = 1$
$x + 2y - z = 5$

4. $2x + 4 < 10$ and $3x - 1 > 5$

5. $|2x - 5| \geq 9$

6. $2x^2 = 7x - 5$

7. $2x^3 + 6x^2 = 20x$

8. Solve for x: $x = \dfrac{ax + b}{c}$.

9. Use function notation to write the equation of the line passing through $(-2, -3)$ and $(2, 5)$.

10. In a campuswide election for student government president, 2800 votes were cast for the two candidates. If the winner had 160 more votes than the loser, how many votes were cast for each candidate?

In Exercises 11–13, graph each equation or inequality in a rectangular coordinate system.

11. $f(x) = -\frac{1}{3}x + 1$

12. $4x - 5y < 20$

13. $y \leq -1$

14. Simplify: $-\dfrac{8x^3y^6}{16x^9y^{-4}}$.

15. Write in scientific notation: 0.0000706.

In Exercises 16–17, perform the indicated operations.

16. $(3x^2 - y)^2$

17. $(3x^2 - y)(3x^2 + y)$

In Exercises 18–20, factor completely.

18. $x^3 - 3x^2 - 9x + 27$

19. $x^6 - x^2$

20. $14x^3y^2 - 28x^4y^2$

CHAPTER 7

Radicals, Radical Functions, and Rational Exponents

Can mathematical models be created for events that appear to involve random behavior, such as stock market fluctuations or air turbulence? Chaos theory, a new frontier of mathematics, offers models and computer-generated images that reveal order and underlying patterns where only the erratic and the unpredictable had been observed. Because most behavior is chaotic, the computer has become a canvas that looks more like the real world than anything previously seen. Magnified portions of these computer images yield repetitions of the original structure, as well as new and unexpected patterns. The computer generates these visualizations of chaos by plotting large numbers of points for functions whose domains are nonreal numbers involving the square root of negative one.

We define $\sqrt{-1}$ in Section 7.7 and hint at chaotic possibilities in the Blitzer Bonus on page 551. If you are intrigued by how the operations of nonreal numbers in Section 7.7 reveal that the world is not random (rather, the underlying patterns are far more intricate than we had previously assumed), we suggest reading *Chaos* by James Gleick, published by Penguin Books.

SECTION 7.1

Radical Expressions and Functions

Objectives

1. Evaluate square roots.
2. Evaluate square root functions.
3. Find the domain of square root functions.
4. Use models that are square root functions.
5. Simplify expressions of the form $\sqrt{a^2}$.
6. Evaluate cube root functions.
7. Simplify expressions of the form $\sqrt[3]{a^3}$.
8. Find even and odd roots.
9. Simplify expressions of the form $\sqrt[n]{a^n}$.

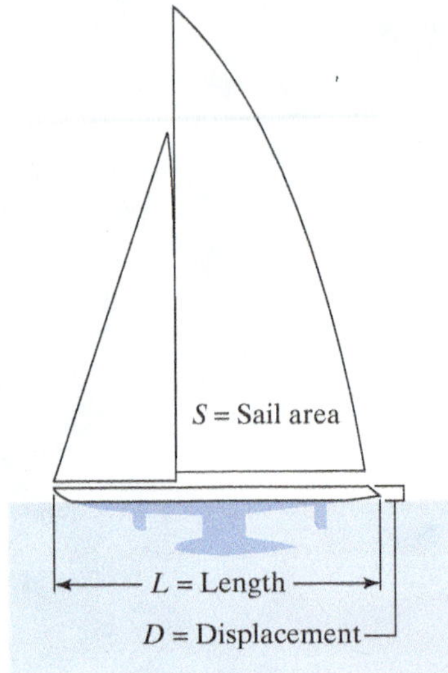

The America's Cup is the supreme event in ocean sailing. Competition is fierce and the costs are huge. Competitors look to mathematics to provide the critical innovation that can make the difference between winning and losing. The basic dimensions of competitors' yachts must satisfy an inequality containing square roots and cube roots:

$$L + 1.25\sqrt{S} - 9.8\sqrt[3]{D} \le 16.296.$$

In the inequality, L is the yacht's length, in meters, S is its sail area, in square meters, and D is its displacement, in cubic meters.

In this section, we introduce a new category of expressions and functions that contain roots. You will see why square root functions are used to describe phenomena that are continuing to grow but whose growth is leveling off.

1 Evaluate square roots.

Square Roots

From our earlier work with exponents, we are aware that the square of 5 and the square of −5 are both 25:

$$5^2 = 25 \quad \text{and} \quad (-5)^2 = 25.$$

The reverse operation of squaring a number is finding the *square root* of the number. For example,

- One square root of 25 is 5 because $5^2 = 25$.
- Another square root of 25 is −5 because $(-5)^2 = 25$.

In general, **if $b^2 = a$, then b is a square root of a.**

The symbol $\sqrt{\ }$ is used to denote the *positive* or *principal square root* of a number. For example,

- $\sqrt{25} = 5$ because $5^2 = 25$ and 5 is positive.
- $\sqrt{100} = 10$ because $10^2 = 100$ and 10 is positive.

Radical sign $\sqrt{a}$ Radicand

Radical expression

The symbol $\sqrt{\ }$ that we use to denote the principal square root is called a **radical sign**. The number under the radical sign is called the **radicand**. Together we refer to the radical sign and its radicand as a **radical expression**.

Definition of the Principal Square Root

If a is a nonnegative real number, the nonnegative number b such that $b^2 = a$, denoted by $b = \sqrt{a}$, is the **principal square root** of a.

The symbol $-\sqrt{}$ is used to denote the negative square root of a number. For example,

- $-\sqrt{25} = -5$ because $(-5)^2 = 25$ and -5 is negative.
- $-\sqrt{100} = -10$ because $(-10)^2 = 100$ and -10 is negative.

EXAMPLE 1 Evaluating Square Roots

Evaluate:

a. $\sqrt{81}$ b. $-\sqrt{9}$ c. $\sqrt{\frac{4}{49}}$

d. $\sqrt{0.0064}$ e. $\sqrt{36 + 64}$ f. $\sqrt{36} + \sqrt{64}$.

Solution

a. $\sqrt{81} = 9$ The principal square root of 81 is 9 because $9^2 = 81$.

b. $-\sqrt{9} = -3$ The negative square root of 9 is -3 because $(-3)^2 = 9$.

c. $\sqrt{\frac{4}{49}} = \frac{2}{7}$ The principal square root of $\frac{4}{49}$ is $\frac{2}{7}$ because $\left(\frac{2}{7}\right)^2 = \frac{4}{49}$.

d. $\sqrt{0.0064} = 0.08$ The principal square root of 0.0064 is 0.08 because $(0.08)^2 = (0.08)(0.08) = 0.0064$.

e. $\sqrt{36 + 64} = \sqrt{100}$ Simplify the radicand.

$= 10$ Take the principal square root of 100, which is 10.

f. $\sqrt{36} + \sqrt{64} = 6 + 8$ $\sqrt{36} = 6$ because $6^2 = 36$. $\sqrt{64} = 8$ because $8^2 = 64$.

$= 14$

Study Tip

In Example 1, parts (e) and (f), observe that $\sqrt{36 + 64}$ is not equal to $\sqrt{36} + \sqrt{64}$. In general,

$$\sqrt{a + b} \neq \sqrt{a} + \sqrt{b}$$

and

$$\sqrt{a - b} \neq \sqrt{a} - \sqrt{b}.$$

✓ CHECK POINT 1 Evaluate:

a. $\sqrt{64}$ b. $-\sqrt{49}$ c. $\sqrt{\frac{16}{25}}$

d. $\sqrt{0.0081}$ e. $\sqrt{9 + 16}$ f. $\sqrt{9} + \sqrt{16}$.

Let's see what happens to the radical expression $\sqrt{x}$ if x is a negative number. Is the square root of a negative number a real number? For example, consider $\sqrt{-25}$. Is there a real number whose square is -25? No. Thus, $\sqrt{-25}$ is not a real number. In general, **a square root of a negative number is not a real number**.

2 Evaluate square root functions.

Square Root Functions

Because each nonnegative real number, x, has precisely one principal square root, $\sqrt{x}$, there is a **square root function** defined by

$$f(x) = \sqrt{x}.$$

The domain of this function is $[0, \infty)$. We can graph $f(x) = \sqrt{x}$ by selecting nonnegative real numbers for x. It is easiest to choose perfect squares, numbers that have rational square roots. **Table 7.1** shows five such choices for x and the calculations for the corresponding outputs. We plot these ordered pairs as points in the rectangular coordinate system and connect the points with a smooth curve. The graph of $f(x) = \sqrt{x}$ is shown in **Figure 7.1**.

Table 7.1

x	$f(x) = \sqrt{x}$	(x, y) or $(x, f(x))$
0	$f(0) = \sqrt{0} = 0$	(0, 0)
1	$f(1) = \sqrt{1} = 1$	(1, 1)
4	$f(4) = \sqrt{4} = 2$	(4, 2)
9	$f(9) = \sqrt{9} = 3$	(9, 3)
16	$f(16) = \sqrt{16} = 4$	(16, 4)

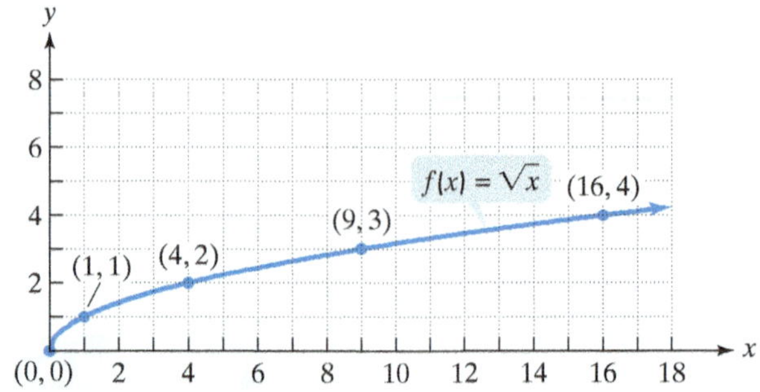

FIGURE 7.1 The graph of the square root function $f(x) = \sqrt{x}$

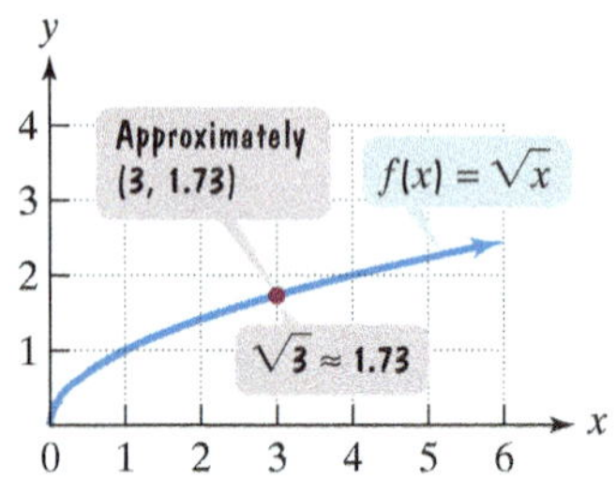

FIGURE 7.2 Visualizing $\sqrt{3}$ as a point on the graph of $f(x) = \sqrt{x}$

Is it possible to choose values of x for **Table 7.1** that are not squares of integers, or perfect squares? Yes. For example, we can let $x = 3$. Thus, $f(3) = \sqrt{3}$. Because 3 is not a perfect square, $\sqrt{3}$ is an irrational number, one that cannot be expressed as a quotient of integers. We can use a calculator to find a decimal approximation of $\sqrt{3}$.

Many Scientific Calculators	**Many Graphing Calculators**
3 [√]	[√] 3 [ENTER]

Rounding the displayed number to two decimal places, $\sqrt{3} \approx 1.73$. This information is shown visually as a point, approximately (3, 1.73), on the graph of $f(x) = \sqrt{x}$ in **Figure 7.2**.

To evaluate a square root function, we use substitution, just as we have done to evaluate other functions.

EXAMPLE 2 Evaluating Square Root Functions

For each function, find the indicated function value:

a. $f(x) = \sqrt{5x - 6}; f(2)$ **b.** $g(x) = -\sqrt{64 - 8x}; g(-3)$.

Solution

a. $f(2) = \sqrt{5 \cdot 2 - 6}$ — Substitute 2 for x in $f(x) = \sqrt{5x - 6}$.

$= \sqrt{4} = 2$ — Simplify the radicand and take the square root.

b. $g(-3) = -\sqrt{64 - 8(-3)}$ — Substitute −3 for x in $g(x) = -\sqrt{64 - 8x}$.

$= -\sqrt{88} \approx -9.38$ — Simplify the radicand: $64 - 8(-3) = 64 - (-24) = 64 + 24 = 88$. Then use a calculator to approximate $\sqrt{88}$.

✓ CHECK POINT 2 For each function, find the indicated function value:

a. $f(x) = \sqrt{12x - 20}; f(3)$

b. $g(x) = -\sqrt{9 - 3x}; g(-5)$.

3 Find the domain of square root functions.

We have seen that the domain of a function f is the largest set of real numbers for which the value of $f(x)$ is a real number. Because only nonnegative numbers have real square roots, the domain of a square root function is the set of real numbers for which the radicand is nonnegative.

EXAMPLE 3 Finding the Domain of a Square Root Function

Find the domain of

$$f(x) = \sqrt{3x + 12}.$$

Solution The domain is the set of real numbers, x, for which the radicand, $3x + 12$, is nonnegative. We set the radicand greater than or equal to 0 and solve the resulting inequality.

$$3x + 12 \geq 0$$
$$3x \geq -12$$
$$x \geq -4$$

The domain of f is $\{x|x \geq -4\}$ or $[-4, \infty)$.

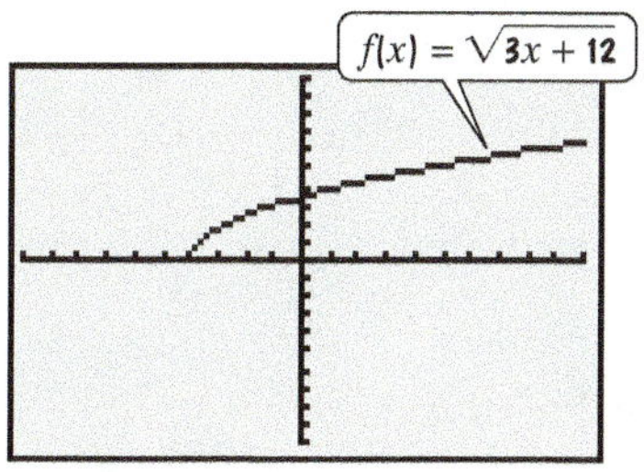

FIGURE 7.3

Figure 7.3 shows the graph of $f(x) = \sqrt{3x + 12}$ in a $[-10, 10, 1]$ by $[-10, 10, 1]$ viewing rectangle. The graph appears only for $x \geq -4$, verifying $[-4, \infty)$ as the domain. Can you see how the graph also illustrates this square root function's range? The graph only appears for nonnegative values of y. Thus, the range is $\{y|y \geq 0\}$ or $[0, \infty)$.

CHECK POINT 3 Find the domain of

$$f(x) = \sqrt{9x - 27}.$$

4 Use models that are square root functions.

The graph of the square root function $f(x) = \sqrt{x}$ is increasing from left to right. However, the rate of increase is slowing down as the graph moves to the right. This is why square root functions are often used to model growing phenomena with growth that is leveling off.

EXAMPLE 4 Modeling with a Square Root Function

By 2005, the amount of "clutter," including commercials and plugs for other shows, had increased to the point where an "hour-long" drama on cable TV was 45.4 minutes. The graph in **Figure 7.4** shows the average number of nonprogram minutes in an hour of prime-time cable television. Although the minutes of clutter grew from 1996 through 2005, the growth was leveling off. The data can be modeled by the function

$$M(x) = 0.7\sqrt{x} + 12.5,$$

where $M(x)$ is the average number of nonprogram minutes in an hour of prime-time cable x years after 1996. According to the model, in 2002, how many cluttered minutes disrupted cable TV action in an hour? Round to the nearest tenth of a minute. What is the difference between the actual data and the number of minutes that you obtained?

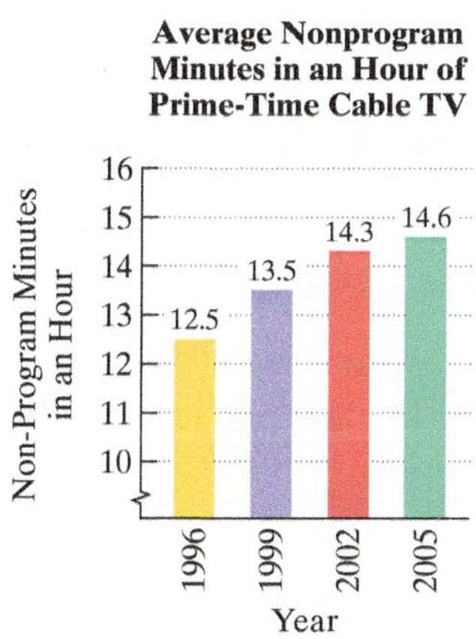

FIGURE 7.4

Source: Nielsen Monitor-Plus

Solution Because 2002 is 6 years after 1996, we substitute 6 for x and evaluate the function at 6.

$$M(x) = 0.7\sqrt{x} + 12.5 \quad \text{Use the given function.}$$
$$M(6) = 0.7\sqrt{6} + 12.5 \quad \text{Substitute 6 for x.}$$
$$\approx 14.2 \quad \text{Use a calculator.}$$

The model indicates that there were approximately 14.2 nonprogram minutes in an hour of prime-time cable in 2002. **Figure 7.4** shows 14.3 minutes, so the difference is $14.3 - 14.2$, or 0.1 minute.

CHECK POINT 4 If the trend from 1996 through 2005 continues, use the square root function in Example 4 to predict how many cluttered minutes, rounded to the nearest tenth, there will be in an hour in 2010.

5 Simplify expressions of the form $\sqrt{a^2}$.

Simplifying Expressions of the Form $\sqrt{a^2}$

You may think that $\sqrt{a^2} = a$. However, this is not necessarily true. Consider the following examples:

$$\sqrt{4^2} = \sqrt{16} = 4$$
$$\sqrt{(-4)^2} = \sqrt{16} = 4.$$

The result is not −4, but rather the absolute value of −4, or 4.

Here is a rule for simplifying expressions of the form $\sqrt{a^2}$:

Simplifying $\sqrt{a^2}$

For any real number a,

$$\sqrt{a^2} = |a|.$$

In words, the principal square root of a^2 is the absolute value of a.

Using Technology

Graphic Connections

The graphs of $f(x) = \sqrt{x^2}$ and $g(x) = |x|$ are shown in a $[-10, 10, 1]$ by $[-2, 10, 1]$ viewing rectangle. The graphs are the same. Thus,

$$\sqrt{x^2} = |x|.$$

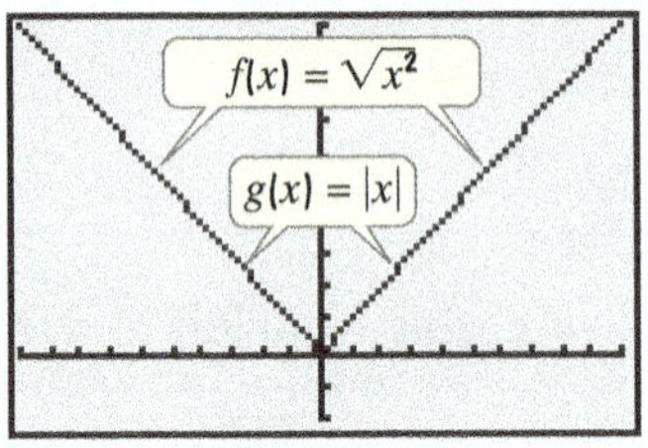

EXAMPLE 5 Simplifying Radical Expressions

Simplify each expression:

a. $\sqrt{(-6)^2}$ **b.** $\sqrt{(x+5)^2}$ **c.** $\sqrt{25x^6}$ **d.** $\sqrt{x^2 - 4x + 4}$.

Solution The principal square root of an expression squared is the absolute value of that expression. In parts (a) and (b), we are given squared radicands. In parts (c) and (d), it will first be necessary to express the radicand as an expression that is squared.

a. $\sqrt{(-6)^2} = |-6| = 6$

b. $\sqrt{(x+5)^2} = |x+5|$

c. To simplify $\sqrt{25x^6}$, first write $25x^6$ as an expression that is squared: $25x^6 = (5x^3)^2$. Then simplify.

$$\sqrt{25x^6} = \sqrt{(5x^3)^2} = |5x^3| \text{ or } 5|x^3|$$

d. To simplify $\sqrt{x^2 - 4x + 4}$, first write $x^2 - 4x + 4$ as an expression that is squared by factoring the perfect square trinomial: $x^2 - 4x + 4 = (x-2)^2$. Then simplify.

$$\sqrt{x^2 - 4x + 4} = \sqrt{(x-2)^2} = |x-2|$$

CHECK POINT 5 Simplify each expression:

a. $\sqrt{(-7)^2}$

b. $\sqrt{(x+8)^2}$

c. $\sqrt{49x^{10}}$

d. $\sqrt{x^2 - 6x + 9}$.

In some situations, we are told that no radicands involve negative quantities raised to even powers. When the expression being squared is nonnegative, it is not necessary to use absolute value when simplifying $\sqrt{a^2}$. For example, assuming that no radicands contain negative quantities that are squared,

$$\sqrt{x^6} = \sqrt{(x^3)^2} = x^3$$
$$\sqrt{25x^2 + 10x + 1} = \sqrt{(5x+1)^2} = 5x + 1.$$

6 Evaluate cube root functions.

Cube Roots and Cube Root Functions

Finding the square root of a number reverses the process of squaring a number. Similarly, finding the cube root of a number reverses the process of cubing a number. For example, $2^3 = 8$, and so the cube root of 8 is 2. The notation that we use is $\sqrt[3]{8} = 2$.

Definition of the Cube Root of a Number

The **cube root** of a real number a is written $\sqrt[3]{a}$.

$$\sqrt[3]{a} = b \quad \text{means that} \quad b^3 = a.$$

For example,

$$\sqrt[3]{64} = 4 \quad \text{because} \quad 4^3 = 64.$$
$$\sqrt[3]{-27} = -3 \quad \text{because} \quad (-3)^3 = -27.$$

In contrast to square roots, the cube root of a negative number is a real number. All real numbers have cube roots. The cube root of a positive number is positive. The cube root of a negative number is negative.

Because every real number, x, has precisely one cube root, $\sqrt[3]{x}$, there is a **cube root function** defined by

$$f(x) = \sqrt[3]{x}.$$

Study Tip

Some cube roots occur so frequently that you might want to memorize them.

$$\sqrt[3]{1} = 1$$
$$\sqrt[3]{8} = 2$$
$$\sqrt[3]{27} = 3$$
$$\sqrt[3]{64} = 4$$
$$\sqrt[3]{125} = 5$$
$$\sqrt[3]{216} = 6$$
$$\sqrt[3]{1000} = 10$$

The domain of this function is the set of all real numbers. We can graph $f(x) = \sqrt[3]{x}$ by selecting perfect cubes, numbers that have rational cube roots, for x. **Table 7.2** shows five such choices for x and the calculations for the corresponding outputs. We plot these ordered pairs as points in the rectangular coordinate system and connect the points with a smooth curve. The graph of $f(x) = \sqrt[3]{x}$ is shown in **Figure 7.5**.

Table 7.2

x	$f(x) = \sqrt[3]{x}$	(x, y) or $(x, f(x))$
-8	$f(-8) = \sqrt[3]{-8} = -2$	$(-8, -2)$
-1	$f(-1) = \sqrt[3]{-1} = -1$	$(-1, -1)$
0	$f(0) = \sqrt[3]{0} = 0$	$(0, 0)$
1	$f(1) = \sqrt[3]{1} = 1$	$(1, 1)$
8	$f(8) = \sqrt[3]{8} = 2$	$(8, 2)$

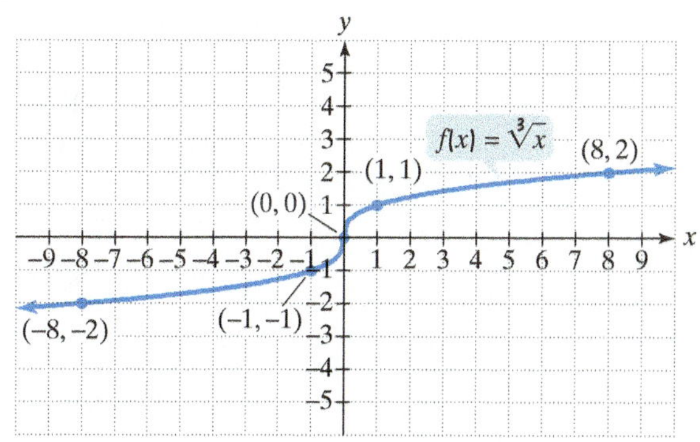

FIGURE 7.5 The graph of the cube root function $f(x) = \sqrt[3]{x}$

Notice that both the domain and the range of $f(x) = \sqrt[3]{x}$ are the set of all real numbers.

EXAMPLE 6 Evaluating Cube Root Functions

For each function, find the indicated function value:

a. $f(x) = \sqrt[3]{x - 2}$; $f(127)$ **b.** $g(x) = \sqrt[3]{8x - 8}$; $g(-7)$.

Solution

a. $f(x) = \sqrt[3]{x - 2}$ — This is the given function.

$f(127) = \sqrt[3]{127 - 2}$ — Substitute 127 for x.

$= \sqrt[3]{125}$ — Simplify the radicand.

$= 5$ — $\sqrt[3]{125} = 5$ because $5^3 = 125$.

b. $g(x) = \sqrt[3]{8x - 8}$ — This is the given function.

$g(-7) = \sqrt[3]{8(-7) - 8}$ — Substitute −7 for x.

$= \sqrt[3]{-64}$ — Simplify the radicand: $8(-7) - 8 = -56 - 8 = -64$.

$= -4$ — $\sqrt[3]{-64} = -4$ because $(-4)^3 = -64$.

CHECK POINT 6 For each function, find the indicated function value:

a. $f(x) = \sqrt[3]{x - 6}$; $f(33)$

b. $g(x) = \sqrt[3]{2x + 2}$; $g(-5)$.

7 Simplify expressions of the form $\sqrt[3]{a^3}$.

Because the cube root of a positive number is positive and the cube root of a negative number is negative, absolute value is not needed to simplify expressions of the form $\sqrt[3]{a^3}$.

Simplifying $\sqrt[3]{a^3}$

For any real number a,

$$\sqrt[3]{a^3} = a.$$

In words, the cube root of any expression cubed is that expression.

EXAMPLE 7 Simplifying a Cube Root

Simplify: $\sqrt[3]{-64x^3}$.

Solution Begin by expressing the radicand as an expression that is cubed: $-64x^3 = (-4x)^3$. Then simplify.

$$\sqrt[3]{-64x^3} = \sqrt[3]{(-4x)^3} = -4x$$

We can check our answer by cubing $-4x$:

$$(-4x)^3 = (-4)^3x^3 = -64x^3.$$

By obtaining the original radicand, we know that our simplification is correct.

CHECK POINT 7 Simplify: $\sqrt[3]{-27x^3}$.

8 Find even and odd roots.

Even and Odd *n*th Roots

Up to this point, we have focused on square roots and cube roots. Other radical expressions have different roots. For example, the fifth root of a, written $\sqrt[5]{a}$, is the number b for which $b^5 = a$. Thus,

$$\sqrt[5]{32} = 2 \qquad \text{because} \qquad 2^5 = 2 \cdot 2 \cdot 2 \cdot 2 \cdot 2 = 32.$$

The radical expression $\sqrt[n]{a}$ represents the ***n*th root** of a. The number n is called the **index**. An index of 2 represents a square root and is not written. An index of 3 represents a cube root.

If the index n in $\sqrt[n]{a}$ is an odd number, a root is said to be an **odd root**. A cube root is an odd root. Other odd roots have the same characteristics as cube roots.

- Every real number has exactly one real root when n is odd. An odd root of a positive number is positive and an odd root of a negative number is negative.

$$3^5 = 3\cdot3\cdot3\cdot3\cdot3 = 243, \text{ so the fifth root of 243 is 3.}$$
$$(-3)^5 = (-3)(-3)(-3)(-3)(-3) = -243, \text{ so the fifth root of } -243 \text{ is } -3.$$

- The (odd) nth root of a, $\sqrt[n]{a}$, is the number b for which $b^n = a$.

$$\sqrt[5]{243} = 3 \qquad \sqrt[5]{-243} = -3$$

$3^5 = 243$ $\quad$ $(-3)^5 = -243$

If the index n in $\sqrt[n]{a}$ is an even number, a root is said to be an **even root**. A square root is an even root. Other even roots have the same characteristics as square roots.

- Every positive real number has two real roots when n is even. One root is positive and one is negative.

$$2^4 = 2\cdot2\cdot2\cdot2 = 16 \quad \text{and} \quad (-2)^4 = (-2)(-2)(-2)(-2) = 16,$$

so both 2 and -2 are fourth roots of 16.

- The positive root, called the **principal nth root** and represented by $\sqrt[n]{a}$, is the nonnegative number b for which $b^n = a$. The symbol $-\sqrt[n]{a}$ is used to denote the negative nth root.

$$\sqrt[4]{16} = 2 \qquad -\sqrt[4]{16} = -2$$

$2^4 = 16$ $\quad$ $(-2)^4 = 16$

- **An even root of a negative number is not a real number.**

$$\sqrt[4]{-16} \text{ is not a real number.}$$

Study Tip

Some higher even and odd roots occur so frequently that you might want to memorize them.

Fourth Roots	Fifth Roots
$\sqrt[4]{1} = 1$	$\sqrt[5]{1} = 1$
$\sqrt[4]{16} = 2$	$\sqrt[5]{32} = 2$
$\sqrt[4]{81} = 3$	$\sqrt[5]{243} = 3$
$\sqrt[4]{256} = 4$	
$\sqrt[4]{625} = 5$	

EXAMPLE 8 Finding Even and Odd Roots

Find the indicated root, or state that the expression is not a real number:

a. $\sqrt[4]{81}$ b. $-\sqrt[4]{81}$ c. $\sqrt[4]{-81}$ d. $\sqrt[5]{-32}$.

Solution

a. $\sqrt[4]{81} = 3$ — The principal fourth root of 81 is 3 because $3^4 = 3\cdot3\cdot3\cdot3 = 81$.

b. $-\sqrt[4]{81} = -3$ — The negative fourth root of 81 is -3 because $(-3)^4 = (-3)(-3)(-3)(-3) = 81$.

c. $\sqrt[4]{-81}$ is not a real number because the index, 4, is even and the radicand, -81, is negative. No real number can be raised to the fourth power to give a negative result such as -81. Real numbers to even powers can only result in nonnegative numbers.

d. $\sqrt[5]{-32} = -2$ because $(-2)^5 = (-2)(-2)(-2)(-2)(-2) = -32$. An odd root of a negative real number is always negative.

CHECK POINT 8 Find the indicated root, or state that the expression is not a real number:

a. $\sqrt[4]{16}$ b. $-\sqrt[4]{16}$ c. $\sqrt[4]{-16}$ d. $\sqrt[5]{-1}$.

9 Simplify expressions of the form $\sqrt[n]{a^n}$.

Simplifying Expressions of the Form $\sqrt[n]{a^n}$

We have seen that

$$\sqrt{a^2} = |a| \quad \text{and} \quad \sqrt[3]{a^3} = a.$$

Expressions of the form $\sqrt[n]{a^n}$ can be simplified in the same manner. Unless a is known to be nonnegative, absolute value notation is needed when n is even. When the index is odd, absolute value bars are not used.

Simplifying $\sqrt[n]{a^n}$

For any real number a,

1. If n is even, $\sqrt[n]{a^n} = |a|$.
2. If n is odd, $\sqrt[n]{a^n} = a$.

EXAMPLE 9 Simplifying Radical Expressions

Simplify:

a. $\sqrt[4]{(x-3)^4}$ b. $\sqrt[5]{(2x+7)^5}$ c. $\sqrt[6]{(-5)^6}$.

Solution Each expression involves the nth root of a radicand raised to the nth power. Thus, each radical expression can be simplified. Absolute value bars are necessary in parts (a) and (c) because the index, n, is even.

a. $\sqrt[4]{(x-3)^4} = |x-3|$ $\quad$ $\sqrt[n]{a^n} = |a|$ if n is even.

b. $\sqrt[5]{(2x+7)^5} = 2x+7$ $\quad$ $\sqrt[n]{a^n} = a$ if n is odd.

c. $\sqrt[6]{(-5)^6} = |-5| = 5$ $\quad$ $\sqrt[n]{a^n} = |a|$ if n is even.

CHECK POINT 9 Simplify:

a. $\sqrt[4]{(x+6)^4}$ b. $\sqrt[5]{(3x-2)^5}$ c. $\sqrt[6]{(-8)^6}$.

7.1 EXERCISE SET

MyMathLab PRACTICE WATCH DOWNLOAD READ REVIEW

Practice Exercises

In Exercises 1–20, evaluate each expression, or state that the expression is not a real number.

1. $\sqrt{36}$
2. $\sqrt{16}$
3. $-\sqrt{36}$
4. $-\sqrt{16}$
5. $\sqrt{-36}$
6. $\sqrt{-16}$
7. $\sqrt{\frac{1}{25}}$
8. $\sqrt{\frac{1}{49}}$
9. $-\sqrt{\frac{9}{16}}$
10. $-\sqrt{\frac{4}{25}}$
11. $\sqrt{0.81}$
12. $\sqrt{0.49}$
13. $-\sqrt{0.04}$
14. $-\sqrt{0.64}$
15. $\sqrt{25-16}$
16. $\sqrt{144+25}$
17. $\sqrt{25}-\sqrt{16}$
18. $\sqrt{144}+\sqrt{25}$
19. $\sqrt{16-25}$
20. $\sqrt{25-144}$

In Exercises 21–26, find the indicated function values for each function. If necessary, round to two decimal places. If the function value is not a real number and does not exist, so state.

21. $f(x) = \sqrt{x-2}$; $f(18), f(3), f(2), f(-2)$
22. $f(x) = \sqrt{x-3}$; $f(28), f(4), f(3), f(-1)$
23. $g(x) = -\sqrt{2x+3}$; $g(11), g(1), g(-1), g(-2)$
24. $g(x) = -\sqrt{2x+1}$; $g(4), g(1), g\left(-\frac{1}{2}\right), g(-1)$
25. $h(x) = \sqrt{(x-1)^2}$; $h(5), h(3), h(0), h(-5)$
26. $h(x) = \sqrt{(x-2)^2}$; $h(5), h(3), h(0), h(-5)$

In Exercises 27–32, find the domain of each square root function. Then use the domain to match the radical function with its graph. [The graphs are labeled (a) through (f) and are shown in [−10, 10, 1] by [−10, 10, 1] viewing rectangles on the next page.]

27. $f(x) = \sqrt{x-3}$
28. $f(x) = \sqrt{x+2}$
29. $f(x) = \sqrt{3x+15}$
30. $f(x) = \sqrt{3x-15}$

31. $f(x) = \sqrt{6 - 2x}$

32. $f(x) = \sqrt{8 - 2x}$

a.
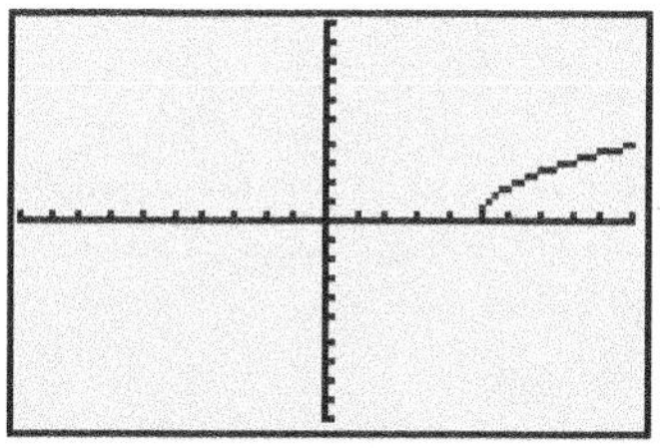

b.
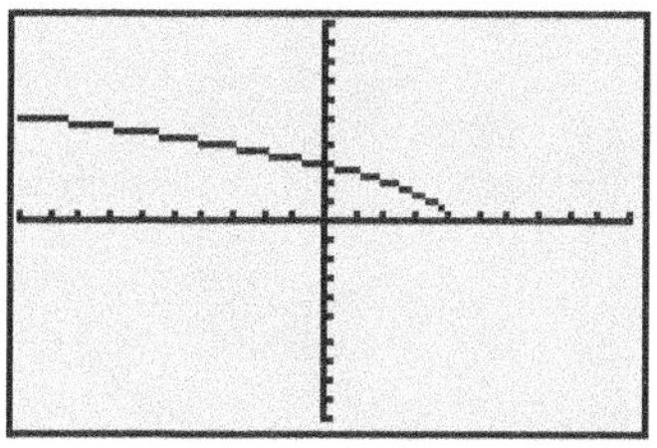

c.
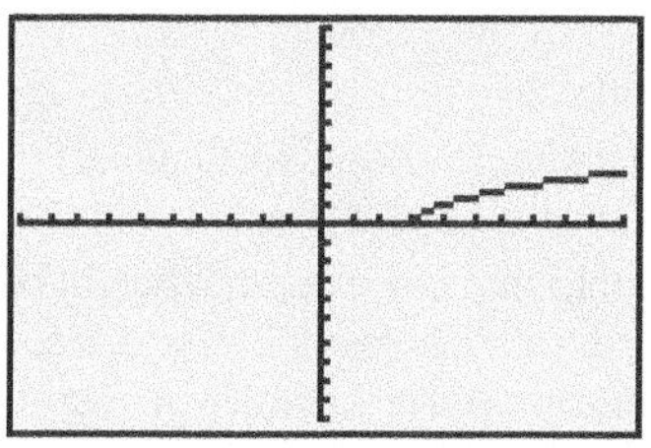

d.
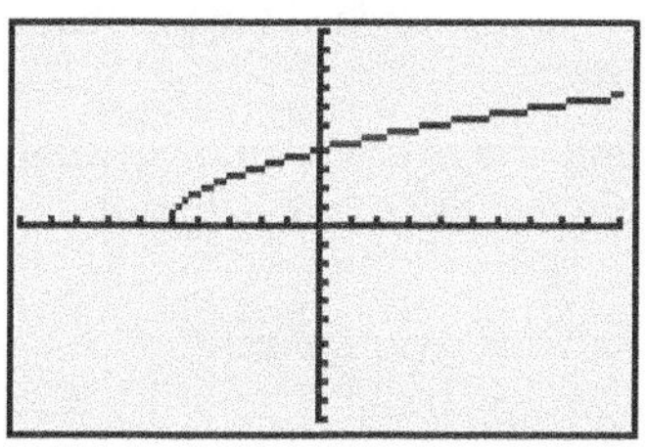

e.
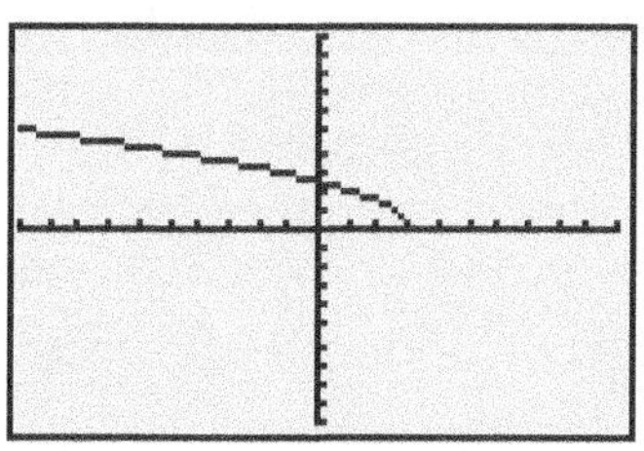

f.
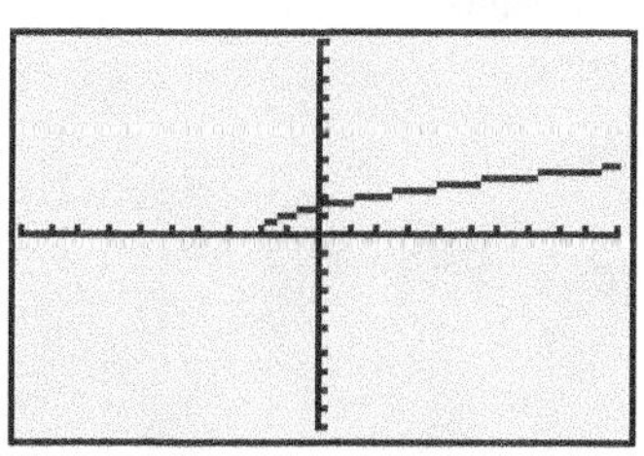

In Exercises 33–46, simplify each expression.

33. $\sqrt{5^2}$ 34. $\sqrt{7^2}$

35. $\sqrt{(-4)^2}$ 36. $\sqrt{(-10)^2}$

37. $\sqrt{(x-1)^2}$ 38. $\sqrt{(x-2)^2}$

39. $\sqrt{36x^4}$ 40. $\sqrt{81x^4}$

41. $-\sqrt{100x^6}$

42. $-\sqrt{49x^6}$

43. $\sqrt{x^2 + 12x + 36}$

44. $\sqrt{x^2 + 14x + 49}$

45. $-\sqrt{x^2 - 8x + 16}$

46. $-\sqrt{x^2 - 10x + 25}$

In Exercises 47–54, find each cube root.

47. $\sqrt[3]{27}$ 48. $\sqrt[3]{64}$

49. $\sqrt[3]{-27}$ 50. $\sqrt[3]{-64}$

51. $\sqrt[3]{\frac{1}{125}}$ 52. $\sqrt[3]{\frac{1}{1000}}$

53. $\sqrt[3]{\frac{-27}{1000}}$ 54. $\sqrt[3]{\frac{-8}{125}}$

In Exercises 55–58, find the indicated function values for each function.

55. $f(x) = \sqrt[3]{x - 1}$; $f(28), f(9), f(0), f(-63)$

56. $f(x) = \sqrt[3]{x - 3}$; $f(30), f(11), f(2), f(-122)$

57. $g(x) = -\sqrt[3]{8x - 8}$; $g(2), g(1), g(0)$

58. $g(x) = -\sqrt[3]{2x + 1}$; $g(13), g(0), g(-63)$

In Exercises 59–76, find the indicated root, or state that the expression is not a real number.

59. $\sqrt[4]{1}$ 60. $\sqrt[5]{1}$

61. $\sqrt[4]{16}$ 62. $\sqrt[4]{81}$

63. $-\sqrt[4]{16}$ 64. $-\sqrt[4]{81}$

65. $\sqrt[4]{-16}$ 66. $\sqrt[4]{-81}$

67. $\sqrt[5]{-1}$ 68. $\sqrt[7]{-1}$

69. $\sqrt[6]{-1}$ 70. $\sqrt[8]{-1}$

71. $-\sqrt[4]{256}$ 72. $-\sqrt[4]{10{,}000}$

73. $\sqrt[6]{64}$ 74. $\sqrt[5]{32}$

75. $-\sqrt[5]{32}$ 76. $-\sqrt[6]{64}$

In Exercises 77–90, simplify each expression. Include absolute value bars where necessary.

77. $\sqrt[3]{x^3}$ 78. $\sqrt[5]{x^5}$

79. $\sqrt[4]{y^4}$ 80. $\sqrt[6]{y^6}$

81. $\sqrt[3]{-8x^3}$ 82. $\sqrt[3]{-125x^3}$

83. $\sqrt[3]{(-5)^3}$ 84. $\sqrt[3]{(-6)^3}$

85. $\sqrt[4]{(-5)^4}$ 86. $\sqrt[6]{(-6)^6}$

87. $\sqrt[4]{(x+3)^4}$ 88. $\sqrt[4]{(x+5)^4}$

89. $\sqrt[5]{-32(x-1)^5}$

90. $\sqrt[5]{-32(x-2)^5}$

Practice PLUS

In Exercises 91–94, complete each table and graph the given function. Identify the function's domain and range.

91. $f(x) = \sqrt{x} + 3$

x	$f(x) = \sqrt{x} + 3$
0	
1	
4	
9	

92. $f(x) = \sqrt{x} - 2$

x	$f(x) = \sqrt{x} - 2$
0	
1	
4	
9	

93. $f(x) = \sqrt{x - 3}$

x	$f(x) = \sqrt{x - 3}$
3	
4	
7	
12	

94. $f(x) = \sqrt{4 - x}$

x	$f(x) = \sqrt{4 - x}$
−5	
0	
3	
4	

In Exercises 95–98, find the domain of each function.

95. $f(x) = \dfrac{\sqrt[3]{x}}{\sqrt{30 - 2x}}$

96. $f(x) = \dfrac{\sqrt[3]{x}}{\sqrt{80 - 5x}}$

97. $f(x) = \dfrac{\sqrt{x - 1}}{\sqrt{3 - x}}$

98. $f(x) = \dfrac{\sqrt{x - 2}}{\sqrt{7 - x}}$

In Exercises 99–100, evaluate each expression.

99. $\sqrt[3]{\sqrt[4]{16} + \sqrt{625}}$

100. $\sqrt[3]{\sqrt{\sqrt{169} + \sqrt{9}} + \sqrt{\sqrt[3]{1000} + \sqrt[3]{216}}}$

Application Exercises

101. The function $f(x) = 2.9\sqrt{x} + 20.1$ models the median height, $f(x)$, in inches, of boys who are x months of age. The graph of f is shown.

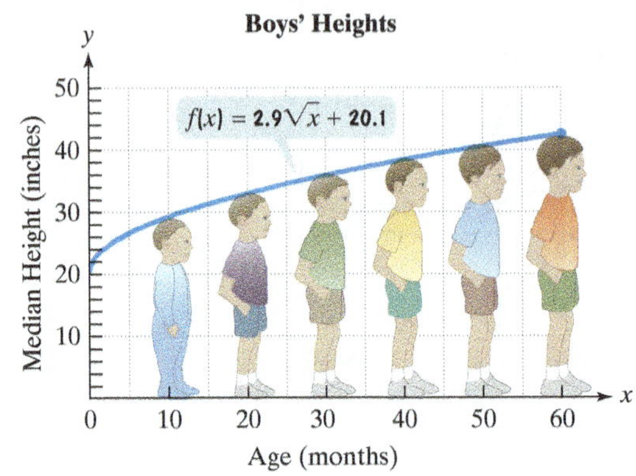

Source: Laura Walther Nathanson, *The Portable Pediatrician for Parents*

a. According to the model, what is the median height of boys who are 48 months, or four years, old? Use a calculator and round to the nearest tenth of an inch. The actual median height for boys at 48 months is 40.8 inches. Does the model overestimate or underestimate the actual height? By how much?

b. Use the model to find the average rate of change, in inches per month, between birth and 10 months. Round to the nearest tenth.

c. Use the model to find the average rate of change, in inches per month, between 50 and 60 months. Round to the nearest tenth. How does this compare with your answer in part (b)? How is this difference shown by the graph?

102. The function $f(x) = 3.1\sqrt{x} + 19$ models the median height, $f(x)$, in inches, of girls who are x months of age. The graph of f is shown.

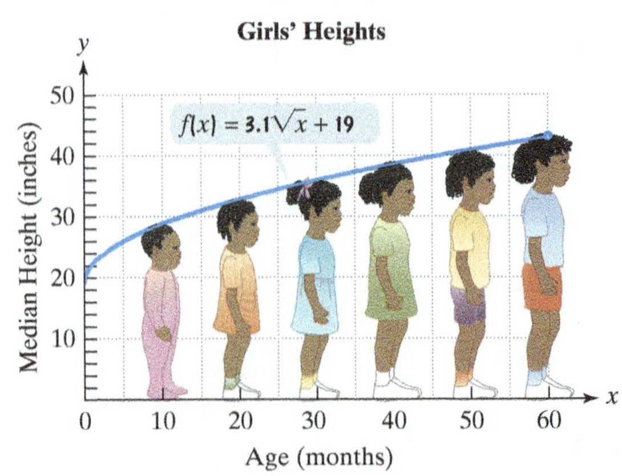

Source: Laura Walther Nathanson, *The Portable Pediatrician for Parents*

a. According to the model, what is the median height of girls who are 48 months, or four years, old? Use a calculator and round to the nearest tenth of an inch. The actual median height for girls at 48 months is 40.2 inches. Does the model overestimate or underestimate the actual height? By how much?

b. Use the model to find the average rate of change, in inches per month, between birth and 10 months. Round to the nearest tenth.

c. Use the model to find the average rate of change, in inches per month, between 50 and 60 months. Round to the nearest tenth. How does this compare with your answer in part (b)? How is this difference shown by the graph?

Police use the function $f(x) = \sqrt{20x}$ to estimate the speed of a car, $f(x)$, in miles per hour, based on the length, x, in feet, of its skid marks upon sudden braking on a dry asphalt road. Use the function to solve Exercises 103–104.

103. A motorist is involved in an accident. A police officer measures the car's skid marks to be 245 feet long. Estimate the speed at which the motorist was traveling before braking. If the posted speed limit is 50 miles per hour and the motorist tells the officer he was not speeding, should the officer believe him? Explain.

104. A motorist is involved in an accident. A police officer measures the car's skid marks to be 45 feet long. Estimate the speed at which the motorist was traveling before braking. If the posted speed limit is 35 miles per hour and the motorist tells the officer she was not speeding, should the officer believe her? Explain.

Writing in Mathematics

105. What are the square roots of 36? Explain why each of these numbers is a square root.

106. What does the symbol $\sqrt{}$ denote? Which of your answers in Exercise 105 is given by this symbol? Write the symbol needed to obtain the other answer.

107. Explain why $\sqrt{-1}$ is not a real number.

108. Explain how to find the domain of a square root function.

109. Explain how to simplify $\sqrt{a^2}$. Give an example with your explanation.

110. Explain why $\sqrt[3]{8}$ is 2. Then describe what is meant by the cube root of a real number.

111. Describe two differences between odd and even roots.

112. Explain how to simplify $\sqrt[n]{a^n}$ if n is even and if n is odd. Give examples with your explanations.

113. Explain the meaning of the words *radical*, *radicand*, and *index*. Give an example with your explanation.

114. Describe the trend in a boy's growth from birth through five years, shown in the graph for Exercise 101. Why is a square root function a useful model for the data?

Technology Exercises

115. Use a graphing utility to graph $y_1 = \sqrt{x}$, $y_2 = \sqrt{x + 4}$, and $y_3 = \sqrt{x - 3}$ in the same $[-5, 10, 1]$ by $[0, 6, 1]$ viewing rectangle. Describe one similarity and one difference that you observe among the graphs. Use the word *shift* in your response.

116. Use a graphing utility to graph $y = \sqrt{x}$, $y = \sqrt{x} + 4$, and $y = \sqrt{x} - 3$ in the same $[-1, 10, 1]$ by $[-10, 10, 1]$ viewing rectangle. Describe one similarity and one difference that you observe among the graphs.

117. Use a graphing utility to graph $f(x) = \sqrt{x}$, $g(x) = -\sqrt{x}$, $h(x) = \sqrt{-x}$, and $k(x) = -\sqrt{-x}$ in the same $[-10, 10, 1]$ by $[-4, 4, 1]$ viewing rectangle. Use the graphs to describe the domains and the ranges of functions f, g, h, and k.

118. Use a graphing utility to graph $y_1 = \sqrt{x^2}$ and $y_2 = -x$ in the same viewing rectangle.

a. For what values of x is $\sqrt{x^2} = -x$?

b. For what values of x is $\sqrt{x^2} \neq -x$?

Critical Thinking Exercises

Make Sense? *In Exercises 119–122, determine whether each statement "makes sense" or "does not make sense" and explain your reasoning.*

119. $\sqrt[4]{(-8)^4}$ cannot be positive 8 because the power and the index cancel each other.

120. If I am given any real number, that number has exactly one odd root and two even roots.

121. I need to restrict the domains of radical functions with even indices, but these restrictions are not necessary when indices are odd.

122. Using my calculator, I determined that $5^5 = 3125$, so 5 must be the fifth root of 3125.

In Exercises 123–126, determine whether each statement is true or false. If the statement is false, make the necessary change(s) to produce a true statement.

123. The domain of $f(x) = \sqrt[3]{x - 4}$ is $[4, \infty)$.

124. If n is odd and b is negative, then $\sqrt[n]{b}$ is not a real number.

125. If $x = -2$, then $\sqrt{x^6} = x^3$.

126. The expression $\sqrt[n]{4}$ represents increasingly larger numbers for $n = 2, 3, 4, 5, 6$, and so on.

127. Write a function whose domain is $(-\infty, 5]$.

128. Let $f(x) = \sqrt{x - 3}$ and $g(x) = \sqrt{x + 1}$. Find the domain of $f + g$ and $\frac{f}{g}$.

129. Simplify: $\sqrt{(2x + 3)^{10}}$.

In Exercises 130–131, graph each function by hand. Then describe the relationship between the function that you graphed and the graph of $f(x) = \sqrt{x}$.

130. $g(x) = \sqrt{x} + 2$

131. $h(x) = \sqrt{x + 3}$

Review Exercises

132. Simplify: $3x - 2[x - 3(x + 5)]$. (Section 1.2, Example 14)

133. Simplify: $(-3x^{-4}y^3)^{-2}$. (Section 1.6, Example 7c)

134. Solve: $|3x - 4| > 11$. (Section 4.3, Example 5 or Example 6b)

Preview Exercises

Exercises 135–137 will help you prepare for the material covered in the next section. In each exercise, use properties of exponents to simplify the expression. Be sure that no negative exponents appear in your simplified expression. (If you have forgotten how to simplify an exponential expression, see the box on page 71.)

135. $(2^3x^5)(2^4x^{-6})$

136. $\frac{32x^2}{16x^5}$

137. $(x^{-2}y^3)^4$

SECTION 7.3

Multiplying and Simplifying Radical Expressions

Objectives

1. Use the product rule to multiply radicals.
2. Use factoring and the product rule to simplify radicals.
3. Multiply radicals and then simplify.

George Tooker (b. 1920) "Mirror II" 1963, egg tempera on gesso panel, 20 × 20 in., 1968.4.

We opened this book with a model that described our improving emotional health as we age. Unfortunately, not everything gets better. The aging process is also accompanied by a number of physical transformations, including changes in vision that require glasses for reading, the onset of wrinkles and sagging skin, and a decrease in heart response. A change in heart response occurs fairly early; after 20, our hearts become less adept at accelerating in response to exercise. In this section's exercise set, you will see how a radical function models changes in heart function throughout the aging process, as we turn to multiplying and simplifying radical expressions.

1 Use the product rule to multiply radicals.

The Product Rule for Radicals

A rule for multiplying radicals can be generalized by comparing $\sqrt{25}\cdot\sqrt{4}$ and $\sqrt{25\cdot4}$. Notice that

$$\sqrt{25}\cdot\sqrt{4} = 5\cdot2 = 10 \quad \text{and} \quad \sqrt{25\cdot4} = \sqrt{100} = 10.$$

Because we obtain 10 in both situations, the original radical expressions must be equal. That is,

$$\sqrt{25}\cdot\sqrt{4} = \sqrt{25\cdot4}.$$

This result is a special case of the **product rule for radicals** that can be generalized as follows:

The Product Rule for Radicals

If $\sqrt[n]{a}$ and $\sqrt[n]{b}$ are real numbers, then

$$\sqrt[n]{a}\cdot\sqrt[n]{b} = \sqrt[n]{ab}.$$

The product of two nth roots is the nth root of the product of the radicands.

Study Tip

The product rule can be used only when the radicals have the same index. If indices differ, rational exponents can be used, as in $\sqrt{x}\cdot\sqrt[3]{x}$, which was Example 7(d) in the previous section.

EXAMPLE 1 Using the Product Rule for Radicals

Multiply:

a. $\sqrt{3}\cdot\sqrt{7}$ b. $\sqrt{x+7}\cdot\sqrt{x-7}$ c. $\sqrt[3]{7}\cdot\sqrt[3]{9}$ d. $\sqrt[8]{10x}\cdot\sqrt[8]{8x^4}$.

Solution In each problem, the indices are the same. Thus, we multiply by multiplying the radicands.

a. $\sqrt{3}\cdot\sqrt{7} = \sqrt{3\cdot7} = \sqrt{21}$

b. $\sqrt{x+7}\cdot\sqrt{x-7} = \sqrt{(x+7)(x-7)} = \sqrt{x^2-49}$

This is not equal to $\sqrt{x^2}-\sqrt{49}$.

c. $\sqrt[3]{7}\cdot\sqrt[3]{9} = \sqrt[3]{7\cdot9} = \sqrt[3]{63}$

d. $\sqrt[8]{10x}\cdot\sqrt[8]{8x^4} = \sqrt[8]{10x\cdot8x^4} = \sqrt[8]{80x^5}$

CHECK POINT 1 Multiply:

a. $\sqrt{5}\cdot\sqrt{11}$ b. $\sqrt{x+4}\cdot\sqrt{x-4}$

c. $\sqrt[3]{6}\cdot\sqrt[3]{10}$ d. $\sqrt[7]{2x}\cdot\sqrt[7]{6x^3}$.

2 Use factoring and the product rule to simplify radicals.

Using Factoring and the Product Rule to Simplify Radicals

In Chapter 5, we saw that a number that is the square of an integer is a **perfect square**. For example, 100 is a perfect square because $100 = 10^2$. A number is a **perfect cube** if it is the cube of an integer. Thus, 125 is a perfect cube because $125 = 5^3$. In general, a number is a **perfect *n*th power** if it is the nth power of an integer. Thus, p is a perfect nth power if there is an integer q such that $p = q^n$.

A radical of index n is **simplified** when its radicand has no factors other than 1 that are perfect nth powers. For example, $\sqrt{300}$ is not simplified because it can be expressed as $\sqrt{100\cdot3}$ and 100 is a perfect square. We can use the product rule in the form

$$\sqrt[n]{ab} = \sqrt[n]{a}\cdot\sqrt[n]{b}$$

to simplify $\sqrt[n]{ab}$ when a or b is a perfect nth power. Consider $\sqrt{300}$. To simplify, we factor 300 so that one of its factors is the greatest perfect square possible.

Using Technology

You can use a calculator to provide numerical support that $\sqrt{300} = 10\sqrt{3}$. First find an approximation for $\sqrt{300}$:

300 [√] ≈ 17.32

or

[√] 300 [ENTER] ≈ 17.32.

Now find an aproximation for $10\sqrt{3}$:

10 [×] 3 [√] ≈ 17.32

or

10 [√] 3 [ENTER] ≈ 17.32.

Correct to two decimal places,

$\sqrt{300} \approx 17.32$ and $10\sqrt{3} \approx 17.32$.

This verifies that

$\sqrt{300} = 10\sqrt{3}$.

Use this technique to support the numerical results for the answers in this section.

Caution: A simplified radical does not mean a decimal approximation.

$\sqrt{300} = \sqrt{100 \cdot 3}$ Factor 300. 100 is the greatest perfect square factor.

$= \sqrt{100} \cdot \sqrt{3}$ Use the product rule: $\sqrt[n]{ab} = \sqrt[n]{a} \cdot \sqrt[n]{b}$.

$= 10\sqrt{3}$ Write $\sqrt{100}$ as 10. We read $10\sqrt{3}$ as "ten times the square root of three."

Simplifying Radical Expressions by Factoring

A radical expression whose index is n is **simplified** when its radicand has no factors that are perfect nth powers. To simplify, use the following procedure:

1. Write the radicand as the product of two factors, one of which is the greatest perfect nth power.
2. Use the product rule to take the nth root of each factor.
3. Find the nth root of the perfect nth power.

EXAMPLE 2 Simplifying Radicals by Factoring

Simplify by factoring:

a. $\sqrt{75}$ b. $\sqrt[3]{54}$ c. $\sqrt[5]{64}$ d. $\sqrt{500xy^2}$.

Solution

a. $\sqrt{75} = \sqrt{25 \cdot 3}$ 25 is the greatest perfect square that is a factor of 75.

$= \sqrt{25} \cdot \sqrt{3}$ Take the square root of each factor: $\sqrt[n]{ab} = \sqrt[n]{a} \cdot \sqrt[n]{b}$.

$= 5\sqrt{3}$ Write $\sqrt{25}$ as 5.

b. $\sqrt[3]{54} = \sqrt[3]{27 \cdot 2}$ 27 is the greatest perfect cube that is a factor of 54: $27 = 3^3$.

$= \sqrt[3]{27} \cdot \sqrt[3]{2}$ Take the cube root of each factor: $\sqrt[n]{ab} = \sqrt[n]{a} \cdot \sqrt[n]{b}$.

$= 3\sqrt[3]{2}$ Write $\sqrt[3]{27}$ as 3.

c. $\sqrt[5]{64} = \sqrt[5]{32 \cdot 2}$ 32 is the greatest perfect fifth power that is a factor of 64: $32 = 2^5$.

$= \sqrt[5]{32} \cdot \sqrt[5]{2}$ Take the fifth root of each factor: $\sqrt[n]{ab} = \sqrt[n]{a} \cdot \sqrt[n]{b}$.

$= 2\sqrt[5]{2}$ Write $\sqrt[5]{32}$ as 2.

d. $\sqrt{500xy^2} = \sqrt{100y^2 \cdot 5x}$ $100y^2$ is the greatest perfect square that is a factor of $500xy^2$: $100y^2 = (10y)^2$.

$= \sqrt{100y^2} \cdot \sqrt{5x}$ Factor into two radicals.

$= 10|y|\sqrt{5x}$ Take the square root of $100y^2$.

✓ CHECK POINT 2 Simplify by factoring:

a. $\sqrt{80}$ b. $\sqrt[3]{40}$

c. $\sqrt[4]{32}$ d. $\sqrt{200x^2y}$.

EXAMPLE 3 Simplifying a Radical Function

If

$$f(x) = \sqrt{2x^2 + 4x + 2},$$

express the function, f, in simplified form.

Solution Begin by factoring the radicand. The GCF is 2. Simplification is possible if we obtain a factor that is a perfect square.

$$\begin{aligned} f(x) &= \sqrt{2x^2 + 4x + 2} && \text{This is the given function.} \\ &= \sqrt{2(x^2 + 2x + 1)} && \text{Factor out the GCF.} \\ &= \sqrt{2(x+1)^2} && \text{Factor the perfect square trinomial: } A^2 + 2AB + B^2 = (A+B)^2. \\ &= \sqrt{2}\cdot\sqrt{(x+1)^2} && \text{Take the square root of each factor. The factor } (x+1)^2 \text{ is a perfect square.} \\ &= \sqrt{2}|x+1| && \text{Take the square root of } (x+1)^2. \end{aligned}$$

In simplified form,

$$f(x) = \sqrt{2}|x + 1|.$$

Using Technology

Graphic Connections

The graphs of

$$f(x) = \sqrt{2x^2 + 4x + 2}, \quad g(x) = \sqrt{2}|x + 1|, \quad \text{and} \quad h(x) = \sqrt{2}(x + 1)$$

are shown in **Figure 7.6** in three separate $[-5, 5, 1]$ by $[-5, 5, 1]$ viewing rectangles. The graphs in **Figure 7.6 (a)** and **(b)** are identical. This verifies that our simplification in Example 3 is correct: $\sqrt{2x^2 + 4x + 2} = \sqrt{2}|x + 1|$. Now compare the graphs in **Figure 7.6 (a)** and **(c)**. Can you see that they are not the same? This illustrates the importance of not leaving out absolute value bars:

$$\sqrt{2x^2 + 4x + 2} \neq \sqrt{2}(x + 1).$$

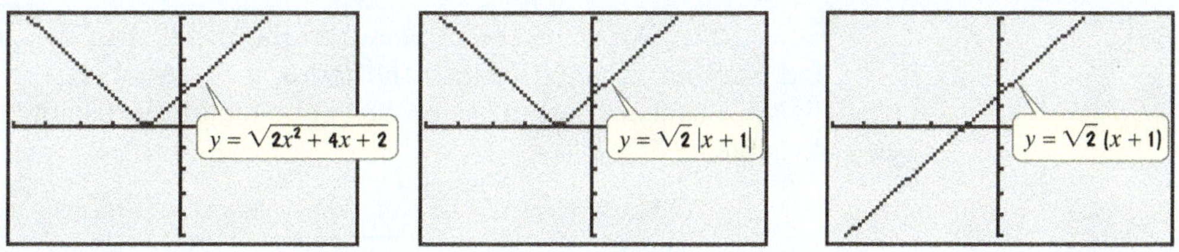

FIGURE 7.6 (a) (b) (c)

✓ **CHECK POINT 3** If $f(x) = \sqrt{3x^2 - 12x + 12}$, express the function, f, in simplified form.

For the remainder of this chapter, in situations that do not involve functions, we will **assume that no radicands involve negative quantities raised to even powers. Based upon this assumption, absolute value bars are not necessary when taking even roots.**

Simplifying When Variables to Even Powers in a Radicand Are Nonnegative Quantities

For any nonnegative real number a,

$$\sqrt[n]{a^n} = a.$$

In simplifying an nth root, how do we find variable factors in the radicand that are perfect nth powers? The **perfect nth powers have exponents that are divisible by n.** Simplification is possible by observation or by using rational exponents. Here are some examples:

- $\sqrt{x^6} = \sqrt{(x^3)^2} = x^3$ or $\sqrt{x^6} = (x^6)^{\frac{1}{2}} = x^3$

  6 is divisible by the index, 2. Thus, x^6 is a perfect square.

- $\sqrt[3]{y^{21}} = \sqrt[3]{(y^7)^3} = y^7$ or $\sqrt[3]{y^{21}} = (y^{21})^{\frac{1}{3}} = y^7$

  21 is divisible by the index, 3. Thus, y^{21} is a perfect cube.

- $\sqrt[6]{z^{24}} = \sqrt[6]{(z^4)^6} = z^4$ or $\sqrt[6]{z^{24}} = (z^{24})^{\frac{1}{6}} = z^4$.

  24 is divisible by the index, 6. Thus, z^{24} is a perfect 6th power.

EXAMPLE 4 Simplifying a Radical by Factoring

Simplify: $\sqrt{x^5y^{13}z^7}$.

Solution We write the radicand as the product of the greatest perfect square factor and another factor. Because the index is 2, variables that have exponents that are divisible by 2 are part of the perfect square factor. We use the greatest exponents that are divisible by 2.

Discover for Yourself

Square the answer in Example 4 and show that it is correct. If it is a square root, you should obtain the given radicand, $x^5y^{13}z^7$.

$$\begin{aligned}\sqrt{x^5y^{13}z^7} &= \sqrt{x^4\cdot x\cdot y^{12}\cdot y\cdot z^6\cdot z} && \text{Use the greatest even power of each variable.}\\ &= \sqrt{(x^4y^{12}z^6)(xyz)} && \text{Group the perfect square factors.}\\ &= \sqrt{x^4y^{12}z^6}\cdot\sqrt{xyz} && \text{Factor into two radicals.}\\ &= x^2y^6z^3\sqrt{xyz} && \sqrt{x^4y^{12}z^6} = (x^4y^{12}z^6)^{\frac{1}{2}} = x^2y^6z^3\end{aligned}$$

CHECK POINT 4 Simplify: $\sqrt{x^9y^{11}z^3}$.

EXAMPLE 5 Simplifying a Radical by Factoring

Simplify: $\sqrt[3]{32x^8y^{16}}$.

Solution We write the radicand as the product of the greatest perfect cube factor and another factor. Because the index is 3, variables that have exponents that are divisible by 3 are part of the perfect cube factor. We use the greatest exponents that are divisible by 3.

$$\begin{aligned}\sqrt[3]{32x^8y^{16}} &= \sqrt[3]{8\cdot 4\cdot x^6\cdot x^2\cdot y^{15}\cdot y} && \text{Identify perfect cube factors.}\\ &= \sqrt[3]{(8x^6y^{15})(4x^2y)} && \text{Group the perfect cube factors.}\\ &= \sqrt[3]{8x^6y^{15}}\cdot\sqrt[3]{4x^2y} && \text{Factor into two radicals.}\\ &= 2x^2y^5\sqrt[3]{4x^2y} && \sqrt[3]{8} = 2 \text{ and } \sqrt[3]{x^6y^{15}} = (x^6y^{15})^{\frac{1}{3}} = x^2y^5.\end{aligned}$$

CHECK POINT 5 Simplify: $\sqrt[3]{40x^{10}y^{14}}$. $2x^3y^4\sqrt[3]{5xy^2}$

EXAMPLE 6 Simplifying a Radical by Factoring

Simplify: $\sqrt[5]{64x^3y^7z^{29}}$.

Solution We write the radicand as the product of the greatest perfect 5th power and another factor. Because the index is 5, variables that have exponents that are divisible by 5 are part of the perfect fifth factor. We use the greatest exponents that are divisible by 5.

$$\begin{aligned}\sqrt[5]{64x^3y^7z^{29}} &= \sqrt[5]{32\cdot 2\cdot x^3\cdot y^5\cdot y^2\cdot z^{25}\cdot z^4} && \text{Identify perfect fifth factors.}\\ &= \sqrt[5]{(32y^5z^{25})(2x^3y^2z^4)} && \text{Group the perfect fifth factors.}\\ &= \sqrt[5]{32y^5z^{25}}\cdot\sqrt[5]{2x^3y^2z^4} && \text{Factor into two radicals.}\\ &= 2yz^5\sqrt[5]{2x^3y^2z^4} && \sqrt[5]{32} = 2 \text{ and } \sqrt[5]{y^5z^{25}} = (y^5z^{25})^{\frac{1}{5}} = yz^5.\end{aligned}$$

CHECK POINT 6 Simplify: $\sqrt[5]{32x^{12}y^2z^8}$.

3 Multiply radicals and then simplify.

Multiplying and Simplifying Radicals

We have seen how to use the product rule when multiplying radicals with the same index. Sometimes after multiplying, we can simplify the resulting radical.

EXAMPLE 7 Multiplying Radicals and Then Simplifying

Multiply and simplify:

a. $\sqrt{15} \cdot \sqrt{3}$ **b.** $7\sqrt[3]{4} \cdot 5\sqrt[3]{6}$ **c.** $\sqrt[4]{8x^3y^2} \cdot \sqrt[4]{8x^5y^3}$.

Solution

a. $\sqrt{15} \cdot \sqrt{3} = \sqrt{15 \cdot 3}$ Use the product rule.

$= \sqrt{45} = \sqrt{9 \cdot 5}$ 9 is the greatest perfect square factor of 45.

$= \sqrt{9} \cdot \sqrt{5} = 3\sqrt{5}$

b. $7\sqrt[3]{4} \cdot 5\sqrt[3]{6} = 35\sqrt[3]{4 \cdot 6}$ Use the product rule.

$= 35\sqrt[3]{24} = 35\sqrt[3]{8 \cdot 3}$ 8 is the greatest perfect cube factor of 24.

$= 35\sqrt[3]{8} \cdot \sqrt[3]{3} = 35 \cdot 2 \cdot \sqrt[3]{3}$

$= 70\sqrt[3]{3}$

c. $\sqrt[4]{8x^3y^2} \cdot \sqrt[4]{8x^5y^3} = \sqrt[4]{8x^3y^2 \cdot 8x^5y^3}$ Use the product rule.

$= \sqrt[4]{64x^8y^5}$ Multiply.

$= \sqrt[4]{16 \cdot 4 \cdot x^8 \cdot y^4 \cdot y}$ Identify perfect fourth factors.

$= \sqrt[4]{(16x^8y^4)(4y)}$ Group the perfect fourth factors.

$= \sqrt[4]{16x^8y^4} \cdot \sqrt[4]{4y}$ Factor into two radicals.

$= 2x^2y\sqrt[4]{4y}$ $\sqrt[4]{16} = 2$ and $\sqrt[4]{x^8y^4} = (x^8y^4)^{\frac{1}{4}} = x^2y$.

Study Tip

Confused about when you should write an expression under one radical and when you should separate the radicals?

- Use $\sqrt[n]{a} \cdot \sqrt[n]{b} = \sqrt[n]{ab}$, writing under one radical, when *multiplying*.
- Use $\sqrt[n]{ab} = \sqrt[n]{a} \cdot \sqrt[n]{b}$, factoring into two radicals, when *simplifying*.

CHECK POINT 7 Multiply and simplify:

a. $\sqrt{6} \cdot \sqrt{2}$

b. $10\sqrt[3]{16} \cdot 5\sqrt[3]{2}$

c. $\sqrt[4]{4x^2y} \cdot \sqrt[4]{8x^6y^3}$.

7.3 EXERCISE SET

MyMathLab MathXL PRACTICE WATCH DOWNLOAD READ REVIEW

Practice Exercises

In Exercises 1–20, use the product rule to multiply.

1. $\sqrt{3} \cdot \sqrt{5}$
2. $\sqrt{7} \cdot \sqrt{5}$
3. $\sqrt[3]{2} \cdot \sqrt[3]{9}$
4. $\sqrt[3]{5} \cdot \sqrt[3]{4}$
5. $\sqrt[4]{11} \cdot \sqrt[4]{3}$
6. $\sqrt[5]{9} \cdot \sqrt[5]{3}$
7. $\sqrt{3x} \cdot \sqrt{11y}$
8. $\sqrt{5x} \cdot \sqrt{11y}$
9. $\sqrt[5]{6x^3} \cdot \sqrt[5]{4x}$
10. $\sqrt[4]{6x^2} \cdot \sqrt[4]{3x}$
11. $\sqrt{x+3} \cdot \sqrt{x-3}$
12. $\sqrt{x+6} \cdot \sqrt{x-6}$
13. $\sqrt[6]{x-4} \cdot \sqrt[6]{(x-4)^4}$
14. $\sqrt[6]{x-5} \cdot \sqrt[6]{(x-5)^4}$
15. $\sqrt{\frac{2x}{3}} \cdot \sqrt{\frac{3}{2}}$
16. $\sqrt{\frac{2x}{5}} \cdot \sqrt{\frac{5}{2}}$
17. $\sqrt[4]{\frac{x}{7}} \cdot \sqrt[4]{\frac{3}{y}}$
18. $\sqrt[4]{\frac{x}{3}} \cdot \sqrt[4]{\frac{7}{y}}$
19. $\sqrt[7]{7x^2y} \cdot \sqrt[7]{11x^3y^2}$
20. $\sqrt[9]{12x^2y^3} \cdot \sqrt[9]{3x^3y^4}$

In Exercises 21–32, simplify by factoring.

21. $\sqrt{50}$
22. $\sqrt{27}$
23. $\sqrt{45}$
24. $\sqrt{28}$
25. $\sqrt{75x}$
26. $\sqrt{40x}$
27. $\sqrt[3]{16}$
28. $\sqrt[3]{54}$
29. $\sqrt[3]{27x^3}$
30. $\sqrt[3]{250x^3}$
31. $\sqrt[3]{-16x^2y^3}$
32. $\sqrt[3]{-32x^2y^3}$

In Exercises 33–38, express the function, f, in simplified form. Assume that x can be any real number.

33. $f(x) = \sqrt{36(x+2)^2}$
34. $f(x) = \sqrt{81(x-2)^2}$
35. $f(x) = \sqrt[3]{32(x+2)^3}$
36. $f(x) = \sqrt[3]{48(x-2)^3}$

37. $f(x) = \sqrt{3x^2 - 6x + 3}$
38. $f(x) = \sqrt{5x^2 - 10x + 5}$

In Exercises 39–60, simplify by factoring. Assume that all variables in a radicand represent positive real numbers and no radicands involve negative quantities raised to even powers.

39. $\sqrt{x^7}$
40. $\sqrt{x^5}$
41. $\sqrt{x^8y^9}$
42. $\sqrt{x^6y^7}$
43. $\sqrt{48x^3}$
44. $\sqrt{40x^3}$
45. $\sqrt[3]{y^8}$
46. $\sqrt[3]{y^{11}}$
47. $\sqrt[3]{x^{14}y^3z}$
48. $\sqrt[3]{x^3y^{17}z^2}$
49. $\sqrt[3]{81x^8y^6}$
50. $\sqrt[3]{32x^9y^{17}}$
51. $\sqrt[3]{(x + y)^5}$
52. $\sqrt[3]{(x + y)^4}$
53. $\sqrt[5]{y^{17}}$
54. $\sqrt[5]{y^{18}}$
55. $\sqrt[5]{64x^6y^{17}}$
56. $\sqrt[5]{64x^7y^{16}}$
57. $\sqrt[4]{80x^{10}}$
58. $\sqrt[4]{96x^{11}}$
59. $\sqrt[4]{(x - 3)^{10}}$
60. $\sqrt[4]{(x - 2)^{14}}$

In Exercises 61–82, multiply and simplify. Assume that all variables in a radicand represent positive real numbers and no radicands involve negative quantities raised to even powers.

61. $\sqrt{12}\cdot\sqrt{2}$
62. $\sqrt{3}\cdot\sqrt{6}$
63. $\sqrt{5x}\cdot\sqrt{10y}$
64. $\sqrt{8x}\cdot\sqrt{10y}$
65. $\sqrt{12x}\cdot\sqrt{3x}$
66. $\sqrt{20x}\cdot\sqrt{5x}$
67. $\sqrt{50xy}\cdot\sqrt{4xy^2}$
68. $\sqrt{5xy}\cdot\sqrt{10xy^2}$
69. $2\sqrt{5}\cdot 3\sqrt{40}$
70. $3\sqrt{15}\cdot 5\sqrt{6}$
71. $\sqrt[3]{12}\cdot\sqrt[3]{4}$
72. $\sqrt[4]{4}\cdot\sqrt[4]{8}$
73. $\sqrt{5x^3}\cdot\sqrt{8x^2}$
74. $\sqrt{2x^7}\cdot\sqrt{12x^4}$
75. $\sqrt[3]{25x^4y^2}\cdot\sqrt[3]{5xy^{12}}$
76. $\sqrt[3]{6x^7y}\cdot\sqrt[3]{9x^4y^{12}}$
77. $\sqrt[4]{8x^2y^3z^6}\cdot\sqrt[4]{2x^4yz}$
78. $\sqrt[4]{4x^2y^3z^3}\cdot\sqrt[4]{8x^4yz^6}$
79. $\sqrt[5]{8x^4y^6z^2}\cdot\sqrt[5]{8xy^7z^4}$
80. $\sqrt[5]{8x^4y^3z^3}\cdot\sqrt[5]{8xy^9z^8}$
81. $\sqrt[3]{x - y}\cdot\sqrt[3]{(x - y)^7}$
82. $\sqrt[3]{x - 6}\cdot\sqrt[3]{(x - 6)^7}$

Practice PLUS

In Exercises 83–90, simplify each expression. Assume that all variables in a radicand represent positive real numbers and no radicands involve negative quantities raised to even powers.

83. $-2x^2y\left(\sqrt[3]{54x^3y^7z^2}\right)$
84. $\dfrac{-x^2y^7}{2}\left(\sqrt[3]{-32x^4y^9z^7}\right)$
85. $-3y\left(\sqrt[5]{64x^3y^6}\right)$
86. $-4x^2y^7\left(\sqrt[5]{-32x^{11}y^{17}}\right)$
87. $\left(-2xy^2\sqrt{3x}\right)\left(xy\sqrt{6x}\right)$
88. $\left(-5x^2y^3z\sqrt{2xyz}\right)\left(-x^4z\sqrt{10xz}\right)$
89. $\left(2x^2y\sqrt[4]{8xy}\right)\left(-3xy^2\sqrt[4]{2x^2y^3}\right)$
90. $\left(5a^2b\sqrt[4]{8a^2b}\right)\left(4ab\sqrt[4]{4a^3b^2}\right)$

Application Exercises

The function

$$d(x) = \sqrt{\frac{3x}{2}}$$

models the distance, d(x), in miles, that a person h feet high can see to the horizon. Use this function to solve Exercises 91–92.

91. The pool deck on a cruise ship is 72 feet above the water. How far can passengers on the pool deck see? Write the answer in simplified radical form. Then use the simplified radical form and a calculator to express the answer to the nearest tenth of a mile.

92. The captain of a cruise ship is on the star deck, which is 120 feet above the water. How far can the captain see? Write the answer in simplified radical form. Then use the simplified radical form and a calculator to express the answer to the nearest tenth of a mile.

Paleontologists use the function

$$W(x) = 4\sqrt{2x}$$

to estimate the walking speed of a dinosaur, W(x), in feet per second, where x is the length, in feet, of the dinosaur's leg. The graph of W is shown in the figure. Use this information to solve Exercises 93–94.

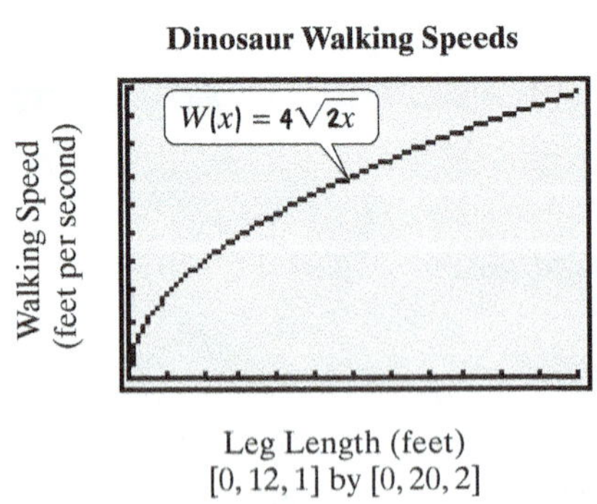

93. What is the walking speed of a dinosaur whose leg length is 6 feet? Use the function's equation and express the answer in simplified radical form. Then use the function's graph to estimate the answer to the nearest foot per second.

94. What is the walking speed of a dinosaur whose leg length is 10 feet? Use the function's equation and express the answer in simplified radical form. Then use the function's graph to estimate the answer to the nearest foot per second.

Your **cardiac index** *is your heart's output, in liters of blood per minute, divided by your body's surface area, in square meters. The cardiac index, $C(x)$, can be modeled by*

$$C(x) = \frac{7.644}{\sqrt[4]{x}}, \quad 10 \le x \le 80,$$

where x is an individual's age, in years. The graph of the function is shown. Use the function to solve Exercises 95–96.

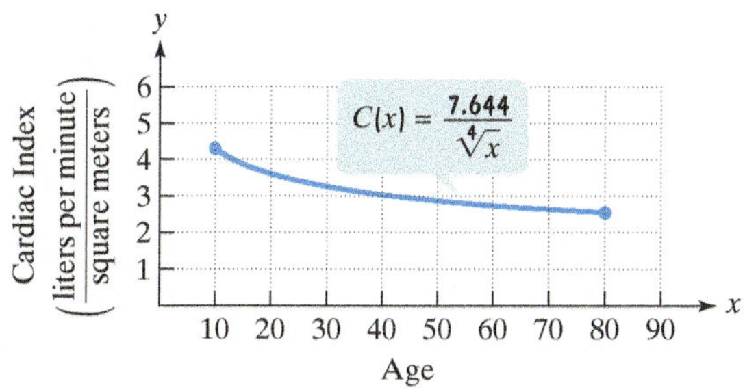

95. a. Find the cardiac index of a 32-year-old. Express the denominator in simplified radical form and reduce the fraction.

b. Use the form of the answer in part (a) and a calculator to express the cardiac index to the nearest hundredth. Identify your solution as a point on the graph.

96. a. Find the cardiac index of an 80-year-old. Express the denominator in simplified radical form and reduce the fraction.

b. Use the form of the answer in part (a) and a calculator to express the cardiac index to the nearest hundredth. Identify your solution as a point on the graph.

Writing in Mathematics

97. What is the product rule for radicals? Give an example to show how it is used.

98. Explain why $\sqrt{50}$ is not simplified. What do we mean when we say a radical expression is simplified?

99. In simplifying an nth root, explain how to find variable factors in the radicand that are perfect nth powers.

100. Without showing all the details, explain how to simplify $\sqrt[3]{16x^{14}}$.

101. As you get older, what would you expect to happen to your heart's output? Explain how this is shown in the graph for Exercises 95–96. Is this trend taking place progressively more rapidly or more slowly over the entire interval? What does this mean about this aspect of aging?

Technology Exercises

102. Use a calculator to provide numerical support for your simplifications in Exercises 21–24 and 27–28. In each case, find a decimal approximation for the given expression. Then find a decimal approximation for your simplified expression. The approximations should be the same.

In Exercises 103–106, determine if each simplification is correct by graphing the function on each side of the equation with your graphing utility. Use the given viewing rectangle. The graphs should be the same. If they are not, correct the right side of the equation and then use your graphing utility to verify the simplification.

103. $\sqrt{x^4} = x^2$; $[0, 5, 1]$ by $[0, 20, 1]$

104. $\sqrt{8x^2} = 4x\sqrt{2}$; $[-5, 5, 1]$ by $[-5, 20, 1]$

105. $\sqrt{3x^2 - 6x + 3} = (x - 1)\sqrt{3}$; $[-5, 5, 1]$ by $[-5, 5, 1]$

106. $\sqrt[3]{2x} \cdot \sqrt[3]{4x^2} = 4x$; $[-10, 10, 1]$ by $[-10, 10, 1]$

Critical Thinking Exercises

Make Sense? *In Exercises 107–110, determine whether each statement "makes sense" or "does not make sense" and explain your reasoning.*

107. Because the product rule for radicals applies when $\sqrt[n]{a}$ and $\sqrt[n]{b}$ are real numbers, I can use it to find $\sqrt[3]{16} \cdot \sqrt[3]{-4}$, but not to find $\sqrt{8} \cdot \sqrt{-2}$.

108. I multiply nth roots by taking the nth root of the product of the radicands.

109. I need to know how to factor a trinomial to simplify $\sqrt{x^2 - 10x + 25}$.

110. I know that I've simplified a radical expression when it contains a single radical.

In Exercises 111–114, determine whether each statement is true or false. If the statement is false, make the necessary change(s) to produce a true statement.

111. $2\sqrt{5} \cdot 6\sqrt{5} = 12\sqrt{5}$

112. $\sqrt[3]{4} \cdot \sqrt[3]{4} = 4$

113. $\sqrt{12} = 2\sqrt{6}$

114. $\sqrt[3]{3^{15}} = 243$

115. If a number is tripled, what happens to its square root?

116. What must be done to a number so that its cube root is tripled?

117. If $f(x) = \sqrt[3]{2x}$ and $(fg)(x) = 2x$, find $g(x)$.

118. Graph $f(x) = \sqrt{(x - 1)^2}$ by hand.

Review Exercises

119. Solve: $2x - 1 \le 21$ and $2x + 2 \ge 12$. (Section 4.2, Example 2)

120. Solve:

$$\begin{aligned} 5x + 2y &= 2 \\ 4x + 3y &= -4. \end{aligned}$$

(Section 3.1, Example 6)

121. Factor: $64x^3 - 27$. (Section 5.5, Example 9)

Preview Exercises

Exercises 122–124 will help you prepare for the material covered in the next section.

122. **a.** Simplify: $21x + 10x$.

b. Simplify: $21\sqrt{2} + 10\sqrt{2}$.

123. **a.** Simplify: $4x - 12x$.

b. Simplify: $4\sqrt[3]{2} - 12\sqrt[3]{2}$.

124. Simplify: $\frac{\sqrt[4]{7y^5}}{\sqrt[4]{x^{12}}}$.

CHAPTER 7 GROUP PROJECT

Group members should consult an almanac, newspaper, magazine, or the Internet and return to the group with as much data as possible that show phenomena that are continuing to grow over time, but whose growth is leveling off. Select the five data sets that you find most intriguing. Let x represent the number of years after the first year in each data set. Model the data by hand using

$$f(x) = a\sqrt{x} + b.$$

Use the first and last data points to find values for a and b. The first data point corresponds to $x = 0$. Its second coordinate gives the value of b. To find a, substitute the second data point into $f(x) = a\sqrt{x} + b$, with the value that you obtained for b. Now solve the equation and find a. Substitute a and b into $f(x) = a\sqrt{x} + b$ to obtain a square root function that models each data set. Then use the function to make predictions about what might occur in the future. Are there circumstances that might affect the accuracy of the predictions? List some of these circumstances.

Chapter 7 Summary

Definitions and Concepts	Examples

Section 7.1 Radical Expressions and Functions

If $b^2 = a$, then b is a square root of a. The principal square root of a, designated $\sqrt{a}$, is the nonnegative number satisfying $b^2 = a$. The negative square root of a is written $-\sqrt{a}$. A square root of a negative number is not a real number.

A radical function in x is a function defined by an expression containing a root of x. The domain of a square root function is the set of real numbers for which the radicand is nonnegative.

Let $f(x) = \sqrt{6 - 2x}$.

$f(-15) = \sqrt{6 - 2(-15)} = \sqrt{6 + 30} = \sqrt{36} = 6$

$f(5) = \sqrt{6 - 2 \cdot 5} = \sqrt{6 - 10} = \sqrt{-4}$, not a real number

Domain of f: Set the radicand greater than or equal to zero.

$$6 - 2x \geq 0$$
$$-2x \geq -6$$
$$x \leq 3$$

Domain of $f = \{x | x \leq 3\}$ or $(-\infty, 3]$

The cube root of a real number a is written $\sqrt[3]{a}$.

$$\sqrt[3]{a} = b \quad \text{means that} \quad b^3 = a.$$

The nth root of a real number a is written $\sqrt[n]{a}$. The number n is the index. Every real number has one root when n is odd. The odd nth root of a, $\sqrt[n]{a}$, is the number b for which $b^n = a$. Every positive real number has two real roots when n is even. An even root of a negative number is not a real number.

If n is even, then $\sqrt[n]{a^n} = |a|$.

If n is odd, then $\sqrt[n]{a^n} = a$.

- $\sqrt[3]{-8} = -2$ because $(-2)^3 = -8$.
- $\sqrt[4]{-16}$ is not a real number.
- $\sqrt{x^2 - 14x + 49} = \sqrt{(x - 7)^2} = |x - 7|$
- $\sqrt[3]{125(x + 6)^3} = 5(x + 6)$

Section 7.2 Rational Exponents

- $a^{\frac{1}{n}} = \sqrt[n]{a}$
- $a^{\frac{m}{n}} = (\sqrt[n]{a})^m$ or $\sqrt[n]{a^m}$
- $a^{-\frac{m}{n}} = \dfrac{1}{a^{\frac{m}{n}}}$

- $121^{\frac{1}{2}} = \sqrt{121} = 11$
- $64^{\frac{1}{3}} = \sqrt[3]{64} = 4$
- $27^{\frac{5}{3}} = (\sqrt[3]{27})^5 = 3^5 = 3 \cdot 3 \cdot 3 \cdot 3 \cdot 3 = 243$
- $16^{-\frac{3}{4}} = \dfrac{1}{16^{\frac{3}{4}}} = \dfrac{1}{(\sqrt[4]{16})^3} = \dfrac{1}{2^3} = \dfrac{1}{8}$
- $(\sqrt[3]{7xy})^4 = (7xy)^{\frac{4}{3}}$

Properties of integer exponents are true for rational exponents. An expression with rational exponents is simplified when no parentheses appear, no powers are raised to powers, each base occurs once, and no negative or zero exponents appear.

Simplify: $\left(8x^{\frac{1}{3}}y^{-\frac{1}{2}}\right)^{\frac{1}{3}}$.

$$= 8^{\frac{1}{3}}\left(x^{\frac{1}{3}}\right)^{\frac{1}{3}}\left(y^{-\frac{1}{2}}\right)^{\frac{1}{3}}$$

$$= 2x^{\frac{1}{9}}y^{-\frac{1}{6}} = \frac{2x^{\frac{1}{9}}}{y^{\frac{1}{6}}}$$

Some radical expressions can be simplifed using rational exponents. Rewrite the expression using rational exponents, simplify, and rewrite in radical notation if rational exponents still appear.

- $\sqrt[9]{x^3} = x^{\frac{3}{9}} = x^{\frac{1}{3}} = \sqrt[3]{x}$
- $\sqrt[5]{x^2} \cdot \sqrt[4]{x} = x^{\frac{2}{5}} \cdot x^{\frac{1}{4}} = x^{\frac{2}{5} + \frac{1}{4}}$

$$= x^{\frac{8}{20} + \frac{5}{20}} = x^{\frac{13}{20}} = \sqrt[20]{x^{13}}$$

Definitions and Concepts | **Examples**

Section 7.3 Multiplying and Simplifying Radical Expressions

The product rule for radicals can be used to multiply radicals

$$\sqrt[n]{a}\cdot\sqrt[n]{b}=\sqrt[n]{ab}.$$

$$\sqrt[3]{7x}\cdot\sqrt[3]{10y^2}=\sqrt[3]{7x\cdot 10y^2}=\sqrt[3]{70xy^2}$$

The product rule for radicals can be used to simplify radicals:

$$\sqrt[n]{ab}=\sqrt[n]{a}\cdot\sqrt[n]{b}.$$

A radical expression with index n is simplified when its radicand has no factors that are perfect nth powers. To simplify, write the radicand as the product of two factors, one of which is the greatest perfect nth power. Then use the product rule to take the nth root of each factor. If all variables in a radicand are positive, then

$$\sqrt[n]{a^n}=a.$$

Some radicals can be simplified after multiplication is performed.

- Simplify: $\sqrt[3]{54x^7y^{11}}$.

$$=\sqrt[3]{27\cdot 2\cdot x^6\cdot x\cdot y^9\cdot y^2}$$
$$=\sqrt[3]{(27x^6y^9)(2xy^2)}$$
$$=\sqrt[3]{27x^6y^9}\cdot\sqrt[3]{2xy^2}=3x^2y^3\sqrt[3]{2xy^2}$$

- Assuming positive variables, multiply and simplify: $\sqrt[4]{4x^2y}\cdot\sqrt[4]{4xy^3}$.

$$=\sqrt[4]{4x^2y\cdot 4xy^3}=\sqrt[4]{16x^3y^4}$$
$$=\sqrt[4]{16y^4}\cdot\sqrt[4]{x^3}=2y\sqrt[4]{x^3}$$

Section 7.4 Adding, Subtracting, and Dividing Radical Expressions

Like radicals have the same indices and radicands. Like radicals can be added or subtracted using the distributive property. In some cases, radicals can be combined once they have been simplified.

$$4\sqrt{18}-6\sqrt{50}$$
$$=4\sqrt{9\cdot 2}-6\sqrt{25\cdot 2}=4\cdot 3\sqrt{2}-6\cdot 5\sqrt{2}$$
$$=12\sqrt{2}-30\sqrt{2}=-18\sqrt{2}$$

The quotient rule for radicals can be used to simplify radicals:

$$\sqrt[n]{\frac{a}{b}}=\frac{\sqrt[n]{a}}{\sqrt[n]{b}}.$$

$$\sqrt[3]{-\frac{8}{x^{12}}}=\frac{\sqrt[3]{-8}}{\sqrt[3]{x^{12}}}=-\frac{2}{x^4}$$

$\sqrt[3]{x^{12}}=(x^{12})^{\frac{1}{3}}=x^4$

The quotient rule for radicals can be used to divide radicals with the same indices:

$$\frac{\sqrt[n]{a}}{\sqrt[n]{b}}=\sqrt[n]{\frac{a}{b}}.$$

Some radicals can be simplified after the division is performed.

Assuming a positive variable, divide and simplify:

$$\frac{\sqrt[4]{64x^5}}{\sqrt[4]{2x^{-2}}}=\sqrt[4]{32x^{5-(-2)}}=\sqrt[4]{32x^7}$$
$$=\sqrt[4]{16\cdot 2\cdot x^4\cdot x^3}=\sqrt[4]{16x^4}\cdot\sqrt[4]{2x^3}$$
$$=2x\sqrt[4]{2x^3}.$$

Section 7.5 Multiplying with More Than One Term and Rationalizing Denominators

Radical expressions with more than one term are multiplied in much the same way that polynomials with more than one term are multiplied.

- $\sqrt{5}(2\sqrt{6}-\sqrt{3})=2\sqrt{30}-\sqrt{15}$
- $(4\sqrt{3}-2\sqrt{2})(\sqrt{3}+\sqrt{2})$

F O I L

$$=4\sqrt{3}\cdot\sqrt{3}+4\sqrt{3}\cdot\sqrt{2}-2\sqrt{2}\cdot\sqrt{3}-2\sqrt{2}\cdot\sqrt{2}$$
$$=4\cdot 3+4\sqrt{6}-2\sqrt{6}-2\cdot 2$$
$$=12+4\sqrt{6}-2\sqrt{6}-4=8+2\sqrt{6}$$

Definitions and Concepts	Examples
Section 7.5 Multiplying with More Than One Term and Rationalizing Denominators (continued)	
Radical expressions that involve the sum and difference of the same two terms are called conjugates. Use $(A + B)(A - B) = A^2 - B^2$ to multiply conjugates.	$(8 + 2\sqrt{5})(8 - 2\sqrt{5}) = 8^2 - (2\sqrt{5})^2 = 64 - 4 \cdot 5 = 64 - 20 = 44$
The process of rewriting a radical expression as an equivalent expression without any radicals in the denominator is called rationalizing the denominator. When the denominator contains a single radical with an *n*th root, multiply the numerator and the denominator by a radical of index *n* that produces a perfect *n*th power in the denominator's radicand.	Rationalize the denominator: $\frac{7}{\sqrt[3]{2x}}$. $= \frac{7}{\sqrt[3]{2x}} \cdot \frac{\sqrt[3]{4x^2}}{\sqrt[3]{4x^2}} = \frac{7\sqrt[3]{4x^2}}{\sqrt[3]{8x^3}} = \frac{7\sqrt[3]{4x^2}}{2x}$
If the denominator contains two terms, rationalize the denominator by multiplying the numerator and the denominator by the conjugate of the denominator.	$\frac{13}{5 - \sqrt{3}} = \frac{13}{5 - \sqrt{3}} \cdot \frac{5 + \sqrt{3}}{5 + \sqrt{3}} = \frac{13(5 + \sqrt{3})}{5^2 - (\sqrt{3})^2} = \frac{13(5 + \sqrt{3})}{25 - 3} = \frac{13(5 + \sqrt{3})}{22}$
Section 7.6 Radical Equations	
A radical equation is an equation in which the variable occurs in a radicand. **Solving Radical Equations Containing *n*th Roots** 1. Isolate one radical on one side of the equation. 2. Raise both sides to the *n*th power. 3. Solve the resulting equation. 4. Check proposed solutions in the original equation. Solutions of an equation to an even power that is radical-free, but not the original equation, are called extraneous solutions.	Solve: $\sqrt{6x + 13} - 2x = 1$. $\sqrt{6x + 13} = 2x + 1$ Isolate the radical. $(\sqrt{6x + 13})^2 = (2x + 1)^2$ Square both sides. $6x + 13 = 4x^2 + 4x + 1$ $(A + B)^2 = A^2 + 2AB + B^2$ $0 = 4x^2 - 2x - 12$ Subtract 6x + 13 from both sides. $0 = 2(2x^2 - x - 6)$ Factor out the GCF. $0 = 2(2x + 3)(x - 2)$ Factor completely. $2x + 3 = 0$ or $x - 2 = 0$ Set variable factors equal to zero. $2x = -3$, $x = -\frac{3}{2}$; $x = 2$ Solve for x. Check both proposed solutions. 2 checks, but $-\frac{3}{2}$ is extraneous. The solution is 2 and the solution set is $\{2\}$.
Section 7.7 Complex Numbers	
The imaginary unit *i* is defined as $i = \sqrt{-1}$, where $i^2 = -1$. The set of numbers in the form $a + bi$ is called the set of complex numbers; *a* is the real part and *b* is the imaginary part. If $b = 0$, the complex number is a real number. If $b \neq 0$, the complex number is an imaginary number.	• $\sqrt{-81} = \sqrt{81(-1)} = \sqrt{81}\sqrt{-1} = 9i$ • $\sqrt{-75} = \sqrt{75(-1)} = \sqrt{25 \cdot 3}\sqrt{-1} = 5i\sqrt{3}$

Definitions and Concepts	Examples

Section 7.7 Complex Numbers (continued)

To add or subtract complex numbers, add or subtract their real parts and add or subtract their imaginary parts.

$$(2 - 4i) - (7 - 10i)$$
$$= 2 - 4i - 7 + 10i$$
$$= (2 - 7) + (-4 + 10)i = -5 + 6i$$

To multiply complex numbers, multiply as if they were polynomials. After completing the multiplication, replace i^2 with -1. When performing operations with square roots of negative numbers, begin by expressing all square roots in terms of i. Then multiply.

F O I L

- $(2 - 3i)(4 + 5i) = 8 + 10i - 12i - 15i^2$
  $= 8 + 10i - 12i - 15(-1)$
  $= 23 - 2i$
- $\sqrt{-36} \cdot \sqrt{-100} = \sqrt{36(-1)} \cdot \sqrt{100(-1)}$
  $= 6i \cdot 10i = 60i^2 = 60(-1) = -60$

The complex numbers $a + bi$ and $a - bi$ are conjugates. Conjugates can be multiplied using the formula

$$(A + B)(A - B) = A^2 - B^2.$$

The multiplication of conjugates results in a real number.

$$(3 + 5i)(3 - 5i) = 3^2 - (5i)^2$$
$$= 9 - 25i^2$$
$$= 9 - 25(-1) = 34$$

To divide complex numbers, multiply the numerator and the denominator by the conjugate of the denominator in order to obtain a real number in the denominator. This real number becomes the denominator of a and b in the quotient $a + bi$.

$$\frac{5 + 2i}{4 - i} = \frac{5 + 2i}{4 - i} \cdot \frac{4 + i}{4 + i} = \frac{20 + 5i + 8i + 2i^2}{16 - i^2}$$
$$= \frac{20 + 13i + 2(-1)}{16 - (-1)}$$
$$= \frac{20 + 13i - 2}{16 + 1}$$
$$= \frac{18 + 13i}{17} = \frac{18}{17} + \frac{13}{17}i$$

To simplify powers of i, rewrite the expression in terms of i^2. Then replace i^2 with -1 and simplify.

Simplify: i^{27}.

$$i^{27} = i^{26} \cdot i = (i^2)^{13} i$$
$$= (-1)^{13} i = (-1)i = -i$$

CHAPTER 7 REVIEW EXERCISES

7.1 *In Exercises 1–5, find the indicated root, or state that the expression is not a real number.*

1. $\sqrt{81}$

2. $-\sqrt{\frac{1}{100}}$

3. $\sqrt[3]{-27}$

4. $\sqrt[4]{-16}$

5. $\sqrt[5]{-32}$

In Exercises 6–7, find the indicated function values for each function. If necessary, round to two decimal places. If the function value is not a real number and does not exist, so state.

6. $f(x) = \sqrt{2x - 5}$; $f(15), f(4), f\left(\frac{5}{2}\right), f(1)$

7. $g(x) = \sqrt[3]{4x - 8}$; $g(4), g(0), g(-14)$

In Exercises 8–9, find the domain of each square root function.

8. $f(x) = \sqrt{x - 2}$

9. $g(x) = \sqrt{100 - 4x}$

In Exercises 10–15, simplify each expression. Assume that each variable can represent any real number, so include absolute value bars where necessary.

10. $\sqrt{25x^2}$

11. $\sqrt{(x + 14)^2}$

12. $\sqrt{x^2 - 8x + 16}$

13. $\sqrt[3]{64x^3}$

14. $\sqrt[4]{16x^4}$

15. $\sqrt[5]{-32(x + 7)^5}$

7.2 *In Exercises 16–18, use radical notation to rewrite each expression. Simplify, if possible.*

16. $(5xy)^{\frac{1}{3}}$

17. $16^{\frac{3}{2}}$

18. $32^{\frac{4}{5}}$

In Exercises 19–20, rewrite each expression with rational exponents.

19. $\sqrt{7x}$

20. $(\sqrt[3]{19xy})^5$

In Exercises 21–22, rewrite each expression with a positive rational exponent. Simplify, if possible.

21. $8^{-\frac{2}{3}}$

22. $3x(ab)^{-\frac{4}{5}}$

In Exercises 23–26, use properties of rational exponents to simplify each expression.

23. $x^{\frac{1}{3}} \cdot x^{\frac{1}{4}}$

24. $\dfrac{5^{\frac{1}{2}}}{5^{\frac{1}{3}}}$

25. $(8x^6y^3)^{\frac{1}{3}}$

25. $\left(x^{-\frac{2}{3}}y^{\frac{1}{4}}\right)^{\frac{1}{2}}$

In Exercises 27–31, use rational exponents to simplify each expression. If rational exponents appear after simplifying, write the answer in radical notation.

27. $\sqrt[3]{x^9y^{12}}$

28. $\sqrt[9]{x^3y^9}$

29. $\sqrt{x} \cdot \sqrt[3]{x}$

30. $\dfrac{\sqrt[3]{x^2}}{\sqrt[4]{x^2}}$

31. $\sqrt[5]{\sqrt[3]{x}}$

32. The function $f(x) = 350x^{\frac{2}{3}}$ models the expenditures, $f(x)$, in millions of dollars, for the U.S. National Park Service x years after 1985. According to this model, what will expenditures be in 2012?

7.3 *In Exercises 33–35, use the product rule to multiply.*

33. $\sqrt{3x} \cdot \sqrt{7y}$

34. $\sqrt[5]{7x^2} \cdot \sqrt[5]{11x}$

35. $\sqrt[6]{x-5} \cdot \sqrt[6]{(x-5)^4}$

36. If $f(x) = \sqrt{7x^2 - 14x + 7}$, express the function, f, in simplified form. Assume that x can be any real number.

In Exercises 37–39, simplify by factoring. Assume that all variables in a radicand represent positive real numbers.

37. $\sqrt{20x^3}$

38. $\sqrt[3]{54x^8y^6}$

39. $\sqrt[4]{32x^3y^{11}z^5}$

In Exercises 40–43, multiply and simplify, if possible. Assume that all variables in a radicand represent positive real numbers.

40. $\sqrt{6x^3} \cdot \sqrt{4x^2}$

41. $\sqrt[3]{4x^2y} \cdot \sqrt[3]{4xy^4}$

42. $\sqrt[5]{2x^4y^3z^4} \cdot \sqrt[5]{8xy^6z^7}$

43. $\sqrt{x+1} \cdot \sqrt{x-1}$

7.4 *Assume that all variables represent positive real numbers. In Exercises 44–47, add or subtract as indicated.*

44. $6\sqrt[3]{3} + 2\sqrt[3]{3}$

45. $5\sqrt{18} - 3\sqrt{8}$

46. $\sqrt[3]{27x^4} + \sqrt[3]{xy^6}$

47. $2\sqrt[3]{6} - 5\sqrt[3]{48}$

In Exercises 48–50, simplify using the quotient rule.

48. $\sqrt[3]{\dfrac{16}{125}}$

49. $\sqrt{\dfrac{x^3}{100y^4}}$

50. $\sqrt[4]{\dfrac{3y^5}{16x^{20}}}$

In Exercises 51–54, divide and, if possible, simplify.

51. $\dfrac{\sqrt{48}}{\sqrt{2}}$

52. $\dfrac{\sqrt[3]{32}}{\sqrt[3]{2}}$

53. $\dfrac{\sqrt[4]{64x^7}}{\sqrt[4]{2x^2}}$

54. $\dfrac{\sqrt{200x^3y^2}}{\sqrt{2x^{-2}y}}$

7.5 *Assume that all variables represent positive real numbers.*

In Exercises 55–62, multiply as indicated. If possible, simplify any radical expressions that appear in the product.

55. $\sqrt{3}(2\sqrt{6} + 4\sqrt{15})$

56. $\sqrt[3]{5}(\sqrt[3]{50} - \sqrt[3]{2})$

57. $(\sqrt{7} - 3\sqrt{5})(\sqrt{7} + 6\sqrt{5})$

58. $(\sqrt{x} - \sqrt{11})(\sqrt{y} - \sqrt{11})$

59. $(\sqrt{5} + \sqrt{8})^2$

60. $(2\sqrt{3} - \sqrt{10})^2$

61. $(\sqrt{7} + \sqrt{13})(\sqrt{7} - \sqrt{13})$

62. $(7 - 3\sqrt{5})(7 + 3\sqrt{5})$

In Exercises 63–75, rationalize each denominator. Simplify, if possible.

63. $\dfrac{4}{\sqrt{6}}$

64. $\sqrt{\dfrac{2}{7}}$

65. $\dfrac{12}{\sqrt[3]{9}}$

66. $\sqrt{\dfrac{2x}{5y}}$

67. $\dfrac{14}{\sqrt[3]{2x^2}}$

68. $\sqrt[4]{\dfrac{7}{3x}}$

69. $\dfrac{5}{\sqrt[5]{32x^4y}}$

70. $\dfrac{6}{\sqrt{3} - 1}$

71. $\dfrac{\sqrt{7}}{\sqrt{5} + \sqrt{3}}$

72. $\dfrac{10}{2\sqrt{5} - 3\sqrt{2}}$

73. $\dfrac{\sqrt{x} + 5}{\sqrt{x} - 3}$

74. $\dfrac{\sqrt{7} + \sqrt{3}}{\sqrt{7} - \sqrt{3}}$

75. $\dfrac{2\sqrt{3} + \sqrt{6}}{2\sqrt{6} + \sqrt{3}}$

In Exercises 76–79, rationalize each numerator. Simplify, if possible.

76. $\sqrt{\dfrac{2}{7}}$

77. $\dfrac{\sqrt[3]{3x}}{\sqrt[3]{y}}$

78. $\dfrac{\sqrt{7}}{\sqrt{5} + \sqrt{3}}$

79. $\dfrac{\sqrt{7} + \sqrt{3}}{\sqrt{7} - \sqrt{3}}$

7.6 *In Exercises 80–84, solve each radical equation.*

80. $\sqrt{2x+4} = 6$

81. $\sqrt{x-5} + 9 = 4$

82. $\sqrt{2x-3} + x = 3$

83. $\sqrt{x-4} + \sqrt{x+1} = 5$

84. $(x^2 + 6x)^{\frac{1}{3}} + 2 = 0$

85. In 2007, state tobacco taxes ranged from \$0.07 per pack in South Carolina to \$2.58 in New Jersey. The graph shows the average state cigarette tax per pack from 2001 through 2007.

Average State Cigarette Tax per Pack

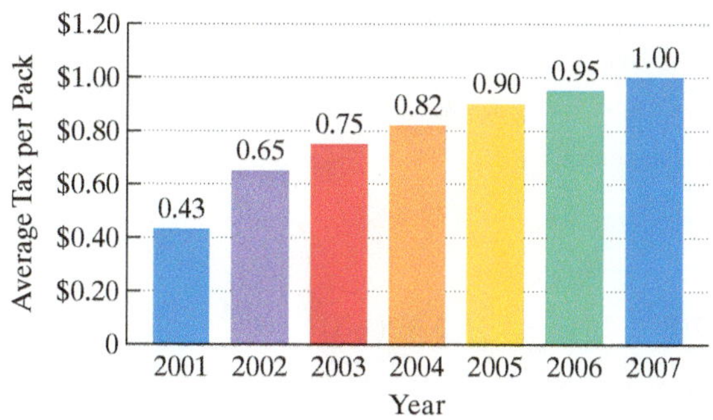

Source: Campaign for Tobacco-Free Kids

The function

$$f(x) = 0.23\sqrt{x} + 0.43$$

models the average state cigarette tax per pack, $f(x)$, x years after 2001.

a. Find and interpret $f(6)$. Round to two decimal places. Does this underestimate or overestimate the tax displayed by the graph? By how much?

b. According to the model, in which year will the average state cigarette tax be \$1.12 per pack?

86. Out of a group of 50,000 births, the number of people, $f(x)$, surviving to age x is modeled by the function

$$f(x) = 5000\sqrt{100 - x}.$$

To what age will 20,000 people in the group survive?

7.7 *In Exercises 87–89, express each number in terms of i and simplify, if possible.*

87. $\sqrt{-81}$

88. $\sqrt{-63}$

89. $-\sqrt{-8}$

In Exercises 90–99, perform the indicated operation. Write the result in the form a + bi.

90. $(7 + 12i) + (5 - 10i)$

91. $(8 - 3i) - (17 - 7i)$

92. $4i(3i - 2)$

93. $(7 - 5i)(2 + 3i)$

94. $(3 - 4i)^2$

95. $(7 + 8i)(7 - 8i)$

96. $\sqrt{-8}\cdot\sqrt{-3}$

97. $\frac{6}{5+i}$

98. $\frac{3+4i}{4-2i}$

99. $\frac{5+i}{3i}$

In Exercises 100–101, simplify each expression.

100. i^{16}

101. i^{23}

CHAPTER 7 TEST

Remember to use your Chapter Test Prep Video CD to see the worked-out solutions to the test questions you want to review.

1. Let $f(x) = \sqrt{8 - 2x}$.

a. Find $f(-14)$.

b. Find the domain of f.

2. Evaluate: $27^{-\frac{4}{3}}$.

3. Simplify: $\left(25x^{-\frac{1}{2}}y^{\frac{1}{4}}\right)^{\frac{1}{2}}$.

In Exercises 4–5, use rational exponents to simplify each expression. If rational exponents appear after simplifying, write the answer in radical notation.

4. $\sqrt[8]{x^4}$

5. $\sqrt[4]{x}\cdot\sqrt[5]{x}$

In Exercises 6–9, simplify each expression. Assume that each variable can represent any real number.

6. $\sqrt{75x^2}$

7. $\sqrt{x^2 - 10x + 25}$

8. $\sqrt[3]{16x^4y^8}$

9. $\sqrt[5]{-\frac{32}{x^{10}}}$

In Exercises 10–17, perform the indicated operation and, if possible, simplify. Assume that all variables represent positive real numbers.

10. $\sqrt[3]{5x^2}\cdot\sqrt[3]{10y}$

11. $\sqrt[4]{8x^3y}\cdot\sqrt[4]{4xy^2}$

12. $3\sqrt{18} - 4\sqrt{32}$

13. $\sqrt[3]{8x^4} + \sqrt[3]{xy^6}$

14. $\frac{\sqrt[3]{16x^8}}{\sqrt[3]{2x^4}}$

15. $\sqrt{3}(4\sqrt{6} - \sqrt{5})$

16. $(5\sqrt{6} - 2\sqrt{2})(\sqrt{6} + \sqrt{2})$

17. $(7 - \sqrt{3})^2$

In Exercises 18–20, rationalize each denominator. Simplify, if possible. Assume all variables represent positive real numbers.

18. $\sqrt{\frac{5}{x}}$

19. $\frac{5}{\sqrt[3]{5x^2}}$

20. $\frac{\sqrt{2} - \sqrt{3}}{\sqrt{2} + \sqrt{3}}$

In Exercises 21–23, solve each radical equation.

21. $3 + \sqrt{2x - 3} = x$
22. $\sqrt{x + 9} - \sqrt{x - 7} = 2$
23. $(11x + 6)^{\frac{1}{3}} + 3 = 0$
24. The function

$$f(x) = 2.9\sqrt{x} + 20.1$$

models the average height, $f(x)$, in inches, of boys who are x months of age, $0 \le x \le 60$. Find the age at which the average height of boys is 40.4 inches.

25. Express in terms of i and simplify: $\sqrt{-75}$.

In Exercises 26–29, perform the indicated operation. Write the result in the form $a + bi$.

26. $(5 - 3i) - (6 - 9i)$
27. $(3 - 4i)(2 + 5i)$
28. $\sqrt{-9} \cdot \sqrt{-4}$
29. $\dfrac{3 + i}{1 - 2i}$
30. Simplify: i^{35}.

CUMULATIVE REVIEW EXERCISES (CHAPTERS 1–7)

In Exercises 1–5, solve each equation, inequality, or system.

1. $$\begin{aligned} 2x - y + z &= -5 \\ x - 2y - 3z &= 6 \\ x + y - 2z &= 1 \end{aligned}$$
2. $3x^2 - 11x = 4$
3. $2(x + 4) < 5x + 3(x + 2)$
4. $\dfrac{1}{x + 2} + \dfrac{15}{x^2 - 4} = \dfrac{5}{x - 2}$
5. $\sqrt{x + 2} - \sqrt{x + 1} = 1$
6. Graph the solution set of the system:

$$\begin{aligned} x + 2y &< 2 \\ 2y - x &> 4. \end{aligned}$$

In Exercises 7–15, perform the indicated operations.

7. $\dfrac{8x^2}{3x^2 - 12} \div \dfrac{40}{x - 2}$
8. $\dfrac{x + \frac{1}{y}}{y + \frac{1}{x}}$
9. $(2x - 3)(4x^2 - 5x - 2)$
10. $\dfrac{7x}{x^2 - 2x - 15} - \dfrac{2}{x - 5}$
11. $7(8 - 10)^3 - 7 + 3 \div (-3)$
12. $\sqrt{80x} - 5\sqrt{20x} + 2\sqrt{45x}$
13. $\dfrac{\sqrt{3} - 2}{2\sqrt{3} + 5}$
14. $(2x^3 - 3x^2 + 3x - 4) \div (x - 2)$
15. $(2\sqrt{3} + 5\sqrt{2})(\sqrt{3} - 4\sqrt{2})$

In Exercises 16–17, factor completely.

16. $24x^2 + 10x - 4$
17. $16x^4 - 1$
18. The amount of light provided by a light bulb varies inversely as the square of the distance from the bulb. The illumination provided is 120 lumens at a distance of 10 feet. How many lumens are provided at a distance of 15 feet?
19. You invested \$6000 in two accounts paying 7% and 9% annual interest. At the end of the year, the total interest from these investments was \$510. How much was invested at each rate?
20. Although there are 2332 students enrolled in the college, this is 12% fewer students than there were enrolled last year. How many students were enrolled last year?

CHAPTER 8

Quadratic Equations and Functions

We are surrounded by evidence that the world is profoundly mathematical. After turning a somersault, a diver's path can be modeled by a quadratic function, $f(x) = ax^2 + bx + c$, as can the path of a football tossed from quarterback to receiver or the path of a flipped coin. Even if you throw an object directly upward, although its path is straight and vertical, its changing height over time is described by a quadratic function. And tailgaters beware: whether you're driving a car, a motorcycle, or a truck on dry or wet roads, an array of quadratic functions that model your required stopping distances at various speeds are available to help you become a safer driver.

The quadratic functions surrounding our long history of throwing things appear throughout the chapter, including Example 6 in Section 8.3 and Example 5 in Section 8.5. Tailgaters should pay close attention to the Section 8.5 opener, Exercises 73–74 and 88–89 in Exercise Set 8.5, and Exercises 30–31 in the Chapter Review Exercises.

SECTION 8.1

The Square Root Property and Completing the Square

Objectives

1. Solve quadratic equations using the square root property.
2. Complete the square of a binomial.
3. Solve quadratic equations by completing the square.
4. Solve problems using the square root property.

"In the future there will be two kinds of corporations; those that go global, and those that go bankrupt."

C. Michael Armstrong, CEO, AT&T

For better or worse, ours is the era of the multinational corporation. New technology that the multinational corporations control is expanding their power. And their numbers are growing fast. There were approximately 300 multinationals in 1900. By 1970, there were close to 7000, and by 1990 the number had swelled to 30,000. In 2001, more than 65,000 multinational corporations enveloped the world.

In this section, you will learn two new methods for solving quadratic equations. These methods are called the *square root method* and *completing the square*. Using these techniques, you will explore the growth of multinational corporations and make predictions about the number of global corporations in the future.

Study Tip

Here is a summary of what we already know about quadratic equations and quadratic functions.

1. A **quadratic equation** in x can be written in the standard form
$$ax^2 + bx + c = 0, \quad a \neq 0.$$
2. Some quadratic equations can be solved by factoring.

Solve: $2x^2 + 7x - 4 = 0.$

$$(2x - 1)(x + 4) = 0$$
$$2x - 1 = 0 \quad \text{or} \quad x + 4 = 0$$
$$2x = 1 \qquad x = -4$$
$$x = \tfrac{1}{2}$$

The solutions are -4 and $\frac{1}{2}$, and the solution set is $\left\{-4, \frac{1}{2}\right\}$.

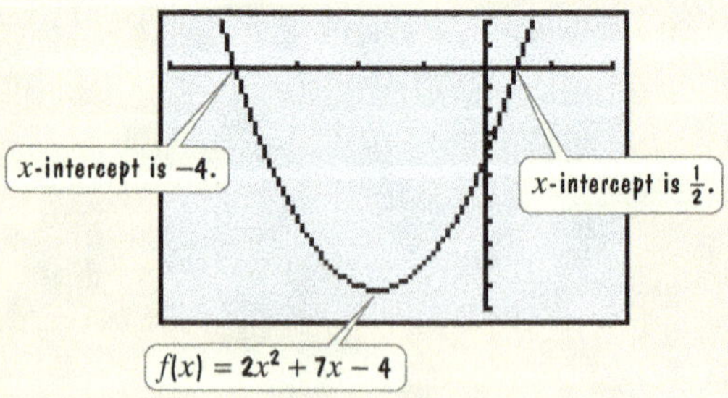

FIGURE 8.1

3. A polynomial function of the form
$$f(x) = ax^2 + bx + c, \quad a \neq 0$$
is a **quadratic function**. Graphs of quadratic functions are shaped like bowls or inverted bowls, with the same behavior at each end.
4. The real solutions of $ax^2 + bx + c = 0$ correspond to the x-intercepts for the graph of the quadratic function $f(x) = ax^2 + bx + c$. For example, the solutions of the equation $2x^2 + 7x - 4 = 0$ are -4 and $\frac{1}{2}$. **Figure 8.1** shows that the solutions appear as x-intercepts on the graph of the quadratic function $f(x) = 2x^2 + 7x - 4$.

Now that we've summarized what we know, let's look at where we go. How do we solve a quadratic equation, $ax^2 + bx + c = 0$, if the trinomial $ax^2 + bx + c$ cannot be factored? Methods other than factoring are needed. In this section, we look at other ways of solving quadratic equations.

1 Solve quadratic equations using the square root property.

The Square Root Property

Let's begin with a relatively simple quadratic equation:

$$x^2 = 9.$$

The value of x must be a number whose square is 9. There are two numbers whose square is 9:

$$x = \sqrt{9} = 3 \quad \text{or} \quad x = -\sqrt{9} = -3.$$

Thus, the solutions of $x^2 = 9$ are 3 and -3. This is an example of the **square root property**.

Discover for Yourself

Solve $x^2 = 9$, or

$$x^2 - 9 = 0,$$

by factoring. What is the advantage of using the square root property?

The Square Root Property

If u is an algebraic expression and d is a nonzero real number, then $u^2 = d$ is equivalent to $u = \sqrt{d}$ or $u = -\sqrt{d}$:

$$\text{If } u^2 = d, \quad \text{then } u = \sqrt{d} \text{ or } u = -\sqrt{d}.$$

Equivalently,

$$\text{If } u^2 = d, \quad \text{then } u = \pm\sqrt{d}.$$

Notice that $u = \pm\sqrt{d}$ is a shorthand notation to indicate that $u = \sqrt{d}$ or $u = -\sqrt{d}$. Although we usually read $u = \pm\sqrt{d}$ as "u equals plus or minus the square root of d," we actually mean that u is the positive square root of d or the negative square root of d.

EXAMPLE 1 Solving a Quadratic Equation by the Square Root Property

Solve: $3x^2 = 18$.

Solution To apply the square root property, we need a squared expression by itself on one side of the equation.

$$3x^2 = 18$$

We want x^2 by itself.

We can get x^2 by itself if we divide both sides by 3.

$$3x^2 = 18 \qquad \text{This is the original equation.}$$

$$\frac{3x^2}{3} = \frac{18}{3} \qquad \text{Divide both sides by 3.}$$

$$x^2 = 6 \qquad \text{Simplify.}$$

$$x = \sqrt{6} \quad \text{or} \quad x = -\sqrt{6} \qquad \text{Apply the square root property.}$$

Now let's check these proposed solutions in the original equation. Because the equation has an x^2-term and no x-term, we can check both values, $\pm\sqrt{6}$, at once.

Check $\sqrt{6}$ and $-\sqrt{6}$:

$$3x^2 = 18 \quad \text{This is the original equation.}$$
$$3(\pm\sqrt{6})^2 \stackrel{?}{=} 18 \quad \text{Substitute the proposed solutions.}$$
$$3 \cdot 6 \stackrel{?}{=} 18 \quad (\pm\sqrt{6})^2 = 6$$
$$18 = 18, \quad \text{true}$$

The solutions are $-\sqrt{6}$ and $\sqrt{6}$. The solution set is $\{-\sqrt{6}, \sqrt{6}\}$ or $\{\pm\sqrt{6}\}$.

CHECK POINT 1 Solve: $4x^2 = 28$.

In this section, we will express irrational solutions in simplified radical form, rationalizing denominators when possible.

EXAMPLE 2 Solving a Quadratic Equation by the Square Root Property

Solve: $2x^2 - 7 = 0$.

Solution To solve by the square root property, we isolate the squared expression on one side of the equation.

$$2x^2 - 7 = 0$$

We want x^2 by itself.

$$2x^2 - 7 = 0 \quad \text{This is the original equation.}$$
$$2x^2 = 7 \quad \text{Add 7 to both sides.}$$
$$x^2 = \frac{7}{2} \quad \text{Divide both sides by 2.}$$
$$x = \sqrt{\frac{7}{2}} \quad \text{or} \quad x = -\sqrt{\frac{7}{2}} \quad \text{Apply the square root property.}$$

Because the proposed solutions are opposites, we can rationalize both denominators at once:

$$\pm\sqrt{\frac{7}{2}} = \pm\frac{\sqrt{7}}{\sqrt{2}} \cdot \frac{\sqrt{2}}{\sqrt{2}} = \pm\frac{\sqrt{14}}{2}.$$

Substitute these values into the original equation and verify that the solutions are $-\frac{\sqrt{14}}{2}$ and $\frac{\sqrt{14}}{2}$. The solution set is $\left\{-\frac{\sqrt{14}}{2}, \frac{\sqrt{14}}{2}\right\}$ or $\left\{\pm\frac{\sqrt{14}}{2}\right\}$.

CHECK POINT 2 Solve: $3x^2 - 11 = 0$.

Some quadratic equations have solutions that are imaginary numbers.

EXAMPLE 3 Solving a Quadratic Equation by the Square Root Property

Solve: $9x^2 + 25 = 0$.

Solution We begin by isolating the squared expression on one side of the equation.

$$9x^2 + 25 = 0$$

We need to isolate x^2.

$$9x^2 + 25 = 0 \quad \text{This is the original equation.}$$

$$9x^2 = -25 \quad \text{Subtract 25 from both sides.}$$

$$x^2 = -\frac{25}{9} \quad \text{Divide both sides by 9.}$$

$$x = \sqrt{-\frac{25}{9}} \quad \text{or} \quad x = -\sqrt{-\frac{25}{9}} \quad \text{Apply the square root property.}$$

$$x = \sqrt{\frac{25}{9}}\sqrt{-1} \qquad x = -\sqrt{\frac{25}{9}}\sqrt{-1}$$

$$x = \frac{5}{3}i \qquad x = -\frac{5}{3}i \qquad \sqrt{-1} = i$$

Using Technology

Graphic Connections

The graph of

$$f(x) = 9x^2 + 25$$

has no x-intercepts. This shows that

$$9x^2 + 25 = 0$$

has no real solutions. Example 3 algebraically establishes that the solutions are imaginary numbers.

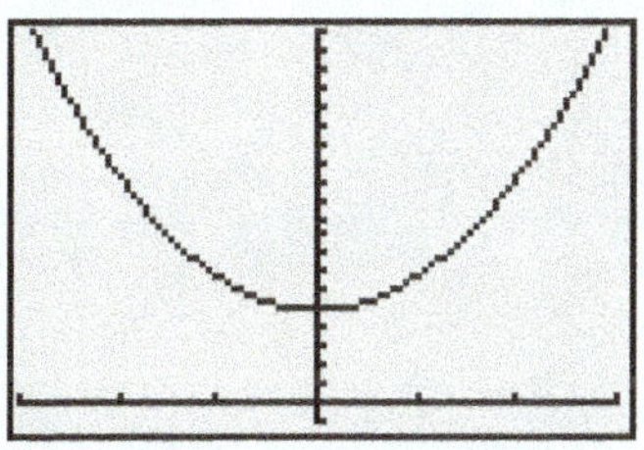

$[-3, 3, 1]$ by $[-5, 100, 5]$

Because the equation has an x^2-term and no x-term, we can check both proposed solutions, $\pm\frac{5}{3}i$, at once.

Check $\frac{5}{3}i$ and $-\frac{5}{3}i$:

$$9x^2 + 25 = 0 \quad \text{This is the original equation.}$$

$$9\left(\pm\frac{5}{3}i\right)^2 + 25 \stackrel{?}{=} 0 \quad \text{Substitute the proposed solutions.}$$

$$9\left(\frac{25}{9}i^2\right) + 25 \stackrel{?}{=} 0 \quad \left(\pm\frac{5}{3}i\right)^2 = \left(\pm\frac{5}{3}\right)^2 i^2 = \frac{25}{9}i^2$$

$$25i^2 + 25 \stackrel{?}{=} 0 \quad 9 \cdot \frac{25}{9} = 25$$

$i^2 = -1$

$$25(-1) + 25 \stackrel{?}{=} 0 \quad \text{Replace } i^2 \text{ with } -1.$$

$$0 = 0, \quad \text{true}$$

The solutions are $-\frac{5}{3}i$ and $\frac{5}{3}i$. The solution set is $\left\{-\frac{5}{3}i, \frac{5}{3}i\right\}$ or $\left\{\pm\frac{5}{3}i\right\}$.

☑ **CHECK POINT 3** Solve: $4x^2 + 9 = 0$.

Can we solve an equation such as $(x - 1)^2 = 5$ using the square root property? Yes. The equation is in the form $u^2 = d$, where u^2, the squared expression, is by itself on the left side.

$$(x - 1)^2 = 5$$

This is u^2 in $u^2 = d$ with $u = x - 1$.

This is d in $u^2 = d$ with $d = 5$.

Discover for Yourself

Try solving

$$(x - 1)^2 = 5$$

by writing the equation in standard form and factoring. What problem do you encounter?

EXAMPLE 4 Solving a Quadratic Equation by the Square Root Property

Solve by the square root property: $(x - 1)^2 = 5$.

Solution

$$(x - 1)^2 = 5 \quad \text{This is the original equation.}$$

$$x - 1 = \sqrt{5} \quad \text{or} \quad x - 1 = -\sqrt{5} \quad \text{Apply the square root property.}$$

$$x = 1 + \sqrt{5} \qquad x = 1 - \sqrt{5} \quad \text{Add 1 to both sides in each equation.}$$

Check $1 + \sqrt{5}$:

$$(x - 1)^2 = 5$$
$$(1 + \sqrt{5} - 1)^2 \stackrel{?}{=} 5$$
$$(\sqrt{5})^2 \stackrel{?}{=} 5$$
$$5 = 5, \quad \text{true}$$

Check $1 - \sqrt{5}$:

$$(x - 1)^2 = 5$$
$$(1 - \sqrt{5} - 1)^2 \stackrel{?}{=} 5$$
$$(-\sqrt{5})^2 \stackrel{?}{=} 5$$
$$5 = 5, \quad \text{true}$$

The solutions are $1 \pm \sqrt{5}$, and the solution set is $\{1 + \sqrt{5}, 1 - \sqrt{5}\}$ or $\{1 \pm \sqrt{5}\}$.

CHECK POINT 4 Solve: $(x - 3)^2 = 10$.

2 Complete the square of a binomial.

Completing the Square

We return to the question that opened this section: How do we solve a quadratic equation, $ax^2 + bx + c = 0$, if the trinomial $ax^2 + bx + c$ cannot be factored? We can convert the equation into an equivalent equation that can be solved using the square root property. This is accomplished by **completing the square**.

Completing the Square

If $x^2 + bx$ is a binomial, then by adding $\left(\frac{b}{2}\right)^2$, which is the square of half the coefficient of x, a perfect square trinomial will result.

The coefficient of x^2 must be 1 to complete the square.

$$x^2 + bx + \left(\frac{b}{2}\right)^2 = \left(x + \frac{b}{2}\right)^2.$$

EXAMPLE 5 Completing the Square

What term should be added to each binomial so that it becomes a perfect square trinomial? Write and factor the trinomial.

a. $x^2 + 8x$ **b.** $x^2 - 7x$ **c.** $x^2 + \frac{3}{5}x$

Solution To complete the square, we must add a term to each binomial. The term that should be added is the square of half the coefficient of x.

$x^2 + 8x$: Add $\left(\frac{8}{2}\right)^2 = 4^2$. Add 16 to complete the square.

$x^2 - 7x$: Add $\left(\frac{-7}{2}\right)^2$, or $\frac{49}{4}$, to complete the square.

$x^2 + \frac{3}{5}x$: Add $\left(\frac{1}{2} \cdot \frac{3}{5}\right)^2 = \left(\frac{3}{10}\right)^2$. Add $\frac{9}{100}$ to complete the square.

a. The coefficient of the x-term in $x^2 + 8x$ is 8. Half of 8 is 4, and $4^2 = 16$. Add 16. The result is a perfect square trinomial.

$$x^2 + 8x + 16 = (x + 4)^2$$

b. The coefficient of the x-term in $x^2 - 7x$ is -7. Half of -7 is $-\frac{7}{2}$, and $\left(-\frac{7}{2}\right)^2 = \frac{49}{4}$. Add $\frac{49}{4}$. The result is a perfect square trinomial.

$$x^2 - 7x + \frac{49}{4} = \left(x - \frac{7}{2}\right)^2$$

c. The coefficient of the x-term in $x^2 + \frac{3}{5}x$ is $\frac{3}{5}$. Half of $\frac{3}{5}$ is $\frac{1}{2}\cdot\frac{3}{5}$, or $\frac{3}{10}$, and $\left(\frac{3}{10}\right)^2 = \frac{9}{100}$. Add $\frac{9}{100}$. The result is a perfect square trinomial.

$$x^2 + \frac{3}{5}x + \frac{9}{100} = \left(x + \frac{3}{10}\right)^2$$

Study Tip

You may not be accustomed to factoring perfect square trinomials in which fractions are involved. The constant in the factorization is always half the coefficient of x.

$$x^2 - 7x + \frac{49}{4} = \left(x - \frac{7}{2}\right)^2$$

Half the coefficient of x, -7, is $-\frac{7}{2}$.

$$x^2 + \frac{3}{5}x + \frac{9}{100} = \left(x + \frac{3}{10}\right)^2$$

Half the coefficient of x, $\frac{3}{5}$, is $\frac{3}{10}$.

✓ **CHECK POINT 5** What term should be added to each binomial so that it becomes a perfect square trinomial? Write and factor the trinomial.

a. $x^2 + 10x$

b. $x^2 - 3x$

c. $x^2 + \frac{3}{4}x$.

3 Solve quadratic equations by completing the square.

Solving Quadratic Equations by Completing the Square

We can solve *any* quadratic equation by completing the square. If the coefficient of the x^2-term is one, we add the square of half the coefficient of x to both sides of the equation. **When you add a constant term to one side of the equation to complete the square, be certain to add the same constant to the other side of the equation.** These ideas are illustrated in Example 6.

EXAMPLE 6 Solving a Quadratic Equation by Completing the Square

Solve by completing the square: $x^2 - 6x + 4 = 0$.

Solution We begin by subtracting 4 from both sides. This is done to isolate the binomial $x^2 - 6x$ so that we can complete the square.

$$x^2 - 6x + 4 = 0 \quad \text{This is the original equation.}$$
$$x^2 - 6x = -4 \quad \text{Subtract 4 from both sides.}$$

Next, we work with $x^2 - 6x = -4$ and complete the square. Find half the coefficient of the x-term and square it. The coefficient of the x-term is -6. Half of -6 is -3 and $(-3)^2 = 9$. Thus, we add 9 to both sides of the equation.

$$x^2 - 6x + 9 = -4 + 9 \quad \text{Add 9 to both sides to complete the square.}$$
$$(x-3)^2 = 5 \quad \text{Factor and simplify.}$$
$$x - 3 = \sqrt{5} \quad \text{or} \quad x - 3 = -\sqrt{5} \quad \text{Apply the square root property.}$$
$$x = 3 + \sqrt{5} \qquad x = 3 - \sqrt{5} \quad \text{Add 3 to both sides in each equation.}$$

The solutions are $3 \pm \sqrt{5}$, and the solution set is $\{3 + \sqrt{5}, 3 - \sqrt{5}\}$ or $\{3 \pm \sqrt{5}\}$.

Study Tip

When you complete the square for the binomial expression $x^2 + bx$, you obtain a different polynomial. When you solve a quadratic equation by completing the square, you obtain an equation with the same solution set because the constant needed to complete the square is added to *both sides*.

If you solve a quadratic equation by completing the square and the solutions are rational numbers, the equation can also be solved by factoring. By contrast, quadratic equations with irrational solutions cannot be solved by factoring. However, all quadratic equations can be solved by completing the square.

✓ CHECK POINT 6 Solve by completing the square: $x^2 + 4x - 1 = 0$.

We have seen that the leading coefficient must be 1 in order to complete the square. If the coefficient of the x^2-term in a quadratic equation is not 1, you must divide each side of the equation by this coefficient before completing the square. For example, to solve $9x^2 - 6x - 4 = 0$ by completing the square, first divide every term by 9:

$$\frac{9x^2}{9} - \frac{6x}{9} - \frac{4}{9} = \frac{0}{9}$$

$$x^2 - \frac{6}{9}x - \frac{4}{9} = 0$$

$$x^2 - \frac{2}{3}x - \frac{4}{9} = 0.$$

Now that the coefficient of the x^2-term is 1, we can solve by completing the square.

EXAMPLE 7 Solving a Quadratic Equation by Completing the Square

Solve by completing the square: $9x^2 - 6x - 4 = 0$.

Solution

$$9x^2 - 6x - 4 = 0$$
This is the original equation.

$$x^2 - \frac{2}{3}x - \frac{4}{9} = 0$$
Divide both sides by 9.

$$x^2 - \frac{2}{3}x = \frac{4}{9}$$
Add $\frac{4}{9}$ to both sides to isolate the binomial.

$$x^2 - \frac{2}{3}x + \frac{1}{9} = \frac{4}{9} + \frac{1}{9}$$
Complete the square: Half of $-\frac{2}{3}$ is $-\frac{2}{6}$, or $-\frac{1}{3}$, and $\left(-\frac{1}{3}\right)^2 = \frac{1}{9}$.

$$\left(x - \frac{1}{3}\right)^2 = \frac{5}{9}$$
Factor and simplify.

$$x - \frac{1}{3} = \sqrt{\frac{5}{9}} \quad \text{or} \quad x - \frac{1}{3} = -\sqrt{\frac{5}{9}}$$
Apply the square root property.

$$x - \frac{1}{3} = \frac{\sqrt{5}}{3} \qquad x - \frac{1}{3} = -\frac{\sqrt{5}}{3}$$
$\sqrt{\frac{5}{9}} = \frac{\sqrt{5}}{\sqrt{9}} = \frac{\sqrt{5}}{3}$

$$x = \frac{1}{3} + \frac{\sqrt{5}}{3} \qquad x = \frac{1}{3} - \frac{\sqrt{5}}{3}$$
Add $\frac{1}{3}$ to both sides and solve for x.

$$x = \frac{1 + \sqrt{5}}{3} \qquad x = \frac{1 - \sqrt{5}}{3}$$
Express solutions with a common denominator.

The solutions are $\frac{1 \pm \sqrt{5}}{3}$ and the solution set is $\left\{\frac{1 \pm \sqrt{5}}{3}\right\}$.

Using Technology

Graphic Connections

Obtain a decimal approximation for each solution of

$$9x^2 - 6x - 4 = 0,$$

the equation in Example 7.

$$\frac{1 + \sqrt{5}}{3} \approx 1.1$$

$$\frac{1 - \sqrt{5}}{3} \approx -0.4$$

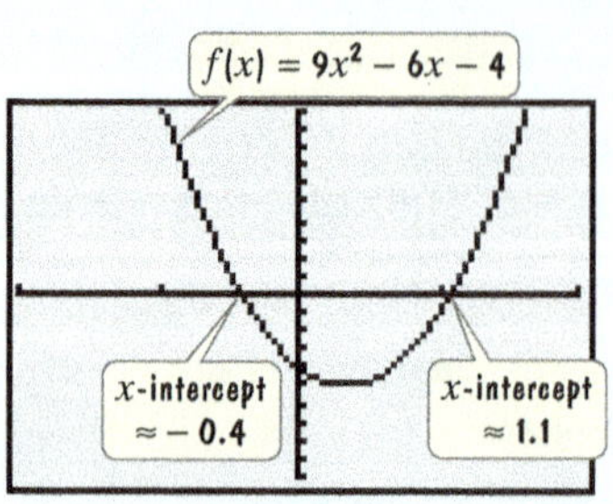

[–2, 2, 1] by [–10, 10, 1]

The x-intercepts of $f(x) = 9x^2 - 6x - 4$ verify the solutions.

✓ CHECK POINT 7 Solve by completing the square: $2x^2 + 3x - 4 = 0$.

4 Solve problems using the square root property.

Applications

We all want a wonderful life with fulfilling work, good health, and loving relationships. And let's be honest: Financial security wouldn't hurt! Achieving this goal depends on understanding how money in a savings account grows in remarkable ways as a result of *compound interest.* **Compound interest** is interest computed on your original investment as well as on any accumulated interest. For example, suppose you deposit \$1000, the principal, in a savings account at a rate of 5%. **Table 8.1** shows how the investment grows if the interest earned is automatically added on to the principal.

Table 8.1 Compound Interest on \$1000

Year	Starting Balance	Interest Earned: $I = Pr$	New Balance
1	\$1000	$\$1000 \times 0.05 = \$50$	\$1050
2	\$1050	$\$1050 \times 0.05 = \$52.50$	\$1102.50
3	\$1102.50	$\$1102.50 \times 0.05 \approx \$55.13$	\$1157.63

A faster way to determine the amount, A, in an account subject to compound interest is to use the following formula:

A Formula for Compound Interest

Suppose that an amount of money, P, is invested at interest rate r, compounded annually. In t years, the amount, A, or balance, in the account is given by the formula

$$A = P(1 + r)^t.$$

Some compound interest problems can be solved using quadratic equations.

EXAMPLE 8 Solving a Compound Interest Problem

You invested \$1000 in an account whose interest is compounded annually. After 2 years, the amount, or balance, in the account is \$1210. Find the annual interest rate.

Solution We are given that

P (the amount invested) = \$1000

t (the time of the investment) = 2 years

A (the amount, or balance, in the account) = \$1210.

We are asked to find the annual interest rate, r. We substitute the three given values into the compound interest formula and solve for r.

$$A = P(1 + r)^t \quad \text{Use the compound interest formula.}$$

$$1210 = 1000(1 + r)^2 \quad \text{Substitute the given values.}$$

$$\frac{1210}{1000} = (1 + r)^2 \quad \text{Divide both sides by 1000.}$$

$$\frac{121}{100} = (1 + r)^2 \quad \text{Simplify the fraction.}$$

$$1 + r = \sqrt{\frac{121}{100}} \quad \text{or} \quad 1 + r = -\sqrt{\frac{121}{100}} \quad \text{Apply the square root property.}$$

$$1 + r = \frac{11}{10} \qquad 1 + r = -\frac{11}{10}$$ $\sqrt{\frac{121}{100}} = \frac{\sqrt{121}}{\sqrt{100}} = \frac{11}{10}$

$$r = \frac{11}{10} - 1 \qquad r = -\frac{11}{10} - 1$$ Subtract 1 from both sides and solve for r.

$$r = \frac{1}{10} \qquad r = -\frac{21}{10}$$ $\frac{11}{10} - 1 = \frac{11}{10} - \frac{10}{10} = \frac{1}{10}$ and $-\frac{11}{10} - 1 = -\frac{11}{10} - \frac{10}{10} = -\frac{21}{10}$.

Because the interest rate cannot be negative, we reject $-\frac{21}{10}$. Thus, the annual interest rate is $\frac{1}{10} = 0.10 = 10\%$.

We can check this answer using the formula $A = P(1 + r)^t$. If \$1000 is invested for 2 years at 10% interest, compounded annually, the balance in the account is

$$A = \$1000(1 + 0.10)^2 = \$1000(1.10)^2 = \$1210.$$

Because this is precisely the amount given by the problem's conditions, the annual interest rate is, indeed, 10% compounded annually.

✓ **CHECK POINT 8** You invested \$3000 in an account whose interest is compounded annually. After 2 years, the amount, or balance, in the account is \$4320. Find the annual interest rate.

In Chapter 5, we solved problems using the Pythagorean Theorem. Recall that in a right triangle, the side opposite the 90° angle is the hypotenuse and the other sides are legs. The Pythagorean Theorem states that the sum of the squares of the lengths of the legs equals the square of the length of the hypotenuse. Some problems that involve the Pythagorean Theorem can be solved using the square root property.

EXAMPLE 9 Using the Pythagorean Theorem and the Square Root Property

a. A wheelchair ramp with a length of 122 inches has a horizontal distance of 120 inches. What is the ramp's vertical distance?

b. Construction laws are very specific when it comes to access ramps for the disabled. Every vertical rise of 1 inch requires a horizontal run of 12 inches. Does this ramp satisfy the requirement?

Solution

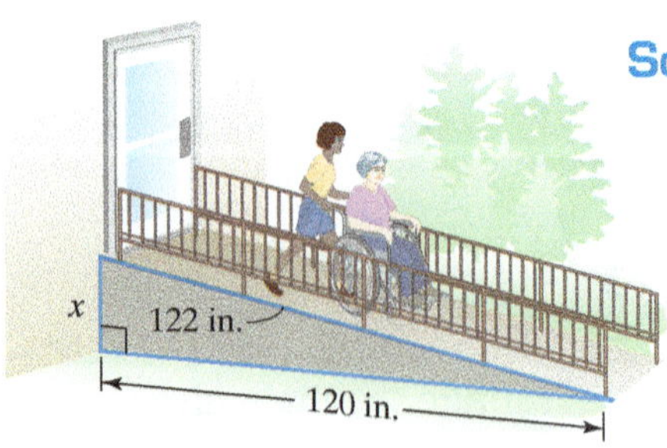

FIGURE 8.2

a. **Figure 8.2** shows the right triangle that is formed by the ramp, the wall, and the ground. We can find x, the ramp's vertical distance, using the Pythagorean Theorem.

(leg)^2	plus	(leg)^2	equals	(hypotenuse)^2
x^2	$+$	120^2	$=$	122^2

We solve this equation using the square root property.

$$x^2 + 120^2 = 122^2$$ This is the equation resulting from the Pythagorean Theorem.

$$x^2 + 14{,}400 = 14{,}884$$ Square 120 and 122.

$$x^2 = 484$$ Isolate x^2 by subtracting 14,400 from both sides.

$$x = \sqrt{484} \quad \text{or} \quad x = -\sqrt{484}$$ Apply the square root property.

$$x = 22 \qquad x = -22$$

Because x represents the ramp's vertical distance, we reject the negative value. Thus, the ramp's vertical distance is 22 inches.

b. Every vertical rise of 1 inch requires a horizontal run of 12 inches. Because the ramp has a vertical distance of 22 inches, it requires a horizontal distance of 22(12) inches, or 264 inches. The horizontal distance is only 120 inches, so this ramp does not satisfy construction laws for access ramps for the disabled.

CHECK POINT 9 A 50-foot supporting wire is to be attached to an antenna. The wire is anchored 20 feet from the base of the antenna. How high up the antenna is the wire attached? Express the answer in simplified radical form. Then find a decimal approximation to the nearest tenth of a foot.

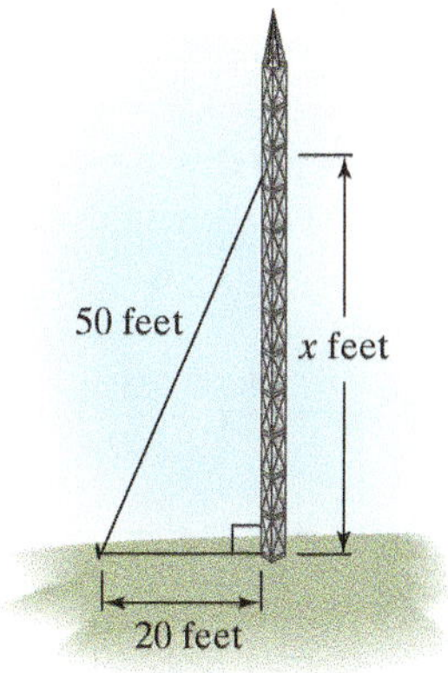

8.1 EXERCISE SET

MyMathLab — PRACTICE | WATCH | DOWNLOAD | READ | REVIEW

Practice Exercises

In Exercises 1–22, solve each equation by the square root property. If possible, simplify radicals or rationalize denominators. Express imaginary solutions in the form $a + bi$.

1. $3x^2 = 75$

2. $5x^2 = 20$

3. $7x^2 = 42$

4. $8x^2 = 40$

5. $16x^2 = 25$

6. $4x^2 = 49$

7. $3x^2 - 2 = 0$

8. $3x^2 - 5 = 0$

9. $25x^2 + 16 = 0$

10. $4x^2 + 49 = 0$

11. $(x + 7)^2 = 9$

12. $(x + 3)^2 = 64$

13. $(x - 3)^2 = 5$

14. $(x - 4)^2 = 3$

15. $2(x + 2)^2 = 16$

16. $3(x + 2)^2 = 36$

17. $(x - 5)^2 = -9$

18. $(x - 5)^2 = -4$

19. $\left(x + \frac{3}{4}\right)^2 = \frac{11}{16}$

20. $\left(x + \frac{2}{5}\right)^2 = \frac{7}{25}$

21. $x^2 - 6x + 9 = 36$

22. $x^2 - 6x + 9 = 49$

In Exercises 23–34, determine the constant that should be added to the binomial so that it becomes a perfect square trinomial. Then write and factor the trinomial.

23. $x^2 + 2x$

24. $x^2 + 4x$

25. $x^2 - 14x$

26. $x^2 - 10x$

27. $x^2 + 7x$

28. $x^2 + 9x$

29. $x^2 - \frac{1}{2}x$

30. $x^2 - \frac{1}{3}x$

31. $x^2 + \frac{4}{3}x$

32. $x^2 + \frac{4}{5}x$

33. $x^2 - \frac{9}{4}x$

34. $x^2 - \frac{9}{5}x$

In Exercises 35–56, solve each quadratic equation by completing the square.

35. $x^2 + 4x = 32$

36. $x^2 + 6x = 7$

37. $x^2 + 6x = -2$

38. $x^2 + 2x = 5$

39. $x^2 - 8x + 1 = 0$

40. $x^2 + 8x - 5 = 0$

41. $x^2 + 2x + 2 = 0$

42. $x^2 - 4x + 8 = 0$

43. $x^2 + 3x - 1 = 0$

44. $x^2 - 3x - 5 = 0$

45. $x^2 + \frac{4}{7}x + \frac{3}{49} = 0$

46. $x^2 + \frac{6}{5}x + \frac{8}{25} = 0$

47. $x^2 + x - 1 = 0$

48. $x^2 - 7x + 3 = 0$

49. $2x^2 + 3x - 5 = 0$

50. $2x^2 + 5x - 3 = 0$

51. $3x^2 + 6x + 1 = 0$

52. $3x^2 - 6x + 2 = 0$

53. $3x^2 - 8x + 1 = 0$

54. $2x^2 + 3x - 4 = 0$

55. $8x^2 - 4x + 1 = 0$

56. $9x^2 - 6x + 5 = 0$

57. If $f(x) = (x - 1)^2$, find all values of x for which $f(x) = 36$.

58. If $f(x) = (x + 2)^2$, find all values of x for which $f(x) = 25$.

59. If $g(x) = \left(x - \frac{2}{5}\right)^2$, find all values of x for which $g(x) = \frac{9}{25}$.

60. If $g(x) = \left(x + \frac{1}{3}\right)^2$, find all values of x for which $g(x) = \frac{4}{9}$.

61. If $h(x) = 5(x + 2)^2$, find all values of x for which $h(x) = -125$.

62. If $h(x) = 3(x - 4)^2$, find all values of x for which $h(x) = -12$.

Practice PLUS

63. Three times the square of the difference between a number and 2 is -12. Find the number(s).

64. Three times the square of the difference between a number and 9 is -27. Find the number(s).

In Exercises 65–68, solve the formula for the specified variable. Because each variable is nonnegative, list only the principal square root. If possible, simplify radicals or rationalize denominators.

65. $h = \frac{v^2}{2g}$ for v

66. $s = \frac{kwd^2}{l}$ for d

67. $A = P(1 + r)^2$ for r

68. $C = \frac{kP_1P_2}{d^2}$ for d

In Exercises 69–72, solve each quadratic equation by completing the square.

69. $\frac{x^2}{3} + \frac{x}{9} - \frac{1}{6} = 0$

70. $\frac{x^2}{2} - \frac{x}{6} - \frac{3}{4} = 0$

71. $x^2 - bx = 2b^2$

72. $x^2 - bx = 6b^2$

73. The ancient Greeks used a geometric method for completing the square in which they literally transformed a figure into a square.

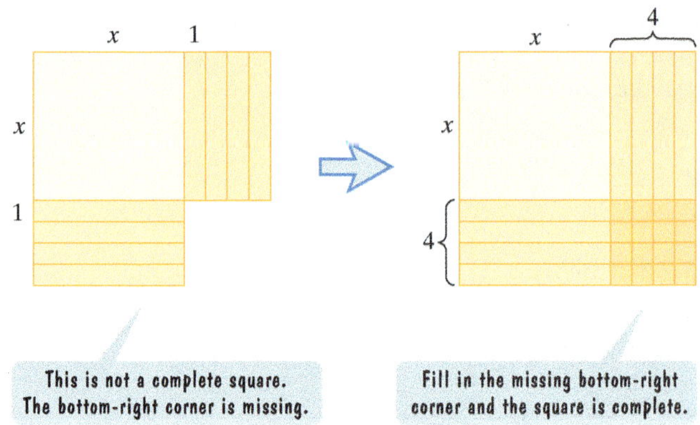

a. Write a binomial in x that represents the combined area of the small square and the eight rectangular stripes that make up the incomplete square on the left.

b. What is the area of the region in the bottom-right corner that literally completes the square?

c. Write a trinomial in x that represents the combined area of the small square, the eight rectangular stripes, and the bottom-right corner that make up the complete square on the right.

d. Use the length of each side of the complete square on the right to express its area as the square of a binomial.

74. An **isosceles right triangle** has legs that are the same length and acute angles each measuring 45°.

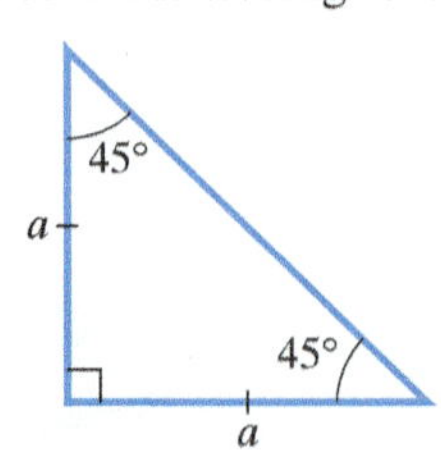

a. Write an expression in terms of a that represents the length of the hypotenuse.
b. Use your result from part (a) to write a sentence that describes the length of the hypotenuse of an isosceles right triangle in terms of the length of a leg.

Application Exercises

In Exercises 75–78, use the compound interest formula

$$A = P(1 + r)^t$$

to find the annual interest rate, r.

75. In 2 years, an investment of \$2000 grows to \$2880.

76. In 2 years, an investment of \$2000 grows to \$2420.

77. In 2 years, an investment of \$1280 grows to \$1445.

78. In 2 years, an investment of \$80,000 grows to \$101,250.

Of the one hundred largest economies in the world, 53 are multinational corporations. In 1970, there were approximately 7000 multinational corporations. By 2001, more than 65,000 corporations enveloped the world. The graph shows this rapid growth from 1970 through 2001, including the starting dates of some notable corporations.

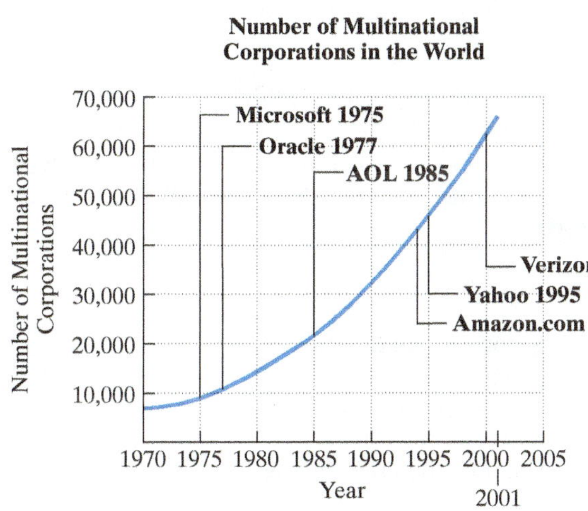

Source: Medard Gabel, *Global Inc*, The New Press, 2003

The data shown can be modeled by the function

$$f(x) = 62.2x^2 + 7000,$$

where $f(x)$ represents the number of multinational corporations in the world x years after 1970. Use this function and the square root property to solve Exercises 79–80.

79. In which year were there 32,000 multinational corporations? How well does the function model the actual number of corporations for that year shown in the graph?

80. In which year were there 46,000 multinational corporations? How well does the function model the actual number of corporations for that year shown in the graph?

The function $s(t) = 16t^2$ models the distance, $s(t)$, in feet, that an object falls in t seconds. Use this function and the square root property to solve Exercises 81–82. Express answers in simplified radical form. Then use your calculator to find a decimal approximation to the nearest tenth of a second.

81. A sky diver jumps from an airplane and falls for 4800 feet before opening a parachute. For how many seconds was the diver in a free fall?

82. A sky diver jumps from an airplane and falls for 3200 feet before opening a parachute. For how many seconds was the diver in a free fall?

Use the Pythagorean Theorem and the square root property to solve Exercises 83–88. Express answers in simplified radical form. Then find a decimal approximation to the nearest tenth.

83. A rectangular park is 6 miles long and 3 miles wide. How long is a pedestrian route that runs diagonally across the park?

84. A rectangular park is 4 miles long and 2 miles wide. How long is a pedestrian route that runs diagonally across the park?

85. The base of a 30-foot ladder is 10 feet from the building. If the ladder reaches the flat roof, how tall is the building?

86. The doorway into a room is 4 feet wide and 8 feet high. What is the length of the longest rectangular panel that can be taken through this doorway diagonally?

87. A supporting wire is to be attached to the top of a 50-foot antenna. If the wire must be anchored 50 feet from the base of the antenna, what length of wire is required?

88. A supporting wire is to be attached to the top of a 70-foot antenna. If the wire must be anchored 70 feet from the base of the antenna, what length of wire is required?

89. A square flower bed is to be enlarged by adding 2 meters on each side. If the larger square has an area of 196 square meters, what is the length of a side of the original square?

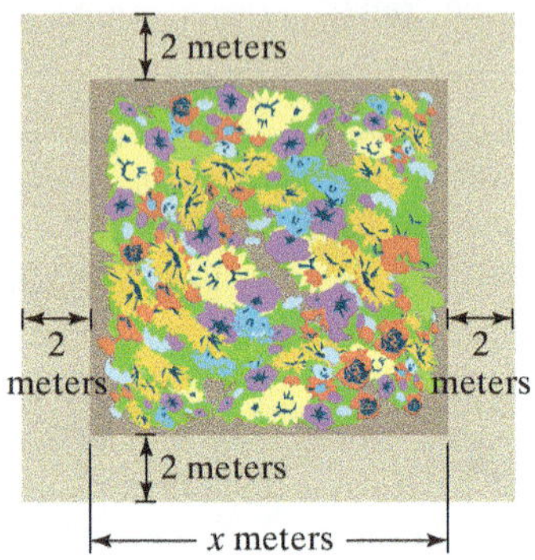

90. A square flower bed is to be enlarged by adding 4 feet on each side. If the larger square has an area of 225 square feet, what is the length of a side of the original square?

Writing in Mathematics

91. What is the square root property?

92. Explain how to solve $(x - 1)^2 = 16$ using the square root property.

93. Explain how to complete the square for a binomial. Use $x^2 + 6x$ to illustrate your explanation.

94. Explain how to solve $x^2 + 6x + 8 = 0$ by completing the square.

95. What is compound interest?

96. In your own words, describe the compound interest formula

$$A = P(1 + r)^t.$$

Technology Exercises

97. Use a graphing utility to solve $4 - (x + 1)^2 = 0$. Graph $y = 4 - (x + 1)^2$ in a $[-5, 5, 1]$ by $[-5, 5, 1]$ viewing rectangle. The equation's solutions are the graph's x-intercepts. Check by substitution in the given equation.

98. Use a graphing utility to solve $(x - 1)^2 - 9 = 0$. Graph $y = (x - 1)^2 - 9$ in a $[-5, 5, 1]$ by $[-9, 3, 1]$ viewing rectangle. The equation's solutions are the graph's x-intercepts. Check by substitution in the given equation.

99. Use a graphing utility and x-intercepts to verify any of the real solutions that you obtained for five of the quadratic equations in Exercises 35–56.

Critical Thinking Exercises

Make Sense? *In Exercises 100–103, determine whether each statement "makes sense" or "does not make sense" and explain your reasoning.*

100. When the coefficient of the x-term in a quadratic equation is negative and I'm solving by completing the square, I add a negative constant to each side of the equation.

101. When I complete the square for the binomial $x^2 + bx$, I obtain a different polynomial, but when I solve a quadratic equation by completing the square, I obtain an equation with the same solution set.

102. When I use the square root property to determine the length of a right triangle's side, I don't even bother to list the negative square root.

103. When I solved $4x^2 + 10x = 0$ by completing the square, I added 25 to both sides of the equation.

In Exercises 104–107, determine whether each statement is true or false. If the statement is false, make the necessary change(s) to produce a true statement.

104. The graph of $y = (x - 2)^2 + 3$ cannot have x-intercepts.

105. The equation $(x - 5)^2 = 12$ is equivalent to $x - 5 = 2\sqrt{3}$.

106. In completing the square for $2x^2 - 6x = 5$, we should add 9 to both sides.

107. Although not every quadratic equation can be solved by completing the square, they can all be solved by factoring.

108. Solve for y: $\dfrac{x^2}{a^2} + \dfrac{y^2}{b^2} = 1$.

109. Solve by completing the square:

$$x^2 + x + c = 0.$$

110. Solve by completing the square:

$$x^2 + bx + c = 0.$$

111. Solve: $x^4 - 8x^2 + 15 = 0$.

Review Exercises

112. Simplify: $4x - 2 - 3[4 - 2(3 - x)]$. (Section 1.2, Example 14)

113. Factor: $1 - 8x^3$. (Section 5.5, Example 9)

114. Divide: $(x^4 - 5x^3 + 2x^2 - 6) \div (x - 3)$. (Section 6.5, Example 1)

Preview Exercises

Exercises 115–117 will help you prepare for the material covered in the next section.

115. **a.** Solve by factoring: $8x^2 + 2x - 1 = 0$.

b. The quadratic equation in part (a) is in the standard form $ax^2 + bx + c = 0$. Compute $b^2 - 4ac$. Is $b^2 - 4ac$ a perfect square?

116. **a.** Solve by factoring: $9x^2 - 6x + 1 = 0$.

b. The quadratic equation in part (a) is in the standard form $ax^2 + bx + c = 0$. Compute $b^2 - 4ac$.

117. **a.** Clear fractions in the following equation and write in the form $ax^2 + bx + c = 0$:

$$3 + \frac{4}{x} = -\frac{2}{x^2}.$$

b. For the equation you wrote in part (a), compute $b^2 - 4ac$.

SECTION 8.2

The Quadratic Formula

Objectives

1. Solve quadratic equations using the quadratic formula.
2. Use the discriminant to determine the number and type of solutions.
3. Determine the most efficient method to use when solving a quadratic equation.
4. Write quadratic equations from solutions.
5. Use the quadratic formula to solve problems.

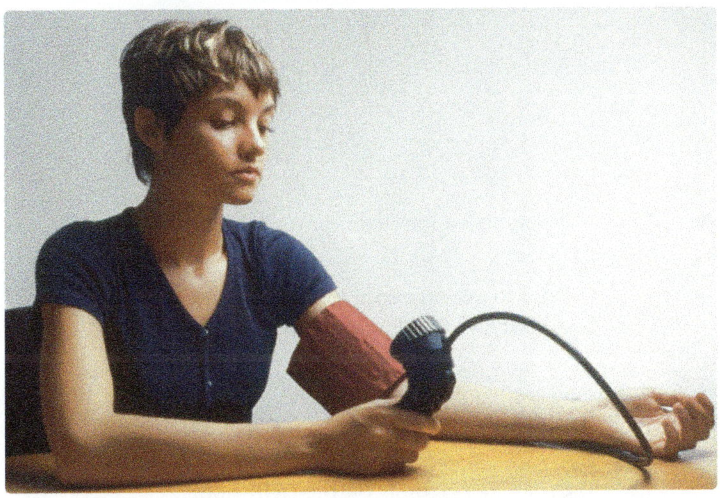

Until fairly recently, many doctors believed that your blood pressure was theirs to know and yours to worry about. Today, however, people are encouraged to find out their blood pressure. That pumped-up cuff that squeezes against your upper arm measures blood pressure in millimeters (mm) of mercury (Hg). Blood pressure is given in two numbers: systolic pressure over diastolic pressure, such as 120 over 80. Systolic pressure is the pressure of blood against the artery walls when the heart contracts. Diastolic pressure is the pressure of blood against the artery walls when the heart is at rest.

In this section, we will derive a formula that will enable you to solve quadratic equations more quickly than the method of completing the square. Using this formula, we will work with functions that model changing systolic pressure for men and women with age.

1 Solve quadratic equations using the quadratic formula.

Solving Quadratic Equations Using the Quadratic Formula

We can use the method of completing the square to derive a formula that can be used to solve all quadratic equations. The derivation given on the next page also shows a particular quadratic equation, $3x^2 - 2x - 4 = 0$, to specifically illustrate each of the steps.

Deriving the Quadratic Formula

Standard Form of a Quadratic Equation	Comment	A Specific Example
$ax^2 + bx + c = 0, a > 0$	This is the given equation.	$3x^2 - 2x - 4 = 0$
$x^2 + \frac{b}{a}x + \frac{c}{a} = 0$	Divide both sides by the coefficient of x^2.	$x^2 - \frac{2}{3}x - \frac{4}{3} = 0$
$x^2 + \frac{b}{a}x = -\frac{c}{a}$	Isolate the binomial by adding $-\frac{c}{a}$ on both sides.	$x^2 - \frac{2}{3}x = \frac{4}{3}$
$x^2 + \frac{b}{a}x + \left(\frac{b}{2a}\right)^2 = -\frac{c}{a} + \left(\frac{b}{2a}\right)^2$ $(\text{half})^2$ $x^2 + \frac{b}{a}x + \frac{b^2}{4a^2} = -\frac{c}{a} + \frac{b^2}{4a^2}$	Complete the square. Add the square of half the coefficient of x to both sides.	$x^2 - \frac{2}{3}x + \left(-\frac{1}{3}\right)^2 = \frac{4}{3} + \left(-\frac{1}{3}\right)^2$ $(\text{half})^2$ $x^2 - \frac{2}{3}x + \frac{1}{9} = \frac{4}{3} + \frac{1}{9}$
$\left(x + \frac{b}{2a}\right)^2 = -\frac{c}{a} \cdot \frac{4a}{4a} + \frac{b^2}{4a^2}$	Factor on the left side and obtain a common denominator on the right side.	$\left(x - \frac{1}{3}\right)^2 = \frac{4}{3} \cdot \frac{3}{3} + \frac{1}{9}$
$\left(x + \frac{b}{2a}\right)^2 = \frac{-4ac + b^2}{4a^2}$ $\left(x + \frac{b}{2a}\right)^2 = \frac{b^2 - 4ac}{4a^2}$	Add fractions on the right side.	$\left(x - \frac{1}{3}\right)^2 = \frac{12 + 1}{9}$ $\left(x - \frac{1}{3}\right)^2 = \frac{13}{9}$
$x + \frac{b}{2a} = \pm\sqrt{\frac{b^2 - 4ac}{4a^2}}$	Apply the square root property.	$x - \frac{1}{3} = \pm\sqrt{\frac{13}{9}}$
$x + \frac{b}{2a} = \pm\frac{\sqrt{b^2 - 4ac}}{2a}$	Take the square root of the quotient, simplifying the denominator.	$x - \frac{1}{3} = \pm\frac{\sqrt{13}}{3}$
$x = \frac{-b}{2a} \pm \frac{\sqrt{b^2 - 4ac}}{2a}$	Solve for x by subtracting $\frac{b}{2a}$ from both sides.	$x = \frac{1}{3} \pm \frac{\sqrt{13}}{3}$
$x = \frac{-b \pm \sqrt{b^2 - 4ac}}{2a}$	Combine fractions on the right side.	$x = \frac{1 \pm \sqrt{13}}{3}$

The formula shown at the bottom of the left column is called the *quadratic formula.* A similar proof shows that the same formula can be used to solve quadratic equations if a, the coefficient of the x^2-term, is negative.

The Quadratic Formula

The solutions of a quadratic equation in standard form $ax^2 + bx + c = 0$, with $a \neq 0$, are given by the **quadratic formula**:

$$x = \frac{-b \pm \sqrt{b^2 - 4ac}}{2a}.$$

x equals negative *b* plus or minus the square root of $b^2 - 4ac$, all divided by $2a$.

To use the quadratic formula, write the quadratic equation in standard form if necessary. Then determine the numerical values for a (the coefficient of the x^2-term), b (the coefficient of the x-term), and c (the constant term). Substitute the values of a, b, and c into the quadratic formula and evaluate the expression. The $\pm$ sign indicates that there are two (not necessarily distinct) solutions of the equation.

EXAMPLE 1 Solving a Quadratic Equation Using the Quadratic Formula

Solve using the quadratic formula: $8x^2 + 2x - 1 = 0.$

Solution The given equation is in standard form. Begin by identifying the values for a, b, and c.

$$8x^2 + 2x - 1 = 0$$

$a = 8$ $b = 2$ $c = -1$

Substituting these values into the quadratic formula and simplifying gives the equation's solutions.

$$x = \frac{-b \pm \sqrt{b^2 - 4ac}}{2a}$$ Use the quadratic formula.

$$x = \frac{-2 \pm \sqrt{2^2 - 4(8)(-1)}}{2(8)}$$ Substitute the values for a, b, and c: $a = 8$, $b = 2$, and $c = -1$.

$$= \frac{-2 \pm \sqrt{4 - (-32)}}{16}$$ $2^2 - 4(8)(-1) = 4 - (-32)$

$$= \frac{-2 \pm \sqrt{36}}{16}$$ $4 - (-32) = 4 + 32 = 36$

$$= \frac{-2 \pm 6}{16}$$ $\sqrt{36} = 6$

Now we will evaluate this expression in two different ways to obtain the two solutions. On the left, we will *add* 6 to −2. On the right, we will *subtract* 6 from −2.

$$x = \frac{-2 + 6}{16} \quad \text{or} \quad x = \frac{-2 - 6}{16}$$

$$= \frac{4}{16} = \frac{1}{4} \qquad = \frac{-8}{16} = -\frac{1}{2}$$

The solutions are $-\frac{1}{2}$ and $\frac{1}{4}$, and the solution set is $\left\{-\frac{1}{2}, \frac{1}{4}\right\}$.

Using Technology

Graphic Connections

The graph of the quadratic function

$$y = 8x^2 + 2x - 1$$

has x-intercepts at $-\frac{1}{2}$ and $\frac{1}{4}$. This verifies that $\left\{-\frac{1}{2}, \frac{1}{4}\right\}$ is the solution set of the quadratic equation

$$8x^2 + 2x - 1 = 0.$$

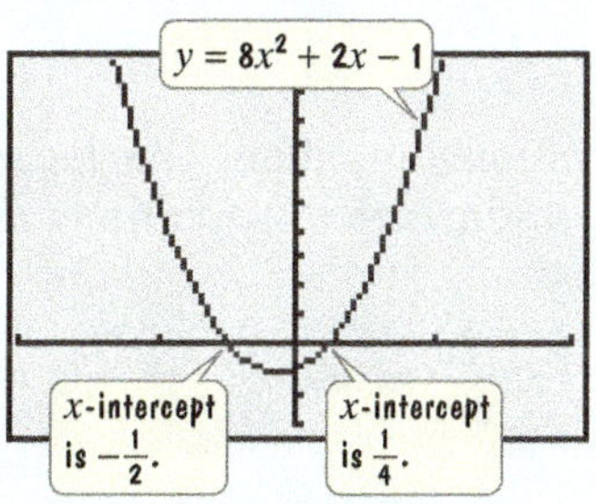

[−2, 2, 1] by [−3, 10, 1]

In Example 1, the solutions of $8x^2 + 2x - 1 = 0$ are rational numbers. This means that the equation can also be solved by factoring. The reason that the solutions are rational numbers is that $b^2 - 4ac$, the radicand in the quadratic formula, is 36, which is a perfect square. If a, b, and c are rational numbers, all quadratic equations for which $b^2 - 4ac$ is a perfect square have rational solutions.

✓ CHECK POINT 1 Solve using the quadratic formula: $2x^2 + 9x - 5 = 0.$

EXAMPLE 2 Solving a Quadratic Equation Using the Quadratic Formula

Solve using the quadratic formula:

$$2x^2 = 4x + 1.$$

Solution The quadratic equation must be in standard form to identify the values for a, b, and c. To move all terms to one side and obtain zero on the right, we subtract $4x + 1$ from both sides. Then we can identify the values for a, b, and c.

$$2x^2 = 4x + 1 \quad \text{This is the given equation.}$$

$$2x^2 - 4x - 1 = 0 \quad \text{Subtract } 4x + 1 \text{ from both sides.}$$

$a = 2$ $b = -4$ $c = -1$

Substituting these values into the quadratic formula and simplifying gives the equation's solutions.

$$x = \frac{-b \pm \sqrt{b^2 - 4ac}}{2a} \quad \text{Use the quadratic formula.}$$

$$x = \frac{-(-4) \pm \sqrt{(-4)^2 - 4(2)(-1)}}{2(2)} \quad \text{Substitute the values for } a, b, \text{ and } c\text{: } a = 2, b = -4, \text{ and } c = -1.$$

$$= \frac{4 \pm \sqrt{16 - (-8)}}{4} \quad (-4)^2 - 4(2)(-1) = 16 - (-8)$$

$$= \frac{4 \pm \sqrt{24}}{4} \quad 16 - (-8) = 16 + 8 = 24$$

$$= \frac{4 \pm 2\sqrt{6}}{4} \quad \sqrt{24} = \sqrt{4 \cdot 6} = \sqrt{4} \cdot \sqrt{6} = 2\sqrt{6}$$

$$= \frac{2(2 \pm \sqrt{6})}{4} \quad \text{Factor out 2 from the numerator.}$$

$$= \frac{2 \pm \sqrt{6}}{2} \quad \text{Divide the numerator and denominator by 2.}$$

The solutions are $\frac{2 \pm \sqrt{6}}{2}$, and the solution set is $\left\{\frac{2 + \sqrt{6}}{2}, \frac{2 - \sqrt{6}}{2}\right\}$ or $\left\{\frac{2 \pm \sqrt{6}}{2}\right\}$.

Using Technology

You can use a graphing utility to verify that the solutions of $2x^2 - 4x - 1 = 0$ are $\frac{2 \pm \sqrt{6}}{2}$. Begin by entering $y_1 = 2x^2 - 4x - 1$ in the Y= screen. Then evaluate this function at each of the proposed solutions.

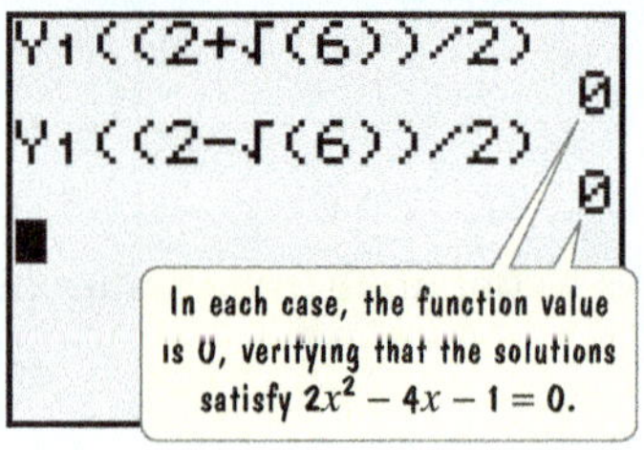

In each case, the function value is 0, verifying that the solutions satisfy $2x^2 - 4x - 1 = 0$.

In Example 2, the solutions of $2x^2 = 4x + 1$ are irrational numbers. This means that the equation cannot be solved by factoring. The reason that the solutions are irrational numbers is that $b^2 - 4ac$, the radicand in the quadratic formula, is 24, which is not a perfect square. Notice, too, that the solutions, $\frac{2 + \sqrt{6}}{2}$ and $\frac{2 - \sqrt{6}}{2}$, are conjugates.

Study Tip

Many students use the quadratic formula correctly until the last step, where they make an error in simplifying the solutions. Be sure to factor the numerator before dividing the numerator and the denominator by the greatest common factor.

$$\frac{4 \pm 2\sqrt{6}}{4} = \frac{2(2 \pm \sqrt{6})}{4} = \frac{\overset{1}{\cancel{2}}(2 \pm \sqrt{6})}{\underset{2}{\cancel{4}}} = \frac{2 \pm \sqrt{6}}{2}$$

You cannot divide just one term in the numerator and the denominator by their greatest common factor.

Incorrect!

$$\frac{\overset{1}{\cancel{4}} \pm 2\sqrt{6}}{\underset{1}{\cancel{4}}} = 1 \pm 2\sqrt{6} \qquad \frac{4 \pm \overset{1}{\cancel{2}}\sqrt{6}}{\underset{2}{\cancel{4}}} = \frac{4 \pm \sqrt{6}}{2}$$

Can all irrational solutions of quadratic equations be simplified? No. The following solutions cannot be simplified:

$$\frac{5 \pm 2\sqrt{7}}{2} \qquad \frac{-4 \pm 3\sqrt{7}}{2}.$$

Other than 1, terms in each numerator have no common factor.

CHECK POINT 2 Solve using the quadratic formula: $2x^2 = 6x - 1$.

EXAMPLE 3 Solving a Quadratic Equation Using the Quadratic Formula

Solve using the quadratic formula:

$$3x^2 + 2 = -4x.$$

Solution Begin by writing the quadratic equation in standard form.

$$3x^2 + 2 = -4x \quad \text{This is the given equation.}$$

$$3x^2 + 4x + 2 = 0 \quad \text{Add 4x to both sides.}$$

$a = 3$ $b = 4$ $c = 2$

Substituting these values into the quadratic formula and simplifying gives the equation's solutions.

$$x = \frac{-b \pm \sqrt{b^2 - 4ac}}{2a} \quad \text{Use the quadratic formula.}$$

$$x = \frac{-4 \pm \sqrt{4^2 - 4 \cdot 3 \cdot 2}}{2 \cdot 3} \quad \text{Substitute the values for a, b, and c: a = 3, b = 4, and c = 2.}$$

$$= \frac{-4 \pm \sqrt{16 - 24}}{6} \quad \text{Multiply under the radical.}$$

$$= \frac{-4 \pm \sqrt{-8}}{6} \quad \text{Subtract under the radical.}$$

$$= \frac{-4 \pm 2i\sqrt{2}}{6} \quad \sqrt{-8} = \sqrt{8(-1)} = \sqrt{8}\sqrt{-1} = i\sqrt{8} = i\sqrt{4 \cdot 2} = 2i\sqrt{2}$$

$$= \frac{2(-2 \pm i\sqrt{2})}{6} \quad \text{Factor out 2 from the numerator.}$$

$$= \frac{-2 \pm i\sqrt{2}}{3} \quad \text{Divide the numerator and denominator by 2.}$$

$$= -\frac{2}{3} \pm i\frac{\sqrt{2}}{3} \quad \text{Express in the form } a + bi\text{, writing } i \text{ before the square root.}$$

The solutions are $-\frac{2}{3} \pm i\frac{\sqrt{2}}{3}$, and the solution set is $\left\{-\frac{2}{3} + i\frac{\sqrt{2}}{3}, -\frac{2}{3} - i\frac{\sqrt{2}}{3}\right\}$ or $\left\{-\frac{2}{3} \pm i\frac{\sqrt{2}}{3}\right\}$.

In Example 3, the solutions of $3x^2 + 2 = -4x$ are imaginary numbers. This means that the equation cannot be solved using factoring. The reason that the solutions are imaginary numbers is that $b^2 - 4ac$, the radicand in the quadratic formula, is -8, which is negative. Notice, too, that the solutions are complex conjugates.

Using Technology

Graphic Connections

The graph of the quadratic function

$$y = 3x^2 + 4x + 2$$

has no x-intercepts. This verifies that the equation in Example 3

$$3x^2 + 2 = -4x, \quad \text{or}$$
$$3x^2 + 4x + 2 = 0$$

has imaginary solutions.

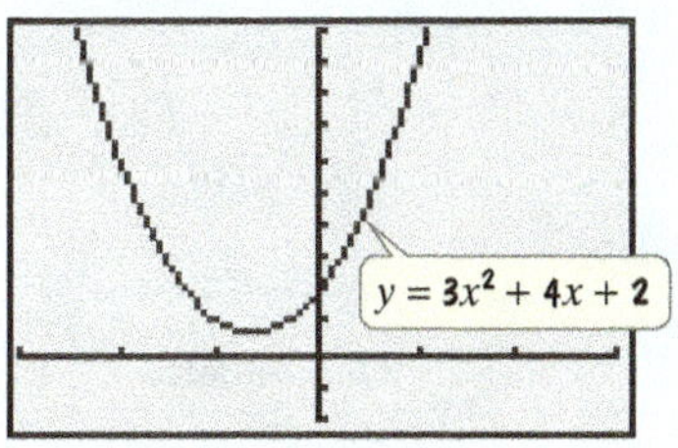

[-3, 3, 1] by [-2, 10, 1]

CHECK POINT 3 Solve using the quadratic formula: $3x^2 + 5 = -6x$.

Some rational equations can be solved using the quadratic formula. For example, consider the equation

$$3 + \frac{4}{x} = -\frac{2}{x^2}.$$

The denominators are x and x^2. The least common denominator is x^2. We clear fractions by multiplying both sides of the equation by x^2. Notice that x cannot equal zero.

$$x^2\left(3 + \frac{4}{x}\right) = x^2\left(-\frac{2}{x^2}\right),\quad x \neq 0$$

$$3x^2 + \frac{4}{x} \cdot x^2 = x^2\left(-\frac{2}{x^2}\right)$$ Use the distributive property.

$$3x^2 + 4x = -2$$ Simplify.

By adding 2 to both sides of $3x^2 + 4x = -2$, we obtain the standard form of the quadratic equation:

$$3x^2 + 4x + 2 = 0.$$

This is the equation that we solved in Example 3. The two imaginary solutions are not part of the restriction that $x \neq 0$.

2 Use the discriminant to determine the number and type of solutions.

The Discriminant

The quantity $b^2 - 4ac$, which appears under the radical sign in the quadratic formula, is called the **discriminant**. **Table 8.2** shows how the discriminant of the quadratic equation $ax^2 + bx + c = 0$ determines the number and type of solutions.

Study Tip

Checking irrational and imaginary solutions can be time-consuming. The solutions given by the quadratic formula are always correct, unless you have made a careless error. Checking for computational errors or errors in simplification is sufficient.

Table 8.2 The Discriminant and the Kinds of Solutions to $ax^2 + bx + c = 0$

Discriminant $b^2 - 4ac$	Kinds of Solutions to $ax^2 + bx + c = 0$	Graph of $y = ax^2 + bx + c$
$b^2 - 4ac > 0$	**Two unequal real solutions:** If a, b, and c are rational numbers and the discriminant is a perfect square, the solutions are *rational*. If the discriminant is not a perfect square, the solutions are *irrational* conjugates.	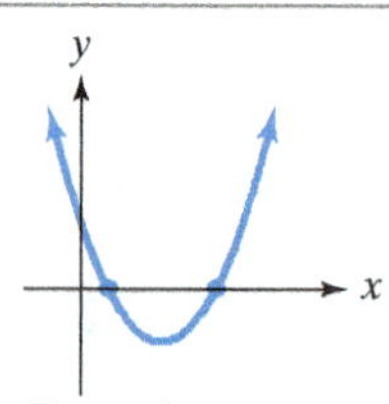 Two x-intercepts
$b^2 - 4ac = 0$	**One solution (a repeated solution) that is a real numbers:** If a, b, and c are rational numbers, the repeated solution is also a rational number.	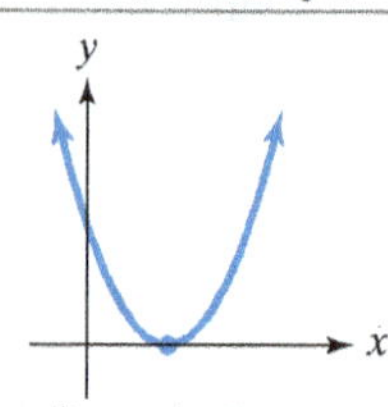 One x-intercept
$b^2 - 4ac < 0$	**No real solution; two imaginary solutions:** The solutions are complex conjugates.	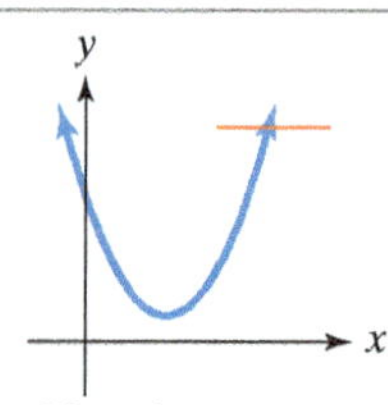 No x-intercepts

EXAMPLE 4 Using the Discriminant

For each equation, compute the discriminant. Then determine the number and type of solutions:

a. $3x^2 + 4x - 5 = 0$ b. $9x^2 - 6x + 1 = 0$ c. $3x^2 - 8x + 7 = 0.$

Study Tip

The discriminant is $b^2 - 4ac$. It is not $\sqrt{b^2 - 4ac}$, so do not give the discriminant as a radical.

Solution Begin by identifying the values for a, b, and c in each equation. Then compute $b^2 - 4ac$, the discriminant.

a. $3x^2 + 4x - 5 = 0$

$a = 3$ $b = 4$ $c = -5$

Substitute and compute the discriminant:

$$b^2 - 4ac = 4^2 - 4 \cdot 3(-5) = 16 - (-60) = 16 + 60 = 76.$$

The discriminant, 76, is a positive number that is not a perfect square. Thus, there are two real irrational solutions. (These solutions are conjugates of each other.)

b. $9x^2 - 6x + 1 = 0$

$a = 9$ $b = -6$ $c = 1$

Substitute and compute the discriminant:

$$b^2 - 4ac = (-6)^2 - 4 \cdot 9 \cdot 1 = 36 - 36 = 0.$$

The discriminant, 0, shows that there is only one real solution. This real solution is a rational number.

c. $3x^2 - 8x + 7 = 0$

$a = 3$ $b = -8$ $c = 7$

$$b^2 - 4ac = (-8)^2 - 4 \cdot 3 \cdot 7 = 64 - 84 = -20$$

The negative discriminant, −20, shows that there are two imaginary solutions. (These solutions are complex conjugates of each other.)

✓ CHECK POINT 4 For each equation, compute the discriminant. Then determine the number and type of solutions:

a. $x^2 + 6x + 9 = 0$
b. $2x^2 - 7x - 4 = 0$
c. $3x^2 - 2x + 4 = 0.$

3 Determine the most efficient method to use when solving a quadratic equation.

Determining Which Method to Use

All quadratic equations can be solved by the quadratic formula. However, if an equation is in the form $u^2 = d$, such as $x^2 = 5$ or $(2x + 3)^2 = 8$, it is faster to use the square root property, taking the square root of both sides. If the equation is not in the form $u^2 = d$, write the quadratic equation in standard form $(ax^2 + bx + c = 0)$. Try to solve the equation by factoring. If $ax^2 + bx + c$ cannot be factored, then solve the quadratic equation by the quadratic formula.

Because we used the method of completing the square to derive the quadratic formula, we no longer need it for solving quadratic equations. However, we will use completing the square in Chapter 10 to help graph certain kinds of equations.

Table 8.3 summarizes our observations about which technique to use when solving a quadratic equation.

Table 8.3 Determining the Most Efficient Technique to Use When Solving a Quadratic Equation

Description and Form of the Quadratic Equation	Most Efficient Solution Method	Example
$ax^2 + bx + c = 0$ and $ax^2 + bx + c$ can be factored easily.	Factor and use the zero-product principle.	$3x^2 + 5x - 2 = 0$ $(3x - 1)(x + 2) = 0$ $3x - 1 = 0$ or $x + 2 = 0$ $x = \frac{1}{3}$ $\quad x = -2$
$ax^2 + c = 0$ The quadratic equation has no x-term. $(b = 0)$	Solve for x^2 and apply the square root property.	$4x^2 - 7 = 0$ $4x^2 = 7$ $x^2 = \frac{7}{4}$ $x = \pm\frac{\sqrt{7}}{2}$
$u^2 = d$; u is a first-degree polynomial.	Use the square root property.	$(x + 4)^2 = 5$ $x + 4 = \pm\sqrt{5}$ $x = -4 \pm \sqrt{5}$
$ax^2 + bx + c = 0$ and $ax^2 + bx + c$ cannot be factored or the factoring is too difficult.	Use the quadratic formula: $x = \frac{-b \pm \sqrt{b^2 - 4ac}}{2a}$.	$x^2 - 2x - 6 = 0$ $a = 1$ $\quad b = -2$ $\quad c = -6$ $x = \frac{-(-2) \pm \sqrt{(-2)^2 - 4(1)(-6)}}{2(1)}$ $= \frac{2 \pm \sqrt{4 - 4(1)(-6)}}{2(1)}$ $= \frac{2 \pm \sqrt{28}}{2} = \frac{2 \pm \sqrt{4}\sqrt{7}}{2}$ $= \frac{2 \pm 2\sqrt{7}}{2} = \frac{2(1 \pm \sqrt{7})}{2}$ $= 1 \pm \sqrt{7}$

4 Write quadratic equations from solutions.

Writing Quadratic Equations from Solutions

Using the zero-product principle, the equation $(x - 3)(x + 5) = 0$ has two solutions, 3 and -5. By applying the zero-product principle in reverse, we can find a quadratic equation that has two given numbers as its solutions.

The Zero-Product Principle in Reverse

If $A = 0$ or $B = 0$, then $AB = 0$.

EXAMPLE 5 Writing Equations from Solutions

Write a quadratic equation with the given solution set:

a. $\left\{-\frac{5}{3}, \frac{1}{2}\right\}$ $\qquad$ b. $\{-5i, 5i\}$.

Jasper Johns, "Zero." © Jasper Johns/Licensed by VAGA, New York, NY.

The special properties of zero make it possible to write a quadratic equation from its solutions.

Solution

a. Because the solution set is $\left\{-\frac{5}{3}, \frac{1}{2}\right\}$, then

$$x = -\frac{5}{3} \quad \text{or} \quad x = \frac{1}{2}.$$

$$x + \frac{5}{3} = 0 \quad \text{or} \quad x - \frac{1}{2} = 0$$ Obtain zero on one side of each equation.

$$3x + 5 = 0 \quad \text{or} \quad 2x - 1 = 0$$ Clear fractions, multiplying by 3 and 2, respectively.

$$(3x + 5)(2x - 1) = 0$$ Use the zero-product principle in reverse: If $A = 0$ or $B = 0$, then $AB = 0$.

$$6x^2 - 3x + 10x - 5 = 0$$ Use the FOIL method to multiply.

$$6x^2 + 7x - 5 = 0$$ Combine like terms.

Thus, one equation is $6x^2 + 7x - 5 = 0$. Many other quadratic equations have $\left\{-\frac{5}{3}, \frac{1}{2}\right\}$ for their solution sets. These equations can be obtained by multiplying both sides of $6x^2 + 7x - 5 = 0$ by any nonzero real number.

b. Because the solution set is $\{-5i, 5i\}$, then

$$x = -5i \quad \text{or} \quad x = 5i.$$

$$x + 5i = 0 \quad \text{or} \quad x - 5i = 0$$ Obtain zero on one side of each equation.

$$(x + 5i)(x - 5i) = 0$$ Use the zero-product principle in reverse: If $A = 0$ or $B = 0$, then $AB = 0$.

$$x^2 - (5i)^2 = 0$$ Multiply conjugates using $(A + B)(A - B) = A^2 - B^2$.

$$x^2 - 25i^2 = 0$$ $(5i)^2 = 5^2i^2 = 25i^2$

$$x^2 - 25(-1) = 0$$ $i^2 = -1$

$$x^2 + 25 = 0$$ This is the required equation.

CHECK POINT 5 Write a quadratic equation with the given solution set:

a. $\left\{-\frac{3}{5}, \frac{1}{4}\right\}$

b. $\{-7i, 7i\}$.

5 Use the quadratic formula to solve problems.

Applications

Quadratic equations can be solved to answer questions about variables contained in quadratic functions.

EXAMPLE 6 Blood Pressure and Age

The graphs in **Figure 8.3** illustrate that a person's normal systolic blood pressure, measured in millimeters of mercury (mm Hg), depends on his or her age. The function

$$P(A) = 0.006A^2 - 0.02A + 120$$

models a man's normal systolic pressure, $P(A)$, at age A.

a. Find the age, to the nearest year, of a man whose normal systolic blood pressure is 125 mm Hg.

b. Use the graphs in **Figure 8.3** to describe the differences between the normal systolic blood pressures of men and women as they age.

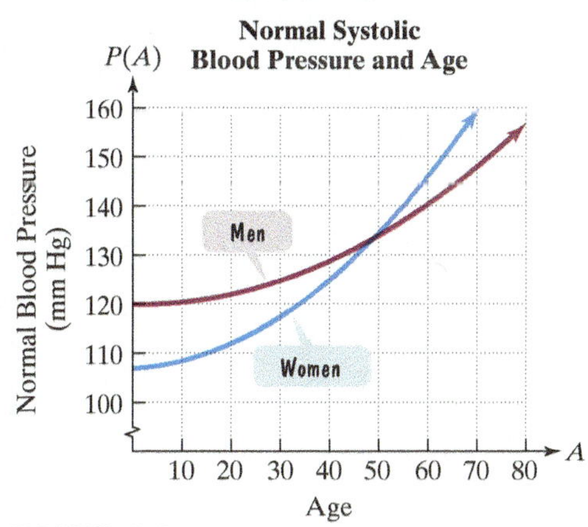

FIGURE 8.3

Solution

a. We are interested in the age of a man with a normal systolic blood pressure of 125 millimeters of mercury. Thus, we substitute 125 for $P(A)$ in the given function for men. Then we solve for A, the man's age.

$$P(A) = 0.006A^2 - 0.02A + 120$$ This is the given function for men.

$$125 = 0.006A^2 - 0.02A + 120$$ Substitute 125 for $P(A)$.

$$0 = 0.006A^2 - 0.02A - 5$$ Subtract 125 from both sides and write the quadratic equation in standard form.

$a = 0.006$ $b = -0.02$ $c = -5$

Because the trinomial on the right side of the equation is prime, we solve using the quadratic formula.

Notice that the variable is A, rather than the usual x.

$$A = \frac{-b \pm \sqrt{b^2 - 4ac}}{2a}$$ Use the quadratic formula.

$$= \frac{-(-0.02) \pm \sqrt{(-0.02)^2 - 4(0.006)(-5)}}{2(0.006)}$$ Substitute the values for a, b, and c: $a = 0.006$, $b = -0.02$, and $c = -5$.

$$= \frac{0.02 \pm \sqrt{0.1204}}{0.012}$$ Use a calculator to simplify the expression under the square root.

$$\approx \frac{0.02 \pm 0.347}{0.012}$$ Use a calculator: $\sqrt{0.1204} \approx 0.347$.

$$A \approx \frac{0.02 + 0.347}{0.012} \quad \text{or} \quad A \approx \frac{0.02 - 0.347}{0.012}$$

$$A \approx 31 \qquad\qquad A \approx -27$$ Use a calculator and round to the nearest integer.

Reject this solution. Age cannot be negative.

Using Technology

On most calculators, here is how to approximate

$$\frac{0.02 + \sqrt{0.1204}}{0.012}.$$

Many Scientific Calculators

[( ] .02 [+] .1204 [√] [ )] [÷] .012 [=]

Many Graphing Calculators

[( ] .02 [+] [√] .1204 [ )] [÷] .012 [ENTER]

If your calculator displays an open parenthesis after √, you'll need to enter another closed parenthesis here.

The positive solution, $A \approx 31$, indicates that 31 is the approximate age of a man whose normal systolic blood pressure is 125 mm Hg. This is illustrated by the black lines with the arrows on the red graph representing men in **Figure 8.4**.

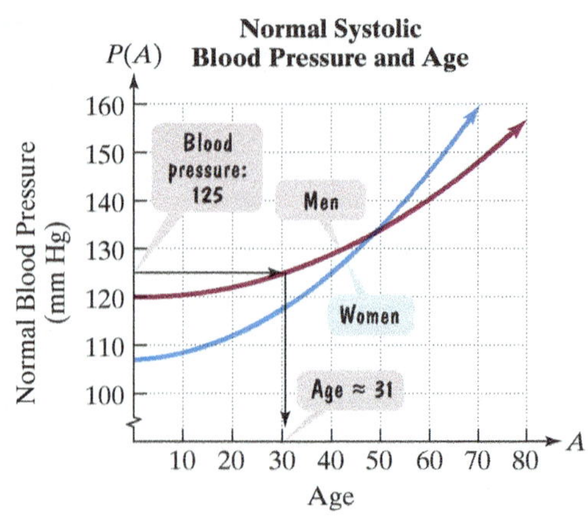

FIGURE 8.4

b. Take a second look at the graphs in **Figure 8.4**. Before approximately age 50, the blue graph representing women's normal systolic blood pressure lies below the red graph representing men's normal systolic blood pressure. Thus, up to age 50, women's normal systolic blood pressure is lower than men's, although it is increasing at a faster rate. After age 50, women's normal systolic blood pressure is higher than men's.

CHECK POINT 6 The function $P(A) = 0.01A^2 + 0.05A + 107$ models a woman's normal systolic blood pressure, $P(A)$, at age A. Use this function to find the age, to the nearest year, of a woman whose normal systolic blood pressure is 115 mm Hg. Use the blue graph in **Figure 8.4** to verify your solution.

8.2 EXERCISE SET

MyMathLab MathXL PRACTICE WATCH DOWNLOAD READ REVIEW

Practice Exercises

In Exercises 1–16, solve each equation using the quadratic formula. Simplify solutions, if possible.

1. $x^2 + 8x + 12 = 0$

2. $x^2 + 8x + 15 = 0$

3. $2x^2 - 7x = -5$

4. $5x^2 + 8x = -3$

5. $x^2 + 3x - 20 = 0$

6. $x^2 + 5x - 10 = 0$

7. $3x^2 - 7x = 3$

8. $4x^2 + 3x = 2$

9. $6x^2 = 2x + 1$

10. $2x^2 = -4x + 5$

11. $4x^2 - 3x = -6$

12. $9x^2 + x = -2$

13. $x^2 - 4x + 8 = 0$

14. $x^2 + 6x + 13 = 0$

15. $3x^2 = 8x - 7$

16. $3x^2 = 4x - 6$

17. $2x(x - 2) = x + 12$

18. $2x(x + 4) = 3x - 3$

In Exercises 19–30, compute the discriminant. Then determine the number and type of solutions for the given equation.

19. $x^2 + 8x + 3 = 0$

20. $x^2 + 7x + 4 = 0$

21. $x^2 + 6x + 8 = 0$

22. $x^2 + 2x - 3 = 0$

23. $2x^2 + x + 3 = 0$

24. $2x^2 - 4x + 3 = 0$

25. $2x^2 + 6x = 0$

26. $3x^2 - 5x = 0$

27. $5x^2 + 3 = 0$

28. $5x^2 + 4 = 0$

29. $9x^2 = 12x - 4$

30. $4x^2 = 20x - 25$

In Exercises 31–46, solve each equation by the method of your choice. Simplify solutions, if possible.

31. $3x^2 - 4x = 4$

32. $2x^2 - x = 1$

33. $x^2 - 2x = 1$

34. $2x^2 + 3x = 1$

35. $(2x - 5)(x + 1) = 2$

36. $(2x + 3)(x + 4) = 1$

37. $(3x - 4)^2 = 16$

38. $(2x + 7)^2 = 25$

39. $\frac{x^2}{2} + 2x + \frac{2}{3} = 0$

40. $\frac{x^2}{3} - x - \frac{1}{6} = 0$

41. $(3x - 2)^2 = 10$

42. $(4x - 1)^2 = 15$

43. $\frac{1}{x} + \frac{1}{x + 2} = \frac{1}{3}$

44. $\frac{1}{x} + \frac{1}{x + 3} = \frac{1}{4}$

45. $(2x - 6)(x + 2) = 5(x - 1) - 12$

46. $7x(x - 2) = 3 - 2(x + 4)$

In Exercises 47–60, write a quadratic equation in standard form with the given solution set.

47. $\{-3, 5\}$

48. $\{-2, 6\}$

49. $\left\{-\frac{2}{3}, \frac{1}{4}\right\}$

50. $\left\{-\frac{5}{6}, \frac{1}{3}\right\}$
51. $\{-6i, 6i\}$
52. $\{-8i, 8i\}$
53. $\{-\sqrt{2}, \sqrt{2}\}$
54. $\{-\sqrt{3}, \sqrt{3}\}$
55. $\{-2\sqrt{5}, 2\sqrt{5}\}$
56. $\{-3\sqrt{5}, 3\sqrt{5}\}$
57. $\{1 + i, 1 - i\}$
58. $\{2 + i, 2 - i\}$
59. $\{1 + \sqrt{2}, 1 - \sqrt{2}\}$
60. $\{1 + \sqrt{3}, 1 - \sqrt{3}\}$

Practice PLUS

Exercises 61–64 describe quadratic equations. Match each description with the graph of the corresponding quadratic function. Each graph is shown in a $[-10, 10, 1]$ *by* $[-10, 10, 1]$ *viewing rectangle.*

61. A quadratic equation whose solution set contains imaginary numbers
62. A quadratic equation whose discriminant is 0
63. A quadratic equation whose solution set is $\{3 \pm \sqrt{2}\}$
64. A quadratic equation whose solution set contains integers

a.
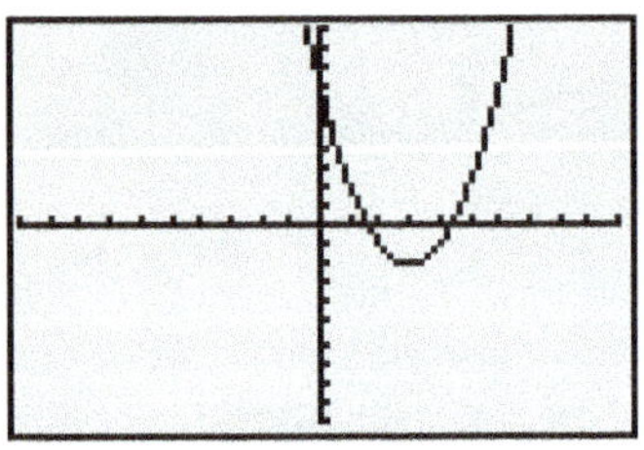

b.
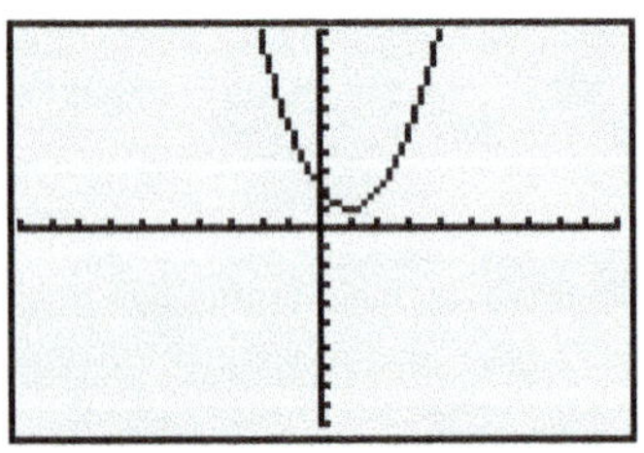

c.
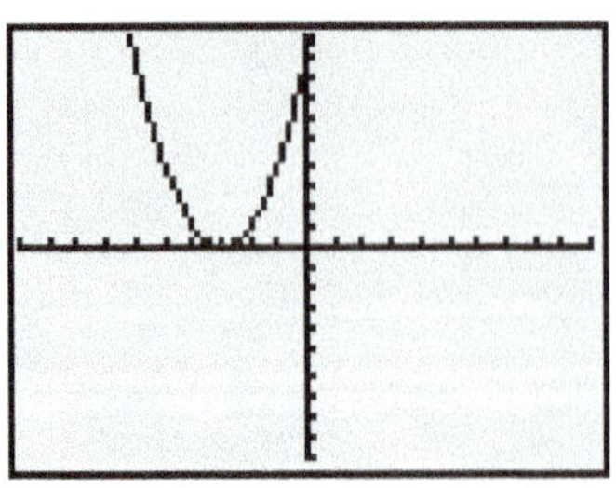

d.
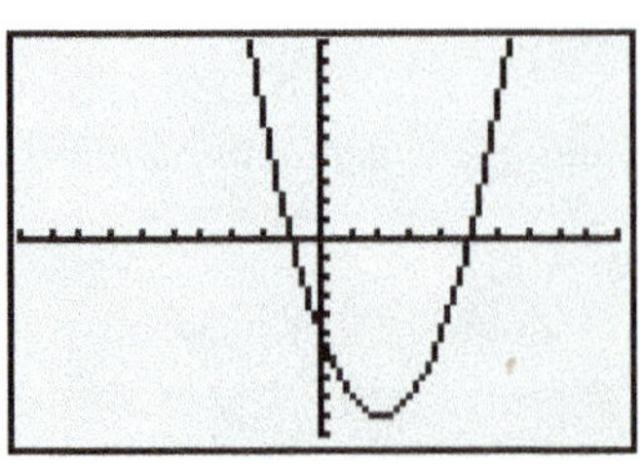

65. When the sum of 6 and twice a positive number is subtracted from the square of the number, 0 results. Find the number.

66. When the sum of 1 and twice a negative number is subtracted from twice the square of the number, 0 results. Find the number.

In Exercises 67–72, solve each equation by the method of your choice.

67. $\frac{1}{x^2 - 3x + 2} = \frac{1}{x + 2} + \frac{5}{x^2 - 4}$

68. $\frac{x - 1}{x - 2} + \frac{x}{x - 3} = \frac{1}{x^2 - 5x + 6}$

69. $\sqrt{2}x^2 + 3x - 2\sqrt{2} = 0$

70. $\sqrt{3}x^2 + 6x + 7\sqrt{3} = 0$

71. $|x^2 + 2x| = 3$

72. $|x^2 + 3x| = 2$

Application Exercises

A driver's age has something to do with his or her chance of getting into a fatal car crash. The bar graph shows the number of fatal vehicle crashes per 100 million miles driven for drivers of various age groups. For example, 25-year-old drivers are involved in 4.1 fatal crashes per 100 million miles driven. Thus, when a group of 25-year-old Americans have driven a total of 100 million miles, approximately 4 have been in accidents in which someone died.

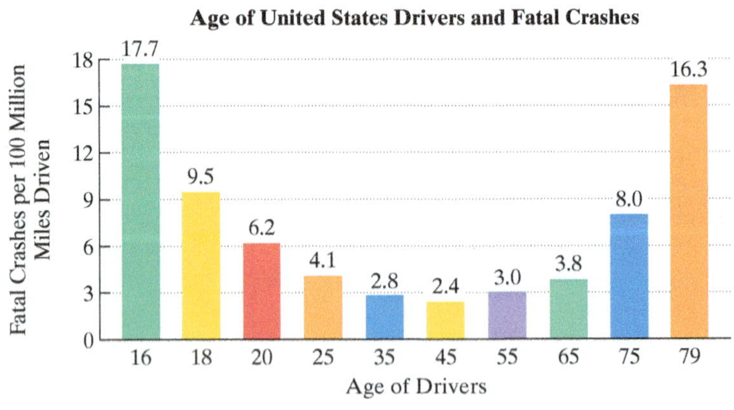

Source: Insurance Institute for Highway Safety

The number of fatal vehicle crashes per 100 million miles, $f(x)$, *for drivers of age* x *can be modeled by the quadratic function*

$$f(x) = 0.013x^2 - 1.19x + 28.24.$$

Use the function to solve Exercises 73–74.

73. What age groups are expected to be involved in 3 fatal crashes per 100 million miles driven? How well does the function model the trend in the actual data shown in the bar graph?

74. What age groups are expected to be involved in 10 fatal crashes per 100 million miles driven? How well does the function model the trend in the actual data shown in the bar graph?

Throwing events in track and field include the shot put, the discus throw, the hammer throw, and the javelin throw. The distance that an athlete can achieve depends on the initial velocity of the object thrown and the angle above the horizontal at which the object leaves the hand.

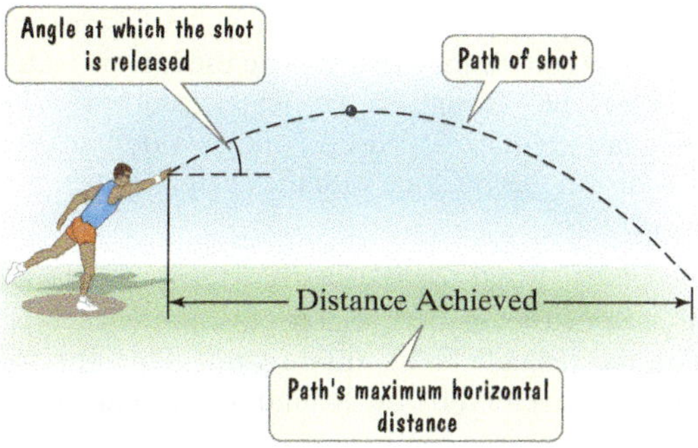

In Exercises 75–76, an athlete whose event is the shot put releases the shot with the same initial velocity, but at different angles.

75. When the shot is released at an angle of 35°, its path can be modeled by the function

$$f(x) = -0.01x^2 + 0.7x + 6.1,$$

in which x is the shot's horizontal distance, in feet, and $f(x)$ is its height, in feet. This function is shown by one of the graphs, (a) or (b), in the figure. Use the function to determine the shot's maximum distance. Use a calculator and round to the nearest tenth of a foot. Which graph, (a) or (b), shows the shot's path?

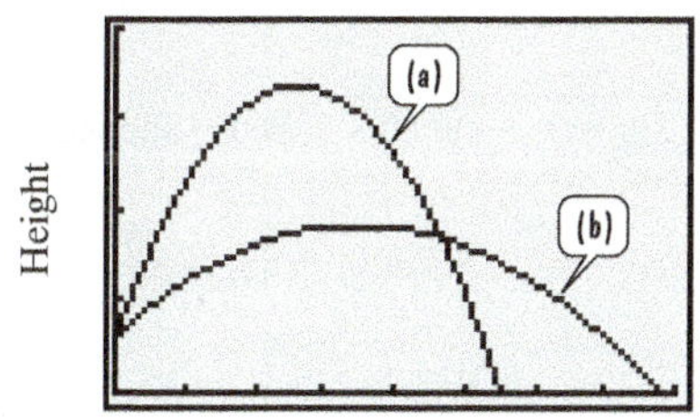

[0, 80, 10] by [0, 40, 10]

76. When the shot is released at an angle of 65°, its path can be modeled by the function

$$f(x) = -0.04x^2 + 2.1x + 6.1,$$

in which x is the shot's horizontal distance, in feet, and $f(x)$ is its height, in feet. This function is shown by one of the graphs, (a) or (b), in the figure above. Use the function to determine the shot's maximum distance. Use a calculator and round to the nearest tenth of a foot. Which graph, (a) or (b), shows the shot's path?

77. The length of a rectangle is 4 meters longer than the width. If the area is 8 square meters, find the rectangle's dimensions. Round to the nearest tenth of a meter.

78. The length of a rectangle exceeds twice its width by 3 inches. If the area is 10 square inches, find the rectangle's dimensions. Round to the nearest tenth of an inch.

79. The longer leg of a right triangle exceeds the shorter leg by 1 inch, and the hypotenuse exceeds the longer leg by 7 inches. Find the lengths of the legs. Round to the nearest tenth of a inch.

80. The hypotenuse of a right triangle is 6 feet long. One leg is 2 feet shorter than the other. Find the lengths of the legs. Round to the nearest tenth of a foot.

81. A rain gutter is made from sheets of aluminum that are 20 inches wide. As shown in the figure, the edges are turned up to form right angles. Determine the depth of the gutter that will allow a cross-sectional area of 13 square inches. Show that there are two different solutions to the problem. Round to the nearest tenth of an inch.

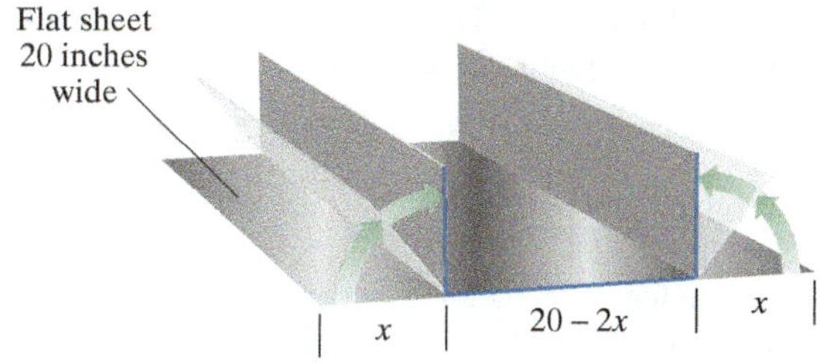

82. A piece of wire is 8 inches long. The wire is cut into two pieces and then each piece is bent into a square. Find the length of each piece if the sum of the areas of these squares is to be 2 square inches.

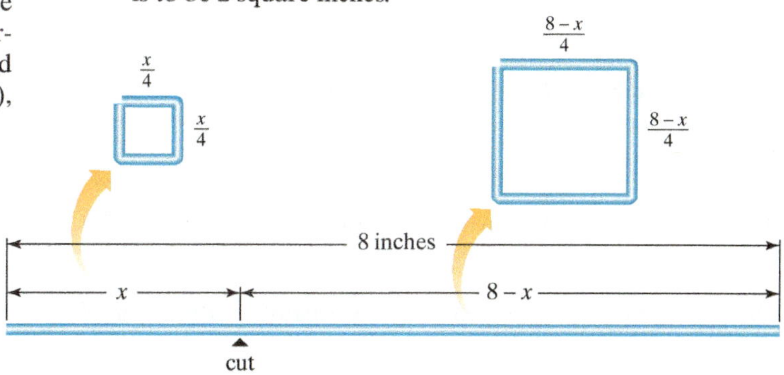

83. Working together, two people can mow a large lawn in 4 hours. One person can do the job alone 1 hour faster than the other person. How long does it take each person working alone to mow the lawn? Round to the nearest tenth of an hour.

84. A pool has an inlet pipe to fill it and an outlet pipe to empty it. It takes 2 hours longer to empty the pool than it does to fill it. The inlet pipe is turned on to fill the pool, but the outlet pipe is accidentally left open. Despite this, the pool fills in 8 hours. How long does it take the outlet pipe to empty the pool? Round to the nearest tenth of an hour.

Writing in Mathematics

85. What is the quadratic formula and why is it useful?

86. Without going into specific details for every step, describe how the quadratic formula is derived.

87. Explain how to solve $x^2 + 6x + 8 = 0$ using the quadratic formula.

88. If a quadratic equation has imaginary solutions, how is this shown on the graph of the corresponding quadratic function?

89. What is the discriminant and what information does it provide about a quadratic equation?

90. If you are given a quadratic equation, how do you determine which method to use to solve it?

91. Explain how to write a quadratic equation from its solution set. Give an example with your explanation.

Technology Exercises

92. Use a graphing utility to graph the quadratic function related to any five of the quadratic equations in Exercises 19–30. How does each graph illustrate what you determined algebraically using the discriminant?

93. Reread Exercise 81. The cross-sectional area of the gutter is given by the quadratic function

$$f(x) = x(20 - 2x).$$

Graph the function in a [0, 10, 1] by [0, 60, 5] viewing rectangle. Then TRACE along the curve or use the maximum function feature to determine the depth of the gutter that will maximize its cross-sectional area and allow the greatest amount of water to flow. What is the maximum area? Does the situation described in Exercise 81 take full advantage of the sheets of aluminum?

Critical Thinking Exercises

Makes Sense? *In Exercises 94–97, determine whether each statement "makes sense" or "does not make sense" and explain your reasoning.*

94. Because I want to solve $25x^2 - 169 = 0$ fairly quickly, I'll use the quadratic formula.

95. I simplified $\frac{3 + 2\sqrt{3}}{2}$ to $3 + \sqrt{3}$ because 2 is a factor of $2\sqrt{3}$.

96. I need to find a square root to determine the discriminant.

97. I obtained -17 for the discriminant, so there are two imaginary irrational solutions.

In Exercises 98–101, determine whether each statement is true or false. If the statement is false, make the necessary change(s) to produce a true statement.

98. Any quadratic equation that can be solved by completing the square can be solved by the quadratic formula.

99. The quadratic formula is developed by applying factoring and the zero-product principle to the quadratic equation $ax^2 + bx + c = 0$.

100. In using the quadratic formula to solve the quadratic equation $5x^2 = 2x - 7$, we have $a = 5, b = 2$, and $c = -7$.

101. The quadratic formula can be used to solve the equation $x^2 - 9 = 0$.

102. Solve for t: $s = -16t^2 + v_0t$.

103. A rectangular swimming pool is 12 meters long and 8 meters wide. A tile border of uniform width is to be built around the pool using 120 square meters of tile. The tile is from a discontinued stock (so no additional materials are available) and all 120 square meters are to be used. How wide should the border be? Round to the nearest tenth of a meter. If zoning laws require at least a 2-meter-wide border around the pool, can this be done with the available tile?

104. The area of the shaded region outside the rectangle and inside the triangle is 10 square yards. Find the triangle's height, represented by $2x$. Round to the nearest tenth of a yard.

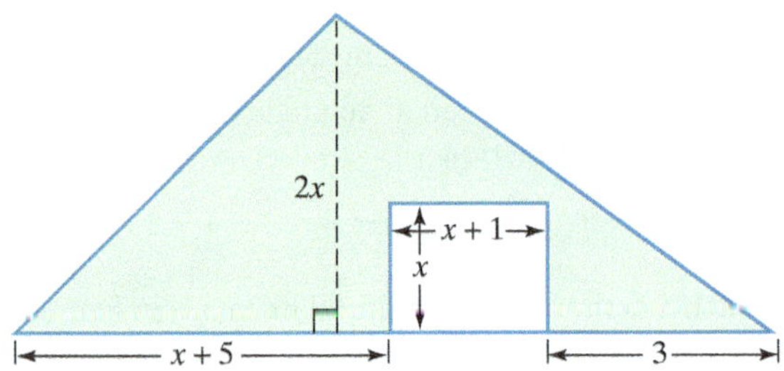

Review Exercises

105. Solve: $|5x + 2| = |4 - 3x|$. (Section 4.3, Example 3)

106. Solve: $\sqrt{2x - 5} - \sqrt{x - 3} = 1$. (Section 7.6, Example 4)

107. Rationalize the denominator: $\frac{5}{\sqrt{3} + x}$. (Section 7.5, Example 5)

Preview Exercises

Exercises 108–110 will help you prepare for the material covered in the next section.

108. Use point plotting to graph $f(x) = x^2$ and $g(x) = x^2 + 2$ in the same rectangular coordinate system.

109. Use point plotting to graph $f(x) = x^2$ and $g(x) = (x + 2)^2$ in the same rectangular coordinate system.

110. Find the x-intercepts for the graph of $f(x) = -2(x - 3)^2 + 8$.

SECTION 8.3

Quadratic Functions and Their Graphs

Objectives

1. Recognize characteristics of parabolas.
2. Graph parabolas in the form $f(x) = a(x - h)^2 + k$.
3. Graph parabolas in the form $f(x) = ax^2 + bx + c$.
4. Determine a quadratic function's minimum or maximum value.
5. Solve problems involving a quadratic function's minimum or maximum value.

We have a long history of throwing things. Before 400 B.C., the Greeks competed in games that included discus throwing. In the seventeenth century, English soldiers organized cannonball-throwing competitions. In 1827, a Yale University student, disappointed over failing an exam, took out his frustrations at the passing of a collection plate in chapel. Seizing the monetary tray, he flung it in the direction of a large open space on campus. Yale students see this act of frustration as the origin of the Frisbee.

In this section, we study quadratic functions and their graphs. By graphing functions that model the paths of the things we throw, you will be able to determine both the maximum height and the distance of these objects.

1 Recognize characteristics of parabolas.

Graphs of Quadratic Functions

The graph of any quadratic function

$$f(x) = ax^2 + bx + c, \quad a \neq 0,$$

is called a **parabola**. Parabolas are shaped like bowls or inverted bowls, as shown in **Figure 8.5**. If the coefficient of x^2 (the value of a in $ax^2 + bx + c$) is positive, the parabola opens upward. If the coefficient of x^2 is negative, the parabola opens downward. The **vertex** (or turning point) of the parabola is the lowest point on the graph when it opens upward and the highest point on the graph when it opens downward.

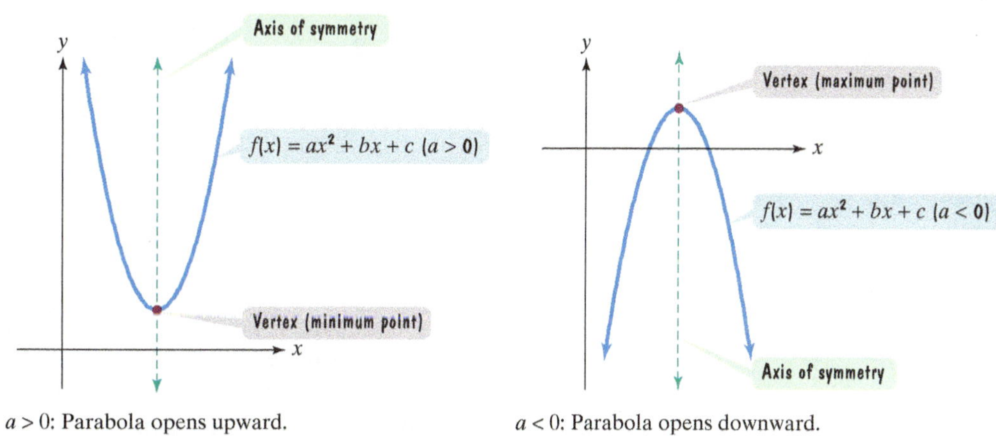

FIGURE 8.5 Characteristics of graphs of quadratic functions

The two halves of a parabola are mirror images of each other. A "mirror line" through the vertex, called the **axis of symmetry**, divides the figure in half. If a parabola is folded along its axis of symmetry, the two halves match exactly.

2 Graph parabolas in the form $f(x) = a(x-h)^2 + k$.

Graphing Quadratic Functions in the Form $f(x) = a(x-h)^2 + k$

One way to obtain the graph of a quadratic function is to use point plotting. Let's begin by graphing the functions $f(x) = x^2$, $g(x) = 2x^2$, and $h(x) = \frac{1}{2}x^2$ in the same rectangular coordinate system. Select integers for x, starting with -3 and ending with 3. A partial table of coordinates for each function is shown below. The three parabolas are shown in **Figure 8.6**.

x	$f(x) = x^2$	(x, y) or $(x, f(x))$
-3	$f(-3) = (-3)^2 = 9$	$(-3, 9)$
-2	$f(-2) = (-2)^2 = 4$	$(-2, 4)$
-1	$f(-1) = (-1)^2 = 1$	$(-1, 1)$
0	$f(0) = 0^2 = 0$	$(0, 0)$
1	$f(1) = 1^2 = 1$	$(1, 1)$
2	$f(2) = 2^2 = 4$	$(2, 4)$
3	$f(3) = 3^2 = 9$	$(3, 9)$

x	$g(x) = 2x^2$	(x, y) or $(x, g(x))$
-3	$g(-3) = 2(-3)^2 = 18$	$(-3, 18)$
-2	$g(-2) = 2(-2)^2 = 8$	$(-2, 8)$
-1	$g(-1) = 2(-1)^2 = 2$	$(-1, 2)$
0	$g(0) = 2 \cdot 0^2 = 0$	$(0, 0)$
1	$g(1) = 2 \cdot 1^2 = 2$	$(1, 2)$
2	$g(2) = 2 \cdot 2^2 = 8$	$(2, 8)$
3	$g(3) = 2 \cdot 3^2 = 18$	$(3, 18)$

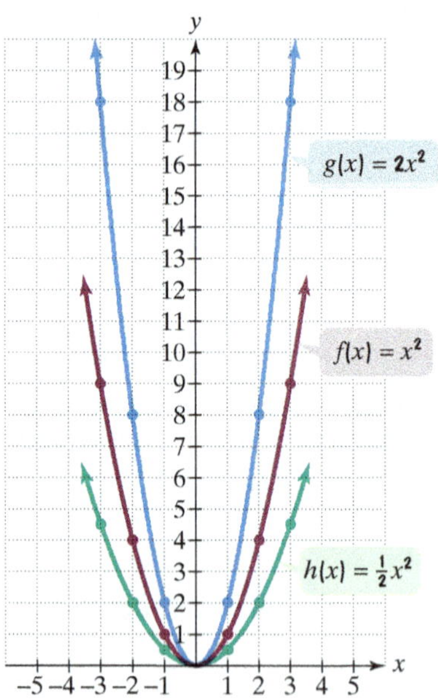

FIGURE 8.6

x	$h(x) = \frac{1}{2}x^2$	(x, y) or $(x, h(x))$
-3	$h(-3) = \frac{1}{2}(-3)^2 = \frac{9}{2}$	$\left(-3, \frac{9}{2}\right)$
-2	$h(-2) = \frac{1}{2}(-2)^2 = 2$	$(-2, 2)$
-1	$h(-1) = \frac{1}{2}(-1)^2 = \frac{1}{2}$	$\left(-1, \frac{1}{2}\right)$
0	$h(0) = \frac{1}{2} \cdot 0^2 = 0$	$(0, 0)$
1	$h(1) = \frac{1}{2} \cdot 1^2 = \frac{1}{2}$	$\left(1, \frac{1}{2}\right)$
2	$h(2) = \frac{1}{2} \cdot 2^2 = 2$	$(2, 2)$
3	$h(3) = \frac{1}{2} \cdot 3^2 = \frac{9}{2}$	$\left(3, \frac{9}{2}\right)$

Can you see that the graphs of f, g, and h all have the same vertex, $(0, 0)$? They also have the same axis of symmetry, the y-axis, or $x = 0$. This is true for all graphs of the form $f(x) = ax^2$. However, the blue graph of $g(x) = 2x^2$ is a narrower parabola than the red graph of $f(x) = x^2$. By contrast, the green graph of $h(x) = \frac{1}{2}x^2$ is a flatter parabola than the red graph of $f(x) = x^2$.

Is there a more efficient method than point plotting to obtain the graph of a quadratic function? The answer is yes. The method is based on comparing graphs of the form $g(x) = a(x-h)^2 + k$ to those of the form $f(x) = ax^2$.

In **Figure 8.7(a)**, the graph of $f(x) = ax^2$ for $a > 0$ is shown in black. The parabola's vertex is (0, 0) and it opens upward. In **Figure 8.7(b)**, the graph of $f(x) = ax^2$ for $a < 0$ is shown in black. The parabola's vertex is (0, 0) and it opens downward.

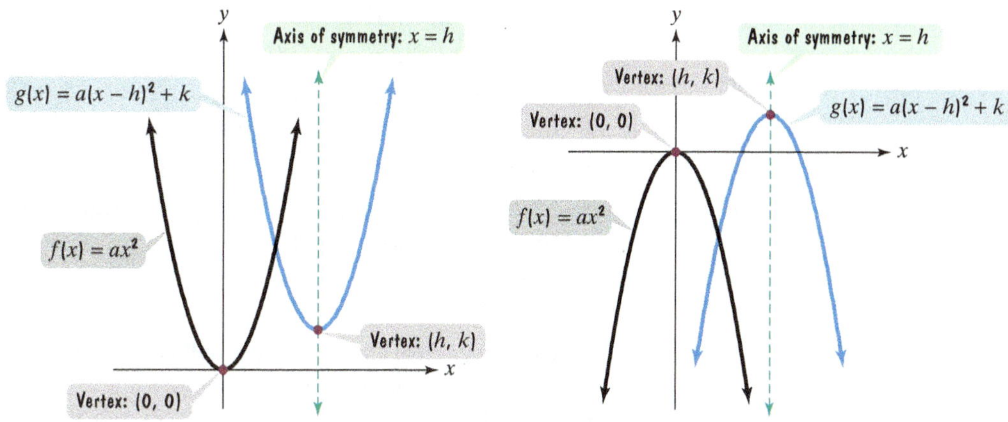

(a) $a > 0$: Parabola opens upward. **(b)** $a < 0$: Parabola opens downward.

FIGURE 8.7 Moving, or shifting, the graph of $f(x) = ax^2$

Figure 8.7(a) and **8.7(b)** also show the graph of $g(x) = a(x - h)^2 + k$ in blue. Compare these graphs to those of $f(x) = ax^2$. Observe that h determines a horizontal move, or shift, and k determines a vertical move, or shift, of the graph of $f(x) = ax^2$:

$$g(x) = a(x - h)^2 + k.$$

If $h > 0$, the graph of $f(x) = ax^2$ is shifted h units to the right.

If $k > 0$, the graph of $y = a(x - h)^2$ is shifted k units up.

Consequently, the vertex (0, 0) on the black graph of $f(x) = ax^2$ moves to the point (h, k) on the blue graph of $g(x) = a(x - h)^2 + k$. The axis of symmetry is the vertical line whose equation is $x = h$.

The form of the expression for g is convenient because it immediately identifies the vertex of the parabola as (h, k).

Quadratic Functions in the Form $f(x) = a(x - h)^2 + k$

The graph of

$$f(x) = a(x - h)^2 + k, \qquad a \neq 0$$

is a parabola whose vertex is the point (h, k). The parabola is symmetric with respect to the line $x = h$. If $a > 0$, the parabola opens upward; if $a < 0$, the parabola opens downward.

The sign of a in $f(x) = a(x - h)^2 + k$ determines whether the parabola opens upward or downward. Furthermore, if $|a|$ is small, the parabola opens more flatly than if $|a|$ is large. On the next page is a general procedure for graphing parabolas whose equations are in this form.

Graphing Quadratic Functions with Equations in the Form $f(x) = a(x - h)^2 + k$

To graph $f(x) = a(x - h)^2 + k$,

1. Determine whether the parabola opens upward or downward. If $a > 0$, it opens upward. If $a < 0$, it opens downward.
2. Determine the vertex of the parabola. The vertex is (h, k).
3. Find any x-intercepts by solving $f(x) = 0$. The equation's real solutions are the x-coordinates of the x-intercepts.
4. Find the y-intercept by computing $f(0)$.
5. Plot the intercepts, the vertex, and additional points as necessary. Connect these points with a smooth curve that is shaped like a bowl or an inverted bowl.

In the graphs that follow, we will show each axis of symmetry as a dashed vertical line. Because this vertical line passes through the vertex, (h, k), its equation is $x = h$. The line is dashed because it is not part of the parabola.

Study Tip

It's easy to make a sign error when finding h, the x-coordinate of the vertex. In

$$f(x) = a(x - h)^2 + k,$$

***h* is the number that follows the subtraction sign.**

- $f(x) = -2(x - 3)^2 + 8$

  The number *after* the subtraction is 3: $h = 3$.

- $f(x) = (x + 3)^2 + 1$
  $= (x - (-3))^2 + 1$

  The number *after* the subtraction is −3: $h = -3$.

EXAMPLE 1 Graphing a Quadratic Function in the Form $f(x) = a(x - h)^2 + k$

Graph the quadratic function $f(x) = -2(x - 3)^2 + 8$.

Solution We can graph this function by following the steps in the preceding box. We begin by identifying values for a, h, and k.

$$f(x) = a(x - h)^2 + k$$

$a = -2$ $h = 3$ $k = 8$

$$f(x) = -2(x - 3)^2 + 8$$

Step 1. Determine how the parabola opens. Note that a, the coefficient of x^2, is −2. Thus, $a < 0$; this negative value tells us that the parabola opens downward.

Step 2. Find the vertex. The vertex of the parabola is at (h, k). Because $h = 3$ and $k = 8$, the parabola has its vertex at $(3, 8)$.

Step 3. Find the x-intercepts by solving $f(x) = 0$. Replace $f(x)$ with 0 in $f(x) = -2(x - 3)^2 + 8$.

$$0 = -2(x - 3)^2 + 8$$ Find x-intercepts, setting $f(x)$ equal to 0.

$$2(x - 3)^2 = 8$$ Solve for x. Add $2(x - 3)^2$ to both sides of the equation.

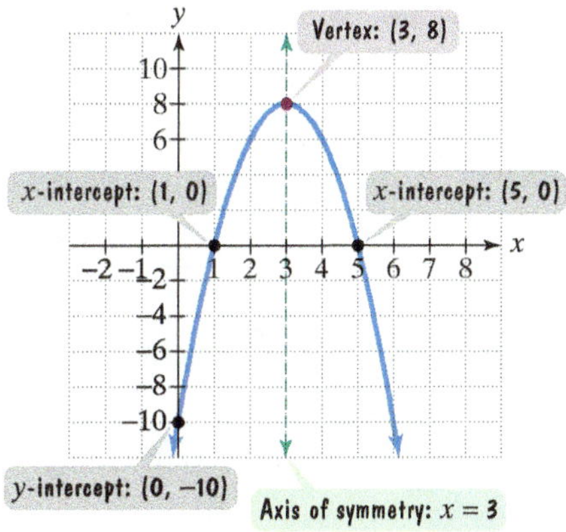

FIGURE 8.8 The graph of $f(x) = -2(x - 3)^2 + 8$

$(x - 3)^2 = 4$	Divide both sides by 2.
$x - 3 = \sqrt{4}$ or $x - 3 = -\sqrt{4}$	Apply the square root property.
$x - 3 = 2$ $\quad$ $x - 3 = -2$	$\sqrt{4} = 2$
$x = 5$ $\quad$ $x = 1$	Add 3 to both sides in each equation.

The x-intercepts are $(5, 0)$ and $(1, 0)$. The parabola passes through $(5, 0)$ and $(1, 0)$.

Step 4. Find the y-intercept by computing $f(0)$. Replace x with 0 in $f(x) = -2(x - 3)^2 + 8$.

$$f(0) = -2(0 - 3)^2 + 8 = -2(-3)^2 + 8 = -2(9) + 8 = -10$$

The y-intercept is $(0, -6.)$ The parabola passes through $(0, -10)$.

Step 5. Graph the parabola. With a vertex at $(3, 8)$, x-intercepts at 5 and 1, and a y-intercept at -10, the graph of f is shown in **Figure 8.8**. The axis of symmetry is the vertical line whose equation is$x = 3$.

✓ CHECK POINT 1 Graph the quadratic function $f(x) = -(x - 1)^2 + 4$.

EXAMPLE 2 Graphing a Quadratic Function in the Form $f(x) = a(x - h)^2 + k$

Graph the quadratic function $f(x) = (x + 3)^2 + 1$.

Solution We begin by finding values for a, h, and k.

$$f(x) = a(x - h)^2 + k \quad \text{Form of quadratic function}$$
$$f(x) = (x + 3)^2 + 1 \quad \text{Given function}$$
$$f(x) = 1(x - (-3))^2 + 1$$

$a = 1$ $\quad$ $h = -3$ $\quad$ $k = 1$

Step 1. Determine how the parabola opens. Note that a, the coefficient of x^2, is 1. Thus, $a > 0$; this positive value tells us that the parabola opens upward.

Step 2. Find the vertex. The vertex of the parabola is at (h, k). Because $h = -3$ and $k = 1$, the parabola has its vertex at $(-3, 1)$.

Step 3. Find the x-intercepts by solving $f(x) = 0$. Replace $f(x)$ with 0 in $f(x) = (x + 3)^2 + 1$. Because the vertex is at $(-3, 1)$, which lies above the x-axis, and the parabola opens upward, it appears that this parabola has no x-intercepts. We can verify this observation algebraically.

$0 = (x + 3)^2 + 1$	Find possible x-intercepts, setting f(x) equal to 0.
$-1 - (x + 3)^2$	Solve for x. Subtract 1 from both sides.
$x + 3 = \sqrt{-1}$ or $x + 3 = -\sqrt{-1}$	Apply the square root property.
$x + 3 = i$ $\quad$ $x + 3 = -i$	$\sqrt{-1} = i$.
$x = -3 + i$ $\quad$ $x = -3 - i$	The solutions are $-3 \pm i$.

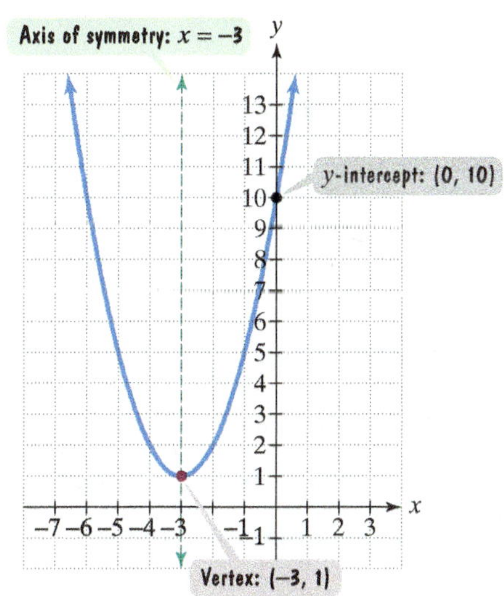

FIGURE 8.9 The graph of $f(x) = (x + 3)^2 + 1$

Because this equation has no real solutions, the parabola has no x-intercepts.

Step 4. Find the y-intercept by computing $f(0)$. Replace x with 0 in $f(x) = (x + 3)^2 + 1$.

$$f(0) = (0 + 3)^2 + 1 = 3^2 + 1 = 9 + 1 = 10$$

The y-intercept is $(0, 10)$. The parabola passes through $(0, 10)$.

Step 5. Graph the parabola. With a vertex at $(-3, 1)$, no x-intercepts, and a y-intercept at $(0, 10)$, the graph of f is shown in **Figure 8.9**. The axis of symmetry is the vertical line whose equation is $x = -3$.

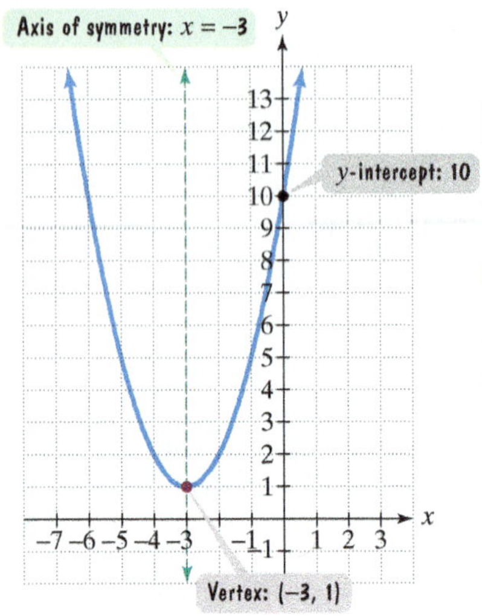

FIGURE 8.9 (repeated)

✓ **CHECK POINT 2** Graph the quadratic function $f(x) = (x-2)^2 + 1$.

Graphing Quadratic Functions in the Form $f(x) = ax^2 + bx + c$

3 Graph parabolas in the form $f(x) = ax^2 + bx + c$.

Quadratic functions are frequently expressed in the form $f(x) = ax^2 + bx + c$. How can we identify the vertex of a parabola whose equation is in this form? Completing the square provides the answer to this question.

$$f(x) = ax^2 + bx + c$$

$$= a\left(x^2 + \frac{b}{a}x\right) + c \quad \text{Factor out } a \text{ from } ax^2 + bx.$$

$$= a\left(x^2 + \frac{b}{a}x + \frac{b^2}{4a^2}\right) + c - a\left(\frac{b^2}{4a^2}\right)$$

Complete the square by adding the square of half the coefficient of x.

By completing the square, we added $a \cdot \frac{b^2}{4a^2}$. To avoid changing the function's equation, we must subtract this term.

$$= a\left(x + \frac{b}{2a}\right)^2 + c - \frac{b^2}{4a}$$

Write the trinomial as the square of a binomial and simplify the constant term.

Now let's compare the form of this equation with a quadratic function in the form $f(x) = a(x-h)^2 + k$.

The form we know how to graph: $f(x) = a(x-h)^2 + k$

$h = -\frac{b}{2a}$ $\quad$ $k = c - \frac{b^2}{4a}$

Equation under discussion: $f(x) = a\left(x - \left(-\frac{b}{2a}\right)\right)^2 + c - \frac{b^2}{4a}$

The important part of this observation is that h, the x-coordinate of the vertex, is $-\frac{b}{2a}$. The y-coordinate can be found by evaluating the function at $-\frac{b}{2a}$.

The Vertex of a Parabola Whose Equation Is $f(x) = ax^2 + bx + c$

Consider the parabola defined by the quadratic function $f(x) = ax^2 + bx + c$. The parabola's vertex is $\left(-\frac{b}{2a}, f\left(-\frac{b}{2a}\right)\right)$. The x-coordinate is $-\frac{b}{2a}$. The y-coordinate is found by substituting the x-coordinate into the parabola's equation and evaluating the function at this value of x.

EXAMPLE 3 Finding a Parabola's Vertex

Find the vertex for the parabola whose equation is $f(x) = 3x^2 + 12x + 8$.

Solution We know that the x-coordinate of the vertex is $x = -\frac{b}{2a}$. Let's identify the numbers a, b, and c in the given equation, which is in the form $f(x) = ax^2 + bx + c$.

$$f(x) = 3x^2 + 12x + 8$$

$a = 3$ $b = 12$ $c = 8$

Substitute the values of a and b into the equation for the x-coordinate:

$$x = -\frac{b}{2a} = -\frac{12}{2 \cdot 3} = -\frac{12}{6} = -2.$$

The x-coordinate of the vertex is -2. We substitute -2 for x into the equation of the function, $f(x) = 3x^2 + 12x + 8$, to find the y-coordinate:

$$f(-2) = 3(-2)^2 + 12(-2) + 8 = 3(4) + 12(-2) + 8 = 12 - 24 + 8 = -4.$$

The vertex is $(-2, -4)$.

☑ **CHECK POINT 3** Find the vertex for the parabola whose equation is $f(x) = 2x^2 + 8x - 1$.

We can apply our five-step procedure and graph parabolas in the form $f(x) = ax^2 + bx + c$.

Graphing Quadratic Functions with Equations in the Form $f(x) = ax^2 + bx + c$

To graph $f(x) = ax^2 + bx + c$,

1. Determine whether the parabola opens upward or downward. If $a > 0$, it opens upward. If $a < 0$, it opens downward.
2. Determine the vertex of the parabola. The vertex is $\left(-\frac{b}{2a}, f\left(-\frac{b}{2a}\right)\right)$.
3. Find any x-intercepts by solving $f(x) = 0$. The real solutions of $ax^2 + bx + c = 0$ are the x-coordinates of the x-intercepts.
4. Find the y-intercept by computing $f(0)$. Because $f(0) = c$ (the constant term in the function's equation), the y-intercept is $(0, c)$.
5. Plot the intercepts, the vertex, and additional points as necessary. Connect these points with a smooth curve.

EXAMPLE 4 Graphing a Quadratic Function in the Form $f(x) = ax^2 + bx + c$

Graph the quadratic function $f(x) = -x^2 - 2x + 1$. Use the graph to identify the function's domain and its range.

Solution

Step 1. Determine how the parabola opens. Note that a, the coefficient of x^2, is -1. Thus, $a < 0$; this negative value tells us that the parabola opens downward.

Step 2. Find the vertex. We know that the x-coordinate of the vertex is $x = -\frac{b}{2a}$. We identify a, b, and c in $f(x) = ax^2 + bx + c$.

$$f(x) = -x^2 - 2x + 1$$

$a = -1 \quad b = -2 \quad c = 1$

Substitute the values of a and b into the equation for the x-coordinate:

$$x = -\frac{b}{2a} = -\frac{-2}{2(-1)} = -\left(\frac{-2}{-2}\right) = -1.$$

The x-coordinate of the vertex is -1. We substitute -1 for x into the equation of the function, $f(x) = -x^2 - 2x + 1$, to find the y-coordinate:

$$f(-1) = -(-1)^2 - 2(-1) + 1 = -1 + 2 + 1 = 2.$$

The vertex is $(-1, 2)$.

Step 3. Find the x-intercepts by solving $f(x) = 0$. Replace $f(x)$ with 0 in $f(x) = -x^2 - 2x + 1$. We obtain $0 = -x^2 - 2x + 1$. This equation cannot be solved by factoring. We will use the quadratic formula to solve it.

$$-x^2 - 2x + 1 = 0$$

$a = -1 \quad b = -2 \quad c = 1$

$$x = \frac{-b \pm \sqrt{b^2 - 4ac}}{2a} = \frac{-(-2) \pm \sqrt{(-2)^2 - 4(-1)(1)}}{2(-1)} = \frac{2 \pm \sqrt{4 - (-4)}}{-2}$$

To locate the x-intercepts, we need decimal approximations. Thus, there is no need to simplify the radical form of the solutions.

$$x = \frac{2 + \sqrt{8}}{-2} \approx -2.4 \quad \text{or} \quad x = \frac{2 - \sqrt{8}}{-2} \approx 0.4$$

The x-intercepts are approximately $(-2.4, 0)$ and $(0.4, 0)$.

Step 4. Find the y-intercept by computing $f(0)$. Replace x with 0 in $f(x) = -x^2 - 2x + 1$.

$$f(0) = -0^2 - 2 \cdot 0 + 1 = 1$$

The y-intercept is (0, 1), which is the constant term in the function's equation. The parabola passes through (0, 1).

Step 5. Graph the parabola. With a vertex at $(-1, 2)$, x-intercepts at $(-2.4, 0)$ and $(0.4, 0)$, and a y-intercept at $(0, 1)$, the graph of f is shown in **Figure 8.10(a)**. The axis of symmetry is the vertical line whose equation is $x = -1$.

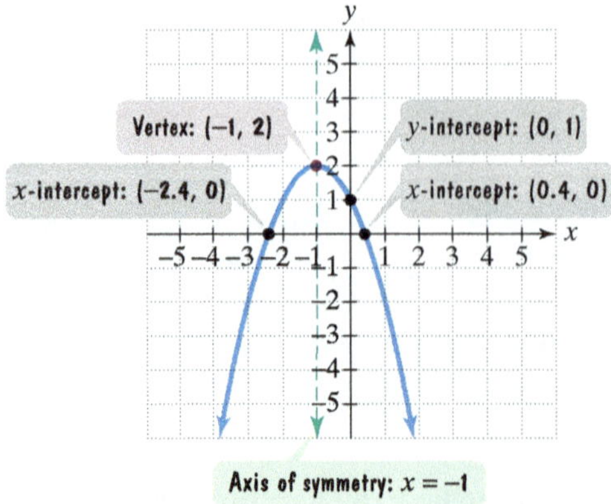

FIGURE 8.10(a) The graph of $f(x) = -x^2 - 2x + 1$

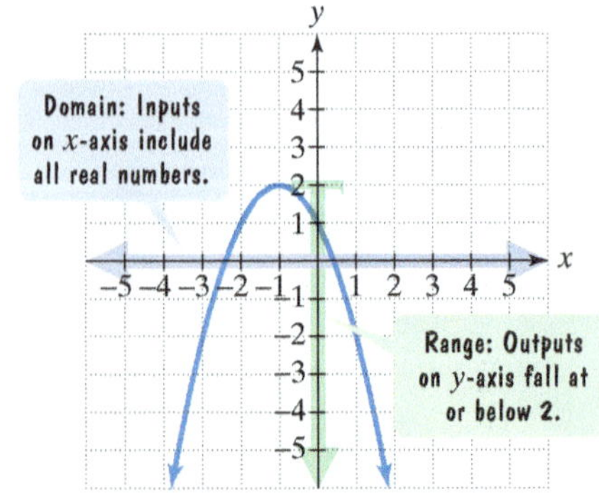

FIGURE 8.10(b) Determining the domain and range of $f(x) = -x^2 - 2x + 1$

Study Tip

The domain of any quadratic function includes all real numbers. If the vertex is the graph's highest point, the range includes all real numbers at or below the y-coordinate of the vertex. If the vertex is the graph's lowest point, the range includes all real numbers at or above the y-coordinate of the vertex.

Now we are ready to determine the domain and range of $f(x) = -x^2 - 2x + 1$. We can use the parabola, shown again in **Figure 8.10(b)**, to do so. To find the domain, look for all the inputs on the x-axis that correspond to points on the graph. As the graph widens and continues to fall at both ends, can you see that these inputs include all real numbers?

$$\text{Domain of } f \text{ is } \{x|x \text{ is a real number}\} \text{ or } (-\infty, \infty).$$

To find the range, look for all the outputs on the y-axis that correspond to points on the graph. **Figure 8.10(b)** shows that the parabola's vertex, $(-1, 2)$, is the highest point on the graph. Because the y-coordinate of the vertex is 2, outputs on the y-axis fall at or below 2.

$$\text{Range of } f \text{ is } \{y|y \leq 2\} \text{ or } (-\infty, 2].$$

CHECK POINT 4 Graph the quadratic function $f(x) = -x^2 + 4x + 1$. Use the graph to identify the function's domain and its range.

4 Determine a quadratic function's minimum or maximum value.

Minimum and Maximum Values of Quadratic Functions

Consider the quadratic function $f(x) = ax^2 + bx + c$. If $a > 0$, the parabola opens upward and the vertex is its lowest point. If $a < 0$, the parabola opens downward and the vertex is its highest point. The x-coordinate of the vertex is $-\frac{b}{2a}$. Thus, we can find the minimum or maximum value of f by evaluating the quadratic function at $x = -\frac{b}{2a}$.

Minimum and Maximum: Quadratic Functions

Consider the quadratic function $f(x) = ax^2 + bx + c$.

1. If $a > 0$, then f has a minimum that occurs at $x = -\frac{b}{2a}$. This minimum value is $f\left(-\frac{b}{2a}\right)$.
2. If $a < 0$, then f has a maximum that occurs at $x = -\frac{b}{2a}$. This maximum value is $f\left(-\frac{b}{2a}\right)$.

In each case, the value of x gives the location of the minimum or maximum value. The value of y, or $f\left(-\frac{b}{2a}\right)$, gives that minimum or maximum value.

EXAMPLE 5 Obtaining Information about a Quadratic Function from Its Equation

Consider the quadratic function $f(x) = -3x^2 + 6x - 13$.

a. Determine, without graphing, whether the function has a minimum value or a maximum value.

b. Find the minimum or maximum value and determine where it occurs.

c. Identify the function's domain and its range.

Solution We begin by identifying a, b, and c in the function's equation:

$$f(x) = -3x^2 + 6x - 13.$$

$a = -3$ $b = 6$ $c = -13$

$f(x) = -3x^2 + 6x - 13$

$a = -3$ $b = 6$ $c = -13$

The given function (repeated)

a. Because $a < 0$, the function has a maximum value.

b. The maximum value occurs at

$$x = -\frac{b}{2a} = -\frac{6}{2(-3)} = -\frac{6}{-6} = -(-1) = 1.$$

The maximum value occurs at $x = 1$ and the maximum value of $f(x) = -3x^2 + 6x - 13$ is

$$f(1) = -3 \cdot 1^2 + 6 \cdot 1 - 13 = -3 + 6 - 13 = -10.$$

We see that the maximum is -10 at $x = 1$.

c. Like all quadratic functions, the domain is $\{x|x \text{ is a real number}\}$ or $(-\infty, \infty)$. Because the function's maximum value is -10, the range includes all real numbers at or below -10. The range is $\{y|y \leq -10\}$ or $(-\infty, -10]$.

We can use the graph of $f(x) = -3x^2 + 6x - 13$ to visualize the results of Example 5. **Figure 8.11** shows the graph in a $[-6, 6, 1]$ by $[-50, 20, 10]$ viewing rectangle. The maximum function feature verifies that the function's maximum is -10 at $x = 1$. Notice that x gives the location of the maximum and y gives the maximum value. Notice, too, that the maximum value is -10 and not the ordered pair $(1, -10)$.

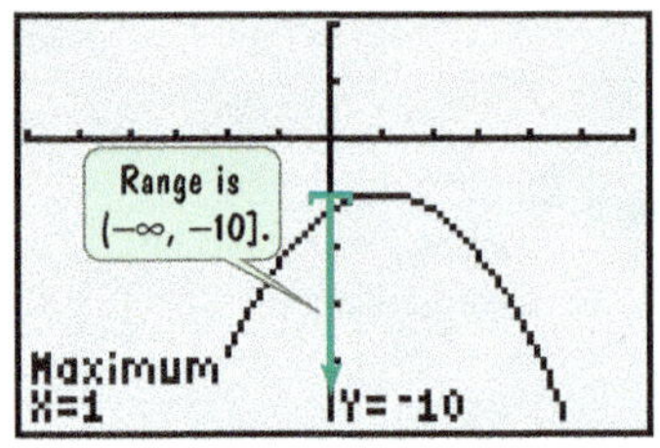

$[-6, 6, 1]$ by $[-50, 20, 10]$

FIGURE 8.11

CHECK POINT 5 Repeat parts (a) through (c) of Example 5 using the quadratic function $f(x) = 4x^2 - 16x + 1000$.

5 Solve problems involving a quadratic function's minimum or maximum value.

Applications of Quadratic Functions

Many applied problems involve finding the maximum or minimum value of a quadratic function, as well as where this value occurs.

EXAMPLE 6 Parabolic Paths of a Shot Put

An athlete whose event is the shot put releases the shot with the same initial velocity, but at different angles. **Figure 8.12** shows the parabolic paths for shots released at angles of 35° and 65°.

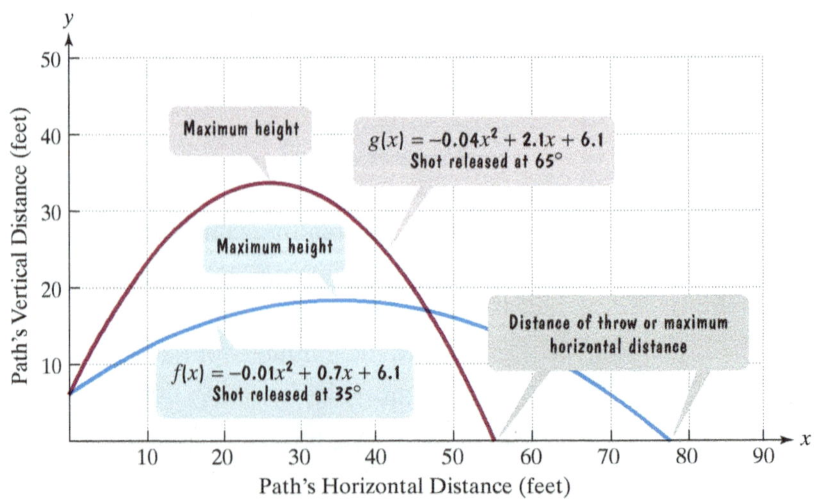

FIGURE 8.12 Two paths of a shot put

When the shot is released at an angle of 35°, its path can be modeled by the function

$$f(x) = -0.01x^2 + 0.7x + 6.1,$$

in which x is the shot's horizontal distance, in feet, and $f(x)$ is its height, in feet. What is the maximum height of this shot's path?

Solution The quadratic function is in the form $f(x) = ax^2 + bx + c$, with $a = -0.01$ and $b = 0.7$. Because $a < 0$, the function has a maximum that occurs at $x = -\frac{b}{2a}$.

$$x = -\frac{b}{2a} = -\frac{0.7}{2(-0.01)} = -(-35) = 35$$

This means that the shot's maximum height occurs when its horizontal distance is 35 feet. Can you see how this is shown by the blue graph of f in **Figure 8.12**? The maximum height of this path is

$$f(35) = -0.01(35)^2 + 0.7(35) + 6.1 = 18.35$$

or 18.35 feet.

CHECK POINT 6 Use function g, whose equation and graph are shown in **Figure 8.12**, to find the maximum height, to the nearest tenth of a foot, when the shot is released at an angle of 65°.

Quadratic functions can also be modeled from verbal conditions. Once we have obtained a quadratic function, we can then use the x-coordinate of the vertex to determine its maximum or minimum value. Here is a step-by-step strategy for solving these kinds of problems:

Strategy for Solving Problems Involving Maximizing or Minimizing Quadratic Functions

1. Read the problem carefully and decide which quantity is to be maximized or minimized.
2. Use the conditions of the problem to express the quantity as a function in one variable.
3. Rewrite the function in the form $f(x) = ax^2 + bx + c$.
4. Calculate $-\frac{b}{2a}$. If $a > 0$, f has a minimum at $x = -\frac{b}{2a}$. This minimum value is $f\left(-\frac{b}{2a}\right)$. If $a < 0$, f has a maximum at $x = -\frac{b}{2a}$. This maximum value is $f\left(-\frac{b}{2a}\right)$.
5. Answer the question posed in the problem.

EXAMPLE 7 Minimizing a Product

Among all pairs of numbers whose difference is 10, find a pair whose product is as small as possible. What is the minimum product?

Solution

Step 1. Decide what must be maximized or minimized. We must minimize the product of two numbers. Calling the numbers x and y, and calling the product P, we must minimize

$$P = xy.$$

Step 2. Express this quantity as a function in one variable. In the formula $P = xy$, P is expressed in terms of two variables, x and y. However, because the difference of the numbers is 10, we can write

$$x - y = 10.$$

We can solve this equation for y in terms of x (or vice versa), substitute the result into $P = xy$, and obtain P as a function of one variable.

$$-y = -x + 10$$ Subtract x from both sides of $x - y = 10$.

$$y = x - 10$$ Multiply both sides of the equation by -1 and solve for y.

Now we substitute $x - 10$ for y in $P = xy$.

$$P = xy = x(x - 10).$$

Because P is now a function of x, we can write

$$P(x) = x(x - 10).$$

Step 3. Write the function in the form $f(x) = ax^2 + bx + c$. We apply the distributive property to obtain

$$P(x) = x(x - 10) = x^2 - 10x.$$

$a = 1$ $b = -10$

Using Technology

Numeric Connections

The TABLE feature of a graphing utility can be used to verify our work in Example 7.

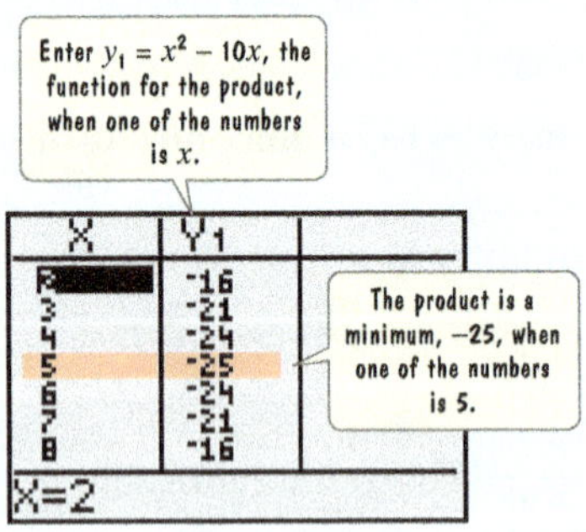

Step 4. Calculate $-\frac{b}{2a}$. If $a > 0$, the function has a minimum at this value. The voice balloons show that $a = 1$ and $b = -10$.

$$x = -\frac{b}{2a} = -\frac{-10}{2(1)} = -(-5) = 5$$

This means that the product, P, of two numbers whose difference is 10 is a minimum when one of the numbers, x, is 5.

Step 5. Answer the question posed by the problem. The problem asks for the two numbers and the minimum product. We found that one of the numbers, x, is 5. Now we must find the second number, y.

$$y = x - 10 = 5 - 10 = -5.$$

The number pair whose difference is 10 and whose product is as small as possible is 5, -5. The minimum product is $5(-5)$, or -25.

✓ **CHECK POINT 7** Among all pairs of numbers whose difference is 8, find a pair whose product is as small as possible. What is the minimum product?

EXAMPLE 8 Maximizing Area

You have 100 yards of fencing to enclose a rectangular region. Find the dimensions of the rectangle that maximize the enclosed area. What is the maximum area?

Solution

Step 1. Decide what must be maximized or minimized. We must maximize area. What we do not know are the rectangle's dimensions, x and y.

Step 2. Express this quantity as a function in one variable. Because we must maximize area, we have $A = xy$. We need to transform this into a function in which A is represented by one variable. Because you have 100 yards of fencing, the perimeter of the rectangle is 100 yards. This means that

$$2x + 2y = 100.$$

We can solve this equation for y in terms of x, substitute the result into $A = xy$, and obtain A as a function in one variable. We begin by solving for y.

$$2y = 100 - 2x \quad \text{Subtract } 2x \text{ from both sides.}$$

$$y = \frac{100 - 2x}{2} \quad \text{Divide both sides by 2.}$$

$$y = 50 - x \quad \text{Divide each term in the numerator by 2.}$$

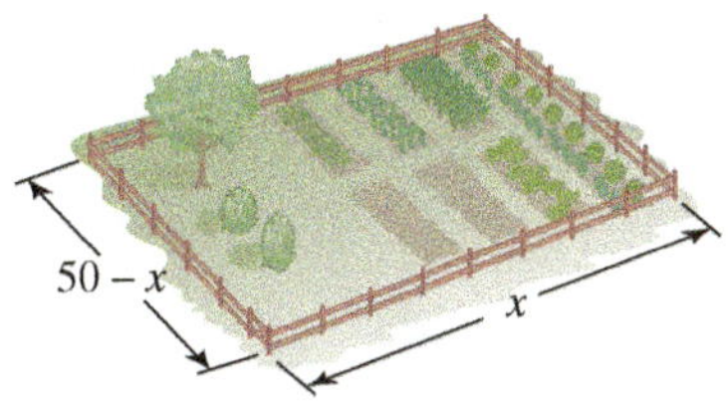

FIGURE 8.13 What value of x will maximize the rectangle's area?

Now we substitute $50 - x$ for y in $A = xy$.

$$A = xy = x(50 - x)$$

The rectangle and its dimensions are illustrated in **Figure 8.13**. Because A is now a function of x, we can write

$$A(x) = x(50 - x).$$

This function models the area, $A(x)$, of any rectangle whose perimeter is 100 yards in terms of one of its dimensions, x.

Step 3. Write the function in the form $f(x) = ax^2 + bx + c$. We apply the distributive property to obtain

$$A(x) = x(50 - x) = 50x - x^2 = -x^2 + 50x.$$

$a = -1$ $b = 50$

Using Technology

Graphic Connections

The graph of the area function

$$A(x) = x(50 - x)$$

was obtained with a graphing utility using a [0, 50, 2] by [0, 700, 25] viewing rectangle. The maximum function feature verifies that a maximum area of 625 square yards occurs when one of the dimensions is 25 yards.

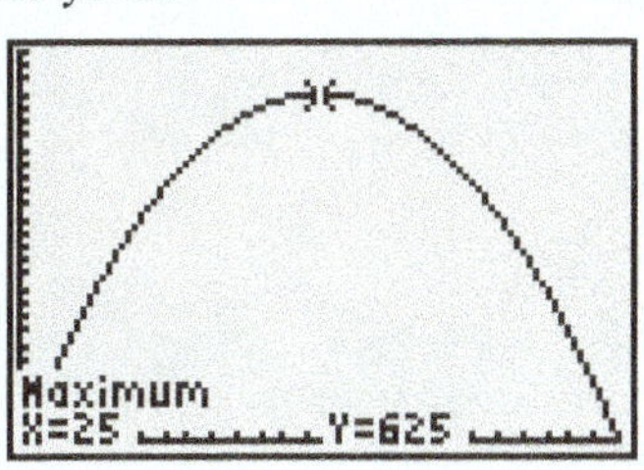

Step 4. Calculate $-\frac{b}{2a}$. If $a < 0$, the function has a maximum at this value. The voice balloons show that $a = -1$ and $b = 50$.

$$x = -\frac{b}{2a} = -\frac{50}{2(-1)} = 25$$

This means that the area, $A(x)$, of a rectangle with perimeter 100 yards is a maximum when one of the rectangle's dimensions, x, is 25 yards.

Step 5. Answer the question posed by the problem. We found that $x = 25$. **Figure 8.13** shows that the rectangle's other dimension is $50 - x = 50 - 25 = 25$. The dimensions of the rectangle that maximize the enclosed area are 25 yards by 25 yards. The rectangle that gives the maximum area is actually a square with an area of 25 yards · 25 yards, or 625 square yards.

CHECK POINT 8 You have 120 feet of fencing to enclose a rectangular region. Find the dimensions of the rectangle that maximize the enclosed area. What is the maximum area?

8.3 EXERCISE SET

MyMathLab — MathXL PRACTICE — WATCH — DOWNLOAD — READ — REVIEW

Practice Exercises

In Exercises 1–4, the graph of a quadratic function is given. Write the function's equation, selecting from the following options:

$$f(x) = (x + 1)^2 - 1,\ g(x) = (x + 1)^2 + 1,$$
$$h(x) = (x - 1)^2 + 1,\ j(x) = (x - 1)^2 - 1.$$

1.

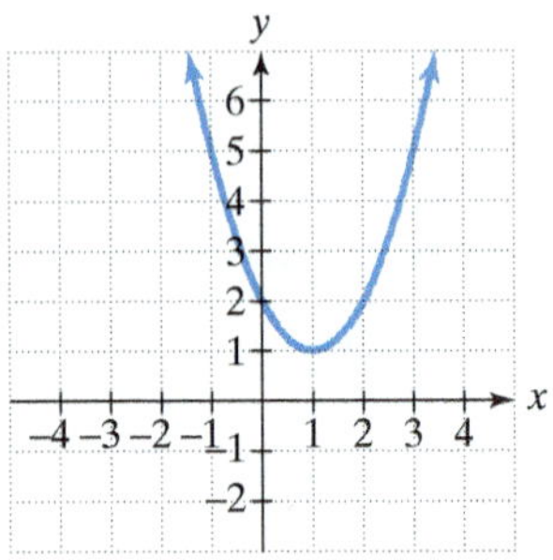

2.

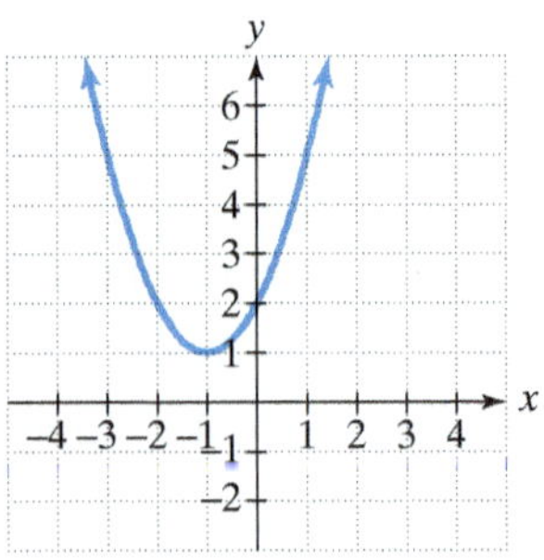

3.

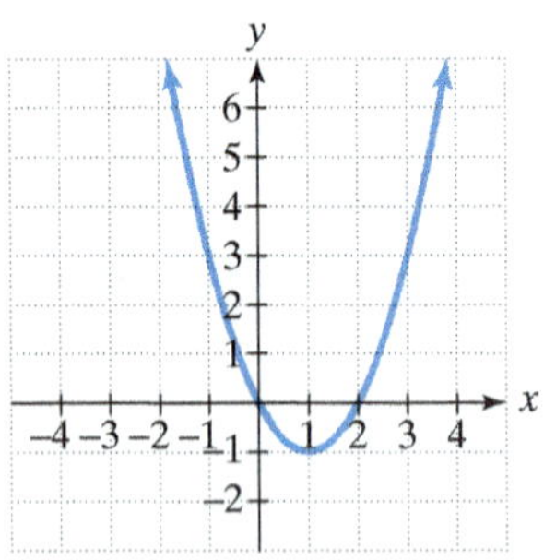

4.

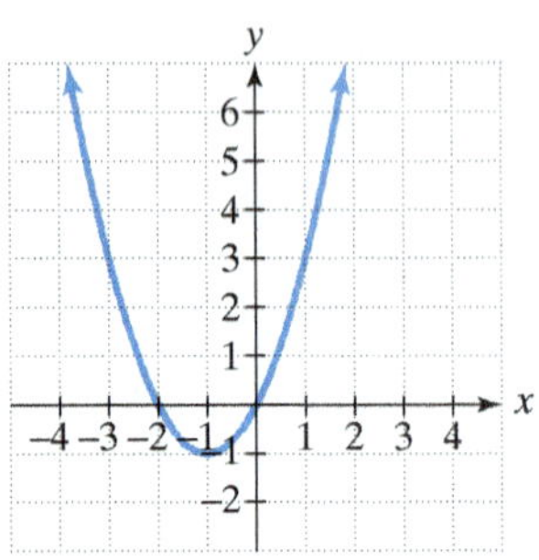

In Exercises 5–8, the graph of a quadratic function is given. Write the function's equation, selecting from the following options:

$$f(x) = x^2 + 2x + 1,\ g(x) = x^2 - 2x + 1,$$
$$h(x) = x^2 - 1,\ j(x) = -x^2 - 1.$$

5.

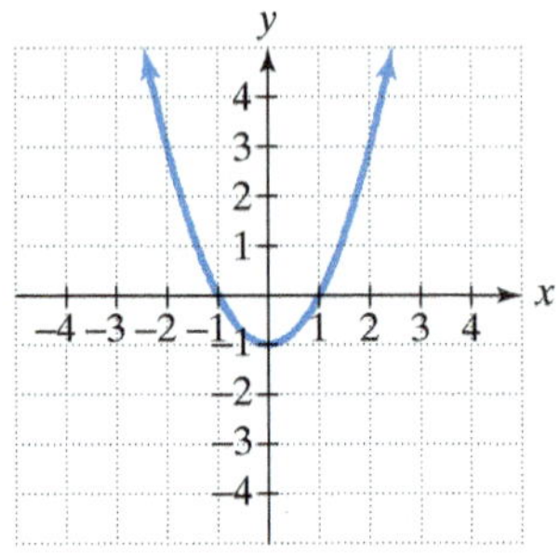

6.

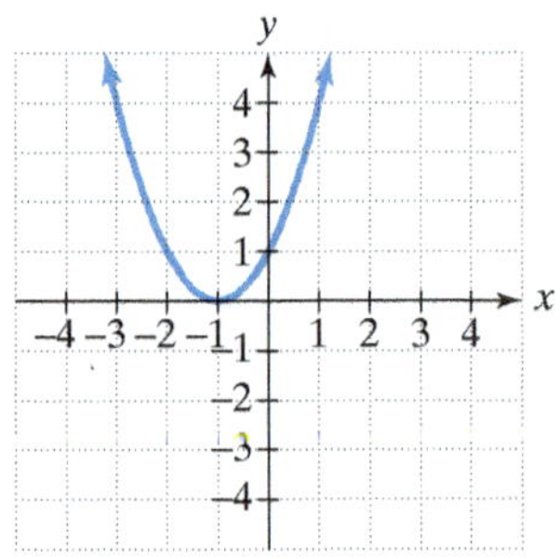

7.

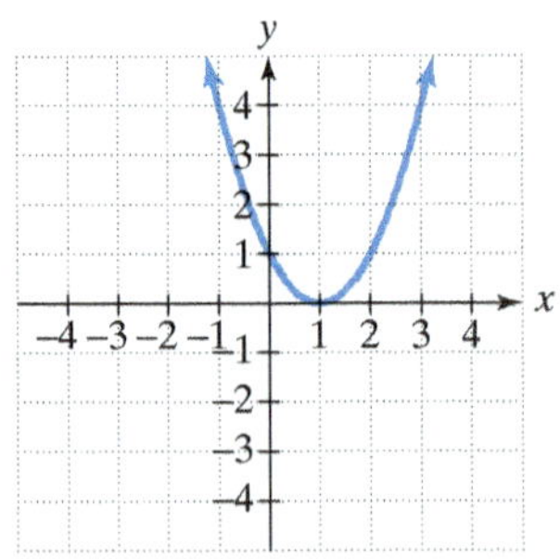

8.

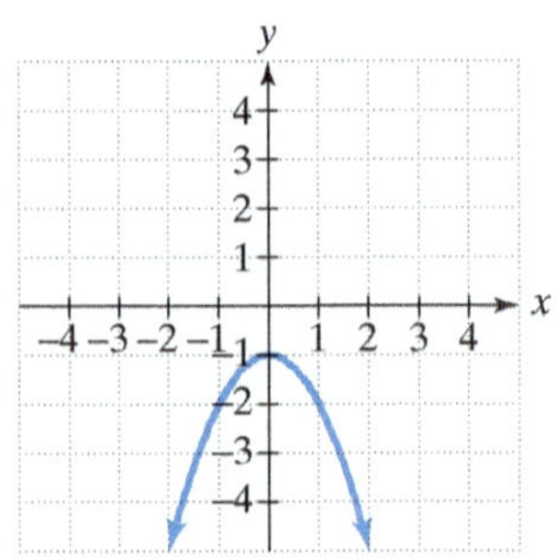

In Exercises 9–16, find the coordinates of the vertex for the parabola defined by the given quadratic function.

9. $f(x) = 2(x - 3)^2 + 1$
10. $f(x) = -3(x - 2)^2 + 12$
11. $f(x) = -2(x + 1)^2 + 5$
12. $f(x) = -2(x + 4)^2 - 8$
13. $f(x) = 2x^2 - 8x + 3$
14. $f(x) = 3x^2 - 12x + 1$
15. $f(x) = -x^2 - 2x + 8$
16. $f(x) = -2x^2 + 8x - 1$

In Exercises 17–38, use the vertex and intercepts to sketch the graph of each quadratic function. Use the graph to identify the function's range.

17. $f(x) = (x - 4)^2 - 1$
18. $f(x) = (x - 1)^2 - 2$
19. $f(x) = (x - 1)^2 + 2$
20. $f(x) = (x - 3)^2 + 2$
21. $y - 1 = (x - 3)^2$
22. $y - 3 = (x - 1)^2$
23. $f(x) = 2(x + 2)^2 - 1$
24. $f(x) = \frac{5}{4} - \left(x - \frac{1}{2}\right)^2$
25. $f(x) = 4 - (x - 1)^2$
26. $f(x) = 1 - (x - 3)^2$
27. $f(x) = x^2 - 2x - 3$
28. $f(x) = x^2 - 2x - 15$
29. $f(x) = x^2 + 3x - 10$
30. $f(x) = 2x^2 - 7x - 4$
31. $f(x) = 2x - x^2 + 3$
32. $f(x) = 5 - 4x - x^2$
33. $f(x) = x^2 + 6x + 3$
34. $f(x) = x^2 + 4x - 1$
35. $f(x) = 2x^2 + 4x - 3$
36. $f(x) = 3x^2 - 2x - 4$
37. $f(x) = 2x - x^2 - 2$
38. $f(x) = 6 - 4x + x^2$

In Exercises 39–44, an equation of a quadratic function is given.

a. *Determine, without graphing, whether the function has a minimum value or a maximum value.*
b. *Find the minimum or maximum value and determine where it occurs.*
c. *Identify the function's domain and its range.*

39. $f(x) = 3x^2 - 12x - 1$
40. $f(x) = 2x^2 - 8x - 3$
41. $f(x) = -4x^2 + 8x - 3$
42. $f(x) = -2x^2 - 12x + 3$
43. $f(x) = 5x^2 - 5x$
44. $f(x) = 6x^2 - 6x$

Practice PLUS

In Exercises 45–48, give the domain and the range of each quadratic function whose graph is described.

45. The vertex is $(-1, -2)$ and the parabola opens up.
46. The vertex is $(-3, -4)$ and the parabola opens down.
47. Maximum $= -6$ at $x = 10$
48. Minimum $= 18$ at $x = -6$

In Exercises 49–52, write an equation of the parabola that has the same shape as the graph of $f(x) = 2x^2$, but with the given point as the vertex.

49. $(5, 3)$
50. $(7, 4)$
51. $(-10, -5)$
52. $(-8, -6)$

In Exercises 53–56, write an equation of the parabola that has the same shape as the graph of $f(x) = 3x^2$ or $g(x) = -3x^2$, but with the given maximum or minimum.

53. Maximum $= 4$ at $x = -2$
54. Maximum $= -7$ at $x = 5$
55. Minimum $= 0$ at $x = 11$
56. Minimum $= 0$ at $x = 9$

Application Exercises

57. The graph shows U.S. adult wine consumption, in gallons per person, for selected years from 1980 through 2005.

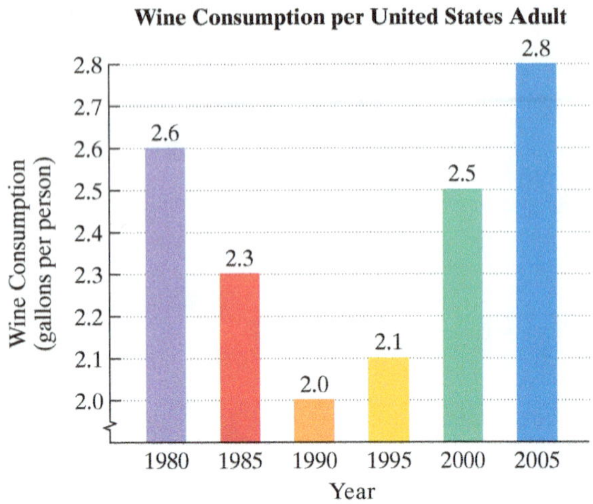

Source: Adams Business Media

The function

$$f(x) = 0.004x^2 - 0.094x + 2.6$$

models U.S. wine consumption, $f(x)$, in gallons per person, x years after 1980.

a. According to this function, what was U.S. adult wine consumption in 2005? Does this overestimate or underestimate the value shown by the graph? By how much?

b. According to this function, in which year was wine consumption at a minimum? Round to the nearest year. What does the function give for per capita consumption for that year? Does this seem reasonable in terms of the data shown by the graph or has model breakdown occurred?

58. The graph shows the number of movie tickets sold in the United States, in billions, from 2000 through 2005.

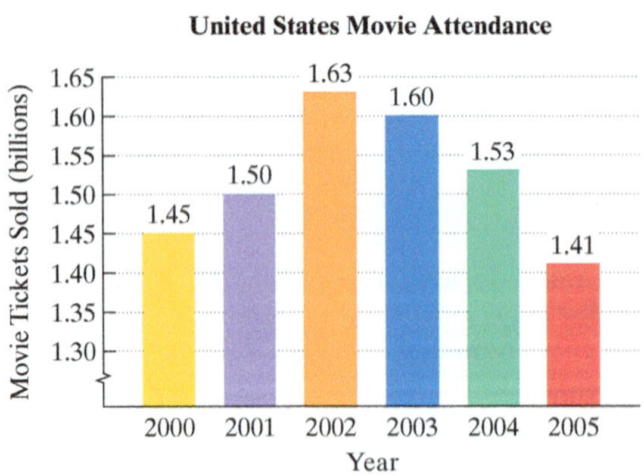

Source: National Association of Theater Owners

The function

$$f(x) = -0.03x^2 + 0.14x + 1.43$$

models U.S. movie attendance, $f(x)$, in billions of tickets sold, x years after 2000.

a. According to this function, how many billions of movie tickets were sold in 2005? Does this overestimate or underestimate the number shown by the graph? By how much?

b. According to this function, in which year was movie attendance at a maximum? Round to the nearest year. What does the function give for the billions of tickets sold for that year? By how much does this differ from the number shown by the graph?

59. A person standing close to the edge on the top of a 160-foot building throws a baseball vertically upward. The quadratic function

$$s(t) = -16t^2 + 64t + 160$$

models the ball's height above the ground, $s(t)$, in feet, t seconds after it was thrown.

a. After how many seconds does the ball reach its maximum height? What is the maximum height?

b. How many seconds does it take until the ball finally hits the ground? Round to the nearest tenth of a second.

c. Find $s(0)$ and describe what this means.

d. Use your results from parts (a) through (c) to graph the quadratic function. Begin the graph with $t = 0$ and end with the value of t for which the ball hits the ground.

60. A person standing close to the edge on the top of a 200-foot building throws a baseball vertically upward. The quadratic function

$$s(t) = -16t^2 + 64t + 200$$

models the ball's height above the ground, $s(t)$, in feet, t seconds after it was thrown.

a. After how many seconds does the ball reach its maximum height? What is the maximum height?

b. How many seconds does it take until the ball finally hits the ground? Round to the nearest tenth of a second.

c. Find $s(0)$ and describe what this means.

d. Use your results from parts (a) through (c) to graph the quadratic function. Begin the graph with $t = 0$ and end with the value of t for which the ball hits the ground.

61. Among all pairs of numbers whose sum is 16, find a pair whose product is as large as possible. What is the maximum product?

62. Among all pairs of numbers whose sum is 20, find a pair whose product is as large as possible. What is the maximum product?

63. Among all pairs of numbers whose difference is 16, find a pair whose product is as small as possible. What is the minimum product?

64. Among all pairs of numbers whose difference is 24, find a pair whose product is as small as possible. What is the minimum product?

65. You have 600 feet of fencing to enclose a rectangular plot that borders on a river. If you do not fence the side along the river, find the length and width of the plot that will maximize the area. What is the largest area that can be enclosed?

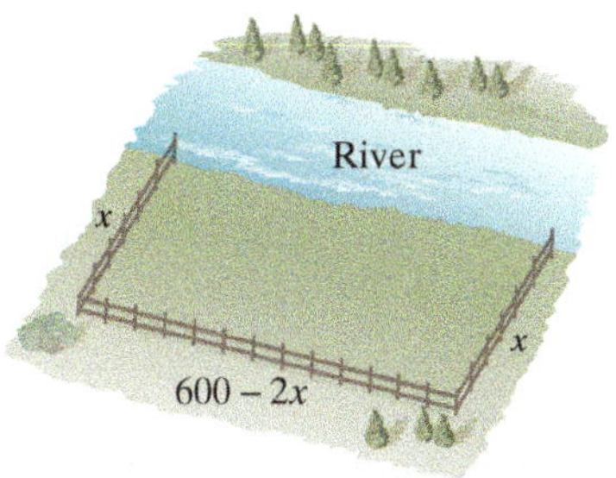

66. You have 200 feet of fencing to enclose a rectangular plot that borders on a river. If you do not fence the side along the river, find the length and width of the plot that will maximize the area. What is the largest area that can be enclosed?

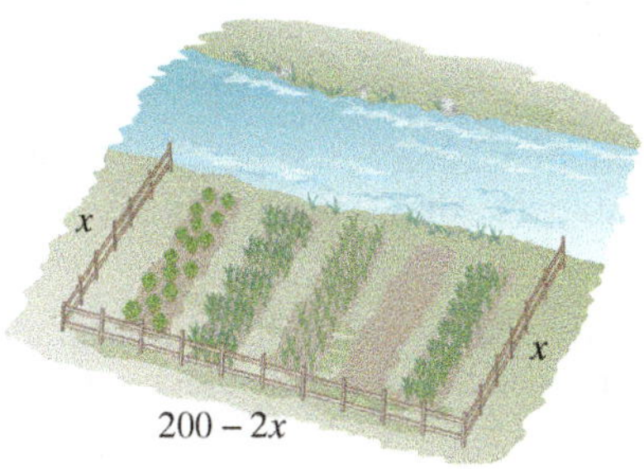

67. You have 50 yards of fencing to enclose a rectangular region. Find the dimensions of the rectangle that maximize the enclosed area. What is the maximum area?

68. You have 80 yards of fencing to enclose a rectangular region. Find the dimensions of the rectangle that maximize the enclosed area. What is the maximum area?

69. A rain gutter is made from sheets of aluminum that are 20 inches wide by turning up the edges to form right angles. Determine the depth of the gutter that will maximize its cross sectional area and allow the greatest amount of water to flow. What is the maximum cross-sectional area?

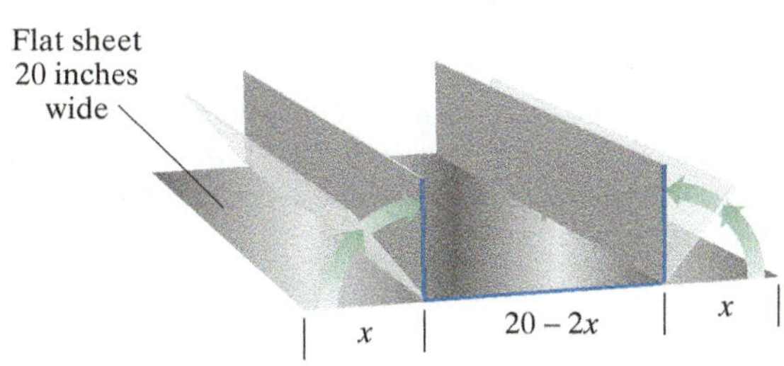

70. A rain gutter is made from sheets of aluminum that are 12 inches wide by turning up the edges to form right angles. Determine the depth of the gutter that will maximize its cross-sectional area and allow the greatest amount of water to flow. What is the maximum cross-sectional area?

In Chapter 3, we saw that the profit, $P(x)$, generated after producing and selling x units of a product is given by the function

$$P(x) = R(x) - C(x),$$

where R and C are the revenue and cost functions, respectively. Use these functions to solve Exercises 71–72.

71. Hunky Beef, a local sandwich store, has a fixed weekly cost of \$525.00, and variable costs for making a roast beef sandwich are \$0.55.

a. Let x represent the number of roast beef sandwiches made and sold each week. Write the weekly cost function, C, for Hunky Beef.

b. The function $R(x) = -0.001x^2 + 3x$ describes the money that Hunky Beef takes in each week from the sale of x roast beef sandwiches. Use this revenue function and the cost function from part (a) to write the store's weekly profit function, P.

c. Use the store's profit function to determine the number of roast beef sandwiches it should make and sell each week to maximize profit. What is the maximum weekly profit?

72. Virtual Fido is a company that makes electronic virtual pets. The fixed weekly cost is \$3000, and variable costs for each pet are \$20.

a. Let x represent the number of virtual pets made and sold each week. Write the weekly cost function, C, for Virtual Fido.

b. The function $R(x) = -x^2 + 1000x$ describes the money that Virtual Fido takes in each week from the sale of x virtual pets. Use this revenue function and the cost function from part (a) to write the weekly profit function, P.

c. Use the profit function to determine the number of virtual pets that should be made and sold each week to maximize profit. What is the maximum weekly profit?

Writing in Mathematics

73. What is a parabola? Describe its shape.

74. Explain how to decide whether a parabola opens upward or downward.

75. Describe how to find a parabola's vertex if its equation is in the form $f(x) = a(x - h)^2 + k$. Give an example.

76. Describe how to find a parabola's vertex if its equation is in the form $f(x) = ax^2 + bx + c$. Use $f(x) = x^2 - 6x + 8$ as an example.

77. A parabola that opens upward has its vertex at (1, 2). Describe as much as you can about the parabola based on this information. Include in your discussion the number of x-intercepts (if any) for the parabola.

Technology Exercises

78. Use a graphing utility to verify any five of your hand-drawn graphs in Exercises 17–38.

79. a. Use a graphing utility to graph $y = 2x^2 - 82x + 720$ in a standard viewing rectangle. What do you observe?

b. Find the coordinates of the vertex for the given quadratic function.

c. The answer to part (b) is (20.5, −120.5). Because the leading coefficient, 2, of the given function is positive, the vertex is a minimum point on the graph. Use this fact to help find a viewing rectangle that will give a relatively complete picture of the parabola. With an axis of symmetry at $x = 20.5$, the setting for x should extend past this, so try Xmin = 0 and Xmax = 30. The setting for y should include (and probably go below) the y-coordinate of the graph's minimum point, so try Ymin = −130. Experiment with Ymax until your utility shows the parabola's major features.

d. In general, explain how knowing the coordinates of a parabola's vertex can help determine a reasonable viewing rectangle on a graphing utility for obtaining a complete picture of the parabola.

In Exercises 80–83, find the vertex for each parabola. Then determine a reasonable viewing rectangle on your graphing utility and use it to graph the parabola.

80. $y = -0.25x^2 + 40x$

81. $y = -4x^2 + 20x + 160$

82. $y = 5x^2 + 40x + 600$

83. $y = 0.01x^2 + 0.6x + 100$

84. The following data show fuel efficiency, in miles per gallon, for all U.S. automobiles in the indicated year.

x (Years after 1940)	y (Average Number of Miles per Gallon for U.S. Automobiles)
1940: 0	14.8
1950: 10	13.9
1960: 20	13.4
1970: 30	13.5
1980: 40	15.9
1990: 50	20.2
2000: 60	22.0

Source: U.S. Department of Transportation

a. Use a graphing utility to draw a scatter plot of the data. Explain why a quadratic function is appropriate for modeling these data.

b. Use the quadratic regression feature to find the quadratic function that best fits the data.

c. Use the equation in part (b) to determine the worst year for automobile fuel efficiency. What was the average number of miles per gallon for that year?

d. Use a graphing utility to draw a scatter plot of the data and graph the quadratic function of best fit on the scatter plot.

Critical Thinking Exercises

Make Sense? *In Exercises 85–88, determine whether each statement "makes sense" or "does not make sense" and explain your reasoning.*

85. Parabolas that open up appear to form smiles ($a > 0$), while parabolas that open down frown ($a < 0$).

86. I must have made an error when graphing this parabola because its axis of symmetry is the y-axis.

87. I like to think of a parabola's vertex as the point where it intersects its axis of symmetry.

88. I threw a baseball vertically upward and its path was a parabola.

In Exercises 89–92, determine whether each statement is true or false. If the statement is false, make the necessary change(s) to produce a true statement.

89. No quadratic functions have a range of $(-\infty, \infty)$.

90. The vertex of the parabola described by $f(x) = 2(x - 5)^2 - 1$ is at (5, 1).

91. The graph of $f(x) = -2(x + 4)^2 - 8$ has one y-intercept and two x-intercepts.

92. The maximum value of y for the quadratic function $f(x) = -x^2 + x + 1$ is 1.

In Exercises 93–94, find the axis of symmetry for each parabola whose equation is given. Use the axis of symmetry to find a second point on the parabola whose y-coordinate is the same as the given point.

93. $f(x) = 3(x + 2)^2 - 5;\quad (-1, -2)$

94. $f(x) = (x - 3)^2 + 2; (6, 11)$

In Exercises 95–96, write the equation of each parabola in $f(x) = a(x - h)^2 + k$ form.

95. Vertex: $(-3, -4)$; The graph passes through the point $(1, 4)$.

96. Vertex: $(-3, -1)$; The graph passes through the point $(-2, -3)$.

97. A rancher has 1000 feet of fencing to construct six corrals, as shown in the figure. Find the dimensions that maximize the enclosed area. What is the maximum area?

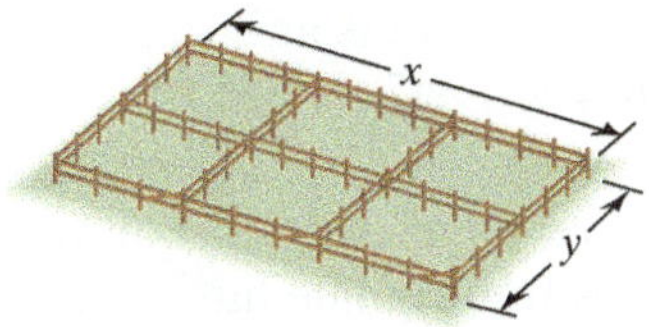

98. The annual yield per lemon tree is fairly constant at 320 pounds when the number of trees per acre is 50 or fewer. For each additional tree over 50, the annual yield per tree for all trees on the acre decreases by 4 pounds due to overcrowding. Find the number of trees that should be planted on an acre to produce the maximum yield. How many pounds is the maximum yield?

Review Exercises

99. Solve: $\dfrac{2}{x+5} + \dfrac{1}{x-5} = \dfrac{16}{x^2 - 25}$. (Section 6.6, Example 5)

100. Simplify: $\dfrac{1 + \dfrac{2}{x}}{1 - \dfrac{4}{x^2}}$. (Section 6.3, Example 1)

101. Solve using determinants (Cramer's Rule):

$$\begin{aligned} 2x + 3y &= 6 \\ x - 4y &= 14. \end{aligned}$$

(Section 3.5, Example 2)

Preview Exercises

Exercises 102–104 will help you prepare for the material covered in the next section.

In Exercises 102–103, solve each quadratic equation for u.

102. $u^2 - 8u - 9 = 0$

103. $2u^2 - u - 10 = 0$

104. If $u = x^{\frac{1}{3}}$, rewrite $5x^{\frac{2}{3}} + 11x^{\frac{1}{3}} + 2 = 0$ as a quadratic equation in u. [*Hint:* $x^{\frac{2}{3}} = \left(x^{\frac{1}{3}}\right)^2$.]

MID-CHAPTER CHECK POINT Section 8.1–Section 8.3

What You Know: We saw that not all quadratic equations can be solved by factoring. We learned three new methods for solving these equations: the square root property, completing the square, and the quadratic formula. We saw that the discriminant of $ax^2 + bx + c = 0$, namely $b^2 - 4ac$, determines the number and type of the equation's solutions. We graphed quadratic functions using vertices, intercepts, and additional points, as necessary. We learned that the vertex of $f(x) = a(x - h)^2 + k$ is (h, k) and the vertex of $f(x) = ax^2 + bx + c$ is $\left(-\dfrac{b}{2a}, f\left(-\dfrac{b}{2a}\right)\right)$. We used the vertex to solve problems that involved minimizing or maximizing quadratic functions.

In Exercises 1–13, solve each equation by the method of your choice. Simplify solutions, if possible.

1. $(3x - 5)^2 = 36$

2. $5x^2 - 2x = 7$

3. $3x^2 - 6x - 2 = 0$

4. $x^2 + 6x = -2$

5. $5x^2 + 1 = 37$

6. $x^2 - 5x + 8 = 0$

7. $2x^2 + 26 = 0$

8. $(2x + 3)(x + 2) = 10$

9. $(x + 3)^2 = 24$

10. $\dfrac{1}{x^2} - \dfrac{4}{x} + 1 = 0$

11. $x(2x - 3) = -4$

12. $\dfrac{x^2}{3} + \dfrac{x}{2} = \dfrac{2}{3}$

13. $\dfrac{2x}{x^2 + 6x + 8} = \dfrac{x}{x+4} - \dfrac{2}{x+2}$

14. Solve by completing the square: $x^2 + 10x - 3 = 0$.

In Exercises 15–18, graph the given quadratic function. Give each function's domain and range.

15. $f(x) = (x - 3)^2 - 4$

16. $g(x) = 5 - (x + 2)^2$

17. $h(x) = -x^2 - 4x + 5$

18. $f(x) = 3x^2 - 6x + 1$

In Exercises 19–20, without solving the equation, determine the number and type of solutions.

19. $2x^2 + 5x + 4 = 0$

20. $10x(x + 4) = 15x - 15$

In Exercises 21–22, write a quadratic equation in standard form with the given solution set.

21. $\left\{-\frac{1}{2}, \frac{3}{4}\right\}$

22. $\{-2\sqrt{3}, 2\sqrt{3}\}$

23. A company manufactures and sells bath cabinets. The function

$$P(x) = -x^2 + 150x - 4425$$

models the company's daily profit, $P(x)$, when x cabinets are manufactured and sold per day. How many cabinets should be manufactured and sold per day to maximize the company's profit? What is the maximum daily profit?

24. Among all pairs of numbers whose sum is -18, find a pair whose product is as large as possible. What is the maximum product?

25. The base of a triangle measures 40 inches minus twice the measure of its height. For what measure of the height does the triangle have a maximum area? What is the maximum area?

GROUP PROJECT

CHAPTER 8

Throughout the chapter, we have considered functions that model the position of free-falling objects. Any object that is falling, or vertically projected into the air, has its height above the ground, $s(t)$, in feet, after t seconds in motion, modeled by the quadratic function

$$s(t) = -16t^2 + v_0 t + s_0,$$

where v_0 is the original velocity (initial velocity) of the object, in feet per second, and s_0 is the original height (initial height) of the object, in feet, above the ground. In this exercise, group members will be working with this position function. The exercise is appropriate for groups of three to five people.

a. Drop a ball from a height of 3 feet, 6 feet, and 12 feet. Record the number of seconds it takes for the ball to hit the ground.

b. For each of the three initial positions, use the position function to determine the time required for the ball to hit the ground.

c. What factors might result in differences between the times that you recorded and the times indicated by the function?

d. What appears to be happening to the time required for a free-falling object to hit the ground as its initial height is doubled? Verify this observation algebraically and with a graphing utility.

e. Repeat part (a) using a sheet of paper rather than a ball. What differences do you observe? What factor seems to be ignored in the position function?

f. What is meant by the acceleration of gravity and how does this number appear in the position function for a free-falling object?

Chapter 8 Summary

Definitions and Concepts	Examples

Section 8.1 The Square Root Property and Completing the Square

The Square Root Property

If u is an algebraic expression and d is a real number, then

$$\text{If } u^2 = d, \quad \text{then} \quad u = \sqrt{d} \quad \text{or} \quad u = -\sqrt{d}.$$

Equivalently,

$$\text{If } u^2 = d, \quad \text{then} \quad u = \pm\sqrt{d}.$$

Solve:

$$\begin{aligned}(x-6)^2 &= 50.\\ x-6 &= \pm\sqrt{50}\\ x-6 &= \pm\sqrt{25\cdot 2}\\ x-6 &= \pm 5\sqrt{2}\\ x &= 6 \pm 5\sqrt{2}\end{aligned}$$

The solutions are $6 \pm 5\sqrt{2}$ and the solution set is $\{6 \pm 5\sqrt{2}\}$.

Completing the Square

If $x^2 + bx$ is a binomial, then by adding $\left(\frac{b}{2}\right)^2$, the square of half the coefficient of x, a perfect square trinomial will result. That is,

$$x^2 + bx + \left(\frac{b}{2}\right)^2 = \left(x + \frac{b}{2}\right)^2.$$

Complete the square:

$$x^2 + \frac{2}{7}x.$$

Half of $\frac{2}{7}$ is $\frac{1}{2}\cdot\frac{2}{7} = \frac{1}{7}$ and $\left(\frac{1}{7}\right)^2 = \frac{1}{49}$.

$$x^2 + \frac{2}{7}x + \frac{1}{49} = \left(x + \frac{1}{7}\right)^2$$

Definitions and Concepts	Examples

Section 8.1 The Square Root Property and Completing the Square (continued)

Solving Quadratic Equations by Completing the Square

1. If the coefficient of x^2 is not 1, divide both sides by this coefficient.
2. Isolate variable terms on one side.
3. Complete the square by adding the square of half the coefficient of x to both sides.
4. Factor the perfect square trinomial.
5. Solve by applying the square root property.

Solve by completing the square:

$$2x^2 + 16x - 6 = 0.$$

$$\frac{2x^2}{2} + \frac{16x}{2} - \frac{6}{2} = \frac{0}{2} \quad \text{Divide by 2.}$$

$$x^2 + 8x - 3 = 0 \quad \text{Simplify.}$$

$$x^2 + 8x = 3 \quad \text{Add 3.}$$

The coefficient of x is 8. Half of 8 is 4 and $4^2 = 16$. Add 16 to both sides.

$$x^2 + 8x + 16 = 3 + 16$$

$$(x + 4)^2 = 19$$

$$x + 4 = \pm\sqrt{19}$$

$$x = -4 \pm \sqrt{19}$$

Section 8.2 The Quadratic Formula

The solutions of a quadratic equation in standard form

$$ax^2 + bx + c = 0, \quad a \neq 0,$$

are given by the quadratic formula

$$x = \frac{-b \pm \sqrt{b^2 - 4ac}}{2a}.$$

Solve using the quadratic formula:

$$2x^2 = 6x - 3.$$

First write the equation in standard form by subtracting $6x$ and adding 3 on both sides.

$$2x^2 - 6x + 3 = 0$$

$a = 2$ $b = -6$ $c = 3$

$$x = \frac{-(-6) \pm \sqrt{(-6)^2 - 4 \cdot 2 \cdot 3}}{2 \cdot 2} = \frac{6 \pm \sqrt{36 - 24}}{4}$$

$$= \frac{6 \pm \sqrt{12}}{4} = \frac{6 \pm \sqrt{4 \cdot 3}}{4} = \frac{6 \pm 2\sqrt{3}}{4}$$

$$= \frac{\cancel{2}(3 \pm \sqrt{3})}{\cancel{2} \cdot 2} = \frac{3 \pm \sqrt{3}}{2}$$

The Discriminant

The discriminant, $b^2 - 4ac$, of the quadratic equation $ax^2 + bx + c = 0$ determines the number and type of solutions.

Discriminant	Solutions
Positive perfect square, with a, b, and c rational numbers	2 real rational solutions
Positive and not a perfect square	2 real irrational solutions
Zero, with a, b, and c rational numbers	1 real rational solution
Negative	2 imaginary solutions

- $2x^2 - 7x - 4 = 0$

$a = 2$ $b = -7$ $c = -4$

$$b^2 - 4ac = (-7)^2 - 4(2)(-4)$$

$$= 49 - (-32) = 49 + 32 = 81$$

Positive perfect square

The equation has 2 real rational solutions.

Definitions and Concepts | Examples

Section 8.2 The Quadratic Formula (continued)

Writing Quadratic Equations from Solutions

The zero-product principle in reverse makes it possible to write a quadratic equation from solutions:

$$\text{If } A = 0 \quad \text{or} \quad B = 0, \quad \text{then} \quad AB = 0.$$

Write a quadratic equation with the solution set $\{-2\sqrt{3}, 2\sqrt{3}\}$.

$$x = -2\sqrt{3} \qquad\qquad x = 2\sqrt{3}$$

$$x + 2\sqrt{3} = 0 \quad \text{or} \quad x - 2\sqrt{3} = 0$$

$$(x + 2\sqrt{3})(x - 2\sqrt{3}) = 0$$

$$x^2 - (2\sqrt{3})^2 = 0$$

$$x^2 - 12 = 0$$

Section 8.3 Quadratic Functions and Their Graphs

The graph of the quadratic function

$$f(x) = a(x - h)^2 + k, \quad a \neq 0,$$

is called a parabola. The vertex, or turning point, is (h, k). The graph opens upward if a is positive and downward if a is negative. The axis of symmetry is a vertical line passing through the vertex. The graph can be obtained using the vertex, x-intercepts, if any, (set $f(x)$ equal to zero and solve), and the y-intercept (set $x = 0$).

Graph: $f(x) = -(x + 3)^2 + 1$.

$$f(x) = -1(x - (-3))^2 + 1$$

$a = -1$ $h = -3$ $k = 1$

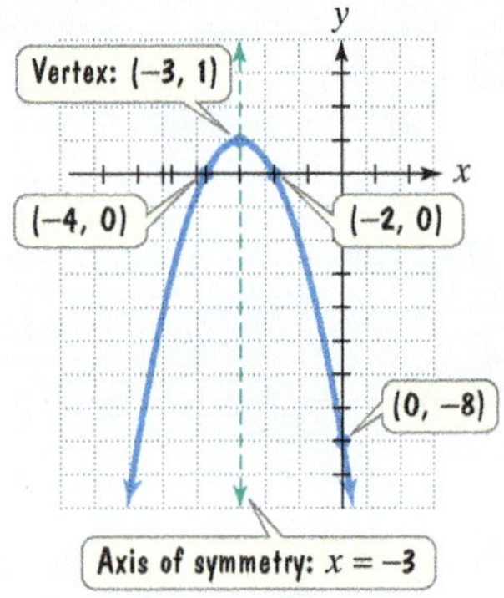

- Vertex: $(h, k) = (-3, 1)$
- Opens downward because $a < 0$
- x-intercepts: Set $f(x) = 0$.

$$0 = -(x + 3)^2 + 1$$

$$(x + 3)^2 = 1$$

$$x + 3 = \pm\sqrt{1}$$

$$x + 3 = 1 \quad \text{or} \quad x + 3 = -1$$

$$x = -2 \qquad\qquad x = -4$$

- y-intercept: Set $x = 0$.

$$f(0) = -(0 + 3)^2 + 1 = -9 + 1 = -8$$

A parabola whose equation is in the form

$$f(x) = ax^2 + bx + c, \quad a \neq 0,$$

has its vertex at

$$\left(-\frac{b}{2a}, f\left(-\frac{b}{2a}\right)\right).$$

The parabola is graphed as described in the left column above. The only difference is how we determine the vertex. If $a > 0$, then f has a minimum that occurs at $x = -\frac{b}{2a}$. This minimum value is $f\left(-\frac{b}{2a}\right)$. If $a < 0$, then f has a maximum that occurs at $x = -\frac{b}{2a}$. This maximum value is $f\left(-\frac{b}{2a}\right)$.

Graph:

$$f(x) = x^2 - 6x + 5.$$

$a = 1$ $b = -6$ $c = 5$

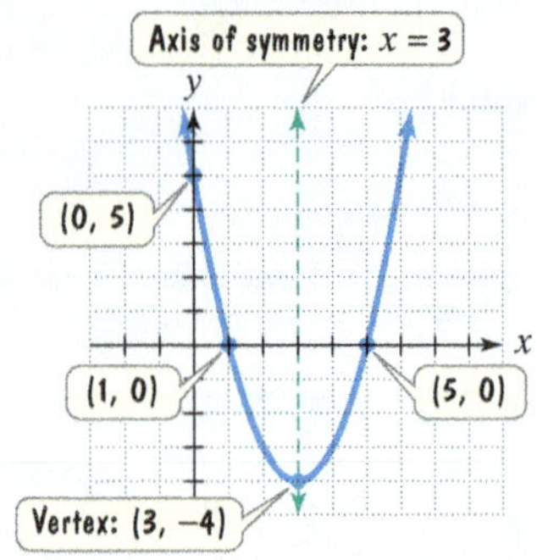

- Vertex:

$$x = -\frac{b}{2a} = -\frac{-6}{2 \cdot 1} = 3$$

$f(3) = 3^2 - 6 \cdot 3 + 5 = -4$

Vertex is at $(3, -4)$.

- Opens upward because $a > 0$.
- x-intercepts: Set $f(x) = 0$.

$$x^2 - 6x + 5 = 0$$

$$(x - 1)(x - 5) = 0$$

$$x = 1 \quad \text{or} \quad x = 5$$

- y-intercept: Set $x = 0$.

$$f(0) = 0^2 - 6 \cdot 0 + 5 = 5$$

Definitions and Concepts | Examples

Section 8.4 Equations Quadratic in Form

An equation that is quadratic in form is one that can be expressed as a quadratic equation using an appropriate substitution. In these equations, the variable factor in one term is the square of the variable factor in the other variable term. Let $u =$ the variable factor that reappears squared. If at any point in the solution process both sides of an equation are raised to an even power, a check is required.

Solve:

$$x^{\frac{2}{3}} - 3x^{\frac{1}{3}} + 2 = 0.$$

$$\left(x^{\frac{1}{3}}\right)^2 - 3x^{\frac{1}{3}} + 2 = 0$$

Let $u = x^{\frac{1}{3}}$.

$$u^2 - 3u + 2 = 0$$

$$(u - 1)(u - 2) = 0$$

$$u - 1 = 0 \quad \text{or} \quad u - 2 = 0$$

$$u = 1 \qquad\qquad u = 2$$

$$x^{\frac{1}{3}} = 1 \qquad\qquad x^{\frac{1}{3}} = 2$$

$$\left(x^{\frac{1}{3}}\right)^3 = 1^3 \qquad\qquad \left(x^{\frac{1}{3}}\right)^3 = 2^3$$

$$x = 1 \qquad\qquad x = 8$$

The solutions are 1 and 8, and the solution set is $\{1, 8\}$.

Section 8.5 Polynomial and Rational Inequalities

Solving Polynomial Inequalities

1. Express the inequality in the form
$$f(x) < 0 \quad \text{or} \quad f(x) > 0,$$
where f is a polynomial function.
2. Solve the equation $f(x) = 0$. The real solutions are the boundary points.
3. Locate these boundary points on a number line, thereby dividing the number line into intervals.
4. Choose one representative number, called a test value, within each interval and evaluate f at that number.
   a. If the value of f is positive, then $f(x) > 0$ for all x in the interval.
   b. If the value of f is negative, then $f(x) < 0$ for all x in the interval.
5. Write the solution set, selecting the interval or intervals that satisfy the given inequality.

This procedure is valid if $<$ is replaced by $\le$ and $>$ is replaced by $\ge$. In these cases, include the boundary points in the solution set.

Solve: $2x^2 + x - 6 > 0$.

The form of the inequality is $f(x) > 0$ with $f(x) = 2x^2 + x - 6$. Solve $f(x) = 0$.

$$2x^2 + x - 6 = 0$$

$$(2x - 3)(x + 2) = 0$$

$$2x - 3 = 0 \quad \text{or} \quad x + 2 = 0$$

$$x = \frac{3}{2} \qquad\qquad x = -2$$

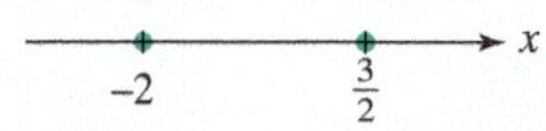

Use −3, 0, and 2 as test values.

$$f(-3) = 2(-3)^2 + (-3) - 6 = 9, \text{ positive}$$
$$f(x) > 0 \text{ for all } x \text{ in } (-\infty, -2).$$

$$f(0) = 2 \cdot 0^2 + 0 - 6 = -6, \text{ negative}$$
$$f(x) < 0 \text{ for all } x \text{ in } \left(-2, \frac{3}{2}\right).$$

$$f(2) = 2 \cdot 2^2 + 2 - 6 = 4, \text{ positive}$$
$$f(x) > 0 \text{ for all } x \text{ in } \left(\frac{3}{2}, \infty\right).$$

The solution set is $\left\{x \mid x < -2 \text{ or } x > \frac{3}{2}\right\}$ or $(-\infty, -2) \cup \left(\frac{3}{2}, \infty\right)$.

Solving Rational Inequalities

1. Express the inequality in the form
$$f(x) < 0 \quad \text{or} \quad f(x) > 0,$$
where f is a rational function.
2. Set the numerator and the denominator of f equal to zero. The real solutions are the boundary points.
3. Locate these boundary points on a number line, thereby dividing the number line into intervals.

Solve: $\dfrac{x}{x + 4} \ge 2.$

$$\frac{x}{x + 4} - \frac{2(x + 4)}{x + 4} \ge 0$$

$$\frac{-x - 8}{x + 4} \ge 0$$

The form of the inequality is $f(x) \ge 0$ with $f(x) = \dfrac{-x - 8}{x + 4}$.

Set the numerator and the denominator equal to zero.

Definitions and Concepts | **Examples**

Section 8.5 Polynomial and Rational Inequalities (continued)

4. Choose one representative number, called a test value, within each interval and evaluate f at that number.
   a. If the value of f is positive, then $f(x) > 0$ for all x in the interval.
   b. If the value of f is negative, then $f(x) < 0$ for all x in the interval.
5. Write the solution set, selecting the interval or intervals that satisfy the given inequality.

This procedure is valid if $<$ is replaced by $\leq$ and $>$ is replaced by $\geq$. In these cases, include any values that make the numerator of f zero. Always exclude any values that make the denominator zero.

$$-x - 8 = 0 \qquad x + 4 = 0$$
$$-8 = x \qquad x = -4$$

Use -9, -7, and -3 as test values.

$$f(-9) = \frac{-(-9) - 8}{-9 + 4} = \frac{1}{-5}, \text{ negative}$$
$f(x) < 0$ for all x in $(-\infty, -8)$.

$$f(-7) = \frac{-(-7) - 8}{-7 + 4} = \frac{-1}{-3} = \frac{1}{3}, \text{ positive}$$
$f(x) > 0$ for all x in $(-8, -4)$.

$$f(-3) = \frac{-(-3) - 8}{-3 + 4} = \frac{-5}{1} = -5, \text{ negative}$$
$f(x) < 0$ for all x in $(-4, \infty)$.

Because of $\geq$, include -8, the value that makes the numerator zero, in the solution set.
The solution set is $\{x|-8 \leq x < -4\}$ or $[-8, -4)$.

CHAPTER 8 REVIEW EXERCISES

8.1 *In Exercises 1–5, solve each equation by the square root property. If possible, simplify radicals or rationalize denominators. Express imaginary solutions in the form $a + bi$.*

1. $2x^2 - 3 = 125$
2. $3x^2 - 150 = 0$
3. $3x^2 - 2 = 0$
4. $(x - 4)^2 = 18$
5. $(x + 7)^2 = -36$

In Exercises 6–7, determine the constant that should be added to the binomial so that it becomes a perfect square trinomial. Then write and factor the trinomial.

6. $x^2 + 20x$
7. $x^2 - 3x$

In Exercises 8–10, solve each quadratic equation by completing the square.

8. $x^2 - 12x + 27 = 0$
9. $x^2 - 7x - 1 = 0$
10. $2x^2 + 3x - 4 = 0$
11. In 2 years, an investment of \$2500 grows to \$2916. Use the compound interest formula
$$A = P(1 + r)^t$$
to find the annual interest rate, r.
12. The function $W(t) = 3t^2$ models the weight of a human fetus, $W(t)$, in grams, after t weeks, where $0 \leq t \leq 39$. After how many weeks does the fetus weigh 588 grams?
13. A building casts a shadow that is double the length of the building's height. If the distance from the end of the shadow to the top of the building is 300 meters, how high is the building? Express the answer in simplified radical form. Then find a decimal approximation to the nearest tenth of a meter.

8.2 *In Exercises 14–16, solve each equation using the quadratic formula. Simplify solutions, if possible.*

14. $x^2 = 2x + 4$
15. $x^2 - 2x + 19 = 0$
16. $2x^2 = 3 - 4x$

In Exercises 17–19, without solving the given quadratic equation, determine the number and type of solutions.

17. $x^2 - 4x + 13 = 0$
18. $9x^2 = 2 - 3x$
19. $2x^2 + 4x = 3$

In Exercises 20–26, solve each equation by the method of your choice. Simplify solutions, if possible.

20. $3x^2 - 10x - 8 = 0$
21. $(2x - 3)(x + 2) = x^2 - 2x + 4$

22. $5x^2 - x - 1 = 0$

23. $x^2 - 16 = 0$

24. $(x - 3)^2 - 8 = 0$

25. $3x^2 - x + 2 = 0$

26. $\frac{5}{x+1} + \frac{x-1}{4} = 2$

In Exercises 27–29, write a quadratic equation in standard form with the given solution set.

27. $\left\{-\frac{1}{3}, \frac{3}{5}\right\}$

28. $\{-9i, 9i\}$

29. $\{-4\sqrt{3}, 4\sqrt{3}\}$

30. The graph shows stopping distances for motorcycles at various speeds on dry roads and on wet roads.

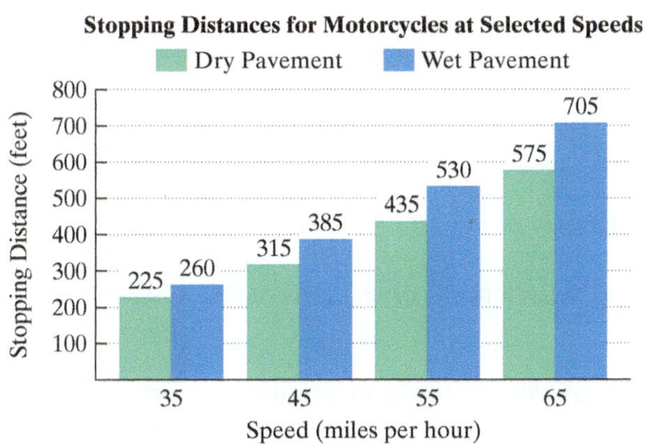

Source: National Highway Traffic Safety Administration

The functions

$$f(x) = 0.125x^2 - 0.8x + 99$$

Dry pavement

and

Wet pavement

$$g(x) = 0.125x^2 + 2.3x + 27$$

model a motorcycle's stopping distance, $f(x)$ or $g(x)$, in feet traveling at x miles per hour. Function f models stopping distance on dry pavement and function g models stopping distance on wet pavement.

a. Use function g to find the stopping distance on wet pavement for a motorcycle traveling at 35 miles per hour. Round to the nearest foot. Does your rounded answer overestimate or underestimate the stopping distance shown by the graph? By how many feet?

b. Use function f to determine a motorcycle's speed requiring a stopping distance on dry pavement of 267 feet.

31. The graphs of the functions in Exercise 30 are shown for $\{x | x \geq 30\}$.

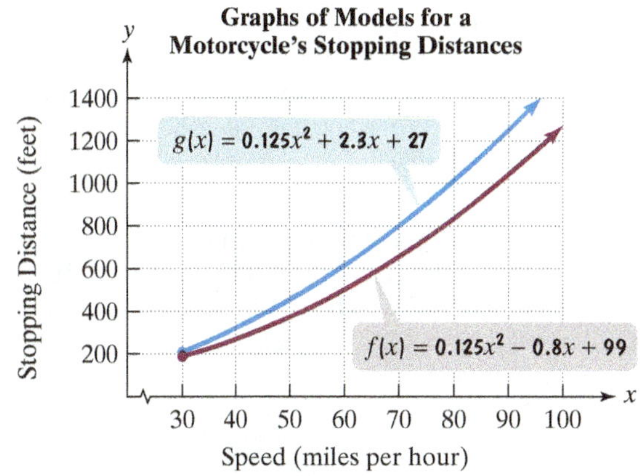

a. How is your answer to Exercise 30(a) shown on the graph of g?

b. How is your answer to Exercise 30(b) shown on the graph of f?

32. A baseball is hit by a batter. The function

$$s(t) = -16t^2 + 140t + 3$$

models the ball's height above the ground, $s(t)$, in feet, t seconds after it is hit. How long will it take for the ball to strike the ground? Round to the nearest tenth of a second.

8.3 *In Exercises 33–36, use the vertex and intercepts to sketch the graph of each quadratic function.*

33. $f(x) = -(x + 1)^2 + 4$

34. $f(x) = (x + 4)^2 - 2$

35. $f(x) = -x^2 + 2x + 3$

36. $f(x) = 2x^2 - 4x - 6$

37. The function

$$f(x) = -0.02x^2 + x + 1$$

models the yearly growth of a young redwood tree, $f(x)$, in inches, with x inches of rainfall per year. How many inches of rainfall per year result in maximum tree growth? What is the maximum yearly growth?

38. A model rocket is launched upward from a platform 40 feet above the ground. The quadratic function

$$s(t) = -16t^2 + 400t + 40$$

models the rocket's height above the ground, $s(t)$, in feet, t seconds after it was launched. After how many seconds does the rocket reach its maximum height? What is the maximum height?

39. The function

$$f(x) = 104.5x^2 - 1501.5x + 6016$$

models the death rate per year per 100,000 males, $f(x)$, for U.S. men who average x hours of sleep each night. How many hours of sleep, to the nearest tenth of an hour, corresponds to the minimum death rate? What is this minimum death rate, to the nearest whole number?

40. A field bordering a straight stream is to be enclosed. The side bordering the stream is not to be fenced. If 1000 yards of fencing material is to be used, what are the dimensions of the largest rectangular field that can be fenced? What is the maximum area?

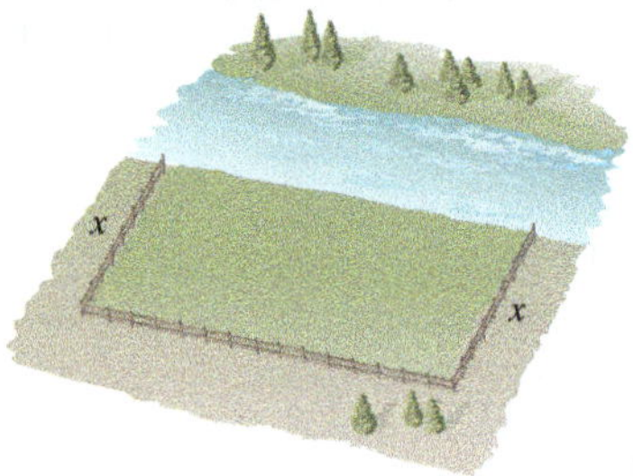

41. Among all pairs of numbers whose difference is 14, find a pair whose product is as small as possible. What is the minimum product?

8.4 *In Exercises 42–47, solve each equation by making an appropriate substitution. When necessary, check proposed solutions.*

42. $x^4 - 6x^2 + 8 = 0$

43. $x + 7\sqrt{x} - 8 = 0$

44. $(x^2 + 2x)^2 - 14(x^2 + 2x) = 15$

45. $x^{-2} + x^{-1} - 56 = 0$

46. $x^{\frac{2}{3}} - x^{\frac{1}{3}} - 12 = 0$

47. $x^{\frac{1}{2}} + 3x^{\frac{1}{4}} - 10 = 0$

8.5 *In Exercises 48–52, solve each inequality and graph the solution set on a real number line.*

48. $2x^2 + 5x - 3 < 0$

49. $2x^2 + 9x + 4 \geq 0$

50. $x^3 + 2x^2 > 3x$

51. $\dfrac{x-6}{x+2} > 0$

52. $\dfrac{x+3}{x-4} \leq 5$

53. A model rocket is launched from ground level. The function

$$s(t) = -16t^2 + 48t$$

models the rocket's height above the ground, $s(t)$, in feet, t seconds after it was launched. During which time period will the rocket's height exceed 32 feet?

54. The function

$$H(x) = \frac{15}{8}x^2 - 30x + 200$$

models heart rate, $H(x)$, in beats per minute, x minutes after a strenuous workout.

a. What is the heart rate immediately following the workout?

b. According to the model, during which intervals of time after the strenuous workout does the heart rate exceed 110 beats per minute? For which of these intervals has model breakdown occurred? Which interval provides a more realistic answer? How did you determine this?

CHAPTER 8 TEST

Remember to use your Chapter Test Prep Video CD to see the worked-out solutions to the test questions you want to review.

Express solutions to all equations in simplified form. Rationalize denominators, if possible.

In Exercises 1–2, solve each equation by the square root property.

1. $2x^2 - 5 = 0$

2. $(x - 3)^2 = 20$

In Exercises 3–4, determine the constant that should be added to the binomial so that it becomes a perfect square trinomial. Then write and factor the trinomial.

3. $x^2 - 16x$

4. $x^2 + \dfrac{2}{5}x$

5. Solve by completing the square: $x^2 - 6x + 7 = 0$.

6. Use the measurements determined by the surveyor to find the width of the pond. Express the answer in simplified radical form.

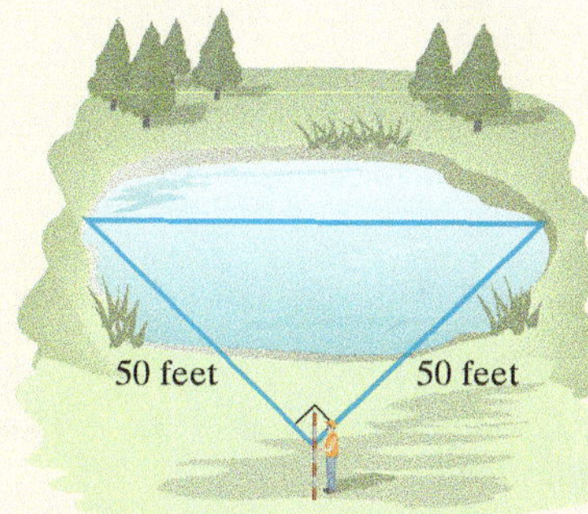

In Exercises 7–8, without solving the given quadratic equation, determine the number and type of solutions.

7. $3x^2 + 4x - 2 = 0$

8. $x^2 = 4x - 8$

In Exercises 9–12, solve each equation by the method of your choice.

9. $2x^2 + 9x = 5$

10. $x^2 + 8x + 5 = 0$

11. $(x + 2)^2 + 25 = 0$

12. $2x^2 - 6x + 5 = 0$

In Exercises 13–14, write a quadratic equation in standard form with the given solution set.

13. $\{-3, 7\}$

14. $\{-10i, 10i\}$

15. By 2007, cellphones with 2-gigabyte storage cards could hold 2000 songs. The bar graph shows the percentage of new cellphones that played music from 2005 through 2007, with projections for 2008 and 2009.

Percentage of New Cellphones That Play Music

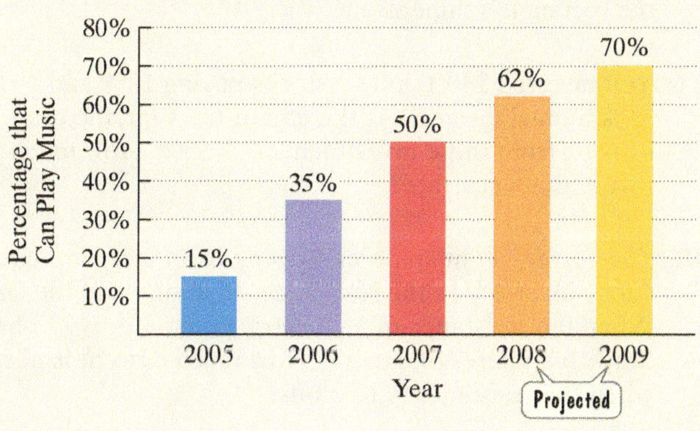

Source: Ovum

The function

$$f(x) = -1.8x^2 + 21x + 15$$

models the percentage of new cellphones that play music, $f(x)$, x years after 2005.

a. Use the function to find the percentage of new cellphones that will play music in 2009. Does this overestimate or underestimate the percentage shown by the graph? By how much?

b. Use the function to determine the first year in which 75% of new cellphones will play music.

In Exercises 16–17, use the vertex and intercepts to sketch the graph of each quadratic function.

16. $f(x) = (x + 1)^2 + 4$

17. $f(x) = x^2 - 2x - 3$

A baseball player hits a pop fly into the air. The function

$$s(t) = -16t^2 + 64t + 5$$

models the ball's height above the ground, $s(t)$, in feet, t seconds after it is hit. Use the function to solve Exercises 18–19.

18. When does the baseball reach its maximum height? What is that height?

19. After how many seconds does the baseball hit the ground? Round to the nearest tenth of a second.

20. The function $f(x) = -x^2 + 46x - 360$ models the daily profit, $f(x)$, in hundreds of dollars, for a company that manufactures x computers daily. How many computers should be manufactured each day to maximize profit? What is the maximum daily profit?

In Exercises 21–23, solve each equation by making an appropriate substitution. When necessary, check proposed solutions.

21. $(2x - 5)^2 + 4(2x - 5) + 3 = 0$

22. $x^4 - 13x^2 + 36 = 0$

23. $x^{\frac{2}{3}} - 9x^{\frac{1}{3}} + 8 = 0$

In Exercises 24–25, solve each inequality and graph the solution set on a real number line.

24. $x^2 - x - 12 < 0$

25. $\dfrac{2x + 1}{x - 3} \le 3$

CUMULATIVE REVIEW EXERCISES (CHAPTERS 1–8)

In Exercises 1–13, solve each equation, inequality, or system.

1. $9(x - 1) = 1 + 3(x - 2)$
2. $3x + 4y = -7$
   $x - 2y = -9$
3. $x - y + 3z = -9$
   $2x + 3y - z = 16$
   $5x + 2y - z = 15$
4. $7x + 18 \le 9x - 2$
5. $4x - 3 < 13$ and $-3x - 4 \ge 8$
6. $2x + 4 > 8$ or $x - 7 \ge 3$
7. $|2x - 1| < 5$
8. $\left|\frac{2}{3}x - 4\right| = 2$
9. $\frac{4}{x - 3} - \frac{6}{x + 3} = \frac{24}{x^2 - 9}$
10. $\sqrt{x + 4} - \sqrt{x - 3} = 1$
11. $2x^2 = 5 - 4x$
12. $x^{\frac{2}{3}} - 5x^{\frac{1}{3}} + 6 = 0$
13. $2x^2 + x - 6 \le 0$

In Exercises 14–17, graph each function, equation, or inequality in a rectangular coordinate system.

14. $x - 3y = 6$
15. $f(x) = \frac{1}{2}x - 1$
16. $3x - 2y > -6$
17. $f(x) = -2(x - 3)^2 + 2$

In Exercises 18–28, perform the indicated operations, and simplify, if possible.

18. $4[2x - 6(x - y)]$
19. $(-5x^3y^2)(4x^4y^{-6})$
20. $(8x^2 - 9xy - 11y^2) - (7x^2 - 4xy + 5y^2)$
21. $(3x - 1)(2x + 5)$
22. $(3x^2 - 4y)^2$
23. $\frac{3x}{x + 5} - \frac{2}{x^2 + 7x + 10}$
24. $\frac{1 - \frac{9}{x^2}}{1 + \frac{3}{x}}$
25. $\frac{x^2 - 6x + 8}{3x + 9} \div \frac{x^2 - 4}{x + 3}$
26. $\sqrt{5xy} \cdot \sqrt{10x^2y}$
27. $4\sqrt{72} - 3\sqrt{50}$
28. $(5 + 3i)(7 - 3i)$

In Exercises 29–31, factor completely.

29. $81x^4 - 1$
30. $24x^3 - 22x^2 + 4x$
31. $x^3 + 27y^3$

In Exercises 32–34, let $f(x) = x^2 + 3x - 15$ and $g(x) = x - 2$. Find each indicated expression.

32. $(f - g)(x)$ and $(f - g)(5)$
33. $\left(\frac{f}{g}\right)(x)$ and the domain of $\frac{f}{g}$
34. $\frac{f(a + h) - f(a)}{h}$
35. Divide using synthetic division:

    $(3x^3 - x^2 + 4x + 8) \div (x + 2)$.
36. Solve for R: $I = \frac{R}{R + r}$.
37. Write the slope-intercept form of the equation of the line through $(-2, 5)$ and parallel to the line whose equation is $3x + y = 9$.
38. Evaluate the determinant: $\begin{vmatrix} -2 & -4 \\ 5 & 7 \end{vmatrix}$.
39. The price of a computer is reduced by 30% to \$434. What was the original price?
40. The area of a rectangle is 52 square yards. The length of the rectangle is 1 yard longer than 3 times its width. Find the rectangle's dimensions.
41. You invested \$4000 in two stocks paying 12% and 14% annual interest. At the end of the year, the total interest from these investments was \$508. How much was invested at each rate?
42. The current, I, in amperes, flowing in an electrical circuit varies inversely as the resistance, R, in ohms, in the circuit. When the resistance of an electric percolator is 22 ohms, it draws 5 amperes of current. How much current is needed when the resistance is 10 ohms?

ANSWERS TO SELECTED EXERCISES

CHAPTER 1

Section 1.1

Check Point Exercises

1. a. $8x + 5$ **b.** $\frac{x}{7} - 2x$ **2.** 21.8; At age 10, the average neurotic level is 21.8. **3.** 608 **4. a.** 33,502 **b.** greater by 612 lobbyists **5. a.** true **b.** true **6. a.** -8 is less than -2.; true **b.** 7 is greater than -3.; true **c.** -1 is less than or equal to -4.; false **d.** 5 is greater than or equal to 5.; true **e.** 2 is greater than or equal to -14.; true

Exercise Set 1.1

1. $x + 5$ **3.** $x - 4$ **5.** $4x$ **7.** $2x + 10$ **9.** $6 - \frac{1}{2}x$ **11.** $\frac{4}{x} - 2$ **13.** $\frac{3}{5 - x}$ **15.** 57 **17.** 10 **19.** $1\frac{1}{9}$ or $\frac{10}{9}$ **21.** 10 **23.** 44 **25.** 46 **27.** $\{1, 2, 3, 4\}$ **29.** $\{-7, -6, -5, -4\}$ **31.** $\{8, 9, 10, \dots\}$ **33.** $\{1, 3, 5, 7, 9\}$ **35.** true **37.** true **39.** false **41.** true **43.** false **45.** true **47.** false **49.** -6 is less than -2.; true **51.** 5 is greater than -7.; true **53.** 0 is less than -4.; false **55.** -4 is less than or equal to 1.; true **57.** -2 is less than or equal to -6.; false **59.** -2 is less than or equal to -2.; true **61.** -2 is greater than or equal to -2.; true **63.** 2 is less than or equal to $-\frac{1}{2}$.; false **65.** true **67.** false **69.** true **71.** false **73.** false **75.** 4.2 **77.** 0.4 **79.** 863; overestimates by 42 **81.** 10°C **83.** 60 ft **103.** does not make sense **105.** does not make sense **107.** false **109.** true **111.** $(2 \cdot 3 + 3) \cdot 5 = 45$ **113.** -6 and -5 **114.** -5 and 5 **115.** 8 **116.** 34; 34

Section 1.2

Check Point Exercises

1. a. 6 **b.** 4.5 **c.** 0 **2. a.** -28 **b.** 0.7 **c.** $-\frac{1}{10}$ **3. a.** 8 **b.** $-\frac{1}{3}$ **4. a.** -3 **b.** 10.5 **c.** $-\frac{3}{5}$ **5. a.** 25 **b.** -25 **c.** -64 **d.** $\frac{81}{625}$ **6. a.** -8 **b.** $\frac{8}{15}$ **7.** 74 **8.** -4 **9.** addition: $9 + 4x$; multiplication: $x \cdot 4 + 9$ **10. a.** $(6 + 12) + x = 18 + x$ **b.** $(-7 \cdot 4)x = -28x$ **11.** $-28x - 8$ **12.** $14x + 15x^2$ or $15x^2 + 14x$ **13.** $12x - 40$ **14.** $42 - 4x$

Exercise Set 1.2

1. 7 **3.** 4 **5.** 7.6 **7.** $\frac{\pi}{2}$ **9.** $\sqrt{2}$ **11.** $-\frac{2}{5}$ **13.** -11 **15.** -4 **17.** -4.5 **19.** $\frac{2}{15}$ **21.** $-\frac{35}{36}$ **23.** -8.2 **25.** -12.4 **27.** 0 **29.** -11 **31.** 5 **33.** 0 **35.** -12 **37.** 18 **39.** -15 **41.** $-\frac{1}{4}$ **43.** 5.5 **45.** $\sqrt{2}$ **47.** -90 **49.** 33 **51.** $-\frac{15}{13}$ **53.** 0 **55.** -8 **57.** 48 **59.** 100 **61.** -100 **63.** -8 **65.** 1 **67.** -1 **69.** $\frac{1}{8}$ **71.** -3 **73.** 45 **75.** 0 **77.** undefined **79.** $\frac{9}{14}$ **81.** -15 **83.** -2 **85.** -24 **87.** 45 **89.** $\frac{1}{121}$ **91.** 14 **93.** $-\frac{8}{3}$ **95.** $-\frac{1}{2}$ **97.** 31 **99.** 37 **101.** addition: $10 + 4x$; multiplication: $x \cdot 4 + 10$ **103.** addition: $-5 + 7x$; multiplication: $x \cdot 7 - 5$ **105.** $(4 + 6) + x = 10 + x$ **107.** $(-7 \cdot 3)x = -21x$ **109.** $\left(-\frac{1}{3} \cdot -3\right)y = y$ **111.** $6x + 15$ **113.** $-14x - 21$ **115.** $-3x + 6$ **117.** $12x$ **119.** $5x^2$ **121.** $10x + 12x^2$ **123.** $18x - 40$ **125.** $8y - 12$ **127.** $16y - 25$ **129.** $12x^2 + 11$ **131.** $x - (x + 4); -4$ **133.** $6(-5x); -30x$ **135.** $5x - 2x; 3x$ **137.** $8x - (3x + 6); 5x - 6$ **139.** \$389 billion **141.** \$25 billion **143.** $-$\$377 billion; deficit **145.** \$552 billion **147.** 38.55%; overestimates by 3.55% **149. a.** $1200 - 0.07x$ **b.** \$780 **169.** makes sense **171.** does not make sense **173.** false **175.** false **177.** true **179.** $\left(2 \cdot 5 - \frac{1}{2} \cdot 10\right) \cdot 9 = 45$ **181.** $\frac{10}{x} - 4x$ **182.** 42 **183.** true **184.** $-5; 0; 3; 4; 3; 0; -5$ **185.** $-8; -3; 0; 1; 0; -3; -8$ **186.** 3; 2; 1; 0; 1; 2; 3

Section 1.3

Check Point Exercises

1.

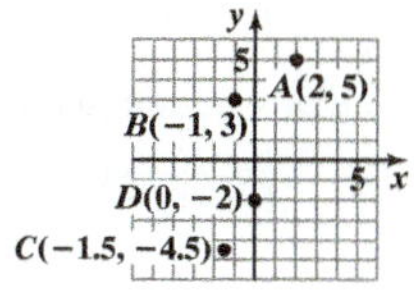

2.

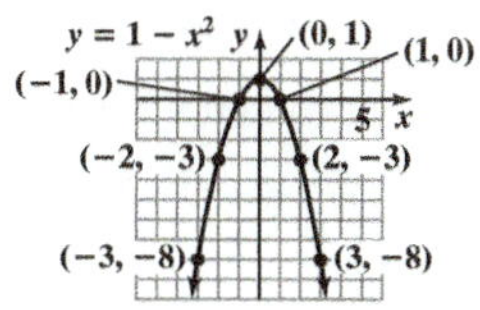

3.

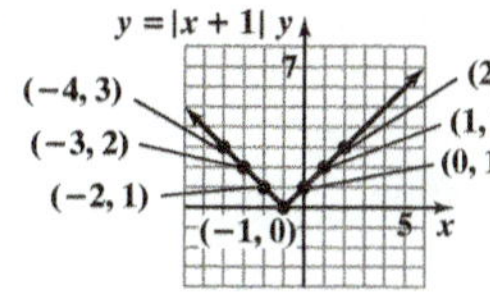

4. a. 0 to 3 hr **b.** 3 to 13 hr **c.** 0.05 mg per 100 ml; after 3 hr **d.** None of the drug is left in the body. **5.** minimum x-value: -100; maximum x-value: 100; distance between tick marks on x-axis: 50; minimum y-value: -100; maximum y-value: 100; distance between tick marks on y-axis: 10

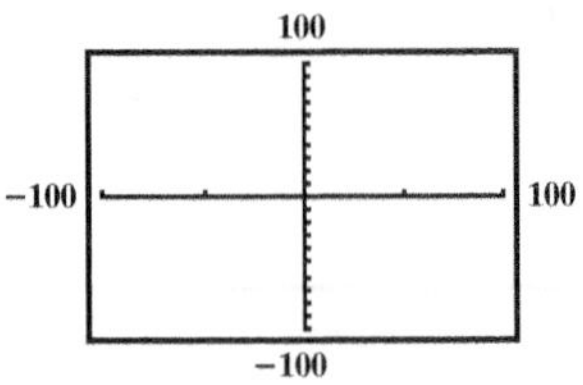

Exercise Set 1.3

1–9.

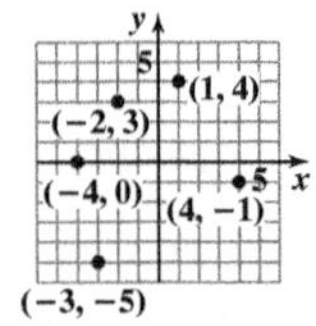

11.

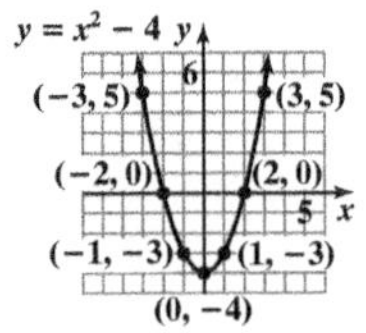

13.

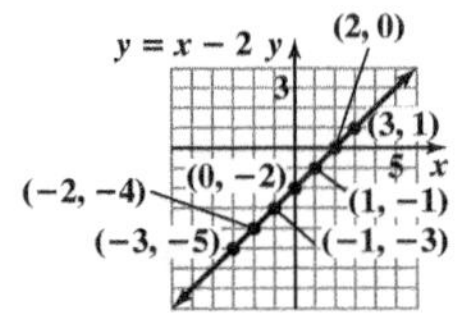

15.

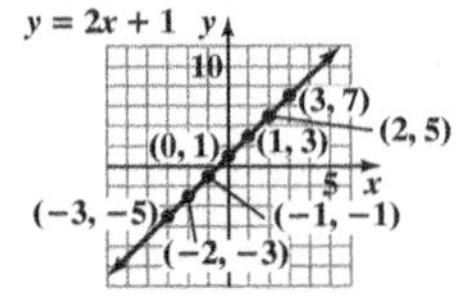

17.

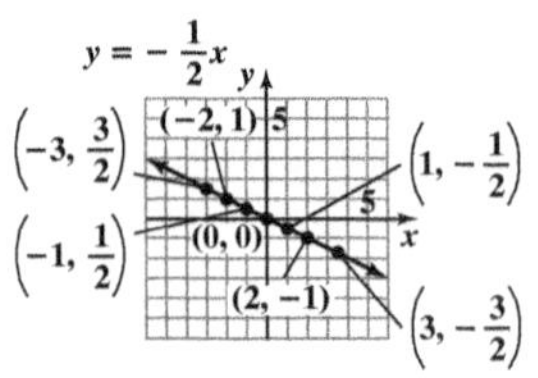

19.

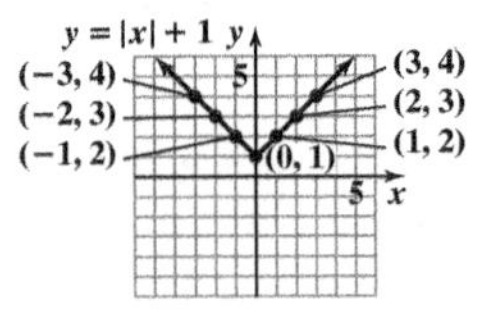

21.

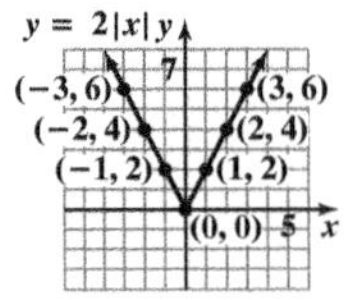

23.

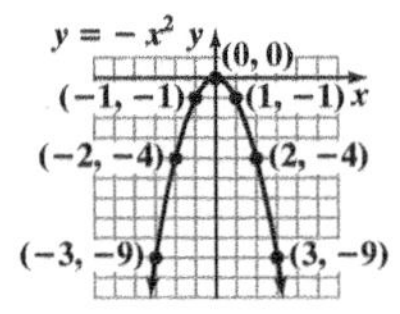

25.

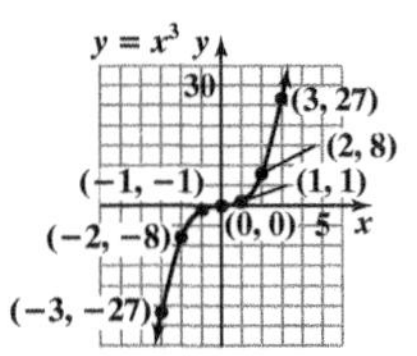

27. c **29.** b **31.** c **33.** no **35.** $(2, 0)$ **37.** $(-2, 4)$ and $(1, 1)$ **39.** $y = 2x + 4$

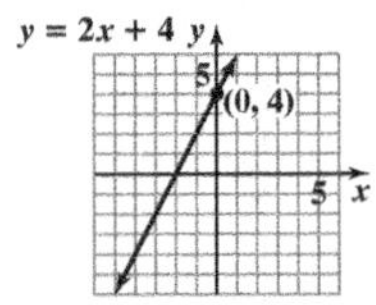

41. $y = 3 - x^2$

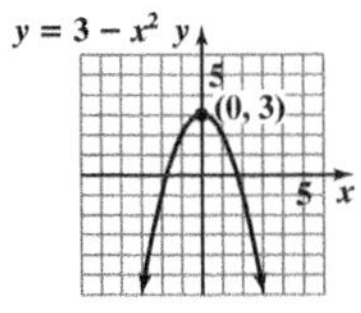

43.

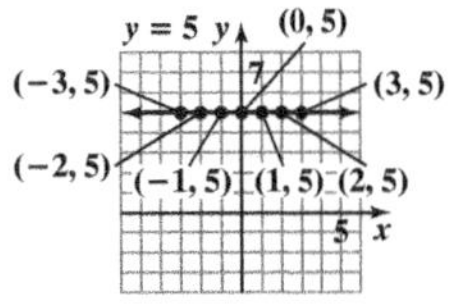

45.

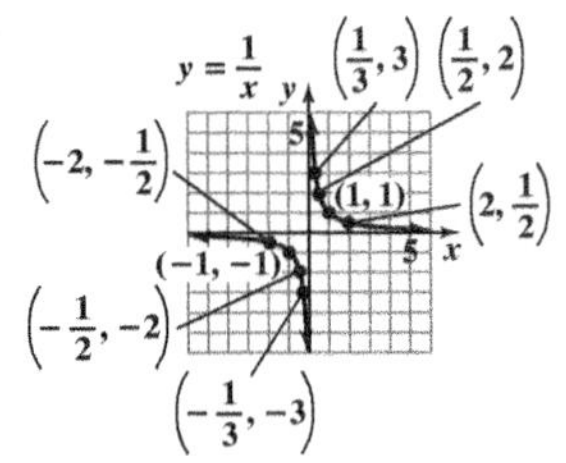

47. 35% **49.** 1945; 94% **51.** 1950–1960; 91% **53.** 8 yr old; 1 awakening **55.** about 1.9 awakenings **57.** a **59.** b **61.** b **63.** c **73.** makes sense **75.** makes sense **77.** false **79.** true **81.** \$15 **83.** 14.3 **84.** 3 **85.** $-14x - 25$ **86.** true **87.** $7 - 3x$ **88.** $15x + 5$

Section 1.4

Check Point Exercises

1. 6 or $\{6\}$ **2.** -1 or $\{-1\}$ **3.** 6 or $\{6\}$ **4.** 1 or $\{1\}$ **5.** no solution or $\varnothing$; inconsistent equation **6.** all real numbers or $\{x \mid x \text{ is a real number}\}$; identity **7.** 2008

Exercise Set 1.4

1. 3 or $\{3\}$ **3.** 11 or $\{11\}$ **5.** 11 or $\{11\}$ **7.** 7 or $\{7\}$ **9.** 13 or $\{13\}$ **11.** 2 or $\{2\}$ **13.** -4 or $\{-4\}$ **15.** 9 or $\{9\}$ **17.** -5 or $\{-5\}$ **19.** 6 or $\{6\}$ **21.** 19 or $\{19\}$ **23.** $\frac{5}{2}$ or $\left\{\frac{5}{2}\right\}$ **25.** 12 or $\{12\}$ **27.** 24 or $\{24\}$ **29.** -15 or $\{-15\}$ **31.** 5 or $\{5\}$ **33.** 13 or $\{13\}$ **35.** -12 or $\{-12\}$ **37.** $\frac{46}{5}$ or $\left\{\frac{46}{5}\right\}$ **39.** all real numbers or $\{x \mid x \text{ is a real number}\}$; identity **41.** no solution or $\varnothing$; inconsistent equation **43.** 0 or $\{0\}$; conditional equation **45.** -10 or $\{-10\}$; conditional equation **47.** no solution or $\varnothing$; inconsistent equation **49.** 0 or $\{0\}$; conditional equation **51.** $3(x - 4) = 3(2 - 2x)$; 2 **53.** $-3(x - 3) = 5(2 - x)$; 0.5 **55.** 2 **57.** -7 **59.** -2 or $\{-2\}$

61. no solution or $\varnothing$ **63.** 10 or $\{10\}$ **65.** -2 or $\{-2\}$ **67. a.** model 1: \$22,228; model 2: \$22,208; Model 1 overestimates the cost by \$10 and model 2 underestimates the cost by \$10. **b.** the school year ending 2012 **69. a.** \$32,000 **b.** \$32,616; \$616 **c.** \$32,597; \$597 **71.** model 1 **73.** 2025 **85.** 3 or $\{3\}$;

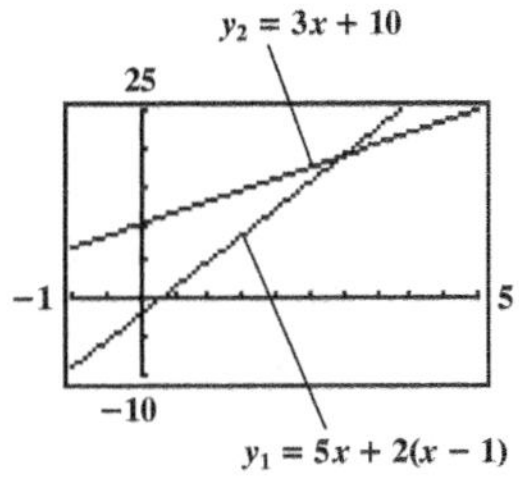

87. -6 or $\{-6\}$;

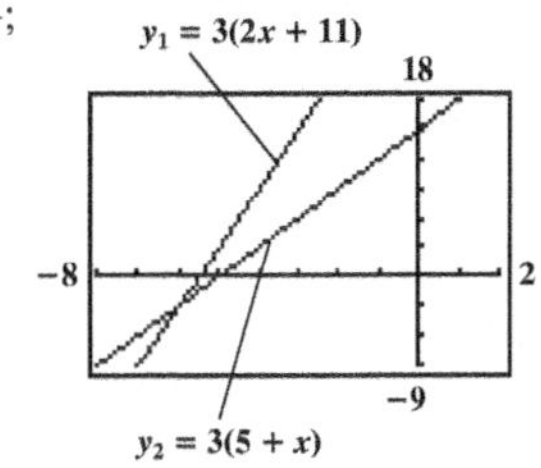

89. makes sense **91.** does not make sense **93.** false **95.** true **97.** $x = \dfrac{c-b}{a}$ **101.** $\dfrac{3}{10}$ **102.** -60 **103.**

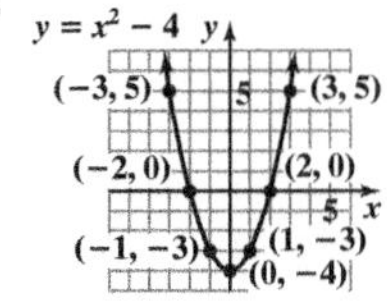

104. a. $3x - 4 = 32$ **b.** 12 **105.** $x + 44$ **106.** $20{,}000 - 2500x$

Mid-Chapter Check Point Exercises

1. $3x + 10$ **2.** 6 or $\{6\}$ **3.** -15 **4.** -7 or $\{-7\}$ **5.** 3 **6.** $13x - 23$ **7.** 0 or $\{0\}$ **8.** $-7x - 34$ **9.** 7 **10.** no solution or $\varnothing$ or inconsistent equation **11.** 3 or $\{3\}$ **12.** -4 **13.** all real numbers or $\{x|x \text{ is a real number}\}$ or identity **14.** 2 **15.**

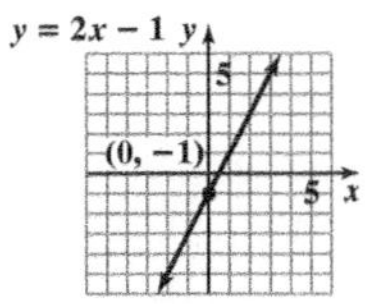

16.

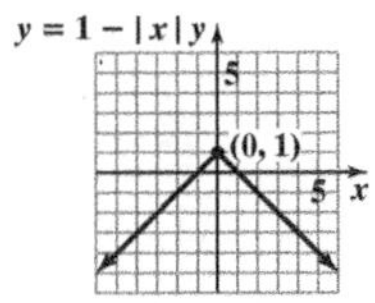

17.

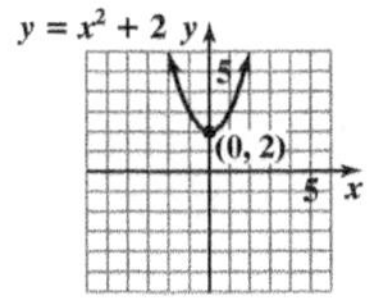

18. true **19.** false **20.** false **21.** true

Section 1.5

Check Point Exercises

1. free: 3; not free: 23; partly free: 20 **2.** 2015 **3.** 300 min **4.** \$1200 **5.** 50 ft by 94 ft **6.** $w = \dfrac{P - 2l}{2}$ **7.** $h = \dfrac{V}{lw}$ **8.** $W = 106 + 6H$ or $W = 6H + 106$ **9.** $C = \dfrac{P}{1 + M}$

Exercise Set 1.5

1. 6 **3.** 25 **5.** 120 **7.** 320 **9.** 19 and 45 **11.** 2 **13.** 8 **15.** all real numbers **17.** sophomore: \$1581; junior: \$2002; senior: \$2846 **19.** 94°, 47°, 39° **21.** 59°, 60°, 61° **23.** 2050 **25.** 2014 **27.** after 5 months; \$165 **29.** 30 times **31. a.** 2014; 22,300 students **b.** $y_1 = 13300 + 1000x$; $y_2 = 26800 - 500x$ **33.** \$420 **35.** \$150 **37.** \$39,000 **39.** \$467.20 **41.** 50 yd by 100 yd **43.** 36 ft by 78 ft **45.** 2 in. **47.** 11 min **49.** $l = \dfrac{A}{w}$ **51.** $b = \dfrac{2A}{h}$ **53.** $P = \dfrac{I}{rt}$ **55.** $p = \dfrac{T - D}{m}$ **57.** $a = \dfrac{2A}{h} - b$ or $a = \dfrac{2A - hb}{h}$ **59.** $h = \dfrac{3V}{\pi r^2}$ **61.** $m = \dfrac{y - y_1}{x - x_1}$ **63.** $d_1 = Vt + d_2$ **65.** $x = \dfrac{C - By}{A}$ **67.** $v = \dfrac{2s - at^2}{2t}$ **69.** $n = \dfrac{L - a}{d} + 1$ **71.** $l = \dfrac{A - 2wh}{2w + 2h}$ **73.** $I = \dfrac{E}{R + r}$ **83. a.** $y = 19.4 + 0.4x$ **85.** makes sense **87.** does not make sense **89.** true **91.** false **93.** 10 problems **95.** 36 plants **97.** -6 is less than or equal to -6.; true **98.** $\dfrac{7}{3}$ **99.** -8 or $\{-8\}$ **100. a.** b^7 **b.** b^{10} **c.** Add the exponents. **101. a.** b^4 **b.** b^6 **c.** Subtract the exponents. **102.** -8

Section 1.6

Check Point Exercises

1. a. b^{11} **b.** $40x^5y^{10}$ **2. a.** $(-3)^3$ or -27 **b.** $9x^{11}y^3$ **3. a.** 1 **b.** 1 **c.** -1 **d.** 10 **e.** 1 **4. a.** $\dfrac{1}{25}$ **b.** $-\dfrac{1}{27}$ **c.** 16 **d.** $\dfrac{3y^4}{x^6}$ **5. a.** $\dfrac{4^3}{7^2} = \dfrac{64}{49}$ **b.** $\dfrac{x^2}{5}$ **6. a.** x^{15} **b.** $\dfrac{1}{y^{14}}$ **c.** b^{12} **7. a.** $16x^4$ **b.** $-27y^6$ **c.** $\dfrac{y^2}{16x^{10}}$ **8. a.** $\dfrac{x^{15}}{64}$ **b.** $\dfrac{16}{x^{12}y^8}$ **c.** $x^{15}y^{20}$ **9. a.** $-12y^9$ **b.** $\dfrac{4y^{14}}{x^6}$ **c.** $\dfrac{64}{x^9y^{15}}$

Exercise Set 1.6

1. b^{11} **3.** x^4 **5.** 32 **7.** $6x^6$ **9.** $20y^{12}$ **11.** $100x^{10}y^{12}$ **13.** $21x^5yz^4$ **15.** b^9 **17.** $5x^5$ **19.** x^5y^5 **21.** $10xy^3$ **23.** $-8a^{11}b^8c^4$ **25.** 1 **27.** 1 **29.** -1 **31.** 13 **33.** 1 **35.** $\dfrac{1}{3^2} = \dfrac{1}{9}$ **37.** $\dfrac{1}{(-5)^2} = \dfrac{1}{25}$ **39.** $-\dfrac{1}{5^2} = -\dfrac{1}{25}$ **41.** $\dfrac{x^2}{y^3}$ **43.** $\dfrac{8y^3}{x^7}$ **45.** $5^3 = 125$ **47.** $(-3)^4 = 81$ **49.** $\dfrac{y^5}{x^2}$ **51.** $\dfrac{b^7c^3}{a^4}$ **53.** x^{60} **55.** $\dfrac{1}{b^{12}}$ **57.** 7^{20} **59.** $64x^3$ **61.** $9x^{14}$ **63.** $8x^3y^6$ **65.** $9x^4y^{10}$ **67.** $-\dfrac{x^6}{27}$ **69.** $\dfrac{y^8}{25x^6}$ **71.** $\dfrac{x^{20}}{16y^{16}z^8}$ **73.** $\dfrac{16}{x^4}$ **75.** $\dfrac{x^6}{25}$ **77.** $\dfrac{81x^4}{y^4}$ **79.** $\dfrac{x^{24}}{y^{12}}$ **81.** x^9y^{12} **83.** a^8b^{12} **85.** $\dfrac{1}{x^6}$

87. $-\frac{4}{x}$ **89.** $\frac{2}{x^7}$ **91.** $\frac{10a^2}{b^5}$ **93.** $\frac{1}{x^9}$ **95.** $-\frac{6}{a^2}$ **97.** $\frac{3y^4z^3}{x^7}$ **99.** $3x^{10}$ **101.** $\frac{1}{x^{10}}$ **103.** $-\frac{5y^8}{x^6}$ **105.** $\frac{8a^9c^{12}}{b}$ **107.** x^{16}

109. $-\frac{27b^{15}}{a^{18}}$ **111.** 1 **113.** $\frac{81x^{20}}{y^{32}}$ **115.** $\frac{1}{100a^4b^{12}c^8}$ **117.** $10x^2y^4$ **119.** $\frac{8}{3xy^{10}}$ **121.** $\frac{y^5}{8x^{14}}$ **123.** $\frac{1}{128x^7y^{16}}$ **125. a.** 1000 aphids

b. 16,000 aphids **c.** 125 aphids **127. a.** one person **b.** 10 people **129. a.** (0, 1) **b.** (4, 10) **131.** d **133.** 0.55 astronomical unit

135. 1.8 astronomical units **147.** makes sense **149.** does not make sense **151.** false **153.** false **155.** false **157.** true

159. x^{9n} **161.** $x^{3n}y^{6n+3}$ **162.**

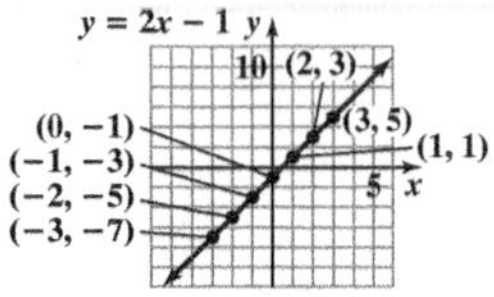

163. $y = \frac{C - Ax}{B}$ **164.** 40 m by 75 m

165. It moves the decimal point three places to the right.

166. It moves the decimal point 2 places to the left.

167. a. 10^5; 100,000 **b.** 10^6; 1,000,000

Section 1.7

Check Point Exercises

1. a. −2,600,000,000 **b.** 0.000003017 **2. a.** 5.21×10^9 **b.** -6.893×10^{-8} **3.** 5.19×10^{11} **4. a.** 3.55×10^{-1} **b.** 4×10^8

5. \$6847 **6.** 3.1×10^7 mi or 31 million mi

Exercise Set 1.7

1. 380 **3.** 0.0006 **5.** −7,160,000 **7.** 1.4 **9.** 0.79 **11.** −0.00415 **13.** −60,000,100,000 **15.** 3.2×10^4 **17.** 6.38×10^{17}

19. -3.17×10^2 **21.** -5.716×10^3 **23.** 2.7×10^{-3} **25.** -5.04×10^{-9} **27.** 7×10^{-3} **29.** 3.14159×10^0 **31.** 6.3×10^7

33. 6.4×10^4 **35.** 1.22×10^{-11} **37.** 2.67×10^{13} **39.** 2.1×10^3 **41.** 4×10^5 **43.** 2×10^{-8} **45.** 5×10^3 **47.** 4×10^{15}

49. 9×10^{-3} **51.** 6×10^{13} **53.** -6.2×10^3 **55.** 1.63×10^{19} **57.** -3.6×10^5 **59.** \$4.65 $\times 10^{10}$ **61.** \$1 $\times 10^8$

63. approximately 67 hot dogs per person **65.** $2.5 \times 10^2 = 250$ chickens **67. a.** \$1.0813 $\times 10^4$; \$10,813 **b.** \$901 **69.** Medicare; \$3242

71. 1.06×10^{-18} g **73.** 3.1536×10^7 **81.** does not make sense **83.** makes sense **85.** false **87.** false **89.** true

91. 1.25×10^{-15} **94.** $85x - 26$ **95.** 4 or {4} **96.** $\frac{y^6}{64x^8}$ **97.** set 1 **98.** −170

99. $5a + 5h + 7$

Review Exercises

1. $2x - 10$ **2.** $6x + 4$ **3.** $\frac{9}{x} + \frac{1}{2}x$ **4.** 34 **5.** 60 **6.** 15 **7.** {1, 2} **8.** {−3, −2, −1, 0, 1} **9.** false **10.** true **11.** true

12. −5 is less than 2.; true **13.** −7 is greater than or equal to −3.; false **14.** −7 is less than or equal to −7.; true **15.** 124 ft **16.** 9.7

17. 5.003 **18.** 0 **19.** −7.6 **20.** −4.4 **21.** 13 **22.** 60 **23.** $-\frac{1}{10}$ **24.** $-\frac{3}{35}$ **25.** −240 **26.** 16 **27.** −32 **28.** $-\frac{5}{12}$

29. 7 **30.** −9.1 **31.** 7 **32.** 9 **33.** −2 **34.** −18 **35.** 55 **36.** 1 **37.** −4 **38.** −13 **39.** $17x - 15$ **40.** $9x^2 + x$

41. $5y - 17$ **42.** $10x$ **43.** $-3x - 8$ **44–46.**

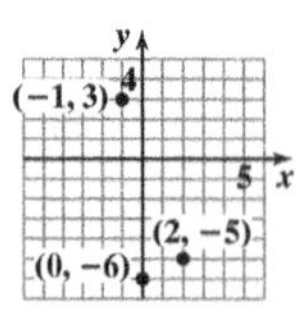

47.

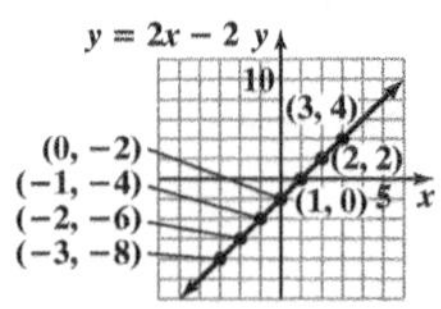

48.

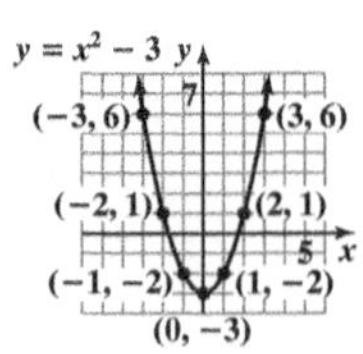

49.

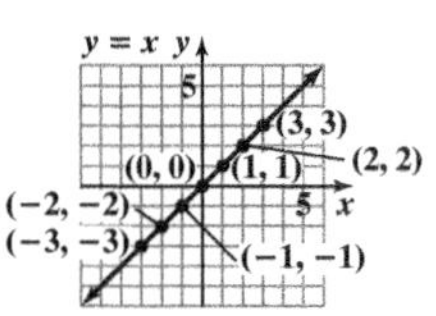

50.

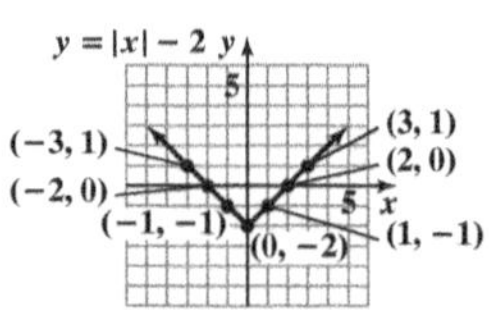

51. minimum x-value: −20; maximum x-value: 40; distance between tick marks on x-axis: 10; minimum y-value: −5; maximum y-value: 5; distance between tick marks on y-axis: 1

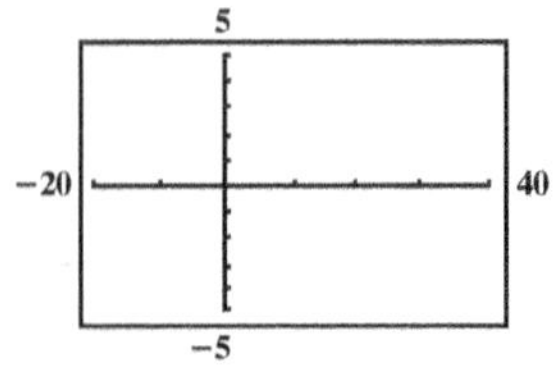

52. 20% **53.** 85 years old **55.** c **56.** 6 or {6} **57.** −10 or {−10} **58.** 5 or {5} **59.** −13 or {−13}

60. -3 or $\{-3\}$ **61.** -1 or $\{-1\}$ **62.** 2 or $\{2\}$ **63.** 2 or $\{2\}$ **64.** $\frac{72}{11}$ or $\left\{\frac{72}{11}\right\}$ **65.** -12 or $\{-12\}$ **66.** $\frac{77}{15}$ or $\left\{\frac{77}{15}\right\}$
67. no solution or $\varnothing$; inconsistent equation **68.** all real numbers or $\{x \mid x \text{ is a real number}\}$; identity **69.** 0 or $\{0\}$; conditional equation
70. $\frac{3}{2}$ or $\left\{\frac{3}{2}\right\}$; conditional equation **71.** no solution or $\varnothing$; inconsistent equation **72. a.** 1997 **b.** overestimates by 2 corporations
73. U2: \$260 million; The Rolling Stones: \$141 million; The Eagles: \$117 million **74.** 25°, 35°, 120° **75. a.** 2018 **b.** \$1177 billion
c. They are shown by the intersection of the graphs at approximately (2018, 1177) **76.** 500 min **77.** \$60 **78.** \$10,000 in sales
79. 44 yd by 126 yd **80. a.** $14{,}100 + 1500x = 41{,}700 - 800x$ **b.** 2017; 32,100 **81.** $h = \frac{3V}{B}$ **82.** $x = \frac{y - y_1}{m} + x_1$
83. $R = \frac{E}{I} - r$ or $R = \frac{E - Ir}{I}$ **84.** $F = \frac{9C + 160}{5}$ or $F = \frac{9}{5}C + 32$ **85.** $g = \frac{s - vt}{t^2}$ **86.** $g = \frac{T}{r + vt}$ **87.** $15x^{13}$ **88.** $\frac{x^2}{y^5}$
89. $\frac{x^4y^7}{9}$ **90.** $\frac{1}{x^{18}}$ **91.** $49x^6y^2$ **92.** $-\frac{8}{y^7}$ **93.** $-\frac{12}{x^7}$ **94.** $3x^{10}$ **95.** $-\frac{a^8}{2b^5}$ **96.** $-24x^7y^4$ **97.** $\frac{3}{4}$ **98.** $\frac{y^{12}}{125x^6}$ **99.** $-\frac{6x^9}{y^5}$
100. $\frac{9x^8y^{14}}{25}$ **101.** $-\frac{x^{21}}{8y^{27}}$ **102.** 7,160,000 **103.** 0.000107 **104.** -4.1×10^{13} **105.** 8.09×10^{-3} **106.** 1.26×10^8 **107.** 2.5×10^1
108. 2.88×10^{13}

Chapter Test

1. $4x - 5$ **2.** 170 **3.** $\{-4, -3, -2, -1\}$ **4.** true **5.** -3 is greater than -1.; false **6.** 259; underestimates by 3 **7.** 17.9
8. -7.6 **9.** $\frac{1}{4}$ **10.** -60 **11.** $\frac{1}{8}$ **12.** 3.1 **13.** -3 **14.** 6 **15.** -4 **16.** $-5x - 18$ **17.** $6y - 27$ **18.** $17x - 22$
19.

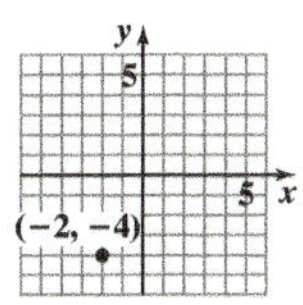

20.

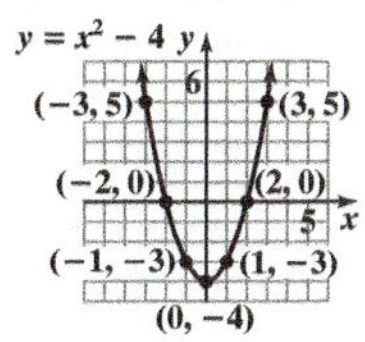

21. 2 or $\{2\}$ **22.** -6 or $\{-6\}$ **23.** no solution or $\varnothing$; inconsistent equation
24. 23 and 49 **25.** 5 yr **26.** 20 prints; \$3.80 **27.** \$50 **28.** 120 yd by 380 yd
29. $h = \frac{3V}{lw}$ **30.** $y = \frac{C - Ax}{B}$ **31.** $-\frac{14}{x^5}$ **32.** $\frac{40}{x^3y^8}$ **33.** $\frac{x^6}{4y^3}$ **34.** $\frac{x^{15}}{64y^6}$
35. $\frac{x^{16}}{9y^{10}}$ **36.** 0.0000038 **37.** 4.07×10^{11} **38.** 5×10^3 **39.** 1.3×10^{10}

CHAPTER 2

Section 2.1

Check Point Exercises

1. domain: {0, 10, 20, 30, 34}; range: {9.1, 6.7, 10.7, 13.2, 15.5} **2. a.** not a function **b.** function **3. a.** 29 **b.** 65 **c.** 46 **d.** $6a + 6h + 9$
4. a. Every element in the domain corresponds to exactly one element in the range. **b.** domain: {0, 1, 2, 3, 4}; range: {3, 0, 1, 2} **c.** 0
d. 2 **e.** $x = 0$ and $x = 4$

Exercise Set 2.1

1. function; domain: {1, 3, 5}; range: {2, 4, 5} **3.** not a function; domain: {3, 4}; range: {4, 5} **5.** function; domain: $\{-3, -2, -1, 0\}$; range: $\{-3, -2, -1, 0\}$ **7.** not a function; domain: {1}; range: {4, 5, 6} **9. a.** 1 **b.** 6 **c.** -7 **d.** $2a + 1$ **e.** $a + 3$ **11. a.** -2
b. -17 **c.** 0 **d.** $12b - 2$ **e.** $3b + 10$ **13. a.** 5 **b.** 8 **c.** 53 **d.** 32 **e.** $48b^2 + 5$ **15. a.** -1 **b.** 26 **c.** 19
d. $2b^2 + 3b - 1$ **e.** $50a^2 + 15a - 1$ **17. a.** $\frac{3}{4}$ **b.** -3 **c.** $\frac{11}{8}$ **d.** $\frac{13}{9}$ **e.** $\frac{2a + 2h - 3}{a + h - 4}$ **f.** Denominator would be zero.
19. a. 6 **b.** 12 **c.** 0 **21. a.** 2 **b.** 1 **c.** -1 and 1 **23.** -2; 10 **25.** -38 **27.** $-2x^3 - 2x$ **29. a.** -1 **b.** 7 **c.** 19
d. 112 **31. a.** {(EL, 1%), (L, 7%), (SL, 11%), (M, 52%), (SC, 13%), (C, 13%), (EC, 3%)} **b.** Yes; Each ideology corresponds to exactly one percentage. **c.** {(1%, EL), (7%, L), (11%, SL), (52%, M), (13%, SC), (13%, C), (3%, EC)} **d.** No; 13% in the domain corresponds to two ideologies in the range, SC and C. **37.** makes sense **39.** makes sense **41.** false **43.** true **45.** true **47.** 3
49. $f(2) = 6$; $f(3) = 9$; $f(4) = 12$; no **50.** 0 **51.** $\frac{y^{10}}{9x^4}$ **52.** $\{-15\}$

53.

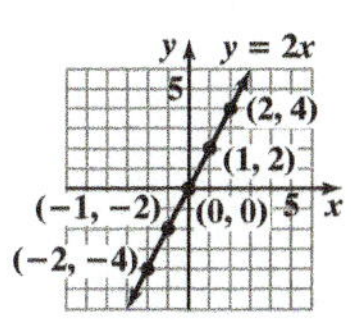

54.

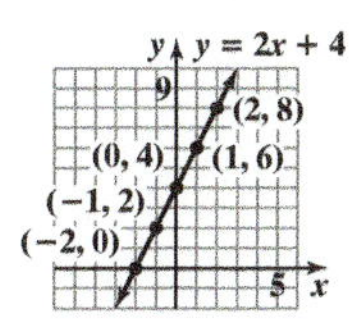

55. a. 3 **b.** -3 and 3 **c.** all real numbers **d.** all real numbers greater than or equal to 1

Section 2.2

Check Point Exercises

1. 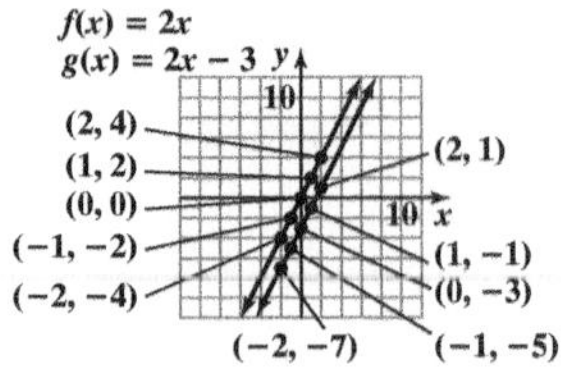

; The graph of g is the graph of f shifted down by 3 units.

2. a. function **b.** function **c.** not a function **3. a.** 400 **b.** 9 **c.** approximately 425
4. a. domain: $\{x|-2 \le x \le 1\}$; range: $\{y|0 \le y \le 3\}$ **b.** domain: $\{x|-2 < x \le 1\}$; range: $\{y|-1 \le y < 2\}$
c. domain: $\{x|-3 \le x < 0\}$; range: $\{y|y = -3, -2, -1\}$

Exercise Set 2.2

1. 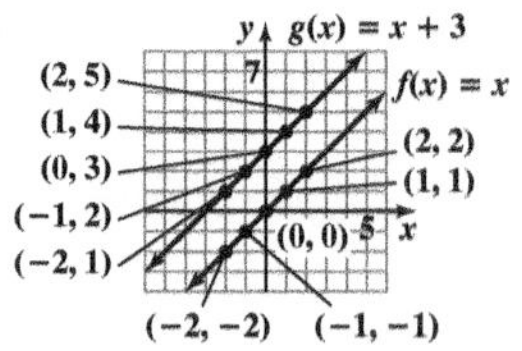

The graph of g is the graph of f shifted up by 3 units.

3. 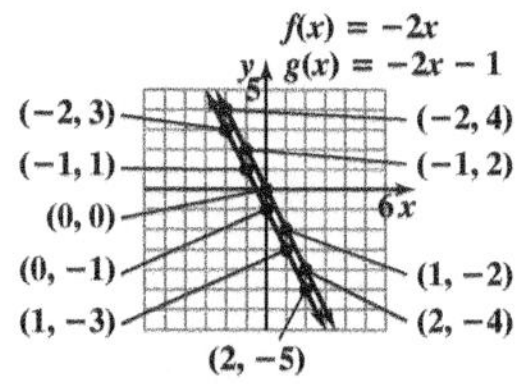

The graph of g is the graph of f shifted down by 1 unit.

5. 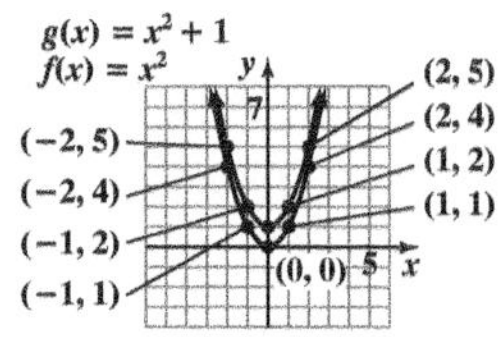

The graph of g is the graph of f shifted up by 1 unit.

7. 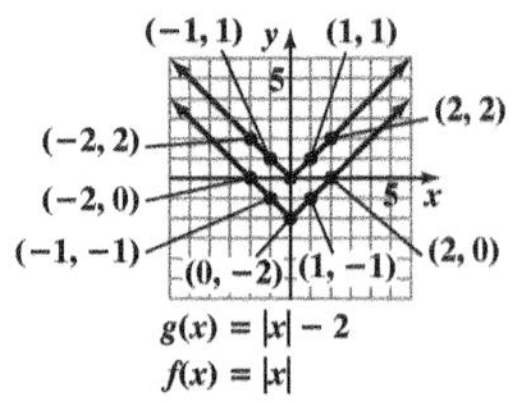

The graph of g is the graph of f shifted down by 2 units.

9. 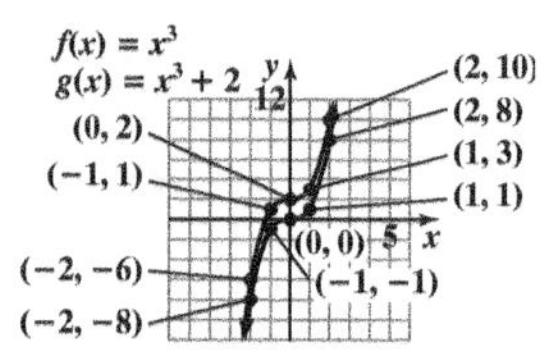

The graph of g is the graph of f shifted up by 2 units.

11. function **13.** not a function **15.** function **17.** not a function **19.** -4 **21.** 4 **23.** 0 **25.** 2 **27.** 2
29. -2 **31.** domain: $\{x|0 \le x < 5\}$; range: $\{y|-1 \le y < 5\}$ **33.** domain: $\{x|x \ge 0\}$; range: $\{y|y \ge 1\}$
35. domain: $\{x|-2 \le x \le 6\}$; range: $\{y|-2 \le y \le 6\}$ **37.** domain: $\{x|x$ is a real number$\}$; range: $\{y|y \le -2\}$
39. domain: $\{x|x = -5, -2, 0, 1, 3\}$; range: $\{y|y = 2\}$ **41. a.** $\{x\,|\,x$ is a real number$\}$ **b.** $\{y\,|\,y \ge -4\}$ **c.** 4 **d.** 2 and 6
e. $(1, 0), (7, 0)$ **f.** $(0, 4)$ **g.** $\{x\,|\,1 < x < 7\}$ **h.** positive **43. a.** 707; Approximately 707,000 bachelor's degrees were awarded to women in 2000.; (20, 707) **b.** overestimates by 2 thousand **45. a.** 68; Approximately 68,000 more bachelor's degrees were awarded to women than to men in 1990.; The points on the graphs with first coordinate 10 are 68 units apart. **b.** overestimates by 13 thousand
47. 440; For 20-year-old drivers, there are 440 accidents per 50 million miles driven.; (20, 440) **49.** $x = 45$; $y = 190$; The minimum number of accidents is 190 per 50 million miles driven and is attributed to 45-year-old drivers. **51.** 3.1; In 1960, Jewish Americans made up about 3.1% of the U.S. population. **53.** 19 and 64; In 1919 and in 1964, Jewish Americans made up about 3% of the U.S. population. **55.** 1940; 3.7%
57. Each year corresponds to only one percentage. **59.** 0.75; It costs \$0.75 to mail a 3-ounce first-class letter. **61.** \$0.58 **69.** makes sense
71. does not make sense **73.** false **75.** true **77.** true **79.** -18 **81.** yes **82.** $\left\{\frac{3}{2}\right\}$ **83.** 76 yd by 236 yd
84. Division by 0 is undefined. **85.** 19 **86.** $10.9x^2 - 35x + 1641$

Section 2.4

Check Point Exercises

1. 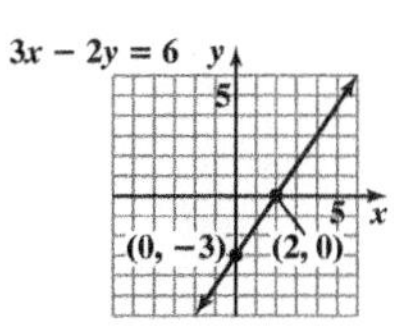

2. **a.** 6 **b.** $-\frac{7}{5}$ **3.** $m = 2; b = -5$

4. 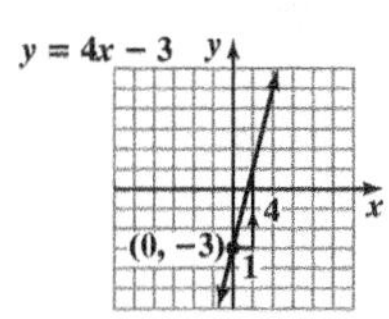

5. 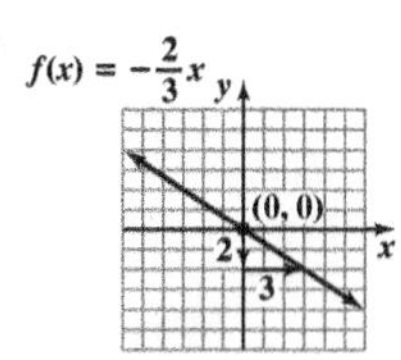

6. 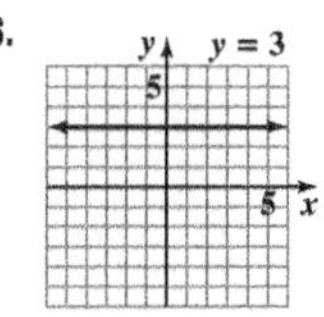

7. 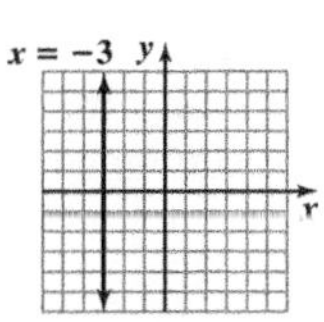

8. 0.57; For the 50–59 age group, the percentage reporting using illegal drugs in the previous month increased by approximately 0.57% each year. **9.** 0.01 mg per 100 mL per hr
10. **a.** $A(x) = 0.11x + 23.9$ **b.** 28.3 yr old

Exercise Set 2.4

1. 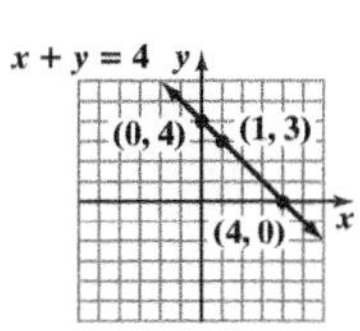

3. 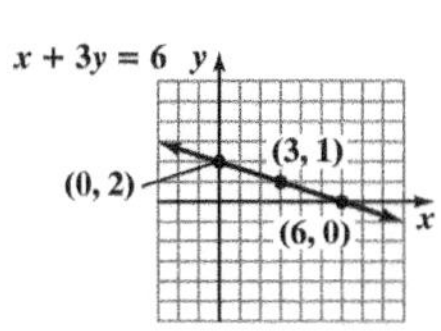

5. 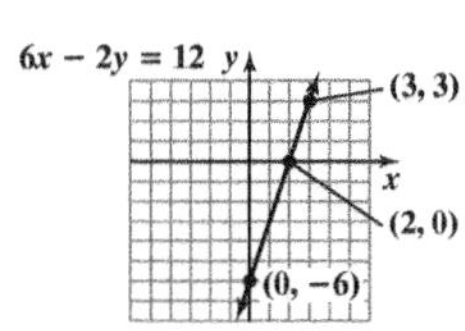

7. 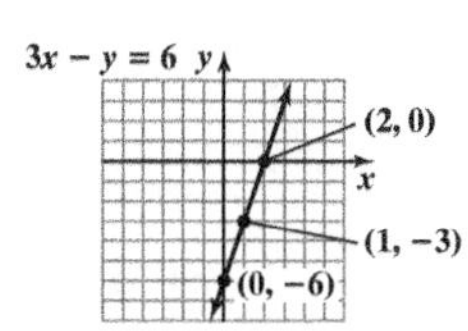

9. 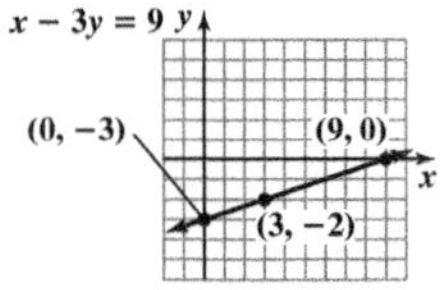

11. 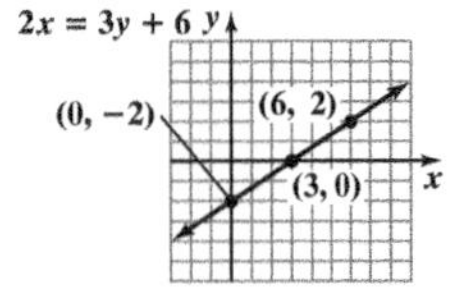

13. 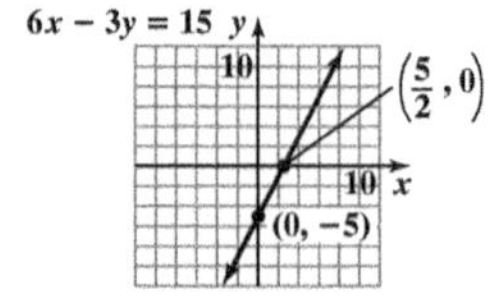

15. $m = 4$; rises **17.** $m = \frac{1}{3}$; rises

19. $m = 0$; horizontal **21.** $m = -\frac{4}{3}$; falls **23.** undefined slope; vertical **25.** L_1: $\frac{2}{3}$; L_2: -2; L_3: $-\frac{1}{2}$

27. $m = 2$; $b = 1$;

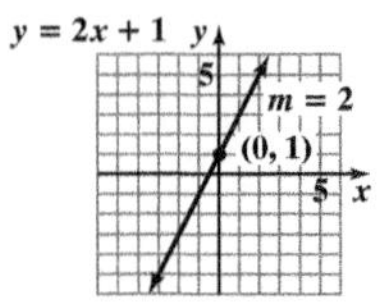

29. $m = -2$; $b = 1$;

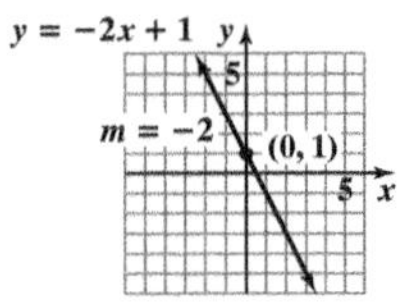

31. $m = \frac{3}{4}$; $b = -2$;

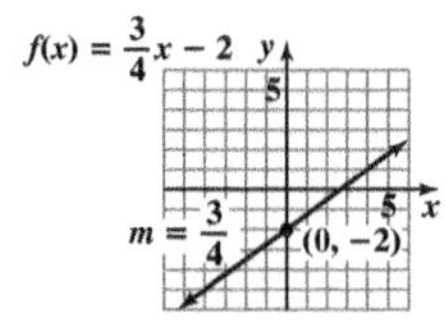

33. $m = -\frac{3}{5}$; $b = 7$;

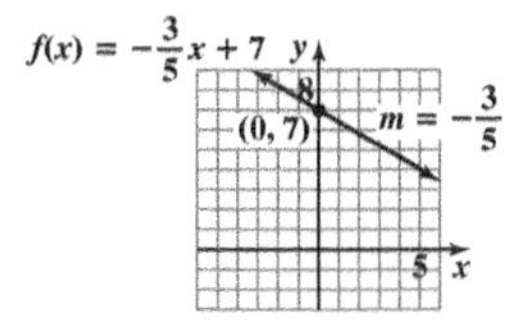

35. $m = -\frac{1}{2}$; $b = 0$;

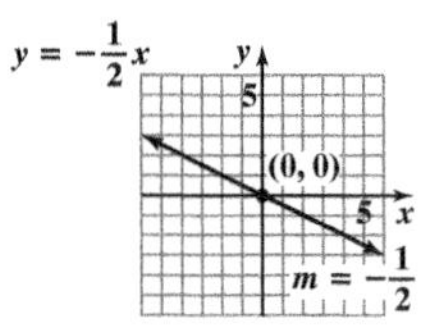

37. $m = 0$; $b = -\frac{1}{2}$;

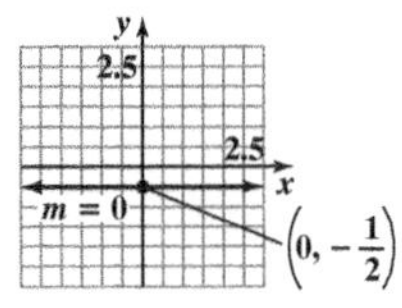

39. a. $y = -2x$

b. $m = -2$; $b = 0$

c.

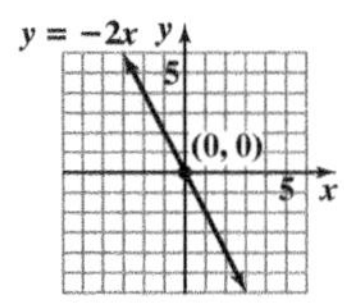

41. a. $y = \frac{4}{5}x$

b. $m = \frac{4}{5}$; $b = 0$

c.

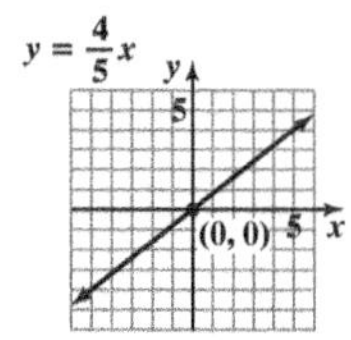

43. a. $y = -3x + 2$

b. $m = -3$; $b = 2$

c.

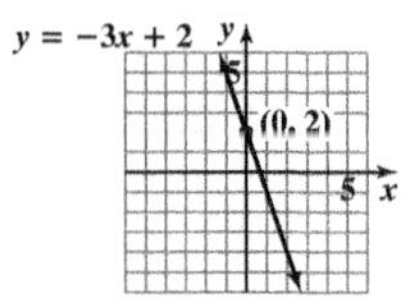

45. a. $y = -\frac{5}{3}x + 5$

b. $m = -\frac{5}{3}$; $b = 5$

c.

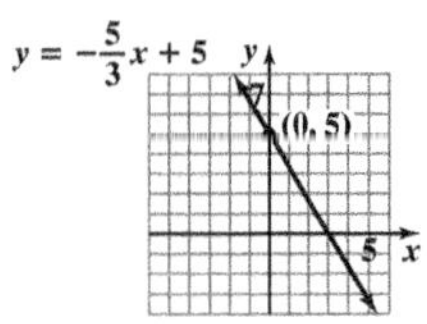

47. 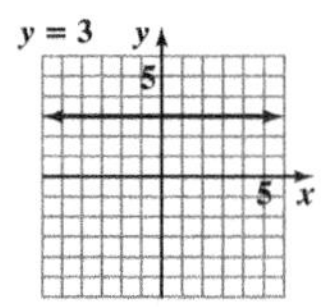

49. 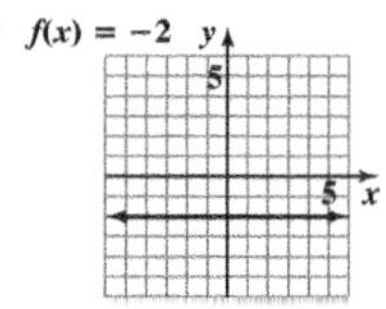

51. 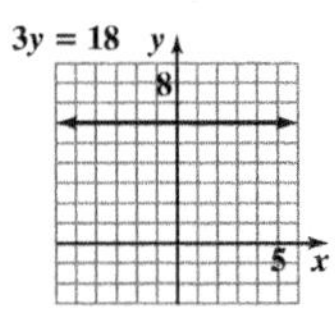

53. 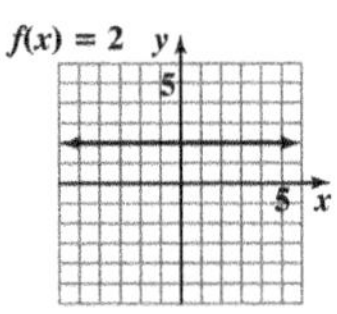

55. 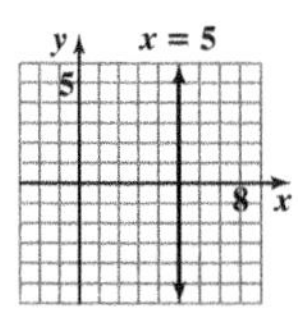

57. 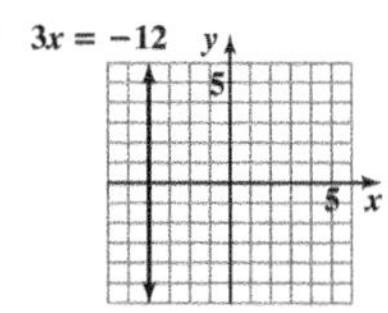

59. 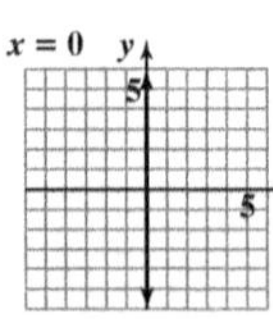

61. $m = -\frac{a}{b}$; falls **63.** undefined slope; vertical **65.** $m = -\frac{A}{B}$; $b = \frac{C}{B}$ **67.** -2

69. 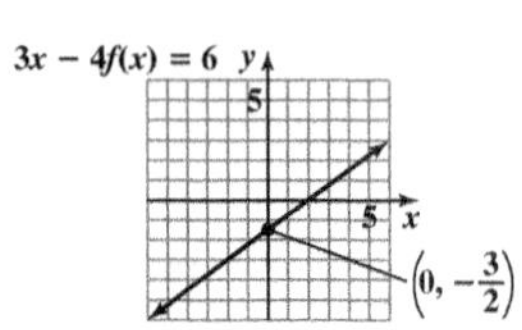

71. 5 **73.** m_1, m_3, m_2, m_4 **75.** $m = 0.01$; The temperature of Earth is increasing by 0.01°F per year. **77.** $m = -0.52$; The percentage of U.S. adults who smoke cigarettes is decreasing by 0.52% each year. **79. a.** 30% **b.** 50% **c.** $m = 4$; average increase of 4% of marriages ending in divorce per year **81. a.** 254; If no women in a country are literate, the mortality rate of children under 5 is 254 per thousand. **b.** -2.4; For each 1% of adult females who are literate, the mortality rate decreases by 2.4 per thousand. **c.** $f(x) = -2.4x + 254$ **d.** 134 per thousand **83.** $P(x) = -1.2x + 47$

103. $m = 2$;

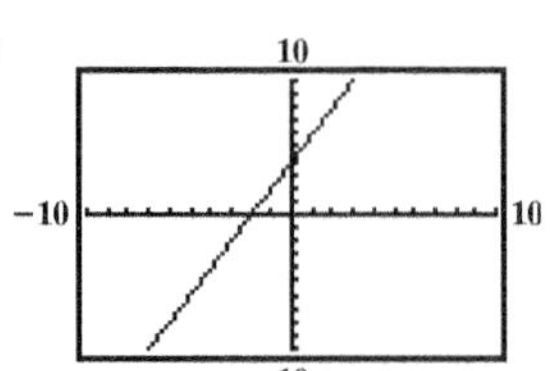

105. $m = -\frac{1}{2}$;

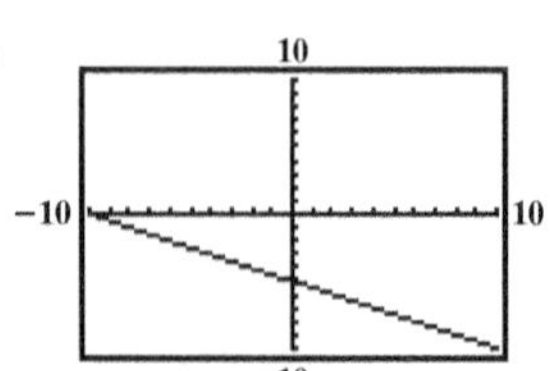

107. does not make sense **109.** does not make sense **111.** false **113.** true **115.** coefficient of x: -6; coefficient of y: 3 **117. a.** $mx_1 + mx_2 + b$ **b.** $mx_1 + mx_2 + 2b$ **c.** no **118.** $16x^4y^6$ **119.** 3.2×10^{-3} **120.** $3x + 17$ **121.** $y = 7x + 33$

122. $y = -\frac{7}{3}x - \frac{2}{3}$ **123. a.** $y = -\frac{1}{4}x + 2; -\frac{1}{4}$ **b.** 4

Section 2.5

Check Point Exercises

1. $y + 3 = -2(x - 4); y = -2x + 5$ or $f(x) = -2x + 5$ **2. a.** $y + 3 = -2(x - 6)$ or $y - 5 = -2(x - 2)$ **b.** $y = -2x + 9$ or $f(x) = -2x + 9$
3. Answers will vary due to rounding.; $f(x) = 0.17x + 72.9$ or $f(x) = 0.17x + 73$; 83.1 yr or 83.2 yr
4. $y - 5 = 3(x + 2); y = 3x + 11$ or $f(x) = 3x + 11$ **5. a.** $m = 3$ **b.** $y + 6 = 3(x + 2); y = 3x$ or $f(x) = 3x$

Exercise Set 2.5

1. $y - 5 = 3(x - 2); f(x) = 3x - 1$ **3.** $y - 6 = 5(x + 2); f(x) = 5x + 16$ **5.** $y + 2 = -4(x + 3); f(x) = -4x - 14$
7. $y - 0 = -5(x + 2); f(x) = -5x - 10$ **9.** $y + \frac{1}{2} = -1(x + 2); f(x) = -x - \frac{5}{2}$ **11.** $y - 0 = \frac{1}{4}(x - 0); f(x) = \frac{1}{4}x$
13. $y + 4 = -\frac{2}{3}(x - 6); f(x) = -\frac{2}{3}x$ **15.** $y - 3 = 1(x - 6)$ or $y - 2 = 1(x - 5); f(x) = x - 3$
17. $y - 0 = 2(x + 2)$ or $y - 4 = 2(x - 0); f(x) = 2x + 4$ **19.** $y - 13 = -2(x + 6)$ or $y - 5 = -2(x + 2); f(x) = -2x + 1$
21. $y - 9 = -\frac{11}{3}(x - 1)$ or $y + 2 = -\frac{11}{3}(x - 4); f(x) = -\frac{11}{3}x + \frac{38}{3}$ **23.** $y + 5 = 0(x + 2)$ or $y + 5 = 0(x - 3); f(x) = -5$
25. $y - 8 = 2(x - 7)$ or $y - 0 = 2(x - 3); f(x) = 2x - 6$ **27.** $y - 0 = \frac{1}{2}(x - 2)$ or $y + 1 = \frac{1}{2}(x - 0); f(x) = \frac{1}{2}x - 1$
29. a. 5 **b.** $-\frac{1}{5}$ **31. a.** -7 **b.** $\frac{1}{7}$ **33. a.** $\frac{1}{2}$ **b.** -2 **35. a.** $-\frac{2}{5}$ **b.** $\frac{5}{2}$ **37. a.** -4 **b.** $\frac{1}{4}$ **39. a.** $-\frac{1}{2}$ **b.** 2
41. a. $\frac{2}{3}$ **b.** $-\frac{3}{2}$ **43. a.** undefined **b.** 0 **45.** $y - 2 = 2(x - 4); y = 2x - 6$ or $f(x) = 2x - 6$
47. $y - 4 = -\frac{1}{2}(x - 2); y = -\frac{1}{2}x + 5$ or $f(x) = -\frac{1}{2}x + 5$ **49.** $y + 10 = -4(x + 8); y = -4x - 42$ or $f(x) = -4x - 42$
51. $y + 3 = -5(x - 2); y = -5x + 7$ or $f(x) = -5x + 7$ **53.** $y - 2 = \frac{2}{3}(x + 2); y = \frac{2}{3}x + \frac{10}{3}$ or $f(x) = \frac{2}{3}x + \frac{10}{3}$
55. $y + 7 = -2(x - 4); y = -2x + 1$ or $f(x) = -2x + 1$ **57.** $f(x) = 5$ **59.** $f(x) = -\frac{1}{2}x + 1$ **61.** $f(x) = -\frac{2}{3}x - 2$
63. $f(x) = 4x - 5$ **65.** $-\frac{A}{B}$ **67. a.** $y - 31.1 = 0.78(x - 10)$ or $y - 38.9 = 0.78(x - 20)$ **b.** $f(x) = 0.78x + 23.3$ **c.** 54.5%
69. a. & b.

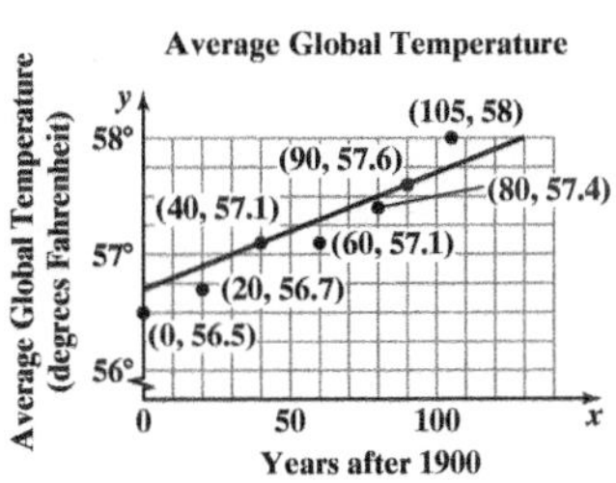

b. $y - 57.1 = 0.01(x - 40)$ or $y - 57.6 = 0.01(x - 90); f(x) = 0.01x + 56.7$ **c.** 58.2°F

71. a. $m \approx 43.1$; The cost of Social Security is projected to increase at a rate of approximately $43.1 billion per year. **b.** $m \approx 51.4$; The cost of Medicare is projected to increase at a rate of approximately $51.4 billion per year. **c.** no; The cost of Medicare is projected to increase at a faster rate than the cost of Social Security.

81. a.

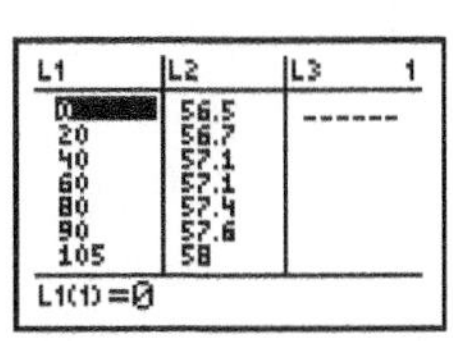

b. & d.

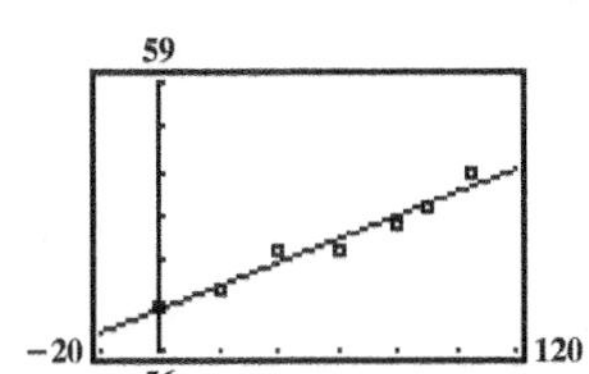

c.

83. makes sense **85.** makes sense **87.** true **89.** true **91.** 4 **93.** (−40, 74) and (97, −200) **95.** 33 **96.** −56
97. 40°, 60°, and 80° **98. a.** yes **b.** yes **99.** (3, −4);

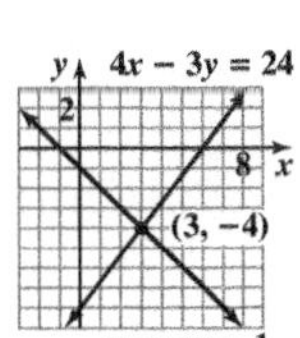

100. {1}

Review Exercises

1. function; domain: {3, 4, 5}; range: {10} **2.** function; domain: {1, 2, 3, 4}; range: {12, 100, π, −6} **3.** not a function; domain: {13, 15}; range: {14, 16, 17} **4. a.** −5 **b.** 16 **c.** −75 **d.** $14a - 5$ **e.** $7a + 9$ **5. a.** 2 **b.** 52 **c.** 70 **d.** $3b^2 - 5b + 2$ **e.** $48a^2 - 20a + 2$

6.

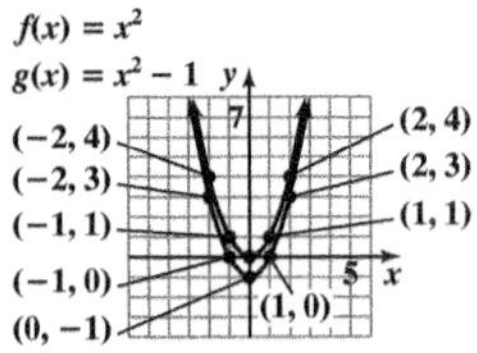

The graph of g is the graph of f shifted down by 1 unit.

7.

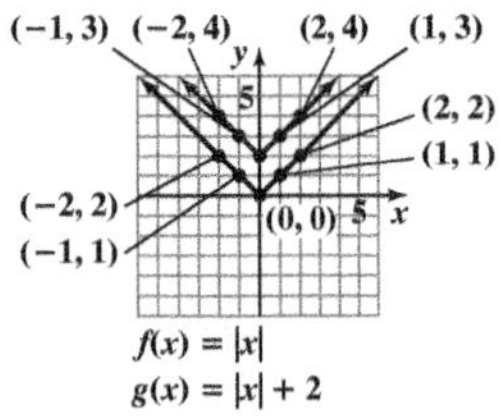

The graph of g is the graph of f shifted up by 2 units.

8. not a function **9.** function **10.** function **11.** not a function **12.** not a function **13.** function **14.** -3 **15.** -2

16. 3 **17.** $\{x|-3 \le x < 5\}$ **18.** $\{y|-5 \le y \le 0\}$

19. a. For each time, there is only one height.
b. 0; The eagle was on the ground after 15 seconds.
c. 45 m
d. 7 and 22; After 7 seconds and after 22 seconds, the eagle's height is 20 meters.
e. Answers will vary.

20. $\{x|x \text{ is a real number}\}$ **21.** $\{x|x \text{ is a real number and } x \neq -8\}$ **22.** $\{x|x \text{ is a real number and } x \neq 5\}$ **23. a.** $6x - 4$ **b.** 14

24. a. $5x^2 + 1$ **b.** 46 **25.** $\{x|x \text{ is a real number and } x \neq 4\}$ **26.** $\{x|x \text{ is a real number and } x \neq -6 \text{ and } x \neq -1\}$ **27.** $x^2 - x - 5; 1$

28. 1 **29.** $x^2 - 3x + 5; 3$ **30.** 9 **31.** -120 **32.** $\dfrac{x^2 - 2x}{x - 5}; -8$ **33.** $\{x|x \text{ is a real number}\}$ **34.** $\{x|x \text{ is a real number and } x \neq 5\}$

35.

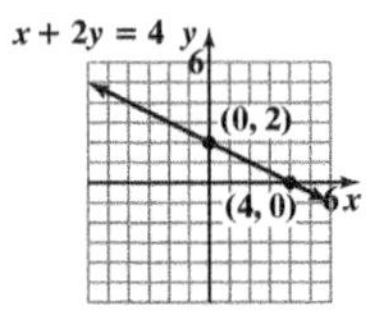

36.

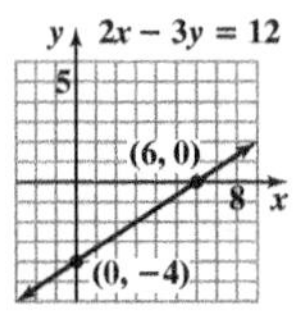

37.

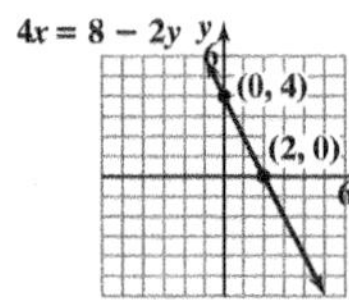

38. 2; rises **39.** $-\frac{2}{3}$; falls

40. undefined; vertical

41. 0; horizontal

42. $m = 2; b = -1;$

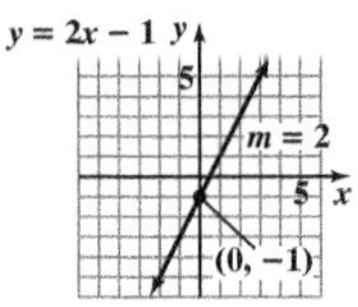

43. $m = -\frac{1}{2}; b = 4;$

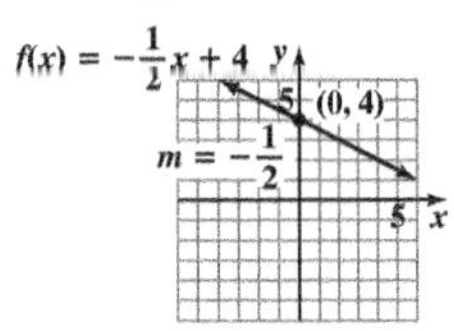

44. $m = \frac{2}{3}; b = 0;$

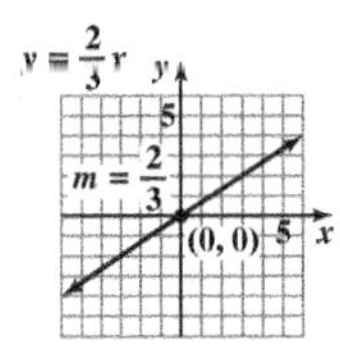

45. $y = -2x + 4; m = -2; b = 4$ **46.** $y = -\frac{5}{3}x; m = -\frac{5}{3}; b = 0$ **47.** $y = -\frac{5}{3}x + 2; m = -\frac{5}{3}; b = 2$

48.

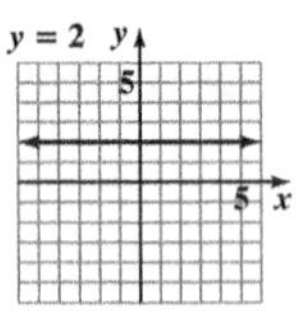

49.

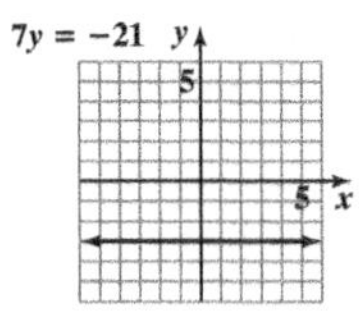

50.

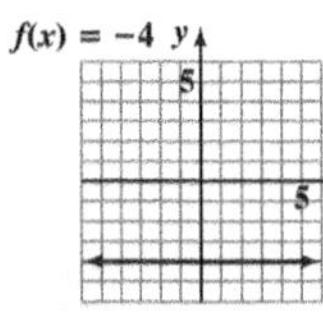

51.

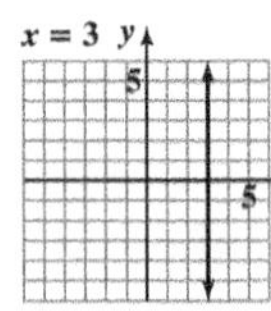

52.

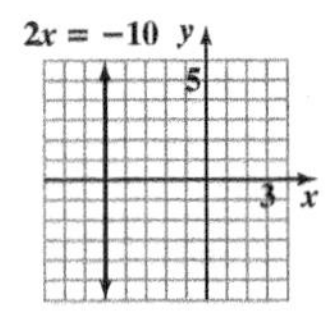

53. -0.27; Record time has been decreasing at a rate of 0.27 second per year since 1900. **54. a.** 137; There was an average increase of approximately 137 discharges per year. **b.** -202; There was an average decrease of approximately 202 discharges per year.

55. a. $F = \frac{9}{5}C + 32$ **b.** 86° **56.** $y - 2 = -6(x + 3); y = -6x - 16$ or $f(x) = -6x - 16$ **57.** $y - 6 = 2(x - 1)$ or $y - 2 = 2(x + 1)$; $y = 2x + 4$ or $f(x) = 2x + 4$ **58.** $y + 7 = -3(x - 4); y = -3x + 5$ or $f(x) = -3x + 5$ **59.** $y - 6 = -3(x + 2); y = -3x$ or $f(x) = -3x$

60. a. $y - 34.6 = 0.8(x - 2)$ or $y - 37.0 = 0.8(x - 5)$ **b.** $f(x) = 0.8x + 33$ **c.** 41 million

Chapter Test

1. function; domain: {1, 3, 5, 6}; range: {2, 4, 6} **2.** not a function; domain: {2, 4, 6}; range: {1, 3, 5, 6} **3.** $3a + 10$ **4.** 28

5.

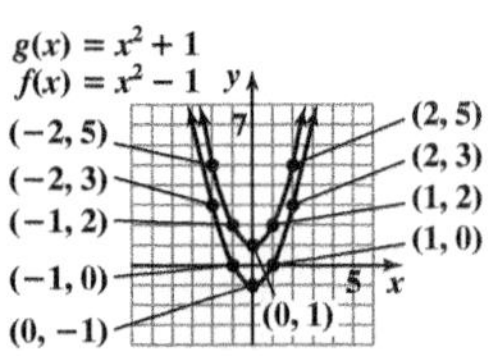

The graph of g is the graph of f shifted up by 2 units.

6. function **7.** not a function **8.** -3 **9.** -2 and 3 **10.** $\{x|x$ is a real number$\}$ **11.** $\{y|y \le 3\}$ **12.** $\{x|x$ is a real number and $x \ne 10\}$

13. $x^2 + 5x + 2$; 26 **14.** $x^2 + 3x - 2$; -4 **15.** -15 **16.** $\frac{x^2 + 4x}{x + 2}$; 3 **17.** $\{x|x$ is a real number and $x \ne -2\}$

18.
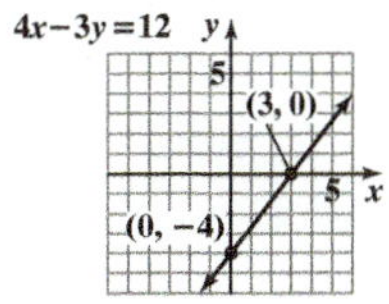

19.
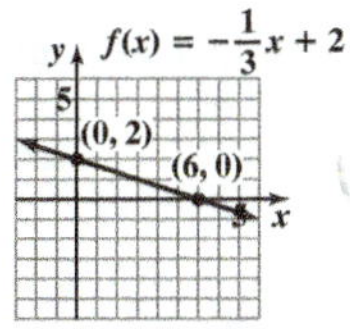

20.
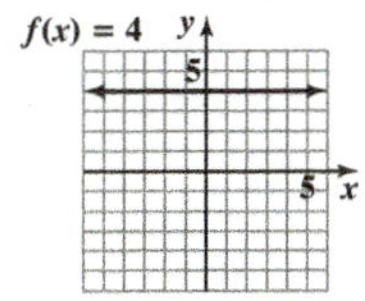

21. $-\frac{1}{2}$; falls **22.** undefined; vertical **23.** 176; In 2005, the number of Super Bowl viewers was 176 million. **24.** 3.6; The number of Super Bowl viewers is increasing at a rate of 3.6 million per year. **25.** $y + 3 = 1(x + 1)$ or $y - 2 = 1(x - 4)$; $y = x - 2$ or $f(x) = x - 2$

26. $y - 3 = 2(x + 2)$; $y = 2x + 7$ or $f(x) = 2x + 7$ **27.** $y + 4 = -\frac{1}{2}(x - 6)$; $y = -\frac{1}{2}x - 1$ or $f(x) = -\frac{1}{2}x - 1$

28. a. $y - 476 = 5(x - 2)$ or $y - 486 = 5(x - 4)$ **b.** $f(x) = 5x + 466$ **c.** 516 per 100,000 residents

Cumulative Review Exercises

1. $\{0, 1, 2, 3\}$ **2.** false **3.** 7 **4.** 15 **5.** $7 + 3x$ or $3x + 7$ **6.** $\{-4\}$ **7.** $\{x|x$ is a real number$\}$; identity **8.** $\{-6\}$ **9.** \$2250

10. $t = \frac{A - p}{pr}$ **11.** $\frac{y^{10}}{9x^8}$ **12.** $\frac{9x^{10}}{y^{12}}$ **13.** 2.1×10^{-5} **14.** function; domain: $\{1, 2, 3, 4, 6\}$; range: $\{5\}$

15.
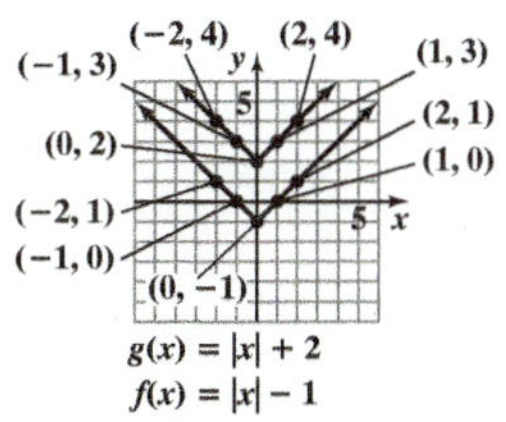

The graph of g is the graph of f shifted up by 3 units.

16. $\{x|x$ is a real number and $x \ne 15\}$ **17.** $2x^2 + x + 5$; 6

18.
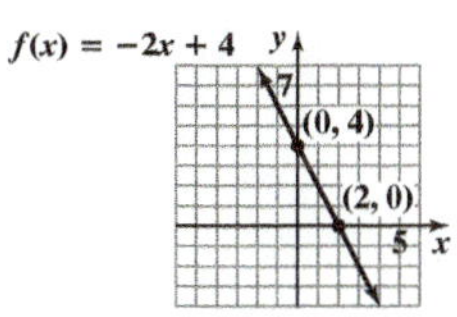

19.
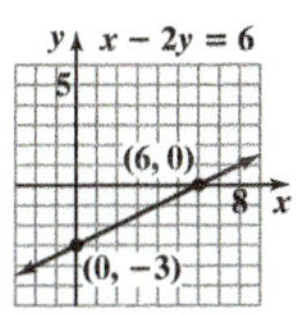

20. $y + 5 = 4(x - 3)$; $y = 4x - 17$ or $f(x) = 4x - 17$

CHAPTER 3

Section 3.1

Check Point Exercises

1. a. not a solution **b.** solution **2.** $(1, 4)$ or $\{(1, 4)\}$ **3.** $(6, 11)$ or $\{(6, 11)\}$ **4.** $(-2, 5)$ or $\{(-2, 5)\}$ **5.** $\left(-\frac{1}{2}, 2\right)$ or $\left\{\left(-\frac{1}{2}, 2\right)\right\}$

6. $(2, -1)$ or $\{(2, -1)\}$ **7.** $\left(\frac{37}{7}, \frac{19}{7}\right)$ or $\left\{\left(\frac{37}{7}, \frac{19}{7}\right)\right\}$ **8.** no solution or $\varnothing$ **9.** $\{(x, y)|x = 4y - 8\}$ or $\{(x, y)|5x - 20y = -40\}$

Exercise Set 3.1

1. solution **3.** not a solution **5.** solution **7.** $\{(3, 1)\}$ **9.** $\left\{\left(\frac{1}{2}, 3\right)\right\}$ **11.** $\{(4, 3)\}$ **13.** $\{(x, y)|2x + 3y = 6\}$ or $\{(x, y)|4x = -6y + 12\}$

15. $\{(1, 0)\}$ **17.** $\varnothing$ **19.** $\{(1, 2)\}$ **21.** $\{(3, 1)\}$ **23.** $\varnothing$ **25.** $\{(2, 4)\}$ **27.** $\{(3, 1)\}$ **29.** $\{(2, 1)\}$ **31.** $\{(2, -3)\}$ **33.** $\varnothing$

35. $\{(3, -2)\}$ **37.** $\varnothing$ **39.** $\{(-5, -1)\}$ **41.** $\left\{(x, y)|y = \frac{2}{5}x - 2\right\}$ or $\{(x, y)|2x - 5y = 10\}$ **43.** $\{(5, 2)\}$ **45.** $\{(2, -3)\}$

47. $\{(-1, 1)\}$ **49.** $\{(-2, -7)\}$ **51.** $\{(7, 2)\}$ **53.** $\{(4, -1)\}$ **55.** $\{(x, y)|2x + 6y = 8\}$ or $\{(x, y)|3x + 9y = 12\}$ **57.** $\left\{\left(\frac{29}{22}, -\frac{5}{11}\right)\right\}$

59. $\{(-2, -1)\}$ **61.** $\{(1, -3)\}$ **63.** $\{(1, -3)\}$ **65.** $\{(4, 3)\}$ **67.** $\{(x, y)|x = 3y - 1\}$ or $\{(x, y)|2x - 6y = -2\}$ **69.** $\varnothing$ **71.** $\{(5, 1)\}$

73. $\left\{\left(\frac{32}{7}, -\frac{20}{7}\right)\right\}$ **75.** $\{(-5, 7)\}$ **77.** $\varnothing$ **79.** $\{(x, y)|x + 2y - 3 = 0\}$ or $\{(x, y)|12 = 8y + 4x\}$ **81.** $\{(0, 0)\}$ **83.** $\{(6, -1)\}$

85. $\left\{\left(\frac{1}{a}, 3\right)\right\}$ **87.** $m = -4, b = 3$ **89.** $y = x - 4$; $y = -\frac{1}{3}x + 4$ **91. a.** approximately (1975, 18); 1975; 18% **b.** 1975; 18%

c. quite well or extremely well **93. a.** $y = 0.04x + 5.48$ **b.** $y = 0.17x + 1.84$ **c.** 2028; 6.6%; Medicare **95. a.** $y = -0.54x + 38$

b. $y = -0.79x + 40$ **c.** 1993; 33.68% **97. a.** 150 sold; 300 supplied **b.** \$100; 250 **109.** makes sense **111.** makes sense

113. false **115.** false **117.** $a = 3, b = 2$ **119.** $\left\{\left(\frac{b_2c_1 - b_1c_2}{a_1b_2 - a_2b_1}, \frac{a_1c_2 - a_2c_1}{a_1b_2 - a_2b_1}\right)\right\}$ **120.** $\left\{\frac{10}{9}\right\}$ **121.** $-128x^{19}y^8$ **122.** 11

123. $0.15x + 0.07y$ **124.** 15 mL **125.** $80x$

Section 3.2

Check Point Exercises

1. hamburger and fries: 1240; fettuccine Alfredo: 1500 **2.** \$3150 at 9%; \$1850 at 11% **3.** 12% solution: 100 oz; 20% solution: 60 oz **4.** boat: 35 mph; current: 7 mph **5. a.** $C(x) = 300{,}000 + 30x$ **b.** $R(x) = 80x$ **c.** (6000, 480,000); The company will break even when it produces and sells 6000 pairs of shoes. At this level, both revenue and cost are \$480,000. **6.** $P(x) = 50x - 300{,}000$

Exercise Set 3.2

1. 3 and 4 **3.** first number: 2; second number: 5 **5. a.** 1500 units; \$48,000 **b.** $P(x) = 17x - 25{,}500$ **7. a.** 500 units; \$122,500 **b.** $P(x) = 140x - 70{,}000$ **9.** after completing college: 41%; after completing high school: 7% **11.** 22 computers and 14 hard drives **13.** \$2000 at 6% and \$5000 at 8% **15.** first fund: \$8000; second fund: \$6000 **17.** \$17,000 at 12%; \$3000 at a 5% loss **19.** California: 100 gal; French: 100 gal **21.** 18-karat gold: 96 g; 12-karat gold: 204 g **23.** cheaper candy: 30 lb; more expensive candy: 45 lb **25.** 8 nickels and 7 dimes **27.** plane: 130 mph; wind: 30 mph **29.** crew: 6 km/hr; current: 2 km/hr **31.** in still water: 4.5 mph; current: 1.5 mph **33.** 86 and 74 **35.** 80°, 50°, 50° **37.** 70 ft by 40 ft **39.** two-seat tables: 6; four-seat tables: 11 **41.** 500 radios **43.** −6000; When the company produces and sells 200 radios, the loss is \$6000. **45. a.** $P(x) = 20x - 10{,}000$ **b.** \$190,000 **47. a.** $C(x) = 18{,}000 + 20x$ **b.** $R(x) = 80x$ **c.** (300, 24,000); When 300 canoes are produced and sold, both revenue and cost are \$24,000. **49. a.** $C(x) = 30{,}000 + 2500x$ **b.** $R(x) = 3125x$ **c.** (48, 150,000); For 48 sold-out performances, both cost and revenue are \$150,000.

59. (6, 300);

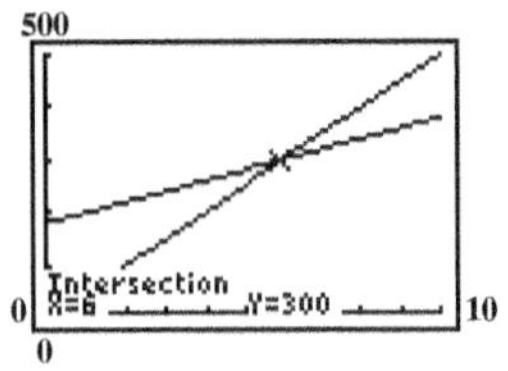

63. does not make sense **65.** makes sense **67.** yes, 8 hexagons and 4 squares **69.** 95 **71.** $y - 5 = -2(x + 2)$ or $y - 13 = -2(x + 6)$; $y = -2x + 1$ or $f(x) = -2x + 1$ **72.** $y - 0 = 1(x + 3)$; $y = x + 3$ or $f(x) = x + 3$ **73.** $\{x|x$ is a real number and $x \neq 3\}$ **74.** yes **75.** $11x + 4y = -3$ **76.** $16a + 4b + c = 1682$

Review Exercises

1. not a solution **2.** solution **3.** $\{(2, 3)\}$ **4.** $\{(x, y) | 3x - 2y = 6\}$ or $\{(x, y) | 6x - 4y = 12\}$ **5.** $\{(-5, -6)\}$ **6.** $\varnothing$ **7.** $\{(3, 4)\}$
8. $\{(23, -43)\}$ **9.** $\{(-4, 2)\}$ **10.** $\left\{\left(3, \frac{1}{2}\right)\right\}$ **11.** $\{(x, y) | y = 4 - x\}$ or $\{(x, y) | 3x + 3y = 12\}$ **12.** $\left\{\left(3, \frac{8}{3}\right)\right\}$ **13.** $\varnothing$
14. TV: \$350; stereo: \$370 **15.** \$2500 at 4%; \$6500 at 7% **16.** 10 ml of 34%; 90 ml of 4% **17.** plane: 630 mph; wind: 90 mph
18. 12 ft by 5 ft **19.** loss of \$4500 **20.** (500, 42,500); When 500 calculators are produced and sold, both cost and revenue are \$42,500.
21. $P(x) = 45x - 22{,}500$ **22. a.** $C(x) = 60{,}000 + 200x$ **b.** $R(x) = 450x$ **c.** (240, 108,000); When 240 desks are produced and sold, both cost and revenue are \$108,000. **23.** no **24.** $\{(0, 1, 2)\}$ **25.** $\{(2, 1, -1)\}$ **26.** infinitely many solutions; dependent equations
27. $y = 3x^2 - 4x + 5$ **28.** 18–29: \$8300; 30–39: \$16,400; 40–49: \$19,500 **29.** $\left[\begin{array}{cc|c} 1 & -8 & 3 \\ 0 & 1 & -2 \end{array}\right]$ **30.** $\left[\begin{array}{cc|c} 1 & -3 & 1 \\ 0 & 7 & -7 \end{array}\right]$ **31.** $\left[\begin{array}{ccc|c} 1 & -1 & \frac{1}{2} & -\frac{1}{2} \\ 1 & 2 & -1 & 2 \\ 6 & 4 & 3 & 5 \end{array}\right]$
32. $\left[\begin{array}{ccc|c} 1 & 2 & 2 & 2 \\ 0 & 1 & -1 & 2 \\ 0 & 0 & 9 & -9 \end{array}\right]$ **33.** $\{(-5, 3)\}$ **34.** no solution or $\varnothing$ **35.** $\{(1, 3, -4)\}$ **36.** $\{(-2, -1, 0)\}$ **37.** 17 **38.** 4
39. -86 **40.** -236 **41.** $\left\{\left(\frac{7}{4}, -\frac{25}{8}\right)\right\}$ **42.** $\{(2, -7)\}$ **43.** $\{(23, -12, 3)\}$ **44.** $\{(-3, 2, 1)\}$ **45.** $y = \frac{5}{8}x^2 - 50x + 1150$; 30-year-old drivers are involved in 212.5 accidents daily and 50-year-old drivers are involved in 212.5 accidents daily.

Chapter Test

1. $\{(2, 4)\}$ **2.** $\{(6, -5)\}$ **3.** $\{(1, -3)\}$ **4.** $\{(x, y)|4x = 2y + 6\}$ or $\{(x, y)|y = 2x - 3\}$ **5.** one-bedroom: 15 units; two-bedroom: 35 units **6.** \$2000 at 6% and \$7000 at 7% **7.** 6% solution: 12 oz; 9% solution: 24 oz **8.** boat: 14 mph; current: 2 mph **9.** $C(x) = 360{,}000 + 850x$ **10.** $R(x) = 1150x$ **11.** (1200, 1,380,000); When 1200 computers are produced and sold, both cost and revenue are \$1,380,000. **12.** $P(x) = 85x - 350{,}000$ **13.** $\{(1, 3, 2)\}$ **14.** $\left[\begin{array}{ccc|c} 1 & 0 & -4 & 5 \\ 0 & -1 & 26 & -20 \\ 2 & -1 & 4 & -3 \end{array}\right]$ **15.** $\{(4, -2)\}$ **16.** $\{(-1, 2, 2)\}$ **17.** 17 **18.** -10 **19.** $\{(-1, -6)\}$ **20.** $\{(4, -3, 3)\}$

Cumulative Review Exercises

1. 1 **2.** $15x - 7$ **3.** $\{-6\}$ **4.** $\{-15\}$ **5.** no solution or $\varnothing$ **6.** \$2000 **7.** $-\frac{x^8}{4y^{30}}$ **8.** $-4a - 3$ **9.** $\{x|x \text{ is a real number and } x \neq -3\}$ **10.** $x^2 - 3x - 1; -1$

11.
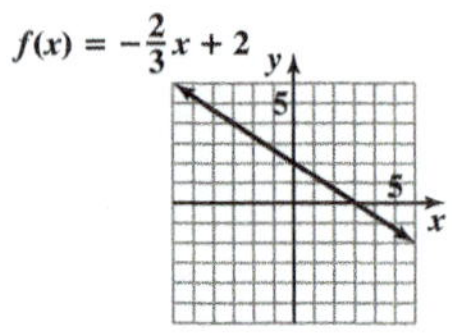

12.
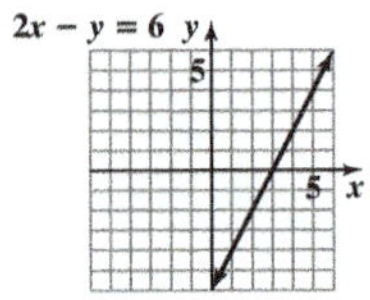

13. $y - 4 = -3(x - 2)$ or $y + 2 = -3(x - 4)$; $y = -3x + 10$ or $f(x) = -3x + 10$ **14.** $y - 0 = -3(x + 1)$; $y = -3x - 3$ or $f(x) = -3x - 3$ **15.** $\left\{\left(7, \frac{1}{3}\right)\right\}$ **16.** $\{(3, 2, 4)\}$ **17.** pad: \$0.80; pen: \$0.20 **18.** 23 **19.** $\{(3, -2, 1)\}$ **20.** $\{(-3, 2)\}$

CHAPTER 4

Section 4.1

Check Point Exercises

1. a. $\{x|-2 \le x < 5\}$
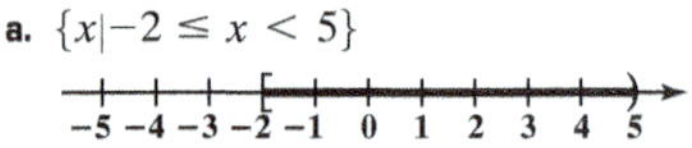

b. $\{x|1 \le x \le 3.5\}$
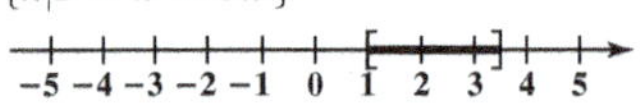

c. $\{x|x < -1\}$
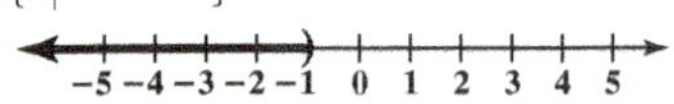

2. $\{x|x > -5\}$ or $(-5, \infty)$
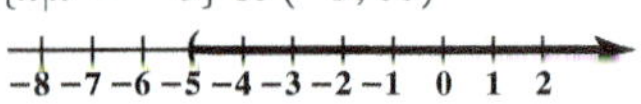

3. $\{x|x < 4\}$ or $(-\infty, 4)$
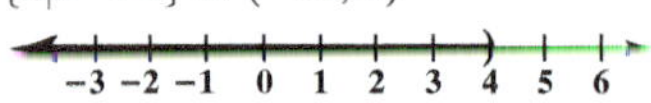

4. $\{x|x \ge 13\}$ or $[13, \infty)$
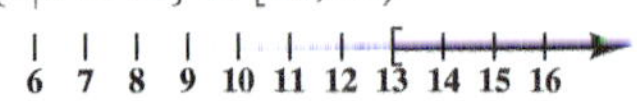

5. a. $\{x|x \text{ is a real number}\}$ or $\mathbb{R}$ or $(-\infty, \infty)$ **b.** $\varnothing$ **6.** more than 720 miles

Exercise Set 4.1

1. $\{x|1 < x \le 6\}$
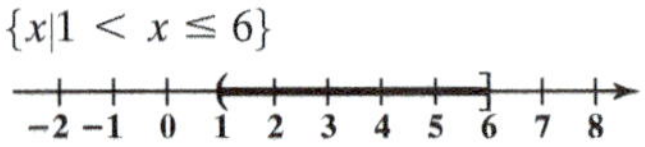

3. $\{x|-5 \le x < 2\}$
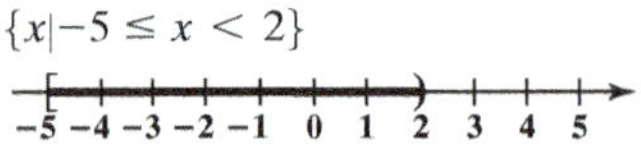

5. $\{x|-3 \le x \le 1\}$
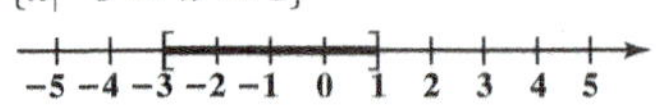

7. $\{x|x > 2\}$
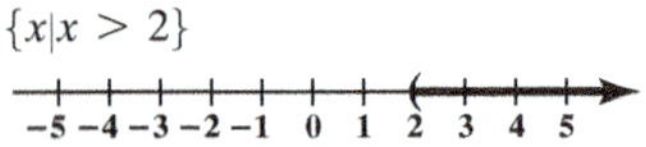

9. $\{x|x \ge -3\}$
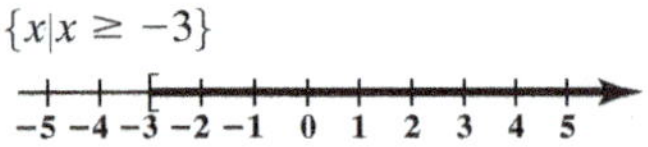

11. $\{x|x < 3\}$

13. $\{x|x < 5.5\}$
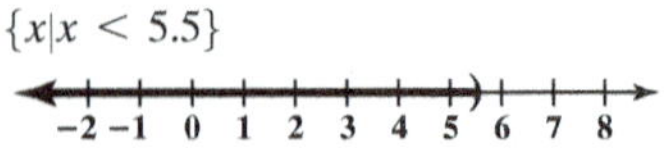

15. $\{x|x < 3\}$ or $(-\infty, 3)$
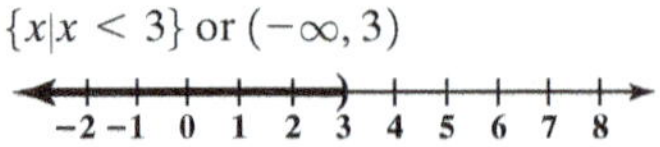

17. $\{x|x \ge 7\}$ or $[7, \infty)$
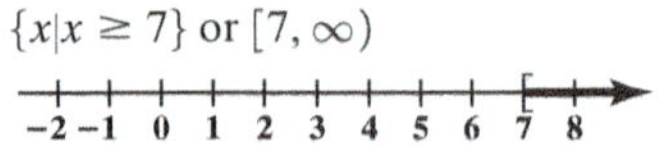

19. $\{x|x \le -4\}$ or $(-\infty, -4]$
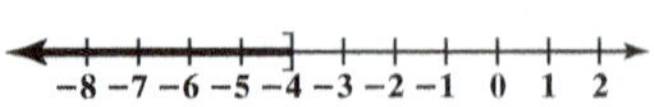

21. $\left\{x \middle| x \le -\frac{2}{5}\right\}$ or $\left(-\infty, -\frac{2}{5}\right]$
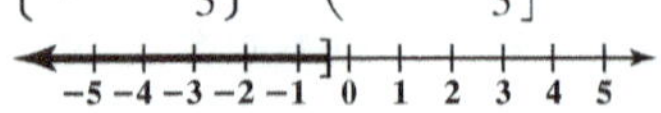

23. $\{x|x \ge 0\}$ or $[0, \infty)$

25. $\{x|x < 1\}$ or $(-\infty, 1)$
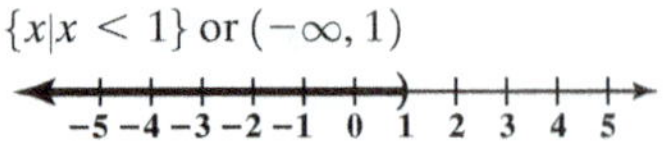

27. $\{x|x \ge 6\}$ or $[6, \infty)$
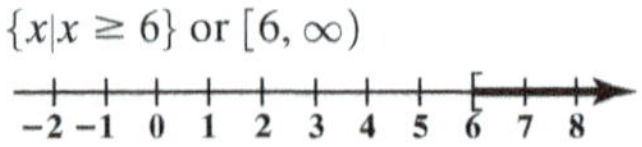

29. $\{x|x \ge -6\}$ or $[-6, \infty)$
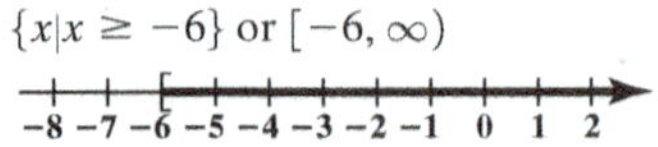

31. $\{x|x < -6\}$ or $(-\infty, -6)$
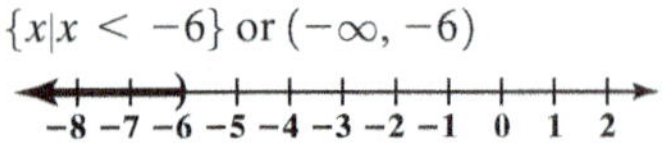

33. $\{x|x \ge 13\}$ or $[13, \infty)$
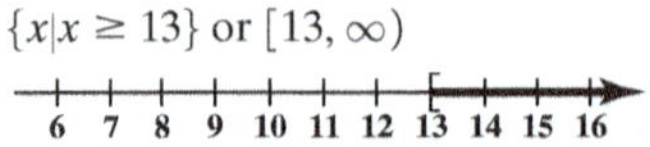

35. $\{x|x \text{ is a real number}\}$ or $\mathbb{R}$ or $(-\infty, \infty)$
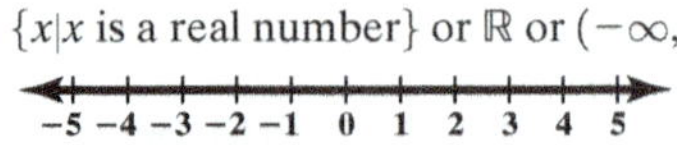

37. $\{x|x \text{ is a real number}\}$ or $\mathbb{R}$ or $(-\infty, \infty)$
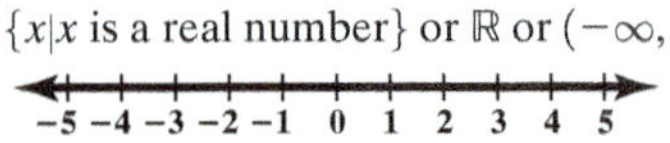

39. $\varnothing$

41. $\{x|x > -6\}$ or $(-6, \infty)$
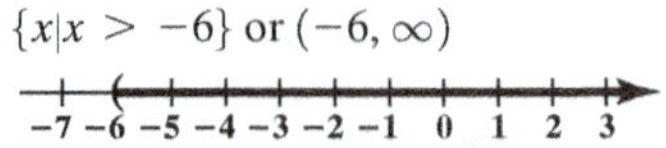

43. $\{x|x \ge -1\}$ or $[-1, \infty)$
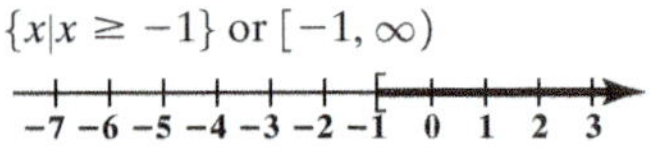

45. $\{x|x < -2\}$ or $(-\infty, -2)$
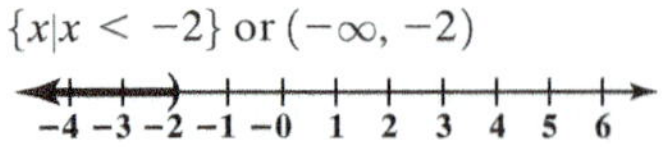

47. $\{x|x < 5\}$ or $(-\infty, 5)$

49. $\left\{x \middle| x \geq -\frac{4}{7}\right\}$ or $\left[-\frac{4}{7}, \infty\right)$ **51.** $\{x|x \geq 6\}$ or $[6, \infty)$ **53.** $\{x|x < 2\}$ or $(-\infty, 2)$

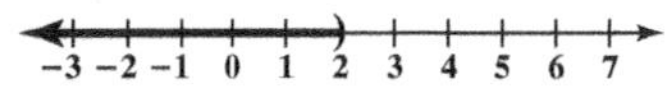

55. $x < \frac{c-b}{a}$ **57.** $\{x|x \leq -3\}$ or $(-\infty, -3]$ **59.** $\{x|x > -1.4\}$ or $(-1.4, \infty)$ **61.** $(0, 4)$ **63.** intimacy ≥ passion or passion ≤ intimacy **65.** commitment > passion or passion < commitment **67.** 9; after 3 years **69.** voting years after 2006 **71.** $\{t|t > 175\}$ or $(175, \infty)$; The women's speed skating times will be less than the men's after the year 2075. **73.** more than 100 miles per day **75.** greater than \$32,000 **77.** more than 6250 tapes **79.** 40 bags or fewer

89.

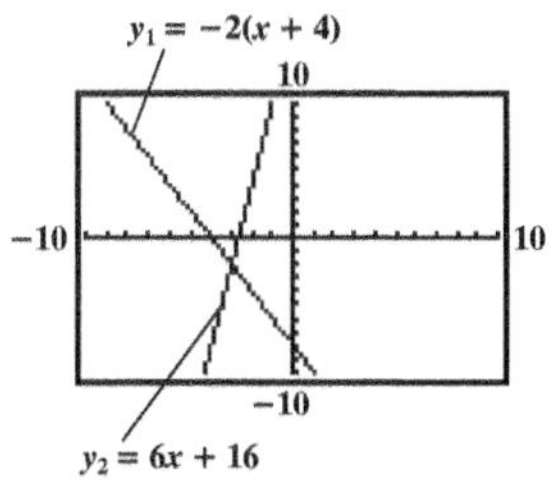

$\{x|x < -3\}$ or $(-\infty, -3)$

91. $\varnothing$

93. a. plan A: $4 + 0.10x$; plan B: $2 + 0.15x$

b.

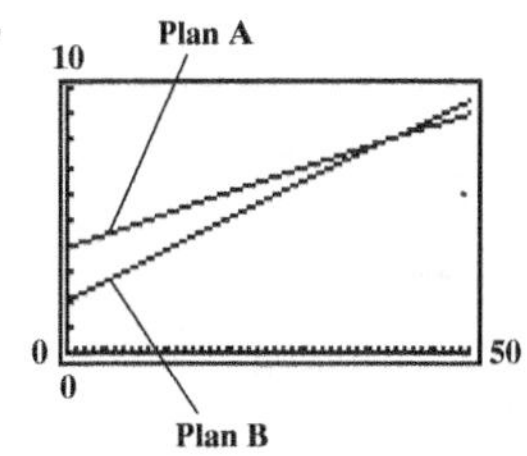

c–d. more than 40 checks per month

95. makes sense **97.** makes sense **99.** false **101.** true **103.** Since $x > y$, $y - x < 0$. Thus, when both sides were multiplied by $y - x$, the sense of the inequality should have been changed. **104.** 29 **105.** $\{(-1, -1, 2)\}$ **106.** $\frac{x^9}{8y^{15}}$ **107. a.** $\{3, 4\}$ **b.** $\{1, 2, 3, 4, 5, 6, 7\}$ **108. a.** $\{x|x < 8\}$ or $(-\infty, 8)$ **b.** $\{x|x < 5\}$ or $(-\infty, 5)$ **c.** any number less than 5 **d.** any number in $[5, 8)$ **109. a.** $\{x|x \geq 1\}$ or $[1, \infty)$ **b.** $\{x|x \geq 3\}$ or $[3, \infty)$ **c.** any number greater than or equal to 3 **d.** any number in $[1, 3)$

Section 4.2

Check Point Exercises

1. $\{3, 7\}$ **2.** $\{x|x < 1\}$ or $(-\infty, 1)$ **3.** $\varnothing$ **4.** $\{x|-1 \leq x < 4\}$ or $[-1, 4)$;

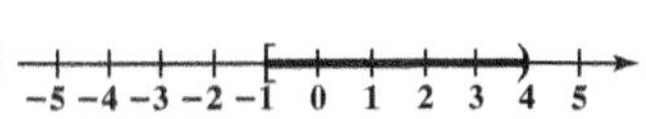

5. $\{3, 4, 5, 6, 7, 8, 9\}$ **6.** $\{x|x \leq 1 \text{ or } x > 3\}$ or $(-\infty, 1] \cup (3, \infty)$ **7.** $\{x|x \text{ is a real number}\}$ or $\mathbb{R}$ or $(-\infty, \infty)$

Exercise Set 4.2

1. $\{2, 4\}$ **3.** $\varnothing$ **5.** $\varnothing$

7. $\{x|x > 6\}$; $(6, \infty)$

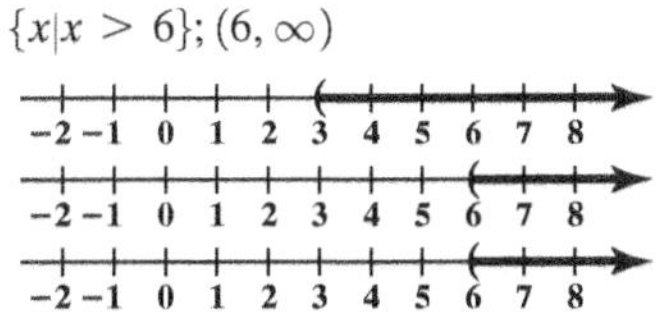

9. $\{x|x \leq 1\}$; $(-\infty, 1]$

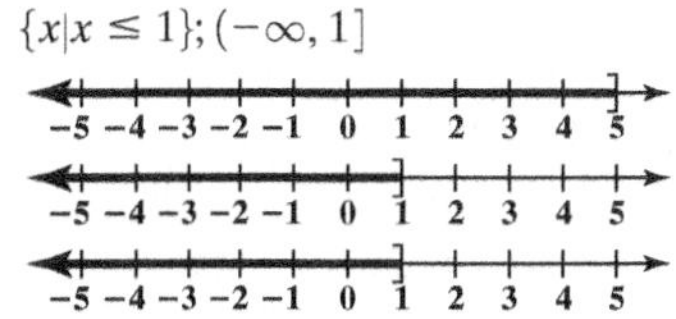

11. $\{x|-1 \leq x < 2\}$; $[-1, 2)$

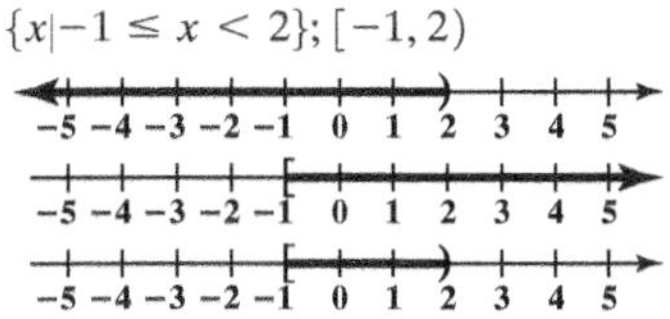

13. $\varnothing$

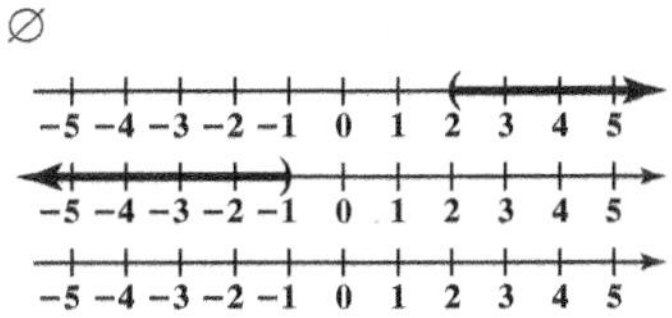

15. $\{x|-6 < x < -4\}$; $(-6, -4)$

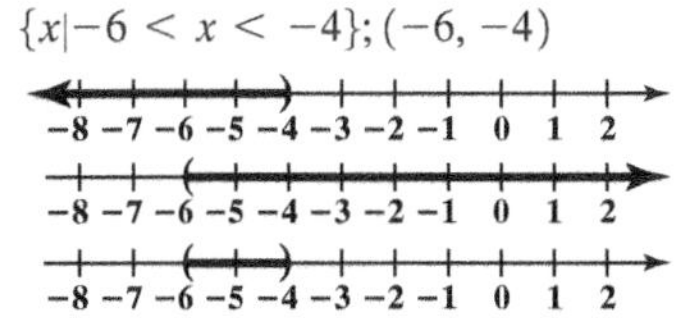

17. $\{x|-3 < x \leq 6\}$; $(-3, 6]$

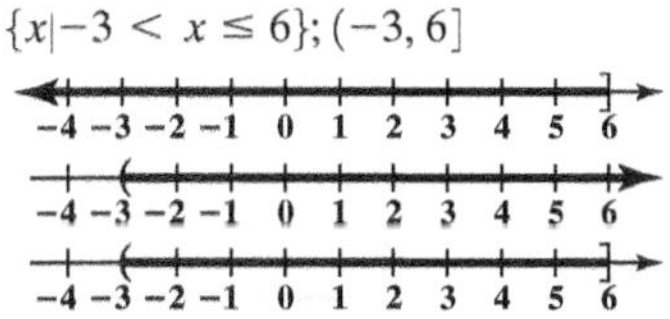

19. $\{x|2 < x < 5\}$; $(2, 5)$

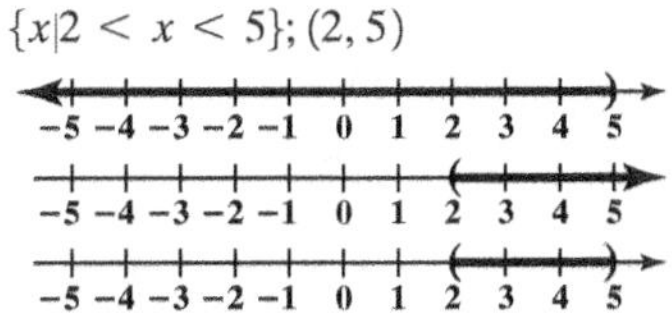

21. $\varnothing$

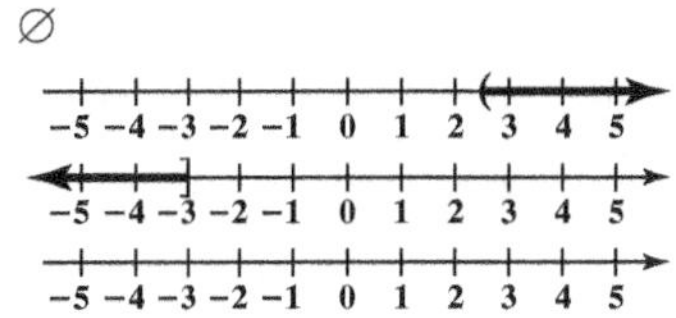

23. $\{x|0 \leq x < 2\}$; $[0, 2)$

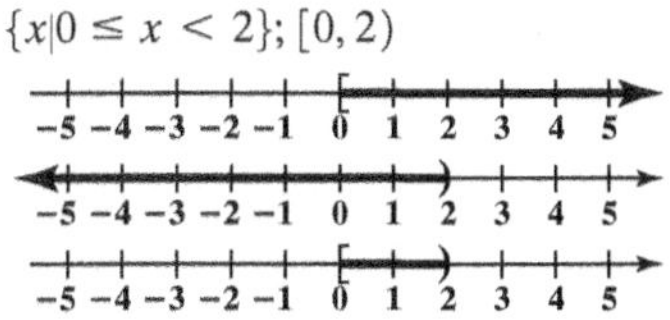

25. $\{x|3 < x < 5\}$; $(3, 5)$

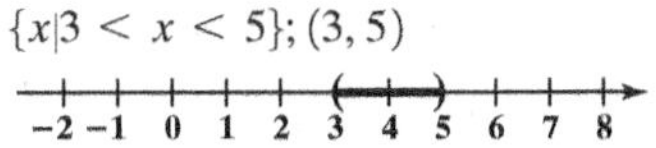

27. $\{x|-1 \leq x < 3\}$; $[-1, 3)$

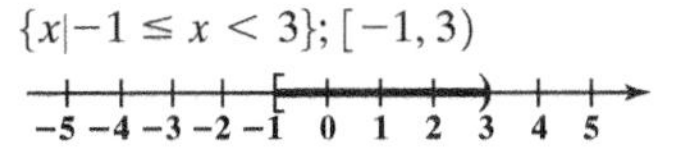

29. $\{x|-5 < x \leq -2\}$; $(-5, -2]$

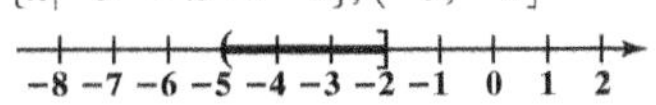

31. $\{x|3 \leq x < 6\}$; $[3, 6)$

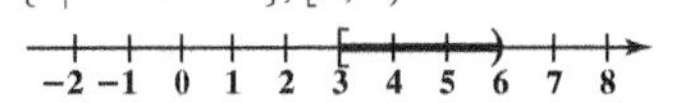

33. $\{1, 2, 3, 4, 5\}$ **35.** $\{1, 2, 3, 4, 5, 6, 7, 8, 10\}$ **37.** $\{a, e, i, o, u\}$

39. $\{x|x > 3\}$; $(3, \infty)$

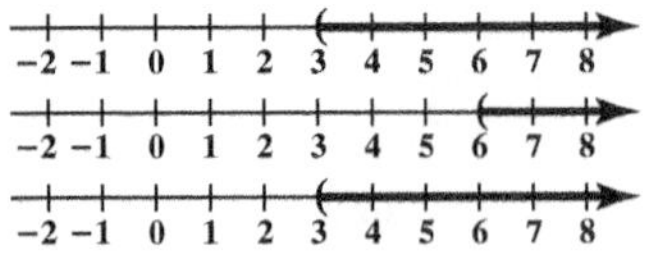

41. $\{x|x \le 5\}$; $(-\infty, 5]$

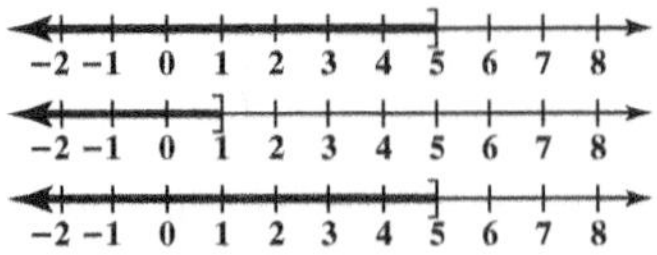

43. $\{x|x$ is a real number$\}$; $(-\infty, \infty)$

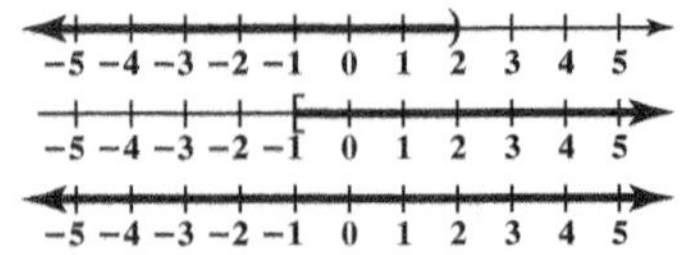

45. $\{x|x < -1 \text{ or } x \ge 2\}$; $(-\infty, -1) \cup [2, \infty)$

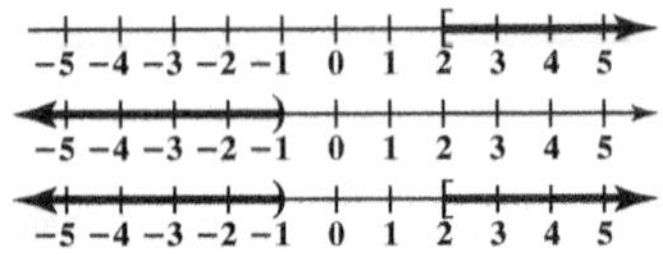

47. $\{x|x < -3 \text{ or } x > 4\}$; $(-\infty, -3) \cup (4, \infty)$

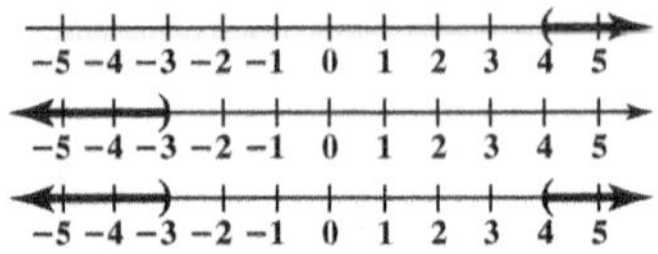

49. $\{x|x \le 1 \text{ or } x \ge 3\}$; $(-\infty, 1] \cup [3, \infty)$

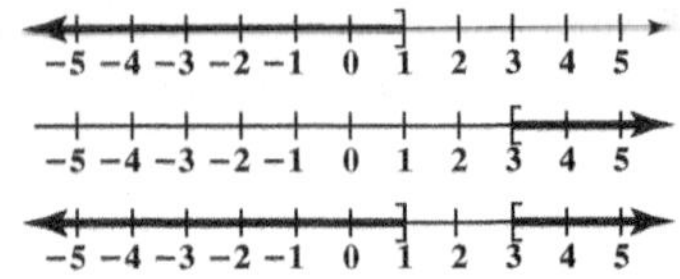

51. $\{x|x$ is a real number$\}$; $(-\infty, \infty)$

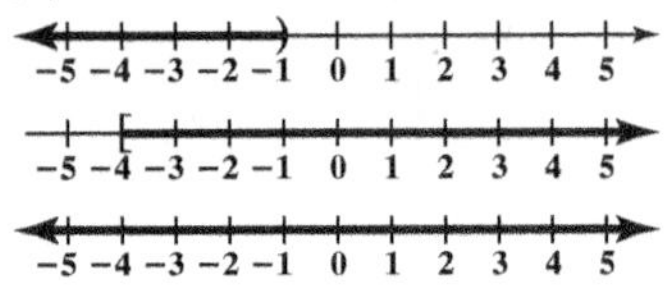

53. $\{x|x < 2\}$; $(-\infty, 2)$

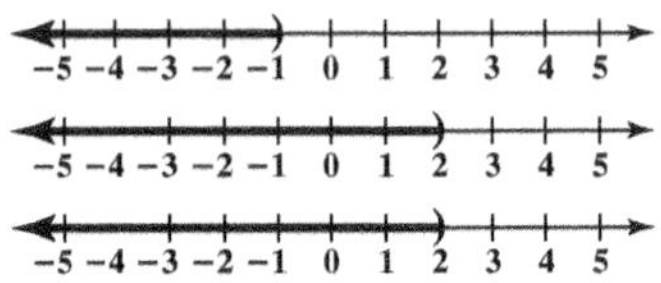

55. $\{x|x > 4\}$ or $(4, \infty)$ **57.** $\{x|x < 0 \text{ or } x > 6\}$ or $(-\infty, 0) \cup (6, \infty)$ **59.** $\frac{b-c}{a} < x < \frac{b+c}{a}$ **61.** $\{x|-1 \le x \le 3\}$ or $[-1, 3]$
63. $\{x|-1 < x < 3\}$ or $(-1, 3)$ **65.** $\{x|-1 \le x < 2\}$ or $[-1, 2)$ **67.** $\{-3, -2, -1\}$ **69.** $\{1974, 1980, 1994\}$
71. $\{1962, 1974, 1980, 1994, 2002\}$ **73.** $\{1962, 2002\}$ **75.** $\varnothing$ **77.** $\{1962, 1974\}$ **79.** between 80 and 110 minutes, inclusive
81. $[76, 126)$; If the highest grade is 100, then $[76, 100]$. **83.** more than 3 and less than 15 crossings per 3-month period

91. $\{x|-2 < x < 6\}$ or $(-2, 6)$;

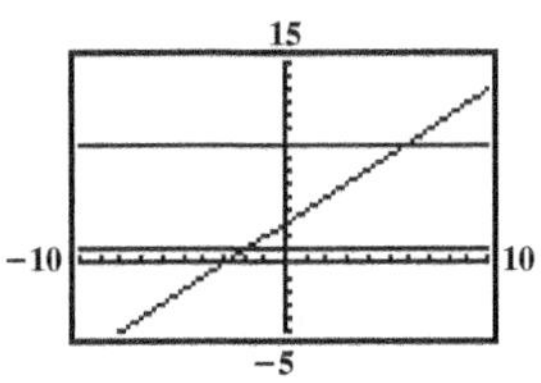

93. $\left\{x \middle| 2 \le x \le \frac{5}{2}\right\}$ or $\left[2, \frac{5}{2}\right]$;

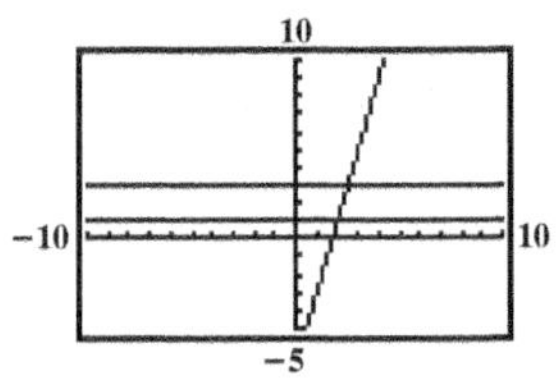

95. Exercise 91:

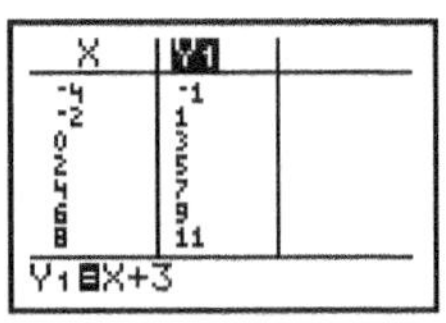

Exercise 93:

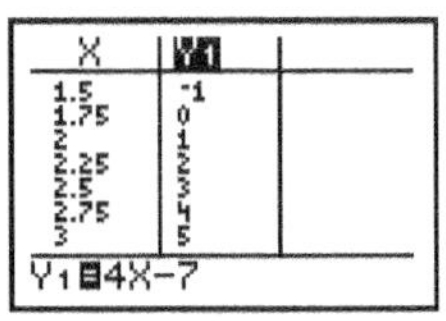

97. makes sense **99.** makes sense **101.** false **103.** false **105.** $(-\infty, 4]$ **107.** $[-1, 4]$ **109.** least: 4 nickels; greatest: 7 nickels
110. $-x^2 + 5x - 9$; -15 **111.** $f(x) = -\frac{1}{2}x + 4$ **112.** $17 - 2x$ **113.** $-\frac{1}{2}$ and 1 **114.** -1 and 3 **115. a.** -5 satisfies the inequality.
b. no

Section 4.3

Check Point Exercises

1. -2 and 3, or $\{-2, 3\}$ **2.** $-\frac{13}{3}$ and 5, or $\left\{-\frac{13}{3}, 5\right\}$ **3.** $\frac{4}{3}$ and 10, or $\left\{\frac{4}{3}, 10\right\}$

4. $\{x|-3 < x < 7\}$ or $(-3, 7)$

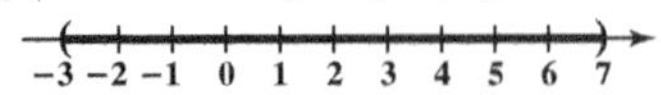

5. $\{x|x \le 1 \text{ or } x \ge 4\}$ or $(-\infty, 1] \cup [4, \infty)$

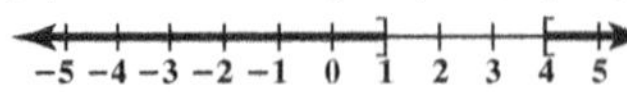

6. a. $\{x|-3 < x < 7\}$ or $(-3, 7)$ **b.** $\{x|x \le 1 \text{ or } x \ge 4\}$ or $(-\infty, 1] \cup [4, \infty)$ **7.** $\left\{x \middle| -\frac{11}{5} \le x \le 3\right\}$ or $\left[-\frac{11}{5}, 3\right]$;

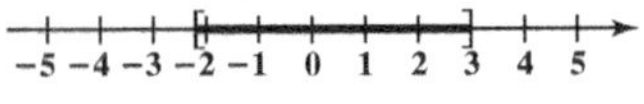

8. $\{x|8.1 \le x \le 13.9\}$ or $[8.1, 13.9]$; The percentage of children in the population who think that not being able to do everything they want is a bad thing is between a low of 8.1% and a high of 13.9%.

Exercise Set 4.3

1. $\{-8, 8\}$ **3.** $\{-5, 9\}$ **5.** $\{-3, 4\}$ **7.** $\{-1, 2\}$ **9.** $\varnothing$ **11.** $\{-3\}$ **13.** $\{-11, -1\}$ **15.** $\{-3, 4\}$ **17.** $\left\{-\frac{13}{3}, 5\right\}$
19. $\left\{-\frac{2}{5}, \frac{2}{5}\right\}$ **21.** $\varnothing$ **23.** $\varnothing$ **25.** $\left\{\frac{1}{2}\right\}$ **27.** $\left\{\frac{3}{4}, 5\right\}$ **29.** $\left\{\frac{5}{3}, 3\right\}$ **31.** $\{0\}$ **33.** $\{4\}$ **35.** $\{4\}$ **37.** $\{-1, 15\}$

39. $\{x|-3 < x < 3\}$ or $(-3, 3)$

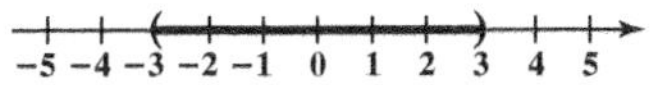

41. $\{x|1 < x < 3\}$ or $(1, 3)$

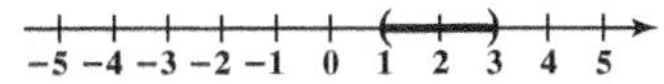

43. $\{x|-3 \le x \le -1\}$ or $[-3, -1]$

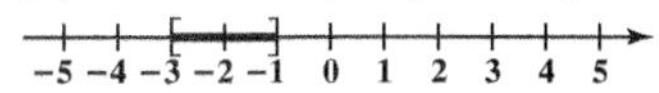

45. $\{x|-1 < x < 7\}$ or $(-1, 7)$

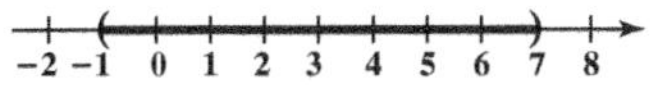

47. $\{x|x < -3 \text{ or } x > 3\}$ or $(-\infty, -3) \cup (3, \infty)$

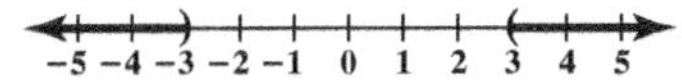

49. $\{x|x < -4 \text{ or } x > -2\}$ or $(-\infty, -4) \cup (-2, \infty)$

51. $\{x|x \le 2 \text{ or } x \ge 6\}$ or $(-\infty, 2] \cup [6, \infty)$

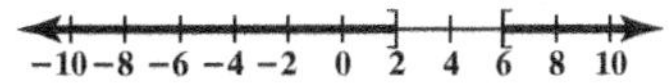

53. $\left\{x \middle| x < \frac{1}{3} \text{ or } x > 5\right\}$ or $\left(-\infty, \frac{1}{3}\right) \cup (5, \infty)$

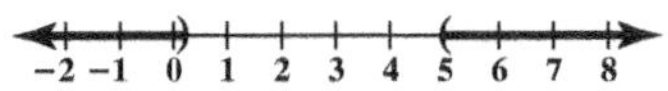

55. $\{x|-5 \le x \le 3\}$ or $[-5, 3]$

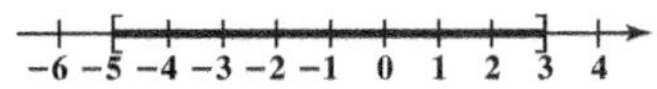

57. $\{x|-6 < x < 0\}$ or $(-6, 0)$

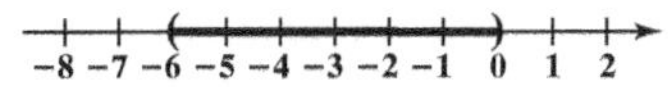

59. $\{x|x \le -5 \text{ or } x \ge 3\}$ or $(-\infty, -5] \cup [3, \infty)$

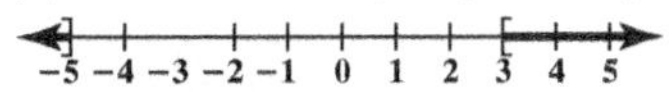

61. $\{x|x < -3 \text{ or } x > 12\}$ or $(-\infty, -3) \cup (12, \infty)$

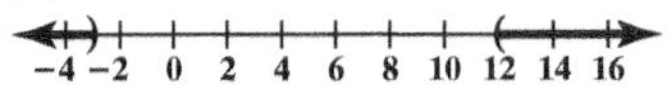

63. $\varnothing$

65. $\{x|x \text{ is a real number}\}$ or $(-\infty, \infty)$

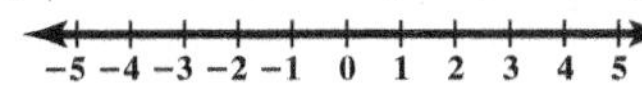

67. $\{x|-9 \le x \le 5\}$ or $[-9, 5]$

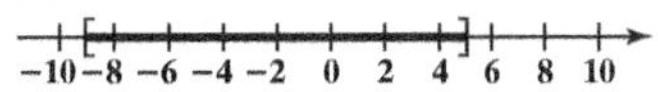

69. $\{x|x < 1 \text{ or } x > 2\}$ or $(-\infty, 1) \cup (2, \infty)$

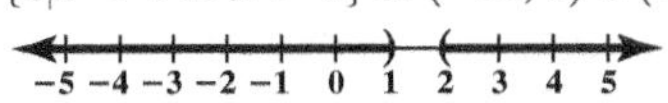

71. $\{x|x < -3 \text{ or } x > 5\}$ or $(-\infty, -3) \cup (5, \infty)$

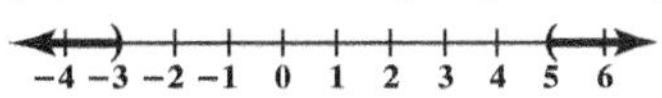

73. $\{x|x \le -1 \text{ or } x \ge 2\}$ or $(-\infty, -1] \cup [2, \infty)$

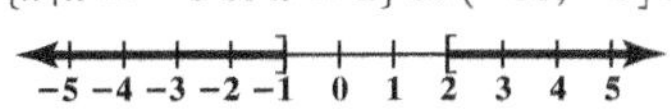

75. $-\frac{3}{2}$ and 4 **77.** -7 and -2 **79.** $\left\{x \middle| -\frac{7}{3} \le x \le 1\right\}$ or $\left[-\frac{7}{3}, 1\right]$ **81.** $\{x|x < -1 \text{ or } x > 4\}$ or $(-\infty, -1) \cup (4, \infty)$

83. $\left(-\infty, -\frac{1}{3}\right] \cup [3, \infty)$ **85.** $\left\{x \middle| \frac{-c-b}{a} < x < \frac{c-b}{a}\right\}$ **87.** $\{3, 5\}$ **89.** $\{x|-2 \le x \le 1\}$ or $[-2, 1]$

91. $\{x|58.6 \le x \le 61.8\}$ or $[58.6, 61.8]$; The percentage of the U.S. population that watched M*A*S*H is between a low of 58.6% and a high of 61.8%.; 1.6% **93.** $\{T|50 \le T \le 64\}$ or $[50, 64]$; The monthly average temperature for San Francisco, CA is between a low of 50°F and a high of 64°F.

95. $\{x|8.59 \le x \le 8.61\}$ or $[8.59, 8.61]$; A machine part that is supposed to be 8.6 cm is acceptable between a low of 8.59 and a high of 8.61 cm.

97. If the number of outcomes that result in heads is 41 or less or 59 or more, then the coin is unfair.

105. $\{-6, 4\}$;

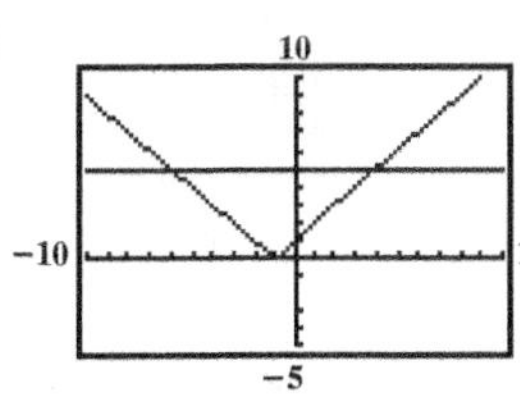

107. $\{2, 3\}$;

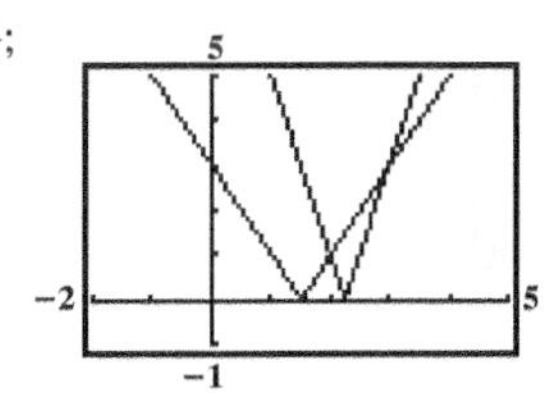

109. $\{x|-2 < x < 3\}$ or $(-2, 3)$;

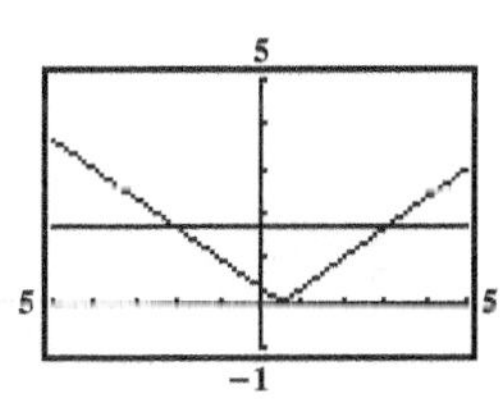

111. $\{x|x < -3 \text{ or } x > 4\}$ or $(-\infty, -3) \cup (4, \infty)$;

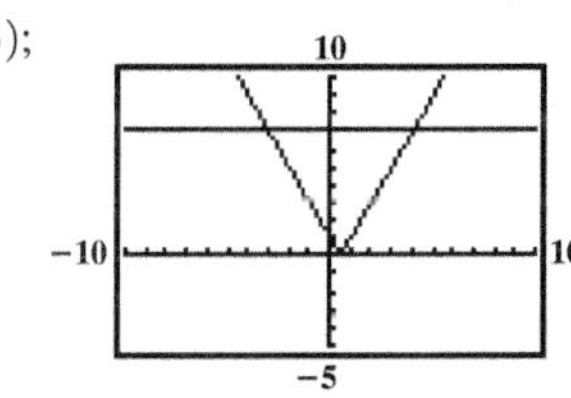

113. $\{x|x \text{ is a real number}\}$ or $(-\infty, \infty)$;

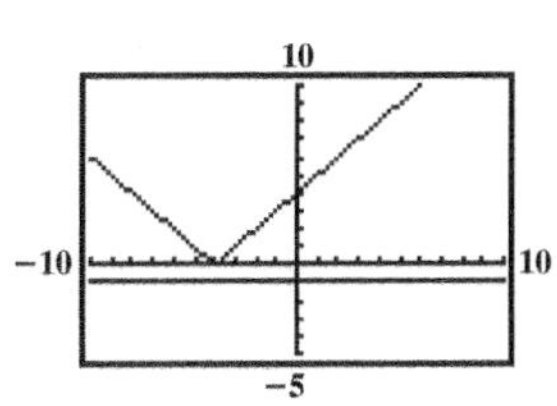

115. makes sense **117.** does not make sense **119.** false **121.** true

123. a. $|x - 4| < 3$ **b.** $|x - 4| \ge 3$ **125.** $\{1\}$

126.

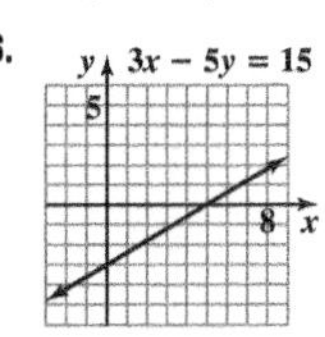

127.

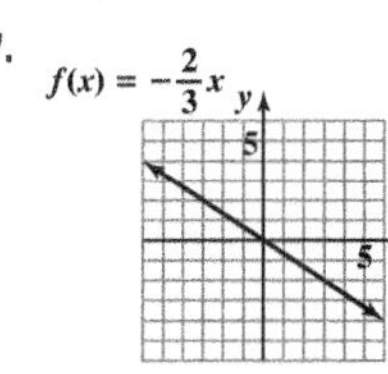

128.

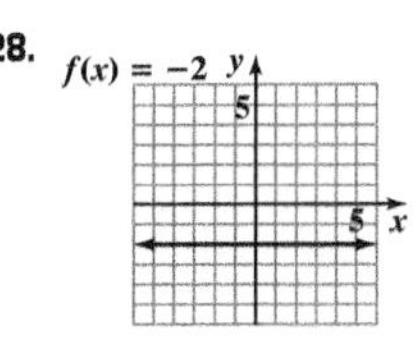

Mid-Chapter Check Point Exercises

1. $\{x|x \le -4\}$ or $(-\infty, -4]$ **2.** $\{x|3 \le x < 5\}$ or $[3, 5)$ **3.** $\left\{\frac{1}{2}, 3\right\}$ **4.** $\{x|x < -1\}$ or $(-\infty, -1)$
5. $\{x|x < -9 \text{ or } x > -5\}$ or $(-\infty, -9) \cup (-5, \infty)$ **6.** $\left\{x \middle| -\frac{2}{3} \le x \le 2\right\}$ or $\left[-\frac{2}{3}, 2\right]$ **7.** $\left\{\frac{1}{2}, \frac{13}{4}\right\}$ **8.** $\{x|-4 < x \le -2\}$ or $(-4, -2]$
9. $\{x|x \text{ is a real number}\}$ or $(-\infty, \infty)$ **10.** $\{x|x \le -3\}$ or $(-\infty, -3]$ **11.** $\{x|x < -3\}$ or $(-\infty, -3)$
12. $\left\{x|x < -1 \text{ or } x > -\frac{1}{5}\right\}$ or $(-\infty, -1) \cup \left(-\frac{1}{5}, \infty\right)$ **13.** $\{x|x \le -10 \text{ or } x \ge 2\}$ or $(-\infty, -10] \cup [2, \infty)$ **14.** $\varnothing$
15. $\left\{x|x < -\frac{5}{3}\right\}$ or $\left(-\infty, -\frac{5}{3}\right)$ **16.** $\{x|x > 4\}$ or $(4, \infty)$ **17.** $\{-2, 7\}$ **18.** $\varnothing$ **19.** no more than 80 miles per day **20.** $[49\%, 99\%)$
21. at least \$120,000 **22.** at least 750,000 discs each month

Section 4.4

Check Point Exercises

1. $4x - 2y \ge 8$
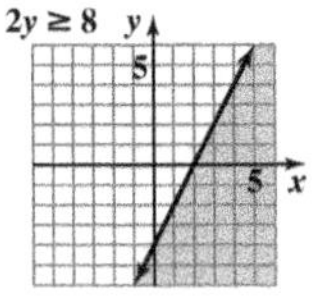

2. $y > -\frac{3}{4}x$
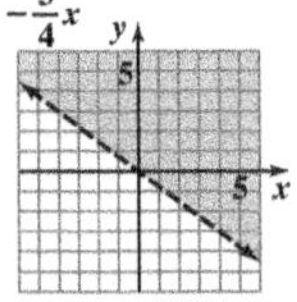

3. a. $y > 1$
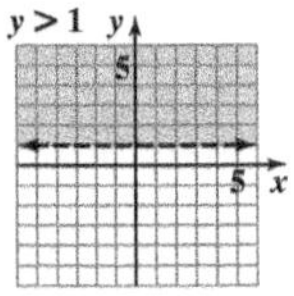

b. $x \le -2$
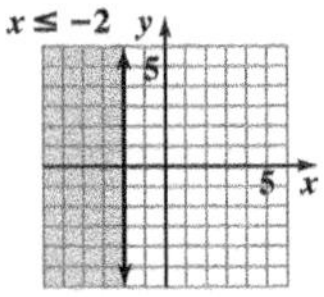

4. $B = (60, 20)$; Using $T = 60$ and $P = 20$, each of the three inequalities for grasslands is true: $60 \ge 35$, true; $5(60) - 7(20) \ge 70$, true; $3(60) - 35(20) \le -140$, true.

5. $x - 3y < 6$, $2x + 3y \ge -6$
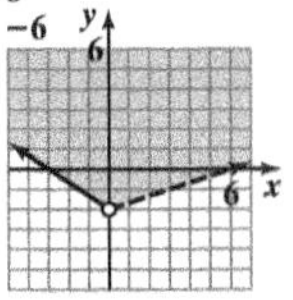

6. $x + y < 2$, $-2 \le x < 1$, $y > -3$
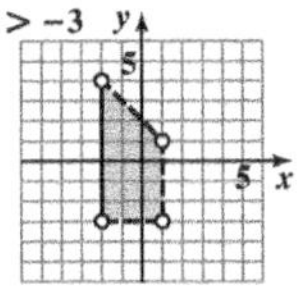

Exercise Set 4.4

1. $x + y \ge 3$
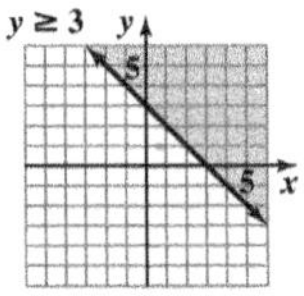

3. $x - y < 5$
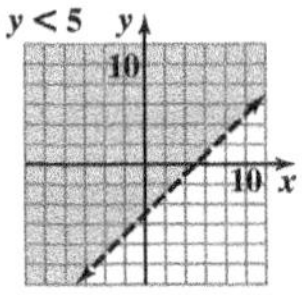

5. $x + 2y > 4$
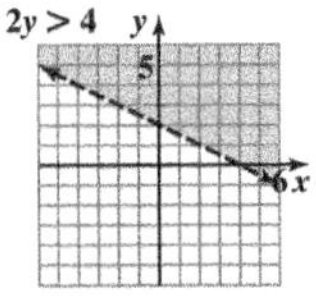

7. $3x - y \le 6$
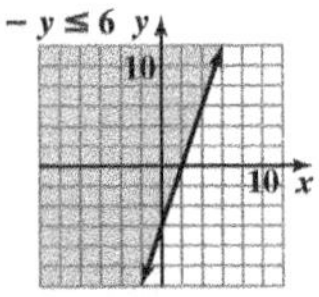

9. $\frac{x}{2} + \frac{y}{3} < 1$
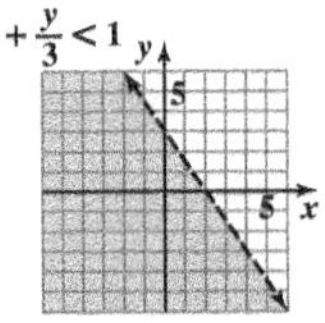

11. $y > \frac{1}{3}x$
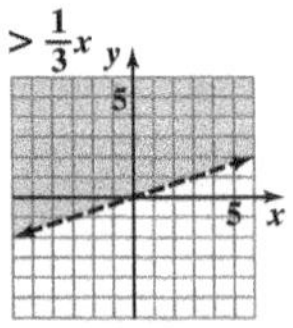

13. $y \le 3x + 2$
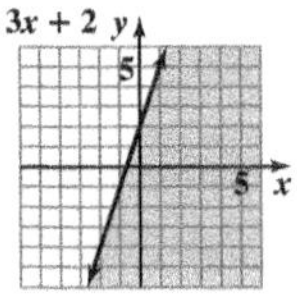

15. $y < -\frac{1}{4}x$
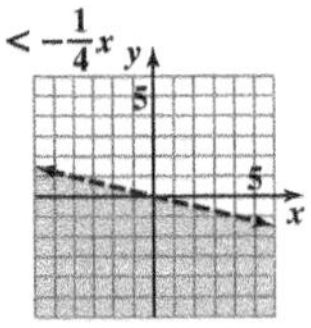

17. $x \le 2$
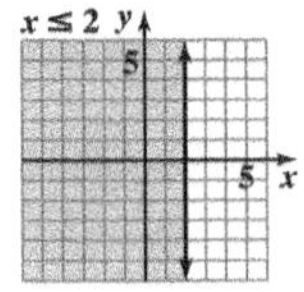

19. $y > -4$
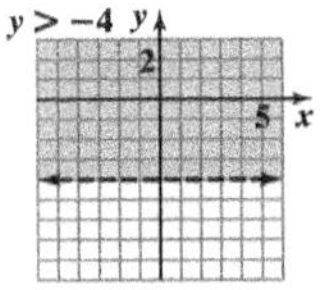

21. $y \ge 0$
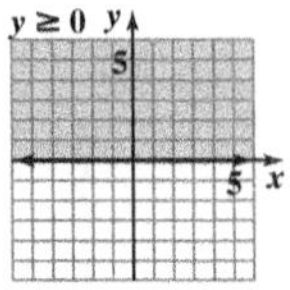

23. $3x + 6y \le 6$, $2x + y \le 8$
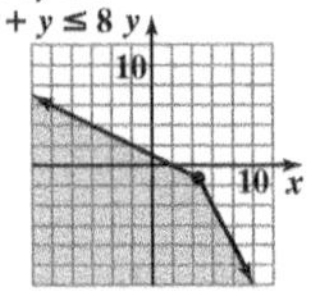

25. $2x - 5y \le 10$, $3x - 2y > 6$
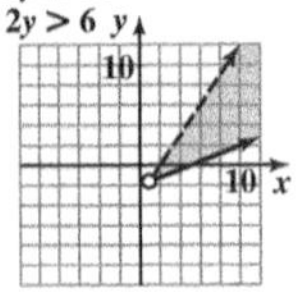

27. $y > 2x - 3$, $y < -x + 6$
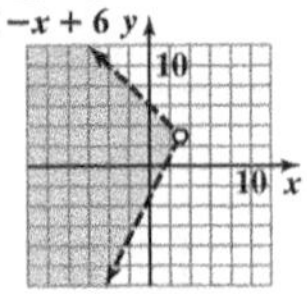

29. $x + 2y \le 4$, $y \ge x - 3$
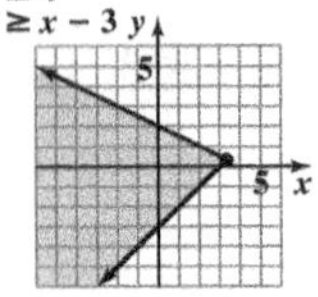

31. $x \le 2$, $y \ge -1$
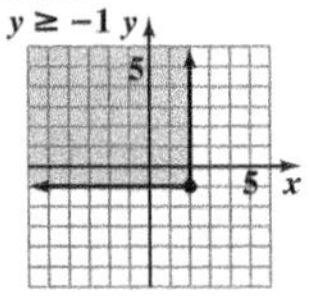

33. $-2 \le x < 5$
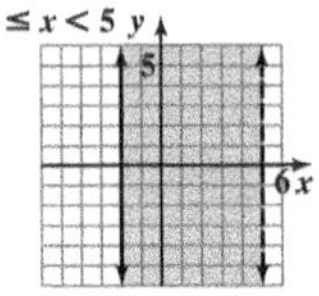

35. $x - y \le 1$, $x \ge 2$
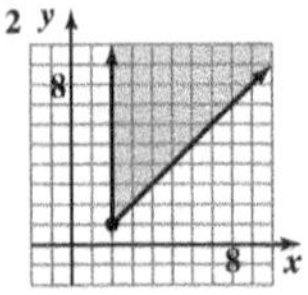

37. $\varnothing$

39. $x + y > 4$, $x + y > -1$
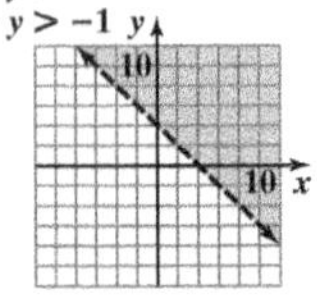

41. 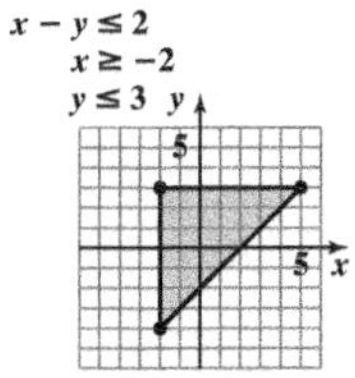

43. 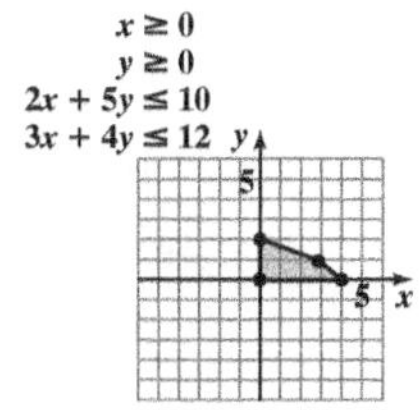

45. 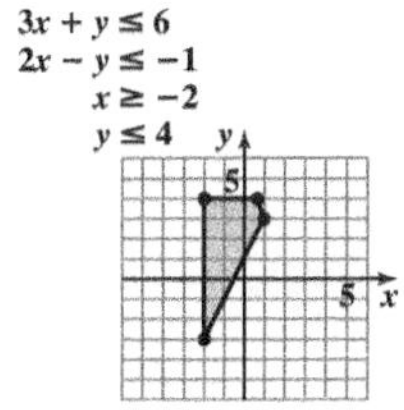

47. $y \ge -2x + 4$
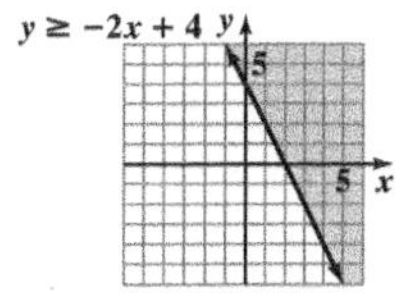

49. $x + y \le 4$ and $3x + y \le 6$
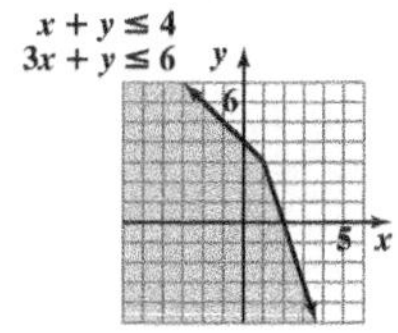

51. $-2 \le x \le 2$; $-3 \le y \le 3$
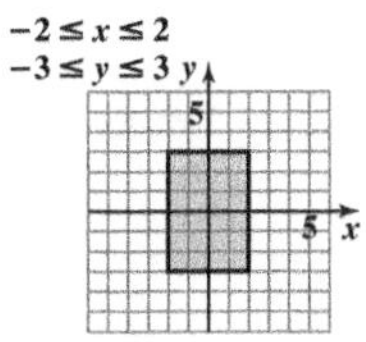

53. 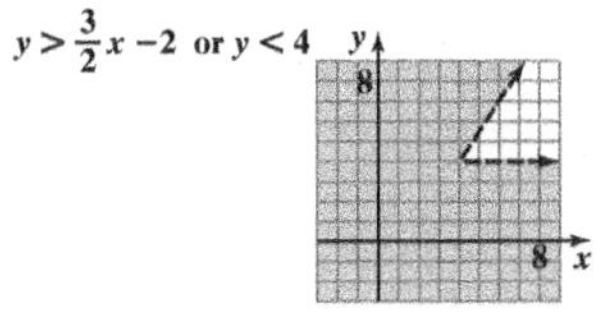

55. no solution **57.** infinitely many solutions **59. a.** $A = (20, 150)$; A 20-year-old with a heart rate of 150 beats per minute is within the target range. **b.** $10 \le 20 \le 70$, true; $150 \ge 0.7(220 - 20)$, true; $150 \le 0.8(220 - 20)$, true **61.** $10 \le a \le 70$; $H \ge 0.6(220 - a)$; $H \le 0.7(220 - a)$

63. a. $y \ge 0$; $x + y \ge 5$; $x \ge 1$; $200x + 100y \le 700$
b. 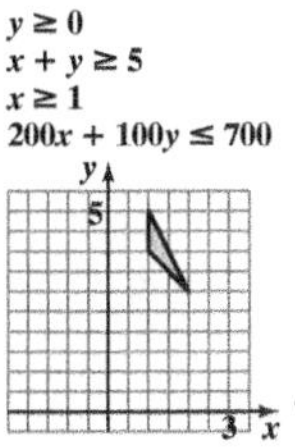

c. 2 nights

75. 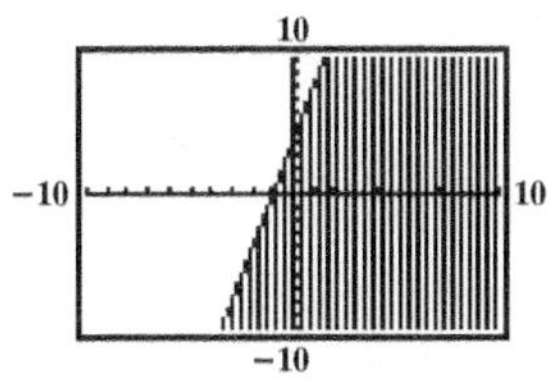

77. 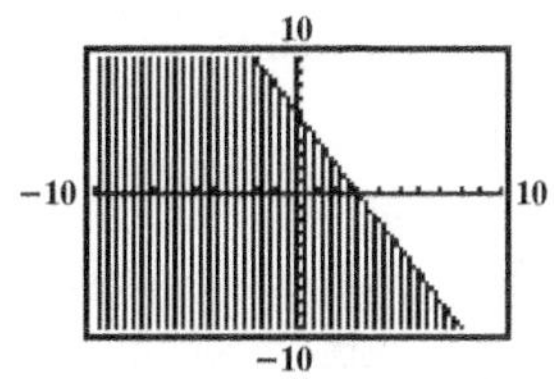

83. makes sense **85.** makes sense **87.** false **89.** true **91.** $x \ge -2, y > -1$ **93.** $y \ge 2x - 2$

95. 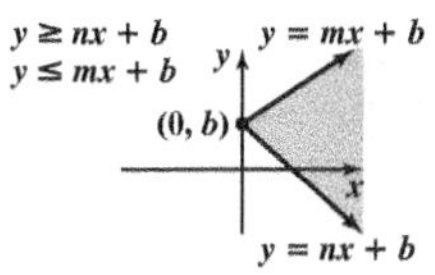

96. $\{(3, 1)\}$ **97.** $\{(2, 4)\}$ **98.** 165

99. a. 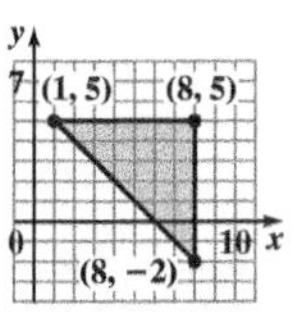

b. (1, 5), (8, 5), (8, −2) **c.** at (1, 5): 13; at (8, 5): 34; at (8, −2): 20

100. a. 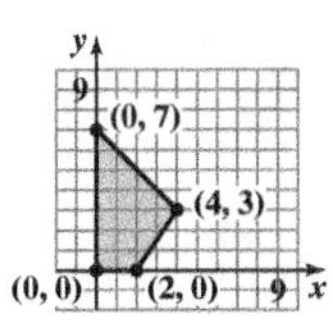

b. (0, 0), (2, 0), (4, 3), (0, 7) **c.** at (0, 0): 0; at (2, 0): 4; at (4, 3): 23; at (0, 7): 35

101. $20x + 10y \le 80{,}000$

Section 4.5

Check Point Exercises

1. $z = 25x + 55y$ **2.** $x + y \le 80$ **3.** $30 \le x \le 80$, $10 \le y \le 30$; $z = 25x + 55y$, $x + y \le 80$, $30 \le x \le 80$, $10 \le y \le 30$
4. 50 bookshelves and 30 desks; $2900 **5.** 30

Exercise Set 4.5

1. (1, 2): 17; (2, 10): 70; (7, 5): 65; (8, 3): 58; maximum: 70; minimum: 17 **3.** (0, 0): 0; (0, 8): 400; (4, 9): 610; (8, 0): 320; maximum: 610; minimum: 0

5. a.

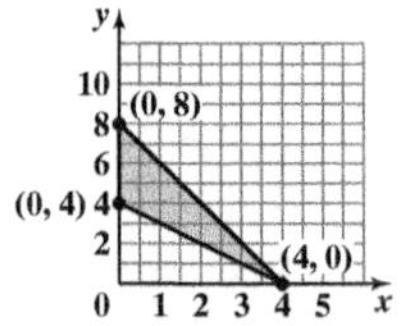

b. (0, 4): 8; (0, 8): 16; (4, 0): 12
c. maximum: 16; at (0, 8)

7. a.

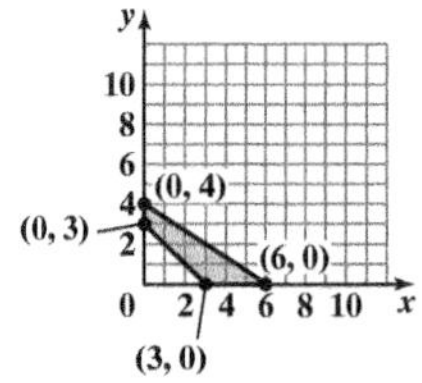

b. (0, 3): 3; (0, 4): 4; (6, 0): 24; (3, 0): 12
c. maximum: 24; at (6, 0)

9. a.

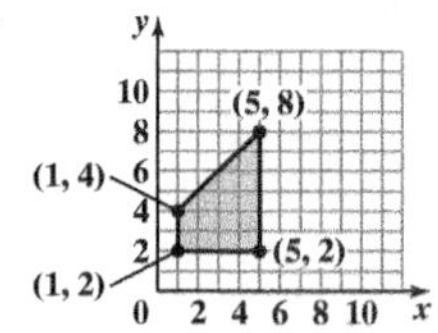

b. (1, 2): −1; (1, 4): −5; (5, 8): −1; (5, 2): 11
c. maximum: 11; at (5, 2)

11. a.

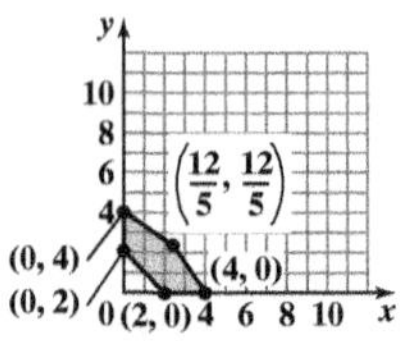

b. (0, 2): 4; (0, 4): 8; $\left(\frac{12}{5}, \frac{12}{5}\right)$: $\frac{72}{5} = 14.4$; (4, 0): 16; (2, 0): 8
c. maximum: 16; at (4, 0)

13. a.

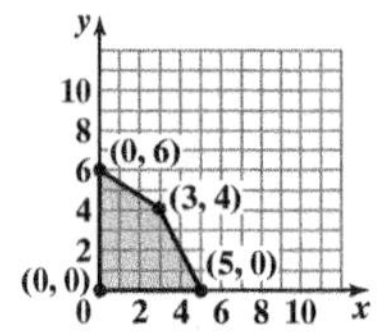

b. (0, 0): 0; (0, 6): 72; (3, 4): 78; (5, 0): 50
c. maximum: 78; at (3, 4)

15. a. $z = 125x + 200y$
b. $x \le 450$; $y \le 200$; $600x + 900y \le 360{,}000$
c.

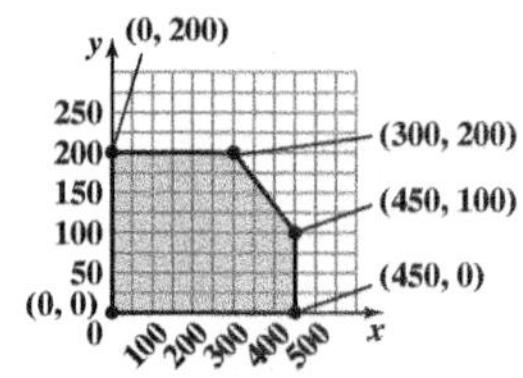

d. 0; 40,000; 77,500; 76,250; 56,250
e. 300; 200; 77,500

17. 40 of model A and 0 of model B **19.** 300 boxes of food and 200 boxes of clothing **21.** 100 parents and 50 students **23.** 10 Boeing 727s and 42 Falcon 20s **29.** does not make sense **31.** makes sense **33.** \$5000 in stocks and \$5000 in bonds **35.** $54x^7y^{15}$ **36.** $L = \frac{12P + W}{2}$ **37.** 10 **38.** $4x^3 + 9x^2 - 13x - 3$ **39.** $5x^3 - 2x^2 + 12x - 15$ **40. a.** g **b.** f

Review Exercises

1. $\{x \mid -2 < x \le 3\}$

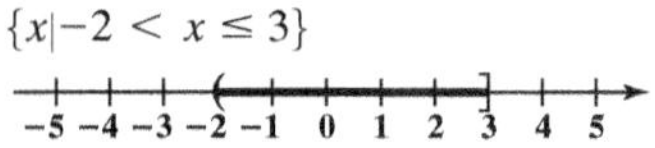

2. $\{x \mid -1.5 \le x \le 2\}$

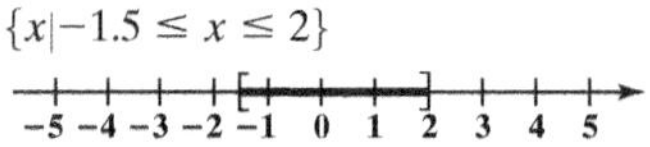

3. $\{x \mid x > -1\}$

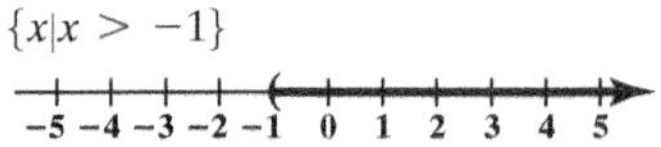

4. $\{x \mid x \ge -2\}$; $[-2, \infty)$

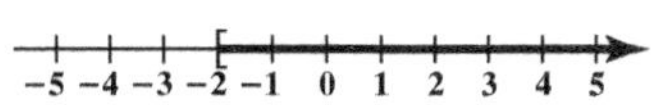

5. $\left\{x \,\middle|\, x \ge \frac{3}{5}\right\}$; $\left[\frac{3}{5}, \infty\right)$

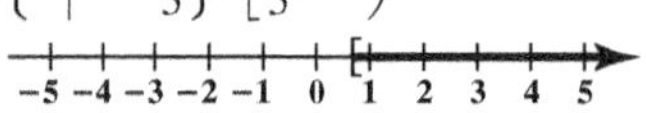

6. $\left\{x \,\middle|\, x < -\frac{21}{2}\right\}$; $\left(-\infty, -\frac{21}{2}\right)$

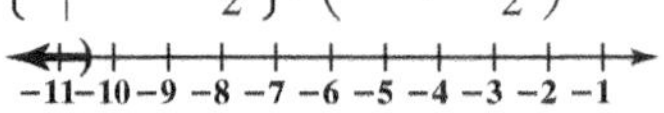

7. $\{x \mid x > -3\}$; $(-3, \infty)$

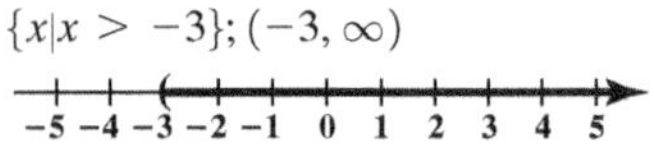

8. $\{x \mid x \le -2\}$; $(-\infty, -2]$

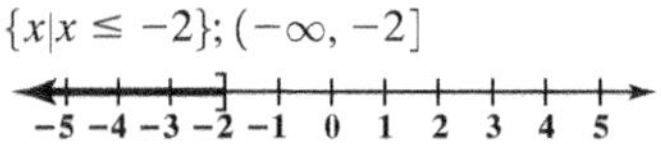

9. ∅ **10.** more than 50 checks **11.** more than \$13,500 in sales **12.** $\{a, c\}$ **13.** $\{a\}$ **14.** $\{a, b, c, d, e\}$ **15.** $\{a, b, c, d, f, g\}$

16. $\{x \mid x \le 3\}$; $(-\infty, 3]$

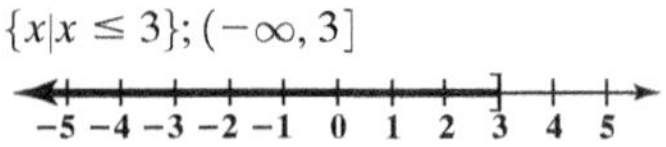

17. $\{x \mid x < 6\}$; $(-\infty, 6)$

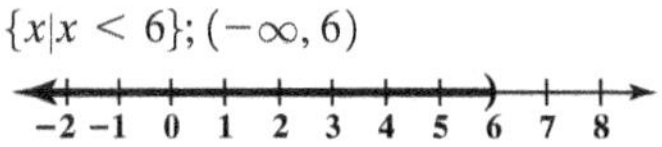

18. $\{x \mid 6 < x < 8\}$; $(6, 8)$

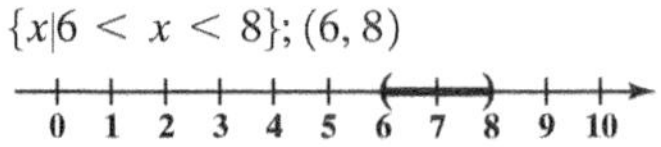

19. $\{x \mid x \le 1\}$; $(-\infty, 1]$

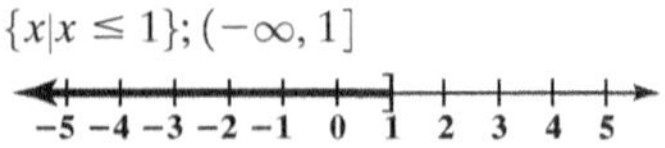

20. ∅

21. $\{x \mid x < 1 \text{ or } x > 2\}$; $(-\infty, 1) \cup (2, \infty)$

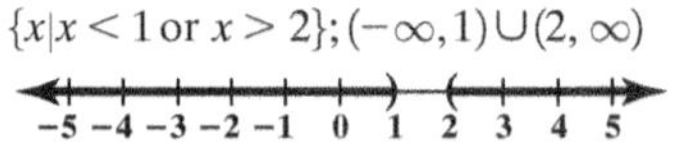

22. $\{x \mid x \le -4 \text{ or } x > 2\}$; $(-\infty, -4] \cup (2, \infty)$

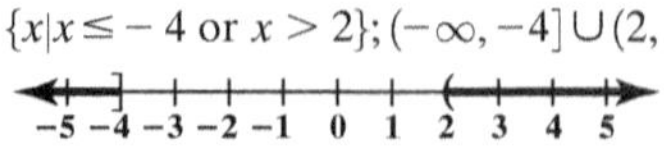

23. $\{x \mid x < -2\}$; $(-\infty, -2)$

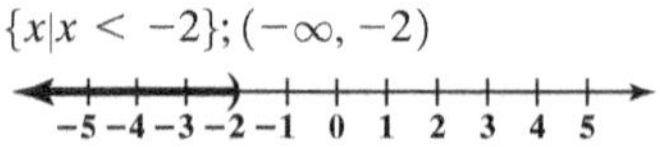

24. $\{x \mid x \text{ is a real number}\}$; $(-\infty, \infty)$

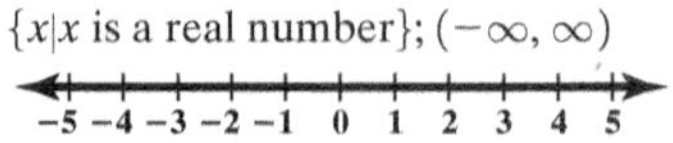

25. $\{x \mid -5 < x \le 2\}$; $(-5, 2]$

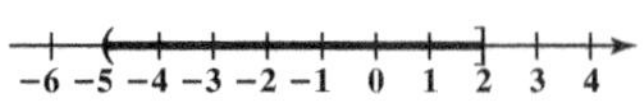

26. $\left\{x \,\middle|\, -\frac{3}{4} \le x \le 1\right\}$; $\left[-\frac{3}{4}, 1\right]$

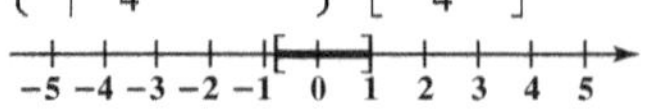

27. [49%, 99%)

28. $\{-4, 3\}$ **29.** ∅ **30.** $\left\{-\frac{11}{2}, \frac{23}{2}\right\}$ **31.** $\left\{-4, -\frac{6}{11}\right\}$

32. $\{x \mid -9 \le x \le 6\}$; $[-9, 6]$

33. $\{x \mid x < -6 \text{ or } x > 0\}$; $(-\infty, -6) \cup (0, \infty)$

34. $\{x \mid -3 < x < -2\}$; $(-3, -2)$

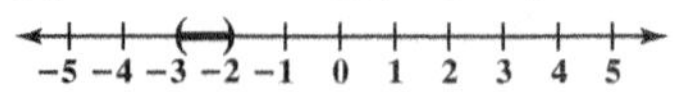

35. $\{x \mid x \le -5 \text{ or } x \ge 1\}$; $(-\infty, -5] \cup [1, \infty)$;

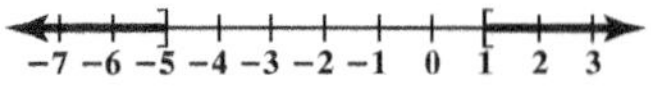

36. ∅ **37.** Approximately 90% of the population sleeps between 5.5 hours and 7.5 hours daily, inclusive.

38. $3x - 4y > 12$

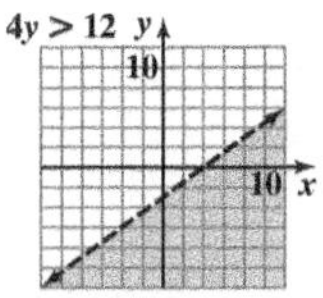

39. $x - 3y \le 6$

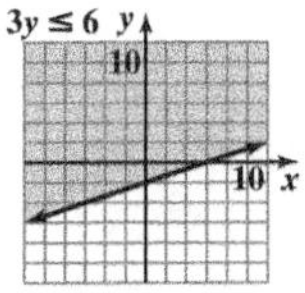

40. $y \le -\frac{1}{2}x + 2$

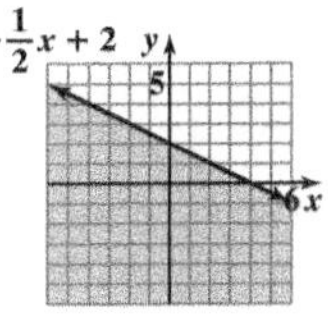

41. $y > \frac{3}{5}x$

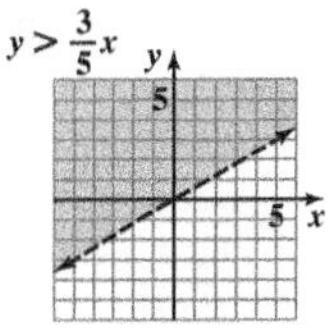

42. $x \le 2$

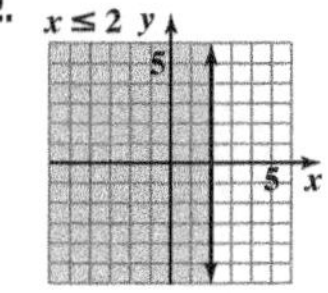

43. $y > -3$

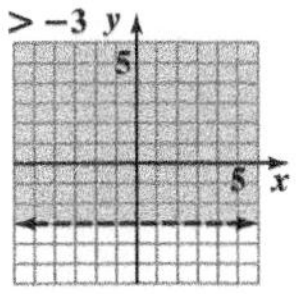

44. $2x - y \le 4$
$x + y \ge 5$

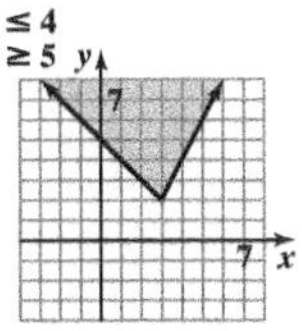

45. $y < -x + 4$
$y > x - 4$

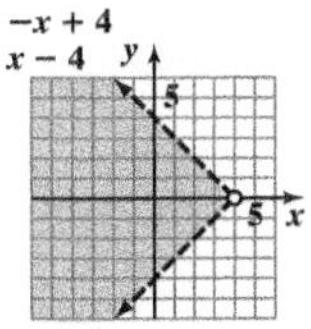

46. $-3 \le x < 5$

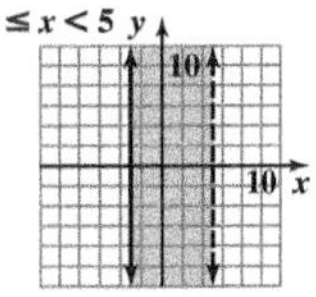

47. $-2 < y \le 6$

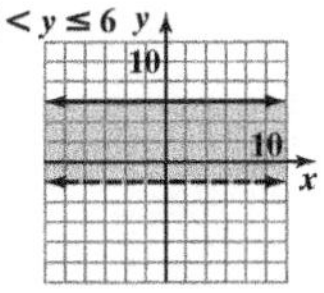

48. $x \ge 3$
$y \le 0$

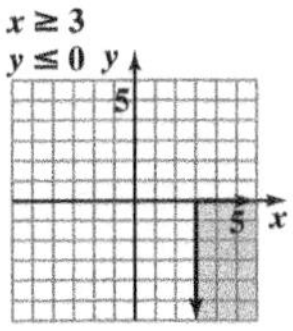

49. $2x - y > -4$
$x \ge 0$

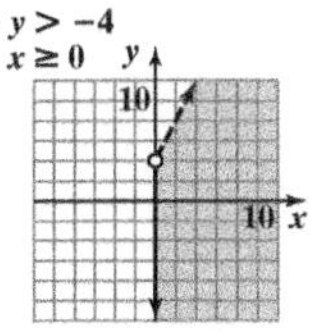

50. $x + y \le 6$
$y \ge 2x - 3$

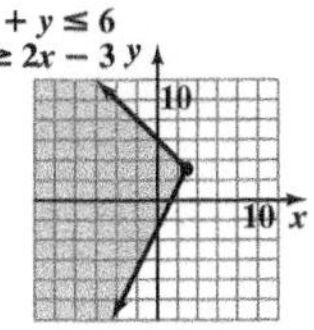

51. $3x + 2y \ge 4$
$x - y \le 3$
$x \ge 0, y \ge 0$

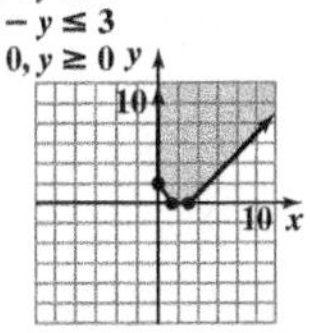

52. $\varnothing$

53. $\left(\frac{1}{2}, \frac{1}{2}\right)$: $\frac{5}{2}$; (2, 2): 10; (4, 0): 8; (1, 0): 2; maximum: 10; minimum: 2

54.

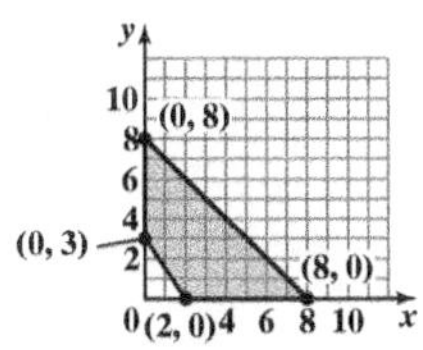

maximum: 24

55.

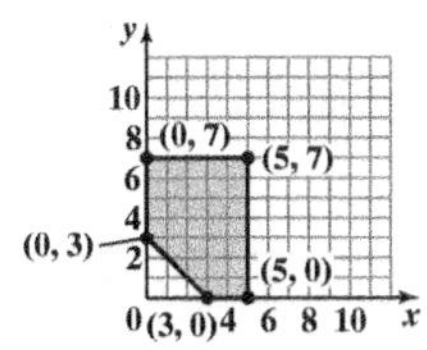

maximum: 33

56.

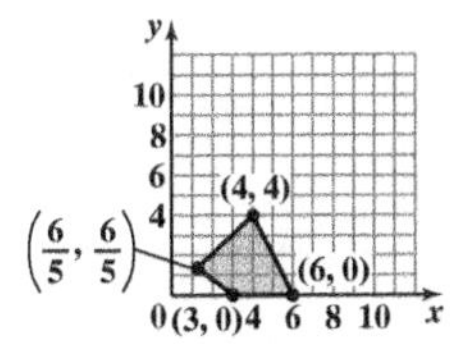

maximum: 44

57. a. $z = 500x + 350y$

b. $x + y \le 200$; $x \ge 10$; $y \ge 80$

c.

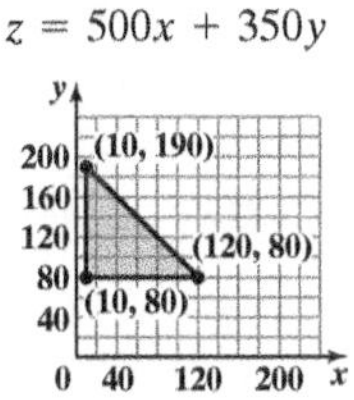

d. (10, 80): 33,000; (10, 190): 71,500; (120, 80): 88,000

e. 120; 80; 88,000

58. 480 of model A and 240 of model B

Chapter Test

1. $\{x|-3 \le x < 2\}$

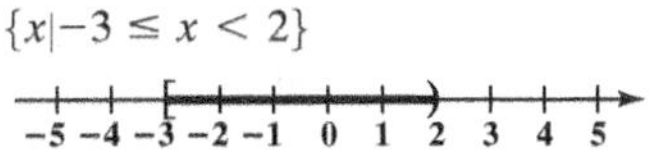

2. $\{x|x \le -1\}$

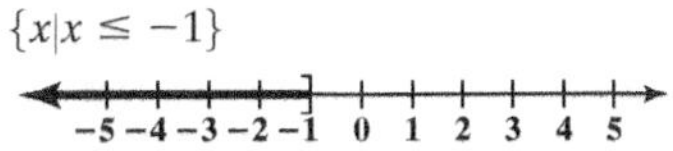

3. $\{x|x \le 12\}$; $(-\infty, 12]$

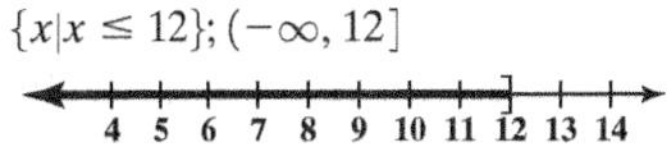

4. $\left\{x \middle| x \ge \frac{21}{8}\right\}$; $\left[\frac{21}{8}, \infty\right)$

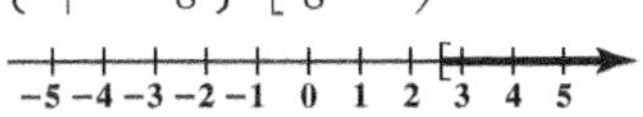

5. more than 200 calls

6. $\{4, 6\}$

7. $\{2, 4, 6, 8, 10, 12, 14\}$

8. $\{x|-2 < x < -1\}$; $(-2, -1)$

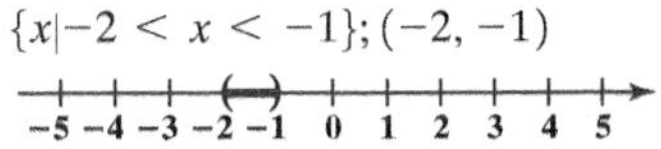

9. $\{x|x \ge -2\}$; $[-2, \infty)$

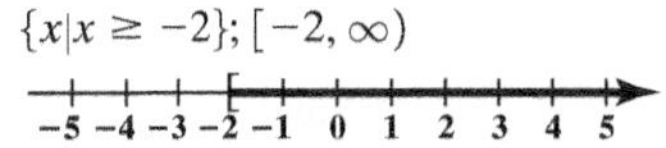

10. $\{x|x < 4\}$; $(-\infty, 4)$

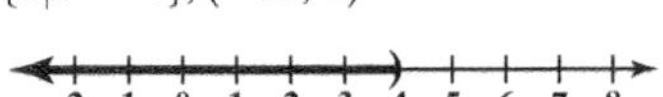

11. $\{x|x \le -4 \text{ or } x > 2\}$; $(-\infty, -4] \cup (2, \infty)$

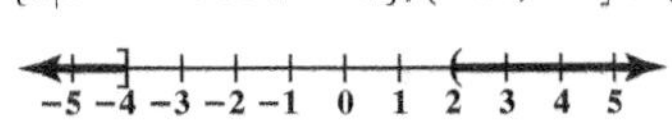

12. $\left\{x \middle| -7 \le x < \frac{13}{2}\right\}$; $\left[-7, \frac{13}{2}\right)$

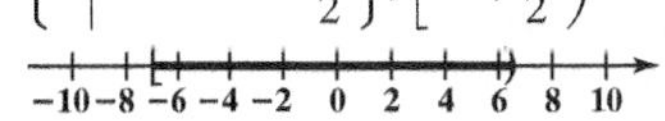

13. $\left\{-2, \frac{4}{5}\right\}$

14. $\left\{-\frac{8}{5}, 7\right\}$

15. $\{x|-3 < x < 4\}$; $(-3, 4)$

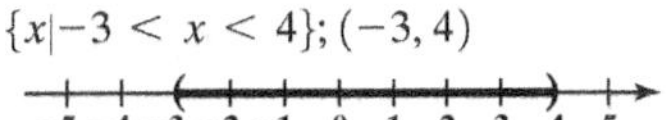

16. $\{x|x \le -1 \text{ or } x \ge 4\}$; $(-\infty, -1] \cup [4, \infty)$

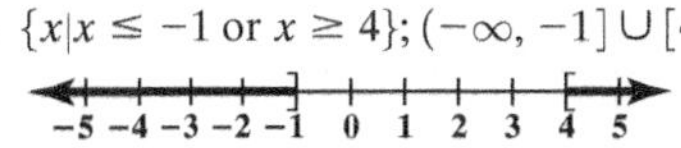

17. $\{b|b < 90.6 \text{ or } b > 106.6\}$ or $(-\infty, 90.6) \cup (106.6, \infty)$; Hypothermia: Body temperature below 90.6°F; Hyperthermia: Body temperature above 106.6°F

18. $3x - 2y < 6$
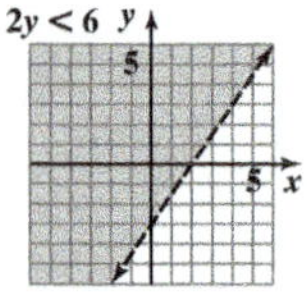

19. $y \geq \frac{1}{2}x - 1$
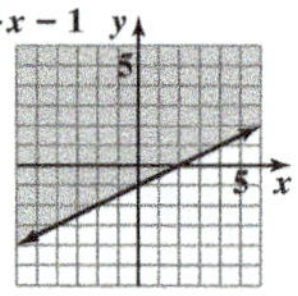

20. $y \leq -1$
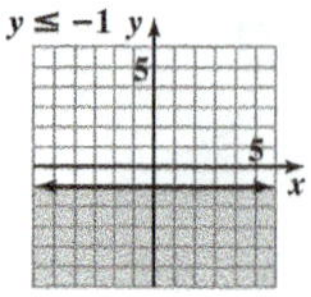

21. $x + y \geq 2$, $x - y \geq 4$
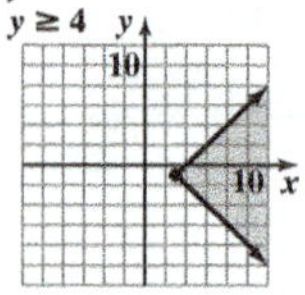

22. $3x + y \leq 9$, $2x + 3y \geq 6$, $x \geq 0$, $y \geq 0$
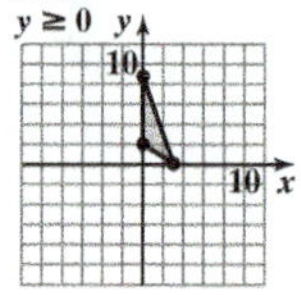

23. $-2 < x \leq 4$
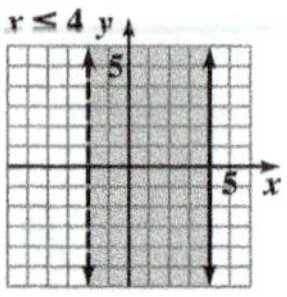

24. maximum: 26 **25.** 50 regular and 100 deluxe; $35,000

Cumulative Review Exercises

1. $\{-1\}$ **2.** $\{8\}$ **3.** $-\frac{2y^7}{3x^5}$ **4.** $22; 4a^2 - 6a + 4$ **5.** $2x^2 + x + 2; 12$ **6.** $f(x) = -\frac{1}{2}x + 4$

7. $f(x) = 2x + 1$
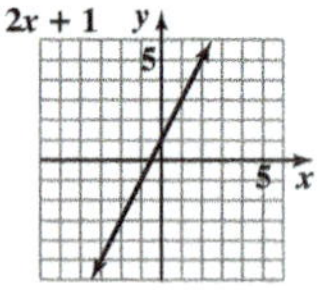

8. $y > 2x$
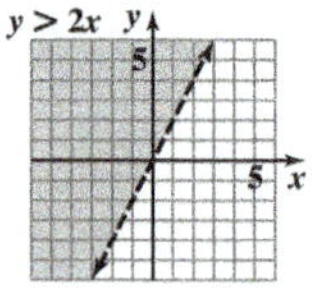

9. $2x - y \geq 6$
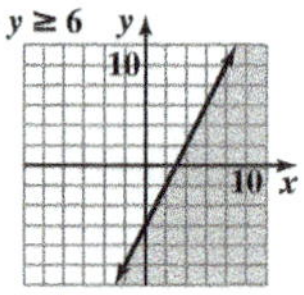

10. $f(x) = -1$
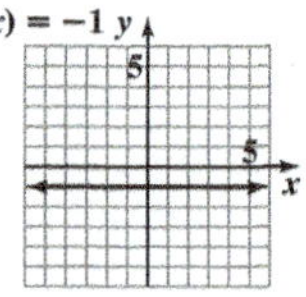

11. $\{(-4, 2, -1)\}$ **12.** $\{(-1, 2)\}$ **13.** -17 **14.** 46 rooms with kitchen facilities and 14 without kitchen facilities **15.** a. and b. are functions.

16. $\{x|x \geq -7\}; [-7, \infty)$
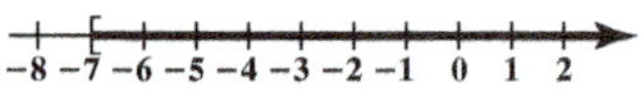

17. $\{x|x < -6\}; (-\infty, -6)$
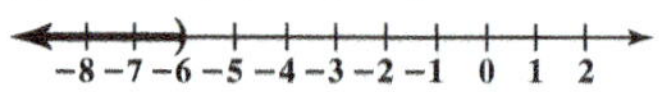

18. $\{x|x \leq 3 \text{ or } x \geq 5\}; (-\infty, 3] \cup [5, \infty)$

19. $\{x|-10 \leq x \leq 7\}; [-10, 7]$

20. $\left\{x \middle| x < \frac{1}{3} \text{ or } x > 5\right\}; \left(-\infty, \frac{1}{3}\right) \cup (5, \infty)$
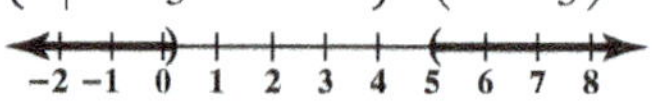

CHAPTER 5

Section 5.1

Check Point Exercises

1.

Term	Coefficient	Degree
$8x^4y^5$	8	9
$-7x^3y^2$	-7	5
$-x^2y$	-1	3
$-5x$	-5	1
11	11	0

The degree of the polynomial is 9, the leading term is $8x^4y^5$, and the leading coefficient is 8.

2. 16 **3.** The graph rises to the left and to the right. **4.** This would not be appropriate over long time periods. Since the graph falls to the right, at some point the ratio would be negative, which is not possible. **5.** The graph does not show the end behavior of the function. The graph should fall to the left. **6.** $-3x^3 + 10x^2 - 10$ **7.** $9xy^3 + 3xy^2 - 15y - 9$ **8.** $10x^3 - 2x^2 + 8x - 10$ **9.** $13x^2y^5 + 2xy^3 - 10$

Exercise Set 5.1

1.

Term	Coefficient	Degree
$-x^4$	-1	4
x^2	1	2

The degree of the polynomial is 4, the leading term is $-x^4$, and the leading coefficient is -1.

3.

Term	Coefficient	Degree
$5x^3$	5	3
$7x^2$	7	2
$-x$	-1	1
9	9	0

The degree of the polynomial is 3, the leading term is $5x^3$, and the leading coefficient is 5.

5.

Term	Coefficient	Degree
$3x^2$	3	2
$-7x^4$	-7	4
$-x$	-1	1
6	6	0

The degree of the polynomial is 4, the leading term is $-7x^4$, and the leading coefficient is -7.

7.

Term	Coefficient	Degree
x^3y^2	1	5
$-5x^2y^7$	-5	9
$6y^2$	6	2
-3	-3	0

The degree of the polynomial is 9, the leading term is $-5x^2y^7$, and the leading coefficient is -5.

9.

Term	Coefficient	Degree
x^5	1	5
$3x^2y^4$	3	6
$7xy$	7	2
$9x$	9	1
-2	-2	0

The degree of the polynomial is 6, the leading term is $3x^2y^4$, and the leading coefficient is 3.

11. 0 **13.** 12 **15.** 56 **17.** -29 **19.** -1
21-23. Graph #23 is not that of a polynomial function.
25. falls to the left and falls to the right; graph (b)
27. rises to the left and rises to the right; graph (a)
29. $11x^3 + 7x^2 - 12x - 4$ **31.** $-\frac{2}{5}x^4 + x^3 + \frac{3}{8}x^2$
33. $9x^2y - 6xy$ **35.** $2x^2y + 15xy + 15$
37. $-9x^4y^2 - 6x^2y^2 - 5x^2y + 2xy$ **39.** $5x^{2n} - 2x^n - 6$
41. $12x^3 + 4x^2 + 12x - 14$ **43.** $22y^5 + 9y^4 + 7y^3 - 13y^2 + 3y - 5$
45. $-5x^3 + 8xy - 9y^2$ **47.** $x^4y^2 + 8x^3y + y - 6x$ **49.** $y^{2n} + 2y^n - 3$ **51.** $8a^2b^4 + 3ab^2 + 8ab$ **53.** $5x^3 + 3x^2y - xy^2 - 4y^3$
55. $5x^4 - x^3 + 5x^2 - 5x + 2$ **57.** $-10x^2y^2 + 4x^2 + 3$ **59.** $-4x^3 - x^2 + 4x + 8; 7$ **61.** $-8x^2 - 2x - 1; -29$ **63.** $-9x^3 - x^2 + 13x + 20$
65. 476,398; In 2000, the cumulative number of AIDS deaths in the U.S. was 476,398. **67.** (10, 476,398)
69. 513,931; 514,071; g **71.** falls to the right; no; Cumulative number of deaths cannot decrease. **73.** no; The graph falls to the right, so eventually there will be a negative number of thefts, which is not possible. **87.** Answers will vary; an example is $f(x) = x^2 - x + 1$.
89. Answers will vary; an example is $f(x) = -x^3 + x + 1$.

91.

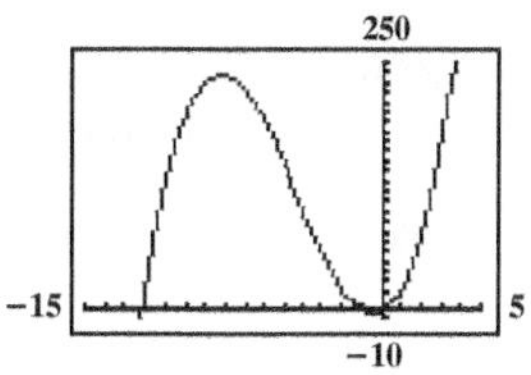

93.

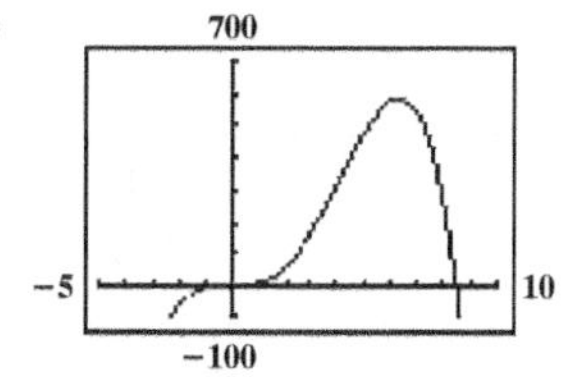

95.

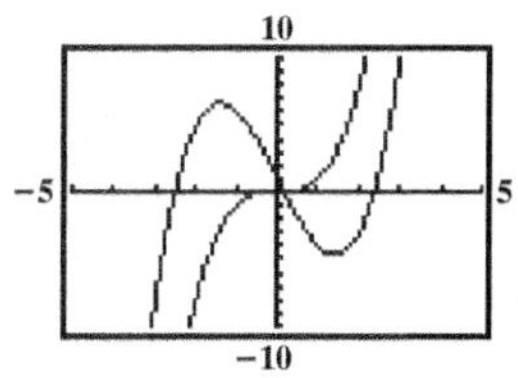

97. makes sense **99.** makes sense **101.** false **103.** false **105.** $3x^{2n} - x^n + 4$ **107.** $9x^2 - 3x + 5$ **108.** $\left\{\frac{2}{3}\right\}$

109.

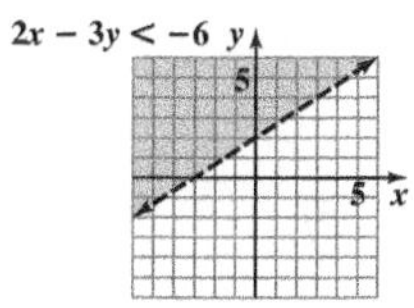

110. $y - 5 = 3(x + 2)$; $y = 3x + 11$ or $f(x) = 3x + 11$ **111.** $10x^7y^9$ **112.** $16x^8 + 6x^5$
113. $3x^3 + 19x^2 + 43x + 35$

Section 5.2

Check Point Exercises

1. a. $-18x^7y^{11}$ **b.** $30x^{10}y^6z^8$ **2. a.** $12x^9 - 18x^6 + 24x^4$ **b.** $10x^5y^9 - 8x^7y^7 - 10x^4y^3$ **3.** $6x^3 - 2x^2 - x + 2$
4. $12x^2y^6 - 8x^2y^4 + 6xy^5 + 2y^2$ **5. a.** $x^2 + 8x + 15$ **b.** $14x^2 + xy - 4y^2$ **c.** $4x^6 - 12x^4 - 5x^3 + 15x$ **6. a.** $x^2 + 16x + 64$
b. $16x^2 + 40xy + 25y^2$ **7. a.** $x^2 - 10x + 25$ **b.** $4x^2 - 24xy^4 + 36y^8$ **8. a.** $x^2 - 9$ **b.** $25x^2 - 49y^2$ **c.** $25a^2b^4 - 16a^2$
9. a. $9x^2 + 12x + 4 - 25y^2$ **b.** $4x^2 + 4xy + y^2 + 12x + 6y + 9$ **10. a.** $x^2 - 10x + 21$ **b.** 5 **11. a.** $a^2 + a - 2$ **b.** $2ah + h^2 - 5h$

Exercise Set 5.2

1. $15x^6$ **3.** $15x^3y^{11}$ **5.** $-6x^2y^9z^9$ **7.** $2x^{3n}y^{n-2}$ **9.** $12x^3 + 8x^2$ **11.** $2y^3 - 10y^2$ **13.** $10x^8 - 20x^5 + 45x^3$ **15.** $28x^2y + 12xy^2$
17. $18a^3b^5 + 15a^2b^3$ **19.** $-12x^6y^3 + 28x^3y^4 - 24x^2y$ **21.** $-12x^{3n} + 20x^{2n} - 2x^{n+1}$ **23.** $x^3 - x^2 - x - 15$ **25.** $x^3 - 1$ **27.** $a^3 - b^3$
29. $x^4 + 5x^3 + x^2 - 11x + 4$ **31.** $x^3 - 4x^2y + 4xy^2 - y^3$ **33.** $x^3y^3 + 8$ **35.** $x^2 + 11x + 28$ **37.** $y^2 - y - 30$ **39.** $10x^2 + 11x + 3$
41. $6y^2 - 11y + 4$ **43.** $15x^2 - 22x + 8$ **45.** $2x^2 + xy - 21y^2$ **47.** $14x^2y^2 - 19xy - 3$ **49.** $x^3 - 4x^2 - 5x + 20$
51. $8x^5 - 40x^3 + 3x^2 - 15$ **53.** $3x^{2n} + 5x^ny^n - 2y^{2n}$ **55.** $x^2 + 6x + 9$ **57.** $y^2 - 10y + 25$ **59.** $4x^2 + 4xy + y^2$
61. $25x^2 - 30xy + 9y^2$ **63.** $4x^4 + 12x^2y + 9y^2$ **65.** $16x^2y^4 - 8x^2y^3 + x^2y^2$ **67.** $a^{2n} + 8a^nb^n + 16b^{2n}$ **69.** $x^2 - 16$ **71.** $25x^2 - 9$
73. $16x^2 - 49y^2$ **75.** $y^6 - 4$ **77.** $1 - y^{10}$ **79.** $49x^2y^4 - 100y^2$ **81.** $25a^{2n} - 49$ **83.** $4x^2 + 12x + 9 - 16y^2$
85. $x^2 + 2xy + y^2 - 9$ **87.** $25x^2 + 70xy + 49y^2 - 4$ **89.** $25y^2 - 4x^2 - 12x - 9$ **91.** $x^2 + 2xy + y^2 + 2x + 2y + 1$
93. $x^4 - 1$ **95. a.** $x^2 + 4x - 12$ **b.** -15 **c.** -12 **97. a.** $x^3 - 27$ **b.** -35 **c.** -27 **99. a.** $a^2 + a + 5$ **b.** $2ah + h^2 - 3h$
101. a. $3a^2 + 14a + 15$ **b.** $6ah + 3h^2 + 2h$ **103.** $48xy$ **105.** $-9x^2 + 3x + 9$ **107.** $16x^4 - 625$ **109.** $x^3 - 3x^2 + 3x - 1$
111. $(2x - 7)^2 = 4x^2 - 28x + 49$ **113. a.** $x^2 + 6x + 4x + 24$ **b.** $(x + 6)(x + 4) = x^2 + 10x + 24$ **115. a.** $x^2 + 12x + 27$
b. $x^2 + 6x + 5$ **c.** $6x + 22$ **117. a.** $4x^2 - 36x + 80$ **b.** $4x^3 - 36x^2 + 80x$ **119. a.** $V(x) = -2x^3 + 10x^2 + 300x$ **b.** rises to the left and falls to the right **c.** no; Because the graph falls to the right, volume will eventually be negative, which is not possible. **d.** 2000; Carry-on luggage with a depth of 10 inches has a volume of 2000 cubic inches. **e.** (10, 2000) **f.** $\{x | 0 < x < 15\}$ or (0, 15), although answers may vary.

131.

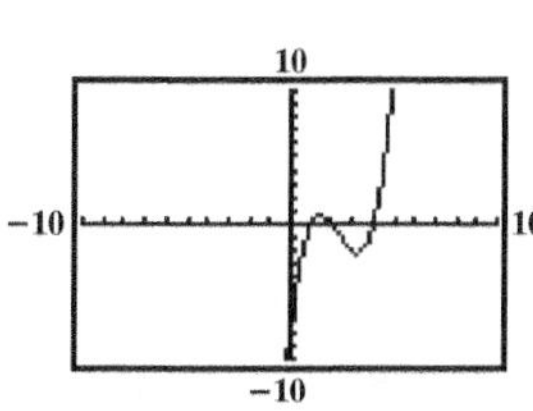

conclusion: $y_1 = y_2$

133.

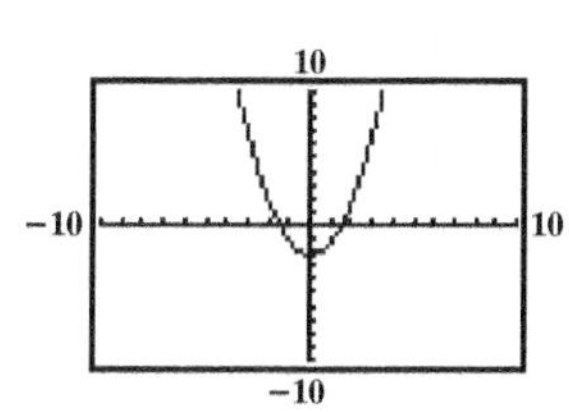

conclusion: $y_1 = y_2$

135. makes sense **137.** makes sense, although answers may vary **139.** false **141.** false **143.** $x^2 + 2x$ **145.** $2x^3 + 12x^2 + 12x + 10$

147. 9 and 11 **148.** $\left\{x \middle| x \le -\frac{14}{3} \text{ or } x \ge 2\right\}$ or $\left(-\infty, -\frac{14}{3}\right] \cup [2, \infty)$ **149.** $\{x|x \ge -3\}$ or $[-3, \infty)$ **150.** 8.034×10^9 **151. a.** $3x^2$ **b.** $6x^2y^2$ **152.** $x^3 - 5x^2 + 3x - 15$ **153.** $3x^3 - 2xy + 12x - 8y$

Section 5.3

Check Point Exercises

1. $10x(2x + 3)$ **2. a.** $3x^2(3x^2 + 7)$ **b.** $5x^3y^2(3 - 5xy)$ **c.** $4x^2y^3(4x^2y^2 - 2xy + 1)$ **3.** $-2x(x^2 - 5x + 3)$ **4. a.** $(x - 4)(3 + 7a)$ **b.** $(a + b)(7x - 1)$ **5.** $(x - 4)(x^2 + 5)$ **6.** $(x + 5)(4x - 3y)$

Exercise Set 5.3

1. $2x(5x + 2)$ **3.** $y(y - 4)$ **5.** $x^2(x + 5)$ **7.** $4x^2(3x^2 - 2)$ **9.** $2x^2(16x^2 + x + 4)$ **11.** $2xy(2xy^2 + 3)$ **13.** $10xy^2(3xy - 1)$ **15.** $2x(6y - 3z + 2w)$ **17.** $3x^2y^4(5xy^2 - 3x^2 + 4y)$ **19.** $5x^2y^4z^2(5xy^2 - 3x^2z^2 + 5yz)$ **21.** $5x^n(3x^n - 5)$ **23.** $-4(x - 3)$ **25.** $-8(x + 6)$ **27.** $-2(x^2 - 3x + 7)$ **29.** $-5(y^2 - 8x)$ **31.** $-4x(x^2 - 8x + 5)$ **33.** $-1(x^2 + 7x - 5)$ **35.** $(x + 3)(4 + a)$ **37.** $(y - 6)(x - 7)$ **39.** $(x + y)(3x - 1)$ **41.** $(3x - 1)(4x^2 + 1)$ **43.** $(x + 3)(2x + 1)$ **45.** $(x + 3)(x + 5)$ **47.** $(x + 7)(x - 4)$ **49.** $(x - 3)(x^2 + 4)$ **51.** $(y - 6)(x + 2)$ **53.** $(y + 1)(x - 7)$ **55.** $(5x - 6y)(2x + 7y)$ **57.** $(4x - 1)(x^2 - 3)$ **59.** $(x - a)(x - b)$ **61.** $(x - 3)(x^2 + 4)$ **63.** $(a - b)(y - x)$ **65.** $(a + 2b)(y^2 - 3x)$ **67.** $(x^n + 1)(y^n + 3)$ **69.** $(a + 1)(b - c)$ **71.** $(x^3 - 5)(1 + 4y)$ **73.** $y^6(3x - 1)^4(6xy - 2y - 7)$ **75.** $(x^2 + 5x - 2)(a + b)$ **77.** $(x + y + z)(a - b + c)$ **79. a.** 16; The ball is 16 feet above the ground after 2 seconds. **b.** 0; The ball is on the ground after 2.5 seconds. **c.** $-8t(2t - 5)$; $f(t) = -8t(2t - 5)$ **d.** 16; 0; yes; no; Answers will vary. **81. a.** $(x - 0.4x)(1 - 0.4) = (0.6x)(0.6) = 0.36x$ **b.** no; 36% **83.** $A = P + Pr + (P + Pr)r = P(1 + r) + Pr(1 + r) = (1 + r)(P + Pr) = P(1 + r)(1 + r) = P(1 + r)^2$ **85.** $A = r(\pi r + 2l)$

93.

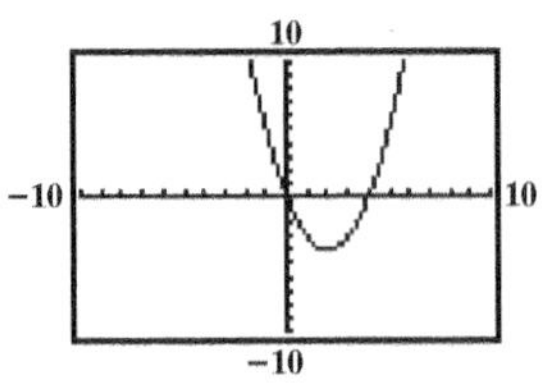

Graphs coincide.; factored correctly

95.

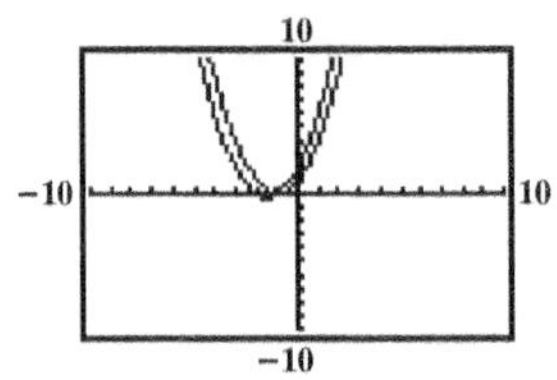

Graphs do not coincide.; $x^2 + 2x + x + 2 = (x + 2)(x + 1)$

97. does not make sense **99.** makes sense **101.** false **103.** true **105.** $x^{2n}(x^{2n} + 1 + x^n)$ **107.** $4y^{2n}(2y^4 + 4y^3 - 3)$ **109.** Answers will vary; an example is $6x^2 - 4x + 9x - 6$. **110.** $\left\{\left(\frac{20}{11}, -\frac{14}{11}\right)\right\}$ **111. a.** function **b.** not a function **112.** length: 8 ft; width: 3 ft **113.** 4 **114.** 2 **115.** 7

Section 5.4

Check Point Exercises

1. $(x + 4)(x + 2)$ or $(x + 2)(x + 4)$ **2.** $(x - 5)(x - 4)$ **3.** $(y + 22)(y - 3)$ **4.** $(x - 3y)(x - 2y)$ **5.** $3x(x - 7)(x + 2)$ **6.** $(x^3 - 5)(x^3 - 2)$ **7.** $(3x - 14)(x - 2)$ **8.** $x^4(3x - 1)(2x + 7)$ **9.** $(2x - y)(x - 3y)$ **10.** $(3y^2 - 2)(y^2 + 4)$ **11.** $(2x - 5)(4x - 1)$

Exercise Set 5.4

1. $(x + 3)(x + 2)$ **3.** $(x + 6)(x + 2)$ **5.** $(x + 5)(x + 4)$ **7.** $(y + 8)(y + 2)$ **9.** $(x - 3)(x - 5)$ **11.** $(y - 2)(y - 10)$ **13.** $(a + 7)(a - 2)$ **15.** $(x + 6)(x - 5)$ **17.** $(x + 4)(x - 7)$ **19.** $(y + 4)(y - 9)$ **21.** prime **23.** $(x - 2y)(x - 7y)$ **25.** $(x + 5y)(x - 6y)$ **27.** prime **29.** $(a - 10b)(a - 8b)$ **31.** $3(x + 3)(x - 2)$ **33.** $2x(x - 3)(x - 4)$ **35.** $3y(y - 2)(y - 3)$ **37.** $2x^2(x + 3)(x - 16)$ **39.** $(x^3 + 2)(x^3 - 3)$ **41.** $(x^2 - 6)(x^2 + 1)$ **43.** $(x + 6)(x + 2)$ **45.** $(3x + 5)(x + 1)$ **47.** $(x + 11)(5x + 1)$ **49.** $(y + 8)(3y - 2)$ **51.** $(y + 2)(4y + 1)$ **53.** $(2x + 3)(5x + 2)$ **55.** $(4x - 3)(2x - 3)$ **57.** $(y - 3)(6y - 5)$ **59.** prime **61.** $(x + y)(3x + y)$ **63.** $(2x + y)(3x - 5y)$ **65.** $(5x - 2y)(3x - 5y)$ **67.** $(3a - 7b)(a + 2b)$ **69.** $5x(3x - 2)(x - 1)$ **71.** $2x^2(3x + 2)(4x - 1)$ **73.** $y^3(5y + 1)(3y - 1)$ **75.** $3(8x + 9y)(x - y)$ **77.** $2b(a + 3)(3a - 10)$ **79.** $2y(2x - y)(3x - 7y)$ **81.** $13x^3y(y + 4)(y - 1)$ **83.** $(2x^2 - 3)(x^2 + 1)$ **85.** $(2x^3 + 5)(x^3 + 3)$ **87.** $(2y^5 + 1)(y^5 + 3)$ **89.** $(5x + 12)(x + 2)$ **91.** $(2x - 13)(x - 2)$ **93.** $(x - 0.3)(x - 0.2)$ **95.** $\left(x + \frac{3}{7}\right)\left(x - \frac{1}{7}\right)$ **97.** $(ax - b)(cx + d)$ **99.** $-x^3y^2(4x - 3y)(x - y)$ **101.** $f(x) = 3x - 13$ and $g(x) = x - 3$, or vice versa **103.** $2x + 1$ by $x + 3$ **105. a.** 32; The diver is 32 feet above the water after 1 second. **b.** 0; The diver hits the water after 2 seconds. **c.** $-16(t - 2)(t + 1)$; $f(t) = -16(t - 2)(t + 1)$ **d.** 32; 0 **107. a.** $x^2 + x + x + x + 1 + 1 = x^2 + 3x + 2$ **b.** $(x + 2)(x + 1)$ **c.** Answers will vary.

117.

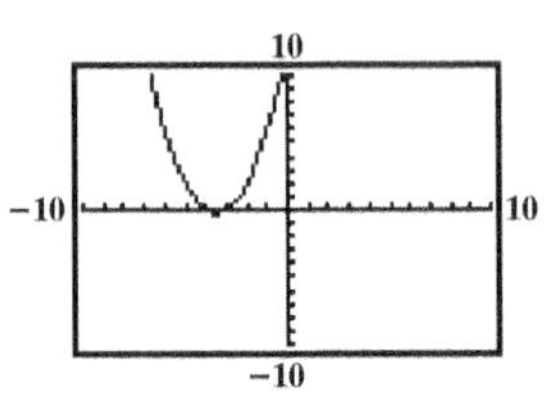

Graphs coincide.; factored correctly

119.

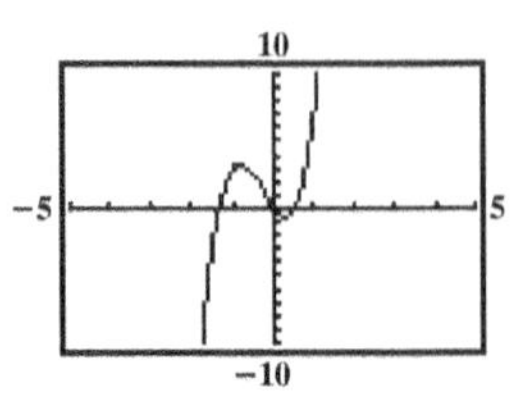

Graphs coincide.; factored correctly

123. makes sense **125.** makes sense **127.** true **129.** false **131.** −16, −8, 8, 16 **133.** $(4x^n - 5)(x^n - 1)$ **135.** $b^2(b^n - 2)(b^n + 5)$ **137.** $d^n(2d - 3)(d - 1)$ **138.** $\{x|x > 5\}$ or $(5, \infty)$ **139.** $\{(2, -1, 3)\}$ **140.** $(x + 2)(4x^2 - 5)$ **141.** $(x + 7)(x + 7)$ or $(x + 7)^2$ **142.** $(x - 4)(x - 4)$ or $(x - 4)^2$ **143.** $(x + 5)(x - 5)$

Mid-Chapter Check Point Exercises

1. $-x^3 + 4x^2 + 6x + 17$ **2.** $-2x^7y^3z^5$ **3.** $30x^5y^3 - 35x^3y^2 - 2x^2y$ **4.** $3x^3 + 4x^2 - 39x + 40$ **5.** $2x^4 - x^3 - 8x^2 + 11x - 4$ **6.** $-x^2 - 5x + 5$ **7.** $-4x^3y - 6x^2y - 7y - 5$ **8.** $8x^2 + 18x - 5$ **9.** $10x^2y^2 - 11xy - 6$ **10.** $9x^2 - 4y^2$ **11.** $6x^3y - 9xy^2 + 2x^2 - 3y$ **12.** $49x^6y^2 - 25x^2$ **13.** $6xh + 3h^2 - 2h$ **14.** $x^4 - 6x^2 + 9$ **15.** $x^5 + 2x^3 + 2x^2 - 15x - 6$ **16.** $4x^2 + 20xy + 25y^2$ **17.** $x^2 + 12x + 36 - 9y^2$ **18.** $x^2 + 2xy + y^2 + 10x + 10y + 25$ **19.** $(x - 8)(x + 3)$ **20.** $(5x + 2)(3y + 1)$ **21.** $(5x - 2)(x + 2)$ **22.** $5(7x^2 + 2x - 10)$ **23.** $9(x - 2)(x + 1)$ **24.** $5x^2y(2xy - 4y + 7)$ **25.** $(3x + 1)(6x + 5)$ **26.** $(4x - 3y)(3x - 4)$ **27.** $(3x - 4)(3x - 1)$ **28.** $(3x^3 + 5)(x^3 + 2)$ **29.** $x(5x - 2)(5x + 7)$ **30.** $(x^3 - 3)(2x - y)$

Section 5.5

Check Point Exercises

1. a. $(4x + 5)(4x - 5)$ **b.** $(10y^3 + 3x^2)(10y^3 - 3x^2)$ **2.** $6y(1 + xy^3)(1 - xy^3)$ **3.** $(4x^2 + 9)(2x + 3)(2x - 3)$ **4.** $(x + 7)(x + 2)(x - 2)$ **5. a.** $(x + 3)^2$ **b.** $(4x + 5y)^2$ **c.** $(2y^2 - 5)^2$ **6.** $(x + 5 + y)(x + 5 - y)$ **7.** $(a + b - 2)(a - b + 2)$ **8. a.** $(x + 3)(x^2 - 3x + 9)$ **b.** $(x^2 + 10y)(x^4 - 10x^2y + 100y^2)$ **9. a.** $(x - 2)(x^2 + 2x + 4)$ **b.** $(1 - 3xy)(1 + 3xy + 9x^2y^2)$

Exercise Set 5.5

1. $(x + 2)(x - 2)$ **3.** $(3x + 5)(3x - 5)$ **5.** $(3 + 5y)(3 - 5y)$ **7.** $(6x + 7y)(6x - 7y)$ **9.** $(xy + 1)(xy - 1)$ **11.** $(3x^2 + 5y^3)(3x^2 - 5y^3)$ **13.** $(x^7 + y^2)(x^7 - y^2)$ **15.** $(x - 3 + y)(x - 3 - y)$ **17.** $(a + b - 2)(a - b + 2)$ **19.** $(x^n + 5)(x^n - 5)$ **21.** $(1 + a^n)(1 - a^n)$ **23.** $2x(x + 2)(x - 2)$ **25.** $2(5 + y)(5 - y)$ **27.** $8(x + y)(x - y)$ **29.** $2xy(x + 3)(x - 3)$ **31.** $a(ab + 7c)(ab - 7c)$ **33.** $5y(1 + xy^3)(1 - xy^3)$ **35.** $8(x^2 + y^2)$ **37.** prime **39.** $(x^2 + 4)(x + 2)(x - 2)$ **41.** $(9x^2 + 1)(3x + 1)(3x - 1)$ **43.** $2x(x^2 + y^2)(x + y)(x - y)$ **45.** $(x + 3)(x + 2)(x - 2)$ **47.** $(x - 7)(x + 1)(x - 1)$ **49.** $(x + 2)^2$ **51.** $(x - 5)^2$ **53.** $(x^2 - 2)^2$ **55.** $(3y + 1)^2$ **57.** $(8y - 1)^2$ **59.** $(x - 6y)^2$ **61.** prime **63.** $(3x + 8y)^2$ **65.** $(x - 3 + y)(x - 3 - y)$ **67.** $(x + 10 + x^2)(x + 10 - x^2)$ **69.** $(3x - 5 + 6y)(3x - 5 - 6y)$ **71.** $(x^2 + x + 1)(x^2 - x - 1)$ **73.** $(z + x - 2y)(z - x + 2y)$ **75.** $(x + 4)(x^2 - 4x + 16)$ **77.** $(x - 3)(x^2 + 3x + 9)$ **79.** $(2y + 1)(4y^2 - 2y + 1)$ **81.** $(5x - 2)(25x^2 + 10x + 4)$ **83.** $(xy + 3)(x^2y^2 - 3xy + 9)$ **85.** $x(4 - x)(16 + 4x + x^2)$ **87.** $(x^2 + 3y)(x^4 - 3x^2y + 9y^2)$ **89.** $(5x^2 - 4y^2)(25x^4 + 20x^2y^2 + 16y^4)$ **91.** $(x + 1)(x^2 - x + 1)(x^6 - x^3 + 1)$ **93.** $(x - 2y)(x^2 - xy + y^2)$ **95.** $(0.2x + 0.3)^2$ or $\frac{1}{100}(2x + 3)^2$ **97.** $x\left(2x - \frac{1}{2}\right)\left(4x^2 + x + \frac{1}{4}\right)$ **99.** $(x - 1)(x^2 + x + 1)(x - 2)(x^2 + 2x + 4)$ **101.** $(x^4 + 1)(x^2 + 4)(x + 2)(x - 2)$ **103.** $(x + 1)(x - 1)(x - 2)(x^2 + 2x + 4)$ **105. a.** $(A + B)^2$ **b.** A^2; AB; AB; B^2 **c.** $A^2 + 2AB + B^2$ **d.** $A^2 + 2AB + B^2 = (A + B)^2$; factoring a perfect square trinomial **107.** $25x^2 - 9 = (5x + 3)(5x - 3)$ **109.** $49x^2 - 36 = (7x + 6)(7x - 6)$ **111.** $3a^3 - 3ab^2 = 3a(a + b)(a - b)$

117.

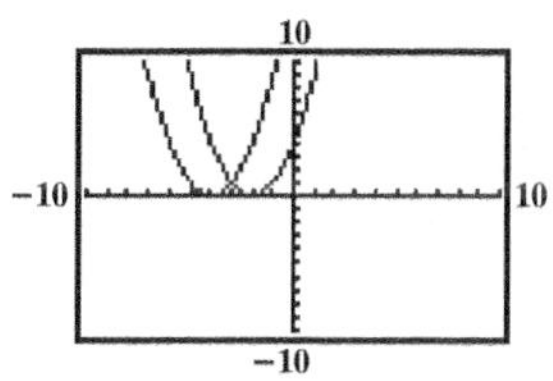

; Graphs do not coincide.; $x^2 + 4x + 4 = (x + 2)^2$

119.

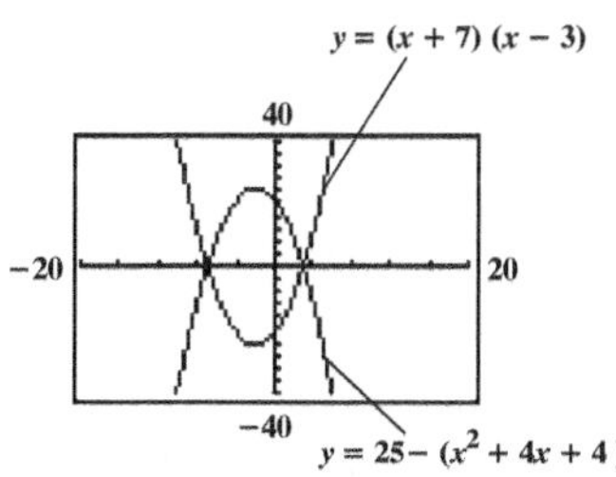

; Graphs do not coincide.; $25 - (x^2 + 4x + 4) = (7 + x)(3 - x)$

121.

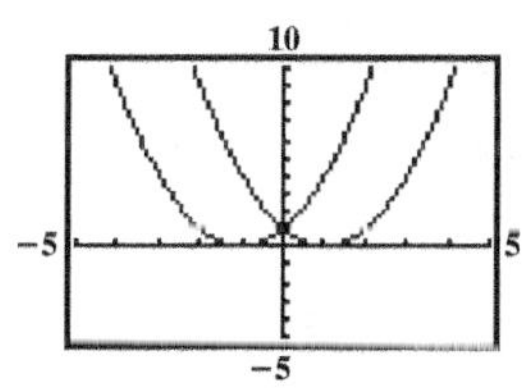

; Graphs do not coincide.; $(x - 3)^2 + 8(x - 3) + 16 = (x + 1)^2$

123.

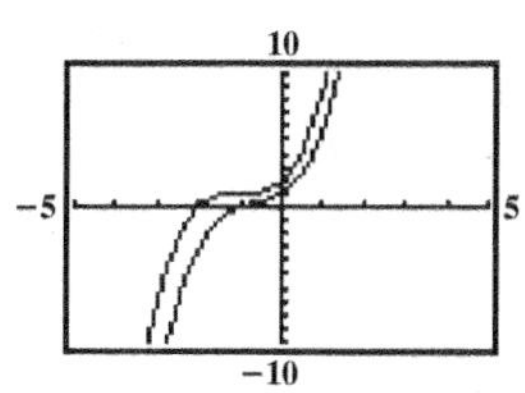

; Graphs do not coincide.; $(x + 1)^3 + 1 = (x + 2)(x^2 + x + 1)$

125. makes sense **127.** does not make sense **129.** false **131.** false **133.** $(y + x)(y^2 - xy + x^2 + 1)$ **135.** $(x^n + y^{4n})(x^{2n} - x^ny^{4n} + y^{8n})$

137. $x^6 - y^6 = (x^3 + y^3)(x^3 - y^3) = (x + y)(x^2 - xy + y^2)(x - y)(x^2 + xy + y^2)$;
$x^6 - y^6 = (x^2 - y^2)(x^4 + x^2y^2 + y^4) = (x + y)(x - y)(x^4 + x^2y^2 + y^4)$; $x^4 + x^2y^2 + y^4 = (x^2 - xy + y^2)(x^2 + xy + y^2)$
139. 1 **140.** $\{x|5 \le x \le 11\}$ or $[5, 11]$ **141.** $\{(-2, 1)\}$ **142.** $(x + 7)(3x - y)$ **143.** $2x(x + 2)^2$ **144.** $5x(x - y)(x - 7y)$
145. $(x + y)(3b + 4)(3b - 4)$

Section 5.6

Check Point Exercises

1. $3x(x - 5)^2$ **2.** $3y(x + 2)(x - 6)$ **3.** $(x + y)(4a + 5)(4a - 5)$
4. $(x + 10 + 6a)(x + 10 - 6a)$ **5.** $x(x + 2)(x^2 - 2x + 4)(x^6 - 8x^3 + 64)$

Exercise Set 5.6

1. $x(x + 4)(x - 4)$ **3.** $3(x + 3)^2$ **5.** $3(3x - 1)(9x^2 + 3x + 1)$ **7.** $(x + 4)(x - 4)(y - 2)$ **9.** $2b(2a + 5)(a - 3)$
11. $(y + 2)(y - 2)(a - 4)$ **13.** $11x(x^2 + y)(x^2 - y)$ **15.** $4x(x^2 + 4)(x + 2)(x - 2)$ **17.** $(x - 4)(x + 3)(x - 3)$
19. $2x^2(x + 3)(x^2 - 3x + 9)$ **21.** $3y(x^2 + 4y^2)(x + 2y)(x - 2y)$ **23.** $3x(2x + 3y)^2$ **25.** $(x - 6 + 7y)(x - 6 - 7y)$ **27.** prime
29. $12xy(x + y)(x - y)$ **31.** $6b(x^2 + y^2)$ **33.** $(x + y)(x - y)(x^2 + xy + y^2)$ **35.** $(x + 6 + 2a)(x + 6 - 2a)$ **37.** $(x^3 - 2)(x^3 + 7)$
39. $(2x - 7)(2 + x^2)$ **41.** $2(3x - 2y)(9x^2 + 6xy + 4y^2)$ **43.** $(x + 5 + y)(x + 5 - y)$ **45.** $(x^4 + y^4)(x^2 + y^2)(x + y)(x - y)$
47. $xy(x + 4y)(x - 4y)$ **49.** $x(1 + 2x)(1 - 2x + 4x^2)$ **51.** $2(4y + 1)(2y - 1)$ **53.** $y(14y^2 + 7y - 10)$ **55.** $3(3x + 2y)^2$
57. $3x(4x^2 + y^2)$ **59.** $x^3y^3(xy - 1)(x^2y^2 + xy + 1)$ **61.** $2(x + 5)(x - 5)$ **63.** $(x - y)(a + 2)(a - 2)$
65. $(c + d - 1)(c + d + 1)[(c + d)^2 + 1]$ **67.** $(p + q)^2(p - q)$ **69.** $(x + 2y)(x - 2y)(x + y)(x - y)$
71. $(x + y + 3)^2$ **73.** $(x - y)^2(x - y + 2)(x - y - 2)$ **75.** $(2x - y^2)(x - 3y^2)$ **77.** $(x - y)(x^2 + xy + y^2 - 1)$
79. $(xy + 1)(x^2y^2 - xy + 1)(x - 2)(x^2 + 2x + 4)$ **81. a.** $x(x + y) - y(x + y)$ **b.** $(x + y)(x - y)$
83. a. $xy + xy + xy + 3x(x) = 3xy + 3x^2$ **b.** $3x(y + x)$ **85. a.** $8x^2 - 2\pi x^2$ **b.** $2x^2(4 - \pi)$

89.

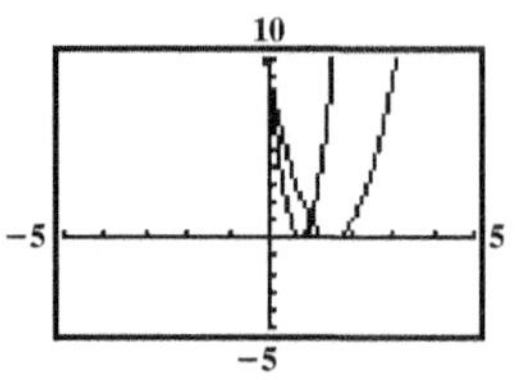

Graphs do not coincide.; $4x^2 - 12x + 9 = (2x - 3)^2$

91.

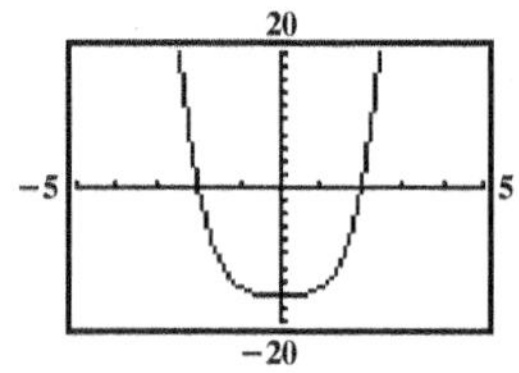

Graphs coincide.; factored correctly

95. makes sense **97.** does not make sense **99.** true **101.** false **103.** $x^n(3x - 1)(x - 4)$ **105.** $(x^2 + 2x + 2)(x^2 - 2x + 2)$
107. $\{-1\}$ **108.** $\frac{8x^9}{y^3}$ **109.** 52 **110.** 0 **111.** 0 **112.** $(x - 5)(x + 3) = 0$

Section 5.7

Check Point Exercises

1. $-\frac{1}{2}$ and 5, or $\left\{-\frac{1}{2}, 5\right\}$ **2. a.** 0 and $\frac{2}{3}$, or $\left\{0, \frac{2}{3}\right\}$ **b.** 5 or $\{5\}$ **c.** -4 and 3, or $\{-4, 3\}$
3. $-2, -\frac{3}{2}$, and 2, or $\left\{-2, -\frac{3}{2}, 2\right\}$ **4.** after 3 seconds; (3, 336) **5.** 2 ft **6.** 5 ft

Exercise Set 5.7

1. $\{-4, 3\}$ **3.** $\{-7, 1\}$ **5.** $\left\{-4, \frac{2}{3}\right\}$ **7.** $\left\{\frac{3}{5}, 1\right\}$ **9.** $\left\{-2, \frac{1}{3}\right\}$ **11.** $\{0, 8\}$ **13.** $\left\{0, \frac{5}{3}\right\}$ **15.** $\{-2\}$ **17.** $\{7\}$ **19.** $\left\{\frac{5}{3}\right\}$
21. $\{-5, 5\}$ **23.** $\left\{-\frac{10}{3}, \frac{10}{3}\right\}$ **25.** $\{-3, 6\}$ **27.** $\{-3, -2\}$ **29.** $\{4\}$ **31.** $\{-2, 8\}$ **33.** $\left\{-1, \frac{2}{5}\right\}$ **35.** $\{-6, -3\}$
37. $\{-5, -4, 5\}$ **39.** $\{-5, 1, 5\}$ **41.** $\{-4, 0, 4\}$ **43.** $\{0, 2\}$ **45.** $\{-5, -3, 0\}$ **47.** 2 and 4; d **49.** -4 and -2; c **51.** $\{-7, -1, 6\}$
53. $\left\{-\frac{4}{3}, 0, \frac{4}{5}, 2\right\}$ **55.** -4 and 8 **57.** $-2, -\frac{1}{2}$, and 2 **59.** -7 and 4 **61.** 2 and 3 **63.** -9 and 5 **65.** 1 second; 0.25, 0.5, 0.75, 1
67. 7 **69.** (7, 21) **71.** length: 9 ft; width: 6 ft **73.** 5 in. **75.** 5 m **77. a.** $4x^2 + 44x$ **b.** 3 ft **79.** length: 10 in.; width: 10 in.
81. length: 12 ft; width: 5 ft **83.** $30\frac{1}{8}$ ft

97.

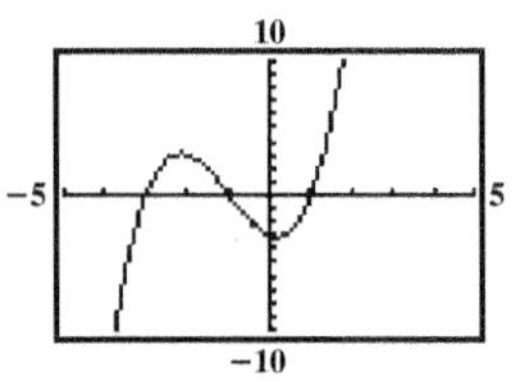

$\{-3, -1, 1\}$

99.

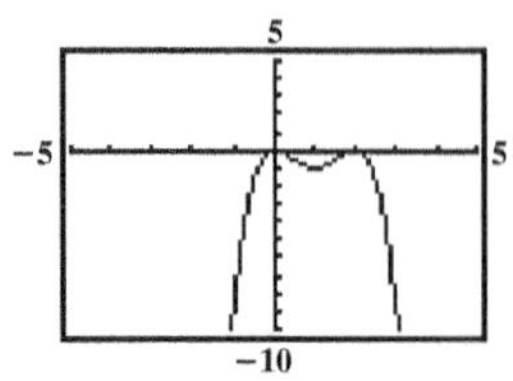

$\{0, 2\}$

101. makes sense **103.** makes sense **105.** false **107.** true **109.** Answers will vary; an example is $x^2 - 4x - 21 = 0$.

111. $\left\{-2, \frac{10}{3}\right\}$ **112.** 12 **113.** \$1700 at 5% and \$1300 at 8% **114.** 30 **115.** $\{x \mid x$ is a real number and $x \neq 2\}$
116. $\frac{(x-9)(x+2)}{(2x-1)(x+2)}; \frac{x-9}{2x-1}$

Review Exercises

1.

Term	Coefficient	Degree
$-5x^3$	-5	3
$7x^2$	7	2
$-\frac{x}{2}$	$-\frac{1}{2}$	1
2	2	0

The degree of the polynomial is 3, the leading term is $-5x^3$, and the leading coefficient is -5.

2.

Term	Coefficient	Degree
$8x^4y^2$	8	6
$-7xy^6$	-7	7
$-x^3y$	-1	4

The degree of the polynomial is 7, the leading term is $-7xy^6$, and the leading coefficient is -7.

3. -31 **4. a.** 380,621; There were 380,621 Americans living with AIDS in 2002. **b.** underestimates by 150 **5.** rises to the left and falls to the right; c **6.** rises to the left and rises to the right; b **7.** falls to the left and rises to the right; a **8.** falls to the left and falls to the right; d **9. a.** 20,787; 23,416; Values are the same as those shown for 1999 and 2000. **b.** The graph of this function falls to the right which indicates that the number of newly declared computer majors would eventually be negative, which is not possible. **10.** $x^3 - 6x^2 - x - 9$ **11.** $12x^3y - 2x^2y - 14y - 17$ **12.** $15x^3 - 10x^2 + 11x - 8$ **13.** $-2x^3y^2 - 4x^3y - 8$ **14.** $3x^3 + 5x^2y - xy^2 - 8y^3$ **15.** $-12x^6y^2z^7$ **16.** $2x^8 - 24x^5 - 12x^3$ **17.** $21x^5y^4 - 35x^2y^3 - 7xy^2$ **18.** $6x^3 + 29x^2 + 27x - 20$ **19.** $x^4 + 4x^3 + 4x^2 - x - 2$ **20.** $12x^2 - 23x + 5$ **21.** $15x^2y^2 + 2xy - 8$ **22.** $9x^2 + 42xy + 49y^2$ **23.** $x^4 - 10x^2y + 25y^2$ **24.** $4x^2 - 49y^2$ **25.** $9x^2y^4 - 16x^2$ **26.** $x^2 + 6x + 9 - 25y^2$ **27.** $x^2 + 2xy + y^2 + 8x + 8y + 16$ **28.** $2x^2 - x - 15$; 21 **29. a.** $a^2 - 9a + 10$ **b.** $2ah + h^2 - 7h$ **30.** $8x^2(2x + 3)$ **31.** $2x(1 - 18x)$ **32.** $7xy(3xy - 2y + 1)$ **33.** $9x^2y(2xy - 3)$ **34.** $-4(3x^2 - 2x + 12)$ **35.** $-1(x^2 + 11x - 14)$ **36.** $(x - 1)(x^2 - 2)$ **37.** $(y - 3)(x - 5)$ **38.** $(x - 3y)(5a + 2b)$ **39.** $(x + 5)(x + 3)$ **40.** $(x + 20)(x - 4)$ **41.** $(x + 17y)(x - y)$ **42.** $3x(x - 1)(x - 11)$ **43.** $(x + 7)(3x + 1)$ **44.** $(3x - 2)(2x - 3)$ **45.** $(5x + 4y)(x - 2y)$ **46.** $x(3x + 4)(2x - 1)$ **47.** $(x + 3)(2x + 5)$ **48.** $(x^3 + 6)(x^3 - 5)$ **49.** $(x^2 + 3)(x^2 - 13)$ **50.** $(x + 11)(x + 9)$ **51.** $(5x^3 + 2)(x^3 + 3)$ **52.** $(2x + 5)(2x - 5)$ **53.** $(1 + 9xy)(1 - 9xy)$ **54.** $(x^4 + y^3)(x^4 - y^3)$ **55.** $(x - 1 + y)(x - 1 - y)$ **56.** $(x + 8)^2$ **57.** $(3x - 1)^2$ **58.** $(5x + 2y)^2$ **59.** prime **60.** $(5x - 4y)^2$ **61.** $(x + 9 + y)(x + 9 - y)$ **62.** $(z + 5x - 1)(z - 5x + 1)$ **63.** $(4x + 3)(16x^2 - 12x + 9)$ **64.** $(5x - 2)(25x^2 + 10x + 4)$ **65.** $(xy + 1)(x^2y^2 - xy + 1)$ **66.** $3x(5x + 1)$ **67.** $3x^2(2x + 1)(2x - 1)$ **68.** $4x(5x^3 - 6x^2 + 7x - 3)$ **69.** $x(x - 2)(x - 13)$ **70.** $-2y^2(y - 3)(y - 9)$ **71.** $(3x - 5)^2$ **72.** $5(x + 3)(x - 3)$ **73.** $(2x - 1)(x + 3)(x - 3)$ **74.** $(3x - y)(2x - 7y)$ **75.** $2y(y + 3)^2$ **76.** $(x + 3 + 2a)(x + 3 - 2a)$ **77.** $(2x - 3)(4x^2 + 6x + 9)$ **78.** $x(x^2 + 1)(x + 1)(x - 1)$ **79.** $(x^2 - 3)^2$ **80.** prime **81.** $4(a + 2)(a^2 - 2a + 4)$ **82.** $(x^2 + 9)(x + 3)(x - 3)$ **83.** $(a + 3b)(x - y)$ **84.** $(3x - 5y)(9x^2 + 15xy + 25y^2)$ **85.** $2xy(x + 3)(5x - 4)$ **86.** $(2x^3 + 5)(3x^3 - 1)$ **87.** $(x + 5)(2 + xy)$ **88.** $(y + 2)(y + 5)(y - 5)$ **89.** $(a^4 + 1)(a^2 + 1)(a + 1)(a - 1)$ **90.** $(x - 4)(3 + y)(3 - y)$ **91. a.** $2xy + 2y^2$ **b.** $2y(x + y)$ **92. a.** $x^2 - 4y^2$ **b.** $(x + 2y)(x - 2y)$ **93.** $\{-5, -1\}$ **94.** $\left\{\frac{1}{3}, 7\right\}$ **95.** $\{-8, 7\}$ **96.** $\{0, 4\}$ **97.** $\{-5, -3, 3\}$ **98.** 9 seconds **99. a.** 20 miles per hour **b.** (20, 40) **c.** As a car's speed increases, its stopping distance gets longer at increasingly greater rates. **100.** length: 9 ft; width: 6 ft **101.** 2 in. **102.** 50 yd, 120 yd, and 130 yd

Chapter Test

1. degree: 3; leading coefficient: -6 **2.** degree: 9; leading coefficient: 7 **3.** 6; 4 **4.** falls to the left and falls to the right **5.** falls to the left and rises to the right **6.** $7x^3y - 18x^2y - y - 9$ **7.** $11x^2 - 13x - 6$ **8.** $35x^7y^3$ **9.** $x^3 - 4x^2y + 2xy^2 + y^3$ **10.** $21x^2 - 20xy - 9y^2$ **11.** $4x^2 - 25y^2$ **12.** $16y^2 - 56y + 49$ **13.** $x^2 + 4x + 4 - 9y^2$ **14.** $3x^2 + x - 10$; 60 **15.** $2ah + h^2 - 5h$ **16.** $x^2(14x - 15)$ **17.** $(9y + 5)(9y - 5)$ **18.** $(x + 3)(x + 5)(x - 5)$ **19.** $(5x - 3)^2$ **20.** $(x + 5 + 3y)(x + 5 - 3y)$ **21.** prime **22.** $(y + 2)(y - 18)$ **23.** $(2x + 5)(7x + 3)$ **24.** $5(x - 1)(x^2 + x + 1)$ **25.** $3(2x + y)(2x - y)$ **26.** $2(3x - 1)(2x - 5)$ **27.** $3(x^2 + 1)(x + 1)(x - 1)$ **28.** $(x^4 + y^4)(x^2 + y^2)(x + y)(x - y)$ **29.** $4x^2y(3y^3 + 2xy - 9)$ **30.** $(x^3 + 2)(x^3 - 14)$ **31.** $(x^2 - 6)(x^2 + 4)$ **32.** $3y(4x - 1)(x - 2)$ **33.** $y(y - 3)(y^2 + 2)$ **34.** $\left\{-\frac{1}{3}, 2\right\}$ **35.** $\left\{-1, \frac{6}{5}\right\}$ **36.** $\left\{0, \frac{1}{3}\right\}$ **37.** $\{-1, 1, 4\}$ **38.** 7 seconds **39.** length: 5 yd; width: 3 yd **40.** 12 units, 9 units, and 15 units

Cumulative Review Exercises

1. $\{4\}$ **2.** $\left\{\left(-4, \frac{1}{2}\right)\right\}$ **3.** $\{(2, 1, -1)\}$ **4.** $\{x \mid 2 < x < 3\}$ or $(2, 3)$ **5.** $\{x \mid x \leq -2$ or $x \geq 7\}$ or $(-\infty, -2] \cup [7, \infty)$ **6.** $\left\{1, \frac{5}{2}\right\}$ **7.** $\{-5, 0, 2\}$ **8.** $x = \frac{b}{c - a}$ **9.** $f(x) = 2x + 1$ **10.** winner: 1480 votes; loser: 1320 votes

11.

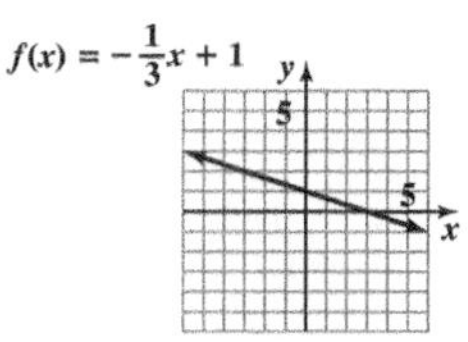

12.

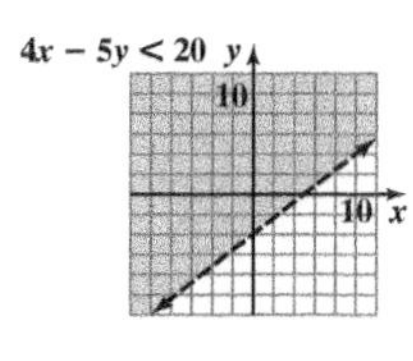

13.

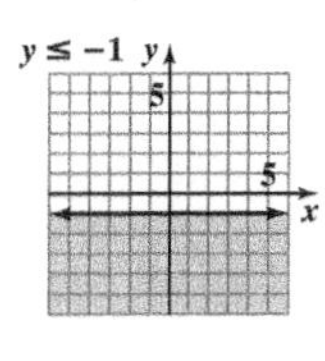

14. $-\frac{y^{10}}{2x^6}$ **15.** 7.06×10^{-5} **16.** $9x^4 - 6x^2y + y^2$ **17.** $9x^4 - y^2$ **18.** $(x + 3)(x - 3)^2$ **19.** $x^2(x^2 + 1)(x + 1)(x - 1)$ **20.** $14x^3y^2(1 - 2x)$

CHAPTER 7

Section 7.1

Check Point Exercises

1. a. 8 **b.** -7 **c.** $\frac{4}{5}$ **d.** 0.09 **e.** 5 **f.** 7 **2. a.** 4 **b.** $-\sqrt{24} \approx -4.90$ **3.** $\{x|x \geq 3\}$ or $[3, \infty)$
4. approximately 15.1 minutes **5. a.** 7 **b.** $|x + 8|$ **c.** $|7x^5|$ or $7|x^5|$ **d.** $|x - 3|$ **6. a.** 3 **b.** -2 **7.** $-3x$ **8. a.** 2 **b.** -2 **c.** not a real number **d.** -1 **9. a.** $|x + 6|$ **b.** $3x - 2$ **c.** 8

Exercise Set 7.1

1. 6 **3.** -6 **5.** not a real number **7.** $\frac{1}{5}$ **9.** $-\frac{3}{4}$ **11.** 0.9 **13.** -0.2 **15.** 3 **17.** 1 **19.** not a real number
21. 4; 1; 0; not a real number **23.** -5; $-\sqrt{5} \approx -2.24$; -1; not a real number **25.** 4; 2; 1; 6 **27.** $\{x|x \geq 3\}$ or $[3, \infty)$; c
29. $\{x|x \geq -5\}$ or $[-5, \infty)$; d **31.** $\{x|x \leq 3\}$ or $(-\infty, 3]$; e **33.** 5 **35.** 4 **37.** $|x - 1|$ **39.** $|6x^2|$ or $6x^2$
41. $-|10x^3|$ or $-10|x^3|$ **43.** $|x + 6|$ **45.** $-|x - 4|$ **47.** 3 **49.** -3 **51.** $\frac{1}{5}$ **53.** $-\frac{3}{10}$ **55.** 3; 2; -1; -4 **57.** -2; 0; 2
59. 1 **61.** 2 **63.** -2 **65.** not a real number **67.** -1 **69.** not a real number **71.** -4 **73.** 2 **75.** -2
77. x **79.** $|y|$ **81.** $-2x$ **83.** -5 **85.** 5 **87.** $|x + 3|$ **89.** $-2(x - 1)$

91.

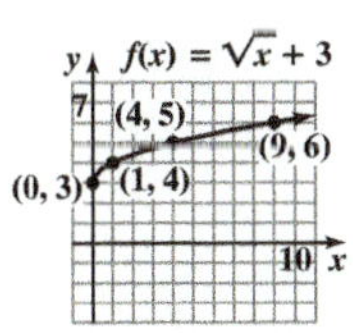

domain: $\{x|x \geq 0\} = [0, \infty)$;
range: $\{y|y \geq 3\} = [3, \infty)$

93.

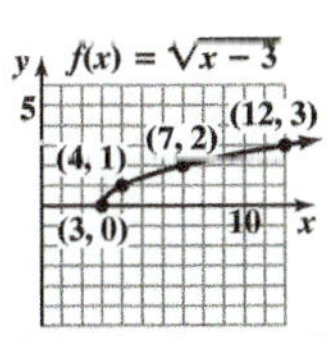

domain: $\{x|x \geq 3\} = [3, \infty)$;
range: $\{y|y \geq 0\} = [0, \infty)$

95. $\{x|x < 15\}$ or $(-\infty, 15)$ **97.** $\{x|1 \leq x < 3\}$ or $[1, 3)$ **99.** 3 **101. a.** 40.2 in.; underestimates by 0.6 in. **b.** 0.9 in. per month **c.** 0.2 in. per month; This is a much smaller rate of change.; The graph is not as steep between 50 and 60 as it is between 0 and 10.
103. 70 mph; The officer should not believe the motorist.; Answers will vary.

115. 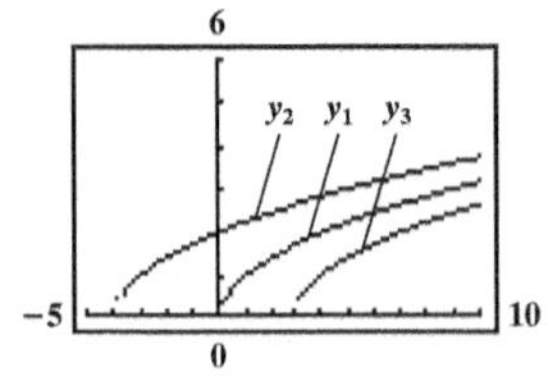

; Answers will vary.

117. 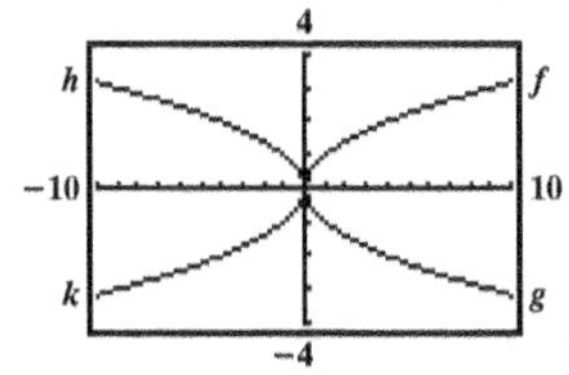

domain of f: $\{x|x \geq 0\}$ or $[0, \infty)$; range of f: $\{y|y \geq 0\}$ or $[0, \infty)$
domain of g: $\{x|x \geq 0\}$ or $[0, \infty)$; range of g: $\{y|y \leq 0\}$ or $(-\infty, 0]$
domain of h: $\{x|x \leq 0\}$ or $(-\infty, 0]$; range of h: $\{y|y \geq 0\}$ or $[0, \infty)$
domain of k: $\{x|x \leq 0\}$ or $(-\infty, 0]$; range of k: $\{y|y \leq 0\}$ or $(-\infty, 0]$

119. does not make sense **121.** makes sense **123.** false **125.** false **127.** Answers will vary; an example is $f(x) = \sqrt{15 - 3x}$.

129. $|(2x + 3)^5|$ **131.** The graph of h is the graph of f shifted left 3 units.

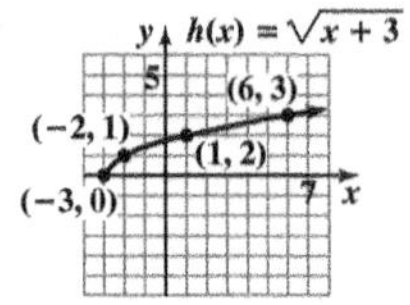

132. $7x + 30$ **133.** $\frac{x^8}{9y^6}$ **134.** $\left\{x \middle| x < -\frac{7}{3} \text{ or } x > 5\right\}$ or $\left(-\infty, -\frac{7}{3}\right) \cup (5, \infty)$ **135.** $\frac{2^7}{x}$ or $\frac{128}{x}$ **136.** $\frac{2}{x^3}$ **137.** $\frac{y^{12}}{x^8}$

Section 7.3

Check Point Exercises

1. a. $\sqrt{55}$ **b.** $\sqrt{x^2 - 16}$ **c.** $\sqrt[3]{60}$ **d.** $\sqrt[7]{12x^4}$ **2. a.** $4\sqrt{5}$ **b.** $2\sqrt[3]{5}$ **c.** $2\sqrt[4]{2}$ **d.** $10|x|\sqrt{2y}$
3. $f(x) = \sqrt{3}|x - 2|$ **4.** $x^4y^5z\sqrt{xyz}$ **5.** $2x^3y^4\sqrt[3]{5xy^2}$ **6.** $2x^2z\sqrt[5]{x^2y^2z^3}$ **7. a.** $2\sqrt{3}$ **b.** $100\sqrt[3]{4}$ **c.** $2x^2y\sqrt[4]{2}$

Exercise Set 7.3

1. $\sqrt{15}$ **3.** $\sqrt[3]{18}$ **5.** $\sqrt[4]{33}$ **7.** $\sqrt{33xy}$ **9.** $\sqrt[5]{24x^4}$ **11.** $\sqrt{x^2 - 9}$ **13.** $\sqrt[6]{(x - 4)^5}$ **15.** $\sqrt{x}$ **17.** $\sqrt[4]{\dfrac{3x}{7y}}$ **19.** $\sqrt[7]{77x^5y^3}$
21. $5\sqrt{2}$ **23.** $3\sqrt{5}$ **25.** $5\sqrt{3x}$ **27.** $2\sqrt[3]{2}$ **29.** $3x$ **31.** $-2y\sqrt[3]{2x^2}$ **33.** $6|x + 2|$ **35.** $2(x + 2)\sqrt[3]{4}$ **37.** $|x - 1|\sqrt{3}$
39. $x^3\sqrt{x}$ **41.** $x^4y^4\sqrt{y}$ **43.** $4x\sqrt{3x}$ **45.** $y^2\sqrt[3]{y^2}$ **47.** $x^4y\sqrt[3]{x^2z}$ **49.** $3x^2y^2\sqrt[3]{3x^2}$ **51.** $(x + y)\sqrt[3]{(x + y)^2}$ **53.** $y^3\sqrt[5]{y^2}$
55. $2xy^3\sqrt[5]{2xy^2}$ **57.** $2x^2\sqrt[4]{5x^2}$ **59.** $(x - 3)^2\sqrt[4]{(x - 3)^2}$ or $(x - 3)^2\sqrt{x - 3}$ **61.** $2\sqrt{6}$ **63.** $5\sqrt{2xy}$ **65.** $6x$ **67.** $10xy\sqrt{2y}$
69. $60\sqrt{2}$ **71.** $2\sqrt[3]{6}$ **73.** $2x^2\sqrt{10x}$ **75.** $5xy^4\sqrt[3]{x^2y^2}$ **77.** $2xyz\sqrt[4]{x^2z^3}$ **79.** $2xy^2z\sqrt[5]{2y^3z}$ **81.** $(x - y)^2\sqrt[3]{(x - y)^2}$
83. $-6x^3y^3\sqrt[3]{2yz^2}$ **85.** $-6y^2\sqrt[5]{2x^3y}$ **87.** $-6x^3y^3\sqrt{2}$ **89.** $-12x^3y^4\sqrt[4]{x^3}$ **91.** $6\sqrt{3}$ miles; 10.4 miles **93.** $8\sqrt{3}$ ft per sec; 14 ft per sec
95. a. $\dfrac{7.644}{2\sqrt[4]{2}} = \dfrac{3.822}{\sqrt[4]{2}}$ **b.** 3.21 liters of blood per minute per square meter; (32, 3.21)

103. Graphs are the same; simplification is correct.

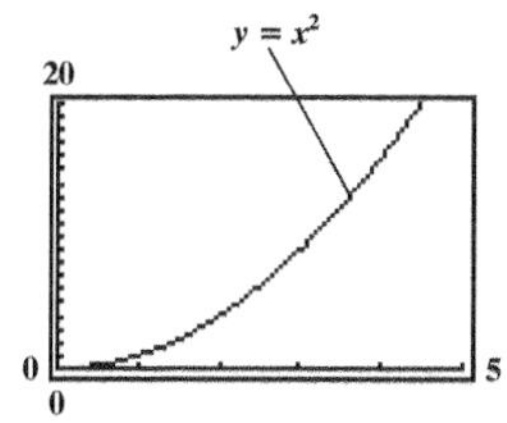

105. Graphs are not the same.; $\sqrt{3x^2 - 6x + 3} = |x - 1|\sqrt{3}$

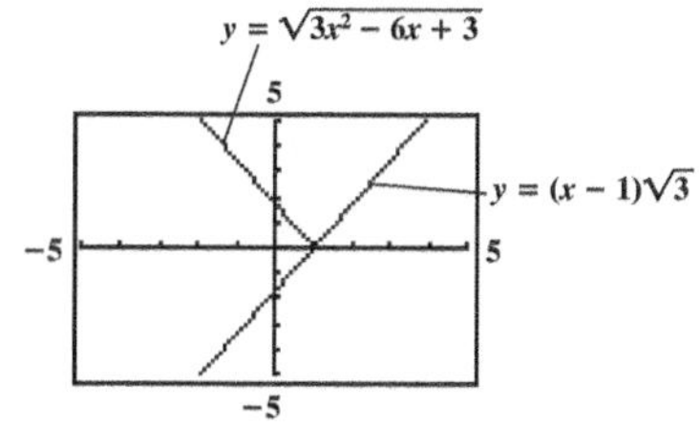

107. makes sense **109.** makes sense **111.** false **113.** false **115.** Its square root is multiplied by $\sqrt{3}$. **117.** $g(x) = \sqrt[3]{4x^2}$
119. $\{x|5 \le x \le 11\}$ or $[5, 11]$ **120.** $\{(2, -4)\}$ **121.** $(4x - 3)(16x^2 + 12x + 9)$ **122. a.** $31x$ **b.** $31\sqrt{2}$ **123. a.** $-8x$
b. $-8\sqrt[3]{2}$ **124.** $\dfrac{y\sqrt[4]{7y}}{x^3}$

Review Exercises

1. 9 **2.** $-\frac{1}{10}$ **3.** -3 **4.** not a real number **5.** -2 **6.** $5; 1.73; 0;$ not a real number **7.** $2; -2; -4$ **8.** $\{x|x \geq 2\}$ or $[2, \infty)$
9. $\{x|x \leq 25\}$ or $(-\infty, 25]$ **10.** $5|x|$ **11.** $|x + 14|$ **12.** $|x - 4|$ **13.** $4x$ **14.** $2|x|$ **15.** $-2(x + 7)$ **16.** $\sqrt[3]{5xy}$
17. $(\sqrt{16})^3 = 64$ **18.** $(\sqrt[5]{32})^4 = 16$ **19.** $(7x)^{1/2}$ **20.** $(19xy)^{5/3}$ **21.** $\frac{1}{8^{2/3}} = \frac{1}{4}$ **22.** $\frac{3x}{a^{4/5}b^{4/5}} = \frac{3x}{\sqrt[5]{a^4b^4}}$ **23.** $x^{7/12}$ **24.** $5^{1/6}$
25. $2x^2y$ **26.** $\frac{y^{1/8}}{x^{1/3}}$ **27.** x^3y^4 **28.** $y\sqrt[3]{x}$ **29.** $\sqrt[6]{x^5}$ **30.** $\sqrt[6]{x}$ **31.** $\sqrt[15]{x}$ **32.** \$3150 million **33.** $\sqrt{21xy}$ **34.** $\sqrt[5]{77x^3}$
35. $\sqrt[6]{(x-5)^5}$ **36.** $f(x) = \sqrt{7}|x - 1|$ **37.** $2x\sqrt{5x}$ **38.** $3x^2y^2\sqrt[3]{2x^2}$ **39.** $2y^2z\sqrt[4]{2x^3y^3z}$ **40.** $2x^2\sqrt{6x}$ **41.** $2xy\sqrt[3]{2y^2}$
42. $xyz^2\sqrt[5]{16y^4z}$ **43.** $\sqrt{x^2 - 1}$ **44.** $8\sqrt[3]{3}$ **45.** $9\sqrt{2}$ **46.** $(3x + y^2)\sqrt[3]{x}$ **47.** $-8\sqrt[3]{6}$ **48.** $\frac{2}{5}\sqrt[3]{2}$ **49.** $\frac{x\sqrt{x}}{10y^2}$ **50.** $\frac{y\sqrt[4]{3y}}{2x^5}$
51. $2\sqrt{6}$ **52.** $2\sqrt[3]{2}$ **53.** $2x\sqrt[4]{2x}$ **54.** $10x^2\sqrt{xy}$ **55.** $6\sqrt{2} + 12\sqrt{5}$ **56.** $5\sqrt[3]{2} - \sqrt[3]{10}$ **57.** $-83 + 3\sqrt{35}$
58. $\sqrt{xy} - \sqrt{11x} - \sqrt{11y} + 11$ **59.** $13 + 4\sqrt{10}$ **60.** $22 - 4\sqrt{30}$ **61.** -6 **62.** 4 **63.** $\frac{2\sqrt{6}}{3}$ **64.** $\frac{\sqrt{14}}{7}$ **65.** $4\sqrt[3]{3}$
66. $\frac{\sqrt{10xy}}{5y}$ **67.** $\frac{7\sqrt[3]{4x}}{x}$ **68.** $\frac{\sqrt[4]{189x^3}}{3x}$ **69.** $\frac{5\sqrt[5]{xy^4}}{2xy}$ **70.** $3\sqrt{3} + 3$ **71.** $\frac{\sqrt{35} - \sqrt{21}}{2}$ **72.** $10\sqrt{5} + 15\sqrt{2}$ **73.** $\frac{x + 8\sqrt{x} + 15}{x - 9}$

74. $\frac{5+\sqrt{21}}{2}$ **75.** $\frac{3\sqrt{2}+2}{7}$ **76.** $\frac{2}{\sqrt{14}}$ **77.** $\frac{3x}{\sqrt[3]{9x^2y}}$ **78.** $\frac{7}{\sqrt{35}+\sqrt{21}}$ **79.** $\frac{2}{5-\sqrt{21}}$ **80.** $\{16\}$ **81.** no solution or $\varnothing$ **82.** $\{2\}$ **83.** $\{8\}$ **84.** $\{-4, -2\}$ **85. a.** 0.99; The average state cigarette tax per pack was approximately \$0.99 six years after 2001, or in 2007.; underestimates by \$0.01 **b.** 9 years after 2001, in 2010 **86.** 84 years old **87.** $9i$ **88.** $3i\sqrt{7}$ **89.** $-2i\sqrt{2}$ **90.** $12+2i$ **91.** $-9+4i$ **92.** $-12-8i$ **93.** $29+11i$ **94.** $-7-24i$ **95.** $113+0i$ or 113 **96.** $-2\sqrt{6}+0i$ or $-2\sqrt{6}$ **97.** $\frac{15}{13}-\frac{3}{13}i$ **98.** $\frac{1}{5}+\frac{11}{10}i$ **99.** $\frac{1}{3}-\frac{5}{3}i$ **100.** 1 **101.** $-i$

Chapter Test

1. a. 6 **b.** $\{x \mid x \le 4\}$ or $(-\infty, 4]$ **2.** $\frac{1}{81}$ **3.** $\frac{5y^{1/8}}{x^{1/4}}$ **4.** $\sqrt{x}$ **5.** $\sqrt[20]{x^9}$ **6.** $5|x|\sqrt{3}$ **7.** $|x-5|$ **8.** $2xy^2\sqrt[3]{xy^2}$ **9.** $-\frac{2}{x^2}$ **10.** $\sqrt{50x^2y}$ **11.** $2x\sqrt[4]{2y^3}$ **12.** $-7\sqrt{2}$ **13.** $(2x+y^2)\sqrt[3]{x}$ **14.** $2x\sqrt[3]{x}$ **15.** $12\sqrt{2}-\sqrt{15}$ **16.** $26+6\sqrt{3}$ **17.** $52-14\sqrt{3}$ **18.** $\frac{\sqrt{5x}}{x}$ **19.** $\frac{\sqrt[3]{25x}}{x}$ **20.** $-5+2\sqrt{6}$ **21.** $\{6\}$ **22.** $\{16\}$ **23.** $\{-3\}$ **24.** 49 months **25.** $5i\sqrt{3}$ **26.** $-1+6i$ **27.** $26+7i$ **28.** $-6+0i$ or -6 **29.** $\frac{1}{5}+\frac{7}{5}i$ **30.** $-i$

Cumulative Review Exercises

1. $\{(-2, -1, -2)\}$ **2.** $\left\{-\frac{1}{3}, 4\right\}$ **3.** $\left\{x \,\middle|\, x > \frac{1}{3}\right\}$ or $\left(\frac{1}{3}, \infty\right)$ **4.** $\left\{\frac{3}{4}\right\}$ **5.** $\{-1\}$

6. $x + 2y < 2$, $2y - x > 4$

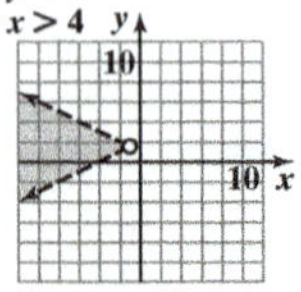

7. $\frac{x^2}{15(x+2)}$ **8.** $\frac{x}{y}$ **9.** $8x^3-22x^2+11x+6$ **10.** $\frac{5x-6}{(x-5)(x+3)}$ **11.** -64 **12.** 0 **13.** $-\frac{16-9\sqrt{3}}{13}$ **14.** $2x^2+x+5+\frac{6}{x-2}$ **15.** $-34-3\sqrt{6}$ **16.** $2(3x+2)(4x-1)$ **17.** $(4x^2+1)(2x+1)(2x-1)$ **18.** about 53 lumens **19.** \$1500 at 7% and \$4500 at 9% **20.** 2650 students

CHAPTER 8

Section 8.1

Check Point Exercises

1. $\pm\sqrt{7}$ or $\{\pm\sqrt{7}\}$ **2.** $\pm\frac{\sqrt{33}}{3}$ or $\left\{\pm\frac{\sqrt{33}}{3}\right\}$ **3.** $\pm\frac{3}{2}i$ or $\left\{\pm\frac{3}{2}i\right\}$ **4.** $3\pm\sqrt{10}$ or $\{3\pm\sqrt{10}\}$ **5. a.** 25; $x^2+10x+25=(x+5)^2$ **b.** $\frac{9}{4}$; $x^2-3x+\frac{9}{4}=\left(x-\frac{3}{2}\right)^2$ **c.** $\frac{9}{64}$; $x^2+\frac{3}{4}x+\frac{9}{64}=\left(x+\frac{3}{8}\right)^2$ **6.** $-2\pm\sqrt{5}$ or $\{-2\pm\sqrt{5}\}$ **7.** $\frac{-3\pm\sqrt{41}}{4}$ or $\left\{\frac{-3\pm\sqrt{41}}{4}\right\}$ **8.** 20% **9.** $10\sqrt{21}$ ft; 45.8 ft

Exercise Set 8.1

1. $\{\pm 5\}$ **3.** $\{\pm\sqrt{6}\}$ **5.** $\left\{\pm\frac{5}{4}\right\}$ **7.** $\left\{\pm\frac{\sqrt{6}}{3}\right\}$ **9.** $\left\{\pm\frac{4}{5}i\right\}$ **11.** $\{-10, -4\}$ **13.** $\{3\pm\sqrt{5}\}$ **15.** $\{-2\pm 2\sqrt{2}\}$ **17.** $\{5\pm 3i\}$ **19.** $\left\{\frac{-3\pm\sqrt{11}}{4}\right\}$ **21.** $\{-3, 9\}$ **23.** 1; $x^2+2x+1=(x+1)^2$ **25.** 49; $x^2-14x+49=(x-7)^2$ **27.** $\frac{49}{4}$; $x^2+7x+\frac{49}{4}=\left(x+\frac{7}{2}\right)^2$ **29.** $\frac{1}{16}$; $x^2-\frac{1}{2}x+\frac{1}{16}=\left(x-\frac{1}{4}\right)^2$ **31.** $\frac{4}{9}$; $x^2+\frac{4}{3}x+\frac{4}{9}=\left(x+\frac{2}{3}\right)^2$ **33.** $\frac{81}{64}$; $x^2-\frac{9}{4}x+\frac{81}{64}=\left(x-\frac{9}{8}\right)^2$ **35.** $\{-8, 4\}$ **37.** $\{-3\pm\sqrt{7}\}$ **39.** $\{4\pm\sqrt{15}\}$ **41.** $\{-1\pm i\}$ **43.** $\left\{\frac{-3\pm\sqrt{13}}{2}\right\}$ **45.** $\left\{-\frac{3}{7}, -\frac{1}{7}\right\}$ **47.** $\left\{\frac{-1\pm\sqrt{5}}{2}\right\}$ **49.** $\left\{-\frac{5}{2}, 1\right\}$ **51.** $\left\{\frac{-3\pm\sqrt{6}}{3}\right\}$ **53.** $\left\{\frac{4\pm\sqrt{13}}{3}\right\}$ **55.** $\left\{\frac{1}{4}\pm\frac{1}{4}i\right\}$ **57.** $-5, 7$ **59.** $-\frac{1}{5}, 1$ **61.** $-2\pm 5i$ **63.** $2\pm 2i$ **65.** $v=\sqrt{2gh}$ **67.** $r=\frac{\sqrt{AP}}{P}-1$ **69.** $\left\{\frac{-1\pm\sqrt{19}}{6}\right\}$ **71.** $\{-b, 2b\}$ **73. a.** x^2+8x **b.** 16 **c.** $x^2+8x+16$ **d.** $(x+4)^2$ **75.** 20% **77.** 6.25% **79.** approximately 1990 **81.** $10\sqrt{3}$ sec; 17.3 sec **83.** $3\sqrt{5}$ mi; 6.7 mi **85.** $20\sqrt{2}$ ft; 28.3 ft **87.** $50\sqrt{2}$ ft; 70.7 ft **89.** 10 m

97.

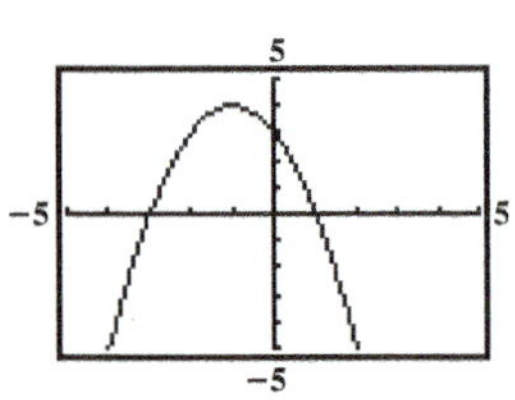

$\{-3, 1\}$

101. makes sense **103.** does not make sense **105.** false **107.** false **109.** $\left\{\frac{-1\pm\sqrt{1-4c}}{2}\right\}$ **111.** $\{\pm\sqrt{5}, \pm\sqrt{3}\}$

112. $4 - 2x$ **113.** $(1 - 2x)(1 + 2x + 4x^2)$ **114.** $x^3 - 2x^2 - 4x - 12 - \frac{42}{x - 3}$ **115. a.** $\left\{-\frac{1}{2}, \frac{1}{4}\right\}$ **b.** 36; yes

116. a. $\left\{\frac{1}{3}\right\}$ **b.** 0 **117. a.** $3x^2 + 4x + 2 = 0$ **b.** -8

Section 8.2

Check Point Exercises

1. -5 and $\frac{1}{2}$, or $\left\{-5, \frac{1}{2}\right\}$ **2.** $\frac{3 \pm \sqrt{7}}{2}$ or $\left\{\frac{3 \pm \sqrt{7}}{2}\right\}$ **3.** $-1 \pm i\frac{\sqrt{6}}{3}$ or $\left\{-1 \pm i\frac{\sqrt{6}}{3}\right\}$ **4. a.** 0; one real rational solution **b.** 81; two real rational solutions **c.** -44; two imaginary solutions that are complex conjugates **5. a.** $20x^2 + 7x - 3 = 0$ **b.** $x^2 + 49 = 0$ **6.** 26 years old; The point (26, 115) lies approximately on the blue graph.

Exercise Set 8.2

1. $\{-6, -2\}$ **3.** $\left\{1, \frac{5}{2}\right\}$ **5.** $\left\{\frac{-3 \pm \sqrt{89}}{2}\right\}$ **7.** $\left\{\frac{7 \pm \sqrt{85}}{6}\right\}$ **9.** $\left\{\frac{1 \pm \sqrt{7}}{6}\right\}$ **11.** $\left\{\frac{3}{8} \pm i\frac{\sqrt{87}}{8}\right\}$ **13.** $\{2 \pm 2i\}$ **15.** $\left\{\frac{4}{3} \pm i\frac{\sqrt{5}}{3}\right\}$ **17.** $\left\{-\frac{3}{2}, 4\right\}$ **19.** 52; two real irrational solutions **21.** 4; two real rational solutions **23.** -23; two imaginary solutions **25.** 36; two real rational solutions **27.** -60; two imaginary solutions **29.** 0; one (repeated) real rational solution **31.** $\left\{-\frac{2}{3}, 2\right\}$ **33.** $\{1 \pm \sqrt{2}\}$ **35.** $\left\{\frac{3 \pm \sqrt{65}}{4}\right\}$ **37.** $\left\{0, \frac{8}{3}\right\}$ **39.** $\left\{\frac{-6 \pm 2\sqrt{6}}{3}\right\}$ **41.** $\left\{\frac{2 \pm \sqrt{10}}{3}\right\}$ **43.** $\{2 \pm \sqrt{10}\}$ **45.** $\left\{1, \frac{5}{2}\right\}$ **47.** $x^2 - 2x - 15 = 0$ **49.** $12x^2 + 5x - 2 = 0$ **51.** $x^2 + 36 = 0$ **53.** $x^2 - 2 = 0$ **55.** $x^2 - 20 = 0$ **57.** $x^2 - 2x + 2 = 0$ **59.** $x^2 - 2x - 1 = 0$ **61.** b **63.** a **65.** $1 + \sqrt{7}$ **67.** $\left\{\frac{-1 \pm \sqrt{21}}{2}\right\}$ **69.** $\left\{-2\sqrt{2}, \frac{\sqrt{2}}{2}\right\}$ **71.** $\{-3, 1, -1 \pm i\sqrt{2}\}$ **73.** 33-year-olds and 58-year-olds; The function models the actual data well. **75.** 77.8 ft; (b) **77.** 5.5 m by 1.5 m **79.** 17.6 in. and 18.6 in. **81.** 9.3 in. and 0.7 in. **83.** 7.5 hr and 8.5 hr

93. 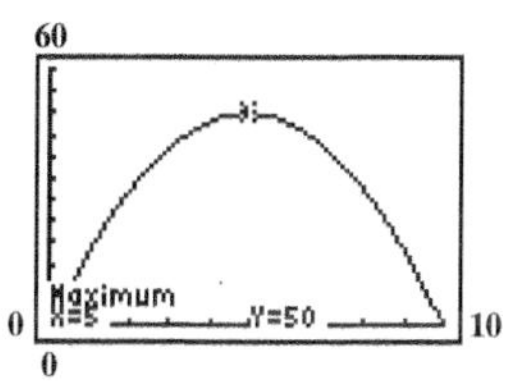

; depth: 5 in.; maximum area: 50 sq in.

95. does not make sense **97.** does not make sense **99.** false **101.** true **103.** 2.4 m; yes **105.** $\left\{-3, \frac{1}{4}\right\}$ **106.** $\{3, 7\}$

107. $\frac{5\sqrt{3} - 5x}{3 - x^2}$ **108.** 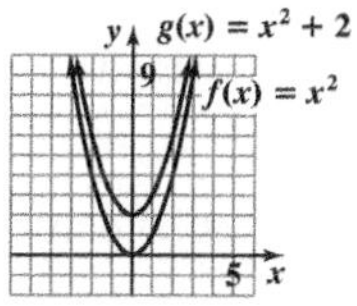

109. 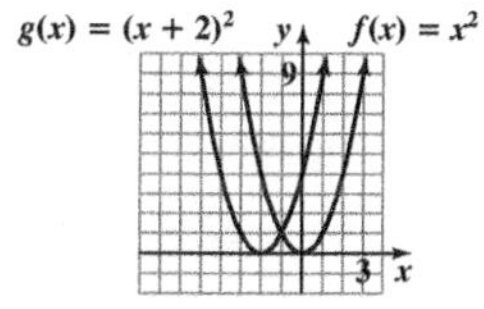

110. 1 and 5

Section 8.3

Check Point Exercises

1. 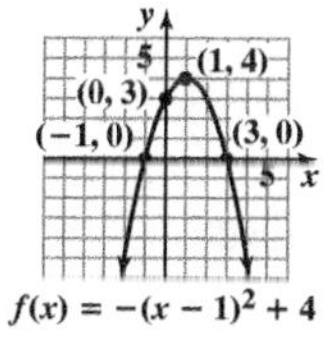

2. 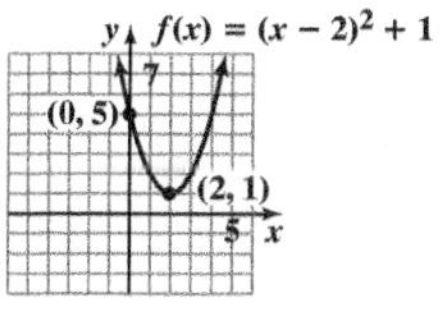

3. $(-2, -9)$

4. 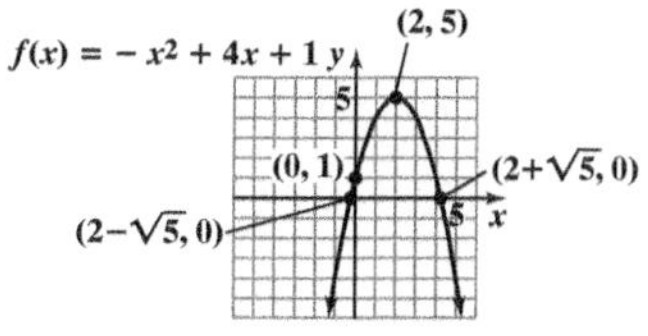

domain: $\{x|x$ is a real number$\}$ or $(-\infty, \infty)$; range: $\{y|y \leq 5\}$ or $(-\infty, 5]$

5. a. minimum **b.** Minimum is 984 at $x = 2$. **c.** domain: $\{x|x$ is a real number$\}$ or $(-\infty, \infty)$; range: $\{y|y \geq 984\}$ or $[984, \infty)$ **6.** 33.7 ft **7.** 4, -4; -16 **8.** 30 ft by 30 ft; 900 sq ft

Exercise Set 8.3

1. $h(x) = (x - 1)^2 + 1$ **3.** $j(x) = (x - 1)^2 - 1$ **5.** $h(x) = x^2 - 1$ **7.** $g(x) = x^2 - 2x + 1$ **9.** $(3, 1)$ **11.** $(-1, 5)$ **13.** $(2, -5)$ **15.** $(-1, 9)$

17.

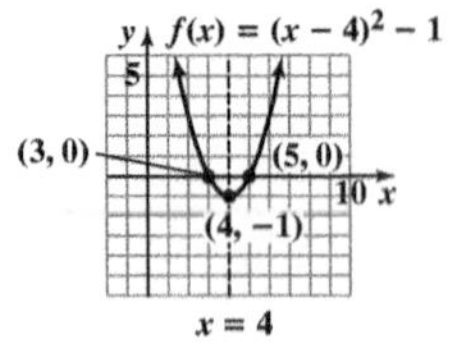

$\{y|y \geq -1\}$ or $[-1, \infty)$

19.

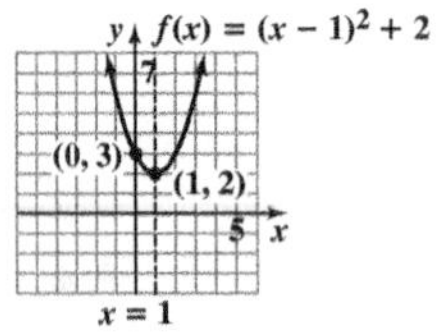

$\{y|y \geq 2\}$ or $[2, \infty)$

21.

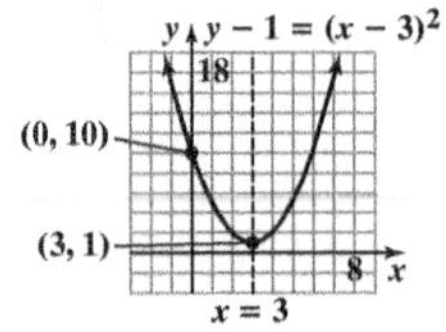

$\{y|y \geq 1\}$ or $[1, \infty)$

23.

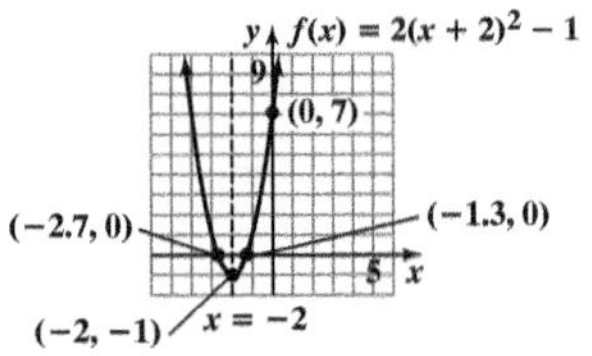

$\{y|y \geq -1\}$ or $[-1, \infty)$

25.

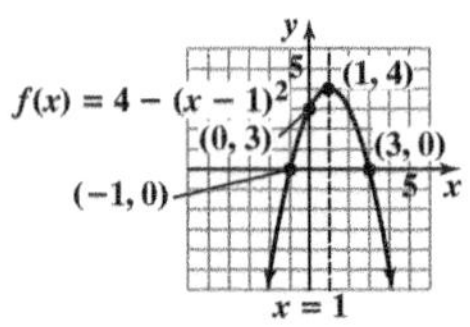

$\{y|y \leq 4\}$ or $(-\infty, 4]$

27.

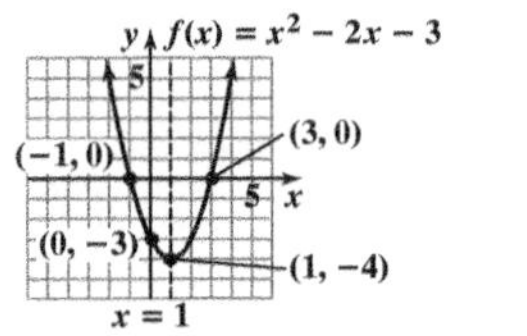

$\{y|y \geq -4\}$ or $[-4, \infty)$

29.

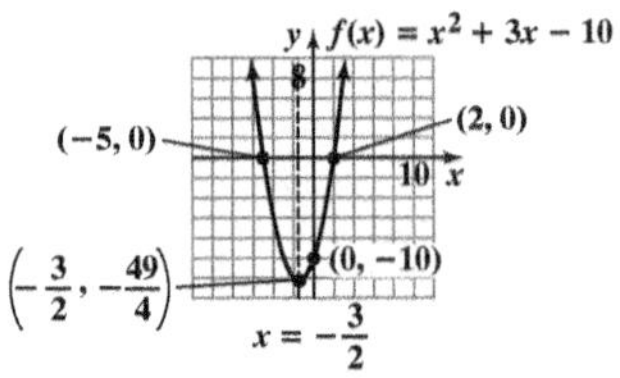

$\left\{y \,\middle|\, y \geq -\frac{49}{4}\right\}$ or $\left[-\frac{49}{4}, \infty\right)$

31.

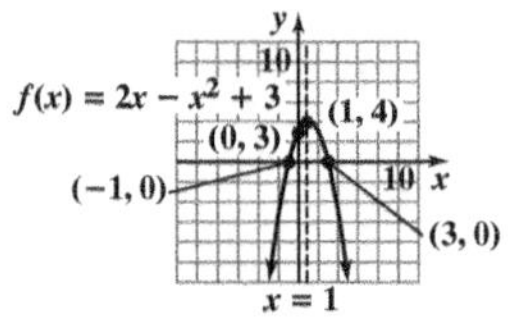

$\{y|y \leq 4\}$ or $(-\infty, 4]$

33.

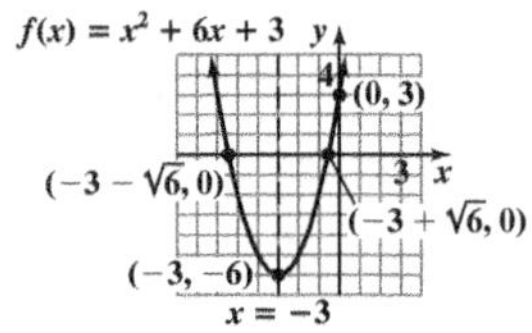

$\{y \,|\, y \geq -6\}$ or $[-6, \infty)$

35.

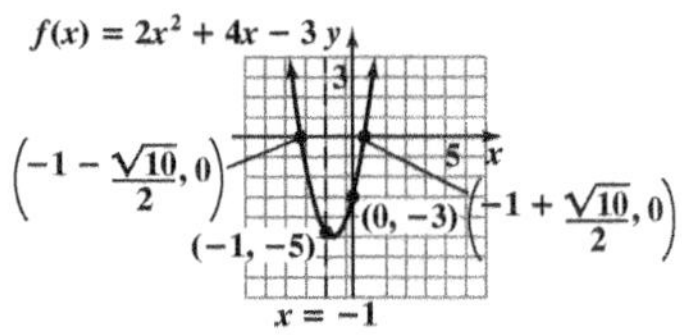

$\{y \,|\, y \geq -5\}$ or $[-5, \infty)$

37.

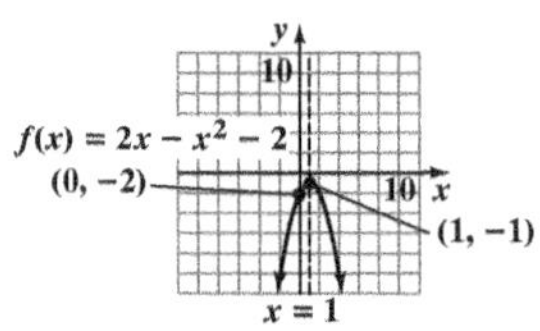

$\{y|y \leq -1\}$ or $(-\infty, -1]$

39. a. minimum **b.** Minimum is -13 at $x = 2$. **c.** domain: $\{x|x \text{ is a real number}\}$ or $(-\infty, \infty)$; range: $\{y|y \geq -13\}$ or $[-13, \infty)$
41. a. maximum **b.** Maximum is 1 at $x = 1$. **c.** domain: $\{x|x \text{ is a real number}\}$ or $(-\infty, \infty)$; range: $\{y|y \leq 1\}$ or $(-\infty, 1]$
43. a. minimum **b.** Minimum is $-\frac{5}{4}$ at $x = \frac{1}{2}$. **c.** domain: $\{x|x \text{ is a real number}\}$ or $(-\infty, \infty)$; range: $\left\{y \,\middle|\, y \geq -\frac{5}{4}\right\}$ or $\left[-\frac{5}{4}, \infty\right)$
45. domain: $\{x|x \text{ is a real number}\}$ or $(-\infty, \infty)$; range: $\{y|y \geq -2\}$ or $[-2, \infty)$ **47.** domain: $\{x|x \text{ is a real number}\}$ or $(-\infty, \infty)$; range: $\{y|y \leq -6\}$ or $(-\infty, -6]$ **49.** $f(x) = 2(x - 5)^2 + 3$ **51.** $f(x) = 2(x + 10)^2 - 5$ **53.** $f(x) = -3(x + 2)^2 + 4$ **55.** $f(x) = 3(x - 11)^2$ **57. a.** 2.75 gallons per person; underestimates by 0.05 gallon **b.** 1992; 2.048 gallons; seems reasonable
59. a. 2 sec; 224 ft **b.** 5.7 sec **c.** 160; 160 feet is the height of the building. **d.**

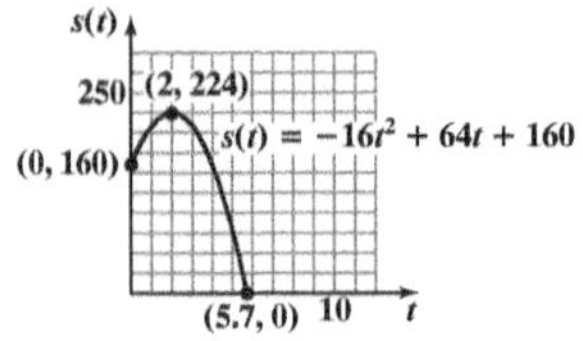

61. 8 and 8; 64 **63.** 8, −8; −64 **65.** length: 300 ft; width: 150 ft; maximum area: 45,000 sq ft **67.** 12.5 yd by 12.5 yd; 156.25 sq yd **69.** 5 in.; 50 sq in. **71. a.** $C(x) = 525 + 0.55x$ **b.** $P(x) = -0.001x^2 + 2.45x - 525$ **c.** 1225 sandwiches; $975.63

79. a.

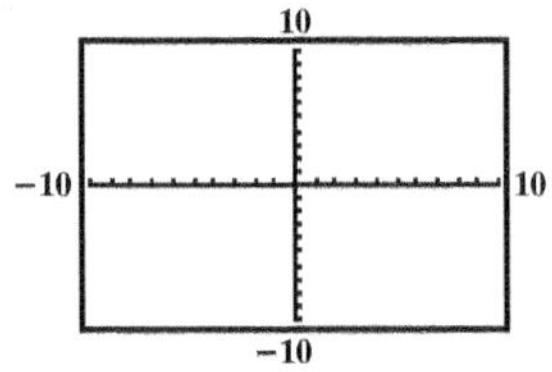

b. $(20.5, -120.5)$ **c.**

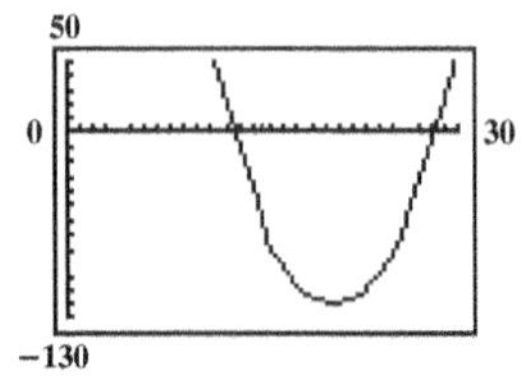

d. Answers will vary.

81. (2.5, 185)

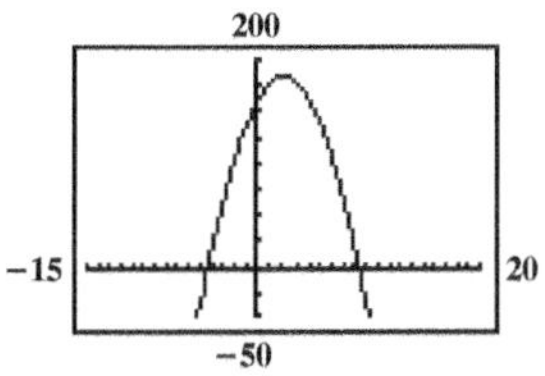

83. (−30, 91)

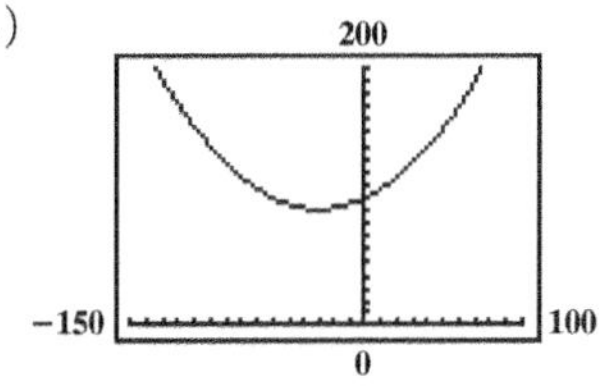

85. makes sense **87.** makes sense **89.** true **91.** false **93.** $x = -2; (-3, -2)$ **95.** $f(x) = \frac{1}{2}(x + 3)^2 - 4$
97. 125 ft by about 166.7 ft; about 20,833 sq ft **99.** $\{7\}$ **100.** $\frac{x}{x-2}$ **101.** $\{(6, -2)\}$
102. $\{-1, 9\}$ **103.** $\left\{-2, \frac{5}{2}\right\}$ **104.** $5u^2 + 11u + 2 = 0$

Mid-Chapter Check Point Exercises

1. $\left\{-\frac{1}{3}, \frac{11}{3}\right\}$ **2.** $\left\{-1, \frac{7}{5}\right\}$ **3.** $\left\{\frac{3 \pm \sqrt{15}}{3}\right\}$ **4.** $\{-3 \pm \sqrt{7}\}$ **5.** $\left\{\pm\frac{6\sqrt{5}}{5}\right\}$ **6.** $\left\{\frac{5}{2} \pm i\frac{\sqrt{7}}{2}\right\}$ **7.** $\{\pm i\sqrt{13}\}$ **8.** $\left\{-4, \frac{1}{2}\right\}$
9. $\{-3 \pm 2\sqrt{6}\}$ **10.** $\{2 \pm \sqrt{3}\}$ **11.** $\left\{\frac{3}{4} \pm i\frac{\sqrt{23}}{4}\right\}$ **12.** $\left\{\frac{-3 \pm \sqrt{41}}{4}\right\}$ **13.** $\{4\}$ **14.** $\{-5 \pm 2\sqrt{7}\}$

15.

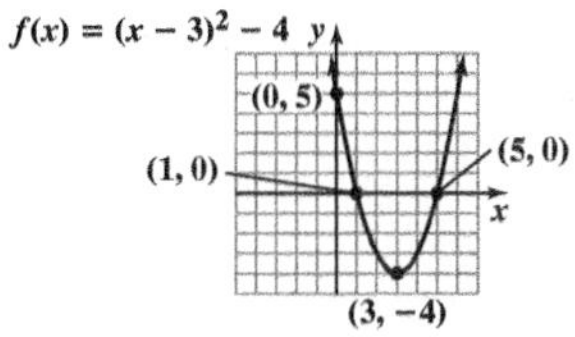

domain: $\{x|x \text{ is a real number}\}$ or $(-\infty, \infty)$; range: $\{y|y \geq -4\}$ or $[-4, \infty)$

16.

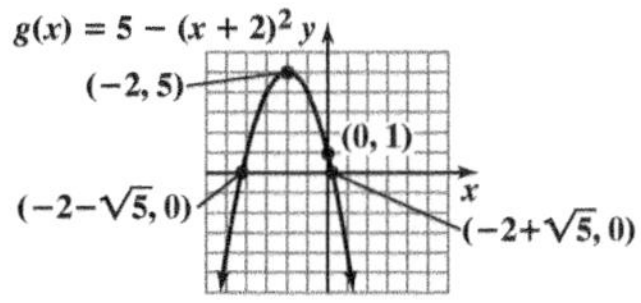

domain: $\{x|x \text{ is a real number}\}$ or $(-\infty, \infty)$; range: $\{y|y \leq 5\}$ or $(-\infty, 5]$

17.

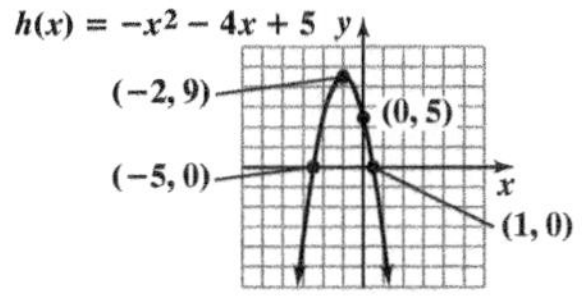

domain: $\{x|x \text{ is a real number}\}$ or $(-\infty, \infty)$; range: $\{y|y \leq 9\}$ or $(-\infty, 9]$

18.

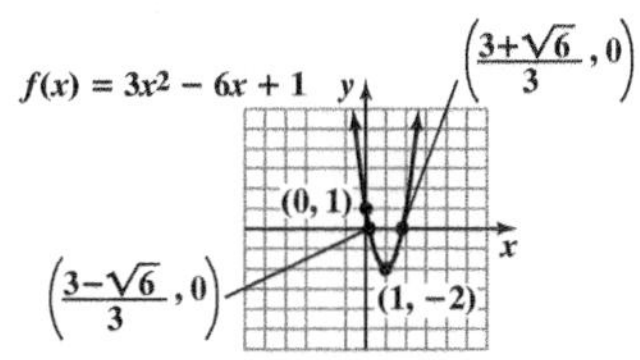

domain: $\{x|x \text{ is a real number}\}$ or $(-\infty, \infty)$; range: $\{y|y \geq -2\}$ or $[-2, \infty)$

19. two imaginary solutions **20.** two real rational solutions **21.** $8x^2 - 2x - 3 = 0$ **22.** $x^2 - 12 = 0$ **23.** 75 cabinets per day; \$1200
24. −9, −9; 81 **25.** 10 in.; 100 sq in.

Section 8.4

Check Point Exercises

1. $-\sqrt{3}, -\sqrt{2}, \sqrt{2}$, and $\sqrt{3}$ or $\{\pm\sqrt{2}, \pm\sqrt{3}\}$ **2.** 16 or $\{16\}$ **3.** $-\sqrt{6}, -1, 1$, and $\sqrt{6}$, or $\{-\sqrt{6}, -1, 1, \sqrt{6}\}$
4. −1 and 2, or $\{-1, 2\}$ **5.** $-\frac{1}{27}$ and 64, or $\left\{-\frac{1}{27}, 64\right\}$

Exercise Set 8.4

1. $\{-2, 2, -1, 1\}$ **3.** $\{-3, 3, -\sqrt{2}, \sqrt{2}\}$ **5.** $\{-2i, 2i, -\sqrt{2}, \sqrt{2}\}$ **7.** $\{1\}$ **9.** $\{49\}$ **11.** $\{25, 64\}$ **13.** $\{2, 12\}$ **15.** $\{-\sqrt{3}, 0, \sqrt{3}\}$
17. $\{-5, -2, -1, 2\}$ **19.** $\left\{-\frac{1}{4}, \frac{1}{5}\right\}$ **21.** $\left\{\frac{1}{3}, 2\right\}$ **23.** $\left\{\frac{-2 \pm \sqrt{7}}{3}\right\}$ **25.** $\{-8, 27\}$ **27.** $\{-243, 32\}$ **29.** $\{1\}$ **31.** $\{-8, -2, 1, 4\}$
33. −2, 2, −1, and 1; c **35.** 1; e **37.** 2 and 3; f **39.** −5, −4, 1, and 2 **41.** $-\frac{3}{2}$ and $-\frac{1}{3}$ **43.** $\frac{64}{15}$ and $\frac{81}{20}$ **45.** $\frac{5}{2}$ and $\frac{25}{6}$
47. ages 20 and 55; The function models the data well. **53.** $\{3, 5\}$ **55.** $\{1\}$ **57.** $\{-1, 4\}$ **59.** $\{1, 8\}$ **61.** makes sense
63. does not make sense **65.** true **67.** false **69.** $\left\{\sqrt[3]{-2}, \frac{\sqrt[3]{225}}{5}\right\}$ **71.** $\frac{1}{5x-1}$ **72.** $\frac{1}{2} + \frac{3}{2}i$ **73.** $\{(4, -2)\}$

Review Exercises

1. $\{\pm 8\}$ **2.** $\{\pm 5\sqrt{2}\}$ **3.** $\left\{\pm\frac{\sqrt{6}}{3}\right\}$ **4.** $\{4 \pm 3\sqrt{2}\}$ **5.** $\{-7 \pm 6i\}$ **6.** 100; $x^2 + 20x + 100 = (x + 10)^2$ **7.** $\frac{9}{4}$; $x^2 - 3x + \frac{9}{4} = \left(x - \frac{3}{2}\right)^2$

8. $\{3, 9\}$ **9.** $\left\{\frac{7 \pm \sqrt{53}}{2}\right\}$ **10.** $\left\{\frac{-3 \pm \sqrt{41}}{4}\right\}$ **11.** 8% **12.** 14 weeks **13.** $60\sqrt{5}$ m; 134.2 m **14.** $\{1 \pm \sqrt{5}\}$ **15.** $\{1 \pm 3i\sqrt{2}\}$

16. $\left\{\frac{-2 \pm \sqrt{10}}{2}\right\}$ **17.** two imaginary solutions **18.** two real rational solutions **19.** two real irrational solutions **20.** $\left\{-\frac{2}{3}, 4\right\}$

21. $\{-5, 2\}$ **22.** $\left\{\frac{1 \pm \sqrt{21}}{10}\right\}$ **23.** $\{-4, 4\}$ **24.** $\{3 \pm 2\sqrt{2}\}$ **25.** $\left\{\frac{1}{6} \pm i\frac{\sqrt{23}}{6}\right\}$ **26.** $\{4 \pm \sqrt{5}\}$ **27.** $15x^2 - 4x - 3 = 0$

28. $x^2 + 81 = 0$ **29.** $x^2 - 48 = 0$ **30. a.** 261 ft; overestimates by 1 ft **b.** 40 mph **31. a.** by the point (35, 261) **b.** by the point (40, 267) **32.** 8.8 sec

33.

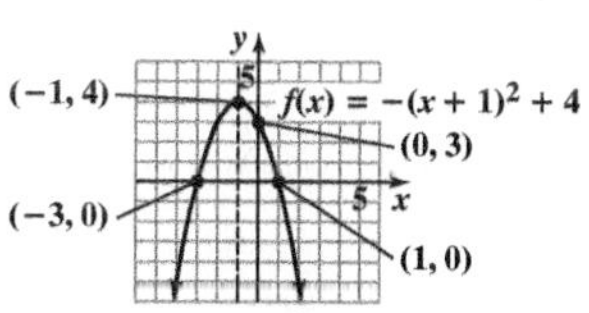

34.

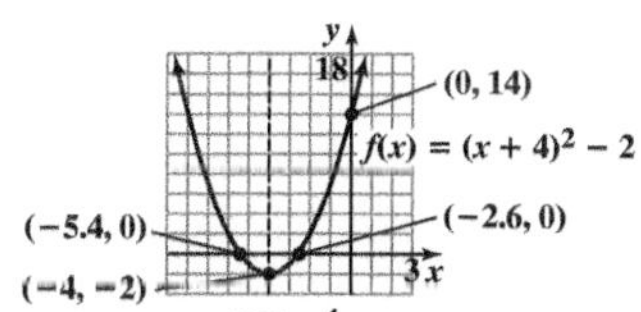

35.

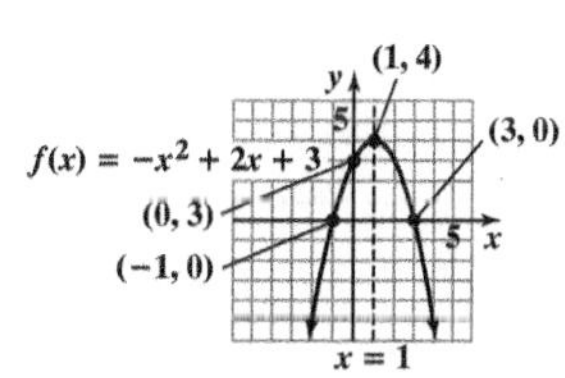

36.

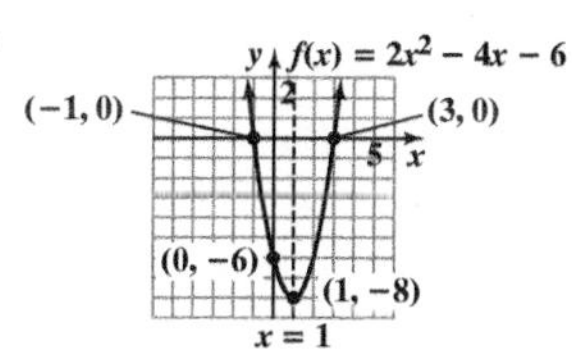

37. 25 in. of rainfall per year; 13.5 in. of growth **38.** 12.5 sec; 2540 feet **39.** 7.2 h; 622 per 100,000 males

40. 250 yd by 500 yd; 125,000 sq yard **41.** −7 and 7; −49 **42.** $\{-\sqrt{2}, \sqrt{2}, -2, 2\}$ **43.** $\{1\}$ **44.** $\{-5, -1, 3\}$ **45.** $\left\{-\frac{1}{8}, \frac{1}{7}\right\}$

46. $\{-27, 64\}$ **47.** $\{16\}$

48. $\left\{x \middle| -3 < x < \frac{1}{2}\right\}$ or $\left(-3, \frac{1}{2}\right)$;

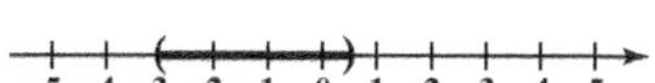

49. $\left\{x \middle| x \le -4 \text{ or } x \ge -\frac{1}{2}\right\}$ or $(-\infty, -4] \cup \left[-\frac{1}{2}, \infty\right)$;

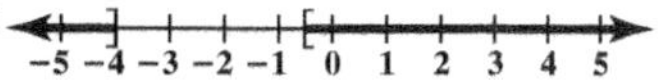

50. $\{x|-3 < x < 0 \text{ or } x > 1\}$ or $(-3, 0) \cup (1, \infty)$; 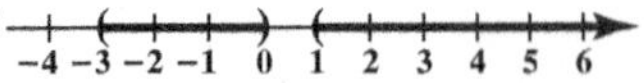

51. $\{x|x < -2 \text{ or } x > 6\}$ or $(-\infty, -2) \cup (6, \infty)$;

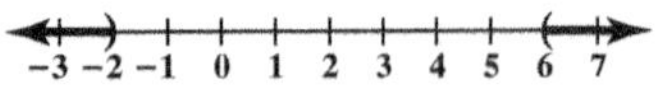

52. $\left\{x \middle| x < 4 \text{ or } x \geq \frac{23}{4}\right\}$ or $(-\infty, 4) \cup \left[\frac{23}{4}, \infty\right)$;

53. between 1 and 2 seconds, excluding $t = 1$ and $t = 2$ **54. a.** 200 beats per minute **b.** between 0 and 4 minutes and more than 12 minutes after the workout; between 0 and 4 minutes; Answers will vary.

Chapter Test

1. $\left\{\pm\frac{\sqrt{10}}{2}\right\}$ **2.** $\{3 \pm 2\sqrt{5}\}$ **3.** $64;\ x^2 - 16x + 64 = (x - 8)^2$ **4.** $\frac{1}{25};\ x^2 + \frac{2}{5}x + \frac{1}{25} = \left(x + \frac{1}{5}\right)^2$ **5.** $\{3 \pm \sqrt{2}\}$ **6.** $50\sqrt{2}$ ft

7. two real irrational solutions **8.** two imaginary solutions **9.** $\left\{-5, \frac{1}{2}\right\}$ **10.** $\{-4 \pm \sqrt{11}\}$ **11.** $\{-2 \pm 5i\}$ **12.** $\left\{\frac{3}{2} \pm \frac{1}{2}i\right\}$

13. $x^2 - 4x - 21 = 0$ **14.** $x^2 + 100 = 0$ **15. a.** 70.2%; overestimates by 0.2% **b.** 2010

16.

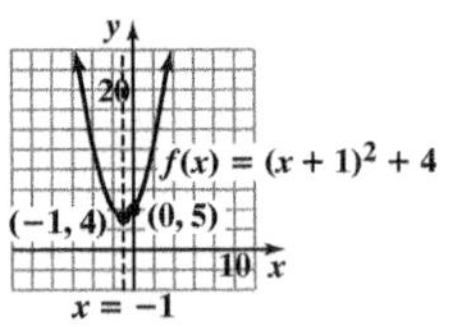

17.

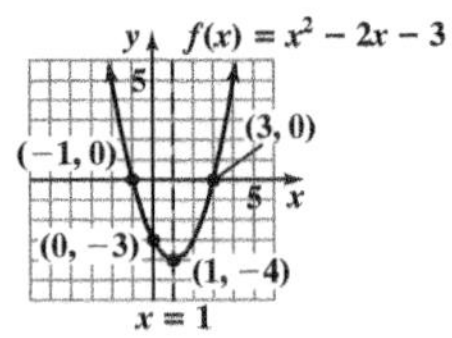

18. after 2 sec; 69 ft **19.** 4.1 sec **20.** 23 computers; \$169 hundreds or \$16,900
21. $\{1, 2\}$ **22.** $\{-3, 3, -2, 2\}$ **23.** $\{1, 512\}$
24. $\{x|-3 < x < 4\}$ or $(-3, 4)$; 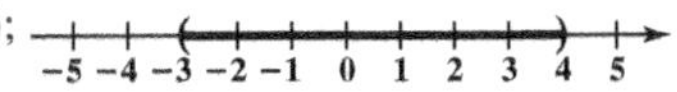

25. $\{x|x < 3 \text{ or } x \geq 10\}$ or $(-\infty, 3) \cup [10, \infty)$; 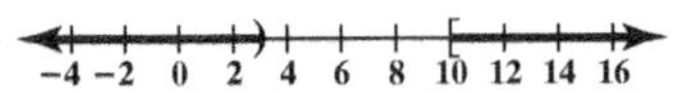

Cumulative Review Exercises

1. $\left\{\frac{2}{3}\right\}$ **2.** $\{(-5, 2)\}$ **3.** $\{(1, 4, -2)\}$ **4.** $\{x|x \geq 10\}$ or $[10, \infty)$ **5.** $\{x|x \leq -4\}$ or $(-\infty, -4]$ **6.** $\{x|x > 2\}$ or $(2, \infty)$

7. $\{x|-2 < x < 3\}$ or $(-2, 3)$ **8.** $\{3, 9\}$ **9.** no solution or $\varnothing$ **10.** $\{12\}$ **11.** $\left\{\frac{-2 \pm \sqrt{14}}{2}\right\}$ **12.** $\{8, 27\}$

13. $\left\{x \middle| -2 \leq x \leq \frac{3}{2}\right\}$ or $\left[-2, \frac{3}{2}\right]$

14.

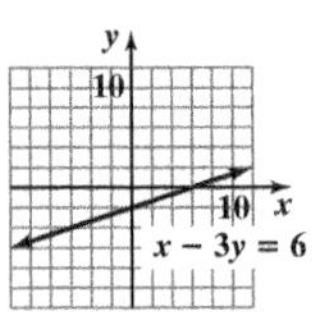

15.

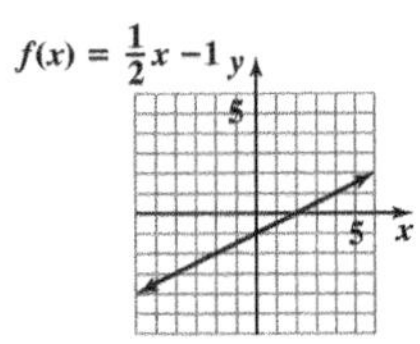

16.

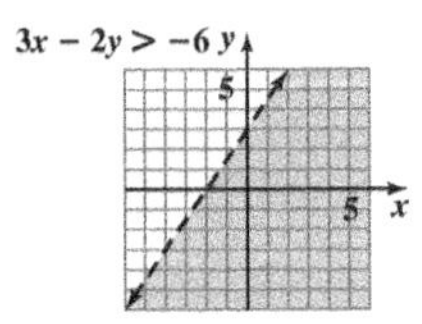

17.

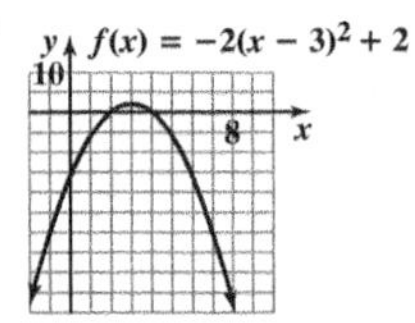

18. $-16x + 24y$ **19.** $-\frac{20x^7}{y^4}$ **20.** $x^2 - 5xy - 16y^2$ **21.** $6x^2 + 13x - 5$ **22.** $9x^4 - 24x^2y + 16y^2$ **23.** $\frac{3x^2 + 6x - 2}{(x + 5)(x + 2)}$

24. $\frac{x - 3}{x}$ **25.** $\frac{x - 4}{3x + 6}$ **26.** $5xy\sqrt{2x}$ **27.** $9\sqrt{2}$ **28.** $44 + 6i$ **29.** $(9x^2 + 1)(3x + 1)(3x - 1)$ **30.** $2x(4x - 1)(3x - 2)$

31. $(x + 3y)(x^2 - 3xy + 9y^2)$ **32.** $x^2 + 2x - 13;\ 22$ **33.** $x + 5;\ \{x|x \text{ is a real number and } x \neq 2\}$ or $(-\infty, 2) \cup (2, \infty)$ **34.** $2a + h + 3$

35. $3x^2 - 7x + 18 - \frac{28}{x + 2}$ **36.** $R = -\frac{Ir}{I - 1}$ or $R = \frac{Ir}{1 - I}$ **37.** $y = -3x - 1$ or $f(x) = -3x - 1$ **38.** 6 **39.** \$620 **40.** 13 yd by 4 yd

41. \$2600 at 12% and \$1400 at 14% **42.** 11 amps

APPLICATIONS INDEX

A

B

R

U

V

W

Y

SUBJECT INDEX

E

M

N

S

T

U

V

Appendix:
From *Introductory Algebra for College Students* by Robert Blitzer

SECTION 10.5

Graphs of Quadratic Equations

Objectives

1. Understand the characteristics of graphs of quadratic equations.
2. Find a parabola's intercepts.
3. Find a parabola's vertex.
4. Graph quadratic equations.
5. Solve problems using a parabola's vertex.

Many sports involve objects that are thrown, kicked, or hit, and then proceed with no additional force of their own. Such objects are called **projectiles**. Paths of projectiles, as well as their heights over time, can be modeled by quadratic equations. In this section, you will learn to use graphs of quadratic equations to gain a visual understanding of the algebra that describes football, baseball, basketball, the shot put, and other projectile sports.

1 Understand the characteristics of graphs of quadratic equations.

Graphs of Quadratic Equations

The graph of any quadratic equation

$$y = ax^2 + bx + c, \quad a \neq 0$$

is called a **parabola**. Parabolas are shaped like bowls or inverted bowls, as shown in **Figure 10.9**. If the coefficient of x^2 (the value of a in $ax^2 + bx + c$) is positive, the parabola opens upward. If the coefficient of x^2 is negative, the parabola opens downward. The **vertex** (or turning point) of the parabola is the lowest point, or minimum point, on the graph when it opens upward and the highest point, or maximum point, on the graph when it opens downward.

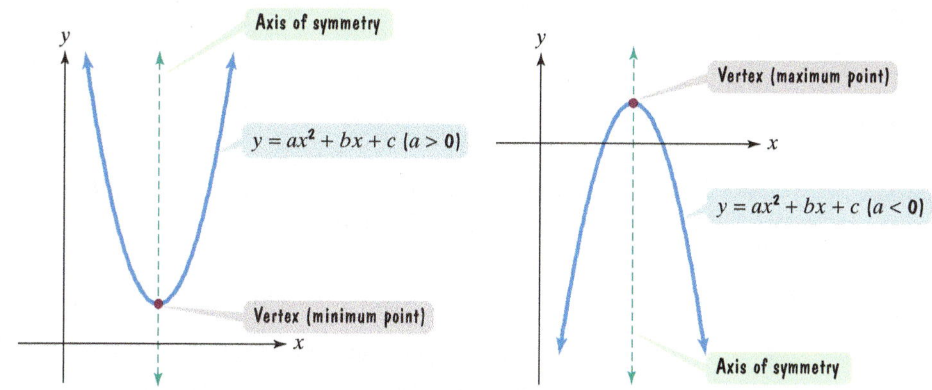

$a > 0$: Parabola opens upward.

$a < 0$: Parabola opens downward.

FIGURE 10.9 Characteristics of graphs of quadratic equations

Look at the unusual image of the word *mirror* shown on the right. The artist, Scott Kim, has created the image so that the two halves of the whole are mirror images of each other. A parabola shares this kind of symmetry. A "mirror line" through the vertex, called the **axis of symmetry**, divides the figure in half. If a parabola is folded along its axis of symmetry, the two halves match exactly.

EXAMPLE 1 Using Point Plotting to Graph a Parabola

Consider the equation $y = x^2 + 4x + 3$.

a. Is the graph a parabola that opens upward or downward?

b. Use point plotting to graph the parabola. Select integers from -5 to 1, inclusive, for x.

Solution

a. To determine whether a parabola opens upward or downward, we begin by identifying a, the coefficient of x^2. The following voice balloons show the values for a, b, and c in $y = x^2 + 4x + 3$. Notice that we wrote x^2 as $1x^2$.

$$y = 1x^2 + 4x + 3$$

a, the coefficient of x^2, is 1.

b, the coefficient of x, is 4.

c, the constant term, is 3.

When a is greater than 0, a parabola opens upward. When a is less than 0, a parabola opens downward. Because $a = 1$, which is greater than 0, the parabola opens upward.

b. To use point plotting to graph the parabola, we first make a table of x- and y-coordinates.

x	$y = x^2 + 4x + 3$	(x, y)
-5	$y = (-5)^2 + 4(-5) + 3 = 8$	$(-5, 8)$
-4	$y = (-4)^2 + 4(-4) + 3 = 3$	$(-4, 3)$
-3	$y = (-3)^2 + 4(-3) + 3 = 0$	$(-3, 0)$
-2	$y = (-2)^2 + 4(-2) + 3 = -1$	$(-2, -1)$
-1	$y = (-1)^2 + 4(-1) + 3 = 0$	$(-1, 0)$
0	$y = 0^2 + 4(0) + 3 = 3$	$(0, 3)$
1	$y = 1^2 + 4(1) + 3 = 8$	$(1, 8)$

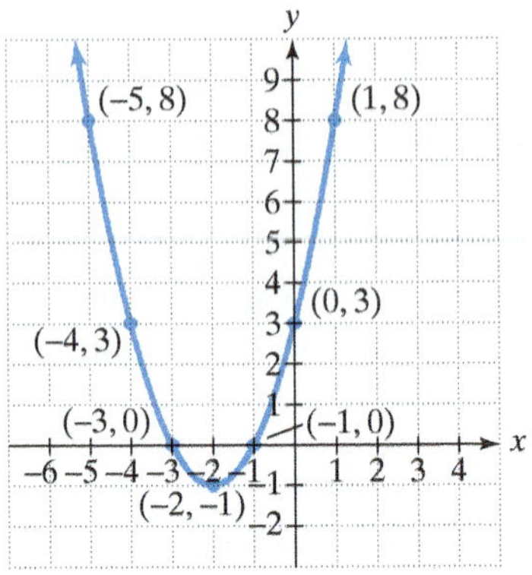

FIGURE 10.10
The graph of $y = x^2 + 4x + 3$

Then we plot the points and connect them with a smooth curve. The graph of $y = x^2 + 4x + 3$ is shown in **Figure 10.10**.

CHECK POINT 1 Consider the equation $y = x^2 - 6x + 8$.

a. Is the graph a parabola that opens upward or downward?

b. Use point plotting to graph the parabola. Select integers from 0 to 6, inclusive, for x.

Several points are important when graphing a quadratic equation. These points, labeled in **Figure 10.11**, are the x-intercepts (although not every parabola has two x-intercepts), the y-intercept, and the vertex. Let's see how we can locate these points.

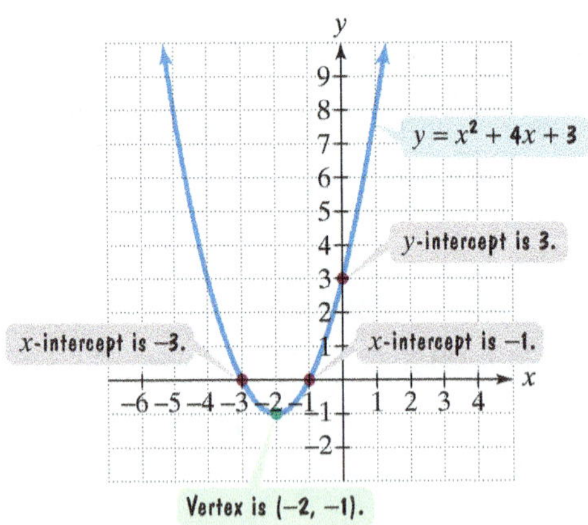

FIGURE 10.11 Useful points in graphing a parabola

2 Find a parabola's intercepts.

Finding a Parabola's x-Intercepts

At each point where a parabola crosses the x-axis, the value of y equals 0. Thus, the x-intercepts can be found by replacing y with 0 in $y = ax^2 + bx + c$. Use factoring or the quadratic formula to solve the resulting quadratic equation for x.

EXAMPLE 2 Finding a Parabola's x-Intercepts

Find the x-intercepts for the parabola whose equation is $y = x^2 + 4x + 3$.

Solution Replace y with 0 in $y = x^2 + 4x + 3$. We obtain $0 = x^2 + 4x + 3$, or $x^2 + 4x + 3 = 0$. We can solve this equation by factoring.

$$x^2 + 4x + 3 = 0$$
$$(x + 3)(x + 1) = 0$$
$$x + 3 = 0 \quad \text{or} \quad x + 1 = 0$$
$$x = -3 \qquad\qquad x = -1$$

Thus, the x-intercepts are -3 and -1. The parabola passes through $(-3, 0)$ and $(-1, 0)$, as shown in **Figure 10.11**.

☑ **CHECK POINT 2** Find the x-intercepts for the parabola whose equation is $y = x^2 - 6x + 8$.

Finding a Parabola's y-Intercept

At the point where a parabola crosses the y-axis, the value of x equals 0. Thus, the y-intercept can be found by replacing x with 0 in $y = ax^2 + bx + c$. Simple arithmetic will produce a value for y, which is the y-intercept.

EXAMPLE 3 Finding a Parabola's y-Intercept

Find the y-intercept for the parabola whose equation is $y = x^2 + 4x + 3$.

Solution Replace x with 0 in $y = x^2 + 4x + 3$.

$$y = 0^2 + 4 \cdot 0 + 3 = 0 + 0 + 3 = 3$$

The y-intercept is 3. The parabola passes through $(0, 3)$, as shown in **Figure 10.11**.

☑ **CHECK POINT 3** Find the y-intercept for the parabola whose equation is $y = x^2 - 6x + 8$.

3 Find a parabola's vertex.

Finding a Parabola's Vertex

Keep in mind that a parabola's vertex is its turning point. The x-coordinate of the vertex for the parabola in **Figure 10.11**, -2, is midway between the x-intercepts, -3 and -1. If a parabola has two x-intercepts, they are found by solving $ax^2 + bx + c = 0$. The solutions of this equation,

$$x = \frac{-b - \sqrt{b^2 - 4ac}}{2a} \quad \text{and} \quad x = \frac{-b + \sqrt{b^2 - 4ac}}{2a},$$

are the x-intercepts. The value of x midway between these intercepts is $x = \frac{-b}{2a}$. This equation can be used to find the x-coordinate of the vertex even when no x-intercepts exist.

The Vertex of a Parabola

For a parabola whose equation is $y = ax^2 + bx + c$,

1. The x-coordinate of the vertex is $\frac{-b}{2a}$.
2. The y-coordinate of the vertex is found by substituting the x-coordinate into the parabola's equation and evaluating.

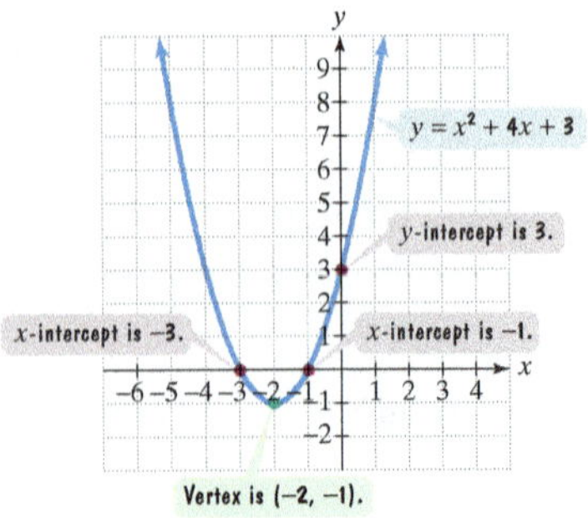

FIGURE 10.11 (repeated)

EXAMPLE 4 Finding a Parabola's Vertex

Find the vertex for the parabola whose equation is $y = x^2 + 4x + 3$.

Solution In the equation $y = x^2 + 4x + 3$, $a = 1$ and $b = 4$.

$$x\text{-coordinate of vertex} = \frac{-b}{2a} = \frac{-4}{2 \cdot 1} = \frac{-4}{2} = -2$$

To find the y-coordinate of the vertex, we substitute -2 for x in $y = x^2 + 4x + 3$ and then evaluate.

$$y\text{-coordinate of vertex} = (-2)^2 + 4(-2) + 3 = 4 + (-8) + 3 = -1$$

The vertex is $(-2, -1)$, as shown in **Figure 10.11**.

CHECK POINT 4 Find the vertex for the parabola whose equation is $y = x^2 - 6x + 8$.

---

4 Graph quadratic equations.

A Strategy for Graphing Quadratic Equations

Here is a procedure to sketch the graph of the quadratic equation, $y = ax^2 + bx + c$:

Graphing Quadratic Equations

The graph of $y = ax^2 + bx + c$, called a parabola, can be graphed using the following steps:

1. Determine whether the parabola opens upward or downward. If $a > 0$, it opens upward. If $a < 0$, it opens downward.
2. Determine the vertex of the parabola. The x-coordinate is $\frac{-b}{2a}$.
   The y-coordinate is found by substituting the x-coordinate into the parabola's equation and evaluating.
3. Find any x-intercepts by replacing y with 0. Solve the resulting quadratic equation for x. The real solutions are the x-intercepts.
4. Find the y-intercept by replacing x with 0. Because $y = a \cdot 0^2 + b \cdot 0 + c$ simplifies to $y = c$, the y-intercept is c, the constant term, and the parabola passes through $(0, c)$.
5. Plot the intercepts and the vertex.
6. Connect these points with a smooth curve.

EXAMPLE 5 Graphing a Parabola

Graph the quadratic equation: $y = x^2 - 2x - 3$.

Solution We can graph this equation by following the steps in the box.

Step 1. Determine how the parabola opens. Note that a, the coefficient of x^2, is 1. Thus, $a > 0$; this positive value tells us that the parabola opens upward.

Step 2. Find the vertex. We know that the x-coordinate of the vertex is $\frac{-b}{2a}$. Let's identify the numbers a, b, and c in the given equation, which is in the form $y = ax^2 + bx + c$.

$$y = x^2 - 2x - 3$$

$a = 1$ $b = -2$ $c = -3$

Now we substitute the values of a and b, $a = 1$ and $b = -2$, into the expression for the x-coordinate:

$$x\text{-coordinate of vertex} = \frac{-b}{2a} = \frac{-(-2)}{2(1)} = \frac{2}{2} = 1.$$

The x-coordinate of the vertex is 1. We can substitute 1 for x in the equation $y = x^2 - 2x - 3$ to find the y-coordinate:

$$y\text{-coordinate of vertex} = 1^2 - 2\cdot 1 - 3 = 1 - 2 - 3 = -4.$$

The vertex is $(1, -4)$.

Step 3. Find the x-intercepts. Replace y with 0 in $y = x^2 - 2x - 3$. We obtain $0 = x^2 - 2x - 3$ or $x^2 - 2x - 3 = 0$. We can solve this equation by factoring.

$$x^2 - 2x - 3 = 0$$
$$(x - 3)(x + 1) = 0$$
$$x - 3 = 0 \quad \text{or} \quad x + 1 = 0$$
$$x = 3 \qquad\qquad x = -1$$

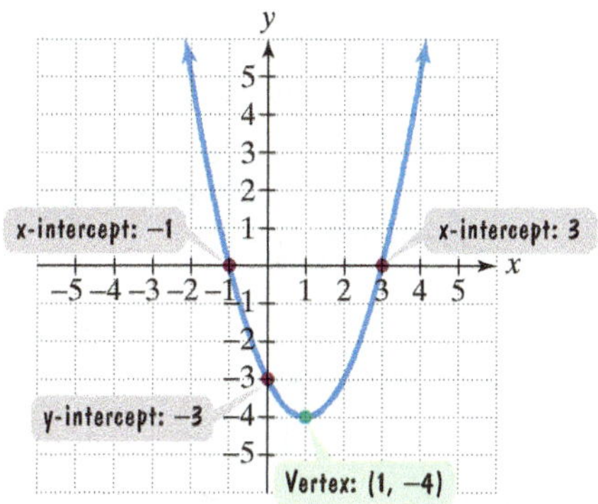

FIGURE 10.12
The graph of $y = x^2 - 2x - 3$

The x-intercepts are 3 and -1. The parabola passes through $(3, 0)$ and $(-1, 0)$.

Step 4. Find the y-intercept. Replace x with 0 in $y = x^2 - 2x - 3$:

$$y = 0^2 - 2\cdot 0 - 3 = 0 - 0 - 3 = -3.$$

The y-intercept is -3. The parabola passes through $(0, -3)$.

Steps 5 and 6. Plot the intercepts and the vertex. Connect these points with a smooth curve. The intercepts and the vertex are shown as the four labeled points in **Figure 10.12**. Also shown is the graph of the quadratic equation, obtained by connecting the points with a smooth curve.

✓ **CHECK POINT 5** Graph the quadratic equation: $y = x^2 + 6x + 5$.

EXAMPLE 6 Graphing a Parabola

Graph the quadratic equation: $y = -x^2 + 4x - 1$.

Solution

Step 1. Determine how the parabola opens. Note that a, the coefficient of x^2, is -1. Thus, $a < 0$; this negative value tells us that the parabola opens downward.

Step 2. Find the vertex. The x-coordinate of the vertex is $\frac{-b}{2a}$.

$$y = -x^2 + 4x - 1$$

$a = -1$ $\quad b = 4$ $\quad c = -1$

$$x\text{-coordinate of vertex} = \frac{-b}{2a} = \frac{-4}{2(-1)} = \frac{-4}{-2} = 2$$

Substitute 2 for x in $y = -x^2 + 4x - 1$ to find the y-coordinate:

$$y\text{-coordinate of vertex} = -2^2 + 4\cdot 2 - 1 = -4 + 8 - 1 = 3.$$

The vertex is $(2, 3)$.

Step 3. Find the x-intercepts. Replace y with 0 in $y = -x^2 + 4x - 1$. We obtain $0 = -x^2 + 4x - 1$ or $-x^2 + 4x - 1 = 0$. This equation cannot be solved by factoring. We will use the quadratic formula to solve it.

$$a = -1, \quad b = 4, \quad c = -1$$

$$x = \frac{-b \pm \sqrt{b^2 - 4ac}}{2a} = \frac{-4 \pm \sqrt{4^2 - 4(-1)(-1)}}{2(-1)} = \frac{-4 \pm \sqrt{16 - 4}}{-2}$$

$$x = \frac{-4 + \sqrt{12}}{-2} \approx 0.3 \quad \text{or} \quad x = \frac{-4 - \sqrt{12}}{-2} \approx 3.7$$

The x-intercepts are approximately 0.3 and 3.7. The parabola passes through (0.3, 0) and (3.7, 0).

Step 4. Find the y-intercept. Replace x with 0 in $y = -x^2 + 4x - 1$:

$$y = -0^2 + 4 \cdot 0 - 1 = -1.$$

The y-intercept is -1. The parabola passes through $(0, -1)$.

Steps 5 and 6. Plot the intercepts and the vertex. Connect these points with a smooth curve. The intercepts and the vertex are shown as the four labeled points in **Figure 10.13**. Also shown is the graph of the quadratic equation, obtained by connecting the points with a smooth curve.

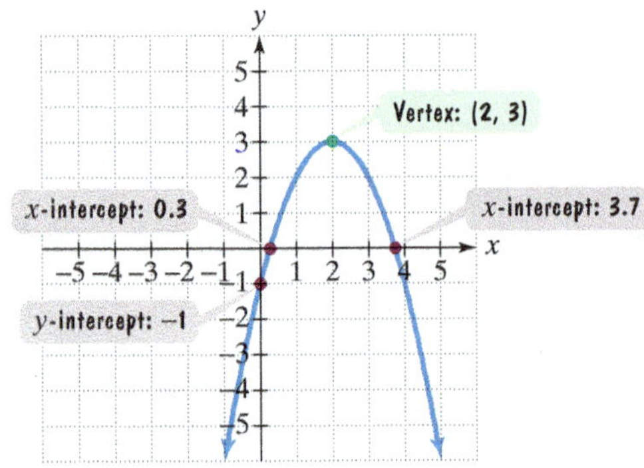

FIGURE 10.13
The graph of $y = -x^2 + 4x - 1$

☑ CHECK POINT 6 Graph the quadratic equation: $y = -x^2 - 2x + 5$.

5 Solve problems using a parabola's vertex.

Applications

Many applied problems involve finding the maximum or minimum value of equations in the form $y = ax^2 + bx + c$. The vertex of the graph is the point of interest. If $a < 0$, the parabola opens downward and the vertex is its highest point. If $a > 0$, the parabola opens upward and the vertex is its lowest point.

EXAMPLE 7 The Parabolic Path of a Punted Football

Figure 10.14 at the top of the next page shows that when a football is kicked, the nearest defensive player is 6 feet from the point of impact with the kicker's foot. The height of the punted football, y, in feet, can be modeled by

$$y = -0.01x^2 + 1.18x + 2,$$

where x is the ball's horizontal distance, in feet, from the point of impact with the kicker's foot.

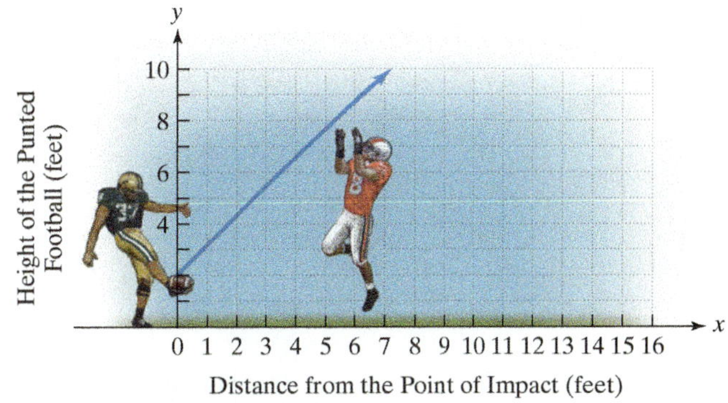

FIGURE 10.14

a. What is the maximum height of the punt and how far from the point of impact does this occur?

b. How far must the nearest defensive player, who is 6 feet from the kicker's point of impact, reach to block the punt?

c. If the ball is not blocked by the defensive player, how far down the field will it go before hitting the ground?

d. Graph the equation that models the football's parabolic path.

Solution

a. We begin by identifying the numbers a, b, and c in the given model.

$$y = -0.01x^2 + 1.18x + 2$$

$a = -0.01$ $b = 1.18$ $c = 2$

Because the coefficient of x^2, -0.01, is negative, the parabola opens downward and the vertex is the highest point on the graph. The y-coordinate of the vertex gives the maximum height of the punt and the x-coordinate reveals how far from the point of impact this occurs.

$$x\text{-coordinate of vertex} = \frac{-b}{2a} = \frac{-1.18}{2(-0.01)} = \frac{-1.18}{-0.02} = 59$$

Substitute 59 for x in $y = -0.01x^2 + 1.18x + 2$ to find the y-coordinate.

$$y\text{-coordinate of vertex} = -0.01(59)^2 + 1.18(59) + 2$$
$$= -34.81 + 69.62 + 2 = 36.81$$

The vertex is (59, 36.81). The maximum height of the punt is 36.81 feet and this occurs 59 feet from the kicker's point of impact.

b. **Figure 10.14** shows that the defensive player is 6 feet from the kicker's point of impact. To block the punt, he must touch the football along its parabolic path. This means that we must find the height of the ball 6 feet from the kicker. Replace x with 6 in the given model, $y = -0.01x^2 + 1.18x + 2$.

$$y = -0.01(6)^2 + 1.18(6) + 2 = -0.36 + 7.08 + 2 = 8.72$$

The defensive player must reach 8.72 feet above the ground to block the punt.

c. Assuming that the ball is not blocked by the defensive player, we are interested in how far down the field it will go before hitting the ground. We are looking for the ball's horizontal distance, x, when its height above the ground, y, is 0 feet. To find this x-intercept, replace y with 0 in $y = -0.01x^2 + 1.18x + 2$. We obtain $0 = -0.01x^2 + 1.18x + 2$, or $-0.01x^2 + 1.18x + 2 = 0$. The equation cannot be solved by factoring. We will use the quadratic formula to solve it.

$-0.01x^2 + 1.18x + 2 = 0$

$a = -0.01$ $b = 1.18$ $c = 2$

The equation for determining the ball's maximum horizontal distance (repeated)

Use a calculator to evaluate the radicand.

$$x = \frac{-b \pm \sqrt{b^2 - 4ac}}{2a} = \frac{-1.18 \pm \sqrt{(1.18)^2 - 4(-0.01)(2)}}{2(-0.01)} = \frac{-1.18 \pm \sqrt{1.4724}}{-0.02}$$

$$x = \frac{-1.18 + \sqrt{1.4724}}{-0.02} \quad \text{or} \quad x = \frac{-1.18 - \sqrt{1.4724}}{-0.02}$$

$$x \approx -1.7 \qquad\qquad x \approx 119.7$$

Use a calculator and round to the nearest tenth.

Reject this value. We are interested in the football's height corresponding to horizontal distances from its point of impact onward, or $x \geq 0$.

If the football is not blocked by the defensive player, it will go approximately 119.7 feet down the field before hitting the ground.

d. In terms of graphing the model for the football's parabolic path, $y = -0.01x^2 + 1.18x + 2$, we have already determined the vertex and the appropriate x-intercept.

vertex: $(59, 36.81)$ — The ball's maximum height, 36.81 feet, occurs at a horizontal distance of 59 feet.

x-intercept: 119.7 — The ball's maximum horizontal distance is approximately 119.7 feet.

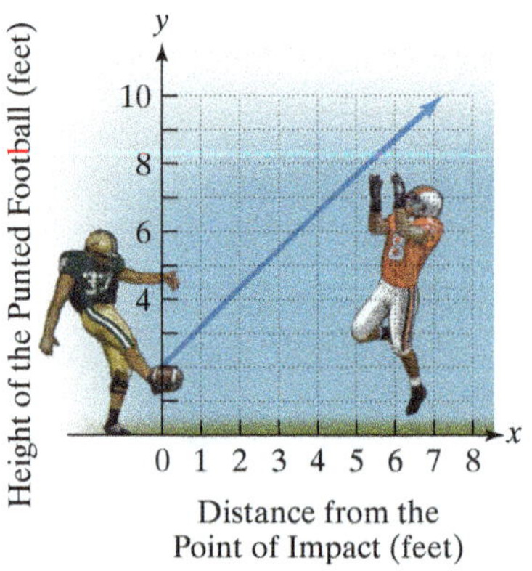

FIGURE 10.14 (repeated)

Figure 10.14 indicates that the y-intercept is 2, meaning that the ball is kicked from a height of 2 feet. Let's verify this value by replacing x with 0 in $y = -0.01x^2 + 1.18x + 2$.

$$y = -0.01 \cdot 0^2 + 1.18 \cdot 0 + 2 = 0 + 0 + 2 = 2$$

Using the vertex, $(59, 36.81)$, the x-intercept, 119.7, and the y-intercept, 2, the graph of the equation that models the football's parabolic path is shown in **Figure 10.15**. The graph is shown only for $x \geq 0$, indicating horizontal distances that begin at the football's impact with the kicker's foot and end with the ball hitting the ground.

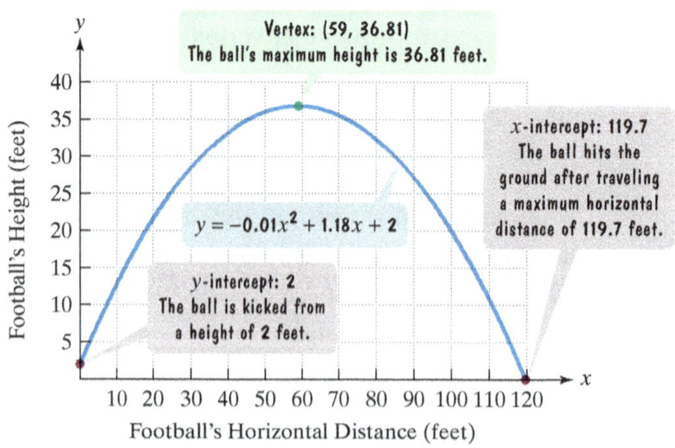

FIGURE 10.15 The parabolic path of a punted football

✓ **CHECK POINT 7** An archer's arrow follows a parabolic path. The height of the arrow, y, in feet, can be modeled by

$$y = -0.005x^2 + 2x + 5,$$

where x is the arrow's horizontal distance, in feet.

a. What is the maximum height of the arrow and how far from its release does this occur?

b. Find the horizontal distance the arrow travels before it hits the ground. Round to the nearest foot.

c. Graph the equation that models the arrow's parabolic path.

10.5 EXERCISE SET

MyMathLab — Math XL PRACTICE · WATCH · DOWNLOAD · READ · REVIEW

Practice Exercises

In Exercises 1–4 determine if the parabola whose equation is given opens upward or downward.

1. $y = x^2 - 4x + 3$

2. $y = x^2 - 6x + 5$

3. $y = -2x^2 + x + 6$

4. $y = -2x^2 - 4x + 6$

In Exercises 5–10, find the x-intercepts for the parabola whose equation is given. If the x-intercepts are irrational numbers, round your answers to the nearest tenth.

5. $y = x^2 - 4x + 3$

6. $y = x^2 - 6x + 5$

7. $y = -x^2 + 8x - 12$

8. $y = -x^2 - 2x + 3$

9. $y = x^2 + 2x - 4$

10. $y = x^2 + 8x + 14$

In Exercises 11–18, find the y-intercept for the parabola whose equation is given.

11. $y = x^2 - 4x + 3$ **12.** $y = x^2 - 6x + 5$

13. $y = -x^2 + 8x - 12$ **14.** $y = -x^2 - 2x + 3$

15. $y = x^2 + 2x - 4$ **16.** $y = x^2 + 8x + 14$

17. $y = x^2 + 6x$ **18.** $y = x^2 + 8x$

In Exercises 19–24, find the vertex for the parabola whose equation is given.

19. $y = x^2 - 4x + 3$ **20.** $y = x^2 - 6x + 5$

21. $y = 2x^2 + 4x - 6$

22. $y = -2x^2 - 4x - 2$

23. $y = x^2 + 6x$ **24.** $y = x^2 + 8x$

In Exercises 25–36, graph the parabola whose equation is given.

25. $y = x^2 + 8x + 7$ **26.** $y = x^2 + 10x + 9$

27. $y = x^2 - 2x - 8$ **28.** $y = x^2 + 4x - 5$

29. $y = -x^2 + 4x - 3$ **30.** $y = -x^2 + 2x + 3$

31. $y = x^2 - 1$ **32.** $y = x^2 - 4$

33. $y = x^2 + 2x + 1$ **34.** $y = x^2 - 2x + 1$

35. $y = -2x^2 + 4x + 5$ **36.** $y = -3x^2 + 6x - 2$

Practice PLUS

In Exercises 37–44, find the vertex for the parabola whose equation is given by first writing the equation in the form $y = ax^2 + bx + c$.

37. $y = (x - 3)^2 + 2$ **38.** $y = (x - 4)^2 + 3$

39. $y = (x + 5)^2 - 4$

40. $y = (x + 6)^2 - 5$

41. $y = 2(x - 1)^2 - 3$

42. $y = 2(x - 1)^2 - 4$

43. $y = -3(x + 2)^2 + 5$

44. $y = -3(x + 4)^2 + 6$

45. Generalize your work in Exercises 37–44 and complete the following statement: For a parabola whose equation is $y = a(x - h)^2 + k$, the vertex is the point _____.

Application Exercises

An athlete whose event is the shot put releases the shot with the same initial velocity, but at different angles. The figure shows the parabolic paths for shots released at angles of 35° and 65°. Exercises 46–47 are based on the equations that model the parabolic paths.

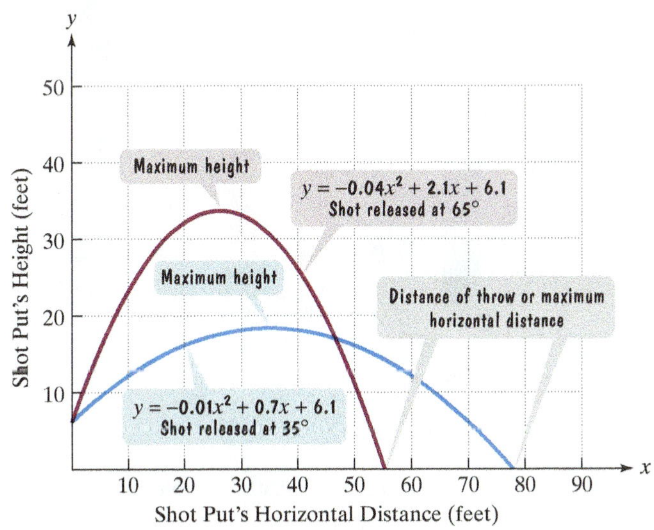

46. When the shot is released at an angle of 65°, its height, y, in feet, can be modeled by

$$y = -0.04x^2 + 2.1x + 6.1,$$

where x is the shot's horizontal distance, in feet, from its point of release. Use this model to solve parts (a) through (c) and verify your answers using the red graph.

a. What is the maximum height, to the nearest tenth of a foot, of the shot and how far from its point of release does this occur?

b. What is the shot's maximum horizontal distance, to the nearest tenth of a foot, or the distance of the throw?

c. From what height was the shot released?

47. When the shot whose path is shown by the blue graph on the previous page is released at an angle of 35°, its height, y, in feet, can be modeled by

$$y = -0.01x^2 + 0.7x + 6.1,$$

where x is the shot's horizontal distance, in feet, from its point of release. Use this model to solve parts (a) through (c) and verify your answers using the blue graph.

a. What is the maximum height of the shot and how far from its point of release does this occur?

b. What is the shot's maximum horizontal distance, to the nearest tenth of a foot, or the distance of the throw?

c. From what height was the shot released?

The Food Stamp Program is the first line of defense against hunger for millions of American families. The program provides benefits for eligible participants to purchase approved food items at food stores. Over half of all participants are children; one out of six is a low-income older adult. The bar graph shows the number of households, in millions, participating in the Food Stamp Program from 1999 through 2004.

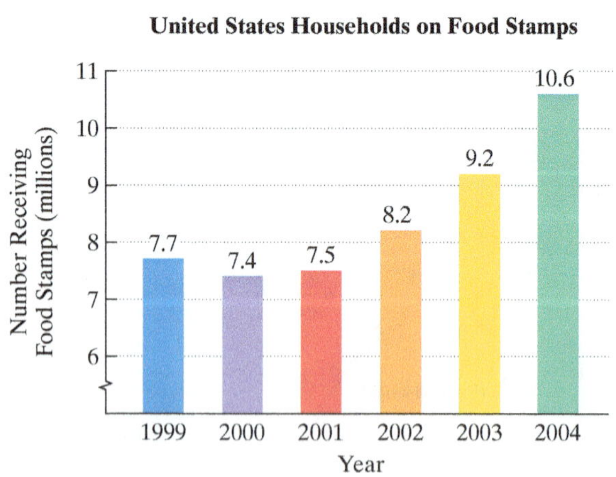

Source: Food Stamp Program

The formula

$$y = 0.22x^2 - 0.50x + 7.68$$

models the number of households, y, in millions, participating in the Food Stamp Program x years after 1999. Use this model to solve Exercises 48–49.

48. a. According to the model, how many households received food stamps in 2004? Does this underestimate or overestimate the actual number shown by the graph? By how much?

b. According to the graph, in which year from 1999 through 2004 was the number of households receiving food stamps at a minimum and how many households received food stamps for that year?

c. According to the model, in which year, to the nearest whole year, was the number of households receiving food stamps at a minimum and how many households received food stamps for that year? How does this compare with the actual data in part (b)?

49. a. According to the model, how many households received food stamps in 1999? Does this underestimate or overestimate the actual number shown by the graph? By how much?

b. According to the graph, in which year from 1999 through 2004 was the number of households receiving food stamps at a minimum and how many households received food stamps for that year?

c. According to the model, in which year, to the nearest whole year, was the number of households receiving food stamps at a minimum and how many households received food stamps for that year? How does this compare with the actual data in part (b)?

50. You have 120 feet of fencing to enclose a rectangular plot that borders on a river. If you do not fence the side along the river, find the length and width of the plot that will maximize the area. What is the largest area that can be enclosed?

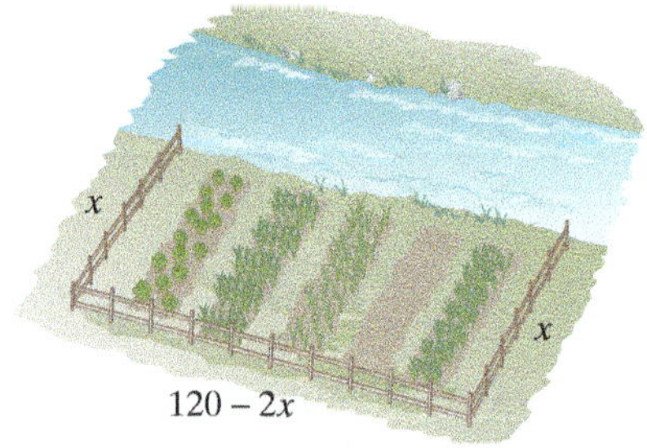

51. The figure shown indicates that you have 100 yards of fencing to enclose a rectangular area. Find the dimensions of the rectangle that maximize the enclosed area. What is the maximum area?

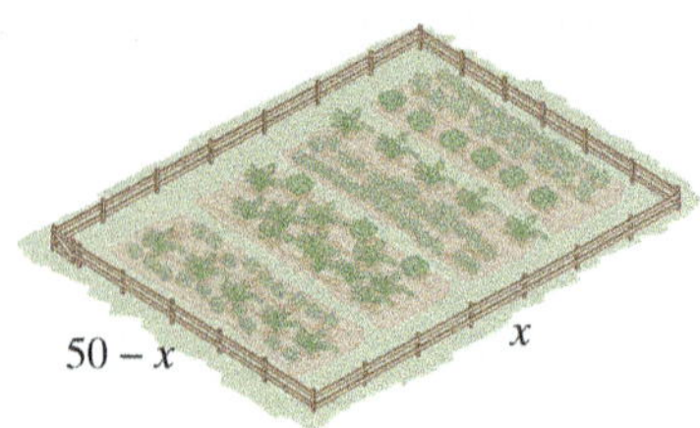

Writing in Mathematics

52. What is a parabola? Describe its shape.

53. Explain how to decide whether a parabola opens upward or downward.

54. If a parabola has two x-intercepts, explain how to find them.

55. Explain how to find a parabola's y-intercept.

56. Describe how to find a parabola's vertex.

57. A parabola that opens upward has its vertex at (1, 2). Describe as much as you can about the parabola based on this information. Include in your discussion the number of x-intercepts, if any, for the parabola.

Critical Thinking Exercises

Make Sense? *In Exercises 58–61, determine whether each statement "makes sense" or "does not make sense" and explain your reasoning.*

58. I must have made an error when graphing this parabola because it is symmetric with respect to the y-axis.

59. Parabolas that open up appear to form smiles ($a > 0$), while parabolas that open down frown ($a < 0$).

60. I threw a baseball vertically upward and its path was a parabola.

61. **Figure 10.14** on page 637 shows that a linear model provides a better description of the football's path than a quadratic model.

In Exercises 62–65, determine whether each statement is true or false. If the statement is false, make the necessary change(s) to produce a true statement.

62. The x-coordinate of the vertex of the parabola whose equation is $y = ax^2 + bx + c$ is $\frac{b}{2a}$.

63. If a parabola has only one x-intercept, then the x-intercept is also the vertex.

64. There is no relationship between the graph of $y = ax^2 + bx + c$ and the number of real solutions of the equation $ax^2 + bx + c = 0$.

65. If $y = 4x^2 - 40x + 4$, then the vertex is the highest point on the graph.

66. Find two numbers whose sum is 200 and whose product is a maximum.

67. Graph $y = 2x^2 - 8$ and $y = -2x^2 + 8$ in the same rectangular coordinate system. What are the coordinates of the points of intersection?

68. A parabola has x-intercepts at 3 and 7, a y-intercept at -21, and (5, 4) for its vertex. Write the parabola's equation.

Technology Exercises

69. Use a graphing utility to verify any five of your hand-drawn graphs in Exercises 25–36.

70. **a.** Use a graphing utility to graph $y = 2x^2 - 82x + 720$ in a standard viewing rectangle. What do you observe?

b. Find the coordinates of the vertex for the given quadratic equation.

c. The answer to part (b) is (20.5, −120.5). Because the leading coefficient, 2 of $y = 2x^2 - 82x + 720$ is positive, the vertex is a minimum point on the graph. Use this fact to help find a viewing rectangle that will give a relatively complete picture of the parabola. With an axis of symmetry at $x = 20.5$, the setting for x should extend past this, so try Xmin = 0 and Xmax = 30. The setting for y should include (and probably go below) the y-coordinate of the graph's minimum point, so try Ymin = −130. Experiment with Ymax until your utility shows the parabola's major features.

d. In general, explain how knowing the coordinates of a parabola's vertex can help determine a reasonable viewing rectangle on a graphing utility for obtaining a complete picture of the parabola.

In Exercises 71–74, find the vertex for each parabola. Then determine a reasonable viewing rectangle on your graphing utility and use it to graph the parabola.

71. $y = -0.25x^2 + 40x$

72. $y = -4x^2 + 20x + 160$

73. $y = 5x^2 + 40x + 600$

74. $y = 0.01x^2 + 0.6x + 100$

Review Exercises

In Exercises 75–77, solve each equation or system of equations.

75. $7(x - 2) = 10 - 2(x + 3)$ (Section 2.3, Example 3)

76. $\frac{7}{x + 2} + \frac{2}{x + 3} = \frac{1}{x^2 + 5x + 6}$ (Section 8.6, Example 4)

77. $5x - 3y = -13$

$x = 2 - 4y$ (Section 5.2, Example 1)

Preview Exercises

Exercises 78–80 will help you prepare for the material covered in the next section.

78. Here are two sets of ordered pairs:

set 1: $\{(1, 5), (2, 5)\}$

set 2: $\{(5, 1), (5, 2)\}$.

In which set is each x-coordinate paired with one and only one y-coordinate?

79. Here are two graphs:

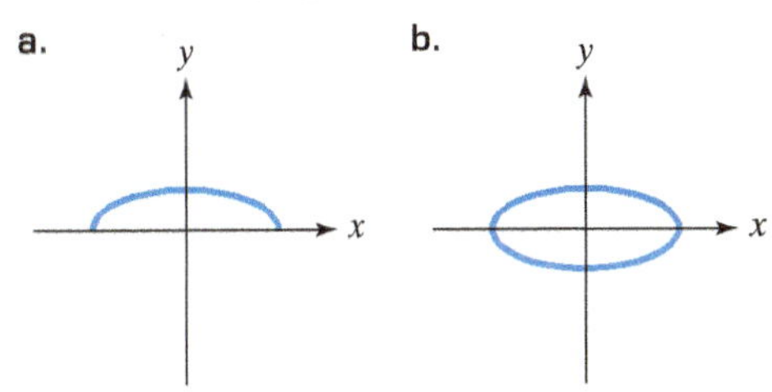

In which graph is each x-coordinate paired with one and only one y-coordinate?

80. Evaluate $x^2 + 3x + 5$ for $x = -3$.

SUPPLEMENT TO §1.8

RULES OF EXPONENTS

i.	$x^a \cdot x^b = x^{a+b}$	(when we multiply like bases raised to powers, we add the powers)
ii.	$(x^a)^b = x^{a \cdot b}$	(when we raise a base-with-a-power to a power, we multiply the powers together)
iii.	$(xy)^a = x^a y^a$	(we distribute a power over multiplication)

Note that one of the most common mistakes is to confuse rules ***i*** and ***ii***. To avoid this mistake, notice that if there are ***two*** factors with like bases, you should use rule ***i***.

EXAMPLE: Simplify the following expressions using the rules of exponents.

a. $-2t^3 \cdot 4t^5$ **b.** $5\left(v^4\right)^2$ **c.** $4(3u)^2$ **d.** x^3y^2

Solutions:

a. $-2t^3 \cdot 4t^5 = -8t^3 \cdot t^5$ (combine the numerical coefficients by multiplication)

$= -8t^{3+5}$ (add the two exponents together using rule ***i***)

$= -8t^8$

b. $5\left(v^4\right)^2 = 5 \cdot v^{4 \cdot 2}$ (apply rule ***ii***; note that the power "2" doesn't affect the coefficient "5")

$= 5v^8$

c. $4(3u)^2 = 4\left(3^2u^2\right)$ (apply rule ***iii***)

$= 4\left(9u^2\right)$ (simplify)

$= 36u^2$ (combine the numerical coeficients by multiplication; note that we don't distribute 4 to 9 *and* u^2 since $9u^2$ is one term)

d. x^3y^2 cannot be simplified further since the bases are different variables.

EXERCISES:

1. Simplify the following expressions using the rules of exponents.

a. $x^5 \cdot x^6$

b. $7y^7 \cdot 3y^3$

c. $2x^2(3x - 2)$

d. $x^5 + x^5$

e. $5t^3 - 2t^3$

f. $9x^5 - x^5$

g. $\left(x^5\right)^6$

h. $2\left(x^5\right)^3$

i. $x\left(x^2\right)^3$

j. $3(2y)^4$

k. $\left(2x^2\right)^3$

l. $(3x)^2$

2. True or False. If a statement is false, change one side of the equation to make it true.

a. $x^2 + x^3 = x^5$

b. $3^2 \cdot 3^4 = 3^6$

c. $3^2 \cdot 2^3 = 6^5$

d. $x^2x^3 = 2x^5$

e. $\left(5^2\right)^3 = 5^5$

f. $x^3 + x^3 = 2x^3$

SUPPLEMENT TO §10.6

We can use ***function notation*** to communicate the information contained in the graph of a function. For example, if the point $(5, 3)$ is on the graph of a function called f, we can write "$f(5) = 3$". This means that if we are asked to find $f(5)$ using the graph of the function f, we would need to search for the y-value that is associated with the x-value 5. Similarly, if we are asked to solve an equation like $f(x) = 4$ using a graph, we would need to search for the x-value(s) associated with the y-value 4.

EXAMPLE: The graph of $y = g(x)$ is given at the right.

a. Find $g(-2)$.

b. Find $g(-3)$.

c. Solve $g(x) = -1$.
(Express your answer in a solution set.)

d. Solve $g(x) = 1$.
(Express your answer in a solution set.)

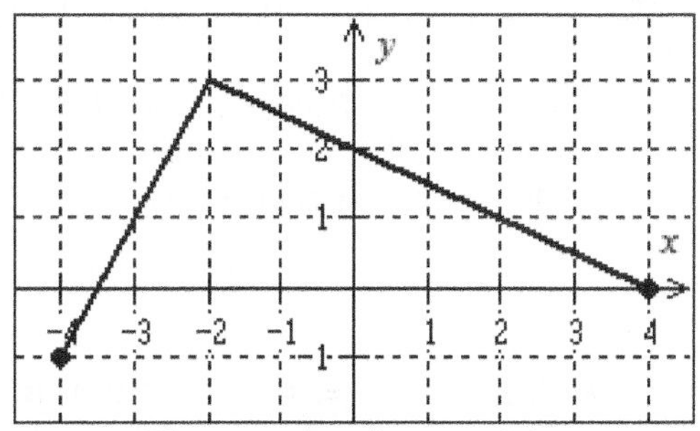

The graph of $y = g(x)$.

Solutions

a. $g(-2) = 3$. (At the x-value -2, the y-value is 3 on the graph.)

b. $g(-3) = 1$. (At the x-value -3, the y-value is 1 on the graph.)

c. The solution set is $\{-4\}$. (At the x-value -4, the y-value is -1 on the graph.)

d. The solution set is $\{-3, 2\}$. (At the x-values -3 and 2, the y-value is 1 on the graph.)

EXERCISES:

1. The graph of $y = G(x)$ is given at the right.

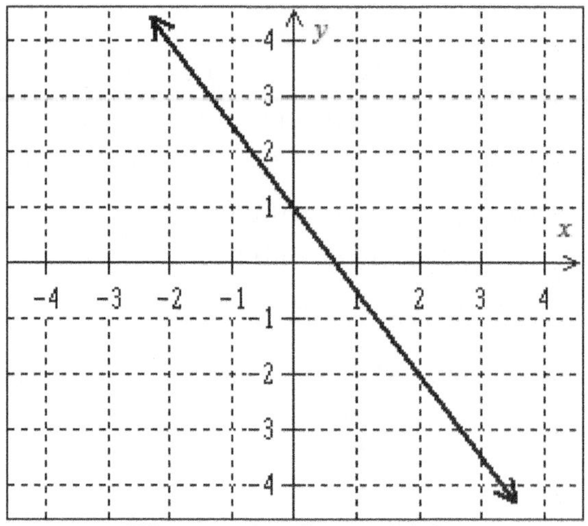

The graph of $y = G(x)$.

a. On the graph, if $x = 0$, then $y =$____.

b. Find $G(-2)$.

c. Find $G(4)$.

d. If $G(x) = -2$, then $x =$____.

e. Solve $G(x) = 4$.

f. Find a formula for $y = G(x)$. [§4.4 review]

g. Use the formula you found in part **f** to find the exact coordinates of the horizontal intercept of the line. [§4.4 review]

2. The graph of $y = f(x)$ is given at the right.

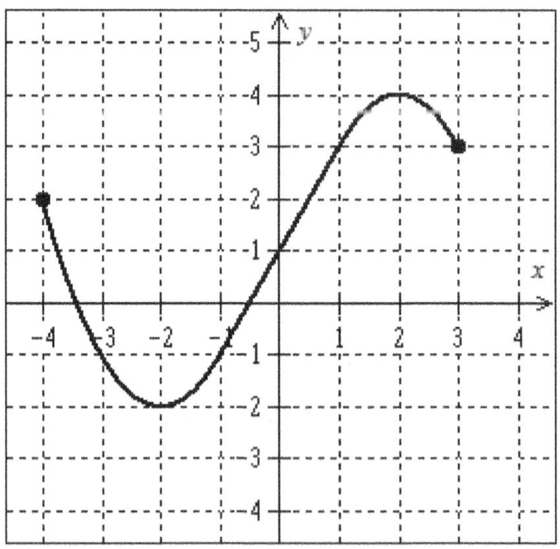

The graph of $y = f(x)$.

a. On the graph, if $x = 1$, then $y =$____.

b. Find $f(0)$.

c. Find $f(-4)$.

d. From the graph we can see that $f(-3) = -1$. What point on the graph is this statement describing?

e. From the graph we can see that $f(2) = 4$. What point on the graph is this statement describing?

f. If $f(x) = 4$, then $x =$____.

g. Solve $f(x) = -1$.

3. Find the following function values if the function H is defined by the set of ordered pairs $\{(2, 5), (7, 10), (-4, 3), (-1, 3), (5, 2)\}$.

a. $H(-4)$ **b.** $H(2)$

4. Billy Jean is driving home from a business meeting. Shown is the graph $y = d(t)$ where d represents her remaining distance (in miles) to home as a function of time, t, (in hours) since she left her meeting.

a. Find and interpret $d(1)$ in the context of the real-world situation. Write in a complete sentence.

b. Find and interpret $d(5)$ in the context of the real-world situation. Write in a complete sentence.

c. Find a formula for $y = d(t)$. [§4.5 review]

d. Using your formula from part **c**, find and interpret the exact value of $d(0)$ in the context of the real-world situation. Write in a complete sentence. [§4.5 review]

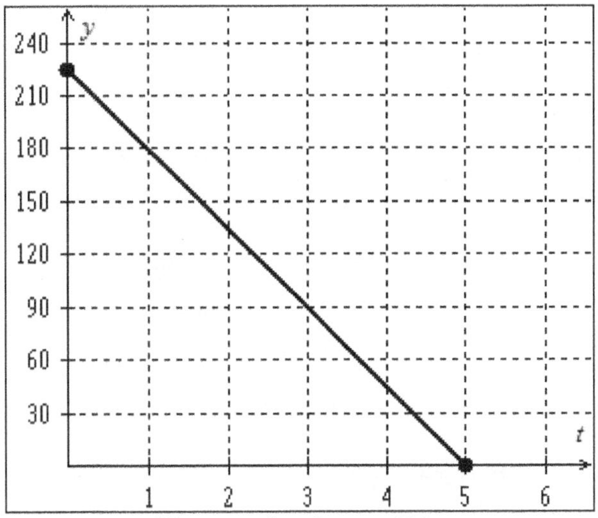

The graph of $y = d(t)$.

5. Instead of using graphs, we can use a *table* to define a function. For example the table below defines the function j. (The top row represents the input values and the bottom row represents the output values, and each column represents an ordered pair that satisfies the function, e.g., the last column implies that the ordered pair $(6, -100)$ satisfies the function.)

x	-3	1	3	6
$j(x)$	170	50	-10	-100

a. Find $j(3)$.

b. When $j(x) = 170$, then $x =$ ____.

c. Is j a linear function? Defend your answer with a strong argument. [§4.3 review]

d. Find a formula for $y = j(x)$ in slope intercept form. [§4.4 review]

e. Use your answer to part **d** to find $j(-8)$.

f. Make a graph of $y = j(x)$ on graph paper to verify that the formula found corresponds to the data in the table. [§4.4 review]

6. An independent tester examined several modern wind turbines (that generate power) to check if they met factory specifications. The Air 403 Wind Turbine came close to its expectations. The amount of power $P(x)$ (in Watts) generated by a wind-speed of x miles per hour blowing on the turbine can be approximated by the formula $P(x) = 20.8x - 262$.

a. Evaluate $P(25)$. Interpret your answer in a complete sentence in the context of the problem.

b. Solve $P(x) = 450$. (Round to the nearest integer). Interpret your answer in a complete sentence in the context of the problem.

c. If you need about 300 Watts to run the power to your heated shed, at what speed does the wind have to be blowing? (Round to the nearest integer.)

To get the turbine to turn at all (and thus generate power), the wind has to be more than just a light breeze. In parts ***d*** *and* ***e****, we will investigate this fact.*

d. Evaluate $P(10)$. Interpret your answer in a complete sentence in the context of the problem.

e. At what speed will the power production be 0 Watts? (Round to the nearest integer.) This speed is called the ***cut-in speed***: if the wind blows even just a tiny amount faster, the turbine will start generating power.

ANSWERS TO THE SUPPLEMENT TO §1.8:

1. **a.** $x^5 \cdot x^6 = x^{11}$

b. $7y^7 \cdot 3y^3 = 21y^{10}$

c. $2x^2(3x - 2) = 6x^3 - 4x^2$

d. $x^5 + x^5 = 2x^5$

e. $5t^3 - 2t^3 = 3t^3$

f. $9x^5 - x^5 = 8x^5$

g. $\left(x^5\right)^6 = x^{30}$

h. $2\left(x^5\right)^3 = 2x^{15}$

i. $x\left(x^2\right)^3 = x \cdot x^6$
$= x^7$

j. $3(2y)^4 = 3\left(16y^4\right)$
$= 48y^4$

k. $\left(2x^2\right)^3 = 8x^6$

l. $(3x)^2 = 9x^2$

2. **a.** $x^2 + x^3 = x^5$ is *false*.

There is no way to combine $x^2 + x^3$ because they are not like terms. However, if we change the left side to multiplication rather than addition, the statement is true: $x^2 \cdot x^3 = x^5$.

b. $3^2 \cdot 3^4 = 3^6$ is *true*.

c. $3^2 \cdot 2^3 = 6^5$ is *false*.
We can only add the exponents when the bases are the same. We can edit the right side of the equation to make it true: $3^2 \cdot 2^3 = 9 \cdot 8$

d. $x^2x^3 = 2x^5$ is *false*.

We can edit the right side of the equation to make the statement true: $x^2x^3 = x^5$.

e. $\left(5^2\right)^3 = 5^5$ is *false*.

We can edit the right side of the equation to make the statement true: $\left(5^2\right)^3 = 5^6$.

f. $x^3 + x^3 = 2x^3$ is *true*.

ANSWERS TO THE SUPPLEMENT TO §10.6:

1. **a.** $y = \underline{1}$

b. $G(-2) = 4$

c. $G(4) = -5$

d. $x = \underline{2}$

e. The solution set is $\{-2\}$.

f. Since this is a graph of a line with slope $m = -\frac{3}{2}$ and vertical intercept $(0, 1)$, the formula for the graph in slope intercept form is $y = -\frac{3}{2}x + 1$, and the formula for the function is $y = G(x) = -\frac{3}{2}x + 1$.

g. The horizontal intercept is the point where $y = 0$. Thus,

$$\begin{aligned} 0 &= -\tfrac{3}{2}x + 1 \\ -1 &= -\tfrac{3}{2}x \\ \tfrac{2}{3} &= x \end{aligned}$$

So the horizontal intercept is $\left(\frac{2}{3}, 0\right)$.

2. **a.** $y = \underline{3}$

b. $f(0) = 1$

c. $f(-4) = 2$

d. the point $(-3, -1)$

e. the point $(2, 4)$

f. $x = \underline{2}$

g. The solution set is $\{-1, 3\}$.

3. **a.** $H(-4) = 3$ **b.** $H(2) = 5$

4. **a.** $d(1) = 180$. This means that one hour after Billy Jean left her meeting, she still had 180 miles left to travel.

b. $d(5) = 0$. This means that 5 hours after Billy Jean left her meeting, she had 0 miles left to travel which implies that she must have arrived at her home.

c. This is a linear graph, so we need to find the slope first. Two points on the graph are $(1, 180)$ and $(5, 0)$. We can use these points to find the slope of the line"

$$\begin{aligned} m &= \frac{y_2 - y_1}{x_2 - x_1} \\ &= \frac{180 - 0}{1 - 5} \\ &= \frac{180}{-4} \\ &= -45 \end{aligned}$$

Now we can plug this slope and the point $(5, 0)$ into the point-slope formula to find the formula for the line:

$$\begin{aligned} (5,0) \Rightarrow \; y &= -45x + b \\ 0 &= -45(5) + b \\ 0 &= -225 + b \\ b &= 225 \end{aligned}$$

Thus, the formula for the line is $y = -45x + 225$ **but** the question asks for "$d(t)$" so t is where the x usually is! This means that we should use t instead of x in our final formula: $y = d(t) = -45t + 225$.

d. $$\begin{aligned} d(0) &= -45(0) + 225 \\ &= 225 \end{aligned}$$

This means that just as Billy Jean was leaving her meeting (i.e., after driving for 0 hours), she had 225 miles left to drive to get home.

5. **a.** $j(3) = -10$.

b. $x = \underline{-3}$

c. If a function is linear, it must have constant slope; thus, we can determine if j is linear by finding the slope between each pair of points and see if the slope is constant. Recall that the slope of the line passing through two points is the ratio of the *rise* to the *run* between the points.

	run	4	2	3
x	-3	1	3	6
$j(x)$	170	50	-10	-100
	rise	-120	-60	-90

- slope between first pair of points: $m = \dfrac{-120}{4} = -30$
- slope between second pair of points: $m = \dfrac{-60}{2} = -30$
- slope between third pair of points: $m = \dfrac{-90}{3} = -30$

Since all the slopes are identical, j is a linear function with slope -30.

d. Since we know that the slope is -30, we know that the formula for the function must look like $y = -30x + b$. Now we have to find b by using one of the points given in the table. Let's use $(1, 50)$:

$$(1, 50) \Rightarrow y = -30x + b$$
$$50 = -30(1) + b$$
$$50 = -30 + b$$
$$b = 80$$

So, the formula must be $y = j(x) = -30x + 80$.

e. $j(-8) = -30(-8) + 80$

$= 240 + 80$

$= 320$

f. Below is the graph of $y = j(x)$. Points from the table are plotted and labeled for clarity.

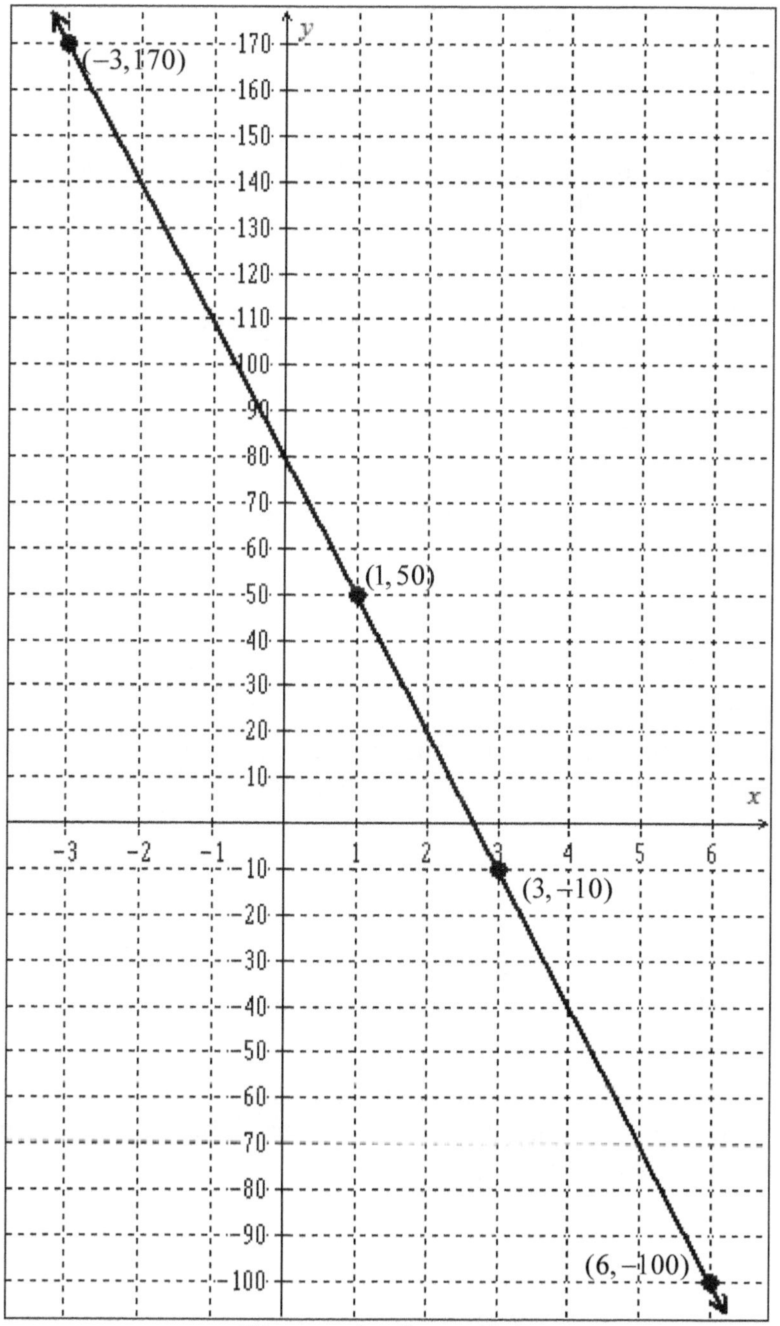

The graph of $y = j(x)$.

6. **a.** $P(25) = 20.8(25) - 262$

$= 258$

This means that when the wind blows at 25 miles per hour, the turbine generates 258 Watts of power.

b.

$$\begin{aligned} P(x) &= 450 \\ 20.8x - 262 &= 450 \\ 20.8x &= 712 \\ x &\approx 34 \end{aligned}$$

This means that in order to generate 450 Watts of power, there needs to be a wind-speed of about 34 miles per hour.

c.

$$\begin{aligned} P(x) &= 300 \\ 20.8x - 262 &= 300 \\ 20.8x &= 562 \\ x &\approx 27 \end{aligned}$$

So the wind needs to blow at about 27 miles per hour in order for the turbine to generate 300 Watts of power.

d. $P(10) = 20.8(10) - 262$

$= -54$

A wind-speed of 10 miles per hour must not be fast enough to make the turbine spin because we cannot generate a *negative* quantity of power.

e.

$$\begin{aligned} P(x) &= 0 \\ 20.8x - 262 &= 0 \\ 20.8x &= 262 \\ x &\approx 13 \end{aligned}$$

This means that when the wind blows at about 13 miles per hour, the turbine just barely starts to spin fast enough to generate power. Thus, the cut-in speed for this turbine is about 13 miles per hour.

Mth70 Required Course Supplement

Supplement to §10.6

1.) Functions f, g, h and j are defined below. Use them to answer the questions.

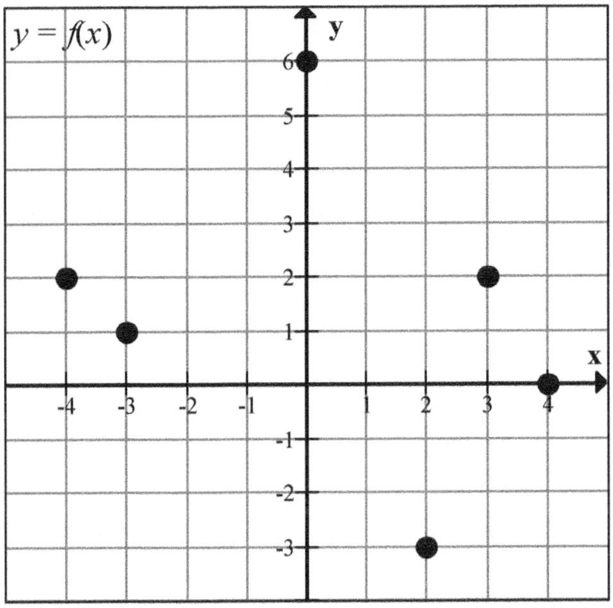

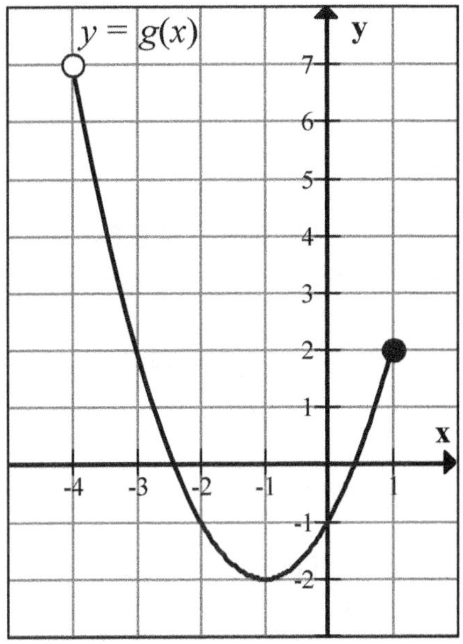

x	$y = h(x)$
-7	-3
0	64
1	64
-5	-4
17	0
20	3

$j(x) = 3x - 7$

a.) Evaluate $f(-3)$.

b.) Evaluate $g(-3)$.

c.) Evaluate $h(0)$.

d.) Evaluate $j(-3)$.

e.) Evaluate $f(1)$.

f.) Evaluate $h(-3)$.

For parts g-o, write your solutions in a solution set.

g.) Solve $j(x) = 11$

h.) Solve $f(x) = 0$.

i.) Solve $h(x) = 64$.

j.) Solve $g(x) = -2$.

k.) Solve $h(x) = 0$.

l.) Solve $g(x) = 2$.

m.) Solve $f(x) = 2$.

n.) Solve $f(x) = 3$.

o.) Estimate the solution: $g(x) = 4$.

p.) State the domain of f.

q.) State the range of f.

r.) State the domain of g.

s.) State the range of g.

t.) State the domain of h.

u.) State the range of h.

v.) State the domain of j.

w.) State the range of j.

2.) Speeding, rapid acceleration and braking wastes gas. These bad habits can lower your gas mileage by 33% at highway speeds and by 5% around town. The graph below gives the Fuel Economy F (in miles per gallon) for the speed driven s (in miles per hour) for a specific model car. Use the graph to answer the questions below. (http://www.fueleconomy.gov/fcg/driveHabits.shtml)

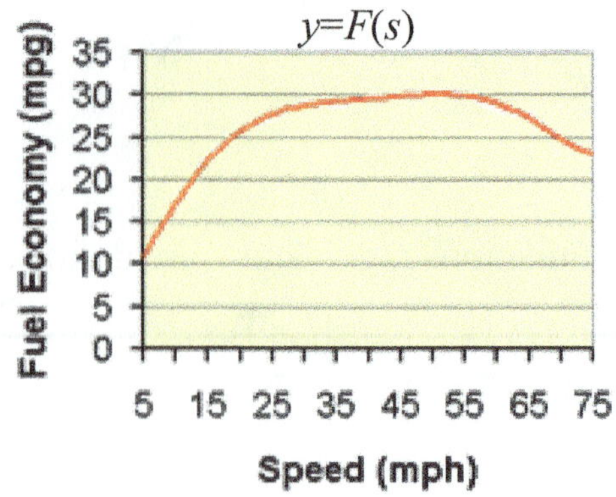

a.) The x value 0 is not shown on the graph above. Can you reason out what $F(0)$ must be? Explain your answer in a complete sentence including units.

b.) Use the graph to estimate $F(25)$ and give the meaning for this situation. Write answer in a complete sentence and use appropriate units.

c.) Use the graph to solve $F(s) = 25$ and give the meaning for this situation.

d.) Find the maximum fuel economy and the corresponding speed using the graph. Write answer in a complete sentence and use appropriate units.

Supplement to §5.3

1.) The graph below shows a system of linear equations. However, the exact solution cannot be determined from the graph. Find the equations of the lines and solve the system algebraically and verify that your result makes sense for the graph.

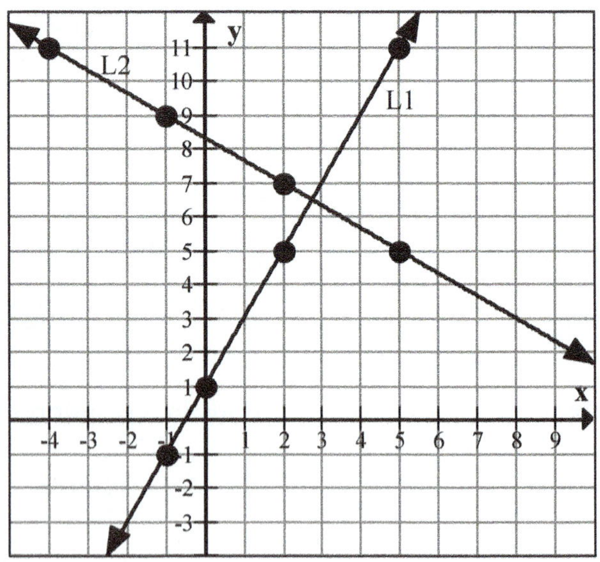

Supplement to §5.4

1.) Hugo Reyes was mixing dried mango slices with dried pineapple to create a delightful island mix. The mango costs \$5.99 per pound and the pineapple costs \$3.99 per pound.

a.) Is it possible for Hugo to mix the fruit in such a way to create a mix that costs a total of \$1.99 per pound?

b.) If Hugo mixed the fruit and calculated that his mix would cost \$4.80 per pound, which ingredient did he add more of?

c.) If Hugo had a 5 pound bag of fruit that he said was worth \$5.99 per pound, how much of each fruit did he add to make it?

d.) If Hugo mixed the fruit and calculated that his mix would cost \$5.00 per pound, which ingredient did he add more of?

2.) Dimensional Analysis practice.

a.) Write 36 minutes in hours as a decimal.

b.) Write 54 minutes in hours as a fraction.

c.) Write 4:20 hours and minutes as minutes.

d.) Write 4.2 hours as hours and minutes.

e.) Find the number of feet (round to the nearest thousandth) in 248 centimeters [2.54 cm = 1 inch; 12 inches = 1 foot].

f.) Find the number of square inches in 4 square feet.

g.) Find the number of square feet in (round to the nearest thousandth) 343 square inches.

Supplement to §6.2

1.) Let $f(x) = x^2 + \frac{1}{3}x + \frac{4}{5}$ and $g(h) = \left(h - \frac{3}{4}\right)(h+8)$. Evaluate and simplify the expressions using correct formatting.

a.) $f(-3)$

b.) $g\left(-\frac{5}{4}\right)$

Supplement to §7.6

1.) Determine if the equation/function is linear, quadratic or something else.

a.) $5x^2 + 2y = 2$

b.) $5x + 2y = 2x$

c.) $y = \sqrt{2}x + 1$

d.) $f(x) = x$

e.) $x = -1$

f.) $\pi x + \sqrt{7}y = 3^2$

g.) $y = 4(x-1) + 2(x-3)$

h.) $y = \frac{3}{x} + 4x$

i.) $h(r) = r^3 + r^2$

j.) $g(x) = (x-1)(x+5)$

k.) $y = \frac{4}{5}$

Supplement to §9.1

If we want to estimate $\sqrt{10}$, we need to find the nearest integers below and above the 10 that are perfect squares (so the square root can be found). These numbers are 9 and 16. So, since 10 is between 9 and 16 (i.e., $9<10<16$) we have that $\sqrt{10}$ will be between $\sqrt{9}$ and $\sqrt{16}$ $\left(\text{i.e., } \sqrt{9}<\sqrt{10}<\sqrt{16}\right)$. We chose perfect squares so we could simplify the radicals. Thus, $\sqrt{10}$ is between 3 and 4 $\left(\text{i.e., } 3<\sqrt{10}<4\right)$, and $\sqrt{10}$ is closer to 3 than 4 since 10 is closer to 9 than 16.

1.) Use your calculator to check this estimate of $\sqrt{10}$.

2.) Fill-in the blanks with integers. Verify using your calculator.

a.) $\sqrt{19}$: Since _____ $<19<$ ______, we know that ______ $<\sqrt{19}<$ ______. Without a calculator, I estimate that $\sqrt{19}\approx$ _________.

b.) $\sqrt{3.2}$: Since _____ $<3.2<$ ______, we know that ______ $<\sqrt{3.2}<$ ______. Without a calculator, I estimate that $\sqrt{3.2}\approx$ _________.

Supplement to §9.4

1.) A student correctly solved an equation and ended up with the solution $\frac{30}{\sqrt{12}}$. However, the answer in the back of the book is $5\sqrt{3}$.

a.) Round both numbers to the nearest tenth.

b.) Show how these answers are the same algebraically.

2.) Solve the equation $\sqrt{3}x+4=10$ algebraically. Rationalize the denominator of your solution.

Supplement to §10.1

1.) Solve the equations below using the square root property, if possible. If not possible, use factoring. Use correct formatting.

a.) $3x^2=48$ b.) $(x-3)^2=4x$ c.) $3x^2-5=-14x$ d.) $3(2x-4)^2-4=20$

2.) The stopping distance of a car is proportional to the square of the its speed before slamming on the brakes. In math words, $D(v)=kv^2$, where D is the stopping distance in feet, v is the speed of the car before braking in miles per hour and k is a number that depends on how good the tires are and the road conditions. For a car with good tires on a good road, $k \approx 0.05$, so $D(v) = 0.05v^2$.

a.) Find $D(60)$ and interpret your answer.

b.) Find $D(70)$ and interpret your answer.

c.) Using your answers above, what is the ratio $\dfrac{D(70)}{D(60)}$ and what does it mean?

d.) A neighbor's basketball rolls out into the road 30ft in front of you. There is no room to swerve to avoid it! What is the maximum speed you could be going and still not hit the ball if you slam on the brakes?

Supplement to §10.3

1.) First, identify the following as being a linear equation, linear expression, quadratic equation or quadratic expression. Then, using correct formatting, for a

i.) linear equation, solve. Write your solution in a solution set and verify your solution.
ii.) linear expression, simplify.
iii.) quadratic equation, solve using the easiest method. Write your solution(s) in a solution set and verify your solution(s).
iv.) quadratic expression, simplify and if possible factor.

a.) $\dfrac{3}{2}x + \dfrac{4}{5} = \dfrac{3}{10}$

Circle: Equation / Expression
Circle: Linear / Quadratic
Solve/simplify/factor as appropriate:
Check solutions, if applicable:

b.) $3x^2 + 11x - 4$

Circle: Equation / Expression
Circle: Linear / Quadratic
Solve/simplify/factor as appropriate:
Check solutions, if applicable:

c.) $2(2x-2)^2 - 10 = 26$

Circle: Equation / Expression
Circle: Linear / Quadratic
Solve/simplify/factor as appropriate:
Check solutions, if applicable:

d.) $(x-4)^2 = x$

Circle: Equation / Expression
Circle: Linear / Quadratic
Solve/simplify/factor as appropriate:
Check solutions, if applicable:

2.) Let $f(x) = (x+6)^2$ and $g(x) = -(10x+1)(x-3)$. Find the missing coordinates in the tables. Note that there may be 2 inputs for some outputs. Approximate any irrational numbers to the tenths. If there is no real number that works, indicate this.

x	$y = f(x)$
-2	
0	
	0
	-2
	9
$\sqrt{5}$	

x	$y = g(x)$
-1	
-4/5	
	0
	24
$-\sqrt{3}$	
	33

3.) Let $f(x) = x^2 - 1$, $g(x) = 3x - 1$ and $h(x) = 2x - 6$. Using correct formatting, find the following:

a.) Find $f(-4)$.

b.) Is "-4" in $f(-4)$ an input or output?

c.) Is "$f(-4)$" an input or output?

d.) If $h(3) = 0$, write a point that you know is on the graph of $y = h(x)$.

e.) If $f(-3) = 8$, write a point that you know is on the graph of $y = f(x)$.

f.) Determine which function would have the point $(2,3)$ on its graph.

g.) Determine which function would have the point $(-2,-7)$ on its graph.

h.) Solve $g(x) = 7$. Write the solution in a solution set.

i.) Evaluate and simplify the expression $g\left(\frac{4}{5}\right) + f\left(\frac{2}{3}\right)$ to simplest form.

j.) Solve $h(x) = 0$. Write solution in a solution set.

k.) Solve $f(x) = 8$. Hint: There are two solutions. Write solutions in a solution set.

l.) Fix and explain what is wrong with the underlined part of each of the following.

i.) $\underline{f(x)} = (3)^2 - 1$

ii.) $g(2) = \underline{3x - 1}$

iii.) $h(a) = \underline{2x - 6}$

4.)a.) If $L(2) = 5$ and $L(-1) = 3$, write two points that you know will be on the graph of L.

b.) Using the points found above, find the formula for the linear function, L, that passes through the two points.

Supplement to §10.5

1. a.) What exactly is the relationship between the equation $y = 2x + 1$ and the picture shown in Figure 1?

b.) What exactly is the relationship between the equation $y = x^2 + 3x + 2$ and the picture shown in Figure 2?

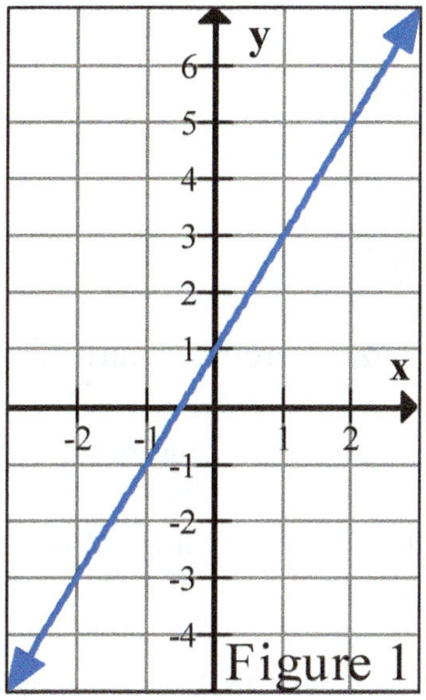

Figure 1

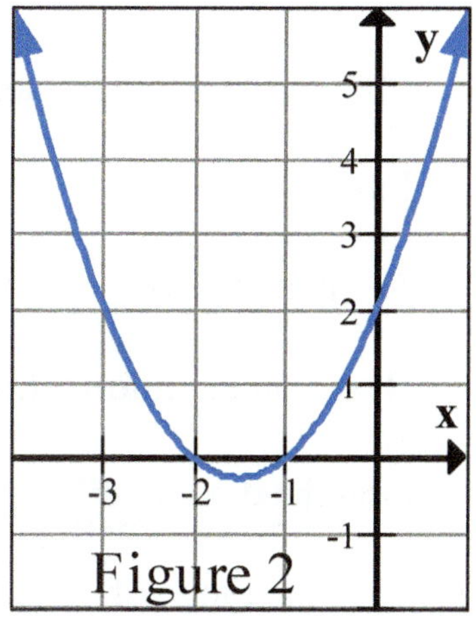

Figure 2

2.) For the function $j(t) = -3t^2 + 8t + 10$,

a.) Find the vertex.

b.) What is the axis of symmetry?

c.) Find the vertical intercept and show work using function notation.

d.) Find the horizontal intercept(s). Round your answers to the tenths with a calculator, if needed.

e.) Based on the information from the previous parts, draw a complete graph of $y = j(t)$.

f.) Find the domain of j and state it in interval notation.

g.) Find the range of j and state it in interval notation.

3.) If the graph of a function is symmetric about the line $x = 5$ and the point $(1, -9)$ is on the graph, what other point must be on the graph? Explain your reasoning.

4.) If the graph of a quadratic function has the points $(1, 6)$ and $(9, 6)$ on it, what is the x-value of the vertex? Explain your reasoning.

5.) For the function $R(x) = x^2 + 4x + 21$,

a.) Find the vertical intercept and show work using function notation.

b.) Find the horizontal intercept(s).

c.) Find the vertex.

d.) What is the axis of symmetry?

e.) Based on the information from the previous parts, draw a complete graph of $y = R(x)$.

f.) Find the domain of R and state it in interval notation.

g.) Find the range of R and state it in interval notation.

6.) If a parabola has vertex (-6, 17), use the information given and symmetry to fill-in the missing value in the table.

x	y
-2	18
	18

7.) An artist, Michael, sells 100 prints over one year for \$10 each. Michael, who is an amateur mathematician and statistician, did some research and found out that for each price increase of \$2 per print, the sales drop by 5 prints per year. The revenue, R, from selling the prints is given by $R(x) = (100 - 5x)(2x + 10)$ where $100 - 5x$ is the number of prints sold and $2x + 10$ is the cost of each print, and x is the number of \$2 price increases he has done.

a.) At what price should Michael sell his prints to maximize the revenue? How many prints will he sell at this price?

b.) Find the horizontal intercepts of $y = R(x)$ and interpret their meaning in the context of the problem.

8.) The river running through L.A., like everything else in L.A., has been paved. This river exists as a meager trickle through the dry months of all-the-time; however, every 17years or so, L.A. receives some rain and everyone's yards turn green for 5 minutes before dying again. The L.A. River collects all this runoff rainwater and channels it out to sea. The depth of water in the L.A. River, D, in ft, is a function of time, t, in hours since a storm ended. The function is given by $D(t) = -t^2 + 2t + 8$.

a.) Evaluate and interpret $D(3)$.

b.) Find and interpret the vertical intercept.

c.) Find and interpret the horizontal intercepts.

d.) At what time will the water be deepest? What is that maximum depth?

e.) Make a graph of $y = D(t)$ on it's domain.

f.) Realistically, what is the range of D and what does it mean?

9.) A trebuchet is a French catapult originally used to launch large projectiles long distances. North of England a man named Hew Kennedy built a full size one to hurl random objects. Suppose that the height of a piano off the ground, h, in feet, is a function of the horizontal distance along the ground in the direction it is thrown, x, also in feet. Given that $h = f(x) = -.002x^2 + .6x + 60$,

a.) Find and interpret the vertical intercept.

b.) Find and interpret the horizontal intercept(s).

c.) Find the vertex. Explain what the vertex means in context.

d.) Find and interpret $f(60)$.

e.) Solve $f(x) = 20$. Explain what this means in context.

f.) If I want to crush a Volkswagen bug with the piano, how far from the trebuchet should I park it?

Mth70 Supplement Key

§10.6

1.) a.) $f(-3)=1$ b.) $g(-3)=2$ c.) $h(0)=64$

d.) $j(-3)=-16$ e.) $f(1)$ does not exist. f.) $h(-3)$ does not exist.

g.) $\{6\}$ h.) $\{4\}$ i.) $\{0,1\}$

j.) $\{-1\}$ k.) $\{17\}$ l.) $\{1,-3\}$

m.) $\{3,-4\}$ n.) $\{\ \}$ o.) $\{-3.5\}$

p.) $\{-4,-3,0,2,3,4\}$ q.) $\{-3,0,1,2,6\}$ r.) $(-4,1]$

s.) $[-2,7)$ t.) $\{-7,-5,0,1,17,20\}$ u.) $\{-4,-3,0,3,64\}$

v.) $\mathbb{R}$ w.) $\mathbb{R}$

2.) a.) $F(0)=0$. This indicates that at a speed of 0 miles per hour, you have a fuel economy of 0 miles per gallon.

b.) $F(25)\approx 27$. This means that when traveling at 25 miles per hour, your fuel economy is about 27 miles per gallon.

c.) The solutions are approximately 20 and 70. This would mean that at the speeds of 20 miles per hour and 70 miles per hour, your fuel economy is about 25 miles per gallon.

d.) The maximum fuel economy of about 30 miles per gallon occurs at a speed of about 50 miles per hour

§5.3

1.) The simplified system is $\begin{cases} y=2x+1 \\ y=-\dfrac{2}{3}x+\dfrac{25}{3} \end{cases}$ The solution to the system is $\left(\dfrac{11}{4},\dfrac{13}{2}\right)$.

§5.4

1.)a.) No, it's not possible. The cheapest mix he could possibly make is made entirely of pineapple and would cost \$3.99 per pound. Even if you add a tiny amount of mango, it would raise the price.

b.) The cost of \$4.80 per pound is closer to \$3.99 than \$5.99. This means that more of the pineapple was added than mango.

c.) He added 5 pounds of mango and no pineapple at all.

d.) The cost of \$5.00 per pound is closer to \$5.99 than \$3.99. This means he added more mango than pineapple.

2.) Dimensional Analysis practice.

a.) $36\min \cdot \frac{1hr}{60\min} = \frac{36}{60}hr = 0.6hr$

b.) $54\min \cdot \frac{1hr}{60\min} = \frac{54}{60}hr = \frac{9}{10}hr$

c.) $4hr \cdot \frac{60\min}{1hr} = 240\min$. So, 4hr20min is 260min.

d.) $0.2hr \cdot \frac{60\min}{1hr} = 12\min$. So 4.20hr is 4hr12min.

e.) $248cm \cdot \frac{1in}{2.54cm} \cdot \frac{1ft}{12in} \approx 8.136ft$

f.) $4ft^2 \cdot \frac{12in}{1ft} \cdot \frac{12in}{1ft} = 576in^2$

g.) $343in^2 \cdot \frac{1ft}{12in} \cdot \frac{1ft}{12in} \approx 2.382ft^2$

§6.2

1.) Let $f(x) = x^2 + \frac{1}{3}x + \frac{4}{5}$ and $g(h) = \left(h - \frac{3}{4}\right)(h+8)$. Evaluate and simplify the expressions using correct formatting.

a.) $f(-3) = \frac{44}{5}$

b.) $g\left(-\frac{5}{4}\right) = -\frac{27}{2}$

§7.6

1.) a.) Quadratic

b.) Linear

c.) Linear

d.) Linear

e.) Linear

f.) Linear

g.) Linear

h.) Something else

i.) Something else

j.) Quadratic

k.) Linear

§9.1

1.) $\sqrt{10} \approx 3.16227766$

2.) a.) $\sqrt{19}$: Since $\underline{16} < 19 < \underline{25}$ (but closer to 16), we know that $\underline{4} < \sqrt{19} < \underline{5}$. Without a calculator, I estimate that $\sqrt{19} \approx$ 4.3 or 4.4. Note: $\sqrt{19} \approx 4.359$.

b.) $\sqrt{3.2}$: Since $\underline{1} < 3.2 < \underline{4}$ (but closer to 4), we know that $\underline{1} < \sqrt{3.2} < 2$. Without a calculator, I estimate that $\sqrt{3.2} \approx 1.7$ or 1.8 Note: $\sqrt{3.2} \approx 1.789$.

§9.4

1.)a.) $\dfrac{30}{\sqrt{12}} \approx 8.7$ and $5\sqrt{3} \approx 8.7$.

b.)

$$\begin{aligned}\frac{30}{\sqrt{12}} &= \frac{30}{\sqrt{4}\sqrt{3}} \\ &= \frac{30}{2\sqrt{3}} \\ &= \frac{15}{\sqrt{3}} \\ &= \frac{15}{\sqrt{3}} \cdot \frac{\sqrt{3}}{\sqrt{3}} \\ &= \frac{15\sqrt{3}}{3} \\ &= 5\sqrt{3}\end{aligned}$$

2.)

$$\begin{aligned}\sqrt{3}x + 4 &= 10 \\ \sqrt{3}x &= 6 \\ x &= \frac{6}{\sqrt{3}} \\ &= \frac{6}{\sqrt{3}} \cdot \frac{\sqrt{3}}{\sqrt{3}} \\ &= \frac{6\sqrt{3}}{3} \\ &= 2\sqrt{3}\end{aligned}$$

The solution set is $\{2\sqrt{3}\}$.

§10.1

1.) The square root property only applies in parts 'a' and 'd'. The square root property can't be used for parts 'b' and 'c' without completing the square (covered in Math 95).

* : The asterisk in the work below indicates the step where the square root property was applied and where the $\pm$ MUST first appear.

a.) Square root property applies

$$3x^2 = 48$$
$$x^2 = 16$$
$$x = \pm\sqrt{16} \quad *$$
$$x = \pm 4$$
$$\{-4, 4\}$$

b.) Square root property does not apply (the 4x on the right-hand side violates the square-root property form)

$$(x-3)^2 = 4x$$
$$(x-3)(x-3) = 4x$$
$$x^2 - 6x + 9 = 4x$$
$$x^2 - 10x + 9 = 0$$
$$(x-1)(x-9) = 0$$
$$x-1=0 \quad or \quad x-9=0$$
$$x=1 \quad or \quad x=9$$
$$\{1, 9\}$$

c.) Square root property does not apply (the variable with the -14 is the fatal error)

$$3x^2 - 5 = -14x$$
$$3x^2 + 14x - 5 = 0$$
$$(x+5)(3x-1) = 0$$
$$x+5=0 \quad or \quad 3x-1=0$$
$$x=-5 \quad or \quad x=\frac{1}{3}$$
$$\left\{-5, \frac{1}{3}\right\}$$

d.) Square root property applies

$$3(2x-4)^2 - 4 = 20$$
$$3(2x-4)^2 = 24$$
$$(2x-4)^2 = 8$$
$$2x-4 = \pm\sqrt{8} \quad *$$
$$2x = 4 \pm \sqrt{8}$$
$$x = \frac{4 \pm \sqrt{8}}{2}$$
$$x = \frac{4 \pm 2\sqrt{2}}{2}$$
$$x = 2 \pm \sqrt{2}$$
$$\{2+\sqrt{2}, 2-\sqrt{2}\} \quad or \quad \{2 \pm \sqrt{2}\}$$

2.) a) $D(60)=180$. This means that at a speed of 60mph, the car will need 180ft to stop.

b.) $D(70)=245$. This means that at a speed of 70mph, the car will need 245ft to stop.

c.) $\frac{D(70)}{D(60)}=\frac{245}{180}\approx 1.36$. This means it takes about 1.36 times as much distance to stop at a speed of 70mph than at 60mph. The stopping distance increased 36% even though the speed only increased by 17%.

d.)
$$\begin{aligned} 30 &= 0.05v^2 \\ 600 &= v^2 \\ \sqrt{600} &= v \\ 24.5 &\approx v \end{aligned}$$

I hope you were driving less than 24.5mph or else you're going to crush a basketball.

§10.3

1.) a.) Equation, Linear, (Solve)

Solution set: $\left\{\frac{-1}{3}\right\}$

$$\begin{aligned} \frac{3}{2}\left(\frac{-1}{3}\right)+\frac{4}{5} &\overset{?}{=} \frac{3}{10} \\ -\frac{1}{2}+\frac{4}{5} &\overset{?}{=} \frac{3}{10} \\ -\frac{5}{10}+\frac{8}{10} &\overset{?}{=} \frac{3}{10} \\ \frac{3}{10} &= \frac{3}{10} \end{aligned}$$

b.) Expression, Quadratic, (Factor)

$(3x-1)(x+4)$

c.) Equation, Quadratic, (Solve)

"Easiest" method to solve:

Zero Factor/Product Property

Solution set: $\left\{\frac{2\pm 3\sqrt{2}}{2}\right\}$ *or* $\left\{1\pm\frac{3\sqrt{2}}{2}\right\}$

Check: $x=\frac{2\pm 3\sqrt{2}}{2}\approx 3.121, -1.121$

c.)
$$\begin{aligned} (2\cdot(3.121)-2)^2-5 &\overset{?}{\approx} 13 \\ 12.995 &\approx 13 \end{aligned}$$

and

$$\begin{aligned} (2\cdot(-1.121)-2)^2-5 &\overset{?}{\approx} 13 \\ 12.995 &\approx 13 \end{aligned}$$

d.) Equation, Quadratic, (Solve)

"Easiest" method to solve:

Quadratic Formula

Solution set: $\left\{\frac{9\pm\sqrt{17}}{2}\right\}$

Check: $x=\frac{9\pm\sqrt{17}}{2}\approx 6.56, 2.44$

$$\begin{aligned} (6.56-4)^2 &\overset{?}{\approx} 6.56 \\ 6.554 &\approx 6.56 \end{aligned}$$

and

$$\begin{aligned} (2.44-4)^2 &\overset{?}{\approx} 2.44 \\ 2.434 &\approx 2.44 \end{aligned}$$

2.) Let $f(x)=(x+6)^2$ and $g(x)=-(10x+1)(x-3)$.

x	$y=f(x)$
-2	16
0	36
-6	0
Non-Real	-2
-3 and -9	9
$\sqrt{5}$	67.8

x	$y=g(x)$
-1	-36
-4/5	-133/5
-1/10 and 3	0
7/5 and 3/2	24
$-\sqrt{3}$	-77.2
Non-Real	33

3.) a.) $f(-4)=15$

b.) input c.) output d.) (3,0) e.) (-3,8)

f.) Function f g.) Function g h.) $\left\{\frac{8}{3}\right\}$ i.) $g\left(\frac{4}{5}\right)+f\left(\frac{2}{3}\right)=\frac{38}{45}$

j.) {3} k.) {-3, 3}

l.) *i.)* $\underline{f(x)}=(3)^2-1$: Here a number has been filled into the variable in function f's definition. This means we are evaluating function f at the number 3. So, this input value should replace the x in $f(x)$ as well: $f(3)=(3)^2-1$.

ii.) $g(2)=\underline{3x-1}$: We are evaluating function g at 2. This means the "2" should replace all the x's in the definition of the function: $g(2)=3\cdot 2-1$
$=5$

iii.) $h(a)=\underline{2x-6}$: We are evaluating function h at "a." which means we need to replace the x's in the function definition with "a." This changes the independent variable in the function definition: $h(a)=2a-6$

4.) a.) $(2,5)$ and $(-1,3)$

b.)

$$m=\frac{5-3}{2-(-1)}=\frac{2}{3}$$

So,

$$y=\frac{2}{3}x+b$$
$$5=\frac{2}{3}(2)+b$$
$$5=\frac{4}{3}+b$$
$$5-\frac{4}{3}=b$$
$$\frac{11}{3}=b$$

So,

$$y=\frac{2}{3}x+\frac{11}{3}$$
$$L(x)=\frac{2}{3}x+\frac{11}{3}$$

§10.5

1. a.) The graph shown is the solution set to the equation $y = 2x + 1$. Every point on the line is a solution.

b.) The graph shown is the solution set to the equation $y = x^2 + 3x + 2$. Every point on the parabola is a solution.

2.) a.) $\left(\frac{4}{3}, \frac{46}{3}\right)$

b.) $t = \frac{4}{3}$

c.) Vertical Intercept: (0, 10)

Work: $j(0) = -3(0)^2 + 8(0) + 10$
$= 10$

d.) $\left(\frac{4-\sqrt{46}}{3}, 0\right), \left(\frac{4+\sqrt{46}}{3}, 0\right)$

e.) Graph shown.

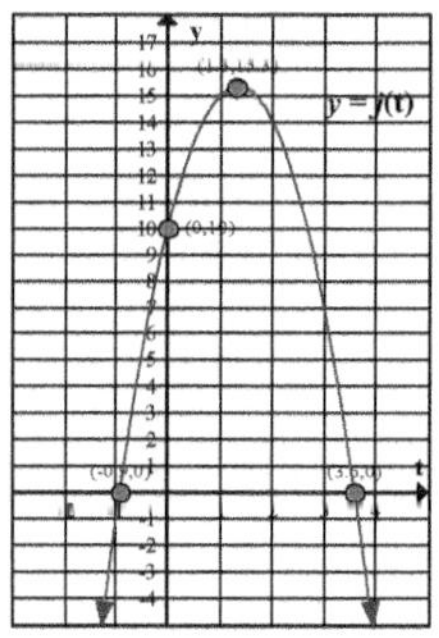

f.) $(-\infty, \infty)$

g.) $\left(-\infty, \frac{46}{3}\right]$

3.) (9, -9) must be on the graph. Explanation: The point $(1, -9)$ is 4 units to the left of the axis of symmetry. Since the parabola is symmetric about its axis of symmetry x = 5, the point that corresponds to $(1, -9)$ will be 4 units to the right of the axis of symmetry and have the same y-coordinate.

4.) The x-value of the vertex is 5. Explanation: The given points $(1, 6)$ and $(9, 6)$ both have the same y-coordinate. This means that the axis of symmetry lies between the two given points and the vertex will lie on the axis of symmetry. Since 5 is halfway between 1 and 9, the equation of the axis of symmetry is $x = 5$ and the x-coordinate of the vertex is 5. Note that we can not determine the y-coordinate of the vertex (the parabola could be "tall" or "short" or have the same shape as $y = x^2$). We can only say that the y-coordinate of the vertex is not 6 because if it was, we would have a line and not a parabola with the original points $(1, 6)$ and $(9, 6)$ a "vertex" of (5,6)!

5.) For the function $R(x) = x^2 + 4x + 21$,

a.) $R(0) = 0^2 + 4(0) + 21$
$= 21$

The vertical intercept is $(0, 21)$

b.) The solutions to $x^2 + 4x + 21 = 0$ are non-Real. This means that there are no (Real) horizontal intercepts.

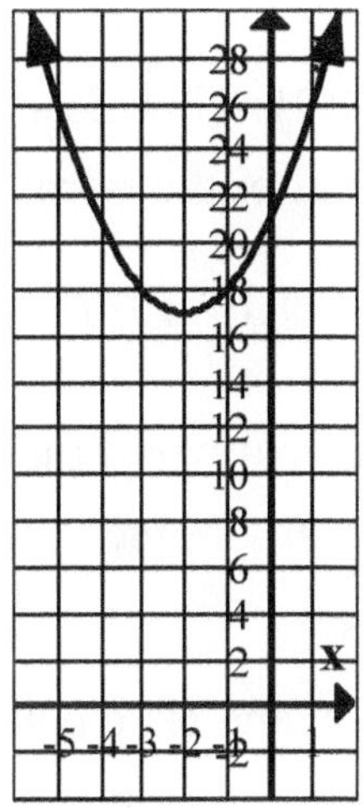

c.) $x = \frac{-b}{2a}$
$= -2$

and $R(-2) = 17$ So, the vertex is $(-2, 17)$.

d.) As above, the axis of symmetry is $x = -2$

e.) Graph shown.

f.) The domain is $(-\infty, \infty)$.

g.) The range is $[17, \infty)$.

6.) Missing value: -10

7.) a.) Vertex of the parabola: (7.5, 1562.5); So to maximize revenue and bring in $1562.50 per year, Michael needs have 7.5 $2 price increases. This means that the cost of each print will be $25 (evaluate $2x + 10$ at $x = 7.5$) and he will need to sell 62.5 prints per year (evaluate 100 - 5x at x = 7.5). Since it is not possible to sell a half of a print, Michael won't be able to earn the maximum possible revenue in one year, but he can aim to sell an average of 62.5 prints per year over several years. Let's take a look at what happens if he sells 62 prints or 63 prints per year: If Michael sells 62 prints per year, the revenue is given by R(7.6) where 7.6 is found by solving 100 – 5x = 62. So, the revenue for selling 62 prints per year is $1562.40 and the price per print is $25.20. If Michael sells 63 prints per year, the revenue is given by R(7.4) where 7.4 is found by solving 100 – 5x = 63. So, the revenue for selling 63 prints per year is $1562.40 and the price per print is $24.80. Can you explain why the revenue is the same for selling 62 and 63 prints?

b.) Horizontal intercepts of $y = R(x)$: (20,0) and (-5,0).
The horizontal intercept (20,0) means that Michael has increased the price of his print by $2, 20 times and he has a revenue of $0. This means that no one feels Michael's prints are worth buying at $50 per print (evaluate 2$x$+10 at x = 20) – poor Michael! The horizontal intercept (-5,0) means that Michael has increased the price of his print by $2, "-5 times" (i.e., he has decreased the price by $2, 5 times) and he has a revenue of $0 because he would be giving his prints away for free (evaluate 2x+10 at x = -5).

8.) a.) $D(3) = 5$. Three hours after the storm ended, the river was 5 ft deep.

b.) The VI is $(0,8)$. When the storm ended, the river was 8ft deep.

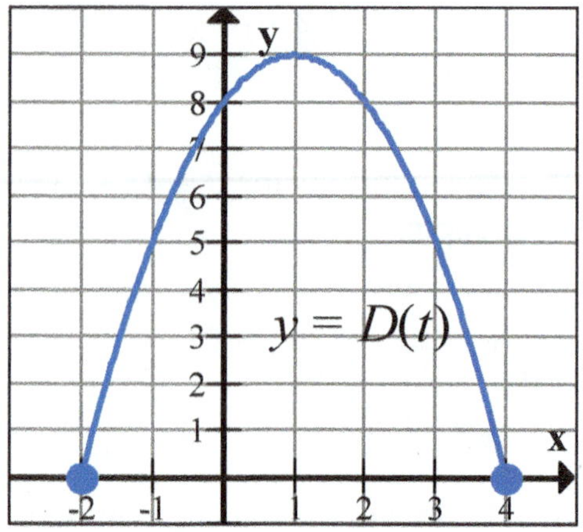

c.) The HI's are $(-2,0)$ and $(4,0)$. Two hours before the storm ended and four hours after the storm ended, the river was dry but in between those times the river was flowing.

d.) The vertex is $(1,9)$. This means that the river had a maximum depth of 9ft 1 hour after the storm ended.

e.) Shown is the graph of $y = D(t)$ on it's domain.

f.) The range of the real world function D is $[0,9]$. This means that the river has a depth somewhere between 0ft and 9ft at all times.

9.) The height of a piano off the ground, h, in feet, is a function of the horizontal distance along the ground in the direction it is thrown, x, also in feet. Given that $h = f(x) = -.002x^2 + .6x + 60$,

a.) The vertical intercept is $(0,60)$. This means that when the piano is thrown, it is launched from an initial height of 60ft above the ground (indicating a very large Trebuchet).

b.) The horizontal intercepts are $(-79.1,0)$ and $(379.1,0)$. The intercept $(379.1,0)$ indicates that the height is 0ft off the ground when the piano has been thrown 379.1ft from the trebuchet, which means that it hit the ground 379.1ft away. The other intercept is a nonsensical answer because the piano did not go backwards from the trebuchet.

c.) The vertex is $(150,105)$ which means that the maximum height the piano reached was 105ft off the ground when it was 150ft horizontally from the trebuchet.

d.) $f(60) = 88.8$. This means that when the piano is 60ft horizontally from the trebuchet, it is 88.8ft above the ground.

e.) The solutions are approximately -56.2 and 356.2. This means that when the piano is 356.2ft horizontally from the trebuchet, the piano is 20ft above the ground. The other solution is a nonsense answer because the piano did not go backwards from the trebuchet.

f.) You should probably park the Volkswagen about 379ft from the trebuchet based on the horizontal intercept. However, if you want to be very accurate, you could estimate that a Bug is about 4ft tall and subsequently solve the equation $f(x) = 4$. The reasonable solution to this is 374.7ft. So if you park the Bug with the middle of the car at about 375ft from the Trebuchet it will accurately land directly on the roof, smashing it to pieces for some reason.

PHOTO CREDITS

CHAPTER 1 CO JFPI Studios, Inc./Corbis/Bettmann **p. 2** (left) Kaz Chiba/Image Bank/Getty Images; (right) Digital Vision Ltd./SuperStock, Inc. **p. 13** Mareschal, Tom/Getty Images Inc.-Image Bank **p. 27** Globe Photos, Inc. **p. 37** Gail Albert Halaban/Corbis/Outline **p. 45** Ron Kimball/Kimball Stock **p. 50** Michael Agliolo/Grant Heilman Photography, Inc. **p. 59** U.S. Air Force/AP Wide World Photos **p. 64** Susumu Nishinaga/Photo Researchers, Inc. **p. 76** SuperStock, Inc.

CHAPTER 2 CO Corbis Royalty Free **p. 96** Columbia Pictures/ZUMA/Corbis/Bettmann **p. 105** National Institute for Biological Standards and Control (U.K.)/Science Photo Library/ Photo Researchers, Inc. **p. 116** (left) Clay McLachlan/IPN/ Aurora & Quanta Productions Inc.; (right) © Robert Llewellyn/ CORBIS All Rights Reserved. **p. 126** David Schmidt/Masterfile Corporation **p. 144** © David Stoecklein/ CORBIS All Rights Reserved. **p. 155** John Serafin

CHAPTER 3 CO TORU/Getty Images **CO inset** Kyodo/Landov LLC **p. 166** UPPA/Topham/The Image Works **p. 182** (left) Esbin-Anderson/Grant Heilman Photography, Inc.; (right) Frazer Harrison/Getty Images **p. 189** AP/Wide World Photos **p. 193** PEANUTS © United Feature Syndicate, Inc. **p. 196** David W. Hamilton/Getty Images, Inc.-Image Bank **p. 208** © Mike Watson/CORBIS All Rights Reserved. **p. 219** David Parker/ Science Museum/Science Photo Library/Photo Researchers, Inc.

CHAPTER 4 CO © Don Mason/CORBIS All Rights Reserved. **p. 240** Phillip & Karen Smith/SuperStock, Inc. **p. 253** Thomas Brummett/Getty Images, Inc.-Photodisc **p. 262** CBS-TV/ Picture Desk, Inc./Kobal Collection **p. 276** Frank Clarkson **p. 286** AP Wide World Photos **p. 287** Pascal Parrot/ Corbis/Sygma

CHAPTER 5 CO AFP/Getty Images, Inc.-Agence France Presse **CO inset** Hugh Threlfall/Alamy Images **p. 304** N. Hashimoto/ Corbis/Sygma **p. 317** Robert F. Blitzer **p. 326** Steve Lyne/ Getty Images **p. 331** © Jasper Johns/Licensed by VAGA, New York, NY. **p. 339** Bob Watkins/Photofusion **p. 353** © Sandro Vannini/CORBIS **p. 362** Phil Jason/Getty Images, Inc.-Taxi **p. 368** SuperStock, Inc.

CHAPTER 7 CO PASIEKA/Photo Researchers, Inc. **p. 486** Reuters/Corbis/Reuters America LLC **p. 498** © Fred Bavendam/Peter Arnold, Inc. **p. 508** © George Tooker **p. 516** Kauko Helavuo/Getty Images, Inc.-Image Bank **p. 524** PEANUTS © United Feature Syndicate, Inc. **p. 533** John G. Ross/Photo Researchers, Inc. **p. 534** Jim West **p. 542** © Steve Chenn/CORBIS **p. 544** © 2005 Roz Chast from Cartoonbank.com. All rights reserved. **p. 551** R. F. Voss "29-Fold M-set Seahorse" computer-generated image. © 1990 R. F. Voss/ IBM Research **p. 553** Robotman reprinted by permission of Newspaper Enterprise Association Inc.

CHAPTER 8 CO Photo courtesy of Dr. Andrew Davidhazy **p. 562** Les Stone/The Image Works **p. 575** Mauro Fermariello/ Photo Researchers, Inc. **p. 583** © Jasper Johns/Licensed by VAGA, New York, NY. **p. 589** Joe McBride/Getty Images, Inc.- Stone Allstock **p. 608** Photonica/Getty Images-Photonica Amana America, Inc. **p. 616** © The New Yorker Collection 1995 Warren Miller from cartoonbank.com. All Rights Reserved.

Definitions, Rules, and Formulas

The Real Numbers

Natural Numbers: $\{1, 2, 3, \ldots\}$
Whole Numbers: $\{0, 1, 2, 3, \ldots\}$
Integers: $\{\ldots, -3, -2, -1, 0, 1, 2, 3, \ldots\}$
Rational Numbers: $\{\frac{a}{b} \mid a \text{ and } b \text{ are integers}, b \neq 0\}$
Irrational Numbers: $\{x \mid x \text{ is real and not rational}\}$

Basic Rules of Algebra

Commutative: $a + b = b + a$; $ab = ba$
Associative: $(a + b) + c = a + (b + c)$; $(ab)c = a(bc)$
Distributive: $a(b + c) = ab + ac$; $a(b - c) = ab - ac$
Identity: $a + 0 = a$; $a \cdot 1 = a$
Inverse: $a + (-a) = 0$; $a \cdot \frac{1}{a} = 1 (a \neq 0)$
Multiplication Properties: $(-1)a = -a$; $(-1)(-a) = a$; $a \cdot 0 = 0$; $(-a)(b) = (a)(-b) = -ab$; $(-a)(-b) = ab$

Set-Builder Notation, Interval Notation, and Graphs

$(a, b) = \{x \mid a < x < b\}$

$[a, b) = \{x \mid a \leq x < b\}$

$(a, b] = \{x \mid a < x \leq b\}$

$[a, b] = \{x \mid a \leq x \leq b\}$

$(-\infty, b) = \{x \mid x < b\}$

$(-\infty, b] = \{x \mid x \leq b\}$

$(a, \infty) = \{x \mid x > a\}$

$[a, \infty) = \{x \mid x \geq a\}$

$(-\infty, \infty) = \{x \mid x \text{ is a real number}\} = \{x \mid x \in R\}$
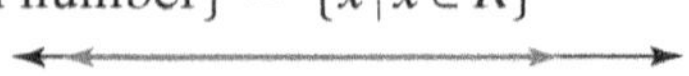

Slope Formula

$$\text{slope } (m) = \frac{\text{change in } y}{\text{change in } x} = \frac{y_2 - y_1}{x_2 - x_1} \quad (x_1 \neq x_2)$$

Equations of Lines

1. *Slope-intercept form:* $y = mx + b$
   m is the line's slope and b is its y-intercept.
2. *Standard form:* $Ax + By = C$
3. *Point-slope form:* $y - y_1 = m(x - x_1)$
   m is the line's slope and (x_1, y_1) is a fixed point on the line.
4. *Horizontal line parallel to the x-axis:* $y = b$
5. *Vertical line parallel to the y-axis:* $x = a$

Systems of Equations

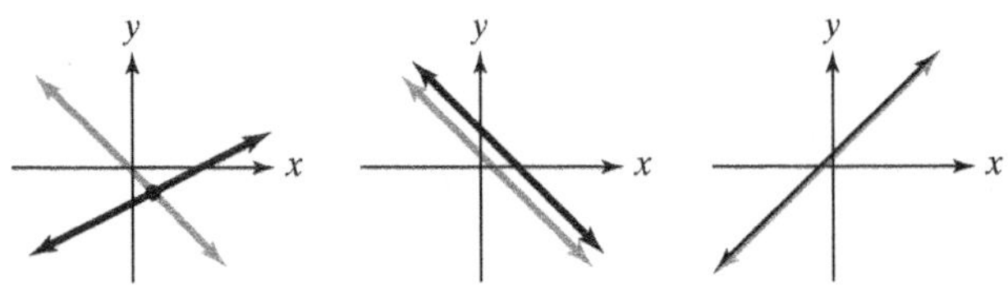

One solution: consistent | No solution: inconsistent | Infinitely many solutions: dependent and consistent

A system of linear equations may be solved: (a)graphically, (b) by the substitution method, (c) by the addition or elimination method, (d) by matrices, or (e) by determinants.

$$\begin{vmatrix} a_1 & b_1 \\ a_2 & b_2 \end{vmatrix} = a_1 b_2 - a_2 b_1$$

Cramer's Rule:

Given a system of a equations of the form

$$\begin{aligned} a_1 x + b_1 y &= c_1 \\ a_2 x + b_2 y &= c_2 \end{aligned}, \quad \text{then } x = \frac{\begin{vmatrix} c_1 & b_1 \\ c_2 & b_2 \end{vmatrix}}{\begin{vmatrix} a_1 & b_1 \\ a_2 & b_2 \end{vmatrix}} \text{ and } y = \frac{\begin{vmatrix} a_1 & c_1 \\ a_2 & c_2 \end{vmatrix}}{\begin{vmatrix} a_1 & b_1 \\ a_2 & b_2 \end{vmatrix}}.$$

Absolute Value

1. $|x| = \begin{cases} x & \text{if } x \geq 0 \\ -x & \text{if } x < 0 \end{cases}$
2. If $|x| = c$, then $x = c$ or $x = -c$. $(c > 0)$
3. If $|x| < c$, then $-c < x < c$. $(c > 0)$
4. If $|x| > c$, then $x < -c$ or $x > c$. $(c > 0)$

Special Factorizations

1. *Difference of two squares:*
   $$A^2 - B^2 = (A + B)(A - B)$$
2. *Perfect square trinomials:*
   $$A^2 + 2AB + B^2 = (A + B)^2$$
   $$A^2 - 2AB + B^2 = (A - B)^2$$

3. *Sum of two cubes:*

$$A^3 + B^3 = (A + B)(A^2 - AB + B^2)$$

4. *Difference of two cubes:*

$$A^3 - B^3 = (A - B)(A^2 + AB + B^2)$$

Variation

English Statement	*Equation*
y varies directly as x.	$y = kx$
y varies directly as x^n.	$y = kx^n$
y varies inversely as x.	$y = \frac{k}{x}$
y varies inversely as x^n.	$y = \frac{k}{x^n}$
y varies jointly as x and z.	$y = kxz$

Exponents

Definitions of Rational Exponents

1. $a^{\frac{1}{n}} = \sqrt[n]{a}$

2. $a^{\frac{m}{n}} = (\sqrt[n]{a})^m$ or $\sqrt[n]{a^m}$

3. $a^{-\frac{m}{n}} = \frac{1}{a^{\frac{m}{n}}}$

Properties of Rational Exponents

If m and n are rational exponents, and a and b are real numbers for which the following expressions are defined, then

1. $b^m \cdot b^n = b^{m+n}$

2. $\frac{b^m}{b^n} = b^{m-n}$

3. $(b^m)^n = b^{mn}$

4. $(ab)^n = a^n b^n$

5. $\left(\frac{a}{b}\right)^n = \frac{a^n}{b^n}$

Radicals

1. If n is even, then $\sqrt[n]{a^n} = |a|$.
2. If n is odd, then $\sqrt[n]{a^n} = a$.
3. The product rule: $\sqrt[n]{a} \cdot \sqrt[n]{b} = \sqrt[n]{ab}$
4. The quotient rule: $\frac{\sqrt[n]{a}}{\sqrt[n]{b}} = \sqrt[n]{\frac{a}{b}}$

Complex Numbers

1. The imaginary unit i is defined as

$$i = \sqrt{-1}, \quad \text{where} \quad i^2 = -1.$$

The set of numbers in the form $a + bi$ is called the set of complex numbers. If $b = 0$, the complex number is a real number. If $b \neq 0$, the complex number is an imaginary number.

2. The complex numbers $a + bi$ and $a - bi$ are conjugates. Conjugates can be multiplied using the formula

$$(A + B)(A - B) = A^2 - B^2.$$

The multiplication of conjugates results in a real number.

3. To simplify powers of i, rewrite the expression in terms of i^2. Then replace i^2 with -1 and simplify.

Quadratic Equations and Functions

1. The solutions of a quadratic equation in standard form

$$ax^2 + bx + c = 0, \quad a \neq 0,$$

are given by the quadratic formula

$$x = \frac{-b \pm \sqrt{b^2 - 4ac}}{2a}.$$

2. The discriminant, $b^2 - 4ac$, of the quadratic equation $ax^2 + bx + c = 0$ determines the number and type of solutions.

Discriminant	**Solutions**
Positive perfect square with a, b, and c rational numbers	2 rational solutions
Positive and not a perfect square	2 irrational solutions
Zero, with a, b, and c rational numbers	1 rational solution
Negative	2 imaginary solutions

3. The graph of the quadratic function

$$f(x) = a(x - h)^2 + k, \quad a \neq 0,$$

is called a parabola. The vertex, or turning point, is (h, k). The graph opens upward if a is positive and downward if a negative. The axis of symmetry is a vertical line passing through the vertex. The graph can be obtained using the vertex, x-intercepts, if any, [set $f(x)$ equal to zero], and the y-intercept (set $x = 0$).

4. A parabola whose equation is in the form

$$f(x) = ax^2 + bx + c, \quad a \neq 0,$$

has its vertex at

$$\left(-\frac{b}{2a}, f\left(-\frac{b}{2a}\right)\right).$$

If $a > 0$, then f has a minimum that occurs at $x = -\frac{b}{2a}$. If $a < 0$, then f has a maximum that occurs at $x = -\frac{b}{2a}$.

Definitions, Rules, and Formulas (continued)

Exponential and Logarithmic Functions

1. Exponential Function: $f(x) = b^x, b > 0, b \neq 1$
Graphs:

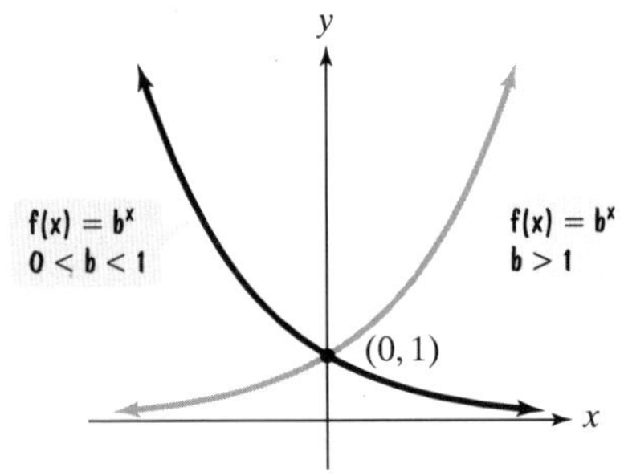

2. Logarithmic Function: $f(x) = \log_b x, b > 0, b \neq 1$
$y = \log_b x$ is equivalent to $x = b^y$.

Graphs:

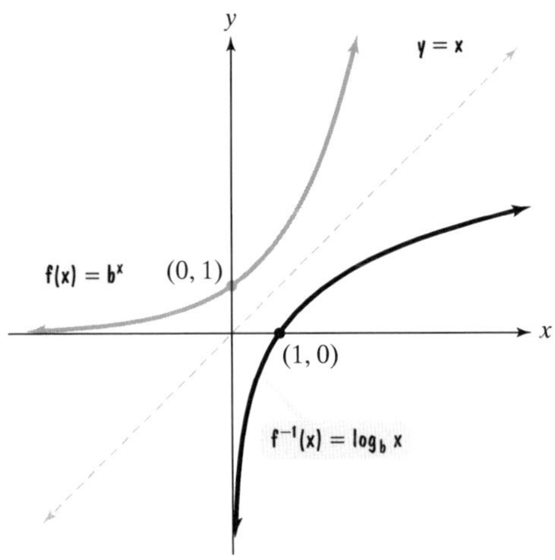

$b > 1$

3. Properties of Logarithms

a. $\log_b(MN) = \log_b M + \log_b N$

b. $\log_b\left(\dfrac{M}{N}\right) = \log_b M - \log_b N$

c. $\log_b M^p = p \log_b M$

d. $\log_b M = \dfrac{\log_a M}{\log_a b} = \dfrac{\ln M}{\ln b} = \dfrac{\log M}{\log b}$

e. $\log_b b^x = x$; $\log 10^x = x$; $\ln e^x = x$

f. $b^{\log_b x} = x$; $10^{\log x} = x$; $e^{\ln x} = x$

Distance and Midpoint Formulas

1. The distance from (x_1, y_1) to (x_2, y_2) is

$$\sqrt{(x_2 - x_1)^2 + (y_2 - y_1)^2}.$$

2. The midpoint of the line segment with endpoints (x_1, y_1) and (x_2, y_2) is

$$\left(\frac{x_1 + x_2}{2}, \frac{y_1 + y_2}{2}\right).$$

Conic Sections

Circle

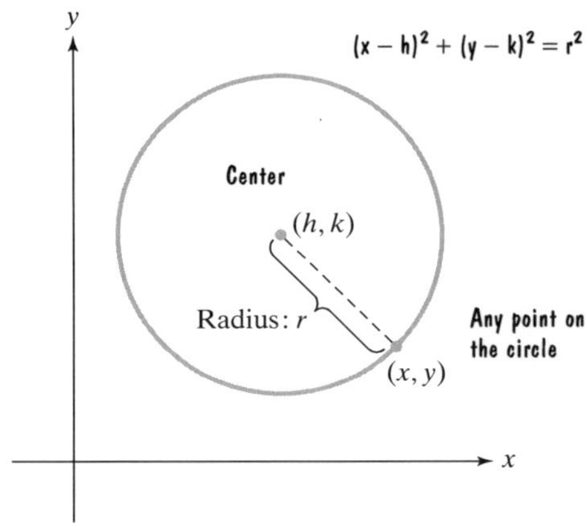

Ellipse

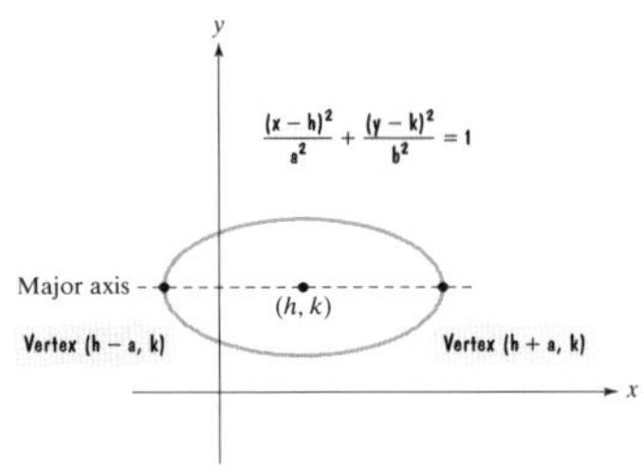

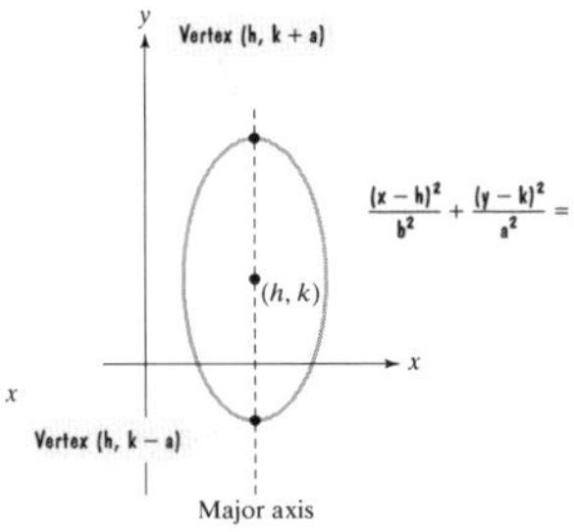

Hyperbola

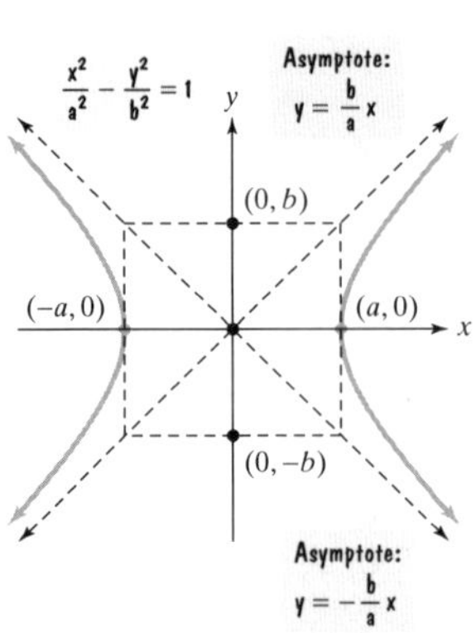

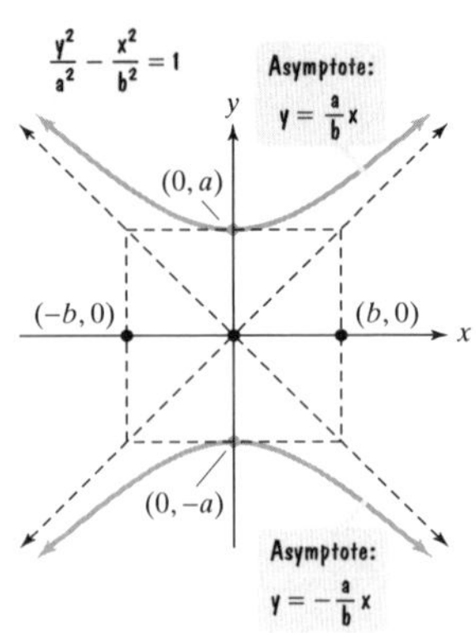